Environmental Physiology
of Animals

Environmental Physiology
of Animals

Pat Willmer

School of Biology
University of St Andrews
St Andrews

Graham Stone

Institute of Cell, Animal and Population Biology
University of Edinburgh
Edinburgh

Ian Johnston

School of Biology
University of St Andrews
St Andrews

**Blackwell
Science**

© 2000
Blackwell Science Ltd
Editorial Offices:
Osney Mead, Oxford OX2 0EL
25 John Street, London WC1N 2BL
23 Ainslie Place, Edinburgh EH3 6AJ
350 Main Street, Malden
  MA 02148 5018, USA
54 University Street, Carlton
  Victoria 3053, Australia
10, rue Casimir Delavigne
  75006 Paris, France

Other Editorial Offices:
Blackwell Wissenschafts-Verlag GmbH
  Kurfürstendamm 57
  10707 Berlin, Germany

Blackwell Science KK
  MG Kodenmacho Building
  7–10 Kodenmacho Nihombashi
  Chuo-ku, Tokyo 104, Japan

First published 2000
Reprinted 2000

Set by Graphicraft Limited, Hong Kong
Printed and bound in the USA

A catalogue record for this title is available
from the British Library

ISBN 0-632-03517-X

Library of Congress
Cataloging-in-publication Data

Willmer, Pat, 1953–
    Environmental physiology of animals/
  P.G. Willmer, G.N. Stone, I.A. Johnston.
        p.   cm.
    Includes bibliographical references and
  index.
    ISBN 0-632-03517-X
    1. Adaptation (Physiology)
    2. Physiology, Comparative.
    3. Ecophysiology. I. Stone, G.
    II. Johnston, Ian A.   III. Title.
  QP82.W48   1999
  571.1—dc21   99–23146
                                   CIP

DISTRIBUTORS

Marston Book Services Ltd
PO Box 269
Abingdon, Oxon OX14 4YN
(Orders: Tel: 01235 465500
         Fax: 01235 465555)

USA
Blackwell Science, Inc.
Commerce Place
350 Main Street
Malden, MA 02148 5018
(Orders: Tel: 800 759 6102
               781 388 8250
         Fax: 781 388 8255)

Canada
Login Brothers Book Company
324 Saulteaux Crescent
Winnipeg, Manitoba R3J 3T2
(Orders: Tel: 204 837 2987)

Australia
Blackwell Science Pty Ltd
54 University Street
Carlton, Victoria 3053
(Orders: Tel: 3 9347 0300
         Fax: 3 9347 5001)

For further information on
Blackwell Science, visit our website:
www.blackwell-science.com

Contents

Preface

There are many books dealing with physiological functioning and the comparative adaptations of animals. Why should one more be added to the list?

We hope that there are three key reasons why this book is different.

1 Above all, we seek to be novel by setting the issues in a strongly **environmental context**; the largest part of this book is devoted to analysing the problems for each kind of environment in turn. In one chapter you can find out how different kinds of seashore animal cope with the ionic problems of salinity change, the thermal problems of both cold sea water and warm drying aerial exposure, and the respiratory hazards of alternate aerial and underwater breathing. Inevitably we include more varied material, from a wider range of animal types; you will find less of a focus on terrestrial reptiles, birds and mammals than in most other books (after all, they represent less than 1% of all animals!).

2 We try to put together issues other than just conventional physiology, to look at **mechanisms and responses** with perspectives from the related and overlapping fields of ecology, behaviour and evolutionary biology. Thus you can find out how the seashore animals also deal with the mechanical problems of tides and waves, how their senses may need to be modified, the difficulties of finding a mate and producing viable offspring, and some of the imposed (anthropogenic) problems that man has introduced to different habitats. The book thus has more environmental biology, more ecology, and a greater evolutionary perspective than many others.

3 Wherever possible we include modern **molecular biological insights** into adaptive problems, and take the level of analysis to the cell, the membrane and the enzyme.

In essence we want to look at how animals cope with particular kinds of environment, putting together the biochemical, physiological, behavioural and ecological adaptations that allow animals to survive there. We believe that this will be a useful contrast to current texts; most of these instead take each system of the animal body in turn (such as ionic and osmotic issues, energy balance, temperature, circulating fluids, nerves, hormones) and then give examples of how they operate in different kinds of animals or in different kinds of habitat. We believe

that our approach will prove useful for a much broader range of **environmental biology** courses, as well as more traditional environmental physiology courses.

Some readers will have little physiological background initially, and will want to understand physiological processes in detail, so in Section B we also offer an updated review of comparative physiology, looking at mechanisms and functions system by system, but incorporating new information on how adaptation is operating at a molecular level. We leave out most of the 'examples' here, and concentrate on principles. This section provides most of the material needed for understanding comparative physiology, though it does not deal with processes largely unaffected by environment (the basic mechanisms of neural, hormonal and muscle physiology, for example, which are broadly conserved in all animals and where excellent texts for specialist courses already exist).

Other readers may already know a good deal about physiological principles, and can use Section B merely as a reference point and reminder when considering particular environments. For them, Section C provides all the key material on environmental adaptation, relevant to a wide range of environmental biology issues. It covers far more than traditional animal physiology textbooks. Extensive cross-references to principles in Section B give the reader a chance to check their understanding of the core physiology.

Thus we hope this book presents material that will be useful to a wide range of readers, in varying kinds of environmental biology courses and at different levels. Above all we hope it will help provide an integrated understanding of environmental adaptation.

Acknowledgements

In this textbook we give further reading sections at the end of each chapter, mostly to modern review material. We have cited sources from which material in the illustrations has been drawn wherever a figure has been used in unmodified form, but we do not cite every source of information in the text. At this level, copious references break up the flow and markedly lengthen the text, making it harder to read. However, we realize that this carries two important risks: firstly for the readers, who might not be able to follow up on points of interest and find out more; and secondly for the scientists who carried out the work and might feel their contributions are going unrecognized. We hope that both of these risks are reduced by modern information technology, making it easier to access literature searches and track down the primary references. Thus we expect that the work we quote from, whether a classic paper or a very recent new insight, can usually be traced quickly and its authorship established. However, we take this opportunity at the outset to offer an enormous general acknowledgement and a debt of thanks to all those whose studies have contributed to this text.

More specifically, we offer personal and very grateful thanks to those who have read and commented on the manuscript for us. Simon Maddrell (University of Cambridge, UK) and Jim Childress (University of California, Santa Barbara, USA) took on the heroic task of reading the whole book, and gave us tremendously helpful inputs. Others read parts of the text as follows: Jon Harrison (Arizona State University, USA) (Chapters 3–8); David Scholnick (University of Colorado at Boulder, USA) (Chapters 3–8); John Spicer (University of Sheffield, UK) (Chapters 4–8); John McLachlan (University of St Andrews, UK) (Chapter 2); Lynn McIlroy (University of St Andrews, UK) (Chapters 4, 5, 6 and 8); Claire Maynard (University of St Andrews, UK) (Chapter 10); Alex Rowe (University of Edinburgh, UK) (Chapter 15); David Paterson (University of St Andrews, UK) (Chapter 10); and Ray Huey (University of Washington, USA) (Chapters 13 & 14).

We also thank those who have provided invaluable assistance in locating sources of both information and illustrations over the course of several years. Here Sandy Edwards has been an invaluable and endlessly patient colleague, and Christina Lamb has also offered enormous help with seeking out references, while Peter Slater, Gordon Cramb, Mike Fedak, Anne Magurran, Su Bryan, Jim Aiton and Susie Whiten have helped out beyond the call of duty. Finally, colleagues at Blackwell Science, especially Ian Sherman and Katrina McCallum, have had a huge and greatly valued impact on the successful production of this book.

To all these, and to many other colleagues and students over the years who have encouraged us in thinking about and writing this book, we are very grateful.

List of Abbreviations

ADH	antidiuretic hormone	HVA	homeoviscous adaptation
AFP	antifreeze protein	HWN	high water, neap (of tide)
AFGP	antifreeze glycoprotein	HWS	high water, spring (of tide)
ALH	ascending loop of Henle (of vertebrate kidney)	ICF	intracellular fluid
BAT	brown adipose tissue ('brown fat')	IL	interleukin
BMR	basal metabolic rate	INA	ice-nucleating agent
BOD	biological oxygen demand	INP	ice-nucleating protein
BP	boiling point	IP_3	inositol triphosphate
CaM	calmodulin	IPP	inositol pentaphosphate
cAMP	cyclic adenosine monophosphate	kDa	kilodalton
CD	collecting duct	LBTS	low body temperature setpoint
CEH	critical equilibrium humidity (above which water vapour is absorbed)	LCT	lower critical temperature
		LDH	lactate dehydrogenase
CEWL	cutaneous evaporative water loss	LH	luteinizing hormone
CPK	creatine phosphokinase	LH	loop of Henle (in vertebrate kidney)
DAG	diacylglycerol	LWN	low water, neap (of tide)
DCT	distal convoluted tubule (in vertebrate kidney)	LWS	low water, spring (of tide)
DLH	descending loop of Henle (in vertebrate kidney)	M_b	body mass
DMSP	dimethylsulphoniopropionate	MDOC	minimum depth of occurrence
DVC	discontinuous ventilation cycle	MFO	mixed function oxidase
E	equilibrium potential across a membrane	MHC	major histocompatibility complex
E_x	equilibrium potential due to a particular ion X	MHWN	median high water, neap (of tide)
ECF	extracellular fluid	MHWS	median high water, spring (of tide)
ECV	extracellular volume	MLWT	median low water (of tide)
EHWN	extreme high water, neap (of tide)	MMR	maximum metabolic rate
EHWS	extreme high water, spring (of tide)	mRNA	messenger RNA
ELWN	extreme low water, neap (of tide)	MWL	median water level (of tide)
ELWS	extreme low water, spring (of tide)	NAD	nicotinamide adenine dinucleotide
ER	endoplasmic reticulum	NIS	non-indigenous species
ETS	electron transfer system (in mitochondria)	NST	non-shivering thermogenesis
EWL	evaporative water loss	OML	oxygen minimum layer
FP	freezing point	p_b	barometric pressure
FPD	freezing point depression	P_c	critical partial pressure
FSH	follicle-stimulating hormone	P_{osm}	osmotic permeability of a membrane
GCM	global circulation model	P_w	permeability to water
GIT	gastrointestinal tract	P_x	permeability for a particular ion X
HBTS	high body temperature setpoint	PAH	*para*-aminohippuric acid
Hif-1	hypoxia-inducible factor 1	PAL	present atmospheric levels
HSP	heat-shock protein	PAr	phosphoarginine (arginine phosphate)

PCr	phosphocreatine (creatine phosphate)		**SR**	sarcoplasmic reticulum
PCT	proximal convoluted tubule (in vertebrate kidney)		**STT**	shivering threshold temperature
PFK	phosphofructokinase		T_a	ambient temperature
PK	protein kinase		T_{abd}	abdominal temperature
PKC	protein kinase C		T_b	body temperature
Q_{10}	temperature coefficient		T_s	surface temperature of body
RBC	red blood cell		T_{pref}	preferred body temperature
REWL	respiratory evaporative water loss		T_{th}	thoracic temperature
RH	relative humidity		**THP**	thermal hysteresis protein
RI	refractive index		**TMAO**	trimethylamine oxide
RPD	redox potential discontinuity		**TNZ**	thermoneutral zone
RQ	respiratory quotient		**TSD**	temperature-dependent sex determination
RVD	regulatory volume decrease		**TTX**	tetrodotoxin (sodium-channel blocker)
SCP	supercooling point		**UCT**	upper critical temperature
SDA	specific dynamic action (heating effect during digestion)		**VMI**	vasomotor index
			VP	vapour pressure

Section A:
Basic Principles

1 The Nature and Levels of Adaptation

1.1 Introduction: comparative, environmental and evolutionary physiology

This book is about how animals cope with the problems posed, and exploit the opportunities offered, by their particular environments. Traditionally the mechanisms for coping with the environment have been treated as issues of **comparative physiology**, which is concerned with investigating both general principles of organismal function (the similarities that exist between all organisms) and the exceptions to the general rules. Comparative physiologists may be interested to look for different ways in which animals solve particular challenges of living in different habitats. Do these different solutions also depend on factors such as animal size or design, or the biological materials used? Much of this traditional physiology relies on laboratory study of either classical laboratory animals (mainly mammals, and certainly with vertebrates very dominant and only a sporadic use of other taxa) or of extremely specialized animals pushed to their extreme performance. This can tell us a great deal about mechanisms, since organisms living in extreme environments illustrate the range of evolutionary possibilities amongst living animals, so that species with extreme development of a particular physiological property are therefore often useful as model systems. But this study of extremes should not make us lose sight of the 'norm' of performance or of more 'generalist' animals. We also need to be wary of the tendency of laboratory-maintained animals to become rather unrepresentative, as there are famous cases where within a few months or a few generations after 'domestication' such animals have dramatically changed their physiological performance.

For these reasons it is very important to study animals in the context of their own habitat and their real needs, so that 'comparative physiology' has tended to be replaced by **environmental physiology** or **ecophysiology** or **physiological ecology**, which in a real sense add traditional natural history to the study of comparative physiology. The primary aim of ecological or environmental physiologists is to understand how animals function in and respond to their natural environments, at all stages of their life cycles. Indeed it quickly becomes apparent that many animals, for most of their lifetime, do not need extreme physiological adaptation and rely instead on behavioural strategies to avoid the worst of their difficulties.

However, ecophysiology has in turn become subject to an increase in conceptual rigour, partly as a result of important critiques of the story-telling nature of the 'adaptationist programme' (the assumption that everything has an adaptive function). We need to move beyond anecdotal science, where after collecting lots of examples of particular features their functions are to be gleaned merely by their apparent correlations with environmental features. Understanding the processes by which particular ecophysiological features arose, and the values of physiological parameters as outcomes of natural selection, may be termed **evolutionary physiology**. Evolutionary physiology is a discipline still somewhat in its infancy, but learning rapidly from other areas of evolutionary biology and from the analytical techniques of population biology and (especially) molecular biology. It involves a more explicit attempt to integrate both short-term and long-term genetic perspectives into physiological ecology. Within species, understanding the genetic basis of physiological traits, and the magnitude and causes of physiological variation, may reveal how they can be shaped in the relatively short term by natural selection. Evolutionary physiology also examines the evolution of traits over longer time periods, across species or higher taxa. Here the crucial point is that the traits we see represent an interaction between ancestral traits and selection, set in the context of ancestral environments. We can only understand this by following the evolution of these traits through a family tree or phylogeny. A phylogenetic analysis allows the evolutionary physiologist to address two crucial issues: firstly, how does the physiology of an ancestral species affect what is possible in its descendants, and secondly, how rapidly can physiological traits evolve? The recent explosion in our knowledge of the relationships between organisms, much of it based on DNA or RNA sequence analysis, provides a wealth of opportunity in this area. Thus, not only can we use explicitly evolutionary analyses to help us understand patterns of mechanistic physiology, but we can now also track the evolution of molecular components of physiological adaptation as a short cut

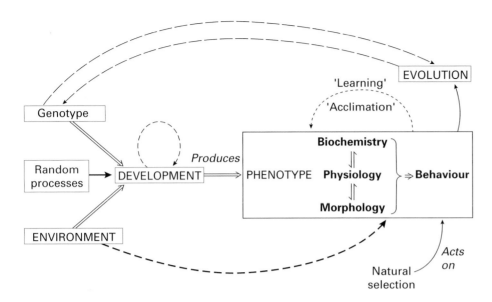

Fig. 1.1 Genotype and environment interact though developmental processes to give a particular phenotype, a suite of biochemical, physiological and morphological traits. The phenotype also includes behaviour, which is limited by all the other phenotypic characters. Selection acts at the whole organism level, and therefore mainly on behaviour. Biochemical and physiological traits are normally only subject to selection if they have detectable and stable effects at the level of behavioural performance (e.g. escape speed, reproductive behaviours, food-gathering efficiency, etc.) that in turn affect reproductive fitness.

for many years of complex laboratory analysis of comparative species differences. Evolutionary physiology may develop all the more quickly set in a strong environmental context, taking both past and present environments into account; for unless there has been substantial climatic or geological upheaval, or migration, animals tend to inherit their environment as well as their genes from their immediate ancestors. Furthermore, organisms from more extreme environments are particularly likely to show clear examples of evolutionary adaptation as a result of intense past selective pressures.

This book is written in the conviction, increasingly common in all kinds of biological literature, that the trend needs to be taken yet further; that physiology cannot be isolated from behavioural strategies and from ecological and life-history strategies, while also increasingly taking full account of evolutionary and molecular studies. Thus, environmental adaptation has to be an interdisciplinary and fully integrated study as biological sciences come of age. A schematic view of the interrelationships of genotype, environment and phenotype is shown in Fig. 1.1, stressing this wider viewpoint; note especially the intervening effect of developmental processes and the key role of behaviour, both as the outcome of phenotypic characters and as the level at which natural selection acts most strongly. Genotypic variation (long-term evolutionary change) can only be subject to selection if it affects reproductive success, mediated by changes in performance and behaviour, though the more immediate changes may concern biochemical, physiological or morphological features which are what we normally see as 'adaptations'. All these issues are discussed further in this chapter.

1.2 The meaning of 'environment'

The two basic concepts in natural selection are fitness and environment. Fitness is strongly linked to adaptedness and therefore to adaptation, and all these concepts are subject to intensive

analysis by ecologists and evolutionary biologists (see section 1.3). However, the concept of the environment is usually largely ignored, perhaps because it is seen as unproblematic. Nevertheless it is worth considering just what we do mean by an animal's environment.

At an obvious level, the environment means the kind of habitat in which an animal lives—the deep sea, a tropical forest, a hot desert or whatever. In this sense an environment is equivalent to the concept of a **biome**, with grossly similar kinds of living space having similar physical characteristics all lumped together. Knowing whether an animal lives in water, in air or amphibiously, or in cold, seasonal or warm latitudes, certainly tells us a lot about the problems it will encounter and the kinds of design and strategy it is likely to show.

Environments are obviously rather more complex than this though, and each species of animal has a more precisely defined environment within a biome, perhaps in the deep-sea benthos, or predominantly arboreal in a forest, or in the litter layer of a freshwater pond. By considering the environment at this level, with some elements of biotic interaction coming in as well as mere physical factors, we gain rather more knowledge about the animal's requirements. Indeed for the parasitic animals considered in Chapter 15, properties of the host organism and of other parasites may be far more important than abiotic factors.

At a third level, each individual animal has its own environment: the totality of all the external factors it experiences, both biotic and physical. This environment is commonly modified by its own behavioural choices and indeed by its very presence. At this level we are really considering microenvironments, or microhabitats, or (on land) microclimates. This is also the level that really matters in terms of environmental and evolutionary physiology. Just because an animal lives in a 'type' of environment as perceived and classified by humans, it does not necessarily experience that environment in the way that we see and experience it. Each animal chooses where to spend its time, to forage or to rest or to seek mates, from a range of possible options. These options change on a very fine spatial scale and a very rapid temporal scale for small animals, but much more coarsely for large animals (for them the environment is sometimes described as 'coarse-grained'). In many cases an animal will choose the least stressful microhabitat, whether in and amongst the ameliorating effects of vegetation, or within a burrow or nest, or merely within the boundary layer of relatively still fluid above a substratum. The presence of the animal will modify local conditions, by adding excretory products (including CO_2) or depleting oxygen or modifying humidity and temperature; an environment containing an animal is no longer the same as the environment before the animal moved in. Thus the environment we want to know about is that of the organism itself and its immediate surroundings, measured on a temporal and spatial scale appropriate to that animal. An animal then becomes part of an interactive system that includes its own internal physico-chemical state and the physical and chemical conditions of its immediate surroundings.

At whatever level we consider them, environments can be helpfully categorized in terms of three important interacting parameters: the basic stress intensity, the magnitude and time-scale of fluctuations, and the energy or resource availability.

1.2.1 Environmental stress

Environments are enormously variable in relation to the stress that they impose on their inhabitants, and this stress may be both **abiotic** (physical and chemical factors) and **biotic** (direct and indirect effects of other organisms, including competition and habitat modification). Since life on Earth evolved in seas that were thermally and osmotically relatively stable, and all cellular machinery was fundamentally selected from its inception to work best in stable rather cool marine conditions, it is often useful to see the abiotic stress as being dependent on how much conditions have diverged from those starting points. Thus life in cool sea water is relatively 'easy', life in a seasonal pond is somewhat tricky, and life in hot desert conditions is spectacularly difficult. Other extreme environments would include polar lands, mountain peaks, hypersaline lakes, hot springs and deep-sea thermal vents. But biotic stress may sometimes work in the opposite direction, in that 'easy' habitats may also have large and diverse populations of other organisms and so impose more competition and predation pressure. Either of these kinds of stress may put an organism at a disadvantage in having to expend more of its own energy to survive, whether in physiological regulation, or avoidance tactics, or competitive or defensive activities.

1.2.2 Magnitude of fluctuations

Environments may be very stable on all timescales relevant to living organisms, the classic case perhaps being deep seas. Or they may vary on an evolutionary and geological timescale of tens or hundreds or thousands of years as land masses move, sea levels rise and fall, materials erode and deposit elsewhere, and rivers change their courses. There may also be changes with a regular annual, lunar or daily cyclicity. Finally, there are changes on a much shorter timescale of hours or minutes or seconds, as the weather changes. Note again that the magnitude of the change is a relative phenomenon, particularly linked to the size of the perceiver. Short-term changes are especially important in relation to very local microenvironments and therefore to very small animals; the difference between the environment above a leaf and the environment below it may be profound, and both may change within seconds in relation to varying solar radiation (insolation), air movements and rainfall. Changeable environments put a high selective premium on versatility or tolerance in animals, rather than on precise adaptations to particular conditions. This may be particularly true where man has intervened in the natural ecosystem to put new stresses on animals, whether from habitat destruction, climate modification, or the introduction of many kinds of toxic chemicals.

1.2.3 Energy or resource availability

Energy is rarely freely available to animals (as it may be to plants), but it is certainly more easily obtained in some habitats than in others. Traditionally it has been thought that where energy is severely restricted, as it may be in deserts and in polar regions, the result is simple communities with short food chains, largely made up of animals with low metabolic demands.

By contrast in rather stable environments with high primary productivity and rapid energy flow (exemplified by tropical rain forests or coral reefs) species may develop more specific habitat preferences, leading to specialization and producing diverse and complex animal communities. The corollary of all this is that stable high-energy communities would favour adaptive radiation, whereas rates of evolutionary change may be rather low in low-energy communities.

Varying resource availability between environments is also important in leading to **resource polymorphism** (or trophic polymorphism), where differences in behaviour, life history strategy or morphology occur within a single species. Striking polymorphisms may occur where interspecific competition is relaxed, and/or there are unfilled niches that can be rapidly exploited, and we will meet many examples of this in Section C.

1.2.4 Selection and the environment

Interactions between all three of these components of an environment tend to determine the kinds and diversity of animals that occur, and the type of selection that operates. Traditionally two main types of selection are recognized, representing either end of a continuum: **r-selection** which occurs in unpredictable environments and **K-selection** which occurs in more predictable environments (Table 1.1). The prefix r refers to the rate of population increase, which tends to be maximized in r-selected

	r-selection	*K*-selection	A-selection
Environment			
Stability	Low	High	High
Abiotic stress	High	Low	High
Energy	Low	High	Low
Individuals			
Body size	Small	Large	Small *or* large
Lifespan	Short	Long	Long
Maturity	Early	Late	Late
Reproduction			
Pattern	Semelparous	Iteroparous	Either
Generation time	Short	Long	Either
Fecundity	High	Low	Low
Offspring	Many, small	Few, large	Either
Parental care	Absent	Common	Possible
Populations			
Density	Fluctuating	High	Low, or fluctuating
Stability	Fluctuating	Steady	Fluctuating
Range	High	Low	Either
Competition	Low	High	Low
Biotic interactions	Few, simple	Many, complex	Few, simple
Overview	Small	Large	Very varied
	Rapid reproductive output	Slow reproductive output	Usually slow
	Colonists	Climax communities	Simple climax
	Generalists	Specialists	Specialists

Table 1.1 Types of selection on animals.

species, while *K* represents the carrying capacity of the environment. Typically, *r*-selected animals are the small, rapidly reproducing, early maturing and short-lived species, producing large numbers of relatively low-investment progeny, many of which will not survive, often reproducing just once and then dying (semelparity) and with potentially wide swings in population size; while *K*-selected animals are large, slowly reproducing and long-lived, producing just a few young and investing heavily in each, often reproducing repeatedly (iteroparity), and with relatively stable population sizes. The *r*-selected animals should live in disturbed habitats as early successional species, have large geographical ranges and relatively non-specialist interactions with other species; they are colonists and opportunists, in environments of high stress and high levels of fluctuation. *K*-selected animals should be more common in climax communities, having complex coevolved relationships with other organisms; they do best with low abiotic environmental stress (though often high biotic stress) and low levels of fluctuation. These designations therefore relate to features both of the environment and of the organisms inhabiting those environments, especially their life-history strategies.

This dichotomy is useful as a way of thinking about different kinds of lifestyle and the kinds of environments and features that might be expected to go together. However, it has many

imperfections, and many animals do not fit even into a continuum between the two extremes, having instead some very *r*-selected features and some very *K*-selected features. Some authors prefer to use a three-way model, where *K*-selection is the norm in predictably favourable habitats but is replaced by **A-selection** (adversity selection) in more extreme environments where conditions are predictably unfavourable. A-selection occurs in habitats that are of high environmental stress but with a low magnitude of fluctuation, and with low energy availability. Animals here have high stress resistance, low fecundity, late maturity and long lifespans, with very low levels of biotic interaction. A-selection might be expected in many kinds of extreme environment; deserts, polar regions, montane habitats, caves, anoxic muds, etc. This may also link to the argument about rates of speciation in different environments. In practice maximum adaptive radiation tends to be exhibited not in thoroughly *K*-selected environments but in areas of abiotic stability where energy is moderately available but *not* unrestricted (e.g. we find tens or hundreds of closely related species in African lakes, in some areas of deep sea and in some patches of tropical forest). Here organisms do not become too specialist, but they can speciate rather rapidly by switching between a wide range of possible resources. The opposite environmental combination of abiotic instability but high resources, which might be exemplified by a temperate estuary, may produce an alternation between periods of relatively low diversity with species coping with fluctuating adverse conditions, and periods of abrupt change when abiotic conditions alter too drastically, perhaps leading to periods of rapid evolution and the generation of whole new groups of related species (cladogenesis).

1.3 The meaning of 'adaptation'

Adaptation is a central concept in biology and one that attracts enormous controversy. It is often used in several different senses, to describe both a pattern and a process; and more often than not it is used rather loosely.

1 Firstly, adaptation is often used as a term for the **characters** or **traits** observed in animals that are the result of selection; for example, the presence of haemoglobin might be said to be an adaptation to allow a greater oxygen carriage in blood.

2 Alternatively, and most 'correctly', adaptation might be defined as a **process**; the means by which natural selection adjusts the frequency of genes that code for traits affecting fitness (most simply, the number of offspring surviving in succeeding generations). For example, increasing haemoglobin concentrations within a taxon might be seen as an adaptation to potentially low oxygen (hypoxic) environments. Evolutionary adaptation then becomes almost synonymous with natural selection itself: necessary attributes for both include variability, repeatability, heritability and differential survivorship of offspring. Adaptation in this sense is a process that normally occurs extremely slowly, over hundreds or thousands of generations, and it is not usually reversible. However, in extreme environments or (as we shall see in Section C) where selective pressures from human interference are strong, it can sometimes occur very quickly.

3 'Adaptation' is also used to describe short-term **compensatory changes** in response to environmental disturbance. This kind of change is the outcome of **phenotypic plasticity**, where pre-existing traits are differentially expressed as appropriate to the local conditions. Here the terms **acclimation** or **acclimatization** are technically more correct (see section 1.5); evolutionary biologists would not use the term adaptation at all in this context.

Great caution must be exercised in using the terms 'adaptation' or 'adaptive', and in the simplistic interpretation of individual traits at the molecular, cellular, tissue or organ levels, in relation to particular environmental factors. A trait should only be considered an adaptation if there is some evidence that it has evolved (has been changed through its evolutionary history) in ways that make it more effective at its task, and thus enhance fitness. In other words, a trait should only be deemed to be an adaptation if it is a consequence of selection for the task it performs; an incidental ability to perform the task is not enough, and nor is mere existence of a good general fit between organism and environment. Evidence that a characteristic is an adaptation is quite hard to come by, and may be of three kinds:

1 Correlation between a character and an environment or use. This is the commonest approach, and where it involves interspecific comparisons of extant species it forms the core of comparative physiology. But it has inherent dangers, which are discussed in more detail in the next section.

2 Comparisons of individual differences within a species. Geographical patterns in the frequencies of different gene variants (alleles) in natural populations can be mapped over environmental gradients.

3 Observation of the effects of altering a character. An organ can be experimentally altered, or a behaviour prevented, and the effects on the efficiency of some particular function in some particular environment observed. In modern physiology we can improve matters by either knocking out or over-expressing a specific gene (rather than a phenotypic character) and observing the effects; indeed, if we know enough about the evolution of a given gene we may be able to piece together the changing physiology that goes with it. This has been possible with a few very well-studied systems such as the lactate dehydrogenase (LDH) gene alleles that are involved in key anaerobic adaptations and that show clines with latitude and temperature (see Chapter 8). But the danger remains of interfering with other processes and thereby breaking the link between cause and effect.

These approaches generally involve correlative evidence, which is never really conclusive. Clear evidence for adaptation is not always possible, and the literature is still dominated by inference; nevertheless this is reasonably well founded by sheer weight of evidence of the first, correlative, kind above. Most of the material in Section C of this book presents adaptations supported by this kind of inference, which may be logically imperfect but may well be correct. In fact, unlike many of the characters that evolutionary biologists deal with, many physiological traits have direct and clearly quantifiable effects in terms of the survival of organisms, and many would argue that it is not necessary to show 'adaptation' (as enhanced survivorship in the field) to make clear inferences about physiological adaptedness.

However, it is important to realize at the outset that not all differences in physiology between species are adaptive. Remember that traits do not evolve 'for a reason'; they evolve entirely by chance and are selected *after* they arise if they enhance the success (fitness) of their possessor. They may persist merely because they do no harm or are linked with some other beneficial trait. They may not have been selected because of the advantage they now seem to endow, but perhaps for some other and now quite unapparent reason; these kinds of traits are sometimes called 'exaptations' rather than adaptations (e.g. Darwin noted that the sutures between bones in mammalian skulls may now be adaptive for allowing the birth of relatively large-brained offspring in some species, but they were selected for long ago in evolutionary history for quite different reasons). In addition, many traits were fixed because of natural selection in an ancestral species but are now present for reasons that have little to do with the current selective regime on the species being considered; this phenomenon is termed 'phylogenetic inertia', and may be attributable to the fundamental conservatism of development, of physiological mechanisms and of underlying genes.

Remember also that natural selection does not necessarily provide maximal or even optimal solutions: biological optimality is usually a relative concept, and a particular design need only be sufficient to do better than the existing alternatives. Therefore it is impossible to say that any one design is 'optimal', or to predict whether a 'better' optimum might exist. The use of optimality in adaptation debates has therefore led to considerable confusion and criticism. It is likely that there will be some sort of match between biological structures and functional requirements, but how good this match should be is not clear. A great many features of animal design, including physiological mechanisms and pathways, have to meet more than one, and sometimes several conflicting, functional demands. They are therefore very likely to represent a working compromise between these competing demands. Adequacy or sufficiency in one function, without unduly disrupting others, may well be the most common evolutionary outcome.

Even where design features appear to serve only one function we should not expect perfect or optimal form or performance. Natural selection filters phenotypic success through genotypic transmission only in relation to *differential* survivorship, relative to other possible phenotypes and genotypes. Optimal construction and performance are therefore not the inevitable outcomes of selection, and the match of optimality to actual design may be largely a reflection of the strength of the selective force. Genetic drift may also tend to reduce the rate of improvement in characters, especially in small populations with low immigration. The potential for optimal adaptation may also be limited by developmental constraints, and by genetic constraints where traits are correlated or linked.

This is why it is so helpful to consider the whole organism and its integrated functioning within its habitat, rather than individual characters or particular physiological systems in isolation. The approach adopted in this book, looking at all aspects of coping with a particular type of environment, may be more useful as a background to modern evolutionary physiology than the 'systems' approach more commonly presented, that deals with each system in turn and only adds in the other bodily systems and the environment as afterthoughts.

1.4 Comparative methods to detect adaptation

Comparisons between and amongst sets of species are the most common ways of looking for adaptations; indeed, it is fair to say that adaptations are essentially 'comparative' phenomena and can only be measured comparatively between or within species. Many studies choose two or more species differing in behaviour or ecology and compare them, to determine whether they show phenotypic differences that could be interpreted as adaptations to their different selective regimes. In a very loose sense, this is called 'the comparative method', and it has served biology well in giving 'evolutionary' explanations of why things are the way they are—in the biological sciences, experiments repeating actual evolutionary events are usually impossible!

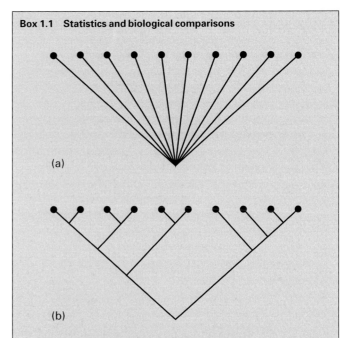

Box 1.1   Statistics and biological comparisons

(a)

(b)

Typically, statistical methods assume that data points are independent, but this would be the case only if we studied 10 'ideal' species that were genuinely unrelated or were absolutely equally related, as shown in the top figure; here, instantaneous speciation resulted in 10 independent lineages leading to 10 living species.

Thus, if we were to test for a correlation between the mean values for these species for two phenotypes (e.g. relative heart mass and maximal oxygen consumption), or perhaps one phenotype and an environmental factor (e.g. blood haemoglobin concentration and altitude), we could claim the nominal $n - 2 = 8$ degrees of freedom for hypothesis testing.

If, instead, the 10 species were actually related as shown in the lower figure, with a hierarchical evolutionary relationship following descent with modification, we would have something fewer than 8 degrees of freedom available for hypothesis testing. The trait of interest might only have arisen or changed once or twice, and then been inherited by several descendant species. These could not be treated as separate examples of adaptive acquisition of that trait. Various analytical methods now exist that explicitly use the phylogenetic topology and branch lengths to allow valid hypothesis testing.

However, there is an inherent problem in looking at patterns across species. As soon as two species have diverged from a common ancestor, differences will appear between them for many traits. Some of these changes will be due to chance shifts in gene frequencies (genetic drift). This is particularly true if one or both species experience low population size (a 'genetic bottleneck'), in which drift can be especially significant. Differences in a given trait therefore do not represent proof of selective modification in either of the species. So what do we need if we want to link differences in environment with differences in physiology? Imagine one species from a warm environment and one from a cold environment, differing in their physiologies. The difference may have nothing to do with environment but could just be due to genetic drift. However, if selection has modified one of the species, how can we tell which has acquired a novel physiology, and thus determine the direction of evolutionary change? We can only work this out if we know the physiology of the common ancestor, usually inferred using a technique called outgroup comparison (see Fig. 1.2c). Then we can say which species has changed, and by how much, giving us a correlation between temperature and physiology.

While one pair of species may show a change in one direction with temperature, another (phylogenetically distinct) pair may show the opposite change; thus we still do not have a general rule. In fact a single two-species comparison never gives sufficient evidence for adaptation (statistical tests of correlation always require a minimum of three data points); but unfortunately such comparisons are remarkably common as a way of proceeding in comparative physiology. However, if many distinct evolutionary radiations show the same change in physiology linked to the same difference in environment then there is evidence for a general correlation between the two. The more radiations showing the pattern, the stronger the case. Crucially, because each radiation starts with a different common ancestor, each represents an *independent* test of whether trait and environment are related. Without a phylogenetic perspective, we make the error of assuming that all species represent statistically independent data points in our analysis, even though we know that the more closely related ones are more likely to share common features and common past environments (see the example in Box 1.1).

Taking a phylogenetic perspective also helps us to avoid two common errors made in comparative physiology. Firstly, knowledge of phylogeny may reveal that we have less support for a relationship than we thought (Fig. 1.2). Secondly, it may show that there are significant associations present, where a naive analysis would miss them (Fig. 1.3).

So, in addition to physiological data, a phylogenetic comparative analysis requires an accurate phylogeny to work with. For a growing diversity of animal groups reasonably reliable phylogenetic trees have been generated, increasingly based on DNA sequence data. Using outgroup comparison we can than infer the phenotypes of ancestors for each radiation— the branch points or 'nodes' on the phylogenetic tree. Once we have worked out the characters at the 'nodes', then the magnitude and direction of changes that have occurred along each branch segment of the phylogenetic tree can be calculated. If we also have independent information on divergence times, available traditionally from the fossil record or more recently by applying the (still debated) idea of a molecular clock to the sequence analyses, then the rates of evolution of characters can also be studied. This approach also permits statistical testing for correlations in the changes of two or more characters, which may allow us to see how the sequence of changes that occurs during the evolution of a complex character may predispose some other trait to change

in a particular direction. If associations between particular characters and particular environmental factors are revealed, then justified and sensible inferences about adaptation are possible.

It must be noted, however, that this is an idealized programme. For the majority of animals we do not yet have broadly agreed phylogenetic trees, nor is there a complete consensus on how these can be achieved. An incorrect phylogeny can lend statistical support to bogus relationships or can obscure genuine ones; thus the comparative method can also appear to support physiological correlations that may yet turn out to be wrong. Phylogenies based on morphological data are particularly prone to errors resulting from convergent evolution, and this may be rife in animals that are adapting to similar environmental constraints, where physiological convergence is especially likely and may be especially revealing from a functional viewpoint. Thus comparative evolutionary physiology has some substantial hurdles to overcome; and the core of our knowledge of physiological function must not be underrated just because it does not measure up to 'phylogenetic correctness'. Nevertheless understanding evolutionary ecophysiology at this kind of level is certainly an outcome much to be desired, and the limitations of more traditional and 'anecdotal' species comparisons should be borne in mind throughout the reading of this book.

1.5 Physiological response on different scales

1.5.1 *Different timescales*

We considered the nature of adaptation in section 1.3 above, but we also need here to distinguish it from other kinds of physiological response: acclimation, acclimatization, and acute adjustments such as changing heart rate. The process of adaptation is usually a long and slow one occurring over generations, and is rarely reversible. In contrast, **acclimatization** is a more rapid phenomenon whereby a physiological or biochemical change occurs within the life of an individual animal, resulting from exposure to new conditions in the animal's environment. Thus migration up a mountain may lead to acclimatization to low oxygen and low pressure; movement southwards out of Arctic areas may force acclimatization to warmer temperatures; and year-long survival in one place may require acclimatization to alternating summer and winter conditions. The term **acclimation** is normally used for similar processes occurring in the laboratory, in response to experimentally imposed changes in conditions. Both acclimatization and acclimation may be reversible. Thus a polar bear is presumed to be *adapted* to polar temperatures; a human polar explorer may become (at least partly) *acclimatized* to them; and a laboratory mouse may be forced to *acclimate* to them. The human and the mouse are both likely to revert gradually to 'normal' when conditions change again.

In practice the terms acclimatization and acclimation are often used interchangeably in modern literature, especially since the term 'acclimatory' is a useful shorthand for a short-term non-genotypic response that has no equivalent derived from the

(a)

Tree unknown. Inference: apparent correlation of Hb presence and inshore life (apparently six independent data points). *But* these species may be linked by evolutionary history and therefore not independent. The inference is untrustworthy

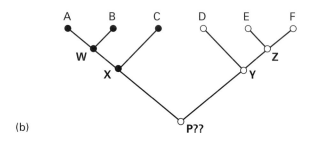

(b)

Tree known. ABC form one group, DEF another. Ancestral species WXYZ are included, but their character states cannot be measured directly, though they may be reconstructed (e.g. using parsimony which assumes that changes in state are rare and minimizes the number of transitions required to give the observed pattern). Here W and X are reconstructed as deep-sea/Hb⁻, and Y and Z as inshore/Hb⁺. This means that the six data points only give *one* real comparison, and no real evidence for a link between Hb and environment. Also since ancestor P is uncertain we do not know which state is ancestral and which is derived

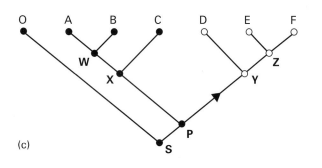

(c)

Tree known *and* condition of an outgroup (O) is known. Relationships and ancestral states as in (b) but now the inference (by 'outgroup comparison') is that deep-sea/Hb⁻ represents the ancestral condition (as in ABC and O). However we still have only one comparison (ABC against DEF) and no real evidence for a link between Hb and environment

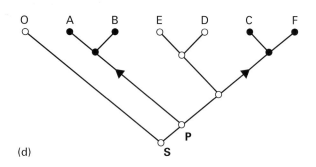

(d)

Tree known and outgroup known, but now suggesting that ancestors S and P were inshore/Hb⁺. Now there are *two* independent transitions from this ancestral state to the derived deep-sea/Hb⁻ state. This strengthens the inference that Hb and environment are causally linked. Including more species in the phylogeny, with more independent transitions, may eventually give statistical support for such a link

Fig. 1.2 The effects of phylogenetic trees on possible interpretations of a single character distribution (presence or absence of haemoglobin, Hb), showing the utility of outgroup comparisons. Solid circles show deep-sea taxa with Hb absent; open circles show inshore taxa with Hb present. Triangles show the sites of changes of state.

word 'acclimatization'. Thus we will often speak of acclimatory responses in animals within their natural environments.

Short-term changes

Very short-term changes in physiological state, such as an increase in heart rate or ventilation or urine flow, are usually acute responses following some behavioural effect such as exercise or a bout of feeding. On a slightly longer timescale, it follows from the discussion above that responses to environmental change occurring within hours or days or weeks are normally acclimatization rather than adaptation. But changes occurring regularly and repeatably, on a seasonal, monthly, or daily basis, are also often acclimatizations rather than adaptations. Remember, though, that these changes in phenotype may be underpinned by genotypic change, in the sense of differential expression of particular genes, even if they are not brought about by any irreversible or heritable change in the genotype itself.

Developmental effects

Embryonic, larval and juvenile phases in a life cycle may occupy

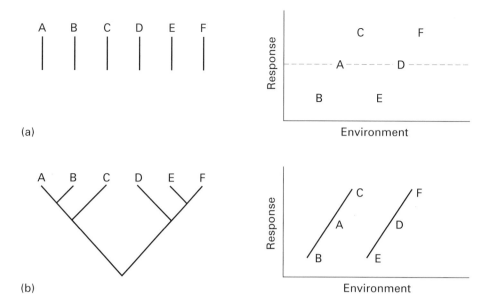

Fig. 1.3 The importance of understanding phylogenetic relationships in making inferences about adaptation and environmental effects on continuous variables. In (a) six species are considered as if they are independent entities; there appears to be no consistent effect of environment. In (b) the phylogenetic tree is known showing that ABC form one radiation and DEF another. Now we can see that *within each radiation* there is a trend for the physiological response to increase in relation to position on the environmental axis. Given more radiations showing similar patterns we could reasonably analyse the effect of the environmental parameter in bringing about a specific adaptive response. (From Huey 1987; courtesy of Cambridge University Press.)

very different environments from adults, and may have very different environmental responses. Such changes of phenotype during development and ontogeny operate on a slightly longer timescale and are more permanent than the acclimatory responses dealt with above. They are of course ubiquitous, and are commonly linked to changes in the environment of the embryo and/or the juvenile that affect its form and function. For example, it is possible to raise littermates of desert rodents either without drinking water or with free access to it; the drought-raised adults have increased relative medullary thickness (the ratio of the cortex to the medulla within the kidney, where it is the medulla that generates the urine-concentrating mechanism) and significantly higher urine concentrations. This kind of phenomenon is usually termed **phenotypic** or **developmental plasticity**, and again operates through differential gene expression.

The environment of an organism interacts with its developmental programme to play a major role in determining the expressed phenotype (see Fig. 1.1). A whole range of environmental factors can act upon development, from the first meiotic division through to the later assumption of juvenile and then adult form. These factors may be abiotic, such as temperature, pressure, pH, humidity, salinity and photoperiod. But there are also biotic factors acting, and these may be external, such as resource availability or population density, or internal, such as hormones. Many of these factors may interact to affect the phenotype produced, usually in a highly complex fashion (Fig. 1.4).

Environmental cues most conspicuously work by activating switches in the patterns of gene expression that make up the developmental programme. This may lead, for example, to the initiation of a developmental arrest, or to the production of alternative phenotypes differing in morphological features, behaviours and life-history characters. Diapause in insects (a resting stage permitting survival through a regularly recurring season

of adverse conditions) provides a good example of a triggered developmental arrest; many other animals in more hostile environments also show periods of arrested development. A striking example of a developmental switch to produce alternate phenotypes is temperature-dependent sex determination (TSD) in reptiles, dealt with in some detail in Chapter 13. Species with TSD are generally found in thermally patchy environments, which allows for the production of both sexes. The sex of the embryo is determined by the cumulative effects of the nest temperature in the period from shortly after egg laying through the first half of embryonic development.

Phenotype variation also arises due to the effects of environmental factors on the rates and degrees of expression of the developmental programme. In contrast to developmental switches, which as we have seen produce discontinuous phenotypes,

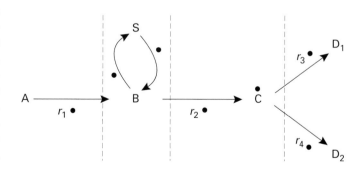

Fig. 1.4 Developmental plasticity and its interaction with the environment. The normal developmental programme of a hypothetical animal runs irreversibly through immature stages A→B→C, and then may switch to alternative adult phenotypes D_1 or D_2. The environment (●) acts at several stages. Firstly, it controls the rates of transition between stages (r_1–r_4). Secondly, it determines whether an optional developmental arrest (S) occurs at stage B. Thirdly, it determines which of the adult phenotypes is produced. (Adapted from Smith-Gill 1983.)

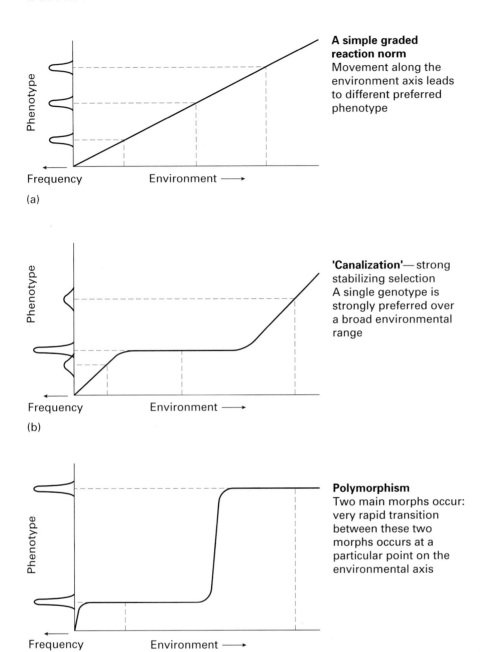

A simple graded reaction norm
Movement along the environment axis leads to different preferred phenotype

(a)

'Canalization'— strong stabilizing selection
A single genotype is strongly preferred over a broad environmental range

(b)

Polymorphism
Two main morphs occur: very rapid transition between these two morphs occurs at a particular point on the environmental axis

(c)

Fig. 1.5 'Reaction norms' as a way of understanding the effects of environment on phenotype.

environmental effects on the rate of development produce a continuum of phenotypes. This kind of phenotypic plasticity has been called 'continuous lability', or 'phenotypic modulation'. The effects of environment on development vary between traits and at different embryological stages. For example, temperatures close to the upper and lower thermal tolerance limits of a species in the first quarter of the developmental period produce a high probability of abnormal phenotypes, but later on may have little effect. Similarly, at any given point during embryogenesis the organs and tissues will have reached different points in their developmental programmes and may therefore exhibit somewhat different environmental sensitivities. In some cases, differential sensitivities of tissues manifest as a change in the relative timing of growth and maturation of different parts of the body, and this is referred to as 'developmental heterochrony'.

The profile of phenotypes produced by a particular genotype in different environments is usually called the 'norm of reaction' or **reaction norm** (Fig. 1.5), and this can clearly be of very variable shape. Morphogenetic processes which show little or no developmental plasticity are said to be 'canalized', i.e. they are fixed at a given level by strong stabilizing selection and produce a very narrow range of phenotypes. Characters that show discontinuous plasticity (and thus appear polymorphic) might be explained by a continuous underlying response to the environment at the cellular or genetic level which has a distinct threshold for phenotypic expression; in this case, the reaction norm would have steeply curved S-shaped regions (Fig. 1.5c).

There is some indication that the phenotypic variation induced by variable environments can be subject to strong selection, i.e. that there may be selection for plasticity itself as an adaptation. Perhaps the most compelling evidence for this comes from studies of temperature effects in amphibian life cycles. Field and laboratory experiments have shown that temperature can account for most of the observed variation in growth and differentiation in populations of the frog *Rana clamitans* along an altitudinal gradient. Growth in montane frog populations has a reduced temperature sensitivity relative to lowland populations, which serves to offset the effect of low temperature. Lowland frogs taken to high altitude have their growth retarded by low temperature to such an extent that the tadpole stage is extended by a whole season relative to the resident montane tadpoles. Such transplant experiments indicate that the phenotypic modulation of this particular trait is under genetic control.

Longer term genotypic/evolutionary effects

Timescales measured in generations are the most important in terms of adaptive effects, and perhaps the only timescales where we should strictly use the term adaptation. Natural selection acts on phenotypes, regardless of their genetic basis, so there can be selection for the results of phenotypic plasticity, as well as for the results of genes producing discrete phenotypes. But the evolutionary response to selection is always at the level of the genotype, with genetic change occurring from one generation to the next. Therefore intraspecific genetic variability and hence heritable variation in fitness are essential prerequisites for long-term evolutionary change. We take a more detailed look at the mechanisms of adaptation at this level in Chapter 2.

1.5.2 Different functional levels

Avoidance, conformity and regulation

When an animal is confronted with changes in its environment, it normally shows one of three categories of response: avoidance, conformity or regulation. Traditionally physiology has largely concerned itself with the last of these, dissecting the mechanisms and the underlying biochemistry of regulation of cells, tissues and the whole body. Thus **homeostasis** (the maintenance of a constant internal environment) always takes an extremely prominent place in physiology textbooks, together with regulatory systems to achieve it. But homeostasis can often be achieved more cheaply by avoidance and behavioural tricks; and it may not need to be achieved at all, with many animals

surviving and flourishing with a conforming lifestyle, involving much less energy and resource expenditure. The distinction of these different kinds of adaptive and acclimatory responses constitutes a key message early on in this book. In the context of environmental adaptation it is appropriate to treat the other levels of response as being at least as important, and homeostasis merely as one of several options.

The strategies accompanying these three responses may be summarized as follows:

1 **Avoiders** have some mechanism for getting away from an environmental problem either in space (e.g. seeking unstressed microhabitats in crevices or burrows, or larger-scale migration) or in time (using torpor or diapause, or producing a resistant egg, pupa or cyst to survive difficult times).

2 **Conformers** undergo changes of internal state similar to the changes of state imposed externally. (They are therefore sometimes termed 'tolerators', although this is a little confusing since regulating animals are also 'tolerating' the external conditions in the sense that they are surviving in them.) Conformers do not attempt to maintain a homeostatic condition for the whole body.

3 **Regulators** maintain some or all of the components of their internal environment close to the original or 'normal' level, irrespective of external conditions.

Figure 1.6 compares the general patterns of conformity and regulation. This kind of analysis may apply to a whole range of environmental and internal state variables, including temperature, osmotic concentration, oxygen levels and pH. Thus we can identify animals that are ionoconformers, osmoconformers, thermoconformers or oxyconformers, and plot similar kinds of diagrams to those in Fig. 1.6(a), and other animals that are osmoregulators, etc. as in Fig. 1.6(b). There will be many such examples later in this book, with hyper-regulation being rather common and hypo-regulation rather rare for most of the key variables.

However, it should be stressed that these categories are not absolute and do merge into each other; commonly we find that there are limits to both regulating and conforming (Fig. 1.7) and that there are no such things as perfect regulators or perfect conformers. For example, osmoconforming animals do tend to show some regulation at extremely low salinities, so that their blood is not so dilute that the cells are irreversibly damaged by excessive swelling; while osmoregulators sometimes lose an ability to regulate at lower salinities and become conformers. Oxyregulators tend to have a critical (species-specific) point beyond which they revert to conforming. Thermoregulators may lose an ability to control their body temperature at both high and low ambient temperatures. Particular species may regulate for some parameters and conform for others.

Behaviour, physiology, biochemistry and morphology

There are four major different but interacting functional levels at which avoidance, conformity or regulation can be effected

(a) External environment (E)

(b) External environment (E)

Fig. 1.6 The relation between the internal environment of an animal and the external conditions; basic principles of (a) conforming (I = E) and (b) regulating (I = constant).

(Box 1.2), corresponding to the four subdivisions of phenotypic traits in Fig. 1.1. These levels are often seen as being hierarchical, though from a modern perspective, in which all changes are viewed as being fundamentally mediated by genes and biochemistry, this hierarchy is somewhat artificial.

Avoidance in space is primarily an attribute brought about by behaviour. For a small animal it might involve a search for an appropriate habitat, using phototactic or chemotactic responses. The 'microhabitat' effect is central here, animals making choices in favour of less stressful local conditions within an apparently harsh macroenvironment. For a larger migratory species the behavioural component might need to be sup-

plemented with physiological adjustments, for example accumulation of food reserves. Avoidance in time may require more complex responses at all levels: an animal entering hibernation (torpor) may accumulate food, construct or find a refuge (thus also avoiding in space), then huddle in a ball to reduce its exposed surface area; it may reduce its core temperature and lower its metabolic rate; it may acquire a thicker insulating layer; and it may mobilize or generate new forms of enzymes and new components in its cell membranes.

Conforming is largely concerned with change at the physiological and biochemical levels. If the internal conditions are allowed to vary markedly, whether in terms of temperature or salinity or oxygen supply, then tissues and cells will need to have biochemical systems in place that can continue to function in the new conditions, especially in terms of appropriate enzymes and stabilized membranes. Just enough must be done to keep the animal functional (though usually at a very low level) in extreme conditions, avoiding potentially irreversible damaging effects of freezing or hypoxia or osmotic water loss (and also the damaging effects of coming out of these states,

Fig. 1.7 Conformity and regulation in the real world; a variety of options for partial conforming and partial regulating.

External environment (E)

- - - - -	'Conformer', but some regulation at extreme low E
··········	'Regulator', but less efficient at extremes
- · - · - ·	Typical 'partial' regulator, conforming in relatively normal conditions but regulating as conditions get more difficult
- - - - -	Essentially a conformer (parallel to E = I line), but internal environment has constant excess of measured variable
— — — -	Regulator but unable to survive too much change (starts to conform and then dies)
——————	Mixed conformer/regulator: regulates (approximately) above some species-specific level

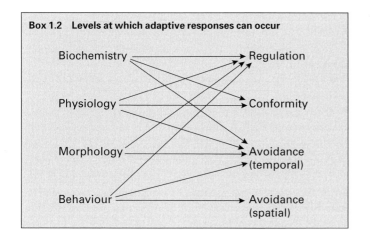

Box 1.2   Levels at which adaptive responses can occur

Biochemistry → Regulation

Physiology → Conformity

Morphology → Avoidance (temporal)

Behaviour → Avoidance (spatial)

since, for example, reoxygenation has hazards of its own). In general the physiological and biochemical changes will be small and cheap to institute, so that conformity (while having the cost of reduced activity or growth or reproduction) may have the benefit of economy.

Regulating, by contrast, may require substantial and expensive change at all these hierarchical levels. Behaviour, as we shall see repeatedly throughout Section C of this book, may remain as the first means of coping; even the best mammalian endotherms continue to use a whole range of behaviours (basking, burrowing, wallowing, huddling, erecting or concealing appendages, etc.) to regulate their temperature. But behaviour will be augmented by substantial physiological and biochemical adjustments; in the case of thermoregulation these may include changes of blood flow or respiratory rate, an increase in shivering or non-shivering thermogenesis, or the production of heat-shock proteins or antifreeze molecules.

It is impossible to give unconditional generalizations about patterns of avoidance, tolerance and regulation across the animal kingdom. But to a first approximation it would be fair to say that different strategies do tend to be found in different phyla, with different body designs, and in different habitats, as follows:

1 Smaller and soft-bodied animals are more likely to be avoiders and conformers. They can use microhabitats more effectively, with concealment in protected crevices or burrows, or on and in other organisms. Because they have a high surface area to volume ratio, they will experience relatively rapid fluxes (whether the flux is of water, ions, thermal energy or respiratory gas) across their surfaces, and working to restore a status quo against these fluxes would be very expensive. They also have little inbuilt protection against swelling and shrinking, and lack the complex outer layers that can be modified to give some insulation or impermeability. In predictably variable habitats, such as estuaries, cyclical avoidance is common (burrowing in mud, hiding in a crevice) and conforming is also often a good solution, with the cyclical environmental changes somewhat smoothed out. But in terrestrial habitats where there is both continuous high environmental stress and high fluctuation,

conforming may not be an option and exceptional strategies for avoidance (torpor, aestivation, encystment, cryptobiosis, etc.) are more common.

2 Animals with hard outer layers (exoskeletons), of small and medium size, may have better options for some regulation and a greater independence of their environments. Arthropods of all kinds are more likely to show partial regulation of osmotic concentrations, their exoskeletons giving them an inevitable 'built-in' resistance to shrinkage or swelling. The outer surfaces can have very much reduced permeability, and may be partly thermally insulated by the addition of fine cuticular hairs, so that all fluxes are slowed and a degree of regulation becomes economically feasible. But behavioural avoidance, aided by the efficient limbs (and sometimes wings) that can be built from an exoskeleton, remains a major part of the overall strategy for coping with environmental change, especially in the more rapidly changing terrestrial habitats.

3 Large animals are much more likely to be regulators in all environments, with the important exception of the relatively equable and unchanging open oceans (where only vertebrates have a serious problem and need to regulate, due to their history of secondary invasion as discussed in Chapter 9). Larger animals operate in a larger scale (coarse-grained) environment, where rapid changes (due to local water or air movements, or localized patches of sunshine) are relatively unimportant. They have lower surface area to volume ratios so that rates of change of state internally are much slower, giving them an 'inertia' effect that smoothes out the fluctuations and gives time for regulatory mechanisms to operate. They may have better opportunities for energy storage (and indeed storage of other resources such as water and even thermal energy). They may also have 'room' for more complex internal regulatory centres, both neural and hormonal. In terrestrial habitats where environmental changes are inherently faster, all of these factors may work together to make regulation the only real option for a large animal. Again, though, remember that regulation does not just mean physiological and biochemical effects; behaviour often still forms the first line of regulatory response.

There are very different costs and benefits of each strategy in terms of energy usage and lifestyle. Avoidance by shutting down in time is cheap but effectively causes the animal to opt out of the race for a while and achieve no growth or reproductive output. Avoidance in space by migration may be transiently very expensive but allows the animal to keep on growing and potentially reproducing in another environment. Avoidance of poor physical environments by either means may give additional benefits, for example by also avoiding predation or competition. Conformity at the extremes of temperature, salinity or hypoxia that are experienced may allow only a minimal 'ticking over' lifestyle, but over a broad range of less extreme variation it is a cheap way of ensuring a reasonably productive lifestyle most of the time. Regulation is usually rather expensive; osmotic regulation underpinned by ionic pumping takes a moderate proportion of the total energy budget of estuarine and fresh-

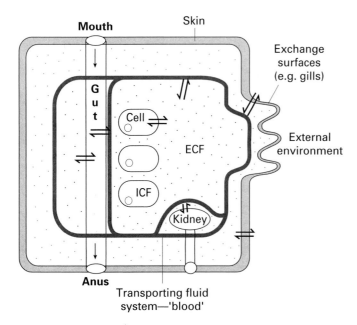

Fig. 1.8 A model of the major body compartments in an animal and exchange routes between them. ECF, extracellular fluid; ICF, intracellular fluid.

water animals, while thermal regulation in terrestrial endotherms may take as much as 90% of the total budget. The pay-off comes primarily in level of performance and greatly extended activity periods, at times and places of environmental adversity; food can be gathered more or less continuously, and all the avoiders and conformers that are relatively inactive become potential prey. With all this extra food, regulators can grow and reproduce faster and/or more reliably and be such successful competitors that despite high costs they may become dominant in many ecosystems.

1.5.3 *Different spatial levels*

Adaptive responses may occur fundamentally at a molecular level, but they 'appear' at various different spatial levels in the whole animal. Some responses are essentially subcellular, and others affect the morphology or activity of whole cells. Yet others manifest as effects on entire tissues or organs, for example changes in muscle size, heart volume or arrangements of vascularization.

However, there is another sense in which different spatial scales of response are important. Animals are made up of several distinct compartments (Fig. 1.8), each of which may show different kinds of adaptive/acclimatory response. The individual cells contain their own fluid environment (**intracellular fluid**; ICF), and they in turn are directly bathed in tissue fluid or **extracellular fluid** (ECF). In many animals the ECF is distinct from a circulating fluid termed **blood** or **haemolymph**, and this may be chemically different from the ECF. (In fact, the relations between these fluids and animal design and body cavities are

quite complex and there are other kinds of possible arrangement; see Box 1.3. But this simple version of a three-compartment system is adequate for the kinds of animals we are normally concerned with.)

In terms of the whole animal, adaptations may therefore occur at several sites:

1 *At the outside surfaces* to maintain differences between the outside world and the circulating blood. Here the adaptive features tend to relate to the 'skin', whether this is a relatively unspecialized epidermal cell layer or a complex multilayered structure with keratinous, chitinous or lipid-containing elements. Sometimes the adaptations relating to exchange processes will be concentrated in or even confined to particular parts of the skin, such as gill surfaces, with other areas of skin relatively inert and impermeable.

2 *Between the circulating fluid and the ECF.* This mainly occurs in vertebrates, where some constituents of blood pass out from capillaries into the ECF and others may be returned to the blood system via the lymph vessels.

3 *Between the ECF and the cells.* Here the adaptive surface is the cell membrane itself, controlling exchanges between ECF and ICF. Total concentrations inside cells must be similar to those outside or the cell will shrink or swell osmotically; cell membranes cannot withstand substantial pressure differences. However, the exact makeup of the cell fluid is very different from that of the ECF, as we shall see in Chapter 4.

4 *Within cells.* Cells themselves are of course strongly compartmentalized and internal membranes may be responsible for regulating exchanges between the cytoplasm and the nucleus, mitochondria or endoplasmic reticulum.

Adaptation to the environment may require modification at all these spatial levels. For many invertebrate marine animals the ECF and blood are effectively identical with sea water and adaptation concerns only levels 3 and 4. But for non-marine animals there is always extensive regulation both at level 1, i.e. the skin of the whole animal, and at the cellular levels.

1.6 Conclusions

Environmental adaptation is a complicated business, integrating all aspects of animal biology. It requires an understanding of animal design and animal physiology above all, but this must be put in context with a detailed understanding of the environment (measured on a suitable temporal and spatial scale), and with an appreciation of ecological and evolutionary mechanisms. There is a need to move away from purely descriptive accounts and over-simplified comparisons, and to put the encyclopedic knowledge already accumulated about what animals can do into a realistic framework of why and how they came to be as they are. Equally there is a need to look beyond the confines of traditional isolated physiological 'systems' (circulation, excretion, respiration, etc.) and see the whole picture of what is needed in order to live in a particular environment; the physiological needs of course, but also the mechanical, sensory,

Box 1.3   Patterns of body fluids and body cavities

(a)   (b)   (c)   (d)

⋰ ECF	▭ Mesodermal tissues	⋎⫽ Blood vessels
⊔⊔ Ectoderm and endoderm (skin, gut)	⋯⋯ Peritoneum	□ 'Body cavity', pseudocoelom, haemocoel or coelom

(a) The extracellular fluid (ECF) bathes all cells directly, sometimes opening out into large cavities or sinuses in some parts of the body. These are technically 'acoelomate' or 'pseudocoelomate' animals, lacking the secondary body cavity of 'higher' animals.
Common—found in platyhelminths, nematodes, other small worms.

(b) The ECF between cells and within tissues may be continuous with a larger open space that forms a body cavity (haemocoel). The fluid in this cavity may be constrained into vessels in some parts of the body, forming a heart and some major arteries, but always opens again into the main cavity; the fluid is often termed haemolymph, and it is normally the same in composition as the ECF. (A coelomic cavity may also be present but is usually small and unimportant.)
Common—found in crustaceans, insects, arachnids, most molluscs.

(c) The ECF is quite distinctly separated from the body cavity fluid (by a complex tissue layer called the peritoneum) and may differ from it chemically. Here the body cavity is usually termed a coelom and the fluid within it is coelomic fluid; the animals are coelomate.
Uncommon—found in sipunculans and other small worm phyla, echinoderms (but in these some coelomic fluid is also moved around in vessels).

(d) The arrangement is coelomate as in (c) but the coelomic fluid and ECF are supplemented with another fluid, contained entirely within vessels, and termed blood. Some leakage from the blood into the ECF occurs and this may become lymph and be recycled back into the blood system.
Common—found in annelids, cephalopod molluscs, all vertebrates.

---

reproductive and life-history adaptations that together make up a successful fully functional animal.

Further reading

Books

Feder, M.E., Bennett, A.F., Burggren, W.W. & Huey, R.B. (1987) *New Directions in Ecological Physiology*. Cambridge University Press, Cambridge.

Harvey, P.H. & Pagel, M.D. (1992) *The Comparative Method in Evolutionary Biology*. Oxford University Press, Oxford.

Louw, G.N. (1993) *Physiological Animal Ecology*. Longman, Harlow.

Sibley, R.M. & Calow, P. (1986) *Physiological Ecology of Animals: an Evolutionary Approach*. Blackwell Scientific Publications, Oxford.

Withers, P.C. (1992) *Comparative Animal Physiology*. Saunders College Publishing.

Reviews and scientific papers

Felsenstein, J. (1985) Phylogenies and the comparative method. *American Naturalist* **125**, 1–15.

Garland, T. & Carter, P.A. (1994) Evolutionary physiology. *Annual Review of Physiology* **56**, 579–621.

Harvey, P.H., Read, A.F. & Nee, S. (1995) Further remarks on the role of phylogeny in comparative ecology. *Journal of Ecology* **83**, 733–734.

Hochachka, P.W. (1998) Is evolutionary physiology useful to mechanistic physiology? The diving response in pinnipeds as a test case. *Zoology: Analysis of Complex Systems* **100**, 328–335.

2 *Fundamental Mechanisms of Adaptation*

2.1 Introduction: adaptation at the molecular and genome level

In Chapter 1 we considered adaptation in relation to selection on phenotypes, as determined by genes and their constituent DNA. But adaptation in practice is essentially a phenomenon associated with proteins. Proteins make up most of the dry matter of cells, and underlie all cellular processes; they form the structural components of the process, they catalyse and control the process, and they are responsible for manufacturing the non-protein components that carry out the process. DNA may store the information needed to make a cell, but it has little other function, and cannot 'do' anything for itself. RNA passes this same information to the cytoplasm outside the nucleus, and has some roles in cell regulation, but in essence its function is to make proteins. Proteins by contrast very obviously 'do things'; they serve as the main agents of molecular recognition, and are precision machines, with moving parts that perform and integrate chemical and mechanical functions. They can transduce motions across membranes, integrate several distinct signals into a single outcome, control rates of reaction of other molecules, and control gene expression itself.

Proteins are the product of the genetic code in the DNA of a cell, and sequence changes in a length of protein-coding DNA often (though not always, due to code redundancy) result in changes in amino-acid sequence. While DNA has very limited structural diversity and an essentially linear character, the resultant proteins have enormous diversity, being made up from assorted chains of 20 different amino acids each of which has its characteristic three-dimensional structure and thus its own biochemical 'personality'.

Genomes are regularly altered by point mutations, especially by single base transformations, deletions and duplications, but they are more substantially remodelled by recombination events, which may produce larger transpositions, deletions and duplications in sections of chromatids and hence changes in 'genetic architecture'. It is these recombinations that are more likely to give a change of physiological performance or of morphology at the cell and the organism level, and thus the phenotypic variation on which selection will act. Adaptation occurs when any change at the DNA level becomes expressed, via protein change, as a trait that is beneficial in a particular environment and so persists due to selective advantage, eventually spreading through a population.

Anything that controls proteins is therefore at the core of adaptation. It is conventional to look at proteins in a 'temporal sequence' of how they are made (transcription, translation), then how they are modified (post-translational changes), and then how their production is controlled by other factors. But since our main perspective here is environmental adaptation, we look first at how the functioning of proteins can be controlled in general terms to alter their function, and then at how proteins are made and destroyed in cells, how processing at the nucleic acid level affects them, and thus how proteins might change through evolutionary time. Then we consider how their functioning may also be controlled at a wider level by signals between cells and within whole organisms. Key phenomena that facilitate protein evolution and organismal adaptation are thus highlighted.

2.2 Controlling protein action

Potential protein activity in cells is set by a balance of synthesis and degradation, processes which are examined later, but achieved protein activity needs more precise spatial and temporal control. The most important system of control is the activity of **ligands**, small molecules that bind onto proteins and alter their conformation, having an **allosteric** effect. A simple example is the binding of glucose onto the enzyme hexokinase, which then facilitates the binding of ATP (resulting in the glucose being transformed into glucose 6-phosphate in the first step of glycolysis, dealt with in Chapter 6). We understand this interaction fairly clearly at the molecular level. The hexokinase consists of two separate but linked structural sections ('domains'—see section 2.3.1), and the binding sites for both glucose and ATP lie in the cleft between the two domains; when glucose binds onto its site, deep in the cleft, it helps to pull the two domains close together and so creates a more precise fit for the ATP site on the edge of the cleft (Fig. 2.1a). Binding of glucose causes a

Open

Domain 1

Glucose

Closed

ATP

Domain 2

(a)

Cooperative allosteric binding
• Binding of A facilitates binding
of B and vice versa

Competitive allosteric binding
• Binding of A inhibited by
presence of B and vice versa

Fig. 2.1 (a) An allosteric effect: binding of glucose between the two domains of hexokinase, allowing ATP to bind more easily. Glucose binding effectively changes the enzyme from an *open* to a *closed* conformation. (b) Models for two-ligand allosteric effects, either cooperative or competitive.

(b)

50-fold increase in the affinity of the hexokinase for ATP. In other proteins the sites at which two ligands bind may not be so close together but they can still affect each other's binding, and are said to be **allosterically coupled** (Fig. 2.1b). Proteins in which conformational changes can affect two widely separated binding sites are likely to have been selected for in evolution because they allow the cell to link together the fate of two molecules. Allosterically controlled proteins are therefore very widespread.

The link between the two ligands may be positive, as in the example of hexokinase. But it may also be negative; here the two ligands bind preferentially to two different conformations of the protein, thus one ligand may effectively 'turn on' or 'turn off' the enzyme in relation to the other ligand. These kinds of allosteric interaction are crucial in regulating metabolic pathways, because it is often the end-product of one pathway that becomes the regulatory ligand for an early step in the path, so providing negative feedback. Allosteric interactions can also be used to give a positive feedback, though, where the ligand is something that accumulates when the cell is short of an end-product of the pathway.

Regulation of enzymes by simple negative feedback may be rather imprecise, and in many cases it is refined by having enzymes made up of several different subunits where the regulatory sites are interactive. The binding of a ligand to a single site can then trigger an allosteric change in other subunits, helping them in their turn to bind the same ligand (Fig. 2.2); a relatively small change in ligand concentration can thus switch the whole protein assembly from inactive to fully active or vice versa. This complex on–off switching is understood in precise molecular detail for some enzymes, with the key amino acids in the polypeptide chains identified.

One of the commonest forms of allosteric control in animal cells is that of reversible **phosphorylation** of the protein. A phosphate group carries three negative charges (PO_4^{3-}) so can cause a structural change in a protein to which it attaches, perhaps by attracting one or several positively charged amino acid side chains. This can in turn alter the protein's configuration elsewhere, allosterically affecting the active binding sites. At any one time at least 10% of the proteins in a cell are phosphorylated, with the phosphate being added from an ATP molecule by a **protein kinase**, and removed by a **protein phosphatase**. These two types of enzyme are therefore key players in the regulation of cell functioning and physiology.

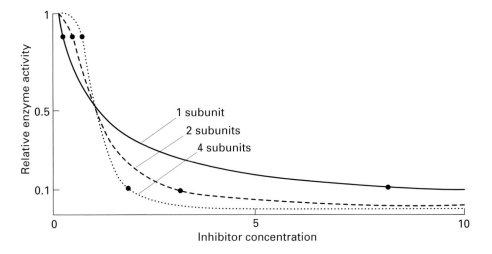

Fig. 2.2 The effects of inhibitory ligand binding on enzyme activity for monomeric and multimeric allosteric enzymes. With only one subunit, effects are relatively slow and high concentrations are needed for 90% inhibition; while for a tetrameric (four subunit) enzyme inhibition is faster and lower concentrations are needed for 90% inhibition. (From Alberts *et al.* 1994, Fig. 5.7.)

Protein kinases (PKs) all belong to a large protein family containing a similar stretch of about 250 amino acids (the 'kinase domain'). Every cell has hundreds of different kinases. There are two main groups: one group work by transferring a phosphate group from ATP onto the hydroxyl group of a serine or threonine side chain in a protein; the other group work by similarly transferring phosphate onto a tyrosine side chain. These reactions are not spontaneously reversible, because of the large amount of chemical energy released when ATP is converted to ADP; but they can be reversed when a protein phosphatase

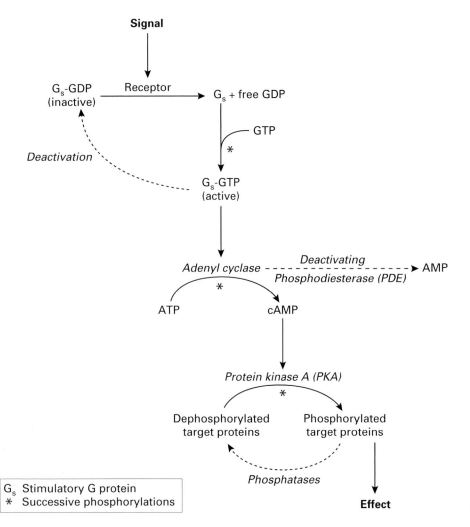

Fig. 2.3 A general model of G-protein cascade effects.

specifically removes the phosphate again, giving a fine-control system. Kinases work in teams, phosphorylating each other in integrated cascade systems where some steps amplify the preceding effect; Fig. 2.3 shows an example for G-protein activation (discussed below). Thus the cell contains a series of **switches**, each of which may itself integrate two or more signals. For example, there are 'cyclin-dependent protein kinases' that only work when another particular protein (cyclin) is present, *and* one specific phosphate binding site is occupied *and* another such site is empty (with phosphate having been removed by a phosphatase); three signals are being integrated. Kinases may also have an indirect switch effect by phosphorylating an inhibitory protein that normally binds to another protein, allowing the last of these to be unmasked and to become active.

There is another group of proteins that are important in regulating cell activity and are active in cascades of cellular protein phosphorylations; these are the **GTP-binding proteins** (or GTPases). The kinases discussed above achieved phosphorylation by catalysing the transfer of a phosphate group from ATP. The GTP-binding proteins instead incorporate the whole nucleotide, GTP, at a binding site within their structure (Fig. 2.3). When this GTP is hydrolysed to GDP the protein becomes inactive, and it is probably so most of the time; this inactivation is brought about by a GTPase-activating protein, which binds to the protein and forces the GTP to hydrolyse to GDP. The now inactive protein must then exchange its GDP for a GTP molecule to become active again, which requires the action of a guanine nucleotide releasing protein, causing release of the GDP. The space is immediately filled by a GTP as this is in excess over GDP in the cell cytoplasm, and the enzyme becomes activated.

This activating effect only occurs in response to various specific extracellular signals. Simple GTP-binding proteins have also been built up through evolution into large versions composed of many subunits (multimeric), the most famous of which are the **G proteins**, which are associated with receptors on cell membranes. These control important biological activities by the hydrolysis of GTP molecules, leading to changes of protein conformation; we will meet them again below.

Proteins that hydrolyse ATP or GTP can be used to perform unusually complex actions. Some of them become directional **motor proteins**, bringing about various kinds of movement. This is achieved because of the high free energy released in the hydrolysis reaction, so that it becomes effectively irreversible. Thus if a protein has three structural conformations and the transitions 1→2, 2→3 and 3→1 are all driven by ATP hydrolysis then this series of changes will proceed automatically, without reversing. This means that a protein undergoing small shape changes can be made to 'move along' the substrate it is attached to (Fig. 2.4). For example, a protein can move along a DNA strand in one direction (allowing 'reading' of the genetic code), or a myosin filament can move along an actin filament (producing muscle contraction); in the latter case the physical movement produced by one ATP-linked conformation change is about 5 nm.

Another complex action that ATP-driven proteins can per-

Fig. 2.4 A model of an allosteric 'motor protein'. The protein has three possible shapes (1, 2 and 3). ATP binding causes the shape change 1→2, and ATP hydrolysis causes the change 2→3, so that the two binding sites effectively 'walk' along a substrate. Release of ADP and phosphate cause the final change 3→1, completing one 'step' or 'stride'. Reversals of movement are normally precluded because one of the stages involves splitting ATP which is essentially energetically irreversible. The protein therefore moves continuously to the right along the substrate. (From Alberts *et al.* 1994, Fig. 5.22.)

form is that of pumping, most commonly pumping ions across a membrane, and these **pump proteins** are particularly important to physiologists. The most famous example is the sodium pump, or Na^+/K^+-ATPase, which normally (as we will see in detail in Chapter 4) moves three sodium ions out of the cell and two potassium ions into the cell for each cycle of conformational changes.

Membrane-bound allosteric pumps that are linked to ATP hydrolysis can also be made to work in reverse; they can employ an existing ionic gradient to synthesize ATP molecules from ADP plus phosphate. The energy from the hydrogen ion (H^+) gradient across the inner mitochondrial membrane is used by such an enzyme, called **ATP synthase**, to make most of the ATP that animals need (see section 6.4.2).

The ability of cells to achieve phosphorylations by harnessing the energy in nucleotide triphosphates (ATP or GTP) to drive allosteric changes in proteins has therefore been central to all evolutionary processes. It provides the fundamental machinery of cells, including switches, motors and transducers (energy converters). Proteins that function in this manner can be assembled into complex arrays that function as multistep machines. Tiny changes in protein structure, produced by alteration of a single key amino acid in the protein chain (and thus the result of as little as one base-pair change in the DNA), can produce greatly amplified effects at the functional level; thus small variation at the molecular level can be strongly selected for or against in the organism. In practice, many of the processes occurring in cells are so intimately linked to each other that there must be strong stabilizing selection to prevent change. However, any small genetic change that is non-lethal but produces protein conformation change can clearly have quite dramatic benefits on a much larger scale. We will meet below a number of reasons why such non-lethal or non-harmful changes are rather likely to occur.

2.3 Control of protein synthesis and degradation

We have seen how protein activity can be controlled by other intracellular molecules (often themselves proteins) that bind to them. However, the availability of proteins to be so controlled is in its turn determined by the more fundamental cellular processes of protein synthesis (from DNA via messenger RNA molecules) and of protein degradation. These too must be subject to elaborate control to produce an appropriate intracellular protein pool.

2.3.1 *Protein synthesis and morphology*

The basic process by which DNA structure is repeatedly transformed into protein sequences, via DNA replication, transcription of DNA into RNA, and translation of the RNA into protein (Fig. 2.5), does not need to be rehearsed here, having become the starting point for nearly all of modern biology. The nucleotide sequence of a gene determines the amino acid sequence of the corresponding protein, via the classic codon–anticodon system in which a triplet of consecutive bases encodes a single amino acid. Remember, though, that DNA is transcribed into single-stranded RNA only in short sections; some of the resultant pieces (the messenger RNAs, or mRNAs) are the protein coders, corresponding to just one or a few proteins, while others are transfer RNAs or ribosomal RNAs that will take part in the translational process.

Proteins once synthesized fold up into their correct configuration. The polypeptide chain automatically forms a complex three-dimensional shape determined by the amino acid side branches and their tendency to form weak bonds with each other. In general the hydrophobic side chains tend to be pushed to the interior of the protein molecule and the hydrophilic side

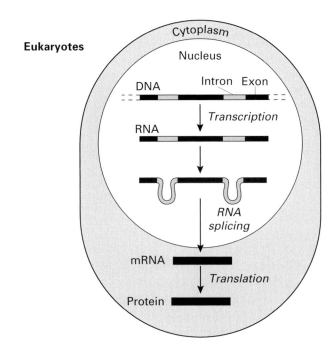

Fig. 2.5 Transfer of information from DNA in the nucleus to protein in the cytoplasm via transcription, splicing and translation. See text for more details. (From Alberts *et al.* 1994, Fig. 3.15.)

chains protrude outwards. This folding is partly controlled by special proteins termed 'molecular chaperones' whose main function is to facilitate protein folding (we will meet these again in their role as the 'heat-shock proteins' in Chapter 8).

It has become clear that there are several structural patterns of folding that occur repeatedly. Two are particularly common: the α-helix and the β-sheet. The α-helix often forms short lengths of protein that cross membranes; the configuration is not very stable in aqueous environments, though two helices may coil together (the 'coiled coil') to give a particularly stable structure common in fibrous proteins (e.g. collagen, keratin). The β-sheet folding system may form large parts of the framework of globular proteins (e.g. lysozyme).

At the next level of complexity, there are only a limited number of ways of combining the α-helices and β-sheets, so certain patterns are commonly repeated and are known as **motifs**. An example is the 'hairpin beta motif', consisting of two antiparallel β-sheets joined at a sharp angle by a loop of polypeptide chain. Combinations of motifs of α-helices and β-sheets may together make up compactly folded globular units within proteins that are known as **domains**, usually containing 50–300 amino acids. A typical domain is therefore made up of a compact core of motifs plus some uniquely shaped linking sections, with the surface covered by protruding loop regions; these loops, or the pits between them, commonly form the binding sites for other molecules.

The number of possible proteins that could theoretically be formed by the normal 20 amino acids transcribed in animals is

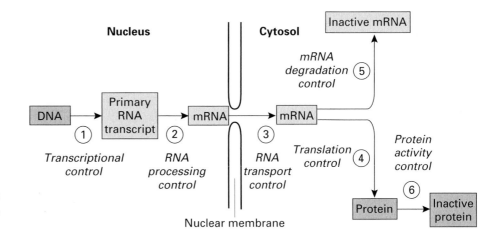

Fig. 2.6 The cluster of genes (*A–E*) in *E. coli* that code for enzymes involved in the manufacture of the amino acid tryptophan. All five are transcribed as a single mRNA molecule controlled by a single promoter. The cluster is called an operon. (From Alberts *et al.* 1994, Fig. 9.24.)

staggering (20^n, where n is the chain length; this gives a value of 160 000 for just four amino acids, and 10^{390} for a more typical chain of 300 amino acids). Yet very few of these possibilities exist naturally. This is probably largely because very few of them would be stable in three dimensions; many would form straggly molecules with many different possible conformations, which would be impossibly difficult to control. The proteins that are selected for in evolution are those that can reliably and repeatably fold up into stable forms, where small conformational changes can be controllably achieved to perform specific structural or catalytic functions.

2.3.2 *Controlling protein synthesis*

In any cell at any particular time some of the DNA is being used to make very large amounts of RNA and other parts are not being transcribed at all. Transcription is controlled by an **RNA polymerase** enzyme, which makes RNA copies of DNA molecules (though in fact the full 'transcription initiation complex' comprises the polymerase and at least seven other proteins). A polymerase appears to act by colliding randomly with DNA molecules but only sticking tightly when it contacts a specific region of DNA termed a **promoter**, which contains the 'start sequence' signalling where RNA synthesis should begin. It opens up the DNA double helix here and then moves along the DNA strand in the 5' to 3' direction, at any one time exposing just a short stretch of unpaired DNA nucleotides. Eventually it meets a stop signal, and releases both the DNA template and the new RNA chain. Figure 2.6 shows the organization of a promoter and

the five genes that it controls, all of them involved in a particular pathway (in this case tryptophan synthesis in the bacterium *Escherichia coli*). This arrangement means that a set of related genes pertinent to a particular metabolic pathway or morphological change can be controlled as one unit.

In animals there are three kinds of RNA polymerase, designated I, II and III, and each is responsible for transcribing different sets of genes. They are structurally similar, all being large multimeric proteins with masses in excess of 500 kDa. RNA polymerase II is responsible for transcribing genes into the RNAs that will make most of the cell's proteins.

Each active gene may make thousands of copies of mRNA in each cell cycle, and the resultant mRNA strand may translate into many thousands of polypeptide chains in the same period, so that amplification factors of millions may be involved. The amount of RNA made from a particular region of DNA must therefore be very tightly controlled, and this is partly determined by the goodness of match between the promoter region sequence and the RNA polymerase. Genes that produce large amounts of mRNA thus have strong promoters and genes that produce mRNA only in small amounts have weak promoters.

Further control of protein synthesis is needed, however, and there are at least five more stages at which this can occur (Fig. 2.7). The first of these, and for many genes the most crucial, is that of transcriptional control. This is the function of a special class of **gene regulatory proteins**, which bind to specific sites on the DNA close to the gene coding sequences. Their manufacture takes up several per cent of the capacity of the genome in higher animals. Their effect is to ensure that the right genes

Fig. 2.7 Six steps at which gene expression can be controlled. (From Alberts *et al.* 1994, Fig. 9.2.)

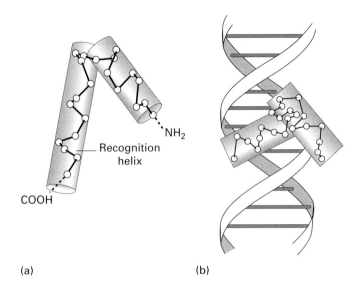

Fig. 2.8 (a) The 'helix-turn-helix' motif in a protein; and (b) the way in which the motif binds to the outside of the DNA double helix. (From Alberts *et al.* 1994, Fig. 9.19.)

are switched on in each cell type at the right time, leading to the differentiation between, for example, nerve cells, gland cells or epidermal cells. All these cells in any one animal contain the same DNA, and all of them have some proteins in common (key metabolic enzymes, cytoskeletal proteins, etc.). However, there are some proteins that are only ever expressed in one cell type (e.g. vertebrates express haemoglobin only in red blood cells), and many other intermediate protein types that have a differential intensity of expression between cell types.

The important fact in gene regulation by proteins is that the DNA helix can be 'read' by proteins from the outside, i.e. without splitting the double helix. There are structural patterns (of the hydrogen bonding sites and hydrophobic patches) inherent to the external grooving on the helix that are unique for each of the base pairs. Thus a protein can recognize and bind to a particular set of nucleotides. Very short DNA sequences, often only 6–15 base pairs long, act as genetic switches; for example the *Drosophila* regulatory protein *bicoid* recognizes the sequence GGGATTAGA (the matching bases are excluded here for simplicity). Thus specific sequences are read by particular motifs on the regulatory proteins. There appear to be rather few such motifs. The most common one is the 'helix-turn-helix' motif (two α-helices separated by a short amino acid chain, creating a fixed angle). The longer carboxyl-terminal α-helix of this motif fits neatly into the groove of one turn of the DNA helix (Fig. 2.8). The now famous 'homeodomain' proteins that serve as vital switches in the development of animals are a subset of the helix-turn-helix regulatory proteins, having one extra helix region added. Another common motif in gene regulatory proteins is the so-called 'zinc finger' motif, where areas of α-helix or β-sheet are linked by a zinc molecule. A third important type is the

'leucine zipper motif', differing in that it works via an interaction between α-helices on two separate monomers of the protein. There are several different monomers, each recognizing a different DNA binding site, and so able to interact with each other to give more sophisticated control (i.e. specific combinations of zipper proteins controlling the gene regulation process). As might be expected, the match required between DNA helix and regulatory protein is very precise; a protein once in place may be making up to 20 bonds of various kinds with the DNA helix structure.

How do regulatory proteins work? At their simplest they recognize a region within the promoter sequence of a DNA strand (see above and Fig. 2.6). This small section within the promoter is termed the **operator**. Promoter and operator regions are arranged so that when the operator is occupied by the regulatory protein the binding of RNA polymerase to the promoter is affected. One possibility is that the polymerase cannot then bind properly to the promoter so transcription cannot occur. This gives a negative control. In the case of the tryptophan synthesis system shown in Fig. 2.6, the regulatory protein is only able to bind to its operator site once it has already bound two molecules of tryptophan. Thus the whole system (one **operon**) can be switched on or off according to the level of free tryptophan in the cell; when there is plenty of tryptophan, a **repressor** protein is activated and therefore bound to the operator, so the RNA polymerase cannot bind to the promoter, and enzymes to produce more tryptophan will not be transcribed. A simple on-off switch with automatic negative feedback is created. Positive control systems are also possible. Gene regulatory proteins that are activating rather than inhibiting generally work by binding to operator regions in such a way that the RNA polymerase is contacted and its own binding to the promoter region improved. The two types of gene regulatory protein, having positive and negative effects, are usually termed **transcriptional activators** and **transcriptional repressors** (or gene activator proteins and gene repressor proteins), respectively. More complex switches can be produced by combining the two, and many genes probably have both repressor and activator proteins, which can be derived from more than one chromosome.

In animal cells additional complexity of control is achieved in a number of ways. One important mechanism is the existence of **general transcription factors**, which must be bound to the promoter site in a complex array before any transcription by RNA polymerase II can occur. Often, one of the components is a protein kinase, which phosphorylates the polymerase and allows it to disengage partly from the transcription factor and begin its own work of transcription. Another important regulatory system involves **enhancers**, sequences of DNA at some distance from the promoter region that can nevertheless influence transcription, seemingly by loops or folds occurring in the actual DNA to allow direct contact of the enhancer at the promoter site. Enhancer regulatory proteins are probably widespread in animals, and most of them exert positive rather than negative control. The 'regulatory complex' of transcription controllers that pertain to any

Fig. 2.9 A model of a single gene (*X*) and all its associated regulatory proteins and transcription factors. The promoter region is next to the gene, where the crucial RNA polymerase and the general transcription factors assemble. The promoter region always contains a short section of TATA repeats, recognized by one key transcription factor. More distantly there are several regions of DNA sequence (arrowed) that are recognized by gene regulatory proteins, usually working in clusters; these are separated by regions of spacer DNA. (From Alberts *et al.* 1994, Fig. 9.34.)

one particular gene can therefore be extremely intricate, and may be spread out over as much as 40 000–50 000 base pairs, including large areas of material that are no more than spacer (Fig. 2.9).

An individual gene regulatory protein may take part in more than one gene regulatory complex, and it might even be an activator in one situation and a repressor in another. Each functions as a regulatory unit that is used to assemble regulatory complexes in different places at different times. This gives an enormous range of possibilities for control of gene transcription. A major way of setting up positional information in animal embryos is by the uneven distribution of gene regulatory proteins along the length of the embryo, so that chromosomes in cells or nuclei at one end of the system are exposed to a different regulatory environment from those at the other end. Where the embryo is fully cellular (that is, in most animals other than insects, which are unusual in having a mosaic embryo with many nuclei sharing a common cytoplasm) setting up these gradients must also require regulation via messengers across cell boundaries, so that the regulatory proteins are themselves properly regulated. Some of the general ways in which proteins can be controlled from outside the cell, discussed in section 2.2 above, must then be operative.

At a yet higher level of organization, there are also some critical gene regulatory proteins (encoded by **master control genes**) that serve as key switches for a whole set of other regulator complexes across a range of cell types. In many cases such genes work via methylation of some part of the genome, or condensation of the chromatin, both of which limit access of the RNA transcribing enzymes. The homeobox genes are pre-eminently of this kind, determining large scale patterning of the developing animal body. Another example relates to the whole process of muscle cell differentiation, which is triggered by a group of 'myogenic regulatory factors' expressed strongly only in muscle cells and integrating the action of many other regulatory systems relating to muscle tissue structure and function. Genes that determine sex could also be included in this category, with a whole complex of characters related to expression of a key control gene.

One final point here is that regulation of gene expression can sometimes be passed on between cells through time as well as through space. Most obviously, once a cell has switched to

become a particular cell type, with the appropriate genes turned on and off, it passes on this suite of gene expression to its daughter cells through mitotic divisions. This **cell memory** allows the formation of tissues.

Clearly the control of gene expression can be extremely complicated, but it is based on simple principles and perhaps on relatively few molecules with complex interactions. Small changes in regulatory proteins are likely to have large effects, yet these will often be well-controlled effects, for example producing positional change without functional changes during development. One important way to get major adaptive change in animals at a physiological level is therefore via mutation in the DNA coding for these regulatory proteins. Another must be the possibility of spatially reshuffling the various regulatory centres on the genome adjacent to the actual gene, thus creating new regulatory circuitry and modifying existing controls.

2.3.3 *Controlling protein degradation*

Another crucial aspect of good cellular functioning is the ability to 'shut off' proteins completely. Intracellular proteolytic mechanisms exist in all animal cells; these mechanisms confer an automatic short half-life on certain normal proteins, as well as recognizing and eliminating abnormal proteins that have not folded up correctly, or disposing of proteins that have been damaged during their working life. Some proteins are degraded continuously, and others may be stable for some time but then get degraded rapidly when a particular point of the cell cycle is reached. Degradation may occur in the cytoplasm or in the endoplasmic reticulum, and in special cases it also occurs in lysosomes.

Cytoplasmic degradation is mainly the work of large protein complexes, sometimes given the special name **proteasomes**. Each of these is shaped like a hollow cylinder formed from many distinct protease enzymes with their active sites facing inwards. The entrance to the core of the cylinder is guarded by another large protein complex, which probably selects the proteins destined for destruction and lets them into the inner chamber. Then the proteases do their work and a multitude of short peptide chains are released.

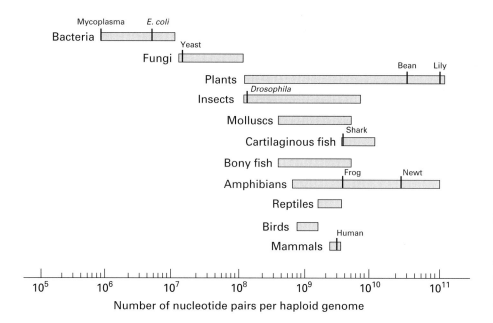

Fig. 2.10 The amount of DNA in the haploid genome of a range of organisms. Note that there is no particular relation with 'complexity' within the eukaryotes: certain amphibians and plants have notoriously large amounts of DNA. (From Alberts *et al.* 1994, Fig. 8.6.)

Proteins due for destruction are recognized by being tagged with a particular marker molecule, **ubiquitin**. This is a small protein that links covalently (thus rather firmly) to the destruction-bound proteins, and serves as a degradation signal. Linking it onto a target protein is achieved by a ubiquitin-conjugating enzyme, which presumably recognizes a signal on a misfolded protein. Recognition of 'normal' proteins destined for a short half-life is a more complicated matter, and seems to reside largely in the nature of the N-terminus of the peptide chain. Certain amino acids confer protection, while others (such as arginine, aspartate and glutamate) provoke proteolytic attack. The sequence of this region can thus be selected over evolutionary time to give an appropriate protein half-life.

2.3.4 *Control via nucleic acid processing systems*

DNA organization

Most of the chromosomal DNA in any animal cell does not code for RNA or for protein; it is 'excess DNA' and does not form genes. The size of the genome, and the proportion of it that codes for RNA, varies between animal groups, and even more widely between the kingdoms of living organisms (Fig. 2.10). The mammalian genome in principle contains enough DNA to make about 3 million average proteins, but probably only constructs about 60 000 proteins in reality. Much of the DNA lies between genes, and a large part of it consists of repeated nucleotide sequences; some of this is tandemly repeated 'satellite DNA' (with very large numbers of repeats of small sequences, perhaps only 2–10 nucleotides in length), and the rest is interspersed repeated DNA. Remember also that even within coded sequences there are introns that are spliced out and discarded after transcription (see below). Thus only about 1% of the DNA sequence

is transcribed into functional RNA sequences. A schematic view of the organization of a chromosome is shown in Fig. 2.11.

Another feature of DNA organization is the existence of **transposable elements** ('jumping genes', or transposons). These are short pieces of DNA (a few hundred to 10 000 nucleotide pairs) that have the remarkable property of being able to move around the genome, either as DNA or via an RNA intermediate. Some merely carry the information for transposition; these may accelerate mutation at the sites where they insert and exit, inducing localized base-pair mutation or some mispairing and thus sequence duplication; they can also multiply and spread within the genome, causing more substantial sequence shifts. Others carry both the instructions for transposition and some other gene as well; the best known cases are genes for antibiotic resistance in bacteria. Thus transposable elements may disproportionately affect the rate of evolutionary change.

DNA repair

DNA and RNA are complex molecules made from simple subunits, and change in DNA is the fundamental mechanism of evolution. Not surprisingly, cells have elaborate mechanisms to limit DNA damage and alteration and so eliminate a high proportion of mutations before they lead to any expression of 'damaged' and useless proteins. Damage arises quite independently of replication for a number of reasons; thermal degradation of purine bases is extremely common, while less frequent changes involve spontaneous deamination of cytosine to uracil, the effects of ultraviolet radiation, and the influence of certain metabolites such as oxygen radicals. Each day thousands of bases may be altered at random in any animal cell. But fewer than one in a thousand alterations causes a heritable mutation; all the rest of them are removed by repair mechanisms

Fig. 2.11 The organization of genes on a typical animal chromosome. The number of genes per chromosome is highly variable, as is the length of each and its exon–intron structure. (From Alberts *et al.* 1994, Fig. 8.7.)

(representing the most fundamental kind of 'stabilizing selection' to prevent inappropriate functional change).

Several different kinds of repair occur, each catalysed by a series of enzymes; all rely on the fact that there are two complementary DNA strands and a damaged strand will not align properly with its undamaged complementary copy. **DNA repair nucleases** recognize altered portions of DNA strands and hydrolyse the phosphodiester bonds that join a damaged nucleotide to its adjacent undamaged nucleotides. **DNA polymerase** then binds to the 3' end of the cut DNA and helps the synthesis of a new copy of the good strand. Finally **DNA ligase** seals the original gap in the damaged strand. Some or all of these repair enzymes can be induced in response to severe DNA damage, providing an important safety mechanism against environmental hazards.

The basic mechanisms of DNA repair are an essential component of the process of genetic recombination in meiosis, where chromosomes undergo crossing-over. Maternal and paternal sequences may be slightly different, so that in the crossing region there will be some mismatched base pairs. This minor problem can be corrected by the DNA repair machinery, giving rise to a major gene conversion where, for example, a copy of the paternal allele has been changed to a copy of the maternal allele (Fig. 2.12). However, where there is a particularly poor match between the recombining strands the pairing is usually aborted and no recombination occurs; excessive scrambling of the genome which might be detrimental to physiological function is thus averted.

RNA processing

Selectivity in synthesis of RNA from DNA occurs at two levels. Only a part of the DNA is transcribed to produce RNA at all; and

only a minor proportion of this RNA survives to be exported into the cytoplasm to take part in protein synthesis (see Fig. 2.6). We dealt with the first of these processes in considering control of protein synthesis, but we still need to consider the second. In eukaryotic cells there is the additional post-transcription stage of **RNA splicing** or editing, where some sections of the transcribed mRNA are removed and take no further part in the translation into protein. **Introns** are the regions of the DNA that are edited out from the primary transcript, and **exons** are the areas that are spliced back together to form the secondary transcript, which is then translated (often with one exon corresponding to one protein domain). RNA processing therefore involves the removal of long nucleotide sequences from the middle of RNA molecules. This RNA splicing of introns occurs in the nucleus out of reach of the ribosomes.

Numbers of introns are very variable (Table 2.1); there may be none at all in some genes, but as many as 75 introns occur within a few mammalian genes. Introns range in size from about 80 to more than 10 000 nucleotides. Their precise sequence is probably irrelevant and they may contain many accumulated mutations; they are sometimes described as 'junk' DNA. The only highly conserved sequences they possess are those required for their removal, at the 5' splice site and the 3' splice site. Splicing is achieved by small complexes of RNA and protein termed **small nuclear ribonucleoproteins**; a number of these together assemble into a **spliceosome**, and collectively they effect removal of the intron.

Single nucleotide changes may be enough to inactivate a splice site, and when this occurs the splicing mechanism may stop or start at the wrong point. Alternative RNA splicing patterns can produce different forms of a protein from the same gene, either at random or more commonly in a regulated pattern. This may involve production of a non-functional protein rather

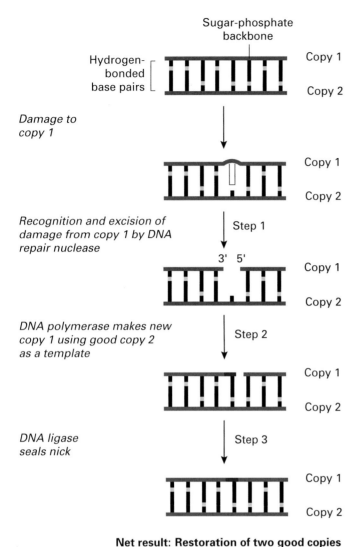

Sugar-phosphate backbone

Hydrogen-bonded base pairs

Copy 1
Copy 2

Damage to copy 1

Copy 1
Copy 2

Recognition and excision of damage from copy 1 by DNA repair nuclease — Step 1

3' 5'

Copy 1
Copy 2

DNA polymerase makes new copy 1 using good copy 2 as a template — Step 2

Copy 1
Copy 2

DNA ligase seals nick — Step 3

Copy 1
Copy 2

Net result: Restoration of two good copies

Fig. 2.12 The role of three kinds of DNA repair enzymes in recognizing and correcting damage (here a single point error) to the DNA double helix. (From Alberts *et al.* 1994, Fig. 6.66.)

Table 2.1 The sizes of genes before and after editing to mRNA, and the numbers of introns they may contain, from the human genome.

Gene	Gene nucleotides	Resultant mRNA nucleotides	Number of introns
β globin	1 500	600	2
Insulin	1 700	400	2
Protein kinase C	11 000	1 400	7
Albumin	25 000	2 100	14
Clotting factor VIII	186 000	9 000	25
Thyroglobulin	300 000	8 700	36

We should also note here that there are some systems where RNA is itself the controlling agent, particularly during development. Non-translated RNA (including in some instances antisense RNA strands and even mitochondrial ribosomal RNA) are known to play some role in particular aspects of patterning (e.g. development of polarity, and induction of muscles) in embryos of *Drosophila* and *Xenopus*, and possibly also in tunicates and mammals. In this context, remember also that the promoter and enhancer sections of DNA are also crucial in development and evolution, though never normally transcribed and translated; it is not *just* proteins that have controlling roles.

2.4 Protein evolution

Point mutation is the most familiar means of genetic change, yet because of the repair processes outlined above only about 1 nucleotide in 200 000 is randomly (heritably) changed by this process every 1000 years. Thus mutation probably serves as no more than 'fine-tuning' for most evolutionary change in proteins. The more important system promoting change is that of recombination, providing the 'scissors and paste' of genetic change, especially the general recombination that occurs between homologous chromosomes during the crossing-over phase of meiosis, giving quite radical genomic change with duplications, deletions or transpositions of whole sections of the nucleotide sequence (Fig. 2.13).

There are several features of protein synthesis and regulation that interact with recombination events to have a major effect on protein evolution:

1 Duplication resulting from recombination is particularly important (Fig. 2.14) because it means there are often multiple copies of genes available. Often these occur as multiple 'tandem repeats' adjacent to each other on a chromosome, and while they stay adjacent there are repair mechanisms that tend to limit mutation in any one copy. Multiple identical copies may be very useful, where large amounts of the transcript are needed; for example, ribosomal RNA genes in vertebrates may occur as hundreds of repeats, many of them adjacent. Heterozygote advantage may also bring about selection for duplications; possession of two copies of the same gene allows an individual to be homozygous for different alleles at each of them but still

than a functional one, or the production of different morphs of the protein in different cell types. RNA processing therefore becomes a major way of producing these alternative protein morphs. There is also some evidence that changes in the splicing patterns could be an important source of variation in the overall evolution of genes and proteins, and thus of organisms.

There are other ways in which RNA is processed and controlled, usually less important in animals but contributing to overall gene expression control. These include systems that: alter the 3' end of the RNA (often by poly A additions); regulate movement from nucleus to cytoplasm; localize mRNAs to certain parts of a cell; or regulate mRNA degradation. These controls are probably mainly effected by further regulatory proteins, and offer yet more scope for relatively controlled and potentially adaptive change.

Fig. 2.14 Where there are tandemly repeated genes in a genome, copies are easily gained or lost due to unequal crossing-over between the sister chromosomes. The long regions of homologous sequence easily align with each other to produce this 'error'. (From Alberts *et al.* 1994, Fig. 8.73.)

Fig. 2.13 A general recombination event causing gene conversion.

effectively heterozygous. But if one copy of a gene is translocated away by further recombination then mutation becomes more likely, and mutational changes in any one copy are not too problematic for the organism. Some altered copies may never be of any use (they are then termed **pseudogenes**), but others may accumulate mutations and eventually acquire new functions. This gradually leads to the evolution of **gene families**, with related proteins having new functions.

2 The intron–exon system is another important factor, since it means that a 'gene' is actually a series of separate bits of nucleotide sequence, including protein-coding sections and regulatory sections. It is presumed to have persisted in eukaryotes because it makes protein synthesis in a multicellular organism much more versatile. Because of RNA processing steps the transcripts of genes can be spliced together in various ways, perhaps with different combinations of regulatory systems, depending on the cell type and developmental stage, so that in practice different but related proteins can be produced from the same gene.

3 The system of incorporating introns must also have led to extra possibilities of non-damaging recombination events between exons, speeding up the process whereby organisms can evolve entirely new proteins with new functions from parts of pre-existing proteins. Further families of genes result. Classic examples of gene groups resulting from exon recombinations are the immunoglobulins and the albumins, discussed in Chapter 15.

4 At least half of all spontaneous mutations observed in *Drosophila* are thought to be due to transposable element effects. Sometimes this is a direct effect on the gene, but often it is because a transposable element affects the timing or the spatial pattern of expression of a nearby gene, having subtle effects on animal development. There is some evidence that transposable elements undergo long periods of quiescence in the genome and then have sudden bursts of activity, known as 'transposition bursts'. These may be triggered by environmental change, giving an obvious link between environment and adaptation. The bursts seem likely to result in increased biological diversity, as the transpositions will potentially bring together two or more new traits that were of little value alone but that become very useful when working together; thus these transposition bursts can produce randomly modified progeny. If it is true that they can do so particularly at times of environmental stress, then the transposable elements may be acting not as mere parasites in the genome but as useful symbiotic factors, generating diversity just when it is most needed.

5 Another possible source of gene families is the joining together by recombination of DNA sequences coding for stable domains, making entirely new proteins but with some shared ancestry; many proteins do show signs of being assembled from shuffling together of pre-existing domains in new combinations.

6 There may be post-translational joining of existing proteins

into even larger 'protein complexes', multimeric forms with molecular weights of 10^5–10^6.

7 Both duplication and recombination may be involved in putting together new genes and new proteins with multiple internally repeated sequences; the collagen family, for example, always has multiple repeats of an ancestral sequence that gives a particularly stable fibrous structure.

8 A different sort of factor that may help explain why new proteins largely evolve by alteration of old ones is the very low probability (referred to earlier) of generating a stable functional new protein from scratch.

For any or all of these reasons, there may be many variants of any one protein, termed **protein isoforms**; these may be the product of different but related genes, or may be the product of one gene modified by RNA processing. Furthermore, genes themselves can clearly be united into gene families and these may show quite high levels of sequence homology, usually also retaining very high levels of three-dimensional structural homology. In most animals multiple variant forms of almost every gene exist; different opsin genes for each kind of photoreceptor, different collagen genes for each kind of connective tissue, or for each stage of the developmental process from embryo to adult, and so on. Families of genes can also be traced between species and often between taxa at much higher levels. Gene families such as globins, collagens and actins occur throughout the animal kingdom in multifarious forms. As an example, the evolutionary history of the globin genes is traced in Fig. 2.15.

Increasingly the evidence points to many multigene families being the results of ancient introns splitting up genes and providing new sites where the genetic material could be shuffled into new arrays. As we have seen, some motifs and domains also occur very widely indeed, shared between otherwise unrelated proteins. Protein evolution can then be viewed largely as a trial-and-error system of recombining separate exons coding for separate protein domains; this will regularly produce new proteins, and there is a high probability of these retaining a good degree of functional integrity but exhibiting slightly altered properties on which selection can act.

The existence of multiple copies of genes means that every animal may have a range of options as to which proteins it uses at any one time or in any particular place. This can happen in two ways:

1 Different variants of a protein (or specifically of an enzyme) occur within one organism, and these are termed **isozymes**; thus we may speak of a muscle dehydrogenase, a liver dehydrogenase, etc., all expressed within one individual. There may also be more than one such variant expressed in one cell type, and this opens up considerable scope for acclimatory change. For example, an animal may express one isozyme in its muscles at low temperatures and switch this off and another on (or at least change their proportionate expression) under different environmental conditions with higher temperatures. One isozyme may have greater thermal stability than another, or better performance at a particular pH, and so on.

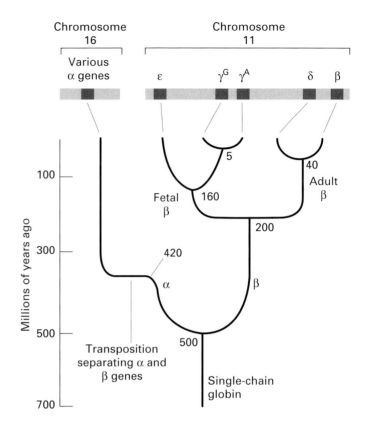

Fig. 2.15 Evolution of the globin gene family in animals. Location of these genes on human chromosomes is also shown. Numbers indicate calculated divergence times (millions of years).

2 Genetically distinct variants (alleles) may occur at a single locus, and where the resultant proteins have different properties they are termed **allozymes**. Populations in different environments commonly have different allele frequencies. Here we have scope for truly adaptive change, with one allozyme being selected in one part of a species' range and another in some other part of its range, leading to gradual diversification and potential speciation. Geographical trends in allele frequencies linked to a selective gradient are termed **clines**.

2.5 Physiological regulation of gene expression

We have considered thus far the ways in which proteins may be synthesized, controlled and degraded within cells largely in relation to proteins as the immediate products of genes. However, proteins are also the main targets of messages from other possible sources, both from within the cell and from the outside environment; this kind of regulation of proteins is the process underlying most of what we conventionally call physiology. The proteins concerned in this control and signalling system may be at the cell surface or intracellular, the latter including agents we have already met (protein kinases, protein phosphatases, GTP-binding proteins) as well as receptor proteins and all the other proteins with which all these key agents interact. There are also some smaller non-protein components of intracellular signalling, however, and these are ubiquitous in all living cells. Together these provide the systems whereby gene expression is communicated to other cells and to all parts of a functioning animal.

2.5.1 Receptors

All signalling molecules, to be effective, have to be recognized by a specific receptor in or on their target cell. The signalling molecules are released by exocytosis or diffusion, and they are extremely varied, including proteins, polypeptides, small peptides or single amino acids, as well as nucleotides, fatty acid derivatives, steroids, and even small molecules such as nitric oxide. But regardless of their chemical nature nearly all these signals act rapidly on their target cells via specific receptor proteins (the exception is nitric oxide which can diffuse through membranes and act directly on target enzymes). The signals are commonly at extracellular concentrations of only 10^{-6}–10^{-9} M and the receptors have very high affinities for them. The signal molecules act as ligands to the receptors so that the receptor protein undergoes a conformational change, usually across the span of the cell membrane, and triggers a series of intracellular signals. We know much more about the nature of receptors since we have been able to target them with recombinant DNA technology and produce large amounts of purified receptor protein.

Receptor proteins are of four main types; one type is intracellular, while the other three main families are cell surface receptors.

Intracellular receptors

Steroid hormones, together with thyroid hormones, vitamin D and the vitamin A-related retinoids, share a hydrophobic character and act by diffusing through the cell membrane to act directly on intracellular receptors. These receptors belong to the steroid-hormone receptor superfamily, characterized by the zinc-finger motif in a key site. Some of these steroid hormones can persist in the blood for hours or days, having long-lasting effects. The receptors to which they link all bind to DNA sequences termed **hormone response elements** to activate gene transcription. Often one or more small groups of genes are activated quickly and the products of these genes in turn activate more genes, giving a long and complex response pattern of gene expression; for example, Fig. 2.16 shows the time course of various egg-white protein mRNAs after the application of oestrogen. At the same time certain other target genes may be repressed by the hormone-receptor complex; for example, prolactin genes are repressed by glucocorticoid action.

Enzyme-linked receptors

These receptors either function directly as enzymes themselves

Fig. 2.16 Time courses of the appearance of various 'egg white' protein mRNAs, after induction by the hormone oestrogen. (Adapted from Palmiter *et al.* 1977.)

or are closely linked to enzymes; mostly they are themselves, or are linked to, a protein kinase. The ligand-binding site is outside the cell and the catalytic site inside.

Ion-channel-linked receptors

These perform the function of both channels and receptors, and are involved in rapid signalling between excitable cells, mediated by fast-acting locally produced neurotransmitters. The signal molecule causes a transient opening or closing of a channel through the protein structure.

G-protein-linked receptors

A G protein is a trimeric GTP-binding regulatory protein, as we met before. G proteins sit in membranes adjacent to both a receptor protein and another target protein, which may belong to either of the two categories already described (channel-linked or enzyme-linked). This is therefore a slightly less direct form of receptor system, but it is much the most common. Its core structure includes a long extracellular binding domain and a chain that folds back and forth across the membrane seven times in all; intracellularly there are G-protein-binding domains and phosphorylation sites (Fig. 2.17).

For all these last three types of cell surface receptors, signals received at the surface must be relayed to the interior of the cell, and often to the nucleus so that expression of particular genes can be affected. Ion-channel-linked systems are a rather special case and not very relevant here, but in most examples

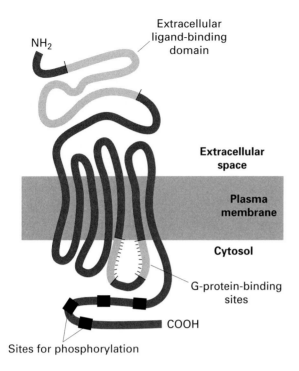

Fig. 2.17 Schematic view of a single G-protein-linked receptor, showing the typical seven folds across the membrane.

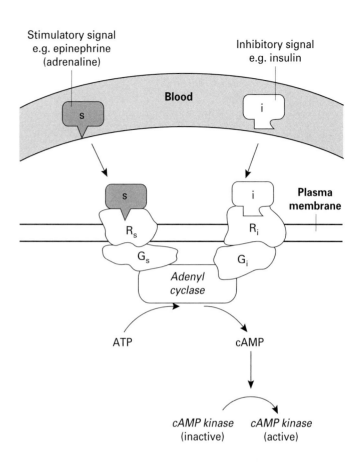

Fig. 2.18 Interaction of two G proteins with a single cAMP-producing adenyl cyclase, giving both stimulatory and inhibitory pathways. G, G proteins; R, receptors.

of the other two categories the immediate result of receptor activation is a cascade of phosphorylations mediated by a second messenger.

2.5.2 *Intracellular mediators and second messengers*

Molecules that help to control major pathways and systems within cells are those that link between protein receptors and their target enzymes or structural proteins or motor proteins or ion channels, or of course their target genes. The receptors are most commonly G-protein-linked receptors, as described above, so the linking stage is usually a G protein in the cell membrane. The messengers produced are often small enough (unlike proteins or polypeptides) to move through gap junctions between adjacent cells, so that signalling can be shared amongst cell communities. They allow extracellular signals to be greatly amplified and also to be integrated together. Two of these 'second messengers' are particularly common—cyclic adenosine monophosphate (**cAMP**) and ionic calcium; these interact with each other and also with other small messenger molecules that are produced in related reactions, as we will see.

Cyclic AMP (cAMP)

In animal cells cAMP is a ubiquitous transmitter between signal receptor sites on the exterior surface of cells and the regulatory enzymes of metabolism. It is synthesized from ATP by a membrane-bound enzyme **adenyl cyclase**, and rapidly destroyed by **cAMP phosphodiesterase** enzymes. When a signal is received that activates the cAMP system the cytoplasmic level of cAMP can increase at least five-fold in a few seconds; this is achieved by modification at the synthesis end (via adenyl cyclase) rather than at the breakdown end.

It works primarily via the amplifier system known as 'cAMP-dependent protein kinase' (or **kinase A**), being linked to the receptor via one or two G proteins (Fig. 2.18). There may be two G proteins (G_s and G_i, indicating stimulatory and inhibitory, respectively) because usually each kinase molecule can be activated or inhibited by ligands binding to separate receptor sites. For example, activation may be by a β-adrenergic agonist or by adrenocorticotropic hormone, and inhibition by an α-adrenergic agonist, muscarine or an opiate. Inside the nucleus, the eventual product of cAMP-induced phosphorylations may bind to a cAMP-response element, adjacent to the transcribed gene, to produce the final response.

The action of 30 or more peptide and catecholamine hormones, as well as some prostaglandins, proceeds via pathways

Fig. 2.19 The roles of inositol triphosphate (IP$_3$) and diacylglycerol (DAG) as intracellular messengers. Both are produced from cleavage of a specific membrane phospholipid following activation of a G protein by an extracellular signal. PIP$_2$, phosphatidylinositol biphosphate. See text for more details. (From Alberts *et al.* 1994, Fig. 15.13.)

involving elevated cAMP concentrations and the activation of kinase A systems. Yet individual hormones trigger distinct and diverse physiological responses in the same cell. How is this achieved? A probable mechanism is that individual kinase pools are compartmentalized at their site of action, with specific anchoring proteins, and only become active when the cAMP in the immediate microenvironment becomes elevated by a particular hormone. In addition, individual hormones activate specific protein kinase subtypes.

Calcium

Calcium is normally regulated by specific ion pumps at very low levels (around 10^{-7} M) intracellularly, but higher (usually $>10^{-3}$ M) extracellularly and in the endoplasmic reticulum. The large transmembrane gradient means that calcium can serve as a messenger simply by opening Ca^{2+} channels and producing an automatic inward flux with a rise of cytoplasmic calcium concentration. When a particular G protein (G$_q$) is activated by an appropriate signal it in turn activates a membrane-bound enzyme, **phospholipase C**. This then hydrolyses a specific phospholipid in the cell membrane to produce a soluble intermediate, **inositol triphosphate** (IP$_3$; Fig. 2.19). The IP$_3$ in its

turn travels through the cytoplasm and releases calcium (via IP$_3$-gated Ca^{2+} channels) from internal membrane stores. These are formed from modified portions of endoplasmic reticulum (ER), which also contain calcium-binding proteins (such as calsequestrin and calreticulin) and pumps to re-sequester the calcium so that it can be used again. Calcium thus released from the ER serves as an intracellular messenger and as such is even more widespread than cAMP. Calcium levels in the cytoplasm of a stimulated cell often show marked oscillations persisting as long as the receptors are activated, suggesting complex positive and negative feedbacks in operation.

Calcium ions may then activate their targets by directly binding to enzymes as an allosteric ligand. Examples of enzymes that directly bind calcium ions are the Ca^{2+}-inhibited guanyl cyclase of rod cells, and protein kinase C. Alternatively, calcium may act indirectly by binding to **calmodulin** (CaM). Calmodulin is a protein that has four Ca^{2+}-binding sites, and a molecular weight of about 16 kDa; it may make up as much as 1% of the total protein content of an animal cell. As more Ca^{2+} is bound so more and more hydrophobic regions are exposed in the molecule. Ca^{2+}-CaM itself acts directly by binding to enzymes, or indirectly through a CaM-dependent multifunctional kinase (PKII or CaM-kinase II); this kinase is particularly important in neural cells, and because it can phosphorylate itself it remains active even after the Ca^{2+} has been sequestered, giving it a role in certain kinds of memory. Examples of enzymes mainly regulated by calmodulin are the Ca^{2+}-ATPase, brain adenyl cyclase, myosin light chain kinase, and phosphorylase kinase.

In all animals calcium is especially important as the immediate trigger for muscle contraction, working via a sarcoplasmic reticulum (SR) receptor similar to that on the endoplasmic reticulum, and called a ryanodine receptor. In some invertebrate muscles and in all vertebrate striated muscles control is exerted via released Ca^{2+} binding to troponin C, which is one of the thin (actin) filament proteins. By contrast, in numerous invertebrate muscles and in vertebrate smooth muscles calcium control of contraction is exercised at the level of the thick (myosin) filament. In most of these cases a cAMP-dependent protein kinase/phosphatase pair is involved, which changes the phosphorylation state of a small subunit of myosin thus turning contraction on and off. Relaxation occurs because calcium is sequestered back into the SR by specific pumps. Thus very localized changes in intracellular Ca^{2+} concentration can have marked physiological effects.

cAMP and calcium interactions

The cAMP and Ca^{2+} intracellular pathways interact in several important ways. Enzymes that influence cAMP levels (adenyl cyclase and phosphodiesterase) are regulated by Ca^{2+}-CaM complexes. Enzymes regulated by cAMP and by Ca^{2+} can also influence each other, or may have interacting effects on shared target molecules. The phosphorylase kinase of skeletal muscle is a classic example, responsible for phosphorylating the key enzyme that mobilizes glycogen.

Phosphorylase kinase has four subunits, one of which is catalytic, one which is regulated by Ca^{2+}-CaM, and two which are regulated by cAMP-mediated systems. Since calcium also controls muscle contraction as we saw above, coordination of crucial ATP-utilizing (myosin ATPase) and ATP-generating (glycolysis) pathways is assured.

Inositol triphosphate, diacylglycerol and protein kinase C

When IP_3 is released by phospholipase C causing hydrolysis of a membrane phospholipid (see above and Fig. 2.19), the other cleavage product is **diacylglycerol** (DAG). The production of these two second messengers is initiated by a variety of signals, including neurotransmitters, hormones and growth factors. IP_3 and DAG have roles in the control of numerous cellular processes, including gametogenesis, fertilization, cell growth, secretion, smooth muscle contraction (as above), sensory perception and neuronal signalling.

IP_3 is at the focal point of two major control systems, one involving G-protein-linked receptors and the other involving tyrosine-linked receptors. These signalling pathways show a diversity of individual components, enabling precise control of the function of a huge variety of specialist cells. Diversity is generated by the stimuli which act through separate receptor types, and by heterogeneity in the G proteins, the phospholipase C, the intracellular IP_3 receptors, and protein kinase C. The latter is a serine/threonine kinase, particularly common in brain tissues; it starts a phosphorylation kinase cascade that leads to at least two important changes in gene regulatory proteins.

DAG has two important roles in cell signalling. Firstly, it directly activates protein kinase C thus interacting with the IP_3 pathways. Secondly, it is a precursor for **arachidonic acid**, which is a messenger in its own right and also serves in turn as the precursor for **eicosanoids** (which include the **prostaglandins** and are important effectors in pain, inflammation and fever responses).

Cyclic GMP (cGMP)

Cyclic GMP is generated directly from an enzyme-linked receptor; this receptor is guanyl cyclase, and there is no intervention of G proteins. The role of cGMP is at present less well understood than that of cAMP, although an increasing number of cGMP-sensitive protein kinases have been characterized. One important role for cGMP is in rod vision in vertebrates. Light energy falling on the receptors in rod cells of the retina is transduced into a chemical change and amplified, closing Na^+ channels in a membrane to induce hyperpolarization, thus initiating a nerve impulse.

Calcium and cGMP are also linked together in some situations by a nitric oxide synthase, where **nitric oxide** (NO) itself is an intermediate signalling molecule. The complex interactions of calcium and cGMP with the nitric oxide system are intricate and somewhat variable between tissues, with effects in neural cells and arterial smooth muscle proving to be particularly important in vertebrates. Further examples of the role of NO in cell–cell interactions are described for parasitic environments in Chapter 15.

The major target and effect of all these second messengers is, ultimately, altered gene expression, giving a host of ways in which genes can be turned on or off as needed in response to signals perceived at the cell surface. Now we need to consider what those signals may be.

2.5.3 *Extracellular control signals— growth factors, hormones, neurotransmitters*

The major extracellular signals for activating intracellular systems are local mediators between cells (paracrine signals), neurotransmitters delivered from nerve cells via chemical synapses, and hormones, which may be either local protein hormones, which we term growth factors, or more conventional hormones delivered at longer range and transmitted through the blood circulation. All of these bind to the receptors, usually in cell membranes, which transduce information into the cell, to start a cascade of intracellular events affecting channels, pumps, enzymes, the cytoskeleton and numerous other proteins and genes.

These signalling molecules have varying half-lives and thus work on very different timescales. In relation to adaptive change, it is the hormones that are major controllers of genetic expression and thus of protein synthesis. They can influence differentiation of dividing cells, and change the patterns of gene expression in already differentiated cells. They may activate master switches that can trigger expression of whole suites of genes, particularly where they are involved in developmental programmes (juvenile hormone in insects is a familiar example) or have trophic effects on the growth of whole tissues. They may mediate exchanges between the whole animal and the outside world, via effects on excretory and secretory processes and on behaviour. But they may also have rather more subtle effects, producing different outcomes on different cells. For example, acetylcholine causes contraction of skeletal muscles, but produces decreased rate and force of contraction in heart muscle cells and causes secretion in yet other non-muscular cells. Dopamine plays a role in numerous aspects of behaviour (sexual responses, hunger, thirst, vigilance) as well as influencing lactation and immune responses. Hormones largely achieve this diversity of response by being effective with several different G-protein-linked receptors; for example, epinephrine (adrenaline) is known to link with at least nine such receptors, and dopamine with at least five.

A good example of multiple hormonal effects is the adrenergic stimulation of a vertebrate heart muscle cell (Fig. 2.20). Norepinephrine (noradrenaline) binds to β-adrenergic receptors on the cell membrane, which in turn interact with G proteins complexed with GDP. The GDP is exchanged for cytoplasmic GTP to produce activated forms of the G proteins. The activated G proteins interact with adenyl cyclase in the membrane and produce an increase in the intracellular cAMP as described above. The cAMP activates protein kinase A, which can then phosphorylate a range of target proteins. It may convert the inactive b dimer of phosphorylase to the active a dimer. At the same time the kinase A activates Ca^{2+} channels leading to an increase in calcium concentration in the cytoplasm; this further activates glycogen phosphorylase, and also overcomes the inhibition of myosin ATPase activity, leading to muscle contraction. There is enormous amplification of the initial signal, and co-ordination of a variety of enzyme sequences that produce and then utilize ATP, giving an immensely complex but organized response to the hormone.

A second example would be the action of glucocorticoids on liver cells. These hormones are released during starvation or prolonged intense exercise, and bind to liver cell membranes. The resultant intracellular signals lead (via pathways similar to those above) to the increased expression of a number of genes, encoding enzymes (e.g. tyrosine aminotransferase) that catalyse the production of glucose from amino acids and other small metabolites. (Note, though, that the same hormone also acts on fat cells, but here the production of some of the same enzymes is reduced.) The overall effect is to increase the supply of glucose in the blood even at the expense of eventual tissue protein degradation. Having common signals producing different effects on different cells again underlines the complexity of the control of gene expression.

As a third example we should consider the action of **growth factors**, small extracellular proteins acting through specific surface receptors and of immense importance in developmental programmes. They have local effects, not involving the circulatory system, and they mediate crucial processes like cell division and cell differentiation. Their action does not involve G proteins or cAMP, but instead a series of intermediates triggered via a tyrosine-specific protein kinase. Under the influence of this kinase, the protein P60 becomes phosphorylated to PP60, and a chain reaction results, the last product acting on gene expression in the nucleus.

Additional sophistication in control systems at the whole organism level (reflecting control at the protein synthesis level) comes about due to the pulsatility of many hormones. Secretion of mammalian growth hormone is cyclical, due to an underlying rhythm in the release of two hypothalamic peptides (growth hormone-releasing hormone, and somatostatin); one or both of these peptides also lead to a cyclicity in release of the main gonadotropic hormones, luteinizing hormone (LH) and follicle-stimulating hormone (FSH) (see Chapter 13). This appears to result in turn from a pulse generator in cells of the basal hypothalamus, possibly due to reciprocal inhibition circuits in two sets of neurones. The frequency of hormonal release can thus be controlled by neural signals related to intracellular signals, giving fine tuning to the system and a new layer of regulation at which adaptation could operate.

2.5.4 Feedback systems and metabolic control

Repeatedly through the discussions in this chapter we have met examples of feedback regulation of one protein or molecule by another. This is a fundamental process in all organismal function, and it is worth dealing with the concept here as it will recur throughout other chapters of this book. The essential principles are shown in Fig. 2.21.

We can most easily examine it by looking at specific cases of enzyme regulation. Enzymes of course are protein catalysts that convert one kind of molecule to another. The flow of carbon through complex metabolic pathways in animals (discussed fully in Chapter 6) is controlled by enzymes, which are subject to all the general kinds of control outlined in the previous sections. The control systems operating are generally reversible, and types with and without some kind of amplification are found.

Where no amplification is present control is usually by a simple feedback; an end-product accumulates or disappears and this provides the **positive feedback** or **negative feedback** by acting allosterically on one of the key enzymes earlier in the pathway to change its activity. Negative feedback is extremely common, and most often involves the final product of a pathway feeding back as an allosteric inhibitor onto the first enzyme in the pathway; thus if large quantities of the product accumulate any wasteful entry of precursors into the path is automatically prevented. An example of the rather less common phenomenon of positive feedback occurs where ADP, the precursor of ATP, is a ligand for several enzymes involved in the pathways of catabolism of foodstuffs that produce ATP. Here ADP is purely regulatory, taking no part in the actual reaction, but serving to ensure an adequate rate of synthesis of ATP. Another example

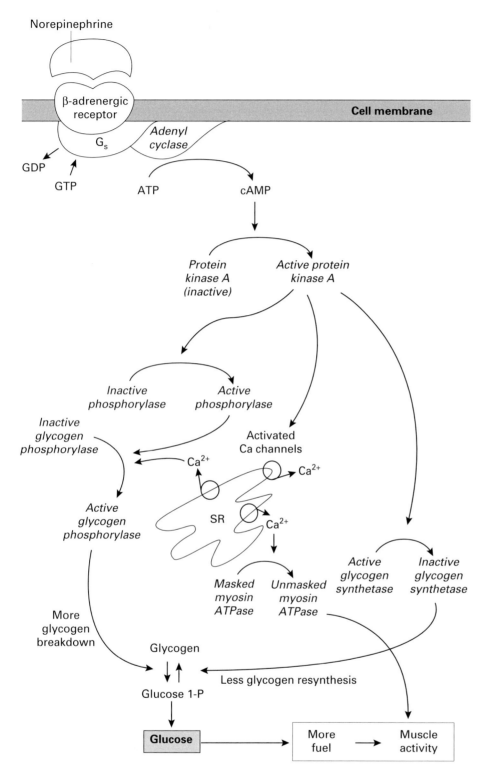

Fig. 2.20 The action of a stimulatory hormone, norepinephrine (noradrenaline), via cell membrane receptors, on various intracellular processes, leading to increased fuel supply and increased muscle activity. SR, sarcoplasmic reticulum.

is that of fructose 1,6-bisphosphate (F-1,6-BP) in the liver; this is a product of an early stage of glycolysis (see Fig. 6.6) which activates the terminal enzyme in glycolysis *before* the 'slug' of metabolites reaches it. This ensures that the central part of the glycolytic chain can work at maximum throughput, limited only by the supply of ADP and phosphate. This control mechanism is stable and works well because as the concentration of F-1,6-BP falls so the later-acting pyruvate kinases also lose activity. Thus,

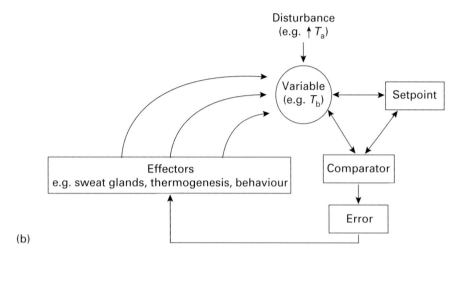

Fig. 2.21 (a) Principles of feedback control. (b) A more complex system where there is a setpoint and a comparator detecting any mismatch with the setpoint and producing an error signal, as in many physiological systems. T_a, ambient temperature; T_b, body temperature. (c) The commonest type of negative feedback at the biochemical level, underlying many enzyme regulation systems; accumulating end-product here serves as the 'error' signal and feeds back to an earlier enzyme as the 'effector'.

although positive feedback sounds potentially risky, stability will always be gained if the response times of an early and a late enzyme in the sequence are different.

In more complex enzyme control systems that have built-in amplification there may be very high gains, i.e. a single signalling molecule can bring about the activation of a large number of enzyme molecules. Systems with built-in amplification typically have a cascade mechanism involving reversible phosphorylations, as in Fig. 2.3. How big the amplification is depends on a number of factors, including the speed of dephosphorylations (related in part to the concentration of the phosphatase enzymes) and the number of steps in the cascade. For example, it has been demonstrated experimentally that for glutamine synthase the amplification is 67 for a single stage, and depending on the conditions it may be 250–1500 for a two-step cascade. Control steps work on different timescales, and this is very important for the integrated functioning of metabolism, as it

provides great flexibility. For example, the immediate conformational responses of enzymes to ligand binding can occur in milliseconds, whereas covalent modifications such as phosphorylation are rarely complete in less than a second and can take many minutes. The effects can be seen in the multistep cascade leading to the activation of glycogen phosphorylase, shown in Fig. 2.20, which has four steps or cycles. This shows the utility of having control steps with different time courses: because the activation of glycogen synthase is much slower than the activation of the phosphorylase, the futile cycling of glucose residues (repeatedly adding and removing phosphate groups) is avoided.

Regulated enzymes are often controlled by more than one feedback mechanism. Usually, the important flux-controlling enzymes have at least two different control systems. For example, the membrane-bound enzyme hydroxymethylglutaryl-CoA reductase is subject to feedback control by cholesterol and to allosteric regulation by other steroid metabolites. Its activity is also regulated by a reductase kinase and by a phosphatase. Furthermore, because the enzyme has a short half-life before being dealt with by proteasomes, changes in concentration also modify its activity. A final control is exerted by a marked diurnal rhythm in its rate of synthesis.

The complex patterns of enzyme regulation observed in present-day animals to some extent reflect a 'fossil record' of the control mechanisms that have evolved. There are probably many examples where the importance of a particular control feature has diminished over time, perhaps due to a changing environment or to some evolutionary innovation. Unless it is positively harmful in the changed circumstances it is likely that the original regulatory domain on the enzyme will remain to supplement the new control system or will gradually assume a new function over time. There may thus be considerable hidden variation in the signalling pathways of cells, and recent evidence has indicated that this can lead to rather rapid change in morphology during periods of environmental stress. Here the heat-shock proteins (HSPs, covered in detail in Chapter 8) are implicated, these being molecules that bind and stabilize proteins. In periods of environmental stress the common form HSP90 is fully occupied with protecting the unfolded proteins in a cell and may therefore be unavailable to monitor and protect the signalling proteins directing development. In *Drosophila*, in the absence of HSP90 a small number of viable mutants therefore appear with a range of developmental malformations that were previously masked. HSP90 thus normally serves as a buffer against existing variation in morphogenesis; environmental stress enables this cryptic variation to be expressed, giving a mechanism (albeit non-heritable) whereby phenotypic change can be accelerated just when it is needed as the environment changes.

The need to have redundancy, cryptic variation or 'fail-safe' mechanisms in important control systems, together with the accumulated variety of past evolutionary experiments, probably accounts for the apparent complexity of the present-day regulation of enzyme activity.

2.6 Conclusions

It is the essential character of proteins that underlies biological adaptation. Small changes in protein shape govern most of what goes on in a cell, by producing switching effects or motor effects or by channelling other molecules across membranes and between cell compartments; these effects can be subject to extremely complex regulations and amplifications to produce controlled effects on whole organisms. Permanent changes in protein structure, brought about by mutations or by recombinational changes in the DNA sequence of a cell, can therefore produce subtle changes in enzymic activity, signalling activity and subcellular morphology, and perhaps, above all, effects on the expression of other proteins; all of which can in turn lead to permanent heritable change in the development of organisms. This is what we see outwardly as change, at a series of levels. At the biochemical level there may be increased thermal tolerance of an enzyme, or expression of a more pH-stable allozyme; morphologically there may be developmental changes of gene expression leading to altered positioning of muscle cells, nerves, blood vessels, even completely modified appendages; physiologically there might be heart rate increments under the influence of calcium and cAMP signalling regimes, or sodium pumping rate changes as more or different channel proteins are synthesized; and at the behavioural level there may be increased attack speed, greater sensitivity to a sex pheromone, or even 'conscious' changes in response (the ultimate adaptive trick). All of these protein-induced changes provide the raw material for adaptation.

However, when all this is said, it remains the case that from a physiologist's viewpoint, the organism is in some senses more than the sum of its molecular or genetic parts. Genes can only function, and can only be selected for, as part of a fully functional reproducing organism. Reducing the scale of analysis to molecules, even when those can now be experimentally manipulated, does not in the end 'explain' what is going on at the whole animal level. That is why the rest of this book concentrates on adaptation at a higher level of organization, from cell and tissue to organism and environment, using a specific molecular understanding of adaptation only in those relatively rare instances where it has been elucidated. We hope, however, that this chapter has set the process of adaptation in a suitable molecular context, so that new information on the molecular interactions and genomic changes underlying ecophysiological modification can be easily assimilated as it becomes available.

Further reading

Books

Alberts, B., Bray, D., Lewis, J., Raff, M., Roberts, K. & Watson, J.D. (1994) *Molecular Biology of the Cell*. Garland, New York.

Hardie, D.G. (1991) *Biochemical Messengers*. Chapman & Hall, London.

Stryer, L. (1988) *Biochemistry*. Freeman, New York.

Watson, J.D., Hopkins, N.H., Roberts, J.W., Steitz, J.A. & Weiner, A.M. (1987) *Molecular Biology of the Gene*. Benjamin-Cummings, Menlo Park, CA.

Wolfe, S.L. (1993) *Molecular and Cellular Biology*. Wadsworth, Belmont, CA.

Reviews and scientific papers

Berridge, M.J. (1993) Inositol triphosphate and calcium signalling. *Nature* **361**, 315–325.

Borgese, F., Sardet, C., Cappadoro, M., Pouyssegur, J. & Motais, R. (1992) Cloning and expression of a cAMP-activated Na+/H+ exchanger; evidence that the cytoplasmic domain mediates hormonal regulation. *Proceedings of the National Academy of Sciences of the United States of America* **89**, 6765–6769.

DeCamilli, P., Emr, S.D., McPherson, P.S. & Novick, P. (1996) Phosphoinositides as regulators in membrane traffic. *Science* **271**, 1533–1539.

Divecha, N. & Irvine, R.F. (1995) Phospholipid signalling. *Cell* **80**, 269–278.

Gilman, A.G. (1987) G proteins: transducers of receptor-generated signals. *Annual Review of Biochemistry* **56**, 615–649.

Tanaka, C. & Nishizuka, Y. (1994) The protein kinase C family for neuronal signalling. *Annual Review of Neuroscience* **17**, 551–567.

Pigliucci, M. (1996) How organisms respond to environmental changes: from phenotypes to molecules (and vice versa). *Trends in Ecology and Evolution* **11**, 168–173.

Rutherford, S.L. & Lindquist, S. (1998) Hsp90 as a capacitor for morphological evolution. *Nature* **396**, 336–342.

Snyder, S.H. (1985) The molecular basis of communication between cells. *Scientific American* **253**, 132–140.

Wurgler-Murphy, S.M. & Saito, H. (1997) Two-component signal transducers and MAPK cascades. *Trends in Biochemical Sciences* **22**, 172–176.

3 The Problems of Size and Scale

3.1 Introduction

Living organisms occur in a wide range of sizes (Table 3.1), spanning 21 orders of magnitude (10^{21}). The smallest living things are the viruses, which contain only genetic material, all their other functions being undertaken by the host cell. Mycoplasmas (also known as pleuropneumonia-like organisms) are only slightly larger, weighing less than 0.1 pg, and are the smallest things that can lead an independent existence and survive and reproduce in an artificial medium. The smallest animals (i.e. multicellular organisms in the kingdom Metazoa) are considerably larger at around 2–10 µg. At the other end of the scale the largest extant animals are mammals: the blue whale weighing in excess of 100 000 kg in the seas, and the African elephant at 5000 kg on land. Each taxon of animals has its own particular size range, largely dependent on inbuilt design constraints; some phyla such as Bryozoa or Platyhelminthes may contain species (both living and extinct) all within two or three orders of magnitude of mass, while others such as Crustacea and Mollusca, or the single class Mammalia, are hugely variable in size. For example, the smallest mammal is a shrew weighing only about 1 g when fully grown, thus being about 100 million times less heavy than the blue whale.

Just the fact of being a particular size has profound consequences for structural and functional relationships in animals. Most structures and physiological variables change in a predictable way with increasing body size. The study of these size-related effects is known as **scaling**. Knut Schmidt-Nielsen defined scaling as dealing with 'the structural and functional consequences of changes in size or scale among otherwise similar organisms'; evolutionary biologists may think of it in terms of a major constraint on the adaptation of certain characters. Mass is most commonly used to measure the size of animals, although length is also useful in a few circumstances, for example in fish swimming. As with man-made structures, three parameters can be changed as the size of an organism is increased, namely the **dimensions**, the **materials** used for its construction and the **design**. Changing dimensions or proportions as size increases is an obvious solution, widely used in

Table 3.1 Ranges of organism and animal sizes.

Organism	Mass	
	Grams	Units to scale
Mycoplasma	10^{-13}	<0.1 pg
Typical bacterium	10^{-10}	0.1 ng
Tetrahymena	10^{-7}	0.1 µg
Amoeba	10^{-4}	0.1 mg
Rotifer	10^{-4}	0.1 mg
Aphid	10^{-3}	1 mg
Bee	10^{-1}	100 mg
Pygmy shrew	10^{0}	1 g
Hamster	10^{2}	100 g
Human	10^{5}	100 kg
Elephant	5×10^{6}	5 000 kg (5 tonnes)
Blue whale	10^{8}	100 000 kg (100 tonnes)

the animal kingdom; a familiar example is that bones are disproportionately thicker in vertebrates as the mass of an animal increases, to support the extra weight and the forces during locomotion without breaking. But the use of new materials, or of an innovative design, are also often selected for to overcome some fundamental constraint associated with size or dimensions. There are many examples of this in the chapters that follow. For example, gas transport by simple diffusion is fast over very short distances but soon becomes limiting as body size increases. The evolutionary solutions to this problem have been mass transport by convection, either on the outside of the animal (ventilation), or inside the animal (circulation), or both. A whole range of specialized ventilated respiratory surfaces have been developed together with pumped circulatory systems to enable animals to overcome the constraints to maximum size posed by diffusion (see section 7.2).

3.2 The principle of similarity: isometric scaling

Relatively simple rules govern the relationships between geometrically similar objects. For example, the linear dimensions of

Fig. 3.1 Isometric triangles, showing that all corresponding linear dimensions, in any isometric figures, always have the same relative proportions.

$$\frac{A_1}{A_2} = \frac{B_1}{B_2} = \frac{C_1}{C_2} = k$$

Fig. 3.2 Double logarithmic plot of the surface area (SA) vs. the volume (V) of a cube (solid line). The regression line has a slope of 0.67. The dashed line shows a plot of the surface area per unit volume against volume, with a slope of −0.33, showing that the relative surface area (the SA/V ratio) decreases with increasing size. (From Schmidt-Nielsen 1984.)

isometric triangles are shown in Fig. 3.1. Geometrically similar objects are sometimes called **isometric**. The surface areas (SA) and volumes (V) of isometric objects are related to their linear dimensions to the second and third power, respectively. Since surface areas are (length)^2 and volumes are (length)^3, it follows that surface area is (volume)$^{2/3}$, or volume to the power 0.67. If the surface area of a cube (unit face area, therefore total SA = 6) is plotted against its volume on logarithmic coordinates a straight line is obtained that corresponds to the equation:

$$\log SA = \log 6 + 0.67 \log V \text{ (Fig. 3.2)}$$

For other shapes the intercept 6 will change but the **isometric scaling exponent** of 0.67 remains the same; in other words the surfaces of isometric bodies are always related to their volumes by the power 0.67. Expressed another way, if the surface area per unit volume is plotted against volume a straight line is produced with a slope of −0.33. Thus smaller bodies have larger surface areas

relative to their volumes than larger objects of the same shape. The surface area to volume (SA/V) ratio of cells, tissues and organs is often a major consideration in physiological comparisons.

In scaling, the units in which measurements are made are not in themselves important; scaling deals with dimensionless ratios. Box 3.1 gives an overview of dimensions and units as used in this book.

Box 3.1   Dimensions and units

If we write an equation using an equals sign, both the magnitude and the dimensions of variables on the two sides of the equation must be identical. The units used to describe the magnitude of a variable must always correctly reflect its dimensions in the three aspects of mass (M), length (L), and time (T).

According to Newton's second law of thermodynamics, **force** is the product of **mass** and **acceleration**:

$F = ma$

Acceleration is the rate of change of velocity, and velocity has the dimensions of length per unit time or LT^{-1}; thus acceleration must have the dimensions LT^{-1}/T, or LT^{-2}. Hence we know that force must have the dimensions MLT^{-2}.

Units represent an arbitrary but self-consistent system to represent the magnitude of a variable, and **SI units** (Système International) are now almost invariably used, where the standard units are **kilograms, metres** and **seconds**. The unit for acceleration is therefore metre per second squared (m s^{-2}), and force is measured in kg m s^{-2}, also known as Newtons and symbolized as N. On average objects fall with an acceleration due to gravity of 9.81 m s^{-2}, and each kilogram of the object is subject to a gravitational force of 9.81 N.

By considering the dimensions of variables we can appreciate that the mass and the weight of an object are not the same thing, and should not be used interchangeably. It is nearly always appropriate to use the mass of biological materials.

The dimensions of some important physical, physiological and ecological variables are as follows:

Variable	Description	Dimensions
Fundamental variables		
Mass		M
Length		L
Time		T
Derived variables		
Density	Mass/volume	ML^{-3}
Velocity	Distance/time	LT^{-1}
Acceleration	Velocity/time	LT^{-2}
Force	Mass × acceleration	MLT^{-2}
Stress	Force/area	ML^{-1}T^{-2}
Work	Force × distance	ML^{2}T^{-2}
Power	Work/time	ML^{2}T^{-3}
Mass specific power	Power/mass	L^{2}T^{-3}
Biomass density	Mass/area	ML^{-2}
Production	Mass/time	MT^{-1}
Productivity	Mass/(time × area)	ML^{-2}T^{-1}

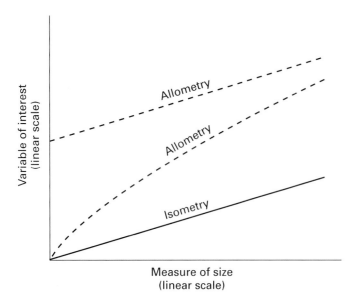

Fig. 3.3 Allometric and isometric variation. *Isometric* variation occurs when a particular variable (*Y*) is related to body size (*X*) by a straight line through the origin, so that the ratio *Y/X* does not vary with changing body size. *Allometric* variation occurs when the same plot intersects the *Y*-axis at some point other than zero, *or* when the plot is a smooth curve; in either case the ratio *Y/X* varies with body size.

3.3 Allometric scaling

If the relationship is plotted between some physiological or eco-logical function (e.g. oxygen consumption, or gonad mass, or reproductive output) and some measure of size (e.g. body mass, or length), then there are three possible outcomes as shown in Fig. 3.3. There may be a straight line relation passing through the origin; then the variable changes in direct proportion or isomet-rically with body size, as outlined above, and the ratio between that variable and body size is always the same. Alternatively there may be a straight line relation that does not pass through the origin, or there may be a curvilinear relation; in either of these cases there is no simple ratio between the two measure-ments (thus a doubling of size does not mean a doubling of oxygen need, or gonad size, etc.) and the function varies **allo-metrically** with body size (from the Greek *allos* meaning differ-ent). In practice, even in animals of the same general body plan, characters do not scale isometrically and biologists must deal with this non-isometric, or allometric scaling.

By far the most common independent variable considered in issues of biological scaling is that of body mass (M_b). A great range of parameters of interest to physiologists has been found to vary with body mass according to allometric equations of the general form:

$$Y = aM_b{}^b$$

In its logarithmic form this becomes:

$$\log Y = \log a + b \log M_b$$

This is described as a 'power formula' because the dependent variable *Y* changes as some power of the independent variable (body mass). Since the logarithmic equation is of the general form for a straight line ($y = mx + c$), it follows that plots of *Y* against M_b on logarithmic axes always do produce a straight line, with a slope of *b* and an intercept of log *a*. The calculated intercept at unity body mass (*a*) is then known as the **pro-portionality constant**, and the slope or gradient *b* is the **mass exponent**. Box 3.2 provides some background on dealing with logarithmic and exponential relationships and on handling the simple mathematics involved.

The proportionality constant can be useful in comparisons between data sets, looking for differences in the level or setting of a particular parameter between animal groups. For example, marsupial mammals have a lower metabolic rate overall (about 30% lower) than do eutherian mammals, with the slope having a lower intercept (lower value of *a*); however, the relation between metabolic rate and size is the same for both groups (i.e. they have the same value of *b*).

The mass exponent *b* is usually the parameter of most interest. If *Y* increases in simple proportion to body mass then the slope *b* will have a value of 1.0. Examples of this are blood volume or lung volume in mammals, both of which are a constant fraction of body mass. If a parameter does not change with body size then the slope *b* will be zero; a good example here is the con-centration of haemoglobin in vertebrate blood, which is broadly invariant across species. If *Y* increases less than would be expected by simple proportionality then *b* will have a positive value of less than 1.0, as with the SA/V ratio already mentioned; here another well-known example is the resting metabolic rate of mammals (see below). If *Y* increases more than expected, *b* will have a positive value greater than 1, as with the bone thickness of vertebrates already mentioned, where for mammals the expon-ent is 1.08. There may even be negative values for *b*, a good example being the relation between mammalian heart rate and size (25 beats per minute in an elephant, compared with up to 1200 per minute in a pygmy shrew), the exponent being −0.25.

The value of *b* provides information on how the variable of interest changes with body size. Similar values of *b* may indicate that similar principles are operating to determine the scaling of two or more variables. If *b* values are the same, then values of *a* are also directly comparable and can be used to make compar-isons between different groups of animals (as above for marsu-pials and placentals). Note that values for the slope and intercept in real populations are not fixed values, but means about which natural variation occurs. This variation is usually quantified using confidence limits (see Fig. 3.4). For a linear relationship the best fit line and the confidence limits about that line are calcu-lated using various possible kinds of regression analysis. Con-fidence limits in *Y* are not a constant value above and below the line, but get wider towards the ends of the sampled variation in *X*. Confidence limits can be calculated for both the slope and the intercept of the best fitted line, and are particularly affected by points with very high or low *X* values (outliers). Notice also that

Box 3.2   Use and handling of logs and exponents

Why use logs?

Taking logs is useful in biological comparisons for several reasons. One is illustrated below: on a log scale it is much easier to compare diagrammatically across a very large size range. With a base 10 scale and exponents, the size axis here can be more easily expressed.

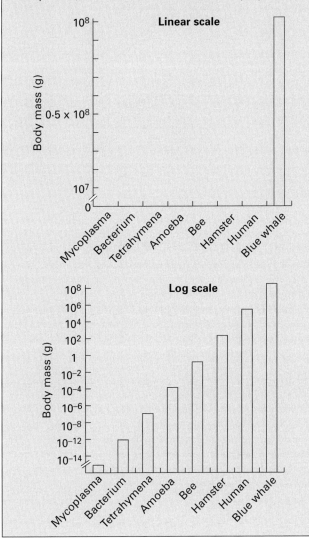

Taking logarithms is also useful because it commonly transforms curvilinear biological relationships into a linear form, making statistical analysis much easier. The curvilinearity very often results from a relationship between two variables that is a power function (see text) and taking logs produces an equation that represents a straight line:

$$\log Y = \log a + b \log X$$

It is tedious to calculate the logs for each data point, but the same effect is achieved by plotting untransformed data on log–log paper. Note that the intercept occurs where $\log X = 0$, i.e. $X = 1$. Linear regression then gives a value of the correlation coefficient r, or more usefully of r^2, which describes the fraction of the variation in Y-values that is explained by the variation in X-values. It also gives the standard errors or the 95% confidence limits for the slope (b) and the intercept (a) of the straight line (see Fig. 3.4).

Handling logs

Logarithms represent the power to which some base must be raised, so a change of one log unit represents a change of an order of magnitude in the antilog. Thus an arithmetic change (e.g. 1 to 2 to 3) in the log corresponds to a multiplicative change in the antilog (e.g. 1 to 10 to 100). Handling logarithms is not difficult—they can be added, subtracted, multiplied and divided according to simple rules:

$$\log (1/X) = -\log X$$
$$\log (XY) = \log X + \log Y$$
$$\log (X/Y) = \log X - \log Y$$
$$\log X^a = a \log X$$
$$\log \sqrt[a]{X} = 1/a \log X$$

Remember also: log 1 in base 10 is zero, and positive numbers below 1 have negative logs (e.g. $\log 0.1 = -1$, $\log 0.001 = -3$); there is no log for zero itself or for negative numbers.

There are also simple rules for operating with powers, analogous to those for logs:

$$X^a X^b = X^{a+b}$$
$$X^a / X^b = X^{a-b}$$
$$(X^a)^b = X^{ab}$$
$$\sqrt[b]{X^a} = X^{a/b}$$
$$X^a Y^a = (XY)^a$$
$$X^a / Y^a = (X/Y)^a$$
$$X^a / X = X^{a-1}$$
$$1/X^a = X^{-a}$$
$$X^0 = 1$$

when comparing two or more good data sets even quite small differences in b are potentially important; remember that the log-log plot inevitably appears to reduce differences that would look substantial on a non-logarithmic graph.

Analysis of scaling can be applied at many levels. Differing allometries may emerge when analysing differences within a species:

1 during the growth of individual animals;
2 across individuals of the same age or developmental stage; or
3 across a range of ages or stages.

It is therefore important to use the right species value when making larger interspecific comparisons. Given this precaution,

variations in physiological parameters can be assessed on the basis of allometric effects at almost any taxonomic level, up to and including the whole animal kingdom. Such analyses have shown that many of the differences between species once attributed to adaptation can largely be explained by scaling effects.

Allometric analysis has had a major impact on environmental physiology; for example, the way in which energy assimilation, respiration and mortality vary with body mass for each species can be described by three allometric equations, and the slopes and intercepts of these have been described as the 'six facts of life', with various combinations of these six facts being possible. However, the very importance of scaling prompts some further

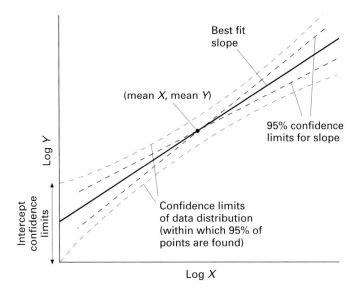

Fig. 3.4 A regression line calculated by the least squares method. Note that the 95% confidence limits for the regression get wider towards the upper and lower extremes of the range. The slope will be affected more by an outlier here than by a point near the middle of the line. 95% confidence limits for the slope make the line more or less steep by rotation about (mean *X*, mean *Y*) (dashed lines).

words of caution. Allometric equations are purely descriptive and do not represent laws; it is therefore important that they should not be used for extrapolating beyond the range of data on which they are based. Allometric relations do not indicate some kind of unbreakable body-weight-related constraint on evolution, since they may be the end-product of several interacting selection pressures. Above all, they do not remove the need for adaptive explanation, as has sometimes been maintained or implied; we may be able to *predict* some character fairly well from a know-ledge of another parameter and of the equation relating the two, but this in itself does not *explain* the character or the selective pressures that gave rise to it. In other words, allometric trends are not 'non-adaptive'. The best known example to throw light on this is probably the story of deer antlers (see Box 3.3).

Allometric relations can also be of use in another way. Single observations may not necessarily fall on the calculated allo-metric regression line, and they may be smaller or larger than expected. Assuming that no errors have been made in making the measurements (or in calculations or statistics) then further insight may be obtained from the data points that significantly deviate from the general allometric line. Comparative biologists are always intrigued by species that do not fit with the pattern of their near relatives, or with individuals that seem to differ from the rest of their species. We noted above that blood volume in mammals is a constant proportion of body mass; but the striking exceptions are the marine diving mammals, many of which have much larger blood volumes than expected, which turns out to be because the blood is used as a significant oxygen store during a

Box 3.3 Problems with scaling and adaptive interpretations

Large male deer have extremely long, heavy and cumbersome antlers; these were up to 3.3 m across in the extinct Irish elk, and many explanations for this have been put forward. But antler size is strongly correlated with body size, and once it was realized that the antlers of the larger species had the same relation to body weight as those of the smaller species some authors maintained that no further search for an adaptive explanation for the huge ornaments on the heads of large deer was necessary.

This cannot be true; the antlers represent a great cost in materials to their possessor (especially as they are renewed every year), and can be a significant handicap in normal daily activities. Deer use their antlers in male/male fighting, and it transpires that the size of the antlers is proportionately greater in species with polygynous mating systems, where male/male competition is greatest. Hence there is clearly selection for larger antler size operating. Probably the two characters of body weight and antler size are under strongly correlated selection pressures, the forces selecting for antler size being the same as those selecting for large bodies.

Hence when we see strong allometric trends we should still be looking for the selective forces responsible for them, just as we would seek for the explanation for any single character.

dive. In the case of metabolic rate, we might also predict some deviations from a simple body size relationship: sloths (if they live up to their reputation) should fall below the line and small rodents and small passerine birds might lie above it. For brain size against body size, we might hope that primates, and *Homo* in particular, would lie well above the line (this does turn out to be true, but is somewhat confounded by an interaction with unusual longevity in some primates and man).

Numerous examples of such scaling are given in this chapter and throughout the book; in particular see section 6.6 on energy balance, sections 7.2, 7.3 and 7.4 on respiration and the circulatory system, and sections 8.7 and 8.8 on temperature regulation. Here we pick out a few examples to show some of the principles and effects of scaling.

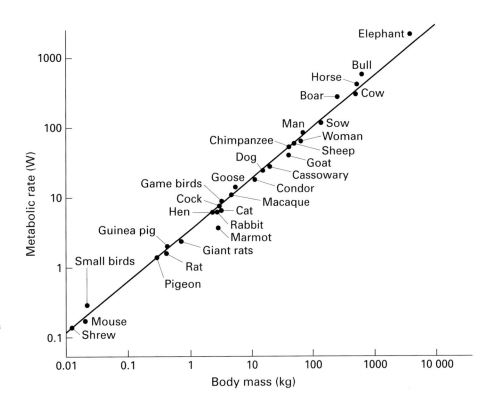

Fig. 3.5 Metabolic rates for a variety of birds and mammals plotted against mass on a double logarithmic scale. (From Schmidt-Nielsen 1984; courtesy of Cambridge University Press.)

3.4 The scaling of metabolic rate

3.4.1 *The allometry of resting metabolism*

The scaling of metabolism is one of the best known and most intensively studied examples. An animal's metabolic rate determines its food intake, growth, excretion and many other functions, which as a consequence will also vary with body size. The terminology of metabolic rates and their measurement is fully explained in Chapter 6.

To compare the effects of some environmental variable on the metabolic rate of different populations or species it is first necessary to take into account that fraction of the variation in the data that can be attributed more simply to differences in the body mass of the groups being compared. The importance of this is illustrated by the following example: in a recent study of the routine oxygen consumption of 143 species of zooplankton covering eight different phyla it was found that as much as 84–96% of the apparently wide variation in metabolic rate could be explained simply by differences in body mass and temperature. Thus it is essential to start by examining the underlying scaling relationships.

Hundreds of scientific papers have been published on the scaling of basal or resting metabolic rate (see Chapter 6) in birds and mammals over the last 60 years. If basal metabolic rates (BMRs) are plotted against body mass on logarithmic coordinates a straight line is obtained, often known as the 'mouse to elephant curve' or the Kleiber relation. Max Kleiber, in 1932, surveyed the metabolic rate of mammals varying in body mass

from 0.15 kg to 679 kg. Metabolic rate (in kcal day^{-1}) was related to body mass (M_b) by the allometric equation:

$$\text{metabolic rate} = 73.3 M_b^{0.74}$$

Statistically the intercept was not different from 70 and the slope was not different from 0.75 (Fig. 3.5), so we can rewrite this as:

$$\text{metabolic rate} = 70 M_b^{0.75}$$

The reasons for this exponent of 0.75 are considered further in Chapter 6, but for the present we need to consider what use can be made of this allometric relation. Within a particular taxonomic group, variation in body mass can in itself explain at least 90% of the variation in metabolic rate. In practice, there are considerable differences in the BMRs of mammals and birds even when they are of the same size; but only when we have factored out the size effect can we begin to examine the causes of the size-independent variation.

This raises a second important issue that has only recently been properly taken account, which is the role of taxonomy. With size taken out, it commonly transpires that the next most important factor is what taxon the animals belong to. For example, we have seen that marsupial mammals generally have about 30% lower BMRs than eutherian mammals of the same size.

Several authors have also suggested that environmental or life-history features affect BMR, for example that BMR may be low in desert species, or low in the various groups of ant-eating mammals. This can only be analysed properly once the effects of scale and of taxonomy have been taken out, and the latter is particularly difficult because species with similar lifestyles are

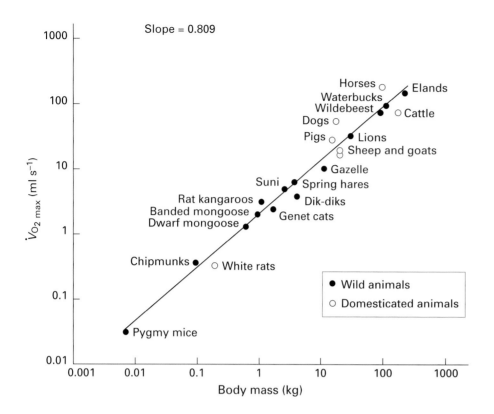

Fig. 3.6 Double logarithmic plot of the maximal rates of oxygen consumption when running on a treadmill, for 21 species of African mammals (mass range 0.007–263 kg). (Reprinted from *Respiration Physiology* **44**, Taylor, C.R. *et al.*, Design of the mammalian respiratory system. III. Scaling of maximal aerobic capacity to body mass: wild and domestic animals, pp. 25–37, copyright 1981, with permission from Elsevier Science.)

often taxonomically related. Some recent studies have begun to disentangle such problems, usually by using the techniques of modern comparative analysis (see Chapter 1). A good example comes from a recent analysis of BMR in muroid rodents (the rats and mice and their relatives). Here the species could be split into three broad strategies of living, the 'vole type', the 'mouse type' and the 'hamster type'. With size and taxonomy taken into account, it turned out that hamster strategy (rich diet; warm, dry open habitat) was associated with the lowest BMR and mouse strategy (rich diet; warm, wet dense habitat) with the highest. This may be partly explained by disproportionately large organs in the high BMR groups, particularly the organs such as heart and kidney that have intrinsically high metabolic demands. These larger organs may in turn be needed to maintain relatively higher daily energy budgets, and the two factors may therefore be 'coevolved' by natural selection in relation to particular species' environments. This kind of analysis allows us to see the real and quantifiable effects of the evolutionary shaping of physiological variables in step with environmental variation—but the first step was to understand the overriding scaling effect.

3.4.2 *The allometry of specific metabolic rate and metabolic scope*

Metabolic rate per unit body mass is referred to as **specific metabolic rate** (see Chapter 6), in other words:

specific metabolic rate = metabolic rate/M_b

Substituting from our earlier equation we get:

specific metabolic rate = $70M_b^{0.75}/M_b$

which can be rearranged as:

specific metabolic rate = $70M_b^{-0.25}$

So specific metabolic rate decreases with increasing body size, the slope of the regression equation being −0.25. This equation tells us that energy metabolism per unit mass is higher in small than in large mammals; in fact small animals like mice degrade chemical energy into heat at about 10 times the rate of an equal mass of much larger mammals like moose. This is at least partly because the smaller mammals have relatively larger surface areas and heat will dissipate more readily from them. If a moose had a specific metabolic rate comparable to that of a mouse, other things being equal it would be liable to cook its own tissues from within!

This kind of analysis also provides information on the constraints on energy metabolism that may be imposed by changes in scale. Again, some species have been able to overcome these constraints following the evolution of different designs. A good example comes from analysis of scaling in mammalian respiration. An important comparison here is the **metabolic scope** of a particular species (this is the ratio of resting rates and maximum rates of metabolism—see Chapter 6). To examine this the oxygen consumptions of animals that have been trained to run on treadmills can be measured. The rate of oxygen consumption ($\dot{V}_{O_2}$) increases with speed of running up to a maximum ($\dot{V}_{O_2\,max}$), after which further increases in speed require anaerobic

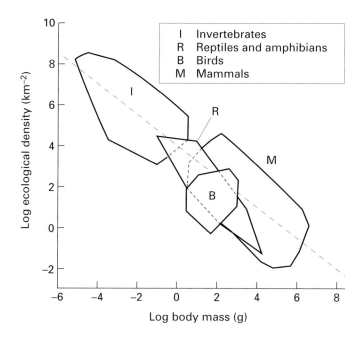

Fig. 3.7 The scaling of ecological densities with body mass for terrestrial animals. Polygons enclose all the data points for particular taxa. Density values for ectotherms were adjusted to correct for the differences in average metabolic rates from endotherms. (Adapted from Damuth 1991.)

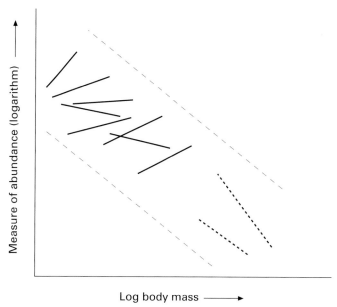

Fig. 3.8 The idealized empirical relationships between abundance and body size at different scales of analysis. For both ecological and regional densities, the overall relationship across a diverse array of species is negative, falling between the dashed lines and giving a slope of about −0.75. For guilds of larger species the individual slopes probably fit well with this, as in the lower right-hand corner (dotted lines). But within guilds of closely related small species (solid lines) the slopes are often shallower than −0.75 or even positive, showing that the larger guild members control more of the energy. (From Damuth 1991.)

metabolism and result in the build-up of lactic acid in the blood (see Chapter 6). Some of the results obtained for wild and domesticated mammals are plotted in Fig. 3.6. For the combined group of 22 species shown, the best fit equation is:

$$\dot{V}o_{2\,max} = 1.92M_b^{0.81}$$

and the 95% confidence limits on the mass exponent b are 0.75–0.87. Other studies have come up with similar values, using a wide range of mammals.

For most mammals, the aerobic metabolic scope values turn out to be around 10, meaning that maximum metabolic rate is about 10 times the resting or basal metabolic rate. But this generalization is again helpful in revealing anomalies for individual species; for example, dogs have aerobic scopes of about 30. Again this raises questions about the particular features of dogs compared with other mammals.

3.4.3 Environmental implications of the scaling of metabolic rate

Because small mammals degrade more energy per unit mass than large mammals, it should follow that a given energy supply will run out more quickly for small animals; or, looked at per unit time, that a given energy supply will support a much smaller biomass of mice than of large moose or elephants. What about actual numbers of animals for a particular energy supply or for

a given area? The abundance of animals in particular habitats is described in terms of population density. Studies spanning many taxa and a large size-range of species have repeatedly shown that the population density for each species decreases approximately as the −0.75 power of body mass (Fig. 3.7). This is the reciprocal of the scaling of individual metabolic requirements discussed above, which means that the overall amount of energy each species uses per unit area of its habitat is roughly independent of body mass.

In practice, guilds of species that include some medium to large mammals usually do show slopes for population density against body mass equal to or more steeply negative than −0.75. This will indeed cancel out the effects of individual metabolic rate (MR) scaling, so that from the point of view of energy control at the ecological level there may be no particular advantage to being a large or a small animal. But a different pattern is found when ecological densities are compared within guilds made up of species of small birds or small mammals that are closely related. In these cases, the slopes on double-logarithmic plots are often less steep than −0.75, which means that the somewhat larger species are better at getting and using energy than smaller species from similar taxa (Fig. 3.8); in ecological terms they are likely to be better competitors. Again the importance of considering taxonomic relations is underlined.

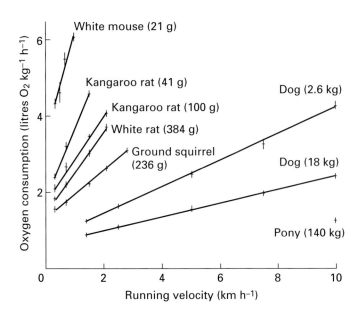

Fig. 3.9 Running velocity plotted against $\dot{V}O_2$ and size, for various running mammals. Oxygen usage increases with running speed, and for each species the increase is linear, but oxygen usage increases more steeply for a small species than for a large one. (Adapted from Taylor *et al.* 1970.)

3.5 Scaling of locomotion

3.5.1 *Scaling of running, swimming and flying*

Locomotion is one of the most obvious activities in which animals use up metabolic substrates. Scaling equations can be very helpful in showing up the relative costs of different kinds of locomotion at different velocities and for different body sizes.

For running animals on land, oxygen consumption increases with speed of running as would be expected, and the slope of this increase is greater for small animals (Fig. 3.9: note that this is

a linear plot, not a logarithmic one). It is more useful for inter-specific comparisons to look at the 'net cost of transport', C, this being the additional energy cost (over and above resting MR) per unit distance moved, or even better to look at the 'specific net transport cost', i.e. the additional energy an animal uses in moving 1 kg of its body mass over a specified distance (C/M_b). The relationship between specific net transport cost and body mass is shown in Fig. 3.10; the mass exponent is −0.28 for the entire data set, telling us that the cost of travelling a unit distance is relatively smaller for larger animals. The figure also illustrates running in reptiles, birds and mammals, and it is evident that birds (mass exponent −0.32) have rather higher transport costs than mammals, and amphibians and reptiles usually have slightly lower costs. For land animals as a whole, this parameter (C/M_b) also turns out to be unaffected by the velocity of movement; in other words, the greater cost of running faster is almost exactly offset by the shorter time needed to cover the distance.

Specific total transport costs are rather higher than the figures so far discussed, because there are other costs involved, such as the need to maintain the correct (usually rather upright) posture while running. These secondary costs are relatively higher for larger animals; for example, because of scaling effects on the skeleton, elephants have to keep their legs rather straight and their body carefully upright to avoid excess stress on their limbs and potential bone breaks. Similarly, large animals require more energy to start and to stop, and take longer to change velocity, because of the greater difficulty of overcoming their own inertia.

The muscles of small animals consume much more energy per gram during steady running than do the muscles of large ones. This is partly due to the general phenomenon of raised specific MR in smaller animals, but also probably because of the increased stride frequency for small animals, so that they must turn their muscles on and off at a higher rate to generate the forces required to move at an 'equivalent speed'. If the cost of each stride in a land mammal is multiplied by the stride period,

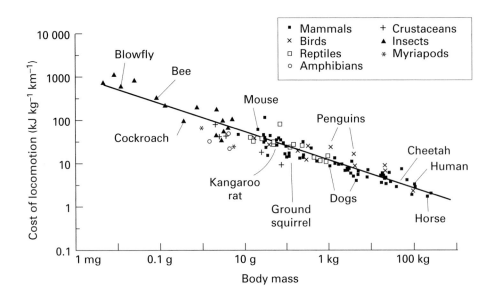

Fig. 3.10 The specific cost of transport in running mode for animals of various kinds, on a double logarithmic plot. To move a unit mass of their tissue a given distance is more expensive for small animals regardless of their taxon or the number of legs they may have. Differences in slope between taxa are slight.

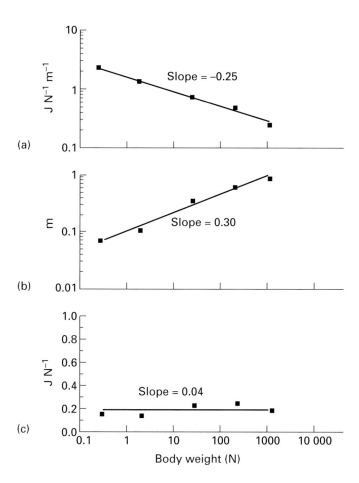

Fig. 3.11 Scaling of strides and the specific cost of transport, based on studies of five species of mammal ranging from horses to kangaroo rats. In (a) the specific cost of transport is shown to decrease with weight; (b) shows step length increasing with weight; and (c) shows the product of these two as an almost constant 'cost coefficient' independent of body size. (Adapted from Kram & Taylor 1990.)

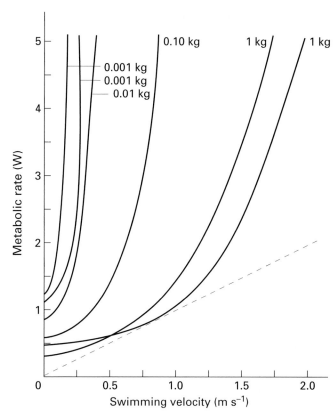

Fig. 3.12 The effects of swimming velocity on metabolic rate for a range of different sizes of fish, showing the sharp rise in costs above a particular size-dependent velocity. For each curve, the intersection with a straight line through the origin (as shown for a 1 kg fish) gives the velocity for minimum total cost of transport. (From Peters 1983; courtesy of Cambridge University Press.)

effectively normalizing for these differences, then the mass-specific cost of locomotion per stride is similar (slope = 0.04) over four orders of body mass (Fig. 3.11).

As a result of the various factors discussed above, almost 85% of the variation in the costs of terrestrial locomotion can be explained by variations in body mass. This turns out to be true not just for mammals but also for a range of other land vertebrates (including even some limbless examples such as snakes), and for many invertebrates with radically different skeletal structures, leg architectures, and numbers of legs.

The discussion so far has dealt with terrestrial locomotion—walking, running, hopping, or even the sinusoidal movements of snakes. When it comes to aquatic and aerial locomotory patterns the scaling relationships become more complex. Swimmers expend little energy in postural control, but do need to do work against the viscous resistance of the medium. The MR of swimming animals increases with speed, but the increase is not linear; above a particular size-dependent velocity, MR increases

sharply (Fig. 3.12). This effectively puts an upper limit on swimming speeds, and the range of achievable speeds is particularly narrow for small animals. Fish do normally swim at the speed at which total transport costs are minimal, but as expected this is more flexible for larger fish than for small ones. Once again, in spite of the different designs and lifestyles among fish, the energetic cost of swimming per unit body mass over unit distance appears to be fairly similar for all species studied, though declining somewhat with increased size.

For flying animals (birds, bats and insects), there is the additional complication that stopping all forward motion does not save energy but rather increases the need for it, as the animal then has to hover. Hovering may be at least twice as expensive as forward flight, and therefore for some birds the MR against velocity plot is strongly curved (Fig. 3.13a) with a clear optimal velocity where costs are minimized. However, the figure also shows that for other fliers, including some birds and all the bats and insects so far tested, the relation between velocity and MR is fairly flat; for example, the bumble-bee *Bombus terrestris* uses about 350 W kg^{-1} at all speeds. Note also (Fig. 3.13b) that the specific costs per unit of body mass decrease with mass,

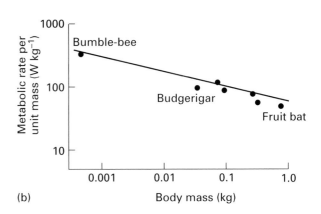

Fig. 3.13 (a) The relation between flight velocity and specific metabolic rate. Initial experiments (as with the budgerigar) showed a strongly U-shaped curve; but not all animals show this, and in many the relation is almost flat. (b) Shows increasing specific metabolic rate with body mass. (Adapted from Alexander 1999.)

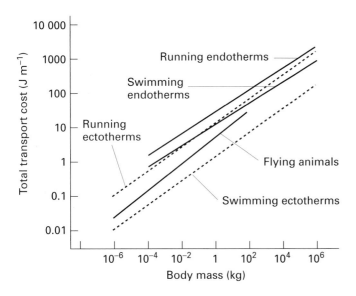

Fig. 3.14 Total transport costs associated with swimming, running and flying in ectotherms (dashed lines) and endotherms (solid lines). (From Peters 1983; courtesy of Cambridge University Press.)

as discussed earlier. However, the minimum total transport costs for flight, like those for running and swimming, increase with increasing body size; Fig. 3.14 shows that for any given body size, the lowest total costs are associated with swimming and the highest costs with walking or running, with flight being intermediate.

The effect of body size on locomotory velocity can also be estimated as shown in Fig. 3.15, and (as expected) maximum speed does increase with increasing size. The exponents are 0.14 for flying, 0.35 for swimming, and 0.38 for running, indicating the lesser sensitivity of flight speeds to body size. For animals of similar size, flying is fastest and swimming slowest. For the largest animals in each category, though, the maximum speeds are not very different for the three locomotory modes, with flight, running and burst swimming all achieving a peak of around 20 m s⁻¹ (calculated for a 1 kg flier, 10 kg runner and 100 kg swimmer, respectively). Measurements of the speeds that animals move at in their normal daily routines indicate that land animals and aquatic animals rarely approach their maximum velocities, whereas flying animals generally do operate closer to their maximum flight velocities (Fig. 3.16).

3.5.2 *Environmental implications of locomotory scaling*

Given the known maximum and average speeds for animals of particular body sizes, it is feasible to estimate the range over which they can operate, either on a daily basis or as distance from a central point such as a nest. In the latter case a 'foraging radius' can be assessed (Fig. 3.16). The resultant foraging area is clearly largest for flying animals, and increases with size within each locomotory mode; this will have implications for territory sizes, and for the habitat (and hence diet) quality required by particular species.

Similarly, different sizes and locomotory modes affect the ability to migrate effectively. For any given migration time, flying animals will be able to complete a much longer distance than walkers or swimmers of similar size; or, to achieve the same distance in the same time, walkers or swimmers would have to be much larger. Global terrestrial or marine migrations are therefore rather slow and prolonged, often taking many months

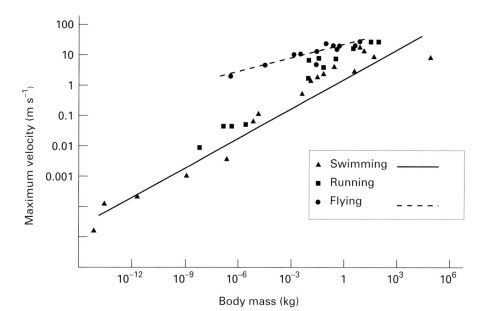

Fig. 3.15 Maximum velocity against body size; the top speed of an animal increases with body size for all locomotory modes. For animals of similar size, flying is the fastest mode and swimming the slowest (though flight velocity is relatively insensitive to size and the *difference* in speed is therefore smaller for large animals). (From Peters 1983; courtesy of Cambridge University Press.)

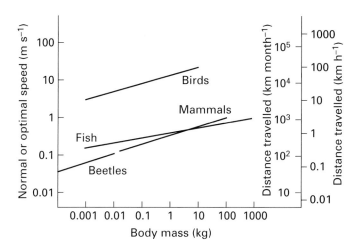

Fig. 3.16 Normal or optimal speeds measured for beetles, fish, birds and mammals, showing that only birds normally approach their maximum velocities of about 20 m s⁻¹. The right-hand axis shows how this influences ecologically relevant factors: the effective foraging radius that results (km travelled in 1 h spent foraging), and the effective migration distance (km travelled in 1 month, moving for 12 h per day). (From Peters 1983; courtesy of Cambridge University Press.)

mental stress; birds are more likely to migrate, whereas terrestrial animals, especially smaller ones, are often more inclined to escape by hibernation and torpor. The smallest migrating mammals have a body mass of about 20 kg (some African antelope), whereas very small birds and even the 1 g monarch butterfly are regular migrators. Many kinds of aquatic animals of all sizes migrate, whether latitudinally or vertically through the water column.

3.6 Conclusions: is there a right size to be?

Scaling and size-dependency are crucial factors in all comparative physiology, and the examples above amply demonstrate the importance of thinking quantitatively and of avoiding oversimplistic interpretations of trends and patterns in animal adaptation. We will meet many other examples, from all environments, where scaling effects must be incorporated in the analysis to understand what is really going on and must be 'taken out' before any other more specific adaptive mechanisms can be properly revealed.

This still leaves us with the question of what really determines the size of a particular species of animal. Clearly there are crucial interacting constraints on size due to:
1 phylogenetic inheritance (e.g. insects are essentially 'small', vertebrates are 'large', and the two groups only just overlap in body mass);
2 basic physiological design (e.g. animals with open blood systems are larger than those with no circulation, and those with closed blood systems may be larger again);
3 basic mechanical design (e.g. animals with hydrostatic skeletons are usually relatively small, as are those with exoskeletons, while those with tubular endoskeletons are relatively large);
4 habitat (e.g. any given design may be larger in aquatic habitats giving inherent support than on land where self-weight is a problem).

(as in salmonid fish), whereas aerial migration by birds can be completed in a matter of days or weeks. This picture is clearly complicated by the need for food, as few animals undertake non-stop migrations relying entirely on their food stores, though many do accumulate fat stores of 25–50% body weight before migrating. Because fish are generally ectothermic, with roughly 10-fold lower MRs than similar sized birds and mammals (see Chapter 6), they can in practice migrate much further without feeding, and have the greatest overall migratory capability.

The scaling effects on migration may have important implications for the 'avoidance strategy' of choice in times of environ-

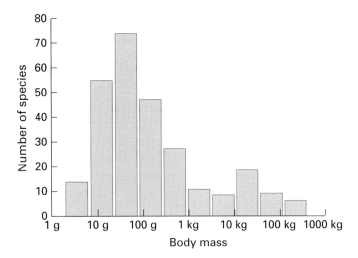

Fig. 3.17 Patterns of body sizes of mammals, for terrestrial Palaearctic species, showing the typical 'right-skewed' distribution and the apparent 'optimum' at about 100 g. (From Purvis & Harvey 1997.)

But within any one design and habitat, what determines species and individual sizes? Again very obvious selective pressures that relate to physiological balances may be invoked, and we discuss in Chapter 14 why either small or large (but not medium) sizes may be favoured in various extreme environments. The best studied examples of size limits and size distributions are the mammals, and this is therefore an instructive example to end with. Mammals show a very clear-cut pattern that also appears in other taxa, that of the 'right-skewed distribution' of body masses so that the modal mass is well below the mean mass (Fig. 3.17); put simply, there are unexpectedly large numbers of rather small species. This observation seems to suggest that there is an optimum body size for a given taxon, which for mammals is set at about 100 g. It may be that there are evolutionary reasons for this, relating to the time of origin of a taxon and a putative tendency to increase in average size through time. Ecological reasons are also feasible. One view might relate to the carrying capacity of habitats for rather a lot of small species but only a few large ones; this is given some support by the observation that within mammals larger land masses support both larger 'large species' and smaller 'small species', giving a greater overall size range. It may also be linked to the ecological pyramid or food web concepts where only a few large (carnivorous) species can be adequately fed by a larger number of small species as prey. One recent analysis suggested yet another reason, which is primarily a function of life histories, with mammals having a maximum 'reproductive power' (the rate at which acquired energy is channelled into offspring) at about 100 g. However, this would suggest that the scaling of reproductive power with size should change sign at this optimum size, and a test of the model using large numbers of species of bat has failed to find any support for this prediction. For both mammals and for other taxa it seems that the jury is still out on just what is the

right size to be, but undoubtedly there is an interaction of physiological, ecological and evolutionary factors here. An interesting consequence may be that interspecific allometries for key factors such as assimilation, respiration and mortality may be 'by-products' of body size optimization, with natural selection acting to switch allocations of resources to production and to growth independently in different species; in that case, many of the interspecific comparisons that produce clear allometric relationships may be somewhat artefactual, and should not be used uncritically in considering the 'big questions' of ecology and life history.

Further reading

Books

Alexander, R.McN. (1996) *Optima for Animals*. Princeton University Press, Princeton, NJ.
Alexander, R.McN. (1999) *Energy for Animal Life*. Oxford University Press, Oxford.
Calder, W.A. (1984) *Size, Function and Life History*. Harvard University Press, Cambridge, MA.
Harvey, P.H. & Pagel, M.D. (1992) *The Comparative Method in Evolutionary Biology*. Oxford University Press, Oxford.
Pennycuick, C.J. (1992) *Newton Rules Biology*. Oxford University Press, Oxford.
Peters, R.H. (1983) *The Ecological Implications of Body Size*. Cambridge University Press, Cambridge.
Schmidt-Nielsen, K. (1984) *Scaling. Why is Animal Size so Important?* Cambridge University Press, Cambridge.

Reviews and scientific papers

Atkinson, D. & Sibly, R.M. (1997) Why are organisms usually bigger in colder environments? Making sense of a life history puzzle. *Trends in Ecology and Evolution* **12**, 235–239.
Biewener, A. (1989) Scaling body support in mammals; limb posture and muscle mechanics. *Science* **245**, 45–48.
Blackburn, T.M. & Gaston, K.J. (1996) On being the right size: different definitions of 'right'. *Oikos* **75**, 551–557.
Chai, P. & Millard, D. (1997) Flight and size constraints: hovering performance of large hummingbirds under maximal loading. *Journal of Experimental Biology* **200**, 2757–2763.
Damuth, J. (1991) Ecology—of size and abundance. *Nature* **351**, 268–269.
Damuth, J. (1993) Copes Rule, the Island Rule, and the scaling of mammalian population density. *Nature* **365**, 748–750.
Damuth, J. (1994) No conflict among abundance rules. *Trends in Ecology and Evolution* **9**, 487–490.
Elgar, M.A. & Harvey, P.H. (1987) Basal metabolic rates in mammals: allometry, phylogeny and ecology. *Functional Ecology* **1**, 25–36.
Full, R.J. & Tu, M.S. (1991) Mechanics of six-legged runners. *Journal of Experimental Biology* **148**, 129–146.
Hedenstrom, A. & Alerstam, T. (1995) Optimal flight speed of birds. *Philosophical Transactions of the Royal Society of London* **348**, 471–487.
Jones, K.E. & Purvis, A. (1997) An optimum body size for mammals? Comparative evidence from bats. *Functional Ecology* **11**, 751–756.
Kozlowski, J. & Weiner, J. (1997) Interspecific allometries are by-products of body size optimisation. *American Naturalist* **149**, 352–380.
Kram, R. & Taylor, C.R. (1990) Energetics of running: a new perspective. *Nature* **346**, 265–267.

Marquet, P.A. & Taper, M.L. (1998) On size and area: patterns of mammalian body size extremes across landmasses. *Evolutionary Ecology* **12**, 127–139.

Packard, G.C. & Boardman, T.J. (1988) The misuse of ratios, indices and percentages in ecophysiological research. *Physiological Zoology* **61**, 1–9.

Pennycuick, C.J. (1997) Actual and 'optimum' flight speeds: field data reassessed. *Journal of Experimental Biology* **200**, 2355–2361.

Purvis, A. & Harvey, P.H. (1997) The right size for a mammal. *Nature* **386**, 332–333.

Ritchie, M.E. & Olff, H. (1999) Spatial scaling laws yield a synthetic theory of biodiversity. *Nature* **400**, 557–560.

Taylor, C.R., Maloiy, G.M.O., Weibel, E.R., Langman, V.A., Kamau, J.M.Z., Seeherman, H.J. & Heglund, N.C. (1981) Design of the mammalian respiratory system III. Scaling maximum aerobic capacity to body mass: wild and domestic mammals. *Respiration Physiology* **44**, 25–37.

West, G.B., Brown, J.H. & Enquist, B.J. (1997) A general model for the origin of allometric scaling laws in biology. *Science* **276**, 122–126.

Section B:
Central Issues in
Comparative Physiology

4 Water, Ions and Osmotic Physiology

4.1 Introduction

The interactions of all life on this planet with the most common molecule found here, water, are probably the most crucial aspects of any review of physiological adaptation. It is a truism to say that water is essential for life, since to a large degree water has been the determinant of life. All living processes on Earth have evolved around the very peculiar properties of this simple molecule, so that retrospectively we are often somewhat mistakenly led to think of water as being serendipitously 'ideal' for supporting life.

All living organisms are in essence a series of interconnected aqueous solutions, contained within bags of phospholipid membranes. Water constitutes 60–90% of the total body mass of animals, this proportion being greater in soft-bodied invertebrates and lower in animals with substantial amounts of stiff skeletal materials such as vertebrates and arthropodan groups. The water is distributed between the cells themselves and the various extracellular fluids (Fig. 4.1). Again, soft- and hard-bodied animals have different patterns, with higher extracellular volumes (ECV) in soft animals arising from their spacious body cavities that

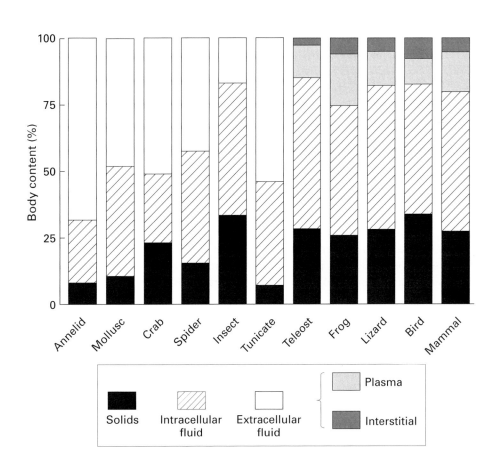

Fig. 4.1 Proportions of body solids, and the distribution of body water, in various fluid compartments for a range of animal taxa. (Adapted from Withers 1992.)

serve a hydrostatic function; for worms and slugs, ECV can be as much as 70% of total water, compared to only 20% in many vertebrates. These volumes must be maintained for the animal to function properly; but even more crucially the concentration of the fluids, particularly intracellularly, has usually to be kept reasonably constant. Yet the concentrations of many components of the internal fluids invariably differ from concentrations in the environment. This chapter is concerned with the mechanisms for keeping volumes and concentrations of biological solutions under control, and thus keeping animal tissues operative, in the face of this fundamental challenge.

4.2 Aqueous solutions

4.2.1 Properties and advantages of water as a basis for life

Water, because of its molecular architecture, has a special set of properties. It consists of two hydrogen atoms covalently bonded to a central oxygen atom; but because the O atom is quite strongly electronegative the bonds have some characteristics of ionic bonds and the water molecule exhibits a dipole effect, with each H atom bearing a slight positive charge and the O atom bearing a double negative charge (Fig. 4.2a). This in turns distorts the molecule away from linearity, giving a bond angle of 104.5°.

The presence of this dipole effect confers on water its ability to form secondary weak bonds (hydrogen bonds) with other molecules (Fig. 4.2b) via its two H atoms (essentially protons, almost bare of electrons since these have been attracted closer to the oxygen atom). Most crucially, one water molecule can bond transiently with two others, the H atoms of one water molecule being attracted to O atoms of two other water molecules; this gives a tetrahedral three-dimensional organization to liquid water. Such organization is very transitory, as it takes only 20 kJ mol^{-1} to break the hydrogen bonds (compared with about 110 kJ mol^{-1} for the H–O covalent bonds within the water molecule), so that each hydrogen bond lasts only about 10^{-10}–10^{-11} s. But at any one moment, a significant proportion of the water molecules are hydrogen bonded together. Therefore the **melting point** and **boiling point** of water are markedly higher than would be expected for similar kinds of molecule (such as NH_3 or H_2S). This means that water will occur both as a solid and as a gas, but by far most commonly as a liquid, at the temperatures experienced at the Earth's surface. As a liquid it has a relatively chaotic structure (Fig. 4.3a), with the hydrogen bonded molecules in rapid motion (described as the 'flickering cluster' structure). Owing to the hydrogen bonding, water also has curious packing properties such that it achieves its **maximum density** at 4°C, a little above its freezing point. Once frozen as ice, the water molecules are packed in one of at least nine possible conformations, the commonest being a regular open hexagonal lattice (Fig. 4.3b), and the material is somewhat less dense. Hence ice forms at the surface of water bodies, and floats on top of such water bodies. If water were a more conventional liquid, with

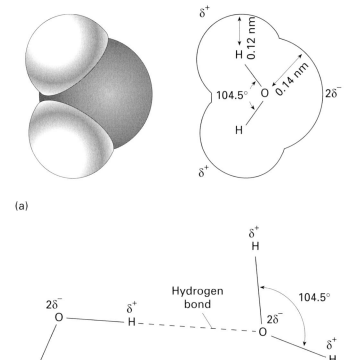

(a)

(b)

Fig. 4.2 (a) The structure of the water (H_2O) molecule, with bond angle and charge distribution. (b) The formation of hydrogen bonds between adjacent water molecules.

maximum density at its freezing point, water bodies would freeze suddenly throughout their volume. In reality, liquid water below ice is insulated from further excessive temperature change, allowing freshwater organisms to survive harsh winters while still living in a fluid medium. Sea water, with its highest density and its freezing point both at about −1.9°C, could in theory freeze to a substantial depth, but in practice it only freezes at the surface when the oceans encounter very low air temperatures, the resulting sea ice having lower density than the sea water and so floating upon it. In fact this freezing usually involves the water component only, leaving salts in solution and thus raising the salinity in the water immediately below the sea ice. However, it should be noted here that most of the ice we see at sea is actually derived from continental and glacial icebergs (a freshwater source) rather than as a result of the sea itself freezing.

The formation of ice crystals requires the presence of a nucleus of molecules onto which the water molecules can condense. As the temperature of fresh water is lowered the probability of this happening increases and at 0°C at atmospheric pressure ice crystals can form and exist in equilibrium with the liquid phase. Almost all water contains impurities (microcrystalline

'Flickering cluster' model

(a)

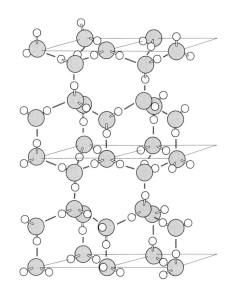

(b)

Fig. 4.3 (a) Diagrammatic representation of the structure of bulk water, bound water and membrane-linked water in a cell (dotted lines are hydrogen bonds). (b) The structure of ice.

dust particles, often called 'motes') which absorb water molecules and act as catalysts for freezing by increasing the surface area for nucleation. It is possible to make ultra-pure water which does not freeze at normal pressures until about −40°C. Water cooled to below 0°C without freezing is referred to as 'under-cooled' or 'supercooled', and many animals rely on supercooling to survive freezing conditions. It is also worth noting that water in animal tissues that is trapped between macromolecules behaves differently from water in the bulk cytoplasm, since it is unable to assume a stable lattice structure and hence is unfreezable; around 35% of the water in a cell is in this so-called 'bound state'.

The hydrogen bonding in water also acts as an energy store (again 20 kJ mol⁻¹), and when ice forms at low ambient temperatures and the crystalline structure of Fig. 4.3(b) results, almost all the water is hydrogen bonded. Hence the melting of ice, which involves breaking a proportion of these bonds, requires a substantial **latent heat of fusion** (6 kJ mol⁻¹). Similarly, vaporizing water involves breaking all the remaining hydrogen bonds, giving an even greater **latent heat of vaporization** (40–44 kJ mol⁻¹ depending on the temperature at which the vapour phase forms; equivalent to about 2.5 kJ g⁻¹ or 580 cal g⁻¹). Many terrestrial animals exploit this property during evaporative cooling, as an excellent means of controlling body temperature.

This internal bonding in water also results in a substantial **surface tension**, exploited by many small animals living at aquatic interfaces. Small animals with hydrophobic cuticles are able to live supported on the water surface of ponds and rockpools, because penetrating the water surface in either direction is quite

difficult. Special arrangements of hydrophobic surfaces or hairs may be required for submerged animals such as fly larvae to breathe air through siphons, or for terrestrial animals such as damselflies to lay eggs below the water surface. Internal bonding in liquid water also results in a very low **compressibility**, so that water is an excellent hydraulic medium to use in hydrostatic skeletons for locomotion in the many groups of worms and other flexible-bodied creatures.

Similarly, water may develop weak bonds with macromolecules and with biological membranes, to give relatively organized **boundary layers** within cells. It may also bind quite strongly to ionic moieties, so that small cations and anions always carry with them a substantial 'shell' of bound water, affecting their transit across membranes and their rates of movement through cytoplasm.

Water also has a small inherent **capacity to ionize**. As the internal hydrogen bonds break and reform every so often three hydrogen atoms will be left associated with one molecule and only one with another, forming H_3O^+ and OH^- ions (hydronium and hydroxyl ions, respectively). The former is often written as H^+, as if it were just a proton, but in fact it always exists as the hydronium ion. Only a very small proportion of water molecules are ionized at any one time, the concentration of the two ions being around 10^{-7} mol l⁻¹. However, an existing hydronium ion can readily donate its spare proton to another water molecule, or displace one of that recipient's protons, and protons may thus appear to 'cascade' through an aqueous system even though any one proton may move only a short distance. Proton cascade conduction through water may play an important part in some biochemical reactions.

The ionization property of water has two other important consequences. Firstly, we can see that technically H_3O^+ is an acid (i.e. it can donate a proton) and OH^- is a base (it can receive a proton). Water can thus act as either an acid or a base, i.e. it is **amphoteric**. It can therefore neutralize charged groups dissolved in it. This is especially important for dissolved amino acids and proteins, as in solution these contain both positively charged amino residues ($-NH_3^+$) and negatively charged carboxyl residues ($-COO^-$); which form predominates will depend upon the acidity of the overall solution, and at normal cellular pH most proteins bear a net negative charge.

The second major consequence of water being ionized is that its **electrical conductivity** is somewhat greater than would be expected, as H_3O^+ and OH^- ions can act as carriers of charge. The conductivity of pure water is fairly low, but when any electrolytes are present it is greatly enhanced.

This brings us to the final point: above all, water acts as a superb **universal solvent**. It can dissolve crystalline ionic compounds such as NaCl, because the dipole on the water molecules can overcome the electrostatic attractions between individual Na^+ and Cl^- ions. Anything ionic will dissolve in water (although certain ions such as carbonate and silicate dissolve only very slightly and are thus often used for making shells and other 'permanent' biological structures). It can also dissolve a range of non-ionic and organic compounds such as sugars and alcohols, as these also have polar properties; and can react with partially polar, technically 'amphipathic', molecules such as soaps (sodium oleate and related compounds), which have hydrophilic 'heads' and hydrophobic 'tails', dispersing the solute into tiny droplets or 'micelles'. Only completely non-polar molecules fail to dissolve in water. In a variety of ways, all of these solvent properties are exploited in living cells.

4.2.2 Concentrations of aqueous solutions; definitions and measurements

Aqueous solutions have to be described in terms of the concentrations of their various solutes, and here there are several options encountered in the biological literature (see Box 4.1). Although **molality** is technically a more 'correct' term, in practice it is most convenient to use **molarity**, as it is more easily measured and molar solutions are easy to prepare volumetrically. The differences between molar and molal measurements in aqueous solutions are nearly always small enough to be ignored.

A 1 molar (1 M) solution contains 1 mole of solute (i.e. its molecular weight in grams) dissolved in 1 litre of solution. Thus a 1 molar solution of sodium chloride contains 58.4 g NaCl per litre of solution. Most biological solutions are 1 molar or less, so that for individual component concentrations it is much more convenient to use the millimolar range. For example the concentration of Na^+ ions in human blood is about 140 mM.

Molarity is a very commonly used measurement in both chemical and biological science; but biological studies also use another kind of concentration measure, that of osmolality or more usefully **osmolarity**, as this is a particularly simple way of expressing the overall concentration of a fluid, and of consider-

---

Box 4.1   Measurements of concentration in biological systems

The chemical properties of a molecule in solution are usually adequately described by its **concentration** (C), although in concentrated media molecular interactions can alter its behaviour so that its **activity** (A) is slightly less than its concentration ($A = \delta C$, where δ is the activity coefficient).

Concentration can be measured in a variety of units:
molarity—moles of solute per litre of solution;
molality—moles of solute per kilogram of solvent.

Osmotic concentration (Π) has the equivalent measures:
osmolarity—total moles of all solutes per litre of solution;
osmolality—total moles of all solutes per kilogram of solvent.

Molal and osmolal solutions in biology always relate to water as the solvent, so that they can be determined as moles per litre of solvent.

In practice when making up artificial solutions the difference is between weighing out the solutes and making them up to a total volume of 1 litre in a volumetric flask, or weighing them out and then adding a premeasured 1 litre of water, giving a somewhat greater final volume.

Chemical potential (μ) can also be used: this is related to concentration in a logarithmic fashion, as follows:

$$\mu = \mu^\circ + RT \ln C$$

μ° is used to specify a standard chemical potential, R is the gas constant and T the absolute temperature. For most biological systems these can be regarded as constants, and μ is taken as proportional to $\ln C$.

However, chemical potential also depends on factors other than concentration. Pressure affects it, as does gravity, and an electrical potential can also have an effect in biological systems. The overall equation becomes:

$$\mu = \mu^\circ + RT \ln C + VP + zFE + mgh$$

where V is the molar volume of the molecule, P the pressure, z the charge on the ion, F the Faraday constant, E the electrical potential, m the mass of the molecule, g the gravitational constant and h the height above ground level.

This more complex equation is normally needed only to compare the chemical potential between two compartments of a system and thus determine which way the molecule of interest will move. Many of the constants therefore cancel out, while P, E and h can all be expressed as differences (ΔP, etc.) between the two regions.

Water potential (ψ) is often particularly useful when looking at movements in different phases (e.g. from liquid to vapour). This is the water chemical potential, as above, divided by the partial molar volume of water, and is given by:

$$\psi = P - \Pi + \rho_w gh$$

where ρ_w is the density of water and overall units are as a pressure (kPa or Torr, etc.). For a liquid it is a composite of osmotic potential, pressure potential and hydrostatic potential.
For water vapour, the equation is:

$$\psi = RT/V \ln (RH/100) + \rho_w gh$$

where RH is relative humidity. This is a composite of a vapour pressure potential and a gravitational potential.

ing how water movements may occur between biological fluid compartments. Osmolarity measures how many solute particles are present in a fluid in total, so that all the constituent solutes contribute to it; for a simple mixture where 100 mM glucose solution is added to 100 mM sucrose solution, the concentration of the resulting solution can be given as 200 milliosmolar or 200 mOsm. Hence the overall concentration of a cellular or extracellular fluid can be given as a single value of osmolarity. This overall **osmotic concentration**, when compared for different adjacent solutions separated by a permeable surface, will indicate which way water will move through the system; it will always flow into the osmotically more concentrated fluid, and out of the more dilute fluid. More concentrated fluids are termed **hyperosmotic**, and more dilute fluids are hypo-osmotic or **hyposmotic**; where two fluids have the same osmotic concentration they are iso-osmotic or **isosmotic**.

Osmolarity is therefore a very useful concept, but it is not especially easy to measure directly. Another way of describing concentration in aqueous solutions is to give a related parameter. Particularly useful here are the **colligative properties**, a series of properties of fluids all of which are related specifically to the number of particles present in a solution but *not* to the nature, size or charge of these particles. Colligative properties include osmotic concentration itself, already referred to, where as we have seen the concentrations of different kinds of molecule can be treated additively irrespective of their nature. But they also include depression of freezing point (FP), raising of boiling point (BP), vapour pressure (VP) reduction, and increasing of refractive index (RI). Any solute particles present in any solution will depress the freezing point and reduce the vapour pressure, while raising the boiling point and the refractive index of that solution, and always in a predictable manner. This means that any of these secondary characteristics of a solution can be used to measure its overall osmotic concentration. Simple refractometers, which measure the bending of light caused by a thin film of the solution introduced between two glass prisms, are a convenient way of checking concentrations in many areas of biological fieldwork. Instruments that measure melting points of fluids (rather than freezing points, which may be affected by supercooling) are even more frequently encountered, and this principle forms the basis of many kinds of laboratory osmometer. Any salt dissolved in fresh water lowers its freezing point, hence the common practice of putting salt on roads during cold nights. Theoretically a 1 molal aqueous solution of an ideal dissolved non-ionic molecule has a freezing point of −1.86°C, and this factor can be used to convert measured freezing point depressions (FPD) into milliosmolar osmotic concentrations.

All that we have said so far is true for non-electrolytes, where molarity and osmolarity are effectively equivalent. But since all the colligative properties depend on the actual number of dissolved particles, electrolytes that by definition dissociate into more than one particle have a much greater effect than their molarity predicts, and this is another reason why the concept

of osmolarity is often particularly useful. Sodium chloride, for example, will dissociate into Na^+ and Cl^- ions, giving *two* moles of particles for each mole of solute, and thus giving roughly twice the freezing point depression expected. Calcium chloride, $CaCl_2$, with three ions when dissolved, will give roughly three times the freezing point depression. Hence freezing point measurements for ionic solutions can still be directly converted into osmotic concentrations and could also be used to calculate molar concentrations. In practice, the conversions do not quite work. This is because solutions of electrolytes do not behave 'ideally' by dissociating fully into their constituent ions; instead, the various ions within the solution show a degree of interaction, cations and anions attracting each other and also attracting the water molecules, so that the solution is incompletely dissociated. Each ionic molecule therefore has its own **osmotic coefficient** (sometimes described as the 'activity coefficient'), expressing the extent to which it does dissociate. To complicate matters further, this coefficient also varies with concentration, with increasing interaction and a lower coefficient as concentration increases. For most biologically important electrolytes the osmotic coefficient is high, usually above 0.8 (80% dissociation) in the normal range of concentrations, so that the osmolarity can be roughly estimated as the expected multiple of the molarity. For example, if a 1 molar aqueous solution of NaCl is made up its osmolarity should be about 2 Osm, and the expected freezing point depression will be about $2 \times 1.86°C$. In fact the measured freezing point is only −3.38°C, indicating an actual osmolarity of 1.82 Osm, so that the osmotic coefficient here is 0.91.

4.3 Passive movements of water and solutes

4.3.1 *Movements through solutions*

The movements of both water and solutes through living tissues and between different compartments of living systems are very complex, but can be seen as following a few relatively simple rules and principles.

The most fundamental rules are those governing **diffusion**, described by Fick's law (Box 4.2) as depending upon the diffusion area, the concentration gradient of the substance moving, and a constant known as the diffusion coefficient (which in practice is often substituted by a parameter such as permeability, *P*). No energy is required or released, the process depending entirely on the inherent kinetic energies of the moving molecules. All diffusional movement therefore usually takes place down a gradient, whether of chemical concentration or heat or electric potential.

But in the special case of ions, diffusive movement is sometimes against a chemical gradient, because ions may also be driven by differences in electrical charge, and most biological membranes do have a resting potential across them, with the cell contents negatively charged relative to the extracellular fluids (see below). Thus positive ions may move into a cell, or negative ions move out of it, against their chemical gradients; but in all

---

Box 4.2   Fick's law of diffusion

Following the argument in Box 4.1 on measures of concentration, we know that the overall **chemical potential** of a molecule in solution is given by:

$$\mu = \mu° + RT \ln C + VP + zFE + mgh$$

A gradient of chemical potential is the driving force for diffusion, so **diffusional flux** (J) over a distance x will be directly proportional to the chemical potential gradient $\Delta\mu/\Delta x$.

Given two regions (concentrations C_1 and C_2) separated by a surface of area A over which exchanges can occur, the flux in one direction can be calculated from the concentration, the gradient of chemical potential, and the ability of the molecule to move (its mobility u):

$$J_1 = uAC_1(-\Delta\mu/\Delta x)$$

The flux in the opposite direction (J_2) can be calculated similarly:

$$J_2 = uAC_2(-\Delta\mu/\Delta x)$$

Since A and $\Delta\mu$ and Δx are the same in both directions, the net flux is therefore:

$$J_{net} = J_2 - J_1 = uA(C_2 - C_1)(-\Delta\mu/\Delta x)$$

which can be simplified (again see Box 4.1) as:

$$J_{net} = -uRTA\,\Delta C/\Delta x$$

The various constants (u, R, T) can be replaced by a single **diffusion coefficient** (D, which varies with temperature and with molecular weight), giving Fick's law:

$$J_{net} = -DA\,\Delta C/\Delta x$$

i.e. net rate of diffusion is proportional to concentration gradient and to exchange area.

This is of central importance in physiology in considering the diffusion of solutes in solution. The same treatment can be extended to describe diffusion of heat through an animal, and diffusion of electric charge in excitable tissue.

---

cases of diffusional passive movement these ions are moving *down* a combined electrochemical gradient.

4.3.2 *Passive movements through membranes*

Membrane structure and properties

The structure and chemistry of cell membranes have been intensively studied since their essential **lipid bilayer** character was established early this century. The bilayer is clearly visible in electron micrographs (Fig. 4.4a), and the familiar Davson–Danielli phospholipid bilayer model of its constitution explains most of its properties. However, the current more complex model, the **fluid mosaic** structure (Fig. 4.4b), proposed by Singer and Nicholson, allows a better understanding of some kinds of membrane permeation. The membrane appears to be a somewhat disorderly lipid bilayer, with fluid properties, having an internal lipid hydrocarbon zone which is made up of the hydrophobic lipid tails, and superficial polar regions where the hydrophilic lipid molecule heads orientate outwards to the surrounding aqueous solution. The lipid components (25–50% of the dry weight) are predominantly phospholipids, especially glycerophosphatides and sphingolipids, with a small percentage of cholesterol in the outer layer acting as a structural stabilizer. Examples of key lipid structures are shown in Fig. 4.5.

There are additional orderly arrays of proteins and lipids in some parts of the membrane surface. Complex proteins straddle the structure completely, forming channels and carriers. Other proteins project only on the inside of the cell, where they form part of internal messenger systems; or only on the outside, many of the latter having short carbohydrate tails that are negatively charged and act to control cell adhesion and separation (forming a 5–7 nm 'thatch' on the outside of the cell), or act as part of external receptor mechanisms. These integral proteins generally have their non-polar portions buried in the hydrocarbon core of the membrane, and their polar regions projecting. They have a globular structure, and can be 'seen' as 5–8 nm particles in electron micrographs of freeze-etched cell membranes (Fig. 4.6); these particles are progressively destroyed when the membranes are treated with proteolytic agents. Electron microscopy has also revealed that the inner side of the membrane is linked to microtubules and microfilaments, which may provide lateral movements and rearrangements of membrane regions according to the cell's needs.

This membrane structure allows considerable possibilities of variation, both within and between cells, so that it is quite incorrect to think of a 'typical' cell membrane as if it covered all possibilities. Within cells, there may be areas of local concentration of particular lipids, creating especially fluid or especially 'solid' zones, or areas where transport proteins are concentrated to create zones of rapid permeation and hence local ion concentration gradients. Even more variation can occur between cells, since specific lipid constituents may be quite different. This can lead to cells with highly insulated and inert surfaces, such as the myelin sheath of some nerve cells; or to specifically active membranes such as those of mitochondria, where protein subunits are tightly packed together, linked to high enzymatic activity. It can also allow the production of membranes that are resistant to environmental change; for example, membranes may be altered at low temperatures or at high pressures to have lipid constituents that are less saturated and/or of shorter chain length, decreasing the melting point and so maintaining membrane fluidity (see Chapter 8). Cholesterol content may also be increased in some cell membranes under various kinds of environmental stress, producing the curious effect of making lipids more mobile when in the solid phase but less so in the liquid phase, so smoothing out the process of lipid transition and preventing sudden changes in membrane performance.

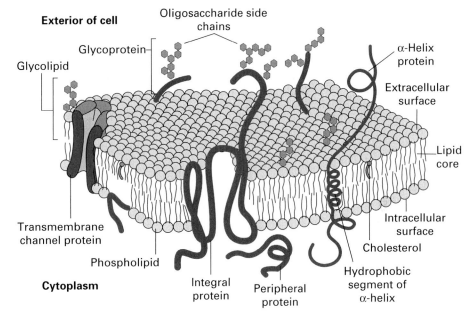

Fig. 4.4 (a) Electron micrograph of a cell membrane indicating the bilayer structure. m, Membrane; is, intercellular space. (From Villee 1977.) (b) Diagrammatic view of the 'fluid mosaic' model of membrane structure, showing the lipid constituents arranged as a bilayer, together with the various protein constituents lying in and crossing the membrane.

Membrane permeation

The permeability of any solute in any biological system, whether a whole animal, a tissue, an epithelial layer or a single cell, depends upon passage through these lipid bilayer membranes. For lipid-soluble solutes, especially those organic molecules with few or no –OH residues (potential hydrogen bonds) and for which there are no special channels or carriers, passage through cell membranes is essentially diffusional and occurs at a rate simply dependent on their lipid solubility. Simple diffusion also occurs for some small water-soluble solutes such as CO_2, which may pass readily through membranes, although water itself may move mainly through small channels. Larger water-soluble molecules need some other way of getting across membranes, and probably use specific larger diameter channels. For example, vertebrate bladder membranes appear to have many small channels for water movement, and also rather fewer larger channels permitting urea movement. Membrane permeability to water can be readily measured using radioisotopes of water (deuterium and tritium oxides), and is usually expressed in $\mu m\ s^{-1}$; the

biological range for water **diffusion permeability** across membranes is around $0.1–1\ \mu m\ s^{-1}$, and rarely up to $10\ \mu m\ s^{-1}$.

The process of **osmosis** can be seen as a special case of diffusion across membranes in which only the solvent (water, in all biological systems) is moving across the membrane (i.e. through the water channels), the solute molecules being restrained by the membrane (Fig. 4.7), which is therefore described as semipermeable. Hence the values of cell water permeability actually measured are often called **osmotic permeability**. This varies dramatically for different biological membranes (Table 4.1); although values for some skins and cuticles are very low, the values for unprotected membranes are clearly generally higher than the diffusional permeability values mentioned in the last paragraph. In some cases this may be partly because water movements are being affected by concurrent salt transport (see below), but it also indicates that actual osmotic flow of water across membranes is *not* purely diffusional. The detailed physics of osmosis remain controversial; put simply, the water molecules in the more dilute (hyposmotic) of two solutions could be thought of as randomly colliding with and passing through the

Saturated fatty acids

e.g. Acetic acid (C_2) CH_3COOH

     Lauric acid (C_{12}) $CH_3(CH_2)_{10}COOH$

     Palmitic acid (C_{16}) $CH_3(CH_2)_{14}COOH$

Unsaturated fatty acids

e.g. Palmitoleic acid (C_{16}) $CH_3(CH_2)_5CH=CH(CH_2)_7COOH$

     Linoleic acid (C_{18}) $CH_3(CH_2)_5CH=CHCH_2CH=CH(CH_2)_7COOH$

Glycerides   **Phospholipid/Glycerophosphatide**

Glycerides

e.g. Glycerol (C_3)
(a carbohydrate)

     H
     |
   HCOH
     |
   HCOH
     |
   HCOH
     |
     H

Triglyceride
(glycerol + three lipid
side chains)

     H
     |
 HCO$-$CO$(CH_2)_n$ CH_3
     |
 HCO$-$CO$(CH_2)_n$ CH_3
     |
 HCO$-$CO$(CH_2)_n$ CH_3
     |
     H

Fig. 4.5 Structures of typical biological lipids, and a pictorial view of a glycerophosphatide, phosphatidyl choline, showing the charges that give the head region its polar character.

membrane more frequently than those on the other side where water concentration is lower, giving rise to a net flow across the membrane from dilute to concentrated solutions. This movement could in theory be through natural transient pores created in the membrane by the random thermal motion of the hydrocarbon 'tails' of the membrane lipids, which might create gaps as large as 0.8 nm, large enough even for hydrated Na^+ and K^+ ions. However, this model does not fully explain the high rate of many osmotic processes, nor does it explain the action of certain hormones that can increase the rate of water flow across membranes (for example, antidiuretic hormone in vertebrate kidneys). Recent work shows that osmosis partly and perhaps predominantly occurs via specific and controllable membrane **water channels** instead, controlled by proteins known as **porins**

having a central water-filled pore. This also explains the fairly common observation of higher osmotic permeability in one direction, usually with inward osmosis happening faster than outward osmotic loss (though linked salt transport may again be confusing the issue here).

Clearly the osmotic movement of water across membranes is not entirely straightforward, and its rate varies depending upon the osmotic gradient created by transport of solutes such as NaCl, a phenomenon often described as 'bulk flow'. It is worth noting that where water is moving fast down an osmotic gradient by this kind of bulk flow system, it may sweep other small solutes along in its path, by a process of **solvent drag**, so speeding up other movements across membranes.

Osmotic movements can be prevented by applying a physical

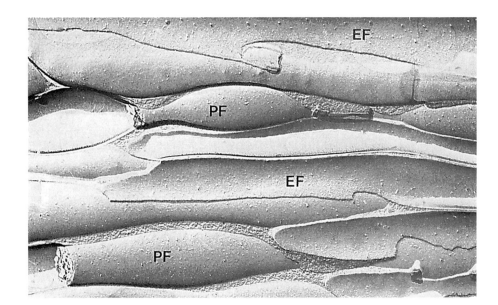

Fig. 4.6 Scanning freeze–fracture electron micrograph of the membrane surfaces of small axons in the neural connective of a mussel. Both the protoplasmic (PF) and the external (EF) faces show protruding membrane proteins as discrete particles 6–10 nm in diameter.

pressure to the more concentrated solution, exactly counter-acting the tendency for the water molecules to move across the membrane into it; this is why an osmotic difference between two solutions is often given as an **osmotic pressure** (Π, see Box 4.6 for more details).

4.3.3 *Electrochemical balance across membranes*

The Donnan equilibrium

A further complication arises in real biological solutions because of the presence of some solutes within cells that cannot pass across membranes but which bear a charge: proteins occurring in the cell in colloidal suspension are the obvious example, commonly bearing a net negative charge at the usual pH of the cell contents. Imagine a cell with a membrane permeable to all ions, where the intracellular solution contains such proteins and all other electrolyte concentrations are equal intracellularly and

Fig. 4.7 A comparison of diffusional permeabilities through biological membranes for key ions and molecules. (From Alberts *et al.* 1994, Fig. 11.2.)

Table 4.1 Osmotic permeabilities for biological membranes, skins and cuticles.

Source/species	P_{osm} (μm s^{-1})	
	Membranes	**Skins and cuticles**
Artificial collodion membrane	1200	
Protozoans		
Marine ciliate	1.0	
Zoothamnium	2.6	
Amoeba	0.5	
Soft-bodied invertebrates		
Asterias (echinoderm)		32.2
Hydra (cnidarian)		9.4
Squid axon membrane	10.8	
Arthropods		
Carcinus (crustacean)		1.4
Potamon (crustacean)		0.6
Astacus (crustacean)		0.3
Uca (crustacean)		0.24
Artemia (crustacean)		0.1
Sialis (insect)		0.05
Aedes (insect)		0.005
Crustacean axon membrane	32.0	
Crustacean muscle membrane	65.4	
Vertebrates		
Salmo (salmon)		0.6
Anguilla (eel) (freshwater)		0.8
(saltwater)		0.2
Rana (frog)		5.6
Bufo (toad)		23.6
Lizard		0.006
Snake		0.000 8
Frog muscle membrane	221.6	
Mammal cell membrane	142.8	

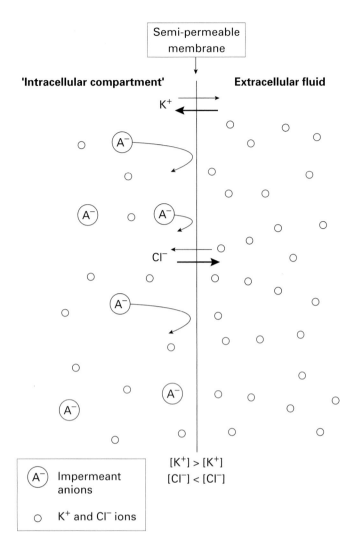

Semi-permeable membrane

'Intracellular compartment'   Extracellular fluid

K^+

A^-

Cl^-

$[K^+] > [K^+]$
$[Cl^-] < [Cl^-]$

A^- Impermeant anions

○ K^+ and Cl^- ions

Fig. 4.8 Osmosis and the Donnan equilibrium across a semi-permeable membrane. With only KCl present and both the anion and cation freely diffusible, both sides of the membrane would equilibrate at identical concentrations. But here, on one side of the membrane (equivalent to the intracellular compartment) there are additional impermeant anions; K^+ diffuses in and Cl^- diffuses out of this compartment until electrochemical equilibrium is reached, when the concentration gradients for the ions (thin arrows) are balanced by an electrical gradient (thick arrows).

Box 4.3   The Nernst equation

The Nernst equation is very important to physiologists because it allows calculation of electric potential of ions from their measured concentrations, or in reverse the calculation of concentration gradients from measured voltage difference across a membrane.

We can start with the formula for chemical potential from Box 4.1:

$$\mu = \mu° + RT \ln C + VP + zFE + mgh$$

Modified to give the difference in chemical potential between two regions this becomes:

$$\mu_1 - \mu_2 = RT \ln (C_1/C_2) + V\Delta P + zF\Delta E + mg\Delta h$$

where ΔP, ΔE and Δh are the pressure difference, electric potential difference and height difference between the two regions.

An ion in equilibrium across a membrane has the same chemical potential on both sides ($\mu_2 = \mu_1$ so $\mu_2 - \mu_1 = 0$), and we can assume the same temperature and pressure and gravitational state on both sides (ΔP and $\Delta h = 0$). So we can rearrange and calculate the difference in electrical potential as:

$$\Delta E = (RT/zF) \ln (C_1/C_2)$$

This is the Nernst equation; it is more familiar and more useful in base 10 log form, where the value of RT/F (for a univalent ion where $z = 1$) is 58 mV. Thus we can write:

$$\Delta E = 58 \log(C_1/C_2)$$

The electrical potential (membrane equilibrium potential, E) produced by a particular ion is therefore directly proportional to the log of the concentration gradient for that ion. An example is given in the text; see also Table 4.3.

---

extracellularly (Fig. 4.8). If the system is allowed to equilibrate, some positive ions (K^+) from the extracellular fluid will move into the cell to counteract the negative charge there, despite the fact that there is no concentration gradient. Similarly, some diffusible anions (Cl–) will leave the cellular compartment because of the electrical gradient, despite this resulting in a slight chloride concentration gradient in the opposite direction. Both anions and cations will continue to redistribute themselves until their overall electrical *and* chemical gradients are in balance and the tendency to move in either direction is matched; at this point

there will be a small net charge across the membrane, inside negative, resisting the tendency of any ions to move back down their concentration gradients. This equilibrium situation, naturally occurring in all cells simply because their membranes are semi-permeable and they contain non-diffusible charged molecules, is called the **Donnan equilibrium**. All cells, even without any of the more complex active transport systems considered later, would thus have an unequal distribution of diffusible ions, and a small resultant electric charge, across their membranes.

When any ion is in equilibrium across a biological membrane, the electrical charge that results across that membrane can be calculated; this most important relationship is described by the **Nernst equation** (Box 4.3). The **electric membrane potential** (E) is related to the logarithm of the concentration gradient. To take a specific example, for the monovalent ion K^+ a mammalian cell might contain a 140 mM solution with the extracellular fluid at 20 mM, so that the membrane potential can be calculated as $58 \times \log (140/20)$, or −49 mV (by convention given as the internal potential relative to the outside, hence inside negative).

The Donnan equilibrium is one of the causes of electrical potentials existing across cell membranes, and in a hypothetical cell that was only permeable to one cation, *or* that was equally

Table 4.2 Hydration states, ionic radii and mobilities for important biological ions, together with relative permeabilities through the Na$^+$ channel and K$^+$ channel in frog axon.

Ion	Atomic mass	Hydration number	Ionic radius (nm) Crystal	Ionic radius (nm) Hydrated	Mobility (μm s^{-1}/volts cm^{-1})	Na$^+$ channel (P/P_{Na})	K$^+$ channel (P/P_{K})
Li$^+$	6.94	>6	0.060	–	4.01	0.930	<0.010
Na$^+$	22.99	4.5	0.095	0.512	5.20	1.000	<0.010
K$^+$	39.10	2.9	0.133	0.396	7.64	0.086	1.000
Ca^{2+}	40.08		0.099	–	–	0	0
NH$_4^+$	18.04		–	–	–	0.160	0.130
Rb$^+$	85.47		0.148	–	–	<0.012	0.910
Cl$^-$	35.45	2.9	0.181	–	7.91	–	–
Br$^-$	79.70	2.4	0.195	–	8.28	–	–

permeable to all ions, it would on its own give a substantial membrane potential. But it usually only contributes a small part of the total potential recorded in real living cells, because several ions can leak across the membrane at different rates. There are in fact two other major reasons why cells have a so-called 'resting potential': the existence of substantial diffusion potentials resulting from differential permeability of the membranes, dealt with next; and the existence of active transport mechanisms across membranes, covered in section 4.4.

Selective permeability and ion channels

All ions have a particular mobility when in solution; some move readily, others rather slowly. A major determinant of this is frictional drag, which in turn is related to ionic size. But it is not necessarily the case that smaller ions are more mobile: as we have seen, ions carry a shell of water molecules bound to them, the 'hydration layer', so that quite small ions may become relatively less mobile, while larger ions may carry a lesser hydration shell (they have a bigger surface area and thus a lower charge density, so bind less tightly to water molecules) and may be quite rapidly mobile. Values of important hydration numbers, and both unhydrated and hydrated ionic radii are given in Table 4.2 (though these values are averages, as the water shell is itself continually exchanging with the surrounding free water). Mobilities are also given, showing that the mobility is greater as the hydration shell gets smaller, and cations of larger atomic mass are in practice *more* mobile.

Differential ionic mobility is one reason for differential ionic permeability in cell membranes. The mechanism of differentiation resides in **ion channels**, though, which are formed by pores within integral proteins in the membrane. All membranes appear to have these channels in them, and the channels are relatively specific for a particular type of ion of a particular size, charge distribution and mobility. Thus we speak of Na$^+$ channels, K$^+$ channels, Ca^{2+} channels and so on. These cationic channels will accept other cations, but will permit their transit across the membrane only slowly. For example, the sodium channel in vertebrate nerve axons is also reasonably permeable to Li$^+$ ions, but rather impermeable to K$^+$ or NH$_4^+$. The K$^+$ channel in the same cell membrane is very impermeable to Na$^+$ ions. Values of relative permeability are also given in Table 4.2; note that they are not directly related to ionic radius, since charge distribution matters as well as size. In certain other cells there are more general cationic or anionic channels, with reasonable permeability to any ion of the correct charge. Most animal cells in their resting state have quite high K$^+$ permeabilities (e.g. P_K 0.001–0.03 μm s^{-1}), moderate Cl$^-$ permeability, and relatively low Na$^+$ and Ca^{2+} permeabilities. (However, it is worth noting that all these values of ion permeability are around 5–8 orders of magnitude lower than the values of cell water permeability, P_{osm}, in Table 4.1, so that any active movement of ions will result in a rapid osmotic water movement, faster than the ion gradient can run down.)

This kind of variable selectivity must be attributable to rather specific pore structures, resulting from membrane proteins with a very precise configuration, and incorporating a 'pore' that discriminates between the hydrated ions according to their size and charge. Investigation of pore structure and selectivity has been greatly aided by the discovery of particular toxins that will block certain kinds of channel and not others. The most famous of these is tetrodotoxin (TTX), derived from the Japanese puffer fish and its relatives, which will quite specifically block the axonal sodium channel when applied to the outside of nerve cells by binding to a site on the outer part of the channel protein. Some local anaesthetics are also specific sodium channel blockers. Other toxins, derived from snakes, scorpions and sea anemones, are known to bind to other parts of the sodium channel structure, causing it to stay abnormally open and thus increasing sodium permeability.

It is worth adding here that certain channels may also include a 'gating' mechanism, the protein undergoing some kind of conformational change to open or close the pore to ion flow. This gating system may be sensitive to ion concentrations, or to certain hormones and intracellular messengers (ligand-gated channels). It may also be voltage sensitive, so that the channel opens or closes as the membrane itself is depolarized or hyperpolarized. This system of altering permeabilities via alterations to the channel protein forms the basis of excitability—the action potential—in nerve and muscle cells.

In recent years channel proteins have begun to be sequenced so that we can recognize families of related membrane channels. The familiar ion channels best known from excitable tissues

	Intracellular fluid (mM)			Extracellular fluid (mM)			Equilibrium potentials (mV)			Resting potential (mV)
	Na	K	Cl	Na	K	Cl	E_K	E_{Na}	E_{Cl}	
Squid nerve	49	410	90	440	22	560	−74	+54	−46	−68
Crab nerve	52	410	26	510	12	540	−87	+57	−76	−80
Insect nerve	86	555		212	124		−37	+23		−25
Frog muscle	10	124	2	110	2	78	−102	+59	−92	−90
Insect muscle	22	86		117	11		−51	+41		−40
Mammal muscle	12	155	4	145	4	123	−98	+67	−91	−90

Table 4.3 Equilibrium potentials in a range of nerve and muscle tissues, with the internal and external ion concentrations. Note that resting membrane potentials are generally close to E_K. (Calculated at 15°C, except for mammalian muscle which was calculated at 37°C.)

probably form one family, and the voltage-gated channels and ligand-gated channels are rather different from these. There is also another category of ion channel, however, strongly selective for sodium and blocked by amiloride, and this has a wide and puzzling distribution including mammalian kidney tubules (where its frequency is increased by the hormone aldosterone), certain snail sensory neurones, some nematode cells that are probably mechanosensors, and parts of the vertebrate brain.

The resting membrane potential

Living animal cells, then, have a membrane that is selectively permeable to a few ions (especially potassium), but rather impermeable to others. This has very important consequences for ionic distributions and electrical potentials developed across the membrane. If we consider an artificial situation, with a membrane that is selectively permeable to potassium and which has a simple concentration gradient across it resulting from 1 M KCl on one side, and 0.1 M KCl on the other, we can visualize what will happen. K⁺ ions will tend to diffuse down their concentration gradient to the more dilute solution, and in doing so will set up a small positive potential on the more dilute 0.1 M side and an equal small negative potential on the 1 M side. However, this process will not go on unchecked until the two sides have equal potassium concentrations, because the electrical potential it creates (as the charged particles move) will lead to an opposing force tending to move potassium ions in the other direction. Therefore diffusion will go on until the electrical forces and the chemical forces balance each other, and at this point there will be a potential difference across the membrane given by the Nernst equation discussed above. This calculated value will be the **potassium equilibrium potential** (E_K). It can be calculated from the Nernst equation for any particular living cell given the internal and external potassium concentrations. Similarly, the **sodium equilibrium potential** (E_{Na}) could be calculated; it would be the achieved membrane potential for a system where sodium was the only mobile ion.

If we look at the actual equilibrium potentials for individual ions in real living cells, with high internal potassium and low sodium and calcium levels, the values obtained are strikingly different from each other. Table 4.3 includes the values for different kinds of axon (including squid axon, the first cell for which these measurements were made) and for various muscle cells. E_K values are large and negative (i.e. potassium drives the cell to a negative potential internally, because positively charged potassium ions tend to move outwards along their concentration gradient); E_{Cl} levels are moderately to highly negative; but E_{Na} (and E_{Ca}) levels are relatively high and positive, because of the tendency of both these ions to move into the cell down their concentration gradients and so produce a small internal positive charge.

But in reality, of course, membranes are exposed to solutions containing complex mixtures of ions, with both cations and anions able to move across the barrier. It is quite easy to extend the Nernst relation to consider several ions at once, and the more complex resultant equation is known as the **Goldman–Hodgkin–Katz relation** (Box 4.4). The only additional parameters needed are the relative permeabilities of each of the ions involved: P_K, P_{Na} and so on. Freely permeable ions will move readily across the membrane and contribute substantially to the membrane potential E, whereas impermeant ions will not move and can have no effect on E. The net contribution of any one ion is therefore the product of its concentration gradient and its permeability. Using the full equation, the value of E for a living cell (sometimes called its **resting potential**) can be calculated; it generally works out at about −30 to −90 mV, in close agreement with the values measured using microelectrodes. In practice the equation is useful in reverse: measured resting potentials can be a tool to calculate ionic permeabilities for particular cells.

In general the Goldman–Hodgkin–Katz equation can be simplified to take account of only sodium and potassium ions (see Box 4.4), since calcium ions have a very low permeability, and chloride ions are nearly always distributed entirely passively across cell membranes. Using this simple version with the squid axon, E is calculated as −73.7 mV, in reasonable agreement with the measured value of −68 mV.

But a quick examination of the actual membrane potential in these cells shows that E is always very close to E_K. For the squid axon E_K is −74 mV with E being −68 mV, whilst for typical vertebrate muscle E_K is −98 mV and E is −90 mV. So it is often quite

In Box 4.3 the derivation of the important Nernst equation was examined, giving the relation between the electrical potential between any two compartments and the concentration difference between them for a particular molecule. This same equation describes the equilibrium potential (E, in millivolts) across a cell membrane due to a Donnan equilibrium (see text), as follows:

$$E = (RT/zF) \ln (C_1/C_2)$$

or:

$$E = 2.303 \, (RT/zF) \log (C_1/C_2)$$

or:

$$E = 58 \log (C_1/C_2)$$

In the simple Donnan equilibrium described there were just two permeant ions, K^+ and Cl^-; in a real cell there may be several. But the principle remains the same: we can calculate the final membrane potential (E) from all the notional contributing potentials *but* we must take into account the relative permeabilities of the ions. The net contribution to E of any ion species by its movement from a compartment across a membrane is the product of its concentration in the original compartment and its permeability. The Goldman–Hodgkin–Katz equation simply puts all these contributions, for each ion in each direction, together:

$$E = \frac{RT}{F} \ln \frac{[K]_o P_K + [Na]_o P_{Na} + [Cl]_i P_{Cl} + [Ca]_o^{1/2} P_{Ca}}{[K]_i P_K + [Na]_i P_{Na} + [Cl]_o P_{Cl} + [Ca]_i^{1/2} P_{Ca}}$$

The extracellular cation concentrations (e.g. for potassium this is $[K]_o$) are placed in the numerator by convention, and the intracellular cation concentrations (e.g. $[K]_i$) in the denominator, so that we calculate the potential of the inside relative to the outside. Note also that the valency or charge term z has been incorporated into the initial constant group as if it were one, which has two consequences: for the only divalent ion (calcium) the terms are raised to the power $1/z$, or $1/2$, and for the only ion where $z = -1$ (chloride) the numerator and denominator must be reversed.

For the squid giant axon, the full equation above gives $E = -73.6$ mV.

In practice, calcium ions are not very permeable across cell membranes, and chloride ions are distributed entirely passively, so the equation can be simplified to contain only the sodium and potassium terms. E is then calculated as -73.7 mV.

Furthermore, as P_K is usually much greater than P_{Na}, we can see from Table 4.3 that the resting potential E for many cells is close to E_K (in this case -74 mV). Only at very low levels of external potassium concentration, in laboratory manipulations with cells, does E differ significantly from E_K, as the contribution of sodium ions becomes more important.

branes, through selectively permeable pores. Some like K^+ and Cl^- move relatively freely having a high permeability, others like Na^+ and Ca^{2+} have much lower permeability. These differences lead to a membrane potential (E) that is always quite close to the potassium equilibrium potential E_K. This small electrical potential across living cell membranes is immensely important for many aspects of cell functioning, from simple cell adhesion and processes of secretion through to more dramatic effects such as excitability and contractility. We know that it is brought about by passive ion movements, determined by the varying concentrations of ions inside and outside cells, and the properties of the membrane and of the ion channels therein. But we also need to know how the cell concentration initially came to be so different from the extracellular fluids, setting up these ionic gradients upon which the membrane potential depends.

This brings us to the topic of active ion movements. There are ion 'pumps' in all cell membranes, which are the major reason for the unequal distribution of Na^+ and K^+ between cells and extracellular fluids; these pumps are continuously moving potassium ions into the cell and sodium ions out to produce the high-potassium cellular milieu. These pumps are therefore, indirectly and in the long term, the single most important reason for the membrane potential, by causing K^+ to be the predominant intracellular cation. In the following section we examine how active pumps work and what other effects they have in cells.

4.4 Non-passive solute movements

Diffusion and osmosis are responsible for the bulk properties and equilibrium compositions of most of the fluids found in the world. The additional type of process that controls the composition of biological aqueous solutions is mediated transport, whereby substances are moved across membranes by interaction with protein **transporters**. This may be against an electrical and/or chemical gradient, or at unexpectedly high rates that are unrelated (or non-linearly related) to concentration differences. Such movements are also characterized by being susceptible to chemical inhibition.

Mediated transport is fundamentally of two kinds. This field has for 30 years been dogged with alternative and conflicting terminologies, but the terms used here are those coming into generally agreed use now. **Primary transport processes** include all those directly dependent on coupling with metabolic reactions in the cell. The best known form of this is primary solute translocation, commonly known simply as **active transport**, where an enzyme within a membrane catalyses the movement of a solute up an electrochemical gradient. Group translocations are a special type of active transport where the transported solute is simultaneously transformed by the transporter enzyme and is released on the other side of the membrane as a different molecular species. Electron translocations are also a form of active transport, where electrons are transported between coupled membrane enzymes; these are sometimes termed 'redox potential pumps'.

acceptable to use a very simple version of the original equation (see Box 4.4) and calculate E simply as being equivalent to E_K. This is because in nearly all cells, the permeability to K^+ ions is much greater than to Na^+ ions (often at least 100-fold greater), and potassium is therefore by far the most important determinant of the resting membrane potential.

To summarize then, many ions move passively across mem-

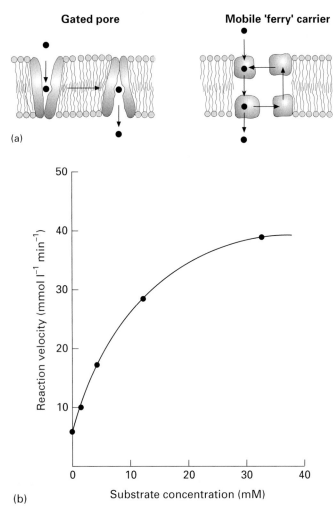

Gated pore

Mobile 'ferry' carrier

(a)

(b)

Fig. 4.9 (a) 'Gate' and 'ferry' models for protein-mediated movements across cell membranes. (b) With such carrier-mediated transport systems there is a non-linear relationship between rate of transport and substrate concentration, giving typical saturation kinetics.

These forms of active transport are all distinct from **secondary transport processes**. These include **symport** and **antiport** (collectively secondary active transport), where two solutes are transported together in the same (symport) or opposite (antiport) directions, with one of the solutes moving up an electrochemical gradient. Secondary transport processes also include the important category of **uniport**, where a single solute is moved down its electrochemical gradient and essentially passively, needing no metabolic energy and no other solute gradient, but with specific mediated interactions with membrane components involved so that the process does not have the kinetics of simple diffusion; this process has traditionally been termed **facilitated diffusion**.

All these processes involve a membrane-bound protein molecule that binds reversibly to the solute molecule at a specific site. The protein–solute complex then undergoes a change in conformation that effectively translocates the solute to the other side of the membrane, where solute and protein disengage. The protein molecule may actually move across the membrane, like a ferry, or it may span the membrane anyway and effectively act as a gate (Fig. 4.9a), guarding a pore in the membrane; both kinds of system are known to occur, and they are quite difficult to distinguish in practice.

4.4.1 Facilitated diffusion (uniport)

Uniporters are the simplest of the membrane transport protein systems, and the term may sometimes be used to include both channels (dealt with above) and facilitated diffusion carriers (or 'facilitated transporters'); here we use it only in the latter sense. The carrier proteins are distinguished from channels by having substrate binding sites that are only accessible from one side of the membrane at any one time.

Uniport facilitated diffusion is thus a protein-mediated but still 'downhill' process. Anything moving across a biological membrane by this means will not obey Fick's law but instead will show saturation kinetics, where transport rate increases hyperbolically with solute concentration and thus approaches saturation, where transport rate cannot increase further (Fig. 4.9b). A classic example is that of glucose uptake into human erythrocytes (red blood cells, or RBCs); the protein carrier for this has a carbohydrate tail on the outside of the cell membrane, but the active site for glucose binding is within the protein and appears to be exposed alternately to the outside and then the inside of the membrane, so shuttling the glucose across. Both sugars and amino acids are moved across eukaryotic cell membranes by uniporter systems, though both are also handled against their concentration gradients by symporters, discussed below. Recently it has also been shown that urea movements in vertebrate kidneys involve uniporters (facilitated transporters), especially in the terminal part of the collecting duct (see Chapter 5).

4.4.2 Primary active transport: membrane ion pumps

Active transport of any solute involves protein-mediated 'uphill' processes, and therefore ultimately requires a continuous expenditure of energy by the splitting of ATP. Energy from a readily reduced, oxidized or hydrolysed chemical bond is converted into the energy of an electrochemical gradient (usually of cations or of protons). In its most familiar form as primary solute translocation, active transport is usually mediated by a membrane protein molecule that is linked directly to ATP splitting (so that the protein carrier is an ATPase enzyme) and that combines reversibly with the substance being moved across the membrane. Active transport by ion pumps is therefore subject to typical enzyme-mediated reaction kinetics (Box 4.5) and the affinity of the pump (its K_m value), and its maximum rate (its V_{max} value), can be calculated in a standard fashion from the Michaelis–Menten equation as shown in the box.

Box 4.5 Enzyme kinetics and the K_m value

Biological reactions commonly occur under the influence of a catalyst in the form of an enzyme (E), with the overall reaction from substrate (S) to product (P) being:

$$E + S \rightleftharpoons ES \rightleftharpoons E + P$$

The rate of the forward reaction (V_f) is then $k_1[S][E]$, where k_1 is the rate constant; for most enzyme-catalysed reactions the reverse reaction (with rate k_2) is negligible and can be ignored. The relationship between reaction velocity and substrate concentration [S] is then hyperbolic as shown below:

The **Michaelis–Menten equation** describes this hyperbolic curve as follows:

$$V_f = V_{max}[S]/(K_m + [S])$$

Maximal reaction velocity (V_{max}) occurs at infinitely high substrate concentrations, so that in practice biochemists normally think in terms of the substrate concentration at half maximal velocity; this is what is termed the **Michaelis–Menten constant, K_m**.

It is difficult to estimate K_m and V_{max} from a hyperbolic plot as above, so reciprocal plots ($1/V$ against $1/[S]$) are normally used instead, given that the equation above can be rearranged as follows:

$$1/V_f = K_m/V_{max}[S] + 1/V_{max}$$

The reciprocal plot (known as a Lineweaver-Burk plot) gives a straight line that has an intercept on the $1/V$ axis at $1/V_{max}$, and an intercept on the $1/[S]$ axis at $-1/K_m$.

Values for K_m and for V_{max} can then be fairly readily found and compared between reactions.

These values are also useful to physiologists in allowing us to compare between tissues and animals such phenomena as rate of ion pumping (equivalent to rate of an ATPase-catalysed reaction).
- K_m values (usually in mM) can be seen as reflecting the 'affinity' of the pump for the ion that it moves; they are often termed K_t values instead ('transport constants').
- V_{max} values (in s^{-1}) indicate the maximum rate of pumping (ion flux) that is possible; again they are often termed J_{max} for pumps and carriers.

At present it is still unclear how many primary ion pumps (ATPases) have evolved in animal systems. Initial work indicated many different kinds operating, but recent studies have tended to reduce this profusion and indicate that relatively few basic mechanisms are at work, though they may handle ions differently in different cells according to circumstance. By far the best studied example of this kind of transport is the **Na⁺/K⁺-ATPase** or **sodium pump**, which is an ATP-splitting enzyme that works only in the presence of both Na⁺ ions in the cell and K⁺ ions outside it. This pump now appears to be common to all animal cells, usually localized in the apical membrane (the side away from the basement membrane on which the epithelium rests) but sometimes in the basal and lateral membranes (e.g. in vertebrate kidney). In many cells this pump exchanges three Na⁺ 'out' for every two K⁺ 'in'. This unequal exchange clearly produces a net movement of charge, and the pump is therefore described as **electrogenic**, or more strictly **electrophoretic**. But sometimes the same pump appears more straightforwardly to exchange one Na⁺ for one K⁺ (or to operate 1 : 1 Na⁺/Na⁺ or K⁺/K⁺ exchanges), becoming **electroneutral**. In many cells the electrogenic effect

of the pump contributes directly to the membrane potential, though this direct effect is usually only a few millivolts. The two roles that the sodium pump plays in setting the resting membrane potential can be demonstrated using inhibitors, since the Na⁺/K⁺-ATPase has a specific reaction to the drug ouabain. Inhibition of a cell by ouabain will often bring about a rapid depolarization of 2–10 mV as the electrogenic 'topping up' effect of the pump is blocked; then over a much longer time course the membrane depolarizes more substantially and the membrane potential, E, declines towards zero, as the ionic gradients upon which the passive equilibrium potential depends are gradually dissipated without the continuing action of the pumped Na⁺/K⁺ exchanges.

The Na⁺/K⁺-ATPase has been extremely well studied from many different tissues, and is now being characterized at the sequence level. It is known to be a four-part protein, or tetramer, with two α polypeptide subunits (molecular weight 110 kDa) that span the cell membrane and contain all the catalytic sites, and two smaller β subunits (35 kDa). At present there are three identified variants (isoforms) of the α unit and two isoforms of

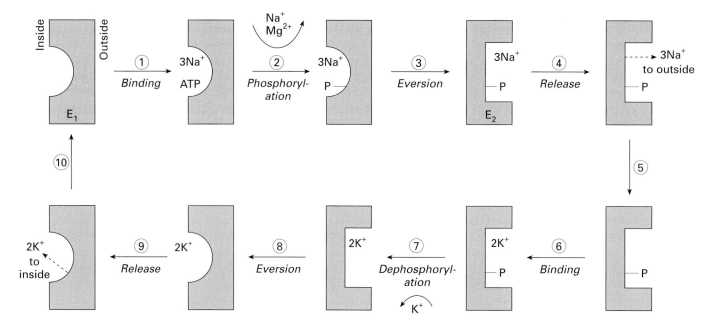

Fig. 4.10 A model for the series of changes taking place when the Na+/K+-ATPase binds with and hydrolyses ATP and simultaneously exchanges Na+ and K+ across the cell membrane.

the β unit, expressed differently according to species, tissue and developmental stage. These polypeptides have distinct regions, in sequence, that are Na+ sensitive, ouabain sensitive, and K+ sensitive, and other zones concerned with phosphorylation; these are arranged in the membrane by complex molecular folding such that there are binding sites for Na+ and for ATP on the inside of the cell, and for K+ and ouabain on the outside. The cycle of events that occurs when the pump is operative and ATP is being hydrolysed is summarized in Fig. 4.10.

A second ion pump that seems to be ubiquitous in animal cells is a calcium-activated enzyme, the Ca2+-ATPase, also usually localized in the apical plasma membrane and found at particularly high concentration in the sarcoplasmic reticulum of striated muscles. Again, it may operate electrogenically, or in partial exchange for H+ ions, according to circumstance. This pump is also made up of α and β dimers (94 kDa and 34 kDa, respectively), and these have quite substantial sequence homology with the Na+/K+-ATPase α and β subunits.

The third type of pump that appears to be universal in cells is a proton (H+) pump. Proton gradients may have been the first system developed in living systems to interconvert chemical energy from energy-rich compounds into electrochemical energy as directional ion gradients; they may date back over 3 billion years. Proton gradients arise in two ways; firstly, from 'redox potential pumps', where the electron transfer systems in cell membranes are coupled to proton translocation, and secondly, from reversible ATP-dependent proton pumps. Both are present in the same membranes, and in mitochondria they both pump protons from the centre of the organelle into the lumen between

the double membrane (see Chapter 6). This mitochondrial proton pump belongs to a gene family known as the **F-ATPases**.

In all eukaryotic cells there are also two additional kinds of proton pumps moving H+ ions out of the cell. One of these is usually an electroneutral H+/K+-ATPase (made up of α and β subunits, and again with strong sequence homology with the Na+/K+-ATPase); it appears to belong to a single gene family together with the Na+/K+-ATPase and the Ca2+-ATPase, and they are collectively designated the **P-ATPase** family. In vertebrates the P-ATPase proton pump operates in the stomach epithelium to acidify the contents in exchange for K+ ions, and is capable of operating against a 4 million-fold proton gradient. The other well-established proton pump is an H-ATPase belonging to a third important gene family, the **V-ATPases**. This pump is often electrogenic and is particularly responsible for acidification in various tissues, for example in urinary compartments such as the reptile bladder and the mammalian nephron; in insects it seems to drive most of the epithelial transport systems. It is worth noting that proton pumps cannot act entirely alone as the proton gradients would be electrically damaging to cells; there is normally some linked movement of an anion as well as a proton, though not always in a 1 : 1 stoichiometric fashion.

The three classes of ATPase have very different structures and operate very differently. The P-ATPases have a phosphorylated enzyme (phosphoenzyme) as an intermediate and are characterized by sensitivity to vanadate; they can exist as a single unit of 94–110 kDa, or may have two kinds of subunit as described for the sodium and calcium pumps above. The V-ATPases and F-ATPases are more like each other; they do not use phosphoenzyme intermediates, and both require several cooperating subunits for full functioning. The V-ATPase may be the primordial proton pump, found in archaebacteria and operating as a hexamer of a single gene product. F-ATPases evolved from this in the eubacteria, as oxygen levels in the Earth's atmosphere increased (see Chapters 6 and 7). Thus modern eukaryotic cells have V-ATPases in their plasma membranes but F-ATPases in their inner mitochondrial membranes (reflecting the descent of these intracellular organelles from eubacterial endosymbionts). Some V-ATPases developed important cation-handling functions in eukaryotes, still seen today in many invertebrates; the family may include sodium-dependent pumps that are insensitive to K^+ (unaffected by ouabain but blocked by the drug ethacrynic acid), and K^+-dependent electrogenic anionic (Cl^-) pumps, which dominate the process of fluid resorption in the insect rectum.

Apart from the V and F families, the P-ATPases developed somewhat later and have proved much more versatile, radiating into many distinct versions and handling a variety of cations, with the Na^+/K^+-ATPase becoming the dominant membrane pump in most animal cells. This is especially the case in vertebrates, where this pump underlies most ionic distributions, with movements of Cl^- and K^+ ions nearly always explained by symport (cotransport) mechanisms.

4.4.3 Secondary active transport; symport and antiport

In many cases of active transport of ions or small solutes, the linkage to ATP splitting is often much less direct than we find with the primary ion pumps, and the active transport is driven instead by the 'energy store' represented by an existing electrochemical gradient, most commonly the Na^+ gradient in vertebrates (though often a proton gradient in prokaryotes, plants and insect epithelia). This may be termed **secondary active transport**, and it is responsible for redistributing actively transported ions and many organic and inorganic solutes such as sugars, amino acids and neurotransmitters in cells. Substances are moved in coupled pairs, so that the system is sometimes called **cotransport**; one of the cotransported substances (A) moves in a direction that dissipates its electrochemical gradient, and the other (B) is thereby moved against its gradient (so the direction of movement can change if the relative gradient changes). If A and B both move in the same direction the process is termed **symport** and the protein carrier is a **symporter** (or simply a **cotransporter**). If B moves in the opposite direction to A then the terms **antiport** and **antiporter** are used (or the carrier may simply be termed an **exchanger**). In either case the cotransport is dependent on the ATP used to maintain the initial ionic gradient. Overall, then, secondary active transport involves a net loss of the energy stored in the gradients set up by primary active transport.

Given the variations in terminology, inevitable confusion arises in the literature, particularly since the genes for individual transporter proteins have become amenable to study. Thus for example we used to speak of a $Na^+/glucose$ cotransporter, and the gene for it became known as SGLT, but we would now call this transporter the glucose-Na^+ symporter (specifying the driven solute first and the driver solute second), and thus the accepted gene abbreviation looks less relevant. Similarly the 'Na^+/H^+ exchanger' in older literature should now ideally be termed the H^+/Na^+ antiporter and its relation to the 'NHE gene' becomes a little mystifying.

Symporters

Some of the clearest examples of symport occur in marine invertebrates (e.g. annelid worms) that can take up amino acids and monosaccharides from sea water against enormous chemical gradients; this uptake occurs in the presence of extracellular sodium, but almost ceases if there is no extracellular sodium. The process is driven by the standard Na^+/K^+-ATPase exchange pump on the basal membrane of the epithelial cells, creating low intracellular sodium and a substantial electrochemical gradient for sodium uptake from the sea water. As a result there is an inward leak of sodium ions, and these are coupled by the specific membrane symporter (protein carrier) to an equal inward movement of the amino acid or sugar (Fig. 4.11). The presence of Na^+ has thus increased the rate of uptake of the required solute, which (continuously driven by the continuously renewed Na^+ gradient) may occur 'uphill' into the animal's cells against 10- or 20-fold concentration gradients. This kind of symport is blocked by application of the Na^+/K^+-ATPase blocker ouabain because of the indirect effect of blocked sodium pumps. It operates in an essentially similar manner in vertebrate intestinal and renal cells, using the symporter products of the SGLT genes mentioned above. Most of the animal Na^+-coupled symporters work with ion-substrate ratios of at least 2 : 1 (e.g. two Na^+ per single sugar molecule moved).

Symport is also enormously important in the flow of anions across membranes. In many systems concerned with salt transport or with volume regulation, chloride ions are handled by a specific symporter system. The apical membrane may bear sodium pumps, and the basal and lateral membranes house the symporter, so that when sodium ions leak in basally along the gradient created by the pump, chloride ions are also moved (see Fig. 4.11). The net result is usually transport of both sodium and chloride across the cell or epithelium. The mechanism is very commonly a $Na^+/K^+/2Cl^-$ symporter, and therefore electroneutral. Some tissues reportedly show different Na^+/Cl^--only, or K^+/Cl^--only systems, though it may be that all these will eventually be revealed as having a similar underlying mechanism. A common feature is that the Cl^- symporters are all sensitive to the diuretic drug furosemide, and all have a very high chloride specificity, handling bromide only partially and ignoring all other chloride-substitutes. Other anionic symporters do exist, however; in vertebrate intestine and kidney there are Na^+-phosphate symporters and Na^+-sulphate symporters, together responsible for a large part of the animal's phosphate and sulphate transmembrane transport.

Antiporters

Antiport is involved in the maintenance of pH in all cells and in

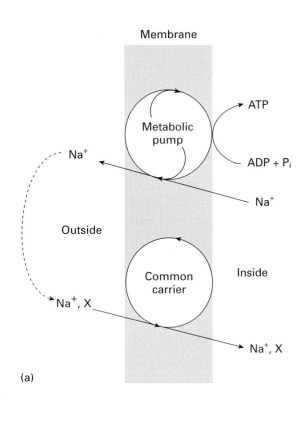

Membrane

Metabolic
pump

ATP

ADP + P$_i$

Na$^+$

Na$^+$

Outside

Na$^+$, X

Common
carrier

Inside

Na$^+$, X

(a)

Lumen   Basal
membrane

Apical
membrane   Blood

Na$^+$
pump

Leak

Na$^+$

Na$^+$–X
symport

Na$^+$
X

X
trans-
porter

X

(b)

[Na$^+$]$_o$ = 140 mM

Intracellular concentration
of alanine

[Na$^+$]$_o$ = 0

(c)   Time

Fig. 4.11 Models of Na$^+$-linked symport of amino acids or sugars (X) across (a) a single membrane or (b) an epithelium. The chemical energy of ATP is used to drive the apical 'sodium pump' so that Na$^+$ leaves the cell. In (a) the Na$^+$ then leaks back in passively and carries X with it; in (b) Na$^+$ leaks in across the other (basal) side of the epithelium and moves X into the cell with it, with a separate X transporter possibly handling movement of X out of the cell apically. In (c) the effects of Na$^+$ concentration on intracellular concentrations of the amino acid alanine are shown.

the maintenance of very low intracellular calcium concentrations in many cells. Na$^+$/H$^+$ antiporters operate to regulate the internal cell pH (electroneutrally in vertebrates, but apparently electrophoretically with a 2 : 1 ratio in most invertebrate epithelia). At the same time, [Ca^{2+}]$_i$ is usually less than 10^{-6} mM, and must be kept at these very low levels by a continuous outward flux (efflux) of calcium. When extracellular Na$^+$ is removed, so that there is no inward sodium leak, efflux of calcium slows very substantially. This suggests that the two ions may be linked via an exchanging carrier (Na$^+$ leaking in and Ca^{2+} leaking out), and this must be a Na$^+$/Ca^{2+} antiporter. The immediate source of energy for the exchange is again the classic sodium pump, as this has set up the sodium gradient on which the Na$^+$/Ca^{2+} exchange depends.

Antiporters also operate in electrogenic K$^+$ transport in some

invertebrate sensilla and in insect midgut, salivary gland and Malpighian tubule. In the lepidopteran midgut K$^+$ is secreted into the lumen from special 'goblet cells' by an electrogenic K$^+$/2H$^+$ antiporter, driven by the gradient resulting from the proton V-ATPase; this generates a high (lumen-positive) electrical gradient, which in turn drives K$^+$ along with amino acids back across the normal lumen cells using a symporter, giving an overall effect of amino acid uptake. In this system, carbonate ions are also secreted into the lumen, making it highly alkaline (pH 8–12).

It is important to remember that water itself is *never* handled by primary or secondary active transport; the process only applies to solutes. In all situations where water may appear to move against the osmotic gradient, closer analysis has always revealed underlying active transport of ions. We will meet many examples of this in ion and water regulatory systems.

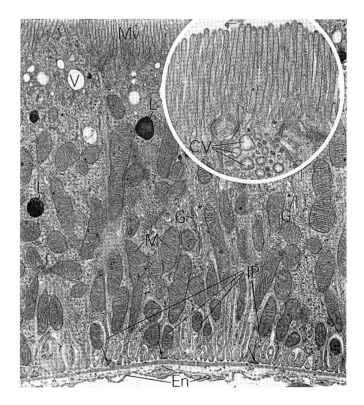

Fig. 4.12 Electron micrograph of a typical salt-transporting epithelial cell (from the proximal convoluted tubule of a vertebrate kidney). Inset shows the inner (luminal) margin. Mv, microvilli (at inner edge, forming a 'brush border'); V, vesicle; M, mitochondria; G, Golgi bodies; IP, interdigitating processes of outer cell margin; L, lysosomes. (From Ross & Romrell 1989.)

4.4.4 *Transepithelial solute movements*

We have seen that active transport mechanisms operating across the cell membrane essentially move a substance either in or out of a particular cell. But because animal cells are usually arrayed in organized and polarized sheets as **epithelia,** with different transport properties at the apical and basal membranes, it is possible to use ATPases, both as primary and secondary active transport, to move substances across cells and across sheets of epithelia, and thus to vary concentrations and conditions between compartments of the animal body. This is achieved by salt-transporting epithelia (arranged in spherical or tubular fashion around a lumen, or as multilobular glands) in which the cells always have a very characteristic appearance (Fig. 4.12). The apical side (often termed luminal when referring to a glandular or tubular system with a clear lumen) has a dense coating of microvilli, giving a very large surface area, while the basal side has less dense but often longer indentations. One or both sides may have heavy endowment with mitochondria, and this serves as a good clue to the site of the transporters. In some transporting tissues the active transporters are apical, and basal movements are passive or involve secondary transport systems, but in others most of the active transporting is basal and apical

flow is passively down the gradient created. Active processing at both membranes is not necessary; moving ions across either one is sufficient to create the gradients that will draw ions passively across the other.

In summary, both organic and inorganic solutes can be transported across biological membranes and thus across cells, against their electrochemical gradients, by coupling to the downhill movements of other molecular species. In animals the coupling is almost always symport or antiport with either sodium or hydrogen ions, the energy being provided by P-ATPase or V-ATPase enzymes and transiently stored as an electrochemical gradient of Na^+ or H^+.

4.4.5 *Other active processes at cell surfaces*

There are several other ways in which materials can be moved across membranes, cells and epithelia faster than by simple diffusion. **Endocytosis** and **exocytosis** are well-known methods for moving small bulk quantities into and out of cells, respectively. Endocytosis involves trapping material lying outside the plasma membrane within a small surface invagination, which is pinched off as an internal vesicle. This is then lysed, to release the contents into the cytoplasm. The process can be split into pinocytosis, where fluid is engulfed, and phagocytosis, where solid matter is involved. The best known mechanism (though there may be several different kinds) is that of **receptor-mediated endocytosis** (Fig. 4.13). Here receptor molecules in the cell

Fig. 4.13 The formation of coated vesicles during receptor-mediated endocytosis. (1) ligand molecules bind to surface receptors in coated pits formed by clathrin molecules bound to the surface membrane; (2) invagination of coated pit so that (3) a coated vesicle is formed; (4) fusion with existing vacuole and (5) further processing, with (6) clathrin and receptor molecules recycled. (Reprinted from *Trends in Biochemical Sciences* 5, Pearse, B., Coated vesicles, pp. 131–134, copyright 1980, with permission from Elsevier Science.)

membrane detect and bind to certain ligands in the extracellular medium; these may be viruses or toxins that the cell will destroy, or plasma proteins, hormones or immunoglobulins to which it will produce a specific response (see Chapter 15). The receptor-ligand complexes can diffuse laterally in the membrane, and once bound they tend to accumulate in specific areas termed 'coated pits', where there are many clathrin molecules; clathrin is a specific protein organized in complex tightly packed mem-

brane arrays. The coated pit invaginates to become a 'coated vesicle', which then fuses with pre-existing vacuoles within the cell. Then the clathrin-coated piece of membrane, with its receptors, is recycled to the surface to attract further ligand-receptor complexes.

Exocytosis is the reverse process, where a small portion of the intracellular medium is organized into a vesicle, which is then extruded at a surface; it is very important in neurosecretory

(a)

(c)

(b)

Fig. 4.14 (a) Electron micrograph of a typical tight junction between epithelial cells. (b) The structure in a freeze–fracture view in epithelial cells. (a & b, From Ross & Romrell 1989.) (c) Diagrammatic view of a tight junction complex. (From Alberts *et al.* 1994, Fig. 19.5.)

and hormonal systems, with calcium ions playing an important controlling role in vesicle release, but it is less involved with cell transport systems in general. However, some cells retain, in reserve just below the transporting apical membrane, a population of vesicles whose membranes contain large quantities of the proteins that are specific ionic channels or pumps. These can be moved into position in the surface membrane using exocytosis controlled by microtubules and microfilaments, to change the pattern or rate of ion movements in response to hormonal triggers; an increase in cell surface area is sometimes detectable. In *Xenopus*, for example, up to 50% of the pumps in developing oocytes are within the cell, and these are continuously exchanged with the surface pumps; as the embryonic blastomeres form, sodium pump distribution becomes polarized and directional Na^+ transport occurs in concert with the formation of the primary embryonic cavity (blastocoel). Pumps and channels can also be mobilized from this vesicle reserve by the action of various hormones.

Another category of movements we should consider briefly are those that occur through membrane junctions. Cells are organized into sheets by junctions; these are of two types, tight junctions and gap junctions. **Tight junctions** (Fig. 4.14) serve to bind cells tightly in place to each other so that they can act as organized epithelia, the outer surface of two neighbouring cells making very close contact and leaving no extracellular space. The linked area may encircle each cell in the epithelium, acting as a barrier against any back leakages *between* cells of substances that have been transported through the cell cytoplasm. Some epithelia may be described as quite 'leaky' and others as quite 'tight' (though the distinction is not very clear-cut). **Gap junctions** are more structured and may have a more active role in cellular transport. They involve a close proximity of two cell membranes, but retain a small intercellular gap of about 2 nm in which a highly organized array of hexagonal units can be detected (Fig. 4.15). Each unit is made largely of protein, with

six subunits structured like a doughnut, leaving a central pore through which the two cells can communicate; molecules as large as certain dyes are able to pass through, and ions, amino acids, nucleotides and sugars can certainly pass. Passage through these cell–cell links can be stopped by varying the intracellular $[Ca^{2+}]$ or $[H^+]$, probably by effects on the spacing of the subunits around each pore.

Finally, we should mention **extracellular pathways** for solute movements. Here we are dealing with transport across an epithelium that occurs *between* the cells rather than through them; the route is sometimes termed **paracellular** to distinguish it from random extracellular diffusion. Clearly paracellular movement could not occur across an epithelium that had non-leaky tight junctions around every cell, but it could nevertheless be an important route in many tissues. Water and various solute molecules could pass between cells down their respective gradients, and in some tissues the lateral gaps between cells could be used specifically to create local high concentrations to facilitate such flow. This is the basis for several models of flow across epithelia to explain rapid transport of water in a variety of tissues. Figure 4.16 shows how such a system might work. Salts (e.g. Na^+, with Cl^- following) are actively transported by classical ion pumps into the intercellular clefts, where the salt concentration builds up rapidly in the restricted volume available; thus there will be a 'standing gradient', with high concentration deep in the cleft and lower concentration luminally, eventually indistinguishable from the actual lumen concentration. Water is therefore drawn into the cleft osmotically from the apical side, either through the leaky junction area (which may be permeable to water but not to ions) or through the cytoplasm, and then fluid moves on down the cleft towards the lumen. The steady pumping of salt at a localized area of membrane thus results in a steady osmotic flow of water across the cell epithelium—so-called 'solute-coupled water transport'.

Fig. 4.15 Structure of a gap junction, where neighbouring cells are coupled through an array of hexagonal subunits on the two closely apposed membranes, forming through-channels. The passage of molecules larger than about 2 nm (such as proteins and nucleic acids) is prevented, but ions, amino acids, sugars and nucleotides can pass from one cell to the other. (From Staehelin 1974.)

Cytoplasm

Gap junction subunits

Cytoplasmic membrane surface

Intercellular space

Lipid bilayer

External membrane surface

Hydrophilic channels

Integral membrane protein

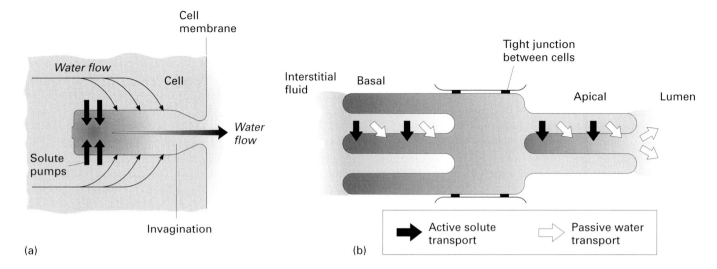

Fig. 4.16 The standing gradient model for solute-linked water transport. (a) In the simple version salts are actively transported out of the cell into the deep clefts (either invaginations into the cell, or clefts between cells), and water follows passively, giving a net flow of fluid. (b) In a more complicated version for a whole epithelium salt is pumped into cell invaginations basally and out to extracellular clefts apically, so that the whole epithelium moves fluid from, for example, its bathing blood to its lumen. The transported solution here is isosmotic, but it can be hyperosmotic if there is solute recycling within the cell (see Chapter 5).

Evidence for this kind of transport is patchy, and quite how important these paracellular routes are remains controversial. In some tissues, such as mammalian gall bladder, there is a detectable distension of the intercellular clefts during rapid secretion, indicating that at least part of the bulk fluid flow is passing by this route; this epithelium is described as 'leaky', as is that of the mammalian small intestine (jejunum) and the insect Malpighian tubule. Such epithelia can produce net fluid transport providing the active transport is fast relative to the 'leak'. But recent evidence has tended to support older ideas that most movement of water in most epithelia is actually transcellular after all, involving simple osmosis and very small concentration gradients at the apical and baso-lateral cell borders, which have a high water permeability. Thus epithelia such as those of the mammalian and insect rectums are 'tight' (in the sense of being non-leaky) and this may well be the norm.

In any event, it is clear that complex patterns of transepithelial transport can be achieved by a combination of the relatively simple subcellular transport systems, and variable permeation mechanisms, that operate in particular localities on cell membranes. An overview of possible transport mechanisms across membranes and epithelia is shown in Fig. 4.17.

4.5 Concentrations of cell contents

We should turn now to look at the actual concentrations of ions and other solutes found inside cells. The earliest living organisms, enclosed by a biological membrane, evolved perhaps 3 billion years ago in the Precambrian, and probably in the fringes of shallow seas. The nature of those seas is somewhat uncertain (see Chapter 9), but we can be reasonably confident that the waters were moderately salty, and there is a reasonable consensus that their composition was not too different from that of modern oceans. It seems very likely that cells evolved in, and have run through almost their entire evolutionary history in, a condition that was essentially marine. It is therefore no surprise to discover that the intracellular contents of the vast majority of cells still reflect this ancestry; most phyla of animals are predominantly marine, and most have intracellular concentrations roughly isosmotic with sea water. It is only the relatively few groups that have successfully invaded freshwater and terrestrial habitats that diverge from this pattern, commonly having both cells and extracellular fluids that are 1/4–1/3 the concentration of sea water. Examples of the ionic and osmotic concentrations of cells from a range of animals are given in Table 4.4.

As is clear from discussions earlier in this chapter, the dominant intracellular cation in animals is always potassium, rather than the sodium that dominates most extracellular fluids and the marine and fresh waters in which animals may live. But even for animals living in very concentrated media such as salt lakes, the total intracellular cation concentration ($[Na^+]_i + [K^+]_i$) does not normally exceed about 300 mM (rarely up to 440 mM as in squid giant axons). This relatively low intracellular ionic concentration is probably related to the destabilizing effects of salts on protein structure, a process sometimes called 'salting in' since it leads to proteins solubilizing into the aqueous phase, and beginning to unfold. Animal cells also differ from their environments in having extremely low calcium concentrations; values given in Table 4.4 include a large proportion of calcium that is effectively bound, and the true free $[Ca^{2+}]_i$ is normally below 10^{-6} mM.

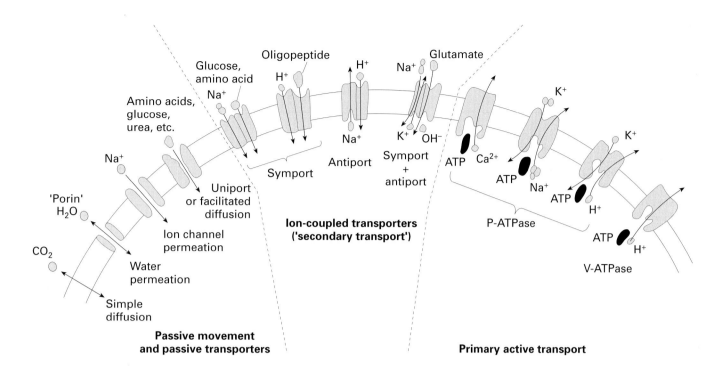

Fig. 4.17 A pictorial summary of the variety of possible methods for moving molecules across membranes.

Hence even the earliest cells may have had in place the transporter mechanisms for pumping cations across the membrane to set up ion gradients, thus benefiting from the various side-effects of these gradients but of course having to pay the energetic costs of continuous ion regulation.

The other major point to notice from Table 4.4 is the important role of small organic molecules inside cells. Most animals have at least 30% of their osmotic concentration provided by amino acids and other small organic molecules, and the figure may be up to 70% in certain cell types. These organic components make up for the low ion levels discussed in the last paragraph, to give an overall osmotic concentration close to that of the extracellular fluid (ECF) and of the environment. These intracellular osmolytes play an important role in osmotic adaptation and volume regulation (see section 4.6), being moved in and out of the cell as its surroundings change. Many of the organic molecules used intracellularly instead of ions are negatively charged at normal pH, and they contribute substantially to the anionic fraction, so that the major inorganic anion, chloride, is also markedly less concentrated inside animal cells than in extracellular fluids, and $[Cl^-]_i$ is usually less than the total internal ionic cation concentration. As mentioned earlier, this chloride distribution arises entirely passively in most cells. Other inorganic anions are numerically relatively insignificant, with sulphate at low concentrations and bicarbonate and phosphate ions often even lower in concentration but functionally very important as buffers of cell pH levels.

4.6 Overall regulation of cell contents

In the previous section we looked at cell contents in a range of animals. Clearly these are average conditions, measured in resting animals under 'normal' conditions, and most of the problems of water and ion balance arise in relation to maintaining these concentrations in the face of environmental challenges. Cells are rather permeable to water, and external changes in medium concentration therefore tend to cause cells to either shrink or swell quite quickly and concentration gradients to run down. The next chapter will deal with the problems of water and ion balance at the level of the whole animal, but we need first to summarize how control of concentration and volume is manifested at the level of individual cells.

Cell concentration regulation depends on uptake and exchange mechanisms whose activity is modified in relation to perceived change, either directly or via intervening sensors and second messengers, and which are crucial to maintaining cell homeostasis. The sodium pump Na^+/K^+-ATPase system, already dealt with, is the single most important effector system for the major cationic Na^+ and K^+ concentrations, keeping internal potassium levels high and sodium levels low. Sodium levels are also in part regulated by pH levels in the cell, via H^+ extrusion and Na^+/H^+ exchange; pH itself depends on a complex acid–base regulation system linked with CO_2 and HCO_3^- levels.

Calcium levels depend on at least two processes already discussed. There are Na^+-Ca^{2+} antiporters, powered by the sodium gradient created by the Na^+/K^+-ATPase system, but this is often the less important system for extruding calcium from cells and maintaining a low $[Ca^{2+}]_i$, because if extracellular Na^+ is entirely removed, calcium efflux persists, albeit at a reduced level. This is due to the specific 'calcium pump' or Ca^{2+}-ATPase membrane

	Solute concentration (mM)							
Animal group	Na+	K+	Cl−	Ca2+*	SO4 2−	PO4 3−	Amino acids	Total NPN†
Marine								
Annelids								
Neanthes (ragworm)	125	195	124	14				412
Molluscs								
Mytilus (mussel)	79	152	94	7	9	39	289	
Sepia (cuttlefish)	31	189	45	2				678
Crustaceans								
Nephrops (prawn)	24	188	53		1	164	476	602
Carcinus (crab)	54	146	53	5				617
Chelicerates								
Limulus (horseshoe crab)	29	129	43	4	1	96	136	
Vertebrates								
Myxine (hagfish)	32	142	43		104		290	
Chimaera (ratfish)	28	120	37	3	189		378	767
Salmo (salmon)	21	264	3		46		49	
Freshwater								
Molluscs								
Anodonta (mussel)	5	21	2	12		20	11	
Crustacea								
Astacus (crayfish)	11	122	14	9				153
Insects								
Sialis (alder fly)	1	135	1					
Land								
Vertebrates								
Rana (frog)	15	126	1	5		10		68
Rattus (rat)	16	152	5	2				70
Homo (human)	14	140	4		4		8	

Table 4.4 Intracellular solute compositions for muscle or nerve tissues from a range of animals from different habitats.

* Values for calcium are total concentrations; most is sequestered and free calcium levels are generally much less than 1 mM, often less than 1 μM.

† For some species total non-protein nitrogen (NPN) is given rather than amino acid concentrations, yielding a higher value.

protein. At normal calcium levels, this pump may be responsible for nearly all the active extrusion of calcium from cells, with the Na+-Ca2+ antiport exchange system only operating when $[Ca^{2+}]_i$ gets abnormally high.

Much of the magnesium present in cells is chelated and bound internally to other molecules; for example, most of the ATPases also have Mg^{2+}-binding sites. Thus free intracellular magnesium is at very low levels. Regulation probably involves symport and antiport with other transport systems, but detailed information is lacking.

Chloride levels are determined largely passively in most vertebrate cells that have been fully analysed, though the anion is symported with Na+ and K+ as described above, often extensively so in invertebrate epithelia. Other anions involve more complex regulatory patterns: phosphate has complex interactions with many other transport systems, and undergoes much binding within cells, and the same is probably true of sulphate and bicarbonate ions.

Overall, for the great majority of cells it would be fair to say that only Na+, K+ and Cl− are really important and variable osmotic components of the cytoplasm; all other ions are tied up in more complex interactions, affecting each other's concentrations in feedbacks and balanced transport systems, with relatively low and relatively constant unbound intracellular levels.

Cell volume regulation depends on all these mechanisms. Cells have ions both inside and outside their membranes, but organic concentrations are much higher inside. In isosmotic conditions the cells would therefore tend to swell, and outward ion pumping is a key factor in counteracting this. The ubiquitous Na+/K+-ATPase is crucial here, since Na+ pumped out will stay out and K+ pumped in will leak back out again as P_K is higher;

Box 4.6 The van't Hoff relation

Cells tend to shrink in hyperosmotic media due to osmotic loss of water and swell up in hyposmotic media due to osmotic water influx. In a simple physical system the relation between cell volume and concentration would be given by the **van't Hoff equation**:

$$\Pi_1(V_1 - b) = \Pi_2(V_2 - b)$$

where Π is the osmotic pressure, V the cell volume and b the total content of solids in the cell (for two sets of conditions 1 and 2, initial and final). Thus $V_1 - b$ is the initial solvent volume and $V_2 - b$ the final solvent volume in the cell.

Plotting cell volume against the ratio of the two osmotic conditions (Π_1/Π_2) should give a straight line with a slope of ($V_1 - b$), *if* the system behaves as a perfect osmometer. This is rarely the case, however, for reasons given in the text.

An indication of the degree of volume regulation can therefore be obtained by comparing the slope value ($V_1 - b$) with the known water content (W) of the cell (or of a whole organism).

This ratio ($V_1 - b$)/W is termed **Ponder's *R*-value**; it is about 0.87 for mammalian kidney cells, 0.94 for vertebrate fibroblasts, and very close to 1 for human red blood cells, which are nearly perfect osmometers. Many invertebrates have values around 0.8.

---

the net effect is sodium extrusion, hence the common name of 'sodium pump', for which prevention of cell swelling could well be the ancestral function in the most primitive of cells.

Cells are many times more permeable to water than they are to any ions, so that the immediate challenge to the cell in a non-isosmotic environment is usually to limit the damage due to water movements—to maintain its volume and avoid swelling or shrinking. In practice, cells do readily act as osmometers, shrinking quite rapidly in hyperosmotic solutions and swelling in hyposmotic solutions (strictly speaking the terms 'hypertonic' and 'hypotonic' are more correct here, referring directly to swelling and shrinking effects, taking into account membrane permeabilities as well as osmotic gradients). A perfect osmometer would behave according to van't Hoff's equation (Box 4.6), so that a graph of cell volume as a function of initial osmotic concentration/experimental osmotic concentration would be a straight line. But although living cells often show an initially straight-line reaction, the response soon deviates from this ideal line and volume changes are less than expected. A number of causes probably underlie this effect. Firstly, a significant portion of internal cell water is bound to membranes, proteins and other constituents and is not free to move. Secondly, neither solutes nor solvents behave 'ideally' except when the solutions are very dilute. Thirdly, some solutes will pass across the membranes too, because real cells are not perfect osmometers. But finally, and most importantly, cells have the specific capacity to regulate their volume, and these regulatory mechanisms will be switched on soon after the initial phase of swelling or shrinking begins. Hence most cells, and indeed many whole animals with soft bodies, when transferred to non-isosmotic media will exhibit the kind of volume response shown in Fig. 4.18, with an initial perturbation of volume and then a return to the original value. Commonly the volume stabilizes within a few minutes or hours for cells, but full regulation may take several days of exposure for whole animals such as marine worms (see Chapters 9 and 10). The capacity to volume regulate can be easily expressed as the ratio of predicted weight (or volume) change to actual weight (or volume) change; this ratio would be 1.0 for a perfect osmometer, and 0 for a perfect volume regulator.

Cells have at least four mechanisms that contribute to volume regulation. Firstly, as we discussed earlier, they may have membranes that are differentially permeable to water movements (inward osmotic gain being faster than outward osmotic loss). Secondly, there may be adaptive reduction in the cell's permeability to water, either as a direct response of the postulated membrane channels or sometimes in response to hormones such as antidiuretic hormone (ADH); this slows water movement though it does not in itself bring about volume regulation. Thirdly, a number of secondary transport systems seem to be switched on in response to cell volume change, such as $Na^+/K^+/2Cl^-$ symport into the cell during shrinkage, or K^+/Cl^- exit during swelling. These secondary transport systems are strongly linked to the fourth mechanism, that of changes in the intracellular solute composition, involving ionic and non-ionic solutes acting as **osmotic effectors**. Such changes may be the most important in longer term volume regulation for the majority of cells. For example, some vertebrate epithelial cells, and many tissue-culture cells, will volume regulate in hyperosmotic media by taking up solutes (especially NaCl), while in hyposmotic media they show loss of solute, in this case mostly KCl. Inevitably the membrane Na^+/K^+-ATPase (sodium pump) is heavily involved in this, so that we should list volume regulation as one of its major functions. A vertebrate red blood cell (RBC) treated with ouabain, or with a metabolic poison that shuts off the ATP supply to the pump, loses most of its ability to control its volume.

But adjustment via ionic solutes does risk undue changes to cell concentrations, with potential effects on many enzyme reaction rates, protein stability, and other aspects of normal cell functioning. Perhaps for this reason, many cells show a different mechanism of long-term volume regulation, by alterations to their non-ionic solutes or **compensatory osmolytes**, which have far fewer effects on enzyme activities. By far the most important compensatory osmolytes in many invertebrate cells are the amino acids, which may make up a large percentage of the normal intracellular osmotic concentration. For example, during acclimation from fresh water to sea water (i.e. to hyperosmotic conditions), most cells from marine invertebrates show a gradual increase of amino-nitrogen in their cells, and an equivalent rise can be detected in the blood of the whole animal (Fig. 4.19a). During hyposmotic adaptation, a decrease in amino acid levels occurs in the cells; the resting level of amino acids closely correlates with the medium salinity (Fig. 4.19b). Both these effects are

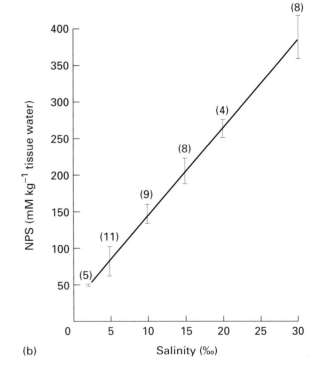

Fig. 4.18 Volume changes in three different invertebrates on transfer to dilute sea water (SW). All show initial volume increments; the sipunculid shows very poor volume regulation, whereas both molluscs are able to regulate their volume back to near normal levels after 5–10 h exposure. (Data from Oglesby 1968; Pierce 1971; Pierce et al. 1983.)

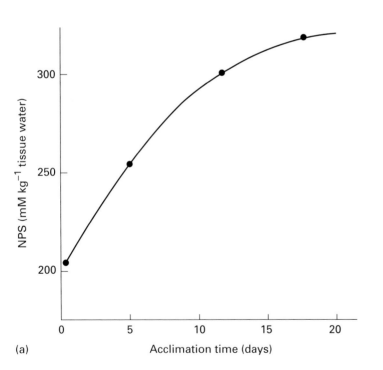

Fig. 4.19 Changes in amino-nitrogen (NPS = ninhydrin positive substances, ninhydrin reacting with amino groups) in tissues of animals exposed to varying osmotic media. (a) The increase in amino acids in *Eriocheir sinensis* (a crab) during acclimation to sea water from fresh water. (Data from Gilles 1979.) (b) The equilibrium levels of amino acids in muscles of *Mya arenaria* (a clam) in relation to salinity of the medium. Numbers in parentheses are sample size, vertical bars show standard error of the mean. (Data from Virkar & Webb 1970.)

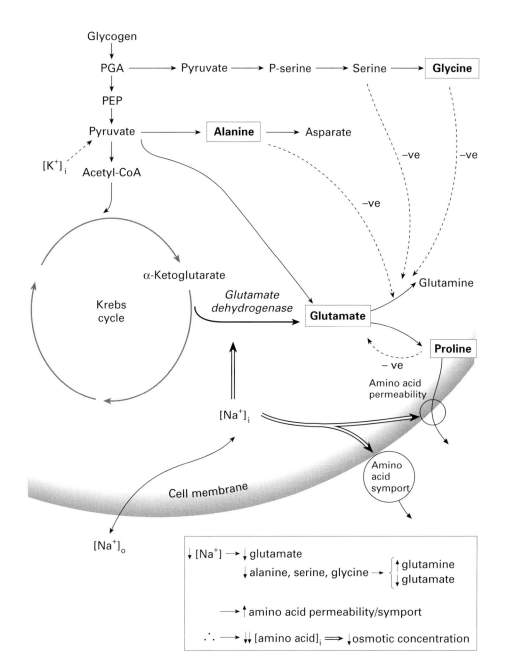

Fig. 4.20 Influences of ion concentrations on key points in the glycolytic and Krebs cycle pathways linking to amino acid synthesis and transaminations, and on transmembrane amino acid movements. Changes in $[Na^+]_o$ can thus bring about changes in intracellular amino acid levels and achieve osmotic compensation. PEP, phosphoenolpyruvate; PGA, phosphoglycerate.

commonly due to changes in the levels of *specific* amino acids, rather than the amino acid pool as a whole. Some amino acids are especially suitable, since they are stabilizing with respect to protein structure and function, opposing the 'salting-out' effect of ions; glycine, alanine and serine are particularly effective here. However, in some animal cells less familiar methylamine compounds such as betaine and taurine are partly responsible, and some insects use proline or carbohydrates like trehalose and the polyol sugars such as sorbitol; all these are described as 'kosmotropic' due to their strong stabilizing influence on membranes via effects on neighbouring water molecules. Thus having relatively high levels of such compounds is both a general protection for macromolecules and membranes during the

initial shock phase of osmotic exposures and ionic changes, and a compensatory protection for the osmotic gap left by maintenance of ion levels in the longer term.

The compensatory changes in these intracellular solutes are brought about by the changing levels of extracellular sodium, causing concomitant small changes in intracellular sodium. In animals acclimating to reduced concentrations (e.g. to lowered salinity), it is the lowered $[Na^+]_i$ that acts as the trigger within the cell, and it may bring about several effects (Fig. 4.20). Firstly, it changes the activity of one or several enzymes (especially glutamate dehydrogenase) that control amino acid synthesis and/or degradation, so leading to altered amino acid levels. Secondly, it alters transmembrane movements of amino acids or sugars, so that there is a rapid efflux and a direct reduction of overall osmotic concentration in the cell. Thirdly, it may work indirectly on gene expression via MAP kinases (see Chapter 2). The resultant changes of concentrations of amino acids or sugars are such as to decrease the osmotic stress on the cell, and so limit water movements and volume changes, without compromising ion levels.

This use of small intracellular organic solutes as osmotic effectors was first studied in marine and estuarine invertebrates (see Chapter 10), but similar phenomena are now well known in many aquatic vertebrates' cells. They may be less common in the cells of terrestrial birds and mammals, which are highly regulated in other ways, but comparable systems certainly operate in the highly stressed renal medulla cells of mammals, where betaine and certain polyols like sorbitol seem to operate to maintain intracellular osmotic balance.

4.7 Conclusions

The movements of water, ions and other small solutes within and between cells, and thus within biological systems in general, can be attributed to the fundamental structure and properties of the cell membrane. The simple characteristics of cell permeation are mediated through a small range of selective ion channels and passive transporters, and of selective ion 'pumps' operating active transport processes that may be coupled with transporters of other ions and small solutes. These processes can be linked together in cells and epithelia, and coupled with osmotically induced water flow. Various combinations of these simple processes turn out to underlie most of the larger scale and apparently complex processes of osmoregulation, salt regulation, excretion and overall animal water balance, which will be covered in the next chapter.

Further reading

Books

Andreoli, T.E., Hoffman, J.F., Fanestil, D.D. & Schultz, S.G. (eds) (1987) *Membrane Physiology*. Plenum Medical, New York.

Bonting, S.L. & de Pont, J.J.H.H.M. (eds) (1981) *Membrane Transport*. Elsevier/North Holland, Amsterdam.

Harvey, W.R. & Nelson, N. (eds) (1994) *Transporters*. Society for Experimental Biology.

Hille, B. (1984) *Ionic Channels of Excitable Membranes*. Sinauer, Sunderland, MA.

Reuss, J.M., Russell, J.M. & Jennings, M.L. (1993) *Molecular Biology and Function of Carrier Proteins*. Rockefeller University Press, New York.

Somero, G.N., Osmond, C.B. & Bolis, C.L. (eds) (1992) *Water and Life*. Springer-Verlag, Berlin.

Stein, W.D. (1986) *Transport and Diffusion across Cell Membranes*. Academic Press, San Diego.

Strange, K. (1994) *Cellular and Molecular Physiology of Cell Volume Regulation*. CRC Press, Boca Raton, FL.

Reviews and scientific papers

Ahearn, G.A., Zhuang, Z., Duerr, J. & Pennington, V. (1994) Role of the invertebrate electrogenic $2Na^+/1H^+$ antiporter in monovalent and divalent cation transport. *Journal of Experimental Biology* **196**, 319–335.

Chamberlin, M.E. & Strange, K. (1989) Anisosmotic cell volume regulation: a comparative view. *American Journal of Physiology* **257**, C159–C173.

Eveloff, J.L. & Warnock, D.G. (1987) Activation of ion transport systems during cell volume regulation. *American Journal of Physiology* **252**, F1–F10.

Galinski, E.A., Stein, M., Amendt, B. & Kinder, M. (1997) The kosmotropic (structure-forming) effect of compensatory solutes. *Comparative Biochemistry and Physiology A* **117**, 357–365.

Gilles, R. (1997) 'Compensatory' organic osmolytes in high osmolarity and dehydration stresses: history and perspectives. *Comparative Biochemistry and Physiology A* **117**, 279–290.

Houssier, C., Gilles, R. & Flock, S. (1997) Effects of compensatory solutes on DNA and chromatin structural organisation in solution. *Comparative Biochemistry and Physiology A* **117**, 313–318.

Kinne, R.K.H. (1993) The role of organic osmolytes in osmoregulation; from bacteria to mammals. *Journal of Experimental Zoology* **265**, 346–355.

Pequeux, A., Gilles, R. & Marshall, W.S. (1988) NaCl transport in gills and related structures. *Advances in Comparative and Environmental Physiology* **1**, 1–73.

Shuttleworth, T.J. (1989) Overview of epithelial ion-transport mechanisms. *Canadian Journal of Zoology* **67**, 3032–3038.

Smith, P.R. & Benos, D.J. (1991) Epithelial Na^+ channels. *Annual Review of Physiology* **53**, 509–530.

5 Animal Water Balance, Osmoregulation and Excretion

5.1 Introduction

5.1.1 The problems of water balance

Maintaining a reasonably constant water balance is a straightforward business for the majority of animals. As we saw in the last chapter, the cells and body fluids of marine animals are in osmotic equilibrium with the surrounding medium, and numerically the marine animals dominate the planet (see Chapter 9). However, when living organisms began to colonize more osmotically challenging habitats—the seashores, estuaries and rivers, and eventually the land—the problems of water balance became paramount. Once living cells moved out of the sea, they faced real problems; water always has a tendency to move into them (from dilute surroundings such as estuaries and fresh water) or move out of them (into the surrounding dry air of terrestrial zones), and neither of these fluxes can be left uncorrected if the cells are to continue functioning.

For non-marine multicellular life, continuing healthy existence depends upon maintaining the cells at a fairly constant water balance, i.e. in reasonable osmotic stasis, and this usually involves using the extracellular fluids to buffer the cells against excessive environmental stress. Controlling this cellular and extracellular osmotic balance ultimately depends on principles outlined in the preceding chapter: the permeation characteristics of membranes and associated surfaces throughout the organism, active processes regulating ion levels, and control of intracellular osmotic effector levels. But for whole animals these processes have additive and interactive effects between different tissues and organs, with particular epithelia carrying out regulatory functions to maintain blood composition and thus provide osmotic stasis for the rest of the body. Therefore the problem of animal water balance can best be examined in terms of the actions, and the control, of particular effector organs.

5.1.2 The components of water balance

Figure 5.1 summarizes the various routes through which water and ions can be gained and lost in idealized animals. For an aquatic animal (Fig. 5.1a), water and ions may be exchanged through the general skin surface and at specialized areas of the skin such as the gills. They may be taken in through the mouth and gut, in the food or via drinking; and they may be lost through the faeces or in urine. For terrestrial animals (Fig. 5.1b), there is normally no opportunity for skin uptake of water or ions (exceptions do occur, and will be covered later), so that the skin and the respiratory system serve only as avenues of ion and water loss; and again losses occur through faecal and urinary deposition. In addition to these routes, all animals may incur small losses of ions and water associated with reproduction, and unpredictable losses from injury, while some may also lose water during moulting processes or from secretions such as slime, mucus, saliva, defensive fluids or toxins. Furthermore, all will have some gain of water from their own internal metabolic oxidative processes, of which water is nearly always a by-product, and this 'metabolic water' may be a very significant component of water balance for small terrestrial animals.

Some of these routes of water and ion gains and losses are unavoidable, continuous and essentially passive, and their effects will be determined by the same factors of concentration gradient, surface area (or surface area to volume ratio) and permeability as were involved in passive movements across cell membranes discussed in Chapter 4. These routes may be partly controlled, via changes in permeability and manipulation of gradients, but movements of water or ions remain passive. Other components of overall water balance are discontinuous routes that may be operative or switched off according to need, and these generally depend upon the principles of active transport also considered in Chapter 4, and therefore depend ultimately on metabolic factors dealt with in Chapter 6.

The principles and mechanisms governing all of these potential gains and losses need to be considered at the level of epithelia, organs and whole animals, in relation to the habitat and design of different animals. First, though, we should look at broad strategies for coping with ionic and osmotic problems and at the water balance problems encountered in different kinds of environment.

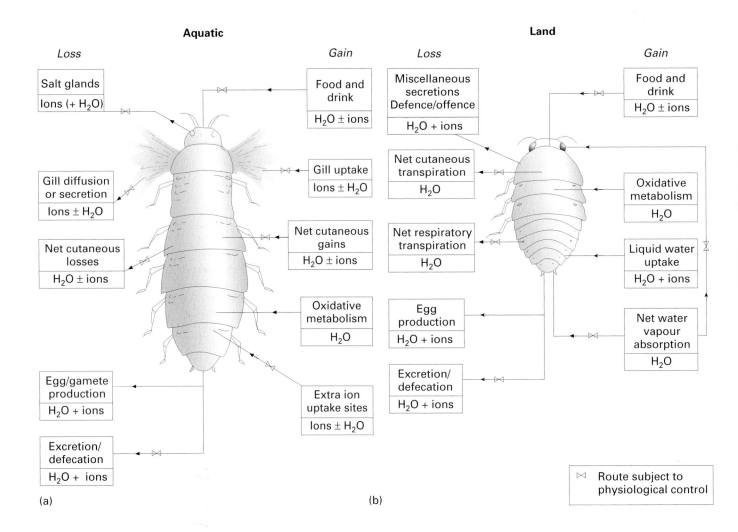

Fig. 5.1 Summary of avenues of water and ion gains and losses for (a) an aquatic animal and (b) a terrestrial animal. (b, Adapted from Edney 1980.)

5.1.3 Avoidance, tolerance, conformity and regulation in osmotic biology

As we discussed in the first chapter of this book, there are distinct general patterns that animals show in dealing with environmental changes. They may **avoid** them by behavioural tricks of various kinds, or they may allow their bodies to be exposed to the changes and thus **tolerate** the conditions. The tolerators may be split into two groups, since some animals may **conform** to changes by letting their own internal conditions follow those of the outside world, while others may **regulate** their own body to maintain reasonable independence of the environment. In the context of water balance, all these options may be available, depending on the habitat, but we also need to distinguish the responses to ionic variance (**ionoconforming** and **ionoregulating**) and to osmotic variance (**osmoconforming** and **osmoregulating**). Table 5.1 summarizes some of the ionic and osmotic

conditions found in the body fluids of animals from different habitats, together with the composition of sea water and fresh water, as a background to what follows.

For marine animals, avoidance is usually neither a need nor an option. Conditions are similar throughout, so that an animal cannot 'avoid' by moving to a less stressful part of the environment. For most marine animals, including virtually all invertebrates and a few vertebrates, conditions are 'easy', since the cells and extracellular fluids (ECF) are in osmotic equilibrium with the medium and the ECF is only marginally different ionically from sea water (see Chapter 9). All these marine animals are therefore conformers, but most of them cannot survive very much change in the concentration of their surroundings; their conforming tolerance *range* is small, and they may be described as **stenohaline**. In practice the lack of environmental variation means that an ability to regulate would be quite unnecessary, so the great majority of animals in the sea are osmoconformers and ionoconformers, and water balance poses no problem.

For many vertebrates living in the sea, different patterns occur because the taxon has an evolutionary history in fresh water,

Table 5.1 Extracellular fluid compositions in a range of animals from different habitats.

Habitat/animal group	Genus	Ionic concentration (mM) Na	K	Ca	Mg	Cl	SO$_4$	HPO$_4$	Osmolarity (mOsm)
Marine									
Seawater		*470*	*10*	*10*	*55*	*570*			*1100*
Cnidaria	*Aurelia*	454	10	10	51	554	15		
Mollusca	*Aplysia*	492	10	13	49	453	28		
	Loligo	419	21	11	52	522	7		
	Sepia	465	22	12	58	591			1160
	Eledone	432	14	11	54	516			1061
	Mytilus	490	13	13	56	573	29		1148
	Strombus	503	11	11	60	577			
	Nerita	484	11	11	55	554			
Annelida	*Arenicola*	459	10	10	52	527	24		
	Neanthes	483	14	13	44	545			1108
Crustacea	*Homarus*	472	10	16	7	470			
	Carcinus	525	13	14	21	502	17		1100
	Pachygrapsus	465	12	11	29				
	Nephrops	512	9	16	10	527			1108
	Ligia	586	14	36	21	596			
Chelicerata	*Limulus*	445	12	10	46	514			1042
Echinodermata	*Asterias*	428	10	12	49	487	27		
	Parastichopsis	473	10	10	53	548			1117
	Echinus	444	10	10	51	519	28		1065
Chordata									
Chondrichthyes	*Scyllium*	269	4	3	1	258	1	1	1075
	Squalus	296	7	3	4	276			1096
Teleostei	*Paralichthys*	180	4	3	1	160	<1		337
	Gadus	174	6	7	3	150			308
Freshwater									
Soft freshwater		*<1*	*<0.01*	*<0.1*	*<0.1*	*<1*			*1–2*
Mollusca	*Anodonta*	16	<1	8	<1	12	<1		66
	Theodoxus	45	2	2	3	33			
	Viviparus	34	1	6	<1	31			
Crustacea	*Cambarus*	146	4	8	4	139			
	Potamon	259	8	13		242			522
	Astacus	208	5	14	1	250			477
	Asellus	137	7			125			
Chordata									
Teleostei	*Carassius*	142	2	6	3	107			
	Salmo	161	5	6	1	120			
Amphibia	*Rana*	92	3	2	2	70			210
Reptilia	*Alligator*	140	4	5	3	111			278
Aves	*Anas*	138	3	2		103		2	294
Terrestrial									
Mollusca	*Poteria*	31	2	5	2	25			
	Pomatias	110	6	16	3	106			
Annelida	*Lumbricus*	76	4	3		43			
Crustacea	*Oniscus*	230	8	17	9	236			
	Porcellio	227	8	15	11	279			
	Gecarcinus	468	12	17	8				
	Holthusiana	270	6	16	5	266			
Insecta	*Locusta*	60	12	17	25				
	Periplaneta	161	8	4	6	144			
Chordata									
Mammalia	*Homo*	142	4	5	2	104	1	2	295
	Rattus	145	6	3	2	116			

Principles

Examples

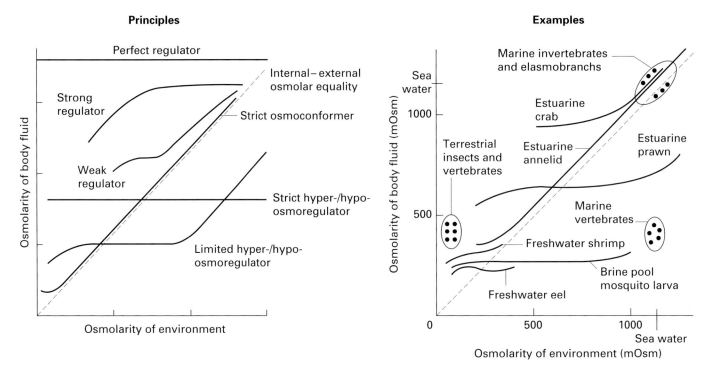

Fig. 5.2 General categories of responses of animal body fluids to variations in external concentrations.

with rather dilute blood, and modern fish and tetrapods are secondary invaders of the oceans. Most of the cartilaginous fish (elasmobranchs or Chondrichthyes) show complete osmoconformity but also strong ionic regulation; their blood is osmotically in equilibrium with sea water due to the presence of various dissolved organic compounds exerting an osmotic effect (**osmolytes**, such as urea and other nitrogenous compounds), while the blood ion levels are kept low, similar to the levels found in freshwater animals. Again this means that water balance is not a significant problem, with no significant osmotic influx or efflux. But in the marine bony fish (Teleosti), and in nearly all the secondarily marine tetrapods such as whales and seals, the blood is both ionically and osmotically dilute, so that water always has a tendency to move out of the animal and ions to move in, and regulation must be continuous. This is by far the most 'difficult' physiological strategy for a marine animal, and it is considered in some detail as a special topic at the end of Chapter 9.

In aquatic habitats other than the sea, osmotic and ionic problems inevitably arise. In estuaries and littoral habitats, many invertebrate animals can operate an avoidance strategy, as conditions follow a tidal cycle and temporary evasion is a good short-term trick when there is a guarantee that conditions will return to the preferred range on a regular basis. Thus aquatic estuarine animals may burrow, or seal their shells, when the tide goes out, while terrestrial animals may climb vegetation, or retreat into sealed crevices and burrows, when the tide comes in.

For those animals that remain active and exposed all the time, though, a broad tolerance range is essential, so these are **euryhaline** animals coping with a broad range of salinities. Many species conform to changes, with the osmotic and ionic concentrations of their ECF following those of the outside fluids. Others are regulators to varying degrees, keeping their body fluids more concentrated (**hyperosmotic regulation**) when the tide goes out and freshwater inputs produce dilution of the habitat. More rarely, some regulators can also keep their body fluids more dilute than their surroundings (**hyposmotic regulation**) by excreting or sequestering osmolytes when they are exposed to salinities at or above that of normal sea water. There are three broad groups in this hypo-regulating category, though numerically they are all insignificant: the extremely tolerant euryhaline animals that can live in salt lakes, such as the brine shrimp *Artemia* and to a lesser extent some larval mosquitoes (see Chapter 12); others like *Palaeomonetes* and *Leander* (two of the common prawns) that may have had a freshwater evolutionary history but now live littorally; and a few animals that live high up on rocky shores where pools may become substantially concentrated by evaporation on hot days. Animals that can achieve hyposmotic regulation can generally also manage hyperosmotic regulation, though the reverse is not true. These patterns are compared in Fig. 5.2 with respect to osmotic concentrations; we will meet many specific examples on similar figures in Chapters 10 and 11. It is worth noting now that there are no *perfect* regulators, and no *perfect* conformers; all animals lie between these two extremes, and all are experiencing some problem with maintenance of water balance. In reality some animals do cope with a very wide range of environmental concentration, while others are relatively restricted either to saline or to fresh waters (the points where a line cease reflect the lower and upper tolerance limits for each species). The level of body fluid concentration maintained in fresh water (itself commonly 0.1–5 mM) is obviously very variable, and shows no very clear phylogenetic pattern. Despite this variation, all freshwater animals (and all terrestrial animals making a secondary return to fresh water) are clearly regulators to some degree, and all have continuous water-balance problems because water will always tend to flow into their bodies from the hyposmotic surrounding fluid.

On land, there is again no alternative to continuous osmotic and ionic regulation, but here maintaining water balance is a struggle to resist water losses to the surrounding drying air. Air does not have an 'osmotic concentration', and the normal terminology cannot be applied; but as a rough guide, all terrestrial animals' body fluids will lose water across any available permeable surface to air at anything less than 99.4% relative humidity—the kind of almost saturated air that can normally only be found in deep, wet caves and crevices, or in air gaps amongst very wet litter and soil. Thus resisting, or compensating for, efflux of water becomes a dominant factor in the environmental physiology of most terrestrial animals.

5.1.4 Tolerance of water loss

There is another aspect that we should consider in relation to overall water balance, and that is the varying tolerance that animals show to actual loss of body water. Uncontrolled dehydration inevitably leads to a proportional increase in osmolyte concentration; if there is no osmoregulation, a loss of $x\%$ of the body water will lead to osmotic concentrations increasing by a factor of $100/(100-x)$, corresponding to a factor of 1.33 for 25% water loss or 2.0 for 50% water loss. All animals are inherently 60–90% water, but this is distributed between the various compartments of the body (see Fig. 4.1), so that many do have inbuilt stores or buffers against this effect; some vertebrates lose water preferentially from the gut when water-stressed, worms lose coelomic fluid from the main body cavity, while insects lose haemolymph water, but nearly all animals have mechanisms to avoid excessive water loss from the cells. Most animals can therefore tolerate overall water loss of a few per cent without undue osmotic problems. However, many animals can tolerate substantially greater loss than this. Clearly for aquatic animals this ability will correlate to a substantial degree with the tolerance of osmotic variation in the body fluids; animals that can cope with high salinities are necessarily tolerating considerable water loss from their bodies. For terrestrial and shore animals, where evaporative water loss is a problem, the tolerance of such loss is extremely variable (Table 5.2). Some slugs and snails, and littoral molluscs such as limpets and chitons, can tolerate 40–75% water loss, causing substantial shrinkage of the body. Amphibious animals such as anurans (frogs and toads) can survive around 30–40% loss, with values up to 48% in the spadefoot toad *Scaphiopus*. Many insects will survive 30–50% loss

Table 5.2 Tolerance of water loss (as maximum % weight loss tolerated) in a variety of animals from terrestrial or semi-terrestrial habitats.

Annelids	
Allolobophora (earthworm)	75
Molluscs	
Patella (limpet)	35–60
Chitons	75
Helix (snail)	45–50
Limax (slug)	80
Sphincterochila (desert snail)	50–55
Crabs	
Gecarcinus	15–18
Uca	18
Insects	
Temperate beetles	25–45
Temperate roaches	25–35
Desert cicada	25
Desert ants, grasshoppers	40–70
Desert tenebrionid beetles	60–75
Frogs	
Rana	28–35
Hyla	35–40
Bufo	42–45
Scaphiopus	45–48
Birds and mammals	
Small birds	4–8
Rat	12–15
Human	10–12
Camel	30

of body water, and those from arid habitats commonly cope with 40–60%. At the other extreme, most terrestrial birds and mammals cannot survive more than a few per cent loss of water; the camel is considered exceptional in tolerating 30%. These vertebrates are very much more dependent on high blood pressure, and constant blood composition, to maintain their respiratory pigments in working order, so ensuring adequate oxygen uptake to support their high metabolic rates. These terrestrial vertebrates need very sophisticated controls to maintain water balance, even though it is the smaller terrestrial animals such as insects, arachnids and land snails that will suffer potentially faster fluxes due to their very high surface area to volume ratio.

5.2 Exchanges occurring at the outer body surface

5.2.1 Passive exchanges

If the surroundings of an animal are at a different osmotic concentration than its interior, gains or losses of water and ions across the animal's outer surface are inevitable; there is no such thing as an 'impermeable' biological surface. We are therefore only discussing *differences* in permeability to water or ions, in

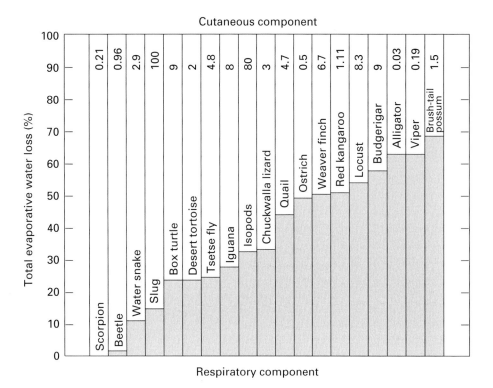

Fig. 5.3 Proportion of total evaporative water loss (EWL) attributable to respiratory loss (REWL—shaded) and to cutaneous loss (CEWL—white) for a variety of animals in their resting state. The total resting EWL is given (mg g^{-1} h^{-1}) for each species. (Adapted from Withers 1992.)

considering how animals may control their cutaneous osmotic fluxes. Water permeability and ion permeability are rather separate issues, just as they were distinct in relation to passage across cell membranes, as discussed in Chapter 4. However, for whole animals the surfaces across which such fluxes occur are usually structurally and chemically much more complicated than simple biological membranes. In some marine animals there may be little more than a surface epithelium, with a thin and loosely organized extracellular coating; at the other extreme, land arthropods are covered in a dense and usually rather rigid cuticle of very complex structure with intricate waterproofing properties, while land tetrapods have a thick outer covering of dead skin cells forming a keratinized layer. Clearly the varying structures and properties of skins will have very different consequences for rates of water and ion flux.

Water permeability

Water permeability in itself is a somewhat complicated issue. For whole animals P_w is commonly measured from rates of *weight* loss (or gain), which are considered equivalent to rates of water loss (or gain) after corrections for any other observed water outputs or inputs such as defecation, urination or feeding. More recently water loss has been assessed using electronic water vapour sensors, or from rates of flux of labelled water (tritiated water, ^3HOH), often expressed as daily turnover of water. Such measures can yield a 'permeability' in units of weight of water lost per unit surface area per unit time, and permeability has therefore been expressed traditionally (but rather loosely) in units such as mg cm^{-2} h^{-1}.

For terrestrial animals, the evaporative water loss (EWL) varies with the dryness (water vapour density) of the surrounding air, and so is more properly expressed as water lost per unit of water vapour pressure deficit, although there are substantial problems in controlling temperature and humidity adequately to ensure correct assessment of this. But this figure of total water loss rate is still potentially misleading for two reasons; firstly, it takes no account of the changing conditions very close to the animal's surface (boundary layers of unstirred, more humid air); and secondly, it almost inevitably contains a substantial component of respiratory evaporative water loss (REWL) in addition to the cutaneous evaporative water loss (CEWL). The first problem is nearly insuperable, as correct modelling of water movements just above and just below an irregular evaporating surface is notoriously complex. The second problem is also difficult to get around. The relationship between the two components (REWL and CEWL) varies with activity, and even in a 'resting' animal is highly variable between taxa (Fig. 5.3), so it is not a simple matter to make corrections to EWL and thereby calculate CEWL. Therefore many recent studies instead express the data on cutaneous loss or gain for all animals as a resistance R, measured in s cm^{-1} (see Box 5.1); on this scale a free water surface has a resistance of zero. Resistance can be determined from experiments on isolated pieces of tissue, removing the respiratory water loss element. But it can also be calculated from whole animal rates of water loss according to the formula given in Box 5.1, and for quiescent animals with enclosed (or artificially blocked) respiratory surfaces the errors will be small. Cuticle permeability is then effectively the reciprocal of resistance ($1/R$), in cm s^{-1}.

Box 5.1 Resistance, permeability and evaporative water loss rates

Evaporative water loss (EWL) depends on the difference in water vapour density χ (in mg water per litre) between the animal and the surrounding air, and by analogy with Fick's law we can derive:

$$EWL = -D \, \Delta\chi/\Delta d$$

where D is the diffusion coefficient for water vapour in air, and d is the diffusion distance (the boundary layer thickness). χ for the animal is taken as the χ for saturated air in equilibrium with the body fluids. In practice the diffusion coefficient and the diffusion distance can be replaced by a single factor, R, the resistance, giving:

$$EWL = -\Delta\chi/R$$

Values of R can be compared between animals, assuming a constant difference in water vapour density applies (roughly true for a particular vapour pressure level, given the similarity of all terrestrial animals' body fluids).

Alternatively, water loss data (EWL) expressed in conventional units as $\mu g \, cm^{-2} \, h^{-1} \, mmHg^{-1}$ can be converted into permeability coefficients (P, with units of $cm \, s^{-1}$) using the equation:

$$EWL \times 2.94 = P \times 10^4$$

This is accurate at 20°C but slightly inacccurate at other temperatures (e.g. the conversion factor is 2.91 at 30°C). This value of permeability P in $cm \, s^{-1}$ is then the reciprocal of resistance (R) in $s \, cm^{-1}$.

Whatever the measurement used, there is a huge range of permeabilities to be found in 'skins'; Table 5.3 shows values of rate of loss (or gain) of water, together with resistance or turnover rate where possible, recorded from a wide spectrum of animals from different habitats. Although the values vary phylogenetically, with some taxa always more permeable than others, there is a clear-cut superimposed habitat effect, with members of the same phylum or class that come from aquatic, mesic and xeric habitats having increasingly impermeable skins. Note that there may also be effects of diet; for example, nectar-feeding birds that get plentiful water supplies are decidedly less impermeable than similar-sized birds that feed on seeds or insects.

Skin surfaces clearly vary from epithelia with nearly unrestricted permeation to very resistant nearly 'waterproof' cuticles. A model of the various layers that may occur in a skin and its cuticular covering is shown in Fig. 5.4. Several features potentially contribute to low P_w. Firstly, the epidermis itself may exert some control over P_w, and hormones may operate here to affect water loss. Secondly, the bulk cuticle material, whatever its constituents (e.g. chitin, proteins, keratin, calcium salts) will usually be inherently relatively impermeable, and its water resistance may alter according to the density of its molecular packing as it becomes more or less hydrated. Thirdly, and usually most important of all, there may be a specific layer conferring even greater impermeability due to a high lipid content; this is commonly at the outer surface, but could be anywhere within the cuticle.

Relatively simple skins are found in many marine soft-bodied animals such as sea anemones, annelid worms and echinoderms;

there is often no secreted layer except perhaps a loose extracellular matrix external to the epidermis, often composed of mucopolysaccharides. Some consolidation of this matrix may occur in essentially 'soft-bodied' animals and may confer a degree of water-flux resistance. Freshwater animals generally have lower permeabilities than related marine species, and brackish-water species may be even less permeable. This may initially seem surprising, but of course most brackish waters change concentration cyclically, and low permeability is a particularly good strategy to smooth out these external changes. In many of the more euryhaline aquatic invertebrates P_w can be acclimated to some degree, over a time course of a few hours or day, according to environmental exposure (see Chapter 10).

The really impermeable surfaces are, not surprisingly, found in animals with complete coverings of exoskeletal materials, especially where the habit is terrestrial. Land animals possessing some kind of relatively impermeable complete outer covering include inactive molluscs, where only the calcareous shell and a small area of mantle tissue are exposed. The other invertebrate groups that show a near-complete and highly impermeable covering are the insects and arachnids from drier habitats. These arthropods provide the very best examples of protective surfaces that minimize CEWL. The main chitin + protein cuticle layers are of relatively low permeability, containing some lipid and with a variable hydration state. But the particular asset of land insects is an additional epicuticular layer of high lipid content. This is made up of C_{20}–C_{37} fatty acid derivatives plus some branched hydrocarbons and some C_{40} wax esters, which may appear as a specific wax layer on the surface of the epicuticle. At one time all these epicuticular lipids were thought to be organized in a single orientated monolayer at the epicuticular boundary, but this theory has now been replaced by a more realistic view (one better able to account for the amount of lipid present and its actual chemistry), which interprets the epicuticular lipid as present in bulk and in a less organized 'liquid crystal' form, able to change phase from more liquid to more pseudocrystalline solid according to conditions (especially temperature). This explains the transition in cuticle permeation (increasing P_w) that often occurs when an insect is warmed (Fig. 5.5), and which is even sharper if related to temperature of the cuticle rather than to air temperature.

In the most xeric insects and arachnids the presence of a waxy epicuticle ensures much slower water loss (R up to $5000 \, s \, cm^{-1}$) than in similarly sized crustaceans ($R = 15$–$75 \, s \, cm^{-1}$), with the same general cuticle design but little or no waxy component. This combination of epidermis, chitin–protein cuticle and waxy epicuticle provides an almost complete body covering that is sufficiently 'waterproof' to allow survival in a desert with no drinking water at all.

Vertebrate integuments also constitute 'complete outer coverings', but their P_w varies quite widely. At one extreme ($R < 50 \, s \, cm^{-1}$) are the soft and non-keratinized skins of amphibians, used for ionic exchanges and even more crucially for a large part of the respiratory exchange of the animal; these skins have

Table 5.3 Comparative measures of 'permeabilities' across the skins of animals.

Habitat	Animal group	Resistance (s cm^{-1})	Water flux (mg cm^{-2} h^{-1})	Transpiration rate (µg cm^{-2} h^{-1} Torr^{-1})	Water turnover (ml g^{-1} day^{-1})	P_{osm} (µm s^{-1})
Marine	Crabs	6–14	14–25			
	Echinoderms					30
	Reptiles	33	0.5			0.002
	Fish	7–35				0.2
Hypersaline	Crustaceans	20–100				
Estuarine/littoral	Amphipods and isopods	3–6		200–300		
	Crabs	4–40		80–200		1.4
Freshwater	Insect larvae	10				0.01
	Insect adults	200				0.05
	Crustaceans	11–46				0.3
	Fish	2–8				
Amphibious	Frogs	3–100	5–40	2400		5–25
	Crocodile	5	70	400		
Terrestrial	Snails active (inactive)	2 (46)		2500 (16)		
	Earthworms	9				
	Isopods	3		80–160		
	Crabs	30–75	1–3			
	Scorpions			8–80		
	Spiders and mites	80		30–60		
	Myriapods		2–4	40–270		
	Collembolans			700		
	Termites			28–37		
	Orthopterans	200	1–2	20–70	0.1–0.4	
	Caterpillars			190		
	Dipteran flies			50–76		
	Beetles			24–50		
	Ants			25		
	Frogs	1.5		2000–2500		
	Lizard	198	0.2			0.005
	Birds	50–70	0.7–2.7			
	Mammals		1–10			
Arid/desert	Isopods			14–30		
	Scorpions, spiders, mites	1300–4000		0.6–2.0		
	Millipedes	430		8		
	Apterygotes			15		
	Hemipteran bugs			12–14		
	Cockroach, cicadas			12–100		
	Beetles	5030	0.1–0.2	3–15	0.05	
	Beetle pupa			1		
	Tsetse fly			8		
	Tsetse fly pupa			0.3		
	Caterpillars			40		
	Ants and wasps			4–26		
	Frog, cocooned	457				
	Tortoises	120	1.6		0.003	
	Birds	158	1.7		0.09	
	Lizards	1360	0.1		0.03	
	Small mammals (rodents)		0.5–0.7		0.03–0.13	
	Large mammals		2–6		0.03–0.09	

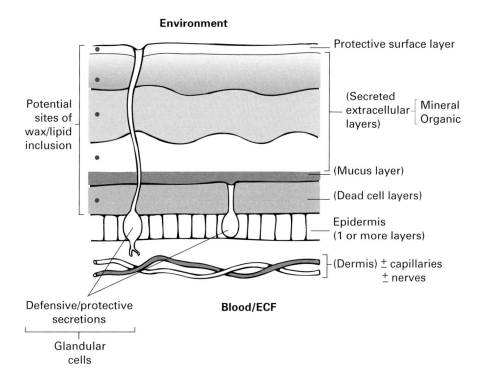

Fig. 5.4 A generalized view of 'skin' and its possible components, in relation to cutaneous water loss. ECF, extracellular fluid.

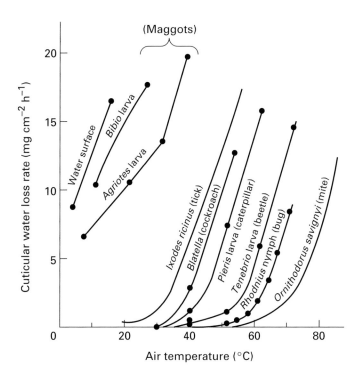

Fig. 5.5 The phenomenon of 'transition temperatures' in a variety of arthropods; cuticular water loss increases with temperature, but begins to increase particularly sharply above some species-specific transition temperature. Note that the transition is not very sharp, and commonly occurs well above normal environmental temperatures. (Data from Wigglesworth 1945; Lees 1947.)

little or no embellishment external to the epidermis itself, apart from secreted mucus. However, in a few relatively xeric amphibian species, such as *Phyllomedusa*, P_w can be further reduced by a waxy secretion from glands that the frog spreads over itself, when R may be raised to 400–500 s cm^{-1}.

Intermediate permeabilities are found in many fish, where keratin scales overlie the softer dermal layers, and the skin itself has little or no respiratory role. As with the invertebrates, species from brackish waters and the very euryhaline species are often the least permeable. Many highly tolerant fish show a hormonally mediated decrease in P_w in response to saltwater exposure.

But the amniotic vertebrates—reptiles, birds and mammals—have distinctly more impermeable skins, with no respiratory function to compromise their osmoregulatory function. The skin is formed largely from layers of dead keratinized cells overlying the living epidermis, often thickened into plates and scales in reptiles, or modified into hair or feathers in the mammals and birds. Lipid material is included in a dispersed fashion in mammalian skin, while in reptile skin the stratum corneum specifically has a rather high phospholipid content. All these animals are distinctly less prone to water loss than most fish and amphibians; aquatic reptiles by only a small degree ($R = 5$–50 s cm^{-1}), but terrestrial reptiles ($R = 200$–500 s cm^{-1}) and all birds and mammals ($R = 50$–200 s cm^{-1}) very substantially. These vertebrates still do not approach the degree of waterproofing found in the arthropodan groups. However, the greater SA/V ratio of desert reptiles and rodents compared with much smaller insects often means that they too lose water so slowly as a percentage of their body weight that they can also survive without access to drinking water.

To complicate the picture, cutaneous water loss in some of these groups can be altered more drastically by controlled and 'planned' losses related to cooling: the phenomenon of sweating, dealt with in detail in Chapter 8. Many mammals have a background low level of water loss from the sweat glands superimposed on the loss through the rest of the skin, with a much faster loss that is turned on during exercise and in hot climates. Similarly, the more waterproof species of frogs effectively sweat through their epidermis when heat-stressed; and a few insects also 'sweat' when overheated.

Ionic permeabilities

The ion permeation properties of skins are due to ion channels in underlying membranes (varying with their number and their properties), and to the structure and electrical charge distributions in overlying extracellular matrices or cuticular layers. Where substantial extracellular material is present the situation is somewhat complicated by ion-binding effects; for example, mucus is often a good ionic binding matrix, and the rate of movement of Na^+ or H^+ ions through it is unexpectedly hard to predict. Certain reagents have traditionally been used to stain sites of high salt permeability in aquatic animals: silver nitrate and potassium permanganate are the most widely used.

Again comparisons are somewhat complicated by varied methods of measurement and different ways of expressing salt movements; some studies determine labelled sodium loss rates directly through pieces of isolated skin or cuticle, while others consider overall sodium loss from whole animals. Data therefore have to be treated with some care. However, for most aquatic invertebrates salts certainly move relatively freely across the surfaces, and again P_{Na} in particular tends to be much lower in brackish and freshwater forms. Specific loss rates of sodium (in $\mu mol\ g^{-1}\ h^{-1}$) when exposed to dilute media vary from over 1000 in fully marine crabs, through 800–900 in marine/littoral crabs, 100–200 in estuarine/freshwater crab, to just 5 in the fully freshwater crayfish. Similar effects can be seen even within a genus, as we will see in Chapter 10.

Unlike P_w, there is little evidence for acclimation or variation of P_{Na} in invertebrates, although it can vary genotypically within a species. However, many aquatic vertebrates can vary their sodium permeability individually according to need, and it is directly under the control of the pituitary hormone prolactin in the skins of fish and frogs.

5.2.2 *Acclimatory changes of permeability*

We have mentioned a number of cases where permeabilities to ions or to water can be altered in the short term—within the life of an individual. How can this be achieved? There are two main possibilities. Firstly, at the level of cell membranes acclimation involves adding or destroying ion channels (and probably also water channels), and there are two possible timescales for this, either recruiting pre-existing channels from vesicles within the cells as described in Chapter 4, or actually synthesizing new channel proteins. Secondly, permeabilities may be altered on a larger scale by alterations to the extracellular components: again this may involve spatial rearrangements, involving phase changes in lipid layers, for example, or it may involve synthesis of new material added to the surface. Insects show these effects particularly clearly, since the permeability of successive exuviae (cast-off skins) can be measured for an individual; they contain lipids of significantly higher melting point when the individual is acclimated to higher temperatures, largely due to incorporation of more straight-chain hydrocarbons relative to branched forms (see Chapter 8).

5.3 Osmoregulation at external surfaces

A substantial proportion of aquatic animals are always either hyperosmotic (most brackish and freshwater animals) or hyposmotic (most marine vertebrates) to the medium in which they live, as we have seen. In both situations, some of the regulation to maintain this situation is done at the external surfaces of the animal, either through the skin as a whole (which thus becomes an important organ actively participating in water balance), or through the more permeable areas such as gills, or via particular modified glands or appendages. External 'skin' surfaces are thus a very important component of overall osmoregulation, and in a few animals they may be the only regulatory system in operation.

5.3.1 *Hyperosmotic regulation and salt uptake*

All the inhabitants of fresh waters, and most of those in brackish waters, are hyperosmotic regulators, continuously having to counter the twin problems of water tending to flow into the animal, and solutes tending to be lost. The water that enters the body has to be excreted, and this process inevitably entails yet more solute lost in the urine. Decreasing the permeability of the skin will help, as discussed in section 5.2, but the respiratory surface of the skin must remain large enough and permeable enough to give adequate gaseous diffusion, so that large gills could be rather an osmotic liability. However, some part of the body must be used to regain solute and redress the balance against the passive fluxes, and the obvious surface to use is that of the permeable respiratory surface itself, localizing all exchange systems there. Hence many aquatic animals, both invertebrates and fish, show **active uptake of ions at the gills**; they may also use their gills for nitrogenous excretion and to regulate acid–base balance. Freshwater animals may take up Na^+ and other ions against at least a 100-fold gradient, for example from a medium at < 5 mM to their own blood at perhaps 500 mM.

The respiratory site is nearly always the site for this uptake, so for animals that normally rely on cutaneous respiration, such as freshwater worms, uptake is directly through the skin. In amphibians salt uptake also occurs through the general skin surface (though anuran amphibians also have lungs), and frog skin is a classical physiological preparation for studying the

Table 5.4 Values of K_m for sodium transport in various aquatic animals.

Habitat/species	Animal group	K_m (mM Na+)
Marine		
Marinogammarus finmarchicus	C	6–10
Carcinus maenas	C	20
Brackish		
Cyathura carinata	C	6
Mesidotea entemon	C	9
Gammarus duebeni	C	1.5–2.0
G. zaddachi	C	1–1.5
Rana cancrivora	V	0.40
Freshwater		
G. lacustris, G. pulex	C	0.1–0.15
Lymnaea stagnalis	M	0.25
Margaritana margaritana	M	0.04
Hirudo medicinalis	A	0.14
Salmo gairdneri	V	0.46
Carassius auratus	V	0.26
Xenopus laevis	V	0.05
Amphibious (freshwater/land)		
Rana pipiens	V	0.20
Bufo boreus	V	0.25
Hyla regilla	V	0.14

A, annelid; C, crustacean; M, mollusc; V, vertebrate.

properties of salt-absorbing epithelia. In most other freshwater animals, salt uptake is localized at the gills. In some larval insects salt uptake occurs in appendages at the tail end, the anal papillae, also inaccurately known as 'anal gills'. But in most molluscs and crustaceans, and in adult insects, this uptake can be shown to be localized in cells of the genuine gill epithelium. Often the whole gill surface is osmoregulatory, but in some animals there is a division of labour; for example, in the mitten crab *Eriocheir* some posterior gills are primarily osmoregulatory while the anterior ones are mainly respiratory.

The standard Na+/K+-ATPase operates at the baso-lateral cell membranes in all these sites, with Na+ symport/antiport systems on the apical membranes. The kinetic properties of the overall sodium transport are highly variable, with K_m values (the concentration for half-maximal transport, so that low K_m indicates high affinity; see Box 4.5) dependent on the medium in which an animal lives (Table 5.4). Thus K_m is 6–20 mM Na+ in marine invertebrates and only 0.05–0.2 mM Na+ in freshwater and amphibious animals, reflecting a high-affinity pump suited to the lower sodium concentrations in their environment. There is still some uncertainty as to whether chloride movements are always dependent on the Na pump, or whether there is a separate and independently active mechanism for handling Cl- ions, as reported for some brackish-water and freshwater invertebrates. Certainly in many epithelia it has been shown that the two

ions appear to move into cells independently of each other, each being coupled to the efflux of an ion of like charge (Na+ for H+ or NH_4^+, Cl- for HCO_3^-; see Chapter 4).

In freshwater teleosts, salt uptake occurs across the integument and gills, as well as in particular cells of the gill epithelium; these cells are conspicuously thicker than the main respiratory epithelium, with the deep basal infoldings and high density of mitochondria characteristic of salt-transporting cells. These so-called **chloride cells** (Fig. 5.6) primarily function to secrete excess salt (NaCl) in marine fish, so their mechanism is covered in the next section; they are less active in freshwater species. However, other ions (especially Ca^{2+}, H+, HCO_3^- and NH_4^+) appear to be handled by these cells according to circumstance, and in fresh water, substantial chloride–bicarbonate exchange is recorded.

5.3.2 *Hyposmotic regulation and salt secretion*

Coping with environments that are more concentrated than the body fluids is a relatively uncommon problem given the numerical dominance of marine invertebrates (i.e. animals isosmotic with the medium), but it is very important because it occurs in all marine teleosts and tetrapods, and also in the invertebrates that can live in hypersaline environments. For all these animals, water will rapidly efflux from the body, potentially causing tissue shrinkage, and salts will tend to leak in.

For invertebrates faced with these problems (see Chapter 12), the drinking rate is increased to acquire the essential water, but this produces a greater salt load, which must be excreted via the gills, excretory organ or rectum.

Hyporegulation in marine vertebrates occurs in lampreys, teleosts and all tetrapods examined to date except the crab-eating frog. In general these animals all have blood concentrations about one-third that of the sea in which they dwell, and their particular physiological problems are dealt with in detail in Chapter 9. Again, drinking sea water is the common response, with water entering the body via the gut. In teleosts, drinking-rate measurements indicate that only a proportion of the salts comes via this route, with substantial amounts also entering through the relatively permeable gills. Monovalent ions are the main input at both sites; divalent ions in the drinking water are largely retained in the gut lumen and excreted with the faeces, while any leaking in at the gills is dealt with by the kidneys. The excess monovalent salt load is secreted largely by the gill epithelia at the **chloride cells** already referred to (Fig. 5.6). These have high levels of chloride ions, especially near their external borders, but present evidence indicates that the main ion pumps are actually, once again, the classic Na+/K+-ATPase 'sodium pumps', since ouabain blocks salt transport (see section 4.4.2). The sodium pumps are basal, and seem to pump in the 'wrong' direction. How can this work? In effect, they produce a steep inward Na+ gradient basally, so that symporters (either Na : Cl or more probably Na : K : 2Cl) can draw ions into the cell from the blood, across the basal membrane. Na+ continues to be pumped out again, and K+ will diffuse back out, so the net effect

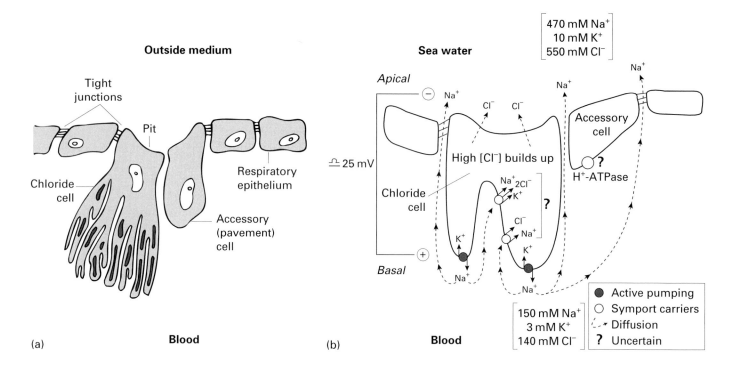

Fig. 5.6 The structure and function of 'chloride cells' occurring in fish gills and salt glands. (a) General structure, thickened cells within the respiratory epithelium. (b) Model of salt transport across the cells in a marine teleost.

is a pile-up of chloride, which then leaves apically via chloride channels, producing a net electrical gradient. This in turn draws sodium across to the apical side, mainly by the paracellular route (chloride probably cannot pass back by this route due to a net negative charge within the cell junctions). Chloride cells are surrounded by accessory 'pavement cells', which may be involved in sodium uptake coupled to an H^+-ATPase; they also exert a special kind of morphological control over the chloride cells, being able to move in and occlude them or move back and expose them as conditions dictate (see Chapter 10).

The gills are thus the main osmoregulatory organs for the major osmotic effectors in a marine teleost. In contrast, for marine elasmobranchs rectal glands are the main site of salt secretion, with a rather similar mechanism to that of the teleost chloride cells.

Other marine vertebrates (the tetrapods) are numerically insignificant in comparison to teleosts, and they do *not* normally use the body surface, or any particular parts thereof, for osmoregulation. Most of them are essentially terrestrial air-breathing animals that spend long periods in and around salt waters, feeding on marine life with a high salt content, and drinking marine waters, so that their water balance problems are not unlike those of the teleosts. However, they are much more physiologically isolated from sea water in that they do not need

large surface areas of permeable gill, since they are periodically breathing air at the surface of the sea; the salt load problem is rather less severe for them. Within this assemblage of tetrapods, two distinct osmoregulatory strategies occur:

1 The reptiles and birds, endowed with kidneys incapable of producing concentrated urine (with the exception of a few birds where urine is slightly hyperosmotic to the blood), swallow some sea water and gain more salt via the general body surface, and then must excrete the excess salt load via structures known as **salt glands**, functioning in similar fashion to the teleost chloride cells by pumping Na^+ in the 'wrong' direction to produce a steep gradient back again. The fluid produced is highly concentrated, made up almost entirely of NaCl at a particular concentration for each species, and it is only produced sporadically when the animal concerned is particularly osmotically stressed. The salt gland system is thus an all-or-nothing osmoregulatory effector (unlike the tetrapod kidney, which can produce varying volumes and concentrations and works all the time). It is also an extremely powerful system, producing higher rates of ion transport per gram of tissue than almost any other regulatory organ (see Chapter 9).

2 Marine mammals are rather different, since like all mammals they can produce hyperosmotic urine, and thus they can regulate using the kidney, leaving even less role for the external surfaces. Whales and seals can produce urine that is hyperosmotic not just to their own blood but also to the surrounding sea water, so ridding themselves of a substantial salt load. Most marine mammals do not drink sea water to any great extent anyway, though they may take in water incidentally when feeding. There are indications that some marine mammals can hold

food in the oesophagus while using a reversed peristalsis to pump sea water back to the mouth where it is spat out.

5.3.3 Terrestrial regulation and water uptake

There are only three possible avenues of water gain in a terrestrial animal: production of oxidative water by metabolic processes, uptake of liquid water, and uptake of water in the vapour phase. All animals inevitably use the first mode, simply by virtue of being alive, but this has only limited scope for regulation; this topic is dealt with in more detail in section 5.8, as it is relevant to the water balance of all animals. Most land animals in practice acquire liquid water by drinking or in combination as food; a few have special modes of acquiring water from puddles and pools, from condensing dew, or from interstitial water held in the substrate. A very small range of animals have special mechanisms to extract water from the vapour phase, from sub-saturated air. Chapter 13 gives a more detailed treatment of these purely terrestrial strategies.

Liquid uptake

For the majority of land animals the most important sources of liquid water are drink and food. However, there are a number of other possibilities used in special circumstances.

1 Use of dew. Dewdrops form by condensation at dawn usually, and may be drunk as they form on plants or on the soil; but some desert species 'collect' dew specifically, on their own bodies or in constructed condensation traps. Spider silk is hygroscopic and dew that collects on webs at dawn or whenever the humidity is high may be drunk by the resident spider.

2 Osmotic uptake of standing water. This is usually only an option for animals that retain fairly permeable skin surfaces. Some crabs can take up water at their gills by immersing themselves in freshwater pools, and some frogs achieve the same thing by squatting in puddles.

3 Osmotic uptake of water from wet soils and surfaces. Quite a range of insects and myriapods (millipedes and centipedes) can use eversible structures at the top of the legs called 'coxal sacs' to absorb water, while some woodlice (isopod crustaceans) use the brood pouch tissues for this purpose. The ventral tube of springtails (collembolan insects) bears eversible vesicles that take up water osmotically, by a process involving active transport of sodium.

4 Suctorial and capillary uptake from soils. Use of interstitial liquid water from less saturated soils generally requires either a very good muscular pump, or structural modifications of the skin surface to produce a strong capillarity that will overcome the capillary attraction of water to soil. Some spiders can extract water from soils by suction, and some ghost crabs can suck water from sand into their gill chambers using muscles acting on the chamber walls. Some littoral amphipod crustaceans have a capillary channel formed by the uropods (terminal posterior

appendages) that can conduct water to the ventral surface and thus to the mouth. Many semi-terrestrial crabs can take up water from sand using a ventral patch of hairy cuticle; these hairs draw up water by capillarity, and conduct it to the leg bases, from where it is drawn into the gill cavity.

Vapour uptake

A small and exclusive range of flightless arthropods has developed the ability to take up water from non-saturated air (Table 5.5); at present the list includes some 70 species, ranging through mites and ticks, some of the apterygote insects, and a range of pterygotes including lice, fleas and orthopterans, plus a

Table 5.5 Critical equilibrium humidity (CEH) values for various non-flying arthropods able to take up water vapour from unsaturated air. Where a range of values is given for a species, the extremes represent variation with temperature.

Taxon	CEH (as % relative humidity)
Arachnida	
Acari—mites and ticks	
Ornithodorus	94
Ixodes	92
Echinolaelaps	90
Acarus	70
Dermatophagoides	52–69
Amblyomma	87–89
Insecta	
Thysanura—apterygotes	
Ctenolepisma	48
Thermobia	44–47
Mallophaga/Siphonaptera—fleas	
Ceratophyllus	82
Goniodes	60
Xenopsylla	50
Liposcelis	54
Phthiraptera—lice	43–52
Psocoptera—booklice	
Liposcellis knulleri	70
L. bostrychophilus	60
L. rufus	58
Orthoptera—roaches + grasshoppers	
Arenivaga	80–83
Chortophaga nymphs	92
Lepidoptera	
Tinea larva	95
Coleoptera—beetles	
Tenebrio larva	88
Lasioderma larva	43
Onymacris larva	84
Crustacea	
Isopoda—woodlice	
Porcellio	87–91

Fig. 5.7 In prepupae of the flea *Xenopsylla*, water vapour uptake (seen as an increase in weight) occurs at humidities above 50% relative humidity (RH) but not at 40% RH, which is below the critical equilibrium humidity (CEH) value for this species. (Adapted from Edney 1947.)

few beetles and caterpillars. Recently, examples have been added from amongst the terrestrial woodlice. Each species has a characteristic **critical equilibrium humidity** (CEH; or sometimes equilibrium relative humidity, ERH), the lowest air humidity at

which it can extract water (see Table 5.5). For most species this value is above 80% relative humidity (RH); in arachnids the total range known is currently 52–95%, and in insects it is 43–95%. An example of the pattern of responses seen during water vapour uptake is shown in Fig. 5.7.

The mechanisms of vapour uptake differ, but always serve to produce a large water activity gradient between the body fluids and the uptake surface. This may be achieved in different sites:
1 In some of the mites and in the desert cockroach, uptake occurs at the mouthparts, in association with salivary glands. Two main principles appear to be involved: the secretion (or perhaps muscular extrusion) of high concentrations of hygroscopic solutes, on which water will condense, and/or the use of structurally complex exposed surfaces to hold this concentrated fluid and draw in fluid by capillarity.
2 In most other cases uptake is via the rectal or anal surfaces, which are already specialized for water absorption. Air is drawn into the anus and exposed to the absorptive surfaces, which in turn may be in close approximation to the ends of the Malpighian tubules where KCl concentrations may reach >4000 mOsm (see below). There may also be special hygroscopic materials, or muscular pumping to create pressure cycles, to facilitate water uptake at such sites in species with very low CEH values; examples are considered in more detail in Chapter 13.

In addition to these localized uptake systems, there are suggestions in the literature of a more general occurrence of surface uptake of water vapour in some insects, including pupal stages where there is no other source of water and where no salivary or anal openings occur; but these have never been fully substantiated and no widely accepted mechanisms have been proposed. The physiology of the uptake of water vapour is an area where

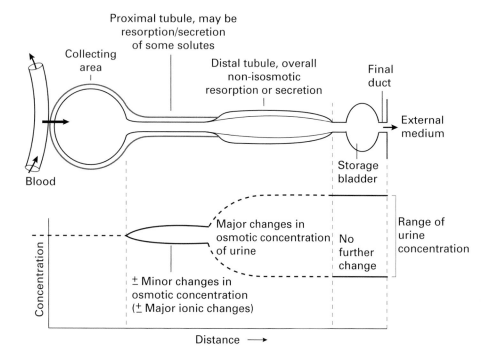

Fig. 5.8 A schematic basic design of osmoregulatory/excretory systems, and the patterns of change of concentration commonly found in each region.

details of the physical and chemical processes remain substantially unresolved.

5.4 Osmoregulatory organs and their excretory products

Some animals can do all of their osmotic regulation and control across all or part of the general body surface, by the mechanisms considered above. However, most animals also have some kind of internal structure that is usually described as their excretory organ, and often the term 'kidney' is used generically to include all these structures, which range from subcellular and unicellular structures such as contractile vacuoles and flame cells, through to complex organs with several different constituent tissues. In practice the primary function of these organelles and organs is almost always that of osmoregulation rather than excretion; the nitrogenous waste from the body that needs to be eliminated is added to the osmoregulated urine almost incidentally (and sometimes not at all). It might therefore be more correct to call these structures osmoregulatory organs, although the distinction becomes somewhat blurred if excretion is allowed a broader definition in terms of removing all excess waste metabolites (thus including excess water or excess ions).

Despite the huge variation in origin, size and apparent complexity in osmoregulatory organs, there are certain common architectural and physiological principles operating in all of them (Fig. 5.8) so that they can be treated together. Nearly all of them are made up of one or many tubular structures, and most include an initial collecting area where the primary urine is formed, then one or more areas where it is modified by addition or removal of particular solutes. Many of the organs include an even more distal area where the urine as a whole can be made more concentrated (hyperosmotic) or more dilute (hyposmotic) than the body fluids.

5.4.1 Production of primary urine

There are only two basic ways of making the initial urinary product in animals (Fig. 5.9). Most taxa use a system based on **ultrafiltration**, where hydrostatic pressure forces blood (or coelomic fluid, or some other component of the body fluids) through a semi-permeable membrane into the lumen of the osmoregulatory organ. The membrane acts as a filter, holding back blood cells, proteins and other large molecules, but allowing water, ions and small solutes such as amino acids and sugars to pass through. The proportion of a solute that passes into the urine (its filtration factor) is roughly determined by its hydrated size; this proportion is generally 1.0 for water, urea, glucose and sucrose, and for mammals it is 0.75 for myoglobin, 0.03 for haemoglobin and less than 0.01 for other serum proteins. The product is thus a simple filtrate of the body fluid, with very little change in osmotic concentration or composition except that cells and proteins are lacking. Most components of the body fluids would therefore be ejected from the body if no further

modification of the urine were to take place. This makes the system ('eject everything and then take back what is wanted') rather safe, in that any small toxic molecules entering the body will be automatically eliminated at the 'kidney' by ultrafiltration. However, it requires large volumes of fluid to be handled; in some animals as much as 99% of the initial filtrate is taken back into the body, involving substantial and expensive active transport.

Ultrafiltration is usually demonstrated by injecting a passive marker unfamiliar to the animal's cells and small enough to go through the filter; its presence can then be detected in the urine. The most common marker is inulin, a soluble polysaccharide that is apparently completely inert in biological systems and thus with no side-effects on cell functioning, and which has a filtration factor in the vertebrate kidney of about 0.98. Presence of inulin in primary urine is a useful (but not infallible) indication of a filtration-based system.

By far the best-known ultrafiltration system is that of the glomerulus in the vertebrate kidney (Fig. 5.10), where a network of blood capillaries is contained within the Bowman's capsule of each nephron and a filtrate of the blood plasma is forced out under pressure through the capillary walls. But filtration systems also occur in most invertebrate excretory organs (Fig. 5.11), including nephridia in worms, the antennal glands of crustaceans, and the molluscan coelomoduct kidney where the filtrate is taken directly from the heart into the pericardial cavity. The closed end of a flame cell in some simple freshwater animals almost certainly operates by ultrafiltration as well (Chapter 11).

A few groups of animals, notably insects but also leeches and some species of teleost fish, use a different system where the primary urine is produced by an **active secretion** process (Fig. 5.9). Membrane pumps move ions from the general body fluids into the lumen of the organ concerned; a model for a single tubule cell (based on the Malpighian tubule in insects) is shown in Fig. 5.12. The membrane pumps will handle most cations, and principally transport whichever is most common (varying with diet). Other oppositely charged ions may then follow down the electrical gradient created, or may be handled by separate pumps, while water follows readily down the osmotic gradient and may carry with it other small molecules in the bulk flow set up through the leaky cell junctions. (Note that these other solutes are effectively filtered across a semi-permeable membrane again, with anything of relative molecular mass less than about 10 000 getting across, so the two methods of urine formation are not quite as distinct as at first appears.) Again, the primary product may well be isosmotic to the blood; but if the pumping of ions is very intense, or if the product is rapidly removed from the lumen to other areas of the organ, or if the membranes where pumps are operating are rather impermeable to water or to particular ions, then the primary urine may potentially differ quite substantially in osmotic and ionic concentration from the main body fluids.

Active secretion of primary urine has the advantage of being

Fig. 5.9 Urine production by (a) ultrafiltration and (b) secretion. Basic principles and examples of mechanisms.

much more controllable; the rate of secretion can be varied according to need and circumstance—for example, being accelerated just after a large fluid meal in sap- or blood-feeders. On the other hand the system is potentially much less 'foolproof' than a filtration-based mechanism, since only molecules that are either specifically pumped, or are of the right size and charge to follow passively, will be passed into the urine; hence the need for a very

'leaky' secretory epithelium (see Chapter 4). Thus inulin injected into an insect does turn up in the urine via the leak route. But there is always a risk that moderately large toxic molecules that cannot leak through and are not 'recognized' by pumps in the tubule walls will be left circulating in the body fluids. Thus it may be that some cold-water marine teleosts, living in a relatively 'safe' and non-variant habitat, can get away with a non-glomerular, secretion-based kidney. However, the argument hardly applies to insects, frequently encountering 'foreign' chemicals in their variable terrestrial world.

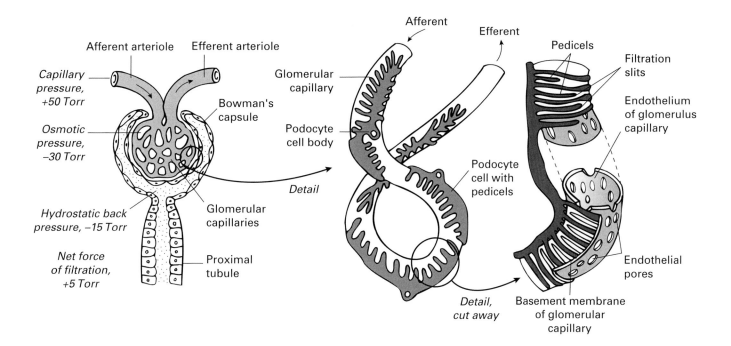

Fig. 5.10 The structure of the vertebrate kidney glomerulus where ultrafiltration of the blood occurs.

5.4.2 *Modification of the urine solutes*

After production of the primary urine, nearly all animals have mechanisms operating further along the tubes or ducts of the osmoregulatory organs that will retrieve particularly useful solutes, or add extra unwanted components, before the urine leaves the body (Fig. 5.8). In particular, glucose and other small sugars, and crucial amino acids, need to be resorbed into the body fluids. In animals with highly regulated body fluids and a strong homeostatic system, especially the terrestrial vertebrates and insects, the excretory system may be responsible for regulating water and certain ion levels; potassium regulation and the control of pH via bicarbonate concentration are common properties of excretory tubules, and in many cases there is also active transport of magnesium into the urine (with sulphate often linked to this) and selective resorption of calcium. In marine elasmobranch fish, where urea is used to increase the osmotic concentration of the blood, there must be a urea-resorbing system in the kidney tubules to recover this compound after it has been ultrafiltered.

The site of these active transport mechanisms is usually at the proximal end of the system, soon after the primary urine is formed. For example, in insects the lower third or so of the Malpighian tubule and the first part of the hindgut (the ileum) often have a resorptive function, taking back KCl and other ions, and leaving a reduced volume of a slightly hyposmotic urine due to a reduced P_w here. Similarly in the vertebrate kidney the volume of the primary urine is reduced greatly in the proximal convoluted tubule (see Fig. 5.16) by resorption of NaCl, with water following; also, all the filtered glucose is actively resorbed here. But regulation of potassium and bicarbonate levels in mammals is a property of the more distal parts of the tubules and the collecting ducts, and in other taxa a degree of resorption and secretion occurs in the distal bladder or the cloaca, just before the urine is excreted (see below).

Certain other molecules are specifically secreted into the urine in many animals. In fact kidneys have a surprising ability to secrete quite 'strange' larger molecules into the lumen, and so protect the body from toxins and drugs; in vertebrates this seems to be largely because the liver initially modifies such molecules, or parts of them, into forms that can be recognized by kidney transport systems. Some molecules have become widely used as markers to study these secretory processes, two of the most useful being *para*-aminohippuric acid (PAH) and the dye phenolsulphonphthalein (phenol red). Both of these are actively secreted into the tubules of a wide range of osmoregulatory organs. PAH is especially useful because it is secreted very rapidly indeed, with none remaining in the efferent blood supply from a healthy mammalian kidney. Phenol red is useful because it colours the urine, and its concentration can be determined colorimetrically. As a result we know that active secretion certainly occurs into the urine in many molluscan kidneys, and in crustacean antennal glands. Active secretion of other toxins such as nicotine, atropine and morphine, often derived from plant diets, has also been identified in insects. Secretion also occurs in the proximal convoluted tubule of the tetrapod kidney, where PAH, certain phenolic compounds, and the antibiotic penicillin are known to be secreted. Again, the distinction between filtration-based and secretion-based kidney systems becomes somewhat blurred.

'Protonephridia' or 'flame cell'

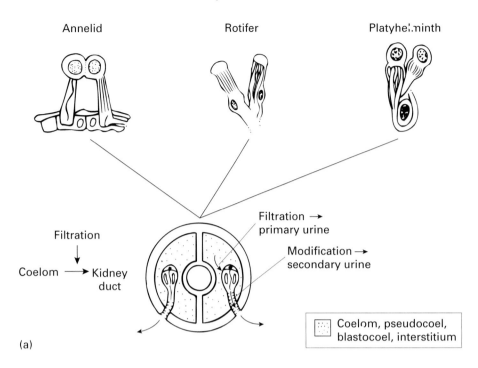

Annelid Rotifer Platyhelminth

Filtration → primary urine

Modification → secondary urine

Filtration
Coelom → Kidney duct

Coelom, pseudocoel, blastocoel, interstitium

(a)

'Metanephridia' or 'coelomoduct'

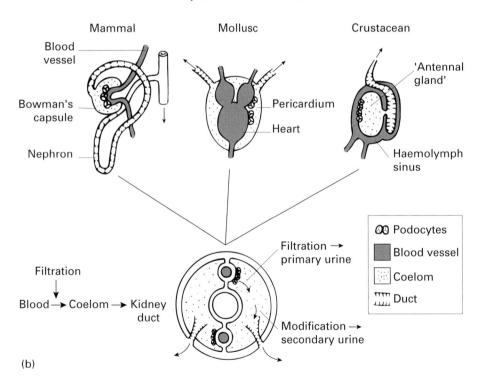

Mammal Mollusc Crustacean

Blood vessel

Bowman's capsule

Nephron

Pericardium

Heart

'Antennal gland'

Haemolymph sinus

Podocytes
Blood vessel
Coelom
Duct

Filtration → primary urine

Modification → secondary urine

Filtration
Blood → Coelom → Kidney duct

(b)

Fig. 5.11 Variety and occurrence of excretory systems based on ultrafiltration: (a) in flame cells; and (b) in more complex nephridial structures. (Adapted from Ruppert & Smith 1988.)

Fig. 5.12 Secretory urine production across cells of Malpighian tubules in insects, driven by an apical cation pump. (Simple model; more details in Chapter 13, see Fig. 13.13.)

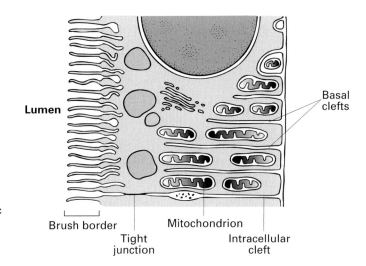

Fig. 5.13 Ultrastructure of a kidney cell from the proximal convoluted tubule where isosmotic resorption occurs. The luminal side has conspicuous projections ('brush border') and the basal side has deep clefts between which mitochondria are tightly packed.

5.4.3 Controlling urine volume and concentration

Even when an animal excretes isosmotic urine, as in most marine invertebrates, some further modification of the primary urine is required before it passes out of the body. Almost always a large part of the volume initially produced must be recovered, and this can be simply achieved by active transport of salts out of the lumen, with water following passively. This allows the waste products and any unfamiliar compounds to be excreted in the smaller volume of (still isosmotic) fluid that remains. The functioning of ion pumps in resorbing ions from the lumen essentially follows the principles outlined in the previous chapter; most commonly, sodium pumps (transmembrane Na^+/K^+-ATPases) operate to extract Na^+ from the urine into the cells, and chloride ions follow. The pump is normally sited on the basal side of classical transporting epithelial cells; the general type was shown in Fig. 4.12, and a specific example from the kidney tubule of a vertebrate is shown in Fig. 5.13. This kind of highly differentiated cell, with luminal microvilli (the 'brush border') giving a vast increase in surface area for water to follow the osmotic gradients created, and with basal indentations housing the pumps and linked with numerous mitochondria, can be found in osmoregulatory organs throughout the animal kingdom.

Hyposmotic urine, which rids the body of excess water, is a characteristic product of freshwater invertebrates and also of most freshwater vertebrates such as frogs or crocodiles. Its production again requires the active resorption of salts from the osmoregulatory tubules, but in this case as ion pumps remove salts from the urine it becomes progressively more dilute, since the P_w of the luminal membranes is kept quite low so that water is less able to follow osmotically. A hyposmotic fluid is either excreted continuously or gathers in a bladder for periodic ejection.

Animals on land generally need to conserve as much water as possible and therefore many of them face the opposite challenge of producing a **hyperosmotic urine**. Secondarily marine vertebrates also need to conserve water, but as we saw earlier (section 5.3.2) only the marine mammals and birds can do this by producing a hyperosmotic excretory fluid. How can animals apparently remove water from their urine, back into their own more dilute body fluid compartment, when we have already established that water itself cannot be actively transported? The mechanisms involved turn out to be various modifications of the simple model shown in Fig. 5.14; a system of cyclical active transport of ions, combined with particular kinds of compartmentalized cellular and tissue architecture in association with differential membrane permeabilities. In some cases this is improved by the addition of a countercurrent flow component.

Concentration by compartmentalization

The model in Fig. 5.14 can operate on a subcellular or on a multicellular level to achieve apparent removal of water from a system and hence produce a hyperosmotic fluid in one compartment by secreting a hyposmotic fluid into another. The cells form part of an epithelium between the lumen of an organ and the blood. Each cell has three essential constituents; a large intracellular compartment (B) that is continuous with the blood side via a relatively broad duct (C), and a series of fine convoluted blind-ending channels (A) near the luminal side of the cell, all of which lead into (B). Cation pumps are concentrated in

Lumen   **Blood**

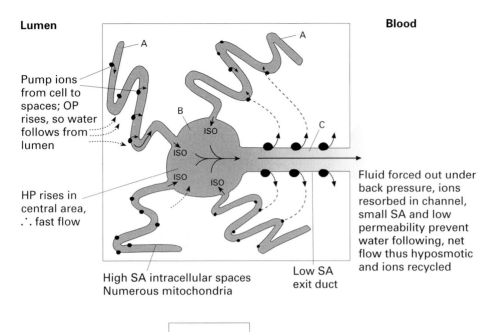

Pump ions from cell to spaces; OP rises, so water follows from lumen

HP rises in central area, ∴ fast flow

High SA intracellular spaces
Numerous mitochondria

Low SA exit duct

Fluid forced out under back pressure, ions resorbed in channel, small SA and low permeability prevent water following, net flow thus hyposmotic and ions recycled

→ Water

← Ions

Fig. 5.14 A model for 'water transport' across a cell within a resorptive epithelium, where solute recycling within the cell occurs and a fluid hyposmotic to the lumen is discharged into the blood, representing net water resorption. HP, hydrostatic pressure; ISO, isosmotic fluid; OP, osmotic pressure (or osmotic concentration); SA, surface area.

the membranes lining these fine channels, and pump Na^+ or K^+ into their narrow lumens, with anions following; water is drawn osmotically out of the intracellular fluid into the channels, since their walls have a high P_w, and ultimately water is drawn down a gradient from the nearby organ lumen contents to balance this ion pumping. This creates an isosmotic fluid in (A) which is continuously forced along the channels to accumulate in the intracellular compartment (B). Both (B) and (C) have much lower P_w values. Fluid accumulating in (B) has nowhere to go except out to the blood via (C), and the flow rate through C will be high

because of hydrostatic pressure resulting from the much larger total surface area of the channels (A) producing all the fluid. As the isosmotic fluid flows down (C) further Na^+ or K^+ ion pumps operate, this time to remove ions from the lumen back into the cell. Because of the high flow, the much reduced surface area, and the relative impermeability to water in this part of the system, water has no chance to follow the ions isosmotically, and so the fluid in (C) becomes progressively more dilute, and by the time it discharges into the body fluid it is very severely depleted of ions. Thus the cell has in effect transported water from the

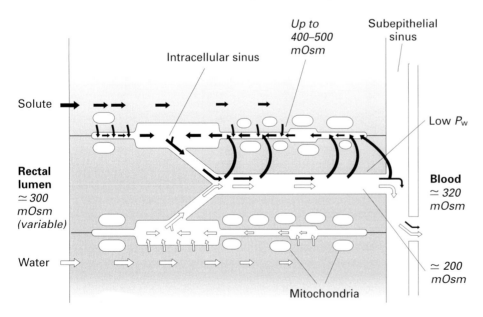

Solute

Intracellular sinus

Up to 400–500 mOsm

Subepithelial sinus

Low P_w

Rectal lumen
≃ 300 mOsm (variable)

Water

Blood
≃ 320 mOsm

≃ 200 mOsm

Mitochondria

Fig. 5.15 A real example of a resorptive epithelium in the insect hindgut (the rectal cells of the cockroach *Periplaneta*) illustrating the principles of intracellular architecture and differential permeability outlined in Fig. 5.14. Note that solute and water movements take place in the *same* intracellular channels in reality. (Adapted from Wall & Oschman 1970.)

Box 5.2 Countercurrent exchangers and countercurrent multipliers

Arranging two flowing parallel systems so that the direction of flow in one is opposite to the direction of flow in the other is the most effective and simple way of ensuring efficient exchange between the two, whether of **water**, **ions**, **gases** or **heat**.

Initially (when flows are first 'turned on') there is a large gradient of component X from fluid A to fluid B at all points in the system; but even after flow has continued for some time there will still be a small gradient from A to B throughout the length of the system. At the left-hand side, B has gained a large concentration of component X but still cannot be as concentrated as the incoming fluid A, and will continue to pick up more of X. Almost complete transfer of X from A to B occurs, at all but the highest of flow rates.

Exchangers

These work 'downhill', acting as passive devices to achieve maximum uptake or exchange. They work for anything diffusible: whatever can pass across the walls will be transferred.

The fluid in the two channels may be:

1 the same, for example cold blood and warm blood, in many rete systems that keep parts of the body cooler or warmer than the rest (e.g. whale flipper, tuna swimming muscle);

2 different, for example oxygenated sea water and deoxygenated blood (many fish gills); oxygenated air and deoxygenated blood (bird lung).

The two channels may be:

1 quite separate, as in the gills and lungs;

2 arranged as a loop, as in the whale flipper, where warm blood moving distally meets cooled blood coming back towards the body in the same continuous flow system.

The exchange may operate:

1 spatially, as in all the above examples with two distinct channels;

2 temporally, where there is just one channel but flow alternates in different directions (e.g. tidal flow in and out of vertebrate lungs). Cool air is warmed as it is drawn in, then loses most of its heat to the cooled passages as it is expired.

Multipliers

These work 'uphill', needing input of energy, acting to concentrate a substance (which may be relatively non-diffusible) in some part of the body.

Therefore they *always* require a loop structure, *always* with the same fluid (blood) in both channels, and they are *always* spatial in character:

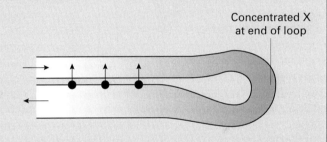

Examples include the loop of Henle in the mammalian kidney, and the gas-accumulating rete of the fish swim-bladder.

---

organ and deposited it in the blood. The organ lumen contents become more and more concentrated. The ions that are pumped into intracellular channels at (A) are recycled back into the cytoplasm at (C) and very little net movement of ions occurs.

A specific example of cellular architectural compartmentalization is shown in Fig. 5.15, based on the resorptive cells found in the rectum of many insects, where the excretory fluid from the Malpighian tubules is finally concentrated. Here the cells are removing water from the gut back into the body fluids, to produce a very concentrated excreta. A similar combination of architectural compartmentalization, where different parts of the system have very different surface areas and very different water permeabilities, can be found on a more complicated multicellular scale in the rectal papillae and pads of some insects, covered in Chapter 13. Mechanisms based on this general pattern

are at work in other systems that appear to transport water 'actively', including parts of the gut of most other terrestrial animals, but at present insect rectal cells are the only system where net cellular production of a hyposmotic fluid is really understood. Clearly water itself is not being actively moved at all; ions are pumped and water follows passively as always.

Additional concentration by countercurrent

A relatively small number of animals are able to achieve very considerable concentration of their urine by supplementing subcellular and cellular architecture with a pattern of gross anatomy whereby two parts of the tubule system run close together and in opposite directions, setting up what is known as a **countercurrent flow** system (Box 5.2). We will meet countercurrent

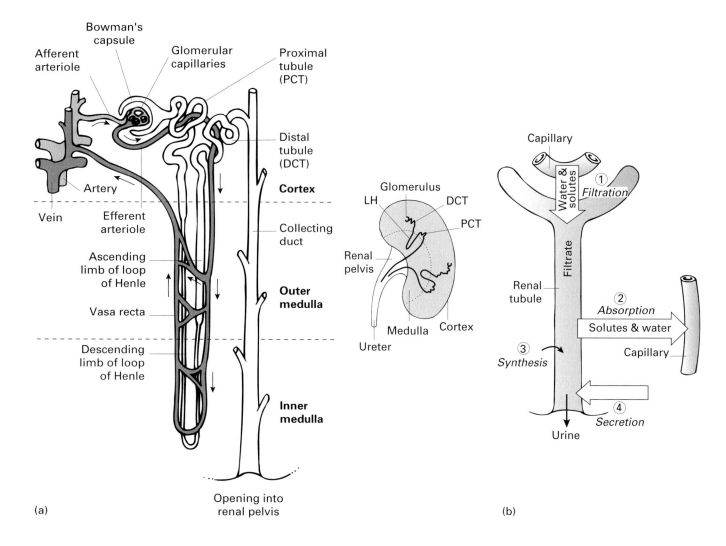

Fig. 5.16 (a) The basic structure of a mammalian nephron; and (b) a simple model of nephron function.

mechanisms regularly in the field of environmental physiology, so it is very important to read this box and get the principles and advantages of **countercurrent exchange** and **countercurrent multiplication** clear from the outset.

In the case of osmoregulation, countercurrent multipliers based on active transport are set up in the excretory/osmoregulatory organs using the constituent tubules to create standing concentration gradients, and so resorb as much water as possible from the urine.

It is worth looking at an example in some detail. By far the best known, architecturally relatively simple but somewhat complex in mechanism, is that of the vertebrate kidney, or rather of the nephrons from which it is made up (Fig. 5.16).

Vertebrate kidney function

First we need to describe the main functions of the various parts of the nephron.

1 The glomerulus in Bowman's capsule produces the ultra-filtrate (collectively filtering up to 25% of the blood per minute in an active mammal).

2 The primary urine is almost immediately reduced in volume by 75% by the proximal convoluted tubule (PCT), which has classical 'leaky' salt-transporting epithelium. This takes out NaCl isosmotically by the usual combination of ion pumping and passive water entrainment, the ions and water passing back into the efferent renal blood system. As it reaches the edge of the cortex the urine is therefore still at about 300 mOsm, isosmotic with the blood. But the 75% reduction in volume has led to a four-fold increase in concentration of relatively impermeant solutes such as urea.

3 The descending loop of Henle (DLH) contains simple epithelial cells with no specializations for transport, having very low P_{Na} and P_{Cl} and P_{urea}, but a high P_w. The first part of the ascending limb likewise has no indications of active transport but the cell membranes have high P_{Na} and P_{Cl}, low P_{urea}, and very low P_w. However, the upper ascending limb is thick-walled with actively transporting cells moving Na and Cl from the tubule lumen to the surrounding interstitial space, in a manner similar to

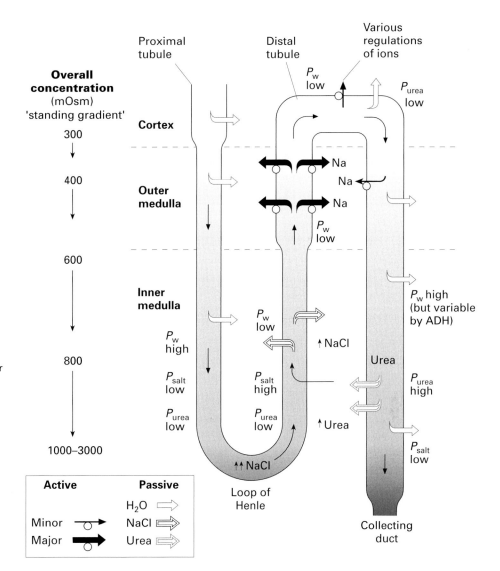

Fig. 5.17 The renal countercurrent multiplier system in mammals; how the standing osmotic gradient across the kidney medulla is set up and maintained by ion fluxes and differential permeabilities. Black arrows indicate active transport and white arrows indicate passive flow. This is the state in a dehydrated mammal. In a hydrated mammal, the supply of antidiuretic hormone (ADH) from the pituitary declines and the distal convoluted tubule (DCT) and collecting duct (CD) become almost impermeable to water so that a copious dilute urine leaves the kidney.

that described for the chloride cells earlier (see section 5.3.2). The effect is to make the interstitial space hyperosmotic. This region also has very low P_w so water does not move rapidly by osmosis to reduce the hyperosmoticity.

4 The distal convoluted tubule (DCT) is rather complex; it is concerned with the detailed modification of ion balance in the urine, usually transporting K^+, H^+ and NH_3 into the lumen and resorbing more Na^+, Cl^- and HCO_3^-. The DCT is under precise endocrine control according to the animal's immediate osmotic and ionic needs; aldosterone from the adrenal glands increases sodium uptake here in particular. Water follows passively here, so again the fluid in the tubule is isosmotic with the surrounding cortex tissue.

5 The collecting duct (CD), into which many nephrons feed, has very low P_{Na} and P_{Cl} but a variable permeability to urea. Some NaCl is actively resorbed here, but by far the most important effect is that as the urine descends through the medullary interstitial space, water leaves the duct osmotically down the steep concentration gradient created through the thickness of the medulla. To allow this the CD normally has a high P_w (although this is under the control of antidiuretic hormone (ADH), discussed below).

This brief review of the sequence of events brings us back to the question of how the medullary concentration gradient was set up. The medulla is made up of all the loops of Henle (which may vary considerably in length) from the thousands or millions of individual nephrons, and their common collecting ducts. In effect it is an assemblage of countercurrent multipliers. Microsampling of both ascending loop of Henle (ALH) and DLH fluids, as well as the interstitial fluids, shows a standing gradient of concentration (Fig. 5.17), increasing from the edge of the cortex towards the inner medulla, where fluids at the hairpin end of the loops can reach 1000 mOsm in normal mammals and more than 3000 mOsm in some species, especially those from arid habitats (see below).

This high concentration in the deeper medulla results from

the permeability properties of the loop together with the ion transport in the ascending limb. Ion pumping from the upper ALH makes the interstitial fluids (ECF) more concentrated and thus also makes the fluid in the descending loop gradually more concentrated as water passes out of its relatively high P_w membranes to the more concentrated interstitial area. The counterflow of the two limbs of the loops ensures that the hairpin area becomes and remains the most concentrated zone; as the very concentrated fluid ascends the loop again it will supply the pumps in the upper ALH with yet more Na+ and Cl- to work with, so that they can pump more ions out, get more water withdrawn, and so on, so that the single effect of the ion pump is multiplied by the anatomy of the system. Fluid ascending the ALH thus becomes gradually less concentrated, and will pass on to the cortex and the DCT at around 300 mOsm again.

This high medullary osmotic concentration is significantly supplemented by urea leaking out of the lower CD cells as these ducts traverse the medulla; until then, urea has been retained within the urine by consistently low P_{urea} values, but the deeper areas of the CDs are at last more permeable to this solute and it follows its concentration gradient into the medullary interstitial spaces. This movement of urea is now known to be facilitated by specific protein uniporters in the CD membranes (see Chapter 4).

The high concentration resulting from the ionic concentration in the loops of Henle, and the urea concentration, inevitably has the desired effect of drawing water out of the urine in the CDs and into the medullary tissue, just before the completed urine is due to leave the kidney, yielding the hyperosmotic product characteristic of most mammals and birds. Furthermore, the extent of this water loss from the CDs can be simply controlled by regulating the P_w of the lower CD cell membranes, via the ADH levels in the blood. The increase in P_w seems to occur both by opening existing water channels and by incorporating subapical cytoplasmic vesicles that are particularly well endowed with water channels into the apical cell membrane; these channels have now been isolated and sequenced. ADH thus keeps the water channels numerous and open; when ADH is withdrawn water cannot leave the duct, and diuresis (copious dilute urine flow) results.

The water that is drawn into the kidney medulla from the CDs is prevented from destroying the crucial concentration gradients by a parallel system of medullary blood vessels (the vasa recta), which are freely permeable to water and solutes and are also arranged as countercurrents (see Fig. 5.16); thus the concentration of the blood follows that of the tissues of the medulla. However, the capillary walls do not let proteins pass, so that the blood (already relatively enriched in protein after passage through the glomerulus) has a colloid osmotic pressure which is sufficient to draw water into the bloodstream.

The net effect of all this structural complexity, ionic pumping and variable permeation is that each nephron produces a small volume of urine at a higher concentration than the body fluids, and often even more concentrated than sea water. At the same time the efferent renal blood has become more dilute than the afferent supply and is taking the recovered water away from the kidney. Once again, water has been removed from a biological fluid without there being any active handling of water itself.

At first sight there is an enormous interspecific diversity in the structure and function of mammalian kidneys, and it would appear that the presence and the length of the loops of Henle in a vertebrate kidney should be related to an ability to produce concentrated urine. The first assumption is broadly true: groups that do not produce a hyperosmotic product usually lack the loops and have much simpler nephron structures than the xeric terrestrial birds and mammals (Fig. 5.18). The second assumption (that length is related to concentration) is not strictly true. Small desert mammals do produce very hyperosmotic urine, approaching 6000 mOsm in some species, and even in excess of 9000 mOsm in the Australian hopping mouse *Notomys*, but they do not have unusually long loops of Henle within the kidney. Filtration pressure in the glomerulus of a mammalian kidney does not increase with increasing size, and total nephron length is also relatively constant (limited by this constant filtration driving force). However, the filtration demand does increase in direct proportion to metabolism, which scales positively with body size. Hence the *number* of nephrons must increase as body mass increases, which results in changes in the geometry of the kidney, increasing the amount of cortex (where most of the nephron is sited) at the expense of the relative thickness of the renal medulla. Urine-concentrating ability should be a function of relative medullary thickness because of the mechanisms already described, and Fig. 5.19 confirms this point; so it might follow that larger mammals would inevitably be incapable of producing urine as hypertonic as that of small mammals. Thus body size alone could have major implications for kidney capability, which may explain why the ubiquitous house mouse, *Mus musculus*, can produce as concentrated a urinary product as some of its desert relatives.

This is a somewhat over-simplified view, because it confuses relative size and absolute size; other things being equal, it should be the absolute length of a tubule loop that matters, not its relative length. *Notomys* has a loop length of only 5.2 mm, whereas a horse has loops seven times longer yet produces urine only up to 1900 mOsm. In fact it turns out that there is no real relationship between absolute loop length and urine concentration for mammals; we need to direct our analysis at a more cellular level to understand urine concentration. Small mammals have higher mass-specific metabolic rates (Chapter 6), and within their kidneys the tubule cells have more densely packed mitochondria, with denser internal cristae, and there are more elaborate lateral membrane foldings; all this indicates that small mammals should have the capacity for much more intensive active transport, on which the urine-concentrating system depends. Thus a mouse kidney could maintain a much stronger osmotic gradient along the loop of Henle than a horse kidney, achieving higher urine concentration with a smaller kidney.

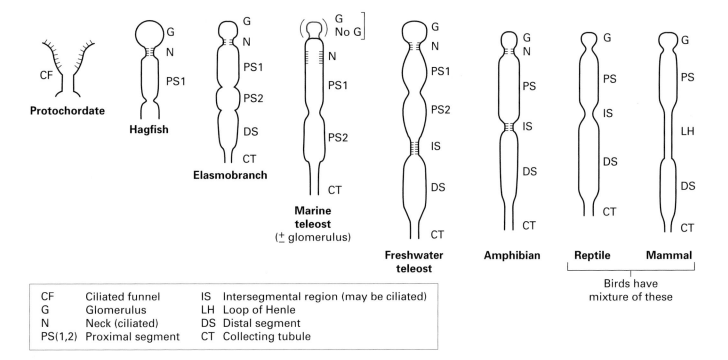

CF	Ciliated funnel	IS	Intersegmental region (may be ciliated)	
G	Glomerulus	LH	Loop of Henle	
N	Neck (ciliated)	DS	Distal segment	
PS(1,2)	Proximal segment	CT	Collecting tubule	

Fig. 5.18 A comparison of the structure of kidney tubules in different taxa of vertebrates. In general, the more 'difficult' the environment the more complex the tubule becomes.

Nevertheless, Fig. 5.19 does reveal that many small rodents living in arid (xeric) habitats do have relatively thicker medullas than other mammals, which correlates with their concentrated urine, and which could be viewed as a 'desert adaptation' super-imposed on a basic body-size-dependent pattern. Desert rodents tend to have almost exclusively long loops and very few short ones, with a high capacity for active transport per unit length, explaining the greater gradients and higher urine concentrations produced. Aquatic mammals have very few long loops, and temperate mammals may have only 25–50% of their loops dipping deep into the medulla. Thus the relations between medulla or loop dimensions, body size and habitat are not nearly as clear-cut as elementary texts would have us believe—this serves as an object lesson in letting enthusiasm for scaling-based explanations (Chapter 3) exceed their legitimate limits.

Rectal countercurrent concentration systems

One further example of concentration of excreted fluids by countercurrent occurs in the cryptonephridial system found in the rectum of certain xeric insects. The system is shown in diagrammatic form in Fig. 5.20. Essentially the Malpighian tubules, which in most insects lie freely in the haemolymph, are here folded back to lie against the rectal walls, with fluid flowing forward in them counter to the rearward fluid flow in the rectum. The whole array is enclosed in a very impermeable 'perinephric membrane'. At a few sites, this membrane is broken by specialized cells (leptophragmata) on which all the KCl transporting mechanism of normal Malpighian tubules is centred, secreting the primary urine constituents into the tubules. But the impermeable perinephric membrane prevents the usual entrained water flow from the haemolymph, and KCl accumulates at high concentrations. Thus it is able to withdraw water osmotically from the rectal contents instead, even when these get quite concentrated, and in effect a standing gradient is set up along the complex such that water is withdrawn along a small osmotic gradient at all points in the rectum; even close to the anus, the highly concentrated fluid in the gut is still losing water to the even more concentrated fluids in the perinephric space and tubule, and ultimately to the blood. More details of this system in particular insects are given in Chapter 13.

5.4.4 *Final urine composition*

After all the processes described above, urine is passed from the body, often via a storage bladder in amphibious and terrestrial animals, partly because for many of these animals intermittent urination takes place only in prescribed places (middens) or as a territory marking system. Occasionally the bladder is used as a site of final modification; for example, acidification in reptile bladder by the unusual V-ATPase H^+ pump (see Chapter 4) is regulated in part by high CO_2 levels, which cause vesicles bearing the ATPase to move to the apical membrane of the bladder epithelium.

$OC_{max} = 837 + 2106RMA$

- ● Rodents
- ○ Bats
- + Other mammals

Cortex
Medulla

Fig. 5.19 The relationship between relative medullary area in the mammalian kidney (taken at the midline in sagittal section) and the maximal urine concentration that can be produced. In general, proportionally larger medullas produce a more concentrated product; note that the relatively largest medullas and very concentrated urines mostly occur in rodents, which are of small body size and common in arid habitats. Insets show diagrammatic cross-sections for three kidneys (not to scale). (From Withers 1992.)

Table 5.6 gives examples of the range of (and/or maximum) urine concentrations achieved in different taxa living in different habitats, including the net urine/plasma (U/P) ratio to clarify the concentration relative to the blood. The best examples of the insect osmoregulatory system are capable of producing urine at around 7000–9000 mOsm; and in some cases where the rectum

is used for vapour uptake it is at least notionally resorbing at around 30 000 mOsm concentrations.

5.4.5 *Nitrogenous excretion in 'excretory' systems*

Comparative biology of nitrogenous excretory products

There is a relatively small range of nitrogen-containing excretory products in animals, with three principal molecules dominating the field. These are **ammonia**, **urea** and **uric acid**. Their chemical formulae and properties, together with those of a few much less common waste products, are shown in Table 5.7.

When proteins and their constituent amino acids are metabolized the amino group that is removed forms ammonia. Ammonia is the ideal waste product for most aquatic animals, with a high nitrogen content making it least wasteful of useful organic carbon; its high toxicity does not matter in aquatic habitats as there is plenty of water available to dissolve it and disperse it away from the body. Often the osmoregulatory organ is not involved at all, with ammonia diffusing away from any permeable surface (very often this will mainly be the gills). Animals that use ammonia as their main excretory product are termed **ammonotelic**. They include most aquatic invertebrates, including the aquatic larval phases of many insects, plus the cyclostome and teleost fish. In many of these animals the ammonia combines with H^+ to form NH_4^+, and this is used as an exchange for Na^+ pumped in at the gills, reducing the work required against an electrical gradient. In many fish up to 90% of the excreted ammonia leaves via the gills rather than the kidneys. Aquatic representatives of other normally terrestrial groups may also excrete a high proportion of ammonia, for example most aquatic insect larvae, the tadpole larvae of frogs and toads, and the crocodiles. Amphibians that remain aquatic throughout their life, such as *Xenopus* and the axolotl, are persistently ammonotelic. Most of the terrestrial woodlice also excrete ammonia, releasing urine from the maxillary glands (where urea is stored as glutamate, which is deaminated just before release

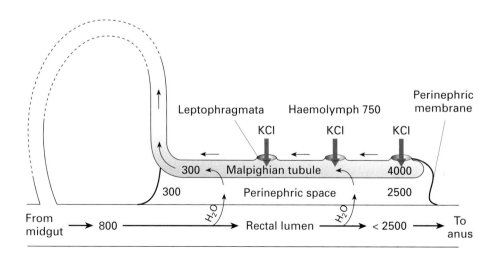

Fig. 5.20 The essential principles of concentration by countercurrent in the cryptonephridial system of certain insects.

Table 5.6 Urine concentrations and urine/plasma (U/P) ratios in animals from different taxa and habitats. For land animals the values are generally maximal rates from dehydrated animals.

Taxon	Habitat	Urine concentration (mOsm)	U/P ratio
Crustaceans			
Lobster	Marine	1050	1.0
Crab	Marine	1020	1.02
Crab	Freshwater	550	1.04
Crab	Littoral	1200	1.17
Molluscs			
Cuttlefish	Marine	1100	1.02
Elasmobranchs			
Dogfish	Marine	800	0.8
Teleosts			
Goby	Marine	406	0.9
Flounder	Marine	375	0.96
Reptiles			
Crocodile	Amphibious	267	0.8
Tuatara	Mesic	270	1.0
Iguana	Mesic/xeric	362	1.0
Desert tortoise	Xeric	337	1.0
Birds			
Pigeon	Mesic	655	1.7
Pelican	Marine/mesic	700	2.0
Ostrich	Xeric	900	2.7
Savanna sparrow	Xeric	2020	5.8
Small mammals			
Rat	Mesic	2900	9
Domestic cat	Mesic	3100	10
Vampire bat	Mesic/xeric	4650	14
Kangaroo rat	Xeric	5500	16
Large mammals			
Beaver	Amphibious	520	1.7
Man	Mesic	1400	4–5
Porpoise	Marine	1800	5
Eland	Xeric	1880	6

to form NH_3) and spreading it over the flattened appendages (pleopods) on the ventral surface of their body, allowing gaseous ammonia to diffuse away. In some species the watery urine is then resorbed in the rectum. Just occasionally, truly terrestrial animals still produce gaseous ammonia: guano bats eliminate a moderate percentage of their excess nitrogen as NH_3 through the lungs, and some hummingbirds can switch to ammonotely in cool conditions when dilute nectar is freely available to give a plentiful water supply.

Urea is something of a compromise product, synthesized from the initial ammonia in combination with carbon dioxide via the urea/ornithine cycle (Fig. 5.21). It has a moderate nitrogen content and intermediate solubility and toxicity, so it can be dissolved at fairly high concentrations and thus requires little water for its excretion. It is commonly used by larger land vertebrates (mammals, amphibians and some reptiles), and by cartilaginous fish, and it occurs sporadically elsewhere in the animal kingdom, for example in intertidal fish like blennies, in some earthworms and land snails, in freshwater turtles, and in some animals from stagnant freshwater habitats. All these animals are described as **ureotelic**. Some of the ureotelic vertebrates use the synthesis of urea as a means of generating protons that contribute to pH homeostasis, and in these cases (notably the ruminant mammals) up to 50% of the liver ATP function may be devoted to urea production.

Uric acid is a purine, synthesized from various amino acids and from the general carbon pool of a cell. It is extremely insoluble and can be produced as an almost solid paste without toxic effects on adjacent living tissues. It is therefore ideal for land animals that are smaller and often particularly water-stressed, including the insects, the terrestrial snails, the birds and most smaller reptiles, and a few arid frogs; these are described as **uricotelic**. Where the excretory product is nominally uric acid, an extra bonus is achieved in that all urate salts are also very insoluble, so that cations can be added to the precipitated paste of excreta and the salts are lost with almost no associated water loss. Some insects exploit this by producing NH_4^+-urate, thus unloading ammonia without any necessary water loss. Terrestrial vertebrates can also lose less water by excreting Na^+ and K^+ as urates from the kidney than they would by using their salt glands. This is true for reptiles; indeed, some xeric reptile species that excrete urates have dispensed with salt glands altogether. However, in birds the usual excretory product is a colloidal gel

Table 5.7 Nitrogenous excretory products and their characteristics.

Excretory product	Formula	M_r	Solubility (mM)	Diffusion coefficient	H/N ratio	C/N ratio	Toxicity
Ammonia	NH_3	17	52.4	High	3	0	High
Urea	$CO(NH_2)_2$	60	39.8	Moderate	2	0.5	Moderate
Allantoin	$C_4H_6O_3N_4$	158	0.016	Low	1.5	1.0	Low
Guanine	$C_4H_5ON_5$	151	0.0013	Low	1.25	0.8	Low
Uric acid	$C_5H_4O_3N_4$	168	0.0015	Low	1.0	1.25	Low

Fig. 5.21 The urea cycle, or ornithine cycle, whereby urea is synthesized from ammonia and carbon dioxide by condensation with ornithine, which is then re-formed by a cyclical process.

$$2NH_4^+ + CO_2 + 4ATP \longrightarrow CO(NH_2)_2 + 2H^+ + H_2O + 4ADP + 4P_i$$

with a rather high water content, in which uric acid represents little saving over the use of urea. Urate is useful for birds at the embryonic stage, however: it is an ideal excretory product for any animals with relatively yolky and impermeable eggs, where a solid paste of excreta can be left behind within the eggshell so that it is not an encumbrance to the hatchling.

The distinction between these three excretory modes is very far from absolute, and many anomalies occur. It is evident from the distribution of these products that the use of particular excretory products is largely related to habitat (water availability) and not to phylogeny. For example, different species of frog can exhibit any of the three modes, with the xeric genera *Chiromantis* and *Phyllomedusa* (both almost independent of water, with special reproductive tricks to avoid the need to return to ponds) producing substantial amounts of uric acid. The same versatility is found in turtles, with the more terrestrial species producing at least 50% uric acid, and with switching possible even within one individual according to its state of hydration. However, even the correlation between habitat and excretory product cannot be taken too far: uric acid turns up in some entirely marine species, for example some prosobranch snails, and also in some freshwater members of the same taxon, whilst some terrestrial snails can excrete gaseous ammonia through their mantle epithelium.

A few other nitrogenous residues occur less commonly in animals' excreta. Some are breakdown products of nucleic acids (the purine and pyrimidine components), rather than of protein. Creatinine is one example, occurring in small quantities in vertebrate urine. Allantoin and allantoic acid also result from purine degradation, and occur in many molluscs and insects, as well as in most mammals except primates. In freshwater insects these products may become the major excretory output, having a greater solubility than uric acid. Another important excretory product is guanine, accounting for up to 90% of the nitrogen excreted in spiders, and occurring in crystalline form in the scales and swim-bladders of some fish, where it reduces the surface permeability. Terrestrial snails also excrete some guanine. Finally, some animals produce small amounts of trimethylamine as a breakdown product of choline metabolism, and trimethylamine oxide (TMAO) is then retained in the blood of some elasmobranch fish (sharks and rays) as an important osmolyte (see Chapter 9).

Excretion of these products

In general, the main nitrogenous excretory product of an animal gets into its urine passively, either in the ultrafiltrate or during bulk flow of water in a secretory urine production system. This is normally true whether that product is ammonia, urea or uric acid (though urates are at least partly actively transported in insect systems). Final concentration of the waste product in the urine is largely determined by the permeability properties of different parts of the tubule system.

However, there are several ways of increasing or of regulating the flow of nitrogenous wastes. One of the most important involves urea; in some animals that can survive prolonged dehydration when urine flow almost ceases, some urea may still be added into the osmoregulatory/excretory tubules by secretion. This is known to occur in some frogs. Shark kidney tubules can also handle urea actively, but in this case it is active resorption that occurs, to keep the high urea blood level used as an osmotic balancing strategy in these fish.

In some snails there is a zone of specialized secretory cells at the distal end of the kidney, and here active uric acid secretion can occur during periods of dehydration. Other water-stressed terrestrial animals, particularly some insects with relatively short adult lives, give up excreting the uric acid at all and resort to depositing it in crystalline form in various parts of their body ('storage excretion'), sometimes using it as a cuticular pigment.

In some animals, tubular filtration and secretion systems for ridding the body of nitrogenous waste are supplemented by tubular synthesis. Some excreted compounds are actually made within the epithelium of the kidney, often by deamination processes resulting in the formation of ammonia (NH_3), which then diffuses into the urine and combines with H^+ to form NH_4^+ ions ready for excretion; this is an important process in both mammals and insects, where it is particularly noticeable in the hindgut of carrion-eating larvae such as the maggots of the flesh-fly *Sarcophaga*.

Sequestration of these products

Some animals, particularly arthropods, are able to cope with varying states of desiccation by regulating the patterns of sequestration of their excretory products. The cockroach *Periplaneta* responds to dehydrating environments by sequestering sodium ions together with uric acid as sodium urate crystals in the fat body, with chloride apparently stored in the hindgut; thus no water need be wasted in excreting these surplus and accumulating osmolytes. Urate crystals are accumulated by a wide range of insects; this strategy is particularly common in juvenile stages, as the excess urate can then be 'dumped' with the old cuticle at the next moult. Some animals also accumulate and store their own urine for prolonged periods; pulmonate snails can do so in the mantle cavity, and some anuran amphibians use the bladder for similar purposes. Such stores may then become sources of water during subsequent drought, particularly in desert species.

5.5 Water regulation via the gut

In those aquatic animals where drinking large volumes of the medium is a major part of their osmoregulatory strategy, the gut clearly becomes a front-line osmoregulatory organ. The same is true, though less immediately obvious, for all non-carnivorous animals; for whereas 'meat' is roughly the same in composition as the tissues of the animal eating it, all other food sources are dissimilar from animal tissue. Since the food intake is therefore not in osmotic and ionic equilibrium with the body tissues, the gut lining may be used as a barrier against excessive ionic stress. Guts may therefore play a crucial role in osmotic and ionic regulation.

In the simplest invertebrate animals, such as sea anemones, hydras and flatworms, the primitive gut (enteron) is the only fluid-filled cavity within the body and inevitably has a vital role in regulation of salt and water balance, especially for those species that live in estuarine or freshwater habitats. The gut is a blind-ending sac enclosed by the body wall, with a single opening; thus there is little opportunity for compartmentalization or specialization. The primary function of digestion is inevitably compounded with that of ion and water regulation, and compromises have to be reached in a multifunctional epithelium.

In all other animal groups a separate anus is added to the system so that the gut becomes tubular, and this opens up new possibilities for modifying the contents differentially in successive areas. It also allows for altering conditions within the gut by adding secretions, so that enzyme functioning can be tuned to local conditions; digestion can occur at a specific pH in certain zones, and ion regulation or water resorption can be localized at a different pH in other areas. In the more complex invertebrates, a variety of caeca and 'digestive glands' are frequently added as side-branches of the gut, which becomes an integrated part of the iono/osmoregulatory system of the whole animal. This is seen most clearly in insect herbivores and in terrestrial herbivorous vertebrates, where the gut becomes extraordinarily complex (see Chapter 13). In all these animals with through-guts, a series of functions are performed from anterior to posterior, by specialized areas that show a degree of similarity throughout the animal kingdom.

5.5.1 Water balance in various gut regions

Salivary glands

In most animals, salivary products serve to lubricate the food as it passes through the mouth and oesophagus, and often begin the process of food breakdown with one or more enzymes. Thus the products of the salivary glands represent no net loss to the animal as they can be recycled through succeeding parts of the gut. However, in some cases the saliva also has an excretory and osmoregulatory role, particularly for animals with a blood-feeding habit. Ixodid ticks, for example, eliminate most of their excretory water, Na^+ and Cl^- by injecting it into the host as they feed; sodium from the blood meal is rapidly taken up from the midgut to the haemolymph, and then pumped into the salivary gland lumen (Fig. 5.22). In other arthropods, water stored in the salivary glands may be used as an emergency resource when an

(a)

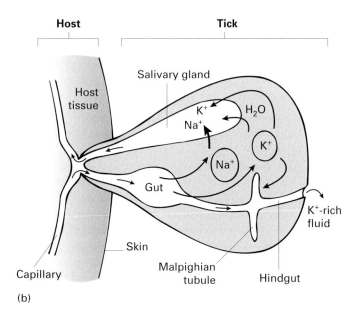

(b)

Fig. 5.22 (a) Structure and (b) function of coxal salivary glands in a tick; about 96% of sodium and 74% of water is excreted by this route during feeding. (a, From Little 1983, courtesy of Cambridge University Press; b, from Kaufmann & Phillips 1973, courtesy of Company of Biologists Ltd.)

individual is water stressed; termites and some fly larvae appear to be able to mobilize water from this source when needed.

Midgut and gut caeca

Midgut epithelia (equivalent in vertebrates to the stomach and a large part of the intestine) are the most important sites of digestive enzyme secretion, and of food absorption, in nearly all animals. But they are also important centres for ion and water absorption, and ion regulation, in many taxa. In most

animals at least 75% of ingested water, and more than 95% of ingested monovalent cations, are taken into the body here.

Complex ion cycling occurs in the midgut, especially in the pterygote insects (Fig. 5.23), with cycles of solute absorption being coupled by symporters to ion movements, and the ions being returned to the system at a different point in the cycle, often in a blind-ending gut caecum. In fact the midgut serves as the main excretory/osmoregulatory organ in many apterygotes; these primitive insects lack Malpighian tubules and use the gut in concert with labial glands opening near the mouthparts and acting as 'kidneys'.

Systems to bypass the midgut absorptive epithelium are often found in animals that are at risk of getting too much water from their diet. This is commonly the case in plant sap-feeders taking in xylem fluids. Many xylem-feeding insects and mites therefore have gut bypass arrangements (Fig. 5.24), where a 'filter chamber' brings about close contact (in countercurrent) between the anterior midgut, Malpighian tubules and anterior hindgut. Much of the water taken into the gut therefore bypasses the main midgut epithelium and passes directly from anterior midgut to rectum, leaving a more concentrated fluid in the rest of the midgut from which digestion and absorption can occur.

Rectum

In most animals, and especially in terrestrial species, the hindgut or rectum becomes a major site of water balance regulation, serving as a site for water resorption from the faeces (and, in many arthropods, from the excretory fluid that has been added to the gut as well). Solute-linked water resorption, described above, is a primary function here in all terrestrial animals. In birds and reptiles, part of the hindgut (the 'integrative segment') is a major resorptive area, so that the resultant precipitation of uric acid further reduces water loss.

In insects even more efficient solute-linked water resorption is possible in the rectum, by mechanisms discussed above. This may be facilitated by the presence of a cuticular lining in the insect hindgut, protecting the rectal cells from the high concentrations of larger faecal toxins that will result from such an extreme degree of concentration as water is removed. In some cases the rectum can even be used for absorbing water vapour from air. However, many terrestrial arthropods have a very different strategy and use their hindguts as a water store for emergency use; in isopods, semi-terrestrial crabs, millipedes and many juvenile and adult insects the gut contents are usually rather dilute, and become significantly more concentrated during times of water stress as the stores are used up.

Integration of gut and excretory functions

The most xeric animals, especially the insects and arachnids from desert habitats, have achieved their success at least partly by integrating most of their salt and water regulation together, with the excretory tubules disgorging their watery product

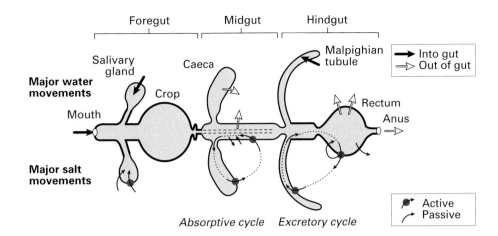

Fig. 5.23 A summary of water (above) and solute (below) movements in the insect gut.

into the gut and letting the hindgut epithelium do most of the resorption. Some relatively xeric earthworms (Australian megascolecids) have independently arrived at the same solution. Most birds and reptiles use their cloacal region (the site where reproductive, rectal and urinary outputs come together) rather similarly to make further changes to the excreta, being able to reflux urine through the cloaca and back into the integrative segment of the hindgut for further modification. In the avian cloaca, ions present as urate salts may be released by the action

of bacteria, with Na^+ and NH_4^+ resorbed into the tissues in exchange for K^+ ions. Remember also the woodlouse trick of allowing urine to flow over the ventral pleopods and then resorbing the fluid via the rectum.

5.6 Regulation of respiratory water exchanges

Respiratory surfaces always have gases diffusing across them, and water therefore inevitably also moves across these surfaces

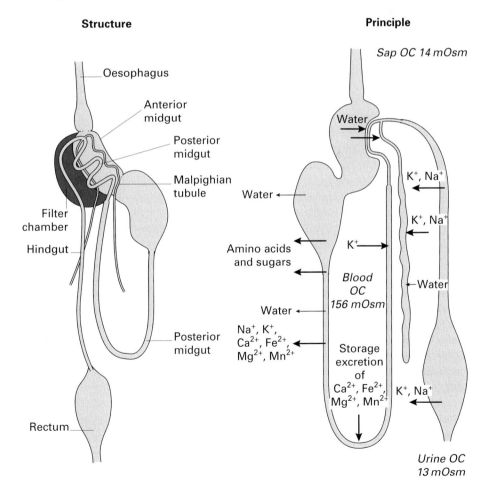

Fig. 5.24 The structure of a 'gut bypass' system in a sap-feeding insect, whereby much of the fluid taken in is filtered directly to the hindgut, bypassing the midgut resorptive areas. OC, osmotic concentration. (Adapted from Cheung & Marshall 1973.)

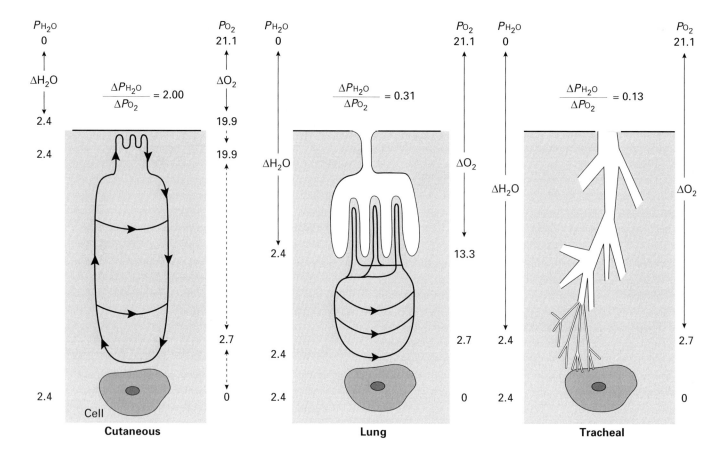

Fig. 5.25 Relationship between diffusive oxygen uptake and water loss across respiratory systems. (Adapted from Withers 1992.)

in any situation where there is a water gradient—i.e. in *all* animals except the isosmotic marine invertebrates. Even in the most xeric animals the respiratory surfaces are therefore always moist; this has sometimes been described as a necessity for gas exchange but actually represents an increased barrier to rapid gas exchange. In non-isosmotic aquatic animals, the solution has been to make the best of the situation and use the permeable gill surfaces as the main site of ion and water regulation, taking up or secreting salts there. The general evolutionary trend in non-aquatic animals has been to replace external and exposed respiratory structures with internalized ones where water balance can be better controlled. In terrestrial animals from very moist habitats (e.g. earthworms and many amphibians) cutaneous respiration is sometimes an adequate solution. But in more xeric forms either gills are concealed within enclosed chambers or they are replaced with either tracheal systems or lungs, entirely within the body and with very restricted openings. Details of these respiratory systems are covered in Chapter 7; here we are concerned only with their role in water loss.

Oxygen uptake, carbon dioxide excretion and evaporative water loss are all essentially diffusive processes. The ratio of water loss to oxygen uptake for any exposed respiratory surface

operating in air can therefore be relatively easily calculated from the ratios of the diffusion coefficients for the two compounds (which works out at about 1.3), multiplied by the ratios of the gradients for the two compounds. Figure 5.25 shows the situation for a cutaneous surface, a lung and a tracheal tube. At a simple cutaneous surface exposed to air the value obtained would be about 2.1 mg H_2O lost per ml O_2 taken up. Any terrestrial system of respiration that can improve on this ratio will be selected for. Invaginated systems achieve this by greatly reducing the partial pressure of oxygen (P_{O_2}) at the final air–fluid interface, so making the gaseous O_2 gradient steeper and increasing the difference between the water gradient (ΔP_{H_2O}) and the oxygen gradient (ΔP_{O_2}).

Lungs and tracheae should therefore both represent a major advantage in terms of water conservation (Fig. 5.25b,c). Most of the simple lungs found in invertebrates achieve some P_{O_2} reduction and thus some water saving, but in vertebrate alveolar lungs this effect is particularly striking, as the P_{O_2} at the internal alveolar surfaces is reduced to around half of its atmospheric level, and the ratio of water loss to oxygen uptake is thus considerably improved compared to cutaneous breathing, to about 0.2–0.3 mg H_2O lost per ml O_2 taken up (a 90% saving). The essential difference between lungs and tracheae is that in the latter the air is delivered directly to the tissues in the gaseous phase, and therefore the P_{O_2} deep within the tracheoles can fall very low, especially when the spiracles are closed, giving a further improvement in the relative water and oxygen gradients (Fig. 5.25c). Theoretically this would allow a water loss as low as 0.1 mg per ml O_2 taken up.

Inevitably, though, water loss from lungs and tracheae is quite high, because large surface areas and slow flow internally are always required and equilibration of air and water occurs at the deep exchange surfaces. Thus moist air, almost invariably saturated, is breathed out. More xeric land animals therefore operate further controls over the respiratory openings, and may add condensers and countercurrent exchange systems to the respiratory flow, to regain as much of the water as possible from the expired air.

The main water-related adaptations of terrestrial vertebrate respiratory systems are clearly concerned with recovering some of the expired water, and the obvious way to do this is to cool the exhalant air. Any cooling of the expired air relative to its temperature at the respiratory surface of the lung will cause it to give up moisture as it leaves the body, since saturated cold air holds less water vapour than saturated warm air. Nasal countercurrents to achieve this can be found in almost all birds and mammals (the rare exceptions are animals with unusually modified 'large' noses such as elephants and pelicans). At their best, where the nasal passage is elongated and the internal anatomy highly specialized, the nasal **respiratory turbinals** of vertebrate endotherms can substantially reduce respiratory water loss. The principle and the results are shown in Fig. 5.26. Incoming air is warmed in the nose before it reaches the alveolar surfaces, and this draws some heat out of the nose, leaving its surfaces cooler. Expired air is subsequently cooled by heat exchange with the cooled nasal mucosa, to an extent depending upon the anatomy of the system, so that the final expired air temperature can be considerably lower than the animal's core body temperature. Hence water condenses in the nose, which is a site of countercurrent recovery both for water and for heat. We will therefore meet this system again in considering thermal biology in Chapter 8, and specific cases of it in operation in terrestrial animals in Chapters 13 and 14.

In tracheal systems air is supplied in the vapour phase via finely divided tubules, almost to the point of uptake in the tissues. However, the final blind endings of the tracheoles are fluid-filled, and at this point gases go into solution and the final uptake of oxygen is largely in the fluid phase.

But there are some more sophisticated features that can be added to improve control and limit water loss from tracheal systems. These include:

1 Directional control of ventilation. Considerable heat is generated in a flying insect, principally in the thorax as this is the site of the main locomotory muscles, so the air there is warmed in passing over thoracic tracheolar surfaces. As it exits rearwards the air will then encounter cooler tissues, so that some water from the tracheal system should be recondensed in the abdomen. This is analogous to the nasal cooling system discussed above for vertebrates, though the evidence for it is less secure.

2 Control of spiracular opening, in response to levels of respiratory gas as a means of regulating for respiratory need, or in response to osmotic concentration or ambient relative humidity, both of which impact more directly on water balance. Spiracular control can be taken to extremes by making breathing discontinuous, with very limited periods of spiracle gaping and thus significant water savings, particularly in resting stages such as pupae.

Examples of these mechanisms are dealt with in detail in Chapter 13.

5.7 Water loss in reproductive systems

Only in terrestrial animals is water loss associated with reproduction a severe problem, and there is a standard solution in all taxa that are 'properly' terrestrial and breed away from water. They all adopt effectively internal fertilization, either using spermatophores or direct internal insemination. In the first case, relatively impermeable spermatophores containing a little fluid and a package of sperm are left by males in some suitable spot and are then picked up by females; or the spermatophore package may be inserted into the female (often into her reproductive system, but sometimes at any point on her body wall), where it breaks down and releases the sperm. In most terrestrial animals the second system is used; there is a switch to full copulation and direct insemination, with a small volume of isosmotic fluid containing the sperm released directly inside the female's vagina.

Eggs and larvae of course have their own water balance problems. In aquatic animals these are essentially an exaggerated version of the adult's problems, compounded by the much larger surface area to volume ratio of the small juvenile stages. Hence the osmoregulatory organs commonly make up an even larger percentage of the body weight in larvae. Many littoral and freshwater animals shorten or entirely suppress the larval stage, and use direct development from well-endowed yolky eggs instead, which reduces these osmotic problems. So too do most of the terrestrial animals, though only the placental mammals and a few isolated examples from other groups have switched to full viviparity, where the offspring's osmotic problems are entirely dealt with by the maternal body. In terrestrial animals that retain egg-laying habits, use of microclimates chosen by the mother in finding an oviposition site very often ameliorates the problems, the extreme case being an endoparasitic habit where the host provides a stable environment. Eggs may be laid in carefully constructed nests in species that use maternal after-care, or in very humid sites on (or inside) a food source, where water and temperature balance are relatively easy to achieve for the egg itself and then for the hatchling.

Nevertheless the egg stage has to undergo a prehatching period of respiration through the surface coating, without losing too much water. This problem is particularly acute for terrestrial insects, reptiles and birds. Insect eggs, like the adults, contain waxy layers lying beneath the chorion of the egg 'shell', and

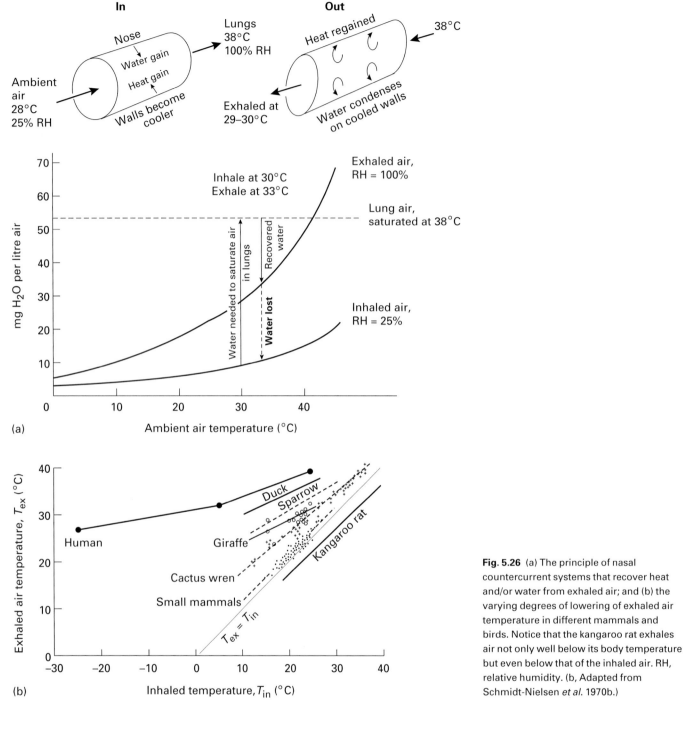

Fig. 5.26 (a) The principle of nasal countercurrent systems that recover heat and/or water from exhaled air; and (b) the varying degrees of lowering of exhaled air temperature in different mammals and birds. Notice that the kangaroo rat exhales air not only well below its body temperature but even below that of the inhaled air. RH, relative humidity. (b, Adapted from Schmidt-Nielsen *et al.* 1970b.)

show temperature transitions in EWL rate similar to those of adult insects. The chorion itself is a complex coating (Fig. 5.27a) with an 'aeropyle' layer of air-filled channels so that air can diffuse into the egg. In some species there is hydrophobic material above the chorion to prevent these air spaces getting flooded, and in others there may be hygroscopic material in parts of the chorion forming a 'hydropyle' layer that draws water into the egg.

The amniotic egg in reptiles and birds has a fairly standard structure but its surface may have very different properties. Some reptiles (lizards and snakes) have leathery eggs only partially impregnated with $CaCO_3$, and these swell and shrink quite substantially, normally increasing in volume as they develop because they are laid in damp soils where their hydration is assured. Hard-shelled eggs occur in crocodiles and most turtles, and these are unable to take up water and change volume. Some

Eggshell of a fly

(a)

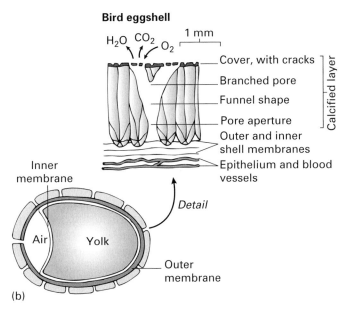

(b)

Fig. 5.27 Water balance in eggs of (a) insects and (b) birds, both having shells with a relatively impermeable inner layer (chorion) to control water loss and above this a surface 'plastron' layer of highly reticulate air spaces (aeropyle) to allow oxygen uptake. The smaller size of insect eggs necessitates a less permeable surface and a more elaborate aeropyle. Many birds' eggs have an internal air store within the egg. (a, Reprinted from Hinton, E.E., *Biology of Insect Eggs*, copyright 1989, with permission from Elsevier Science.)

of the hard-shelled reptilian eggs share with birds' eggs the property of only allowing the passage of water vapour, being nearly impermeable to liquid water. These shells always have pores for respiratory exchange (Fig. 5.27b), whose size and properties are related to the crystalline structure of the mineral shell material and are characteristic for particular species. But in all cases, water is inevitably lost; this can only be controlled after laying by control of the humidity and temperature of the nest site, which is achieved by parental nesting in birds but usually only by initial site selection in most reptiles. Even in a carefully constructed and tended hen's nest, it can be calculated that an egg incubated for a standard 21 days takes up 5 litres of oxygen, gives off 4 litres of carbon dioxide, and loses 10 litres of water vapour (though this is equivalent to just a few grams of water).

5.8 Water gain

5.8.1 Metabolic water

Respiration of foodstuffs, whether carbohydrate, lipid or protein, ultimately yields carbon dioxide and water, so that metabolic water is invariably a component of water gain in animals. Metabolic water yields for particular substrates are shown in Table 5.8, indicating a maximum water output for lipids per gram of substrate, but of glucose per unit of energy yield (because fats require much more oxygen uptake for their complete oxidation). However, for many animals even in relatively dry terrestrial habitats metabolic water is only a small component of their overall water balance compared with inputs from food and drink; and while metabolic water production may be highest in absolute terms in summer when activity is greater, its percentage contribution to overall water inputs may be greater when an animal is in a resting state in winter.

For small xeric animals, though, metabolic water can be the principal contributor to achieving net water balance. Clothes moths (*Tineola*) kept at very low humidities may use as much as 70% of their food intake primarily for water production. Desert tenebrionid beetles in the dry desert winter may get a large input from metabolic water, whereas in summer they are more active and seek out any alternative external water sources.

Can metabolic water be regulated? In animals with simple diffusive respiration, such as many non-flying insects, extra food does seem to be eaten at low humidities, with a rise in metabolic rate, to give extra water; this occurs in flour-moth larvae (*Ephestia*), for example, and can best be interpreted as a means

Table 5.8 Metabolic water yields during the oxidation of different food types.

Substrate	Water formed (g water per g food)	Metabolic energy yield (kJ per g food)	Water/energy (g water per kJ)
Starch	0.56	17.6	0.032
Lipid	1.07	39.3	0.027
Protein (with urea excreted)	0.39	17.0	0.022
Protein (with uric acid excreted)	0.50	18.4	0.027

Table 5.9 Water contents of various types of food.

Food type	Water content (% by weight)
Leaves	75–90
Xylem sap	>95
Fruits	60–90
Most seeds	5–10
Rice grains	12
Nectar	30–90
Milk	87
Meat	70–90
Whole animals	60–80

of improving the animal's water balance. But in animals with ventilated respiration this would not normally work as the extra respiration needed to metabolize the extra food leads to a net water loss; hence of course the old idea that camels could metabolize their fat store to gain water is a fallacy, as more water would be lost (in extra respiration to provide the necessary oxygen) than could be gained by the metabolic processes. However, recent work shows that in rapidly flying insects water is indeed lost more slowly than it is generated; some large bees produce excess water in flight and have to void it.

5.8.2 *Food and drink*

Drinking and eating ('preformed water' sources) are far more commonly the main component of water gain and the main avenue for regulating the input side of water balance. The water content of food (Table 5.9) is high for carnivores and insectivores, but extremely variable for plant-feeders, ranging from dry seeds to exceptionally watery sap (many xylem tubes contain fluid that is 99.8–99.9% water), which may put the consumer into water overload and necessitate gut bypass mechanisms (section 5.5.1). Blood-sucking animals and dilute nectar feeders may have a limited version of the same problem, and some bees will regurgitate nectar to evaporate off excess water from their tongues.

The nature of the food also affects the nature of the animal's salt problems. Eating animal flesh, or blood, naturally yields principally Na, K and Cl in roughly the correct proportions for the consuming animal. But eating plants may give a high-K diet with inadequate sodium (probably because plants have lower ECF volumes), and eating certain plant fluids (nectar, and many xylem and phloem saps) may give almost no salt input.

Many animals therefore have sophisticated systems for regulation of food choice, mediated by chemoreceptors (especially for salts and sugars) on food-handling appendages or in the mouth region. Direct stimulation from such sites is usually integrated with inhibition from gut stretch receptors, to regulate the occurrence of feeding. Control from higher centres (hypothalamic nuclei in vertebrates) has a separate role in determining food preferences. Here nitrogen contents, calorific values or water contents may be key factors in many species, but specific attractants may also operate. Modified amino acids are often the key triggers; for example, glutathione is a potent feeding attractant for many cnidarians (hydras and sea anemones), and taurine has a similar effect in shrimps and some other marine filter-feeders. Many terrestrial insects prefer to feed on plants with specific secondary compounds, often using these same compounds as part of their defensive repertoire (see Chapter 13). Herbivorous mammals are generally less fussy and tend to maximize nutrient intake, though many will select legumes (with higher nitrogen, higher sodium and phosphorus, and lower fibre) in preference to most other plant families. In more difficult environments, however, food choices may be more refined. Many nectar-feeders will only take certain nectar concentrations (rather dilute in desert bees, compared with a fairly broad concentration range preferred in temperate species); and primates, such as baboons, from arid areas become much more selective for plants of high moisture content.

In general, the limited volume of the gut means that eventually feeding may cease even in the presence of preferred foods, and before the needs of water balance are fully met. This is usually brought about by the inhibitory effect of stretch receptors in the gut wall.

5.9 The costs and energetics of regulating water and ion balance

The energy required for ionic and osmotic regulation cannot be assessed accurately just by looking at the oxygen consumption or metabolic rate of an animal when exposed to different media. The analysis of metabolic rates is fraught with problems anyway (see Chapter 6); and in this situation too many factors are changing, and the varying costs of locomotion or buoyancy, or of respiration itself, will be compounded with those of actual osmoregulation. In practice, many euryhaline animals, such as the mitten crab *Eriocheir*, show little or no difference in metabolic rate in sea water or in very dilute brackish waters, and results from fish are confusingly variable. The cost of osmoregulating must therefore be calculated from simple thermodynamic considerations, though this can only give the minimum costs involved. Comparisons for different kinds of animal can then show the relative costs of different osmotic strategies such as high impermeability (hence low urine volume), or of localized ion uptake areas, or of producing isosmotic as against hyposmotic urine. Some examples for invertebrates, including both hyporegulators and weak and strong hyperosmotic regulators, are shown in Table 5.10, and several points are apparent:

1 It may be quite surprisingly 'cheap' to be an osmoregulator. The minimum cost in fresh water or brackish water of hyperosmotic regulation is rarely more than 10% of the total metabolic output for an aquatic animal, and may be estimated as less than 1%. (However, real costs are probably markedly higher than these estimates, given the density of mitochondria found in all osmoregulatory tissues.)

Table 5.10 Examples of the costs of osmoregulating.

| Species | Concentration of medium (mOsm) | Blood concentration (mOsm) | Urine concentration (mOsm) | Osmotic work (J g^{-1} h^{-1}) | | | % of metabolic rate |
				Ion uptake	Urine	Total	
Potamobius (crayfish)	12	840	240			0.002	0.3
Eriocheir (mitten crab)	12	640	640			0.005	0.5
Anodonta (freshwater clam)	12	85	50			0.001	1.3
Nereis (ragworm)	3	200	200	0.110	0.003	0.113	2.7
Carcinus (shore crab)	370	600	600	0.301	0.008	0.309	11
Aedes (mosquito larva)	1000	330	High	0.29	1.07	1.36	22
Artemia (brine shrimp)	6000	350	High	1.92	?	(1.92)	>33

2 The cost of osmoregulation is lower for the freshwater and estuarine crustaceans shown (values for the shore crab *Carcinus* are affected by a low measured metabolic rate (MR)) than for the soft-bodied animals such as the freshwater clam *Anodonta*. The latter is only a very weak regulator, maintaining a very low blood concentration (one-tenth that of the crayfish *Potamobius*), but because it has relatively permeable tissues it takes in water quite rapidly and has a high urine volume. Animals with relatively high permeability are therefore constrained to have very low blood concentrations; if they tried to osmoregulate more strongly, the costs would escalate substantially as the ionic gradients increased. This may explain why costs for regulating fish, also relatively permable, are often in the range 2–4% of total MR. The advantages of the arthropodan design, with 'built-in' low permeability and also a degree of inherent volume control, are evident.

3 There is an inherent advantage to producing hyposmotic urine, since the crayfish and mollusc, which recover up to two-thirds of the ions from their urine, have lower osmotic costs than the mitten crabs and ragworms, which produce isosmotic urine. Thus it would seem to be cheaper to osmoregulate (take up ions) via the kidney rather than via the skin or gills, presumably because the ionic gradients are lower there. It may also be cheaper to produce urine in the kidneys by secretion (as insects do) rather than by ultrafiltration (as most other animals do). The latter system requires a very large volume of fluid to be sent through the kidney, and a great deal of active resorption is therefore needed; probably more active transport is required in total than in a system where the primary urine is made by active salt transport.

4 Hyposmotic regulation in the brine shrimp and the mosquito larva, which involves very hyperosmotic urine production, is relatively much more costly (22–33% of the metabolic budget) than hyperosmotic regulation in any animal yet investigated.

Water vapour uptake, in those few animals that can do it, is certainly thermodynamically difficult to achieve, since coupling energy expenditure to actual water absorption is tricky, but the energetic cost again turns out to be quite low. Ticks and insects that are able to absorb water vapour use only a few per cent of their daily metabolic budgets in this way, and there is no detectable increase in metabolic rate when vapour uptake is occurring.

5.10 Roles of nervous systems and hormones

Many epithelia involved in osmoregulation are under hormonal control, either to regulate permeability to particular ions or water, to control the rate of transport of one or more ions, or even to change the direction of this transport.

Permeability changes are best known in the vertebrate kidney, where aldosterone increases the P_{Na} of the apical membrane in the distal convoluted tubule; in the short term this is by activation of pre-existing quiescent channels from vesicles within the cell (see Chapter 4), but during prolonged exposure a further effect occurs with the synthesis of new channel proteins. In the same organ, ADH also activates some quiescent Na channels, but in addition it causes an increase in epithelial water permeability in the collecting duct, partly by the incorporation of especially 'leaky' cytoplasmic vesicles into the cell membrane, so that water can leave the tubule osmotically and the urine that remains behind is more concentrated. Similar mechanisms operate in invertebrates; in some insects at least seven classes of active regulators have been found to date. For example the peptides known as leucokinins act on some of the upper Malpighian tubule cells (via intracellular calcium) to raise the chloride permeability, while diuretic hormone and its mimic 5-hydroxytryptamine (5HT) cause changes in P_w in the lower Malpighian tubules with effects on V-ATPases, cotransporters and apical P_{Cl}.

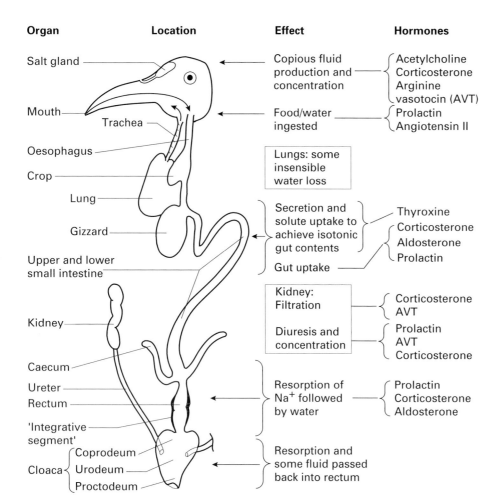

Organ	Location	Effect	Hormones

Fig. 5.28 A summary of the organs involved in osmoregulation and excretion in a terrestrial vertebrate, and the hormones that have a regulatory effect at each site.

Controlling the rates of secretion of ions generally involves activation of secretory tissues, principally via their ion pump ATPases. In insect tubules a small neuropeptide acting via cAMP or cGMP activates the cation-transporting apical V-ATPase. In vertebrates there are often also secondary effects. For example, the rectal salt gland of elasmobranch fish is stimulated by a small peptide hormone from the gut (vasoactive intestinal peptide, or VIP), which stimulates chloride secretion in at least three ways: it activates the Na^+/K^+-ATPase; enhances the symport pathway handling Na^+ and Cl^-; and increases the chloride permeability of the apical cell membrane. VIP is also released as a neurosecretory product of nerves innervating this salt gland, giving even faster control.

Changes of transport direction are well known for NaCl absorption and secretion in mammalian lower gut, in teleost gills, and in the hindgut of larval stages of some euryhaline insects. In at least some of these cases the same cells are involved in both directions.

Mechanisms of hormonal control of salt and water balance in most taxa clearly involve the well-documented actions of cyclic nucleotides (cAMP, cGMP, etc.) and calcium ions, working on protein kinases and leading to phosphorylation of proteins in cell membranes (see Chapter 2). Two distinct protein kinases are

known that inhibit Na^+/K^+-ATPase activity by phosphorylating its α subunit, and these kinases are affected by such hormones as norepinephrine (noradrenaline), angiotensin and dopamine. Insulin also has a direct activating effect both on this pump and on sodium channels in vertebrate striated muscles. A wide range of other molecules are known to affect the P-ATPase family (see Chapter 4), some directly and others (such as fatty acids and prostaglandins) via indirect effects on lipid–protein interactions in the membrane.

Such individual control systems are almost invariably integrated into multiple feedback control systems, especially in insects and mammals. Several hormones interact to control permeability values, whether of skin, gill or renal surfaces, and another suite of hormones act to modulate salt uptake or loss rates and the overall directions of ion movements. A single example may serve to show the complex interactions that can operate; Fig. 5.28 summarizes hormonal control mechanisms in avian water balance.

5.11 Conclusions

Life on Earth has evolved around the properties and peculiarities of water, and all life remains dependent upon water. In aquatic ecosystems, and in most temperate and tropical terrestrial habitats, water is abundant and its conservation within animal bodies is not too difficult, but maintaining a precise osmotic equilibrium may nevertheless be a perpetual task, requiring some proportion of the metabolic efforts of all cells or of some specialized subset of body cells. In hyposmotic habitats it is excess water uptake that must be avoided, and the proportion of the metabolic effort devoted to this may be increased. In many land habitats, dehydration is a permanent danger, and sophisticated structural, physiological and behavioural adaptations are necessary to prevent fatal water losses. It is probably fair to say that controlling water balance is even more essential than controlling temperature stresses, and (together with acquiring a food supply) is the single most important factor in the lives of many animals.

Further reading

Books

Brown, J.A., Balmet, R.J. & Rankin, J.C. (eds) (1993) *New Insights in Vertebrate Kidney Function*. Cambridge University Press, Cambridge.

Dejours, P., Bolis, L., Taylor, C.R. & Weibel, E.R. (eds) (1987) *Comparative Physiology: Life in Water and on Land*. Liviana Press, Padova.

Gilles, R. & Gilles-Baillien, M. (1985) *Transport Processes: Iono- and Osmoregulation*. Springer-Verlag.

Greger, R. (1988) *NaCl Transport in Epithelia*. Advances in Comparative and Environmental Physiology; 1. Springer-Verlag, Berlin.

Hadley, N.F. (1994) *Water Relations of Terrestrial Arthropods*. Academic Press, San Diego, CA.

Kinne, R.K.H. (1990) *Urinary Concentrating Mechanisms*. Karger, Basel.

Somero, G.N., Osmond, C.B. & Bolis, C.L. (eds) (1992) *Water and Life*. Springer-Verlag, Berlin.

Reviews and scientific papers

Atkinson, D.A. (1992) Functional roles of urea synthesis in vertebrates. *Physiological Zoology* **65**, 243–267.

Beuchat, C.A. (1990) Body size, medullary thickness and urine concentrating ability in mammals. *American Journal of Physiology* **258**, R298–R308.

Foskett, J.K. & Scheffey, C. (1982) The chloride cell: definitive identification as the salt-secretory cell in teleosts. *Science* **215**, 164–166.

Gaede, K. & Knülle, W. (1997) On the mechanism of water vapour sorption from unsaturated atmospheres by ticks. *Journal of Experimental Biology* **200**, 1491–1498.

Gibbs, A.G. (1998a) The role of lipid physical properties in lipid barrier functions. *American Zoologist* **38**, 268–279.

Gibbs, A.G. (1998b) Waterproofing properties of cuticular lipids. *American Zoologist* **38**, 471–482.

Gibbs, A.G. & Mousseau, T.A. (1994) Thermal acclimation and genetic variation in cuticular lipids of the lesser migratory grasshopper (*Melanoplus sanguinipes*): effects of lipid composition on biophysical properties. *Physiological Zoology* **67**, 1523–1543.

Gilles, R. (1987) Volume regulation in cells of euryhaline invertebrates. *Current Topics in Membranes and Transport* **30**, 205–247.

Greenwald, L. (1992) Concentrating mammalian urine. *Nature* **356**, 295.

Handler, J.S. & Moo Kwon, H. (1997) Kidney cell survival in high tonicity. *Comparative Biochemistry and Physiology A* **117**, 301–316.

Jørgensen, C.B. (1997) Urea and amphibian water economy. *Comparative Biochemistry and Physiology A* **117**, 161–170.

Karnaky, K.J. (1986) Structure and function of the chloride cells of *Fundulus heteroclitus* and other teleosts. *American Zoologist* **26**, 209–224.

Lockey, K.H. (1988) Lipids of the insect cuticle: origin, composition and function. *Comparative Biochemistry and Physiology* **89B**, 595–645.

Rasmussen, A.D. & Anderson, O. (1996) Apparent water permeability as a physiological parameter in crustaceans. *Journal of Experimental Biology* **199**, 2555–2564.

Ruppert, E.E. & Smith, P.R. (1988) The functional organisation of filtration nephridia. *Biological Reviews* **63**, 231–258.

Schmidt-Nielsen, B. (1988) Excretory mechanisms as examples of the principle 'The whole is greater than the sum of its parts'. *Physiological Zoology* **61**, 312–321.

Shoemaker, V.H. & Nagy, K. (1977) Osmoregulation in amphibians and reptiles. *Annual Review of Physiology* **39**, 449–471.

Treherne, J.E. & Willmer, P.G. (1975) Hormonal control of integumentary water loss: evidence for a novel neuroendocrine system in an insect. *Journal of Experimental Biology* **63**, 142–159.

Wright, P.A. (1995) Nitrogen excretion: three end products, many physiological roles. *Journal of Experimental Biology* **198**, 273–281.

6 Metabolism and Energy Supply

6.1 Introduction

Metabolic reactions within cells are the source of all the macromolecules within a living body, synthesizing these large end-products by **anabolic** processes. At the same time other large macromolecules are being broken down to produce usable energy, by **catabolic** processes. The free energy ($\Delta G°$) produced by catabolism is involved in so many physiological processes that it is essential to understand the source of this energy and to consider the effects of the environmental variables that will influence its rate of production (the 'metabolic rate' of an animal).

The phosphagens are the high-energy phosphate compounds that mediate energy transfers in living organisms. Of these, **adenosine triphosphate (ATP)** is the common currency of energy metabolism for animals, providing the link between energy-yielding and energy-requiring reactions. ATP may be produced independently of oxygen (anaerobic metabolism) or through pathways which require oxygen (aerobic metabolism). Life on Earth certainly evolved in a reducing anaerobic environment, perhaps 3500 million years ago (MYA). Free oxygen would be created in such an environment in small quantities by the photolysis of water, but would have been rapidly taken up by reduced ferrous rocks, which acted as an oxygen sink for at least another 1500 million years. Even after some of the prokaryotic heterotrophs acquired chlorophyll (probably around 3000 mya) and became photosynthetic 'plants' that released oxygen as a by-product, free oxygen levels stayed very low because the oceans and vast beds of reduced minerals acted as sinks. Thus the first prokaryotic organisms were necessarily dependent on anaerobic metabolism; oxygen was toxic to them.

Perhaps around 2300–2000 MYA the ocean and mineral sinks were becoming filled and free oxygen levels began to rise slowly (Fig. 6.1); this change may have taken place in three stages, from total planetary anoxia, through atmospheric and surface ocean oxidization, to deep oceanic oxidization and the formation of the first geological oxidized 'red beds' rich in ferric oxides (dated 2000 MYA). All organisms would have had to develop O_2 detoxification systems to survive, and at this point aerobic metabolism also probably began to evolve. Current thinking suggests that this important evolutionary innovation may have had multiple origins in different groups of prokaryotes, and that it became possible at about 1% of present atmospheric levels (PAL) of oxygen. Gradual further rises in oxygen level accompanied the subsequent evolution of eukaryotic cells and then (perhaps 1000–800 MYA) of the first multicellular plants and animals.

Most metazoan animals now use aerobic metabolism for most of their energy requirements, reaping the benefits of pathways that are many times more efficient (generating 36–38 moles of ATP for each mole of substrate broken down aerobically, instead of just 2–6 with different anaerobic routes). However, many organisms still thrive in totally anaerobic conditions, for example in marine muds and sediments, in bogs and wetlands, or in the guts of other animals. Furthermore, nearly all organisms commonly revert to using anaerobic pathways during temporary periods of reduced oxygen availability (hypoxia) or when their rate of energy utilization is greater than can be met by the slower aerobic pathways, for example during strenuous exercise.

6.2 Metabolic intermediaries

6.2.1 Adenosine triphosphate

ATP is the classic 'high-energy molecule', and interconversions between it and the other adenylates (the diphosphate and monophosphate forms, ADP and AMP) are the key to many metabolic processes. ATP consists of a purine base (adenine) linked to a five-carbon sugar (ribose) to form adenosine, and then three phosphate groups linked on by high-energy esteric bonds. Hydrolysis of the terminal phosphate group (Fig. 6.2) releases about 30.5 kJ mol^{-1} of ATP under normal cell conditions; and remember that 1 mole of ATP is equivalent to 6×10^{23} molecules of ATP (Avogadro's number). ATP acts to transfer

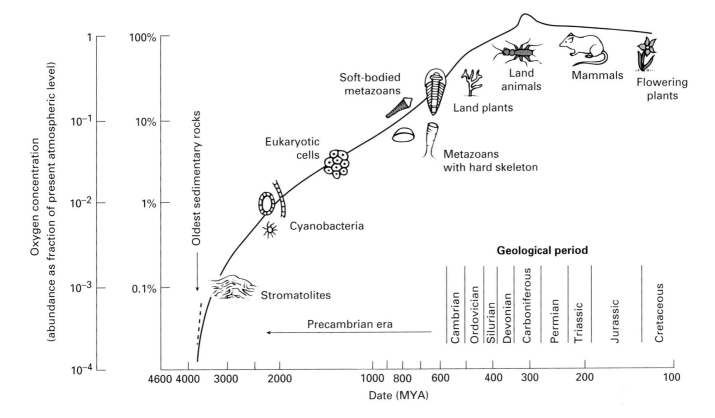

Fig. 6.1 Atmospheric oxygen levels through the history of Earth, with key events shown. (Adapted from Wayne 1991; by permission of Oxford University Press.)

chemical energy to other molecules during energy-requiring interactions, yielding adenosine diphosphate (ADP) and inorganic phosphate (P_i) in the process; this ADP is then recharged with a phosphate group, again by cellular metabolism. As a result energy is constantly cycling through ATP, and the individual molecules have a relatively short lifetime.

Fig. 6.2 Structure and hydrolysis of adenosine triphosphate (ATP).

Around 60% of the energy that is used to recharge ADP appears as heat, due to inefficiencies in the reactions involved. Hence metabolic activity always generates heat, referred to as 'metabolic heat' when considering problems of animal temperature balance (Chapter 8). With the exception of muscles, the rate of ATP turnover is relatively low in most tissues. During the transition from rest to maximal activity, the ATP turnover (or cycling, or flux) in muscle can increase by one or more orders of magnitude, representing a massive heat production in a very active animal. However, the rates of ATP turnover also vary systematically between large and small animals so that mass-specific metabolic rates are higher in smaller animals (see Table 6.1, and Chapter 3).

Adenosine triphosphate (ATP)

Adenosine diphosphate (ADP)

Inorganic phosphate (P_i)

Table 6.1 ATP turnover rates during peak exercise in animals of varying size.

Species	Body mass	Muscle ATP turnover (μmol ATP g^{-1} min^{-1})
Insects		
Locust	1 g	5400
Birds		
Hummingbird	2.5 g	600
Mammals		
Mouse	7.2 g	227
Ferret	500 g	82
Wallaby	4.8 kg	57
Goat	20 kg	48
Human	70 kg	30
Eland	210 kg	32

Fig. 6.3 Structures of the phosphagens (a) phosphocreatine (PCr) and (b) phosphoarginine (PAr).

6.2.2 *Phosphagens—rapidly mobilized fuels*

At any one time an animal does not contain a great amount of ATP, with values of only 2–8 μmol ATP g^{-1} being common; evidently it is not normally used to store much energy. Other **phosphagens** are used for this storage, the commonest being creatine phosphate (PCr) and arginine phosphate (PAr), which are phosphorylated derivatives of guanidinium compounds (Fig. 6.3). These may be up to 10 times as concentrated as ATP in tissues such as muscle and brain, and they are the immediate short-term source of ATP in most animals. They may be used to provide the necessary ATP under both aerobic and anaerobic conditions; thus they act in effect as a 'buffering' system for ATP, ensuring that its concentration is low and the ATP/ADP ratio is always kept favourable for ATP synthesis.

Phosphagens are the simplest and most rapid precursors for generating ATP as only one enzymic step is involved, catalysed by a kinase enzyme (e.g. creatine phosphokinase, CPK). Kinases belong to a class of enzymes that transfer the high-energy phosphate bond of ATP to another molecule:

$$PCr + ADP + H^+ \rightleftharpoons ATP + creatine$$

$$PAr + ADP + H^+ \rightleftharpoons ATP + arginine$$

For a phosphagen to be a useful source of ATP, it must be able to transfer its high-energy phosphate bond (~P) to ADP at high rates and at the correct time. This is achieved in most cells by having large amounts of the right kinase, with the right kinetic properties, present in the right locations in the cytoplasm. These kinases therefore occur in high concentrations in the cell as a whole, and are present as different isozymes (see Chapter 2), close to their sites of utilization and often bound to other proteins; for example, in muscles different isozymes

occur, some bound to myosin ATPase, and others within the sarcolemma and sarcoplasmic reticulum with easy access to Na$^+$/K$^+$-ATPase and Ca^{2+}-ATPase, respectively (Fig. 6.4). The overall concentration of kinases is highest in tissues with high ATP demands, such as the skeletal muscles of animals capable of very rapid movements. These same tissues usually have high concentrations of the phosphagens. Thus the white muscle of the rainbow trout contains around 30 mmol PCr per kg wet mass, and the adductor muscle of the scallop contains about 50 mmol PAr per kg wet mass. Both of these muscles are involved in escape behaviours. In slower muscles and in tissues such as neural ganglia the PCr levels may be only one-tenth of these values.

At equilibrium the enzymes catalysing the reactions shown above are driven to the right (i.e. in the direction of ATP synthesis) by a number of mechanisms, including the phosphagens' high affinity for ADP (at least during the initial stages of work). The equilibrium constants for these kinases at pH 7.0 are of the order 2×10^9, which means that the kinases work extremely effectively and the phosphagen stores can be largely used up without affecting ATP concentrations (Fig. 6.5).

PCr and PAr stores are sufficient to support a modest amount of work lasting just a few seconds (about 2–5 s in mammalian muscle). In fish and many invertebrates the energy required for capturing prey or escaping from predators can often be supplied entirely from the breakdown of phosphagens, which are subsequently recharged using aerobic pathways. However, further amounts of ATP can be obtained by a pathway that results initially in the formation of adenosine monophosphate (AMP):

$$2ADP \rightleftharpoons ATP + AMP$$

This reaction is at equilibrium and can only proceed in the

Fig. 6.4 The role of creatine phosphokinase (CPK) in controlling the flow of adenylates between sites of ATP synthesis and utilization in muscle. Different isozymes of CPK occur in at least five sites where they facilitate the supply of ATP from phosphocreatine (PCr). Three of these (1–3) are associated with ATPases. A fourth (4) is linked to porins (P) in the outer mitochondrial membrane, using the ATP from oxidative phosphorylation, while a fifth (5) occurs in the cytosol and uses ATP from glycolytic processes. (Adapted from Wallimann *et al.* 1992.)

direction of further ATP synthesis if the AMP is removed in another reaction, catalysed by the enzyme AMP deaminase and yielding inosine monophosphate (IMP) and ammonia:

$$AMP \rightarrow IMP + NH_3$$

Molluscs and crustaceans lack the AMP deaminase enzyme and therefore cannot exploit this source of ATP. Those animals that do have it may show a depletion of their total adenylate pool through breakdown of AMP when exercised to exhaustion.

In addition to being fuels in anaerobic metabolism, phosphagens (particularly PCr) play an important part in aerobic pathways, by shuttling inorganic phosphate (P_i) between sites of ADP formation and utilization, and by involvement in the facilitated diffusion of ATP. Hummingbird flight muscles are among the most aerobic tissues known and have very high levels of PCr.

In summary, phosphagens are a useful way of supporting cellular metabolism at high power outputs while buffering the cell's ATP levels, especially at localities close to vitally active ATPase enzymes.

6.3 Anaerobic metabolism

6.3.1 Uses and significance of anaerobic metabolism

Most animals use anaerobic pathways as part of their repertoire in maintaining their energy budgets, and in certain circumstances where oxygen is unavailable the entire metabolism may be dependent on anaerobic processes; we will meet examples in more detail when looking at animals from marine, freshwater and semi-terrestrial sediments, and at parasites living in the vertebrate gut. In many environments the need for anaerobic metabolism is transient, though. This may occur on a regular cycle for animals in intertidal and estuarine environments, for example, where anoxia may persist for hours or even for days according to the state of the tides. Or it may occur more randomly where spurts of exercise are needed (e.g. for prey capture or escape responses) or where an air-breathing diving animal spends time under water, suffering bouts of oxygen shortage. All these animals are very tolerant of hypoxia or anoxia (see Chapter 7). Their metabolic rates usually decline during anaerobic respiration, and in intertidal invertebrates the reduction

Fig. 6.5 The concentration of phosphocreatine (PCr) showing how it is used to maintain ATP levels constant despite varying ATP usage regimes. Total creatine and adenine levels are also shown. (From Woledge *et al.* 1985.)

may be to only a small fraction of the normal resting metabolic rate (see below). In vertebrates tolerance of hypoxia is much less profound, except in a few special cases where, for example, fish can survive in ice-bound ponds for several weeks. Higher vertebrates only use anaerobic pathways to any great extent during short bursts of intense activity. Here the rapidity of ATP generation is an advantage; the reaction is entirely in the cell cytoplasm, and nothing need be transported in or out of the cell to support the catabolic reactions. However, the rapid build-up of end-products soon inhibits further anaerobiosis, and the amount of energy that can be generated before this occurs is severely limited. Anaerobic energy yields are therefore only really important in activities lasting up to a few minutes. Longer term submaximal work by mammalian muscles does involve some anaerobic component, but the end-product (lactate) is removed by the liver and other aerobic tissues at about the same rate that it is formed.

6.3.2 *Basic glycolytic pathways*

The process of **glycolysis** takes place in the cytoplasm and if oxygen is limited it can be an entirely anaerobic process. Written at its simplest it involves the reaction:

$$glucose + 2ADP + 2P_i \rightarrow 2 \ lactate + 2ATP$$

However, the glycolytic pathway is in fact a complex fermentation reaction, consisting of 10 enzyme-catalysed steps (Fig. 6.6).

Because it is a very ancient pathway, many of its features, such as the relative proportions of the different enzymes, are highly conserved between animals, and its dual roles in providing ATP and providing carbon molecules for synthetic (anabolic) reactions are similar throughout the metazoan kingdom. However, as we will see there are a number of ways in which its rate and outcome can be modified according to environmental circumstance.

Glucose usually enters cells by means of specific transport proteins (see Chapter 4). It is initially phosphorylated by ATP to form glucose 6-phosphate, a reaction catalysed by the enzyme hexokinase:

$$glucose + ATP \xrightarrow{\text{hexokinase}} glucose \ 6\text{-phosphate} + ADP$$

Under anaerobic conditions, 1 mol of glucose can be fermented to 1 mol of anaerobic end-product, with a net generation of 2 mol of ATP (Fig. 6.6). If the ATP is coupled to an ATPase, such as the myosin ATPase in muscle, then 2 mol of H+ are also generated, which are accepted by the coenzyme NAD (nicotinamide adenine dinucleotide):

$$glucose + 2ADP + 2P_i + 2NAD$$
$$\rightarrow 2 \ pyruvate \ + 2ATP + 2NADH + H^+$$

In practice at physiological pH the net generation of protons is less than this reaction would suggest, as some are consumed in converting the pyruvate to lactate, a common end-product of glycolysis. Nevertheless, the generation of H+ cannot proceed indefinitely as it leads to acidosis, and this is one factor limiting the duration of anaerobic glycolysis that can be sustained. Protons are buffered or exported from cells to minimize the impact of their accumulation. Lactate concentrations can reach 40 mmol per kg wet tissue mass in mammalian muscle, and 200 mmol per kg wet tissue mass in diving turtles where survival of hypoxia is crucial.

The reducing equivalents (NADH) formed by the action of glyceraldehyde phosphate dehydrogenase during glycolysis (Fig. 6.6) are oxidized to regenerate NAD in the terminal step of the pathway, which is catalysed by another dehydrogenase enzyme. This enzyme is lactate dehydrogenase (LDH) when lactate is the endpoint of glycolysis acting as the terminal electron acceptor (and note that LDH is so ubiquitous and so relatively easy to study and sequence that it has become an important source of comparative and evolutionary studies).

6.3.3 *Alternative endpoints of glycolysis*

Arthropods, many worms and echinoderms all rely mainly on classical glycolysis for anaerobic energy production, resulting in the accumulation of large amounts of lactate and sometimes small amounts of alanine. In most vertebrates too, glycolysis to lactate is the only anaerobic pathway of any real importance, and is used in exercise and also in more chronic anoxia such as during diving in air-breathing tetrapods. But a few invertebrates, especially but not exclusively in parasitic groups, have

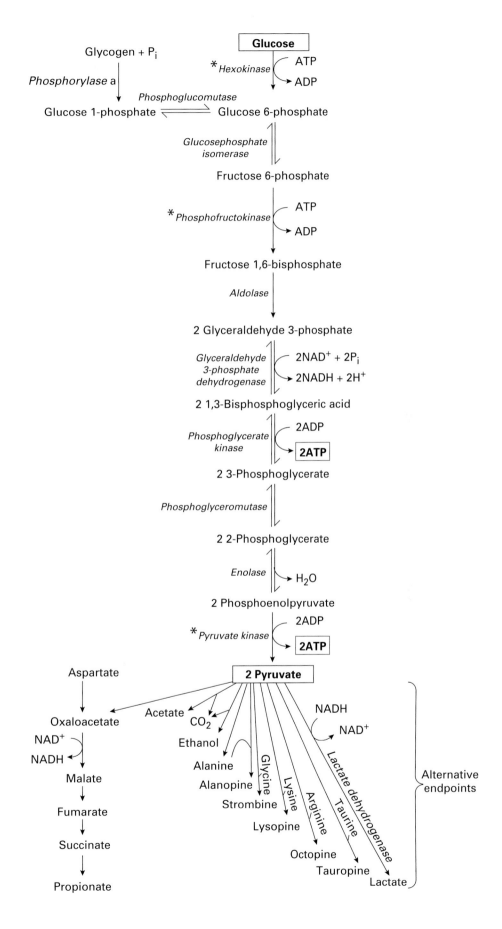

Fig. 6.6 Glycolysis pathways and key side paths, also showing the main points of control (∗).

Table 6.2 ATP yield from different fermentation routes.

Substrate	Endpoints	mol ATP per mol substrate
Glucose	Lactate, octopine, alanopine or strombine	2
Glucose	Succinate	4
Glucose	Propionate	6
Aspartate or glutamate	Succinate	1
Aspartate or glutamate	Propionate	2
Branched-chain fatty acids	Volatile fatty acids	1

ethanol as an endpoint; parasite anaerobic pathways are discussed further in Chapter 15. Ethanol is also an endpoint in some fish, of which the goldfish is the best studied. In these animals a complete absence of oxygen can be tolerated for long periods at low temperatures, with CO_2 and ethanol as end-products. Glycogen is converted to pyruvate by anaerobic glycolysis in the cytoplasm, and pyruvate is then decarboxylated to acetaldehyde in the mitochondria; the acetaldehyde is subsequently reduced to ethanol in the cytoplasm by alcohol dehydrogenase (note that this is the same pathway that is quite common in plants). Ethanol is excreted, allowing glycolysis to proceed without the accumulation of an acid end-product. Goldfish using this pathway can survive through several days of anoxia at 0–4°C because their metabolic rate is reduced by more than 80%, sparing their glycogen stores.

In many invertebrate taxa a different strategy is used, with lactate dehydrogenase replaced by a functionally analogous imino dehydrogenase, having an identical ATP yield per glycosyl unit. Here the final step is the condensation of an amino acid with the pyruvate produced by glycolysis, allowing the regeneration of the electron acceptor coenzyme, NAD:

pyruvate (RCO–R′) + R″–NH$_2$ + NADH + H$^+$
  → RCHR′–NHR″ + H$_2$O + NAD

Particular amino acids such as alanine, glycine, lysine and arginine are used, according to species, to remove the pyruvate; the end-products are the imino acids alanopine, strombine, lysopine and octopine, respectively. In effect, oxygen acts as the terminal electron acceptor to form water. Octopine formation appears to occur when there are high resting levels of phosphoarginine in the muscles, and is common in swimming cephalopods and certain bivalves.

Aspartate and certain branched-chain amino acids can also be fermented directly in some invertebrates (Fig. 6.6); for example, in intertidal molluscs, where hypoxia can be a problem, the aspartate is deaminated and then fed on via parts of the Krebs cycle (see below) to produce a higher ATP yield and a variety of end-products such as succinate and propionate; the amino group is transferred into glutamate and then alanine, so may again lead to the formation of alanopine. Alternatively, in some

metazoan parasites in profoundly hypoxic conditions other glycolytic products occur (see Chapter 15), especially oxaloacetate derived from pyruvate; these feed on into the Krebs cycle, and are again converted through succinate and propionate, sometimes eventually leading to methylated products such as 2-methylbutyrate. Such processes may yield an extra 4–6 ATP per glucose molecule. The ATP yields of the different substrates of fermentation are shown in Table 6.2. Pathways producing succinate or propionate from aspartate or glutamate, or those using branched amino acids, are not favoured for anaerobic muscular work, because they cannot sustain a high enough flux of ATP.

6.3.4 *Control of glycolysis*

Control of glycolysis is achieved in two important ways: firstly, by exogenous activating signals (hormones or neurotransmitters, acting through the G protein/adenyl cyclase/cAMP/protein kinase systems discussed in Chapter 2); and secondly, through the regulatory properties of glycolytic enzymes themselves where products act as negative feedbacks on earlier stages in the process (see Chapter 2).

There are tissue-specific isozymes of almost every glycolytic enzyme, and most of the steps in glycolysis are readily reversible under physiological conditions. The exceptions are those catalysed by isozymes of hexokinase, phosphofructokinase and pyruvate kinase (starred in Fig. 6.6). These three enzymes act in a coordinated fashion to control a major part of the flux through the pathway. Phosphofructokinase (PFK) is the most important control step; this enzyme is inhibited by high levels of ATP and citrate, and is activated by AMP and fructose 2,6-bisphosphate. The presence of fructose 2,6-bisphosphate in the vertebrate liver signals that glucose is abundant. This pattern of regulation ensures that the activity of PFK matches the need either for generating ATP or for providing the building blocks for synthetic reactions. When PFK is inhibited, glucose 6-phosphate accumulates, and inhibits hexokinase, the first enzyme in the pathway (Fig. 6.6). Pyruvate kinase is the last step in the pathway. It is allosterically inhibited by ATP and alanine, and is activated by fructose 1,6-bisphosphate (Fig. 6.6). The feed-forward activation of pyruvate kinase by glycolytic intermediates ensures that its activity increases in parallel with PFK when the demand for ATP is high (see Chapter 2). Pyruvate kinase occurs in a number of different forms (isozymes) providing variation in the control of the terminal step of glycolysis between tissues. The isozyme present in mammalian liver is controlled by a reversible phosphorylation. The hormone glucagon (which in concert with insulin controls blood sugar levels in vertebrates) triggers a cascade that leads to an increase in the less active phosphorylated form of pyruvate kinase. This ensures that the liver does not consume glucose for synthetic reactions when it is more urgently needed for ATP generation in other tissues such as muscles and brain.

Fig. 6.7 Glycogen structure and bonding.

However, a puzzling feature of glycolytic control is that it appears to be 'more than the sum of its parts'; manipulation of gene expression in yeast has shown that changes in overall rates of glycolysis are greater than can be accounted for by changes in individual enzyme steps. There is recent evidence that in muscle some fraction of the key enzymes may be present but inaccessible or unreactive, providing a latent pool that can be called upon to increase rates of flux without substrate concentrations changing much. This may be the main control mechanism, with varying substrate or product or modulator levels providing the fine tuning.

6.3.5 Fuel sources for glycolysis

Glucose is the primary fuel source for glycolysis, but it is stored as the polysaccharide glycogen in many tissues, especially muscle (and the liver of vertebrates). Glycogen can enter glycolysis either via conversion to glucose or indirectly by conversion to glucose 6-phosphate (Fig. 6.6). Glycogen concentrations are particularly high in species that have a well-developed tolerance for hypoxia or anoxia. For example, in species of marine bivalves found in the intertidal zone stored glycogen concentrations can exceed 1% of wet body mass (see Chapter 10).

Glycogen is a large branched polymer where glucose molecules are linked together by α-1,4-glycosidic bonds in the straight parts of the polymer, and the branches are created by α-1,6-glycosidic bonds (Fig. 6.7). Glycogen is often present in the cytoplasm in the form of granules (readily visible with the electron microscope) which also contain the enzymes for glycogen synthesis and degradation as well as some of the enzymes that control these processes. The synthesis and degradation of glycogen may be hormonally regulated via different enzyme pathways involving glycogen phosphorylase and glycogen synthase.

Both of these enzymes exist in active and inactive forms. The reaction cascades that convert the inactive to the active forms are controlled by phosphorylation reactions which have cAMP as a common second messenger (see Chapter 2). The fact that the glycogen polymer is branched increases the solubility of glycogen molecules and creates a large number of terminal residues that are the sites of action for glycogen phosphorylase and synthase. Thus branching increases the maximum rate at which glycogen synthesis and breakdown can occur.

Other sugars, including fructose and galactose, can also enter the glycolytic pathway. Galactose is converted in four steps to glucose 6-phosphate, which can then enter glycolysis. In mammals, the lactose present in milk is hydrolysed to fructose and glucose. Fructose (which is present in many plant foods, and therefore also in honey) can enter the glycolytic pathway of animals via a fructose 1-phosphate pathway, which is particularly important in liver cells. Fructose can also be phosphorylated to fructose 6-phosphate by hexokinase. This reaction only predominates in tissues such as white fat cells, which have higher concentrations of fructose than glucose. In other tissues the formation of fructose 6-phosphate is competitively inhibited because the affinity of hexokinase for glucose is much greater than for fructose.

We saw above (section 6.3.2) that amino acids can also feed into glycolytic pathways. Thus carbohydrates and amino acids from proteins can be used as fuel for glycolysis, giving a range of end-products. However, the one fuel that cannot be used is lipid, which is normally in too reduced a state to be fermented.

6.4 Aerobic metabolism

In fermentation, organic molecules serve as terminal electron and proton acceptors. Since life evolved in an essentially anaerobic atmosphere these pathways must be among the most ancient means of energy production. However, carbon skeletons are only partially degraded in the process, and despite a roughly 35–50% efficient process (depending on the end-products) only

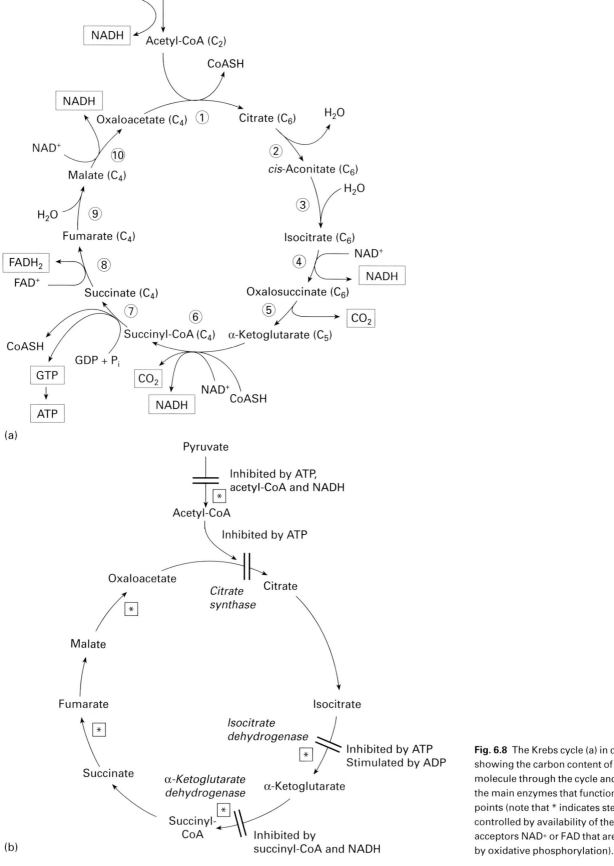

(a)

(b)

Fig. 6.8 The Krebs cycle (a) in detail, showing the carbon content of each molecule through the cycle and (b) showing the main enzymes that function as control points (note that * indicates steps also controlled by availability of the electron acceptors NAD^+ or FAD that are regenerated by oxidative phosphorylation).

about 4.7% of the total energy of the glucose molecule is released in producing pyruvate. Whenever possible then, animals proceed further with the breakdown of glucose, ensuring that it is more completely oxidized with O_2 acting as the terminal electron acceptor. These additional steps represent the evolutionarily 'later' reactions of **aerobic** metabolism. The net reaction for complete glucose oxidation in aerobic conditions can be written as:

$$glucose + 36(ADP + P_i) + 6O_2 \rightarrow 36ATP + 6CO_2 + 6H_2O$$

6.4.1 The Krebs cycle

The classic way of achieving further carbon compound breakdown and better energy yield is to convert the glycolytic end-product pyruvate not into lactate but (by condensation with coenzyme A) into acetyl-CoA. This then serves as the entry substrate into the series of reactions known as the Krebs cycle (also known as the citric acid cycle or tricarboxylic acid cycle), occurring in the mitochondrial matrix (Fig. 6.8). The Krebs cycle is both the major pathway for ATP production and a source of intermediates for the synthesis of a whole range of other molecules, including proteins.

The acetyl-CoA (a two-carbon fragment, C_2) that starts the cycle is combined with oxaloacetate (also C_4) to form citrate (C_6). This then passes on through a series of C_6, C_5 and C_4 intermediates, the last of which is again oxaloacetate. The two carbons that entered as acetyl-CoA are released as $2CO_2$ in each turn of the cycle, at steps 5 and 6 in Fig. 6.8. At the same time, one molecule of the high-energy phosphate compound guanosine triphosphate (GTP) is formed by the succinyl-CoA synthetase reaction (step 7 in Fig. 6.8) per turn of the Krebs cycle; this is then transferred to ATP. Simultaneously, H^+ is transferred to the proton acceptors NAD and FAD (flavin adenine dinucleotide) at four stages in the cycle, with an additional proton transfer to NAD at the pyruvate–acetyl-CoA stage. The net reaction per turn of the cycle is therefore:

$$pyruvate + 4NAD + FAD + GDP + P_i$$
$$\rightarrow 2CO_2 + 4NADH + FADH_2 + GTP$$

(Remember that as each glucose yields two pyruvate molecules to feed into the cycle, the yield per molecule of glucose is twice this.)

Any Krebs cycle intermediates that are drawn off for biosynthesis must be replenished by a series of 'anaplerotic pathways' (from the Greek meaning 'to fill up'). For example, oxaloacetate converted into amino acids for protein synthesis is subsequently replaced by the carboxylation of pyruvate using the enzyme pyruvate carboxylase.

The CO_2 formed in the Krebs cycle is a major waste product of aerobic metabolism, and it is generated by the production of ever shorter carbon fragments, from the carboxyl groups of the cycle intermediates. At the same time molecular oxygen is used up, and this will appear as water after combining with hydrogen molecules during oxidative phosphorylation, so that water becomes another end-product (the so-called 'metabolic water' referred to in considering animal water balance in Chapters 4 and 5).

6.4.2 The electron transfer system

During the Krebs cycle, the two carbon atoms in acetyl-CoA have appeared as CO_2 with the formation of NADH and $FADH_2$. Each glucose molecule produces a total of 10 NADH and two $FADH_2$. These represent the 'fuel' for the major energy-storing process of aerobic respiration, as they are reoxidized via the **electron transfer system** (ETS) located in the inner mitochondrial membranes.

The reduced molecules NADH and $FADH_2$ each contain a pair of electrons with a high transfer potential which can be donated to molecular oxygen via a series of intermediates; in the process free energy is liberated, which can be used to generate ATP by a process called **oxidative phosphorylation** (Fig. 6.9). Oxidative phosphorylation is carried out by molecular complexes termed respiratory assemblies, located on the folded cristae of the inner mitochondrial membranes. NADH itself cannot easily cross the membranes, but electrons are transferred from NADH via carriers and then passed on to O_2 via the respiratory assemblies. These comprise three large protein complexes: NADH-Q reductase, cytochrome reductase and cytochrome oxidase (Fig. 6.9a,c). The electron-carrying groups on these proteins are flavins, iron-sulphur clusters, haems and copper ions. Electrons are carried from NADH-Q reductase to cytochrome reductase by the reduced form of a hydrophobic quinone called ubiquinone (Q in Fig. 6.9). Ubiquinone also functions to carry electrons from $FADH_2$ to cytochrome reductase. The protein cytochrome c then shuttles electrons from cytochrome reductase to cytochrome oxidase (note that cytochrome c has been another good source of comparative sequence studies). Cytochrome oxidase at the end of the ETS chain catalyses the transfer of electrons to molecular oxygen, the final electron acceptor.

The step-by-step transfer of electrons from NADH or $FADH_2$ to O_2 through these carriers does not directly produce any ATP, but results in the pumping of protons out of the mitochondrial matrix to the space between the inner and outer mitochondrial membranes, generating a proton-motive force consisting of a pH gradient and a transmembrane electric potential. ATP is then synthesized when protons flow back to the mitochondrial matrix through an **ATP synthase**, a large protein complex straddling the inner membrane and made up from F_1 and F_0 subunits (Fig. 6.9b). The precise details of this process remain unclear, though it is known that the proton gradient is needed not to form the ATP but to release it from the enzyme. Whatever the detail, it is evident that oxidation and phosphorylation are coupled by the large proton electrochemical gradient across the inner mitochondrial membrane.

As we have seen, the H^+ that were transferred from intermediates to coenzymes in the Krebs cycle produce a total of four reduced coenzyme molecules for each two-carbon fragment entering the Krebs cycle. The three molecules of NADH donate

Fig. 6.9 Oxidative phosphorylations by the electron transfer system (ETS) in mitochondrial membranes. (a) The basic sequence of electron carriers, protons being 'pumped' by the three complexes shown in boxes. (b) The ATP synthase, where a positive membrane potential allows proton flow through the F_0 subunit activating the F_1 ATP-synthesizing unit. (c) The arrangement of these major proteins in the inner mitochondrial membrane and a view of the overall process.

their hydrogen atoms to cytochromes at the beginning of the electron transfer chain forming a total of nine ATP molecules. The reduced coenzyme $FADH_2$ enters at a later point in the ETS chain producing a further two ATP molecules. The net reaction for complete aerobic glucose oxidation produces 36 molecules of ATP (36 mol of ATP per mole glucose) (Table 6.3). Where glycogen is the starting point the net yield is slightly higher. The conversion efficiency is about 41%, with the remaining 59% of available energy released as heat.

6.4.3 Fuels for aerobic metabolism

Glucose and related sugars are not the only fuel sources for aerobic carbon breakdown. In fact the Krebs cycle is extremely flexible in allowing other potential fuels to be fed into the system, some of which need to be considered here as they have particular consequences for the overall energetic gain. Fuels from within the cell may be used for short-term activity, but long-term aerobic exercise requires muscles and other tissues to

Table 6.3 Overview of ATP production in aerobic metabolism.

Reaction sequence	ATP yield per glucose
Glycolysis (in cytoplasm)	
Phosphorylation of glucose	−1
Phosphorylation of fructose 6-phosphate	−1
Dephosphorylation of 2 molecules of 1,3-bisphosphoglycerate	+2
Dephosphorylation of 2 molecules of phosphoenolpyruvate	+2
(2 NADH are formed in the oxidation of 2 molecules of glyceraldehyde 3-phosphate)	
Conversion of pyruvate into acetyl-CoA feeding into Krebs cycle (in mitochondria)	
(2 NADH are formed)	
Krebs cycle (in mitochondria)	
2 molecules of GTP are formed from 2 molecules of succinyl-CoA	+2
(6 NADH are formed in the oxidation of 2 molecules each of isocitrate, α-ketoglutarate and malate)	
(2 FADH$_2$ are formed in the oxidation of 2 molecules of succinate)	
Oxidative phosphorylation (in mitochondria)	
2 NADH formed in glycolysis; each yields 2 ATP	+4
2 NADH formed in the oxidative decarboxylation of pyruvate; each yields 3 ATP	+6
2 FADH$_2$ formed in the Krebs cycle; each yields 2 ATP	+4
6 NADH formed in the Krebs cycle; each yields 3 ATP	+18
Net yield per glucose	+36

be supplied with exogenous fuels, for example from the liver in vertebrates or from the fat body in insects.

Lipids

Lipids in the form of fatty acids are one of the most common inputs to metabolism in animals. In general, the fatty acids present in animal cells contain an even number of carbon atoms, often between 14 and 24 (see Fig. 4.5). The most common fatty acids are C_{16} and C_{18}. The alkyl chain may be saturated, or it may be unsaturated (i.e. containing one or more carbon–carbon double bonds); 16 : 0 denotes a fatty acid with 16 carbon atoms and no double bonds, whereas 16 : 2 signifies the presence of two double bonds.

Note that at physiological pH fatty acids are ionized (giving, for example, palmitate, stearate, oleate). However, the fatty acids are stored in an uncharged form in combination with glycerol as 'triglycerides', more correctly known as triacylglycerols (see Fig. 4.5). The triacylglycerols are highly concentrated fuels because they are both highly reduced and very non-polar, enabling them to be stored in an anhydrous state. One gram of nearly anhydrous triacylglycerol stores more than six times as much energy as 1 g of hydrated glycogen (note that glycogen cannot be stored unhydrated, and 1 g anhydrous glycogen binds about 2 g water). Many animals contain cells that are specialized for the synthesis, storage and mobilization of triacylglycerols, for example the fat body in insects and the white adipose

tissue in vertebrates (Fig. 6.10). In vertebrates triacylglycerol is hydrolysed back to fatty acids and glycerol by lipases; the resultant process is termed **lipolysis** and is up-regulated by a variety of hormones, including epinephrine (adrenaline), norepinephrine (noradrenaline), glucagon and adrenocorticotropic hormone (ACTH). These hormones stimulate adenyl cyclases to cause an increase in cAMP concentration, which then stimulates a protein kinase that phosphorylates the lipase and so activates it. Lipolysis in vertebrates is inhibited by the hormone insulin.

The glycerol produced by lipolysis can be converted to glyceraldehyde 3-phosphate and either metabolized to pyruvate by glycolysis or to glucose by 'gluconeogenic' pathways. Before the fatty acids released by lipolysis can be used as fuels they must first be activated to acyl-CoAs on the outer mitochondrial membrane. Since the inner mitochondrial membrane is relatively impermeable to long-chain acyl-CoA molecules, a special transporting system is needed to get them into the matrix. The acyl-CoA molecules combine with carnitine and are shuttled across the inner mitochondrial membrane by a translocase enzyme (Fig. 6.11). Saturated acyl-CoA is then degraded by the so-called β-oxidation spiral, involving four basic steps generating acetyl-CoA (Fig. 6.12), which is further metabolized by the Krebs cycle and ETS as before (Figs 6.8 & 6.9). With 1 mol of the fatty acid palmitate as a starting substrate, a total of 129 mol ATP is generated; starting with a triacylglycerol (glycerol + 3 palmitates) the yield is 403 mol ATP. These values are far higher than the yields of 39 ATP from glycogen and 18 ATP from lactate (moles

Fig. 6.10 Micrograph of adipose tissue in a mammal. The cytoplasm of the cells is largely occluded by huge lipid deposits (L), with a nucleus (N) sometimes visible. A capillary (C) and a venule (V) are also present. (From Ross & Romrell 1989.)

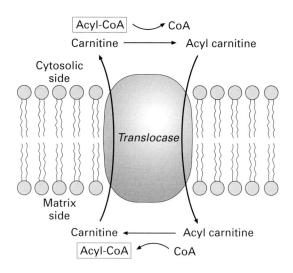

Fig. 6.11 Mechanisms for acyl-CoA crossing a membrane; acyl carnitine is formed and translocated into the mitochondrial matrix, then the carnitine returns to the cytoplasmic side in exchange for the acyl carnitine.

per mole of substrate), underlining the value of fats in energy storage.

Amino acids

Amino acids, ultimately derived from protein breakdown, are also metabolized by pathways feeding into the Krebs cycle at various different points (Fig. 6.13). The ATP yields are relatively modest; with alanine, glutamate and proline as substrates they

Fig. 6.12 The β-oxidation spiral.

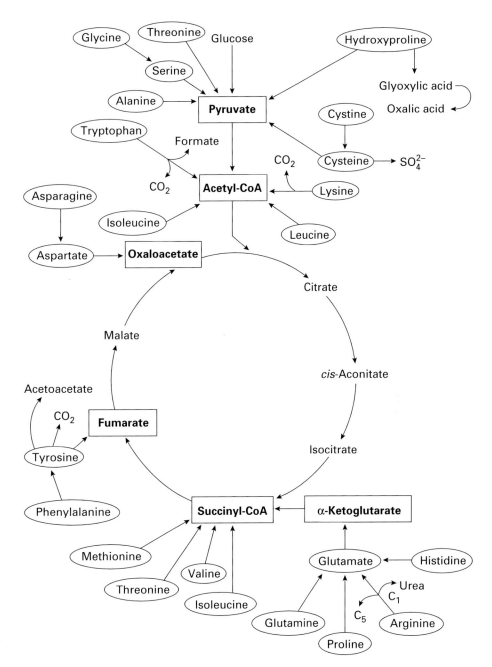

Fig. 6.13 Amino acids feeding into the Krebs cycle as alternative fuels.

are 15, 27 and 30 mol per mole of substrate, respectively. Note that where proteins or amino acids are fed into the oxidative cycle there will also be production of bicarbonate and ammonium ions as wastes.

Other fuels

In the tissues of some animals acetoacetate and 3-hydroxy-butyrate are important alternative fuels. For example, the heart and renal cortex of mammals use acetoacetate in preference to glucose, whilst the brain can adapt to acetoacetate during periods of starvation. The utilization of acetoacetate as a fuel is shown in Fig. 6.14.

Lactic acid

The lactic acid formed in glycolysis (see above) is another important fuel for aerobic metabolism in some animals. In many vertebrates lactate produced in muscle fibres specialized for glycolysis is transported by the circulation to more aerobic tissues, where it is oxidized. Even within mammals the types and sequence of fuels used vary between species. For example, the average human contains sufficient glycogen to sustain a lactate turnover rate of 500 mmol kg^{-1} min^{-1}, which is sufficient for about 3 h of moderate intensity exercise, while in llamas lactate flux only makes a contribution to work for a few minutes at the start of exercise. The shuttling of lactate between sites of

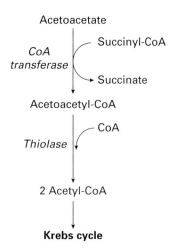

Fig. 6.14 Acetoacetate as a fuel.

formation and utilization is also minimized or absent in fish and insects. Although lactate concentrations can reach 40 mmol kg⁻¹ in the white muscle of salmonid fish following strenuous exercise (and even higher levels in some diving vertebrates), its clearance requires many hours of recovery. Some of the lactate is oxidized and some is converted back to glucose in the kidney and the liver where it is stored as glycogen.

Each of these different potential fuels yields a different net ATP output. These can perhaps best be compared by the amount of each fuel needed for a particular level of ATP generation. Thus an ATP turnover rate of 20 mmol g⁻¹ min⁻¹ could be supported by a palmitate flux (turnover rate) of 0.15 units, a glucose flux of 1.06 units, a lactate flux of 1.2 units and an oxygen flux of 3.3 units. Since the supply of oxygen needed is so high, the maximum rate of aerobic work in many animals may in practice be limited by the capacity of the heart and circulatory system to deliver oxygen-loaded blood to the tissues.

6.4.4 Control of aerobic metabolism; metabolic depression

Most of the enzymes in the Krebs cycle, the β-oxidation spiral and the ETS do not seem to occur as distinctly different isozymes, and control is exerted mainly by enzyme–substrate interactions and feedbacks. The major control points of the Krebs cycle are illustrated in Fig. 6.8. Highly phosphorylated molecules such as ATP can reduce the rate of formation of acetyl-CoA by diminishing the activities of citrate synthase, isocitrate dehydrogenase and α-ketoglutarate dehydrogenase. The activity of the pyruvate dehydrogenase complex is also regulated, by a combination of reversible phosphorylation, feedback regulation by nucleotides, and product inhibition. Enzyme phosphorylation is an important mechanism for coping with hypoxia and anoxia; key glycolytic enzymes can be 'turned down' by phosphorylation, with their substrate and activator affinities altered,

and the effect can be magnified by changing the concentration of the allosteric activators such as fructose 2,6-biphosphate. By such means, a coordinated metabolic depression right through the glycolysis/Krebs cycle/ETS pathway results. This 'hypometabolic state' is used in many animals faced with poor oxygen supply, dehydration, increased acidity or alkalinity, or freezing.

6.4.5 Aerobic metabolism in insects

Insects are a notable exception to some of the patterns of metabolic fuelling outlined above. Insect flight is a very energy-demanding process, and one which is supplied almost entirely by aerobic metabolic pathways. The maximum rate of oxygen consumption increases 50-fold or more between rest and take-off, and magnetic resonance spectroscopy in resting and flying locusts has shown a 600-fold increase in ATP production. However, relatively little energy is stored in the flight muscles themselves. The most important energy store is the fat body, which contains fat (triacylglycerol) and glycogen, and these are broken down and delivered to the muscles via the haemo-lymph. The direct substrates for aerobic respiration are the disaccharide trehalose, the lipid diacylglycerol, and the amino acid proline. Carbohydrate is used exclusively in many Diptera and Hymenoptera, whereas lipids (the fatty acid chains of di-acylglycerol) are the main fuels in Orthoptera and Lepidoptera. Proline is used as a fuel in insects such as the tsetse fly (*Glossina*) and the Colorado potato beetle (*Leptinotarsa*), where it is partly oxidized to alanine (Fig. 6.15). The alanine is transported to the fat body for the re-synthesis of proline, and the acetyl-CoA required is derived from fatty acids. Thus proline-supported insect flight can also deplete lipid stores. Insect flight muscles contain very low concentrations of lactate dehydrogenase, and lactate is of minimal importance as a fuel. The flight muscles are also unusual in containing only low levels of phosphagens such as phosphoarginine. This is because insect flight muscle can shift between its resting and maximal rates of ATP production more rapidly than the tissues of any other animals, due to its radically different oxygen transport system (see Chapter 7). It therefore does not require significant buffering of ATP.

As a rule of thumb insects that are oxidizing exclusively carbohydrates usually only make short flights; for example, honey-bees (*Apis*) rarely travel more than 3 km from their hives. Lipid stores contain more energy per unit mass than carbohydrate stores and are therefore preferable for long flights, as in migrating monarch butterflies. In many cases carbohydrate is used at and soon after take-off, followed by lipid during prolonged flight. For example, during a continuous flight lasting 10 h, which requires a metabolic rate of 280 J per gram body mass per hour, a locust would have to oxidize 500 mg of glycogen, equivalent to 35% of its body mass. In contrast, a similar flight requires the oxidation of only 70 mg lipid. In other words, on average, 1 mg of lipid delivers as much energy as about 8 mg of glycogen. This difference is related to the higher energetic value

Fig. 6.15 Proline as a fuel.

of lipid relative to carbohydrate and to the fact that glycogen is stored in cells in association with water.

The mobilization of fuels in insects is under the control of hormones secreted from the glandular lobe of the corpus cardiacum (part of a neurosecretory complex just behind the brain). These hormones are octapeptides and decapeptides, with sequences that are highly conserved between different insect species. The hormones activate a lipid shuttle whereby high-density lipoproteins are used to transport digested lipid from the gut to tissues and to and from storage sites. In addition octopamine (a short neurosecretory peptide) is probably involved in the rapid mobilization of diacylglycerol during the first few minutes of insect flight. The whole system is supported by oxygen delivery via the tracheoles (see Chapter 7), which in insects is so effective that the *in vivo* cytochrome oxidase turnover rates approach those of

the theoretical maximum (whereas in vertebrates the cytochrome oxidase only works at about 10% of its theoretical maximum).

6.5 Metabolic rates

Metabolic rate is one of the most commonly measured physiological variables, and is a measure of the total energy metabolized by an animal in unit time. It has proved immensely useful in comparative studies of animal adaptation and performance. But despite its wide usage, its value can be determined only indirectly, from any one of four kinds of analysis:
1 the energy value of food taken in minus that of the wastes excreted;
2 the amount of oxygen used up (or carbon dioxide produced);
3 the amount of heat produced;
4 the amount of metabolic water produced.

Of these, the usage of oxygen is most easily measured, and rate of **oxygen consumption** ($\dot{V}o_2$) has therefore been the routine measure of metabolic rate in physiological studies. It does not of course take account of any anaerobic metabolism, so can be very misleading for some kinds of animal. However, oxygen consumption does correlate closely with **heat production**, whatever the kind of foodstuff being metabolized (Table 6.4), with an average value of 20 kJ of heat per litre of oxygen. Conversions between oxygen usage and heat production are therefore relatively simple. Heat production can be assessed by whole animal calorimetry (Fig. 6.16), and calorimeters can be calibrated with some fuel of known energy content (Table 6.4).

Conversions between these two parameters and the first measure given above, the **food value taken in**, are less straightforward because we know that different foodstuffs have different calorific values (see above); fat yields more than twice as much energy per gram used up, in comparison with carbohydrates or proteins (Table 6.4). This makes fat a very good energy store for mobile animals to whom body weight matters, as it provides greater energy per unit weight of body tissue (though of course this stored fat still requires as much oxygen to release the energy as any other store would do). The difficulty of using food intake as a measure of metabolic rate in situations where diet composition may not be known in great detail is partly overcome by use of the **respiratory quotient** (RQ). This is the ratio (fairly simple to measure) of carbon dioxide released to oxygen used up, and it differs for each of the main nutrient categories (Table 6.4). An RQ value close to 0.7 indicates that fat is principally being metabolized, whereas an RQ approaching 1.0 indicates carbohydrate-based metabolism. Intermediate values are more confusing, with several possible combinations of substrate, but even these can be disentangled if the amount of nitrogen excretion (ammonia, urea or uric acid) is also measured, as this gives a value for protein metabolism; with this taken out, the proportions of the remainder that are due to fat and carbohydrate, respectively, can be calculated.

The final technique for assessing metabolic rates, from **metabolic water production**, is in practice rather difficult, since a small amount of metabolic water is diluted into the large water

Nutrient	Heat production		RQ (CO₂ formed /O₂ used)	Metabolic water (g per g food)
	kJ per g consumed	kJ per l O₂ consumed		
Carbohydrates	17.4	20.9	1.00	0.56
Lipids	39.3	19.6	0.71	1.07
Proteins	17.8	18.6	0.80	0.4–0.5

Table 6.4 Energy content, respiratory quotient (RQ), and metabolic water production of foods.

content of the whole animal. However, improvements in isotope trace techniques (e.g. using doubly labelled water) have facilitated these kinds of analyses, and they are increasingly being used. Again the release of water depends on the nature of the substrate being metabolized, so that corrections as outlined above must be made.

In theory then, and to a large extent in practice, metabolic rates can be measured for any aerobically respiring animal. There are, however, a number of problems of definition and measurement inherent in making any comparative analyses of metabolic rates in animals.

6.5.1 Metabolic rate and activity

The first complication in measuring an animal's metabolic rate is that it is substantially affected by levels of rest and activity. An active fish or crustacean may have a metabolic rate several times that of a resting individual, whilst for flying insects the change

Fig. 6.16 Basic whole-animal calorimetric apparatus. (Adapted from Kleiber, M. *The Fire of Life: An Introduction to Animal Energetics*. Copyright © 1961. Reprinted by permission of Wiley-Liss, Inc., a subsidiary of John Wiley & Sons, Inc.)

may be more than 100-fold (partly because body temperature rises sharply during activity). To complicate matters further, in certain kinds of activity such as diving, metabolic rates may actually be temporarily reduced (see Chapter 9).

Therefore several levels of metabolism (and metabolic rates) are commonly recognized:

1 Standard metabolic rate (SMR) is the level required for a minimal resting lifestyle with no spontaneous activity, no digestion of food, and no physical, thermal or psychological stress. In ectotherms this is temperature-dependent. In an endotherm, where this standard metabolism is not varied with temperature (see later), the term **basal metabolic rate** (BMR) may be used instead, and must be measured in the thermoneutral zone (see Chapter 8).

2 Routine metabolic rate (or 'resting metabolic rate') is the level of metabolism for minimal (but normal and unrestrained) activity, and is rather loosely defined in practice.

3 Average daily metabolic rates may be worked out for the routine activities of a 24-h period and this is sometimes termed **field metabolic rate**; or a metabolic rate may be calculated in laboratory studies over a long enough time period to ensure that body mass (M_b) remains constant, food inputs balancing metabolic outputs, giving the **sustained metabolic rate**. Field and sustained metabolic rates are probably the most useful parameters to characterize an animal's normal energy expenditure.

4 Active metabolic rate is the level required for specified levels of exercising activity (forced if necessary), up to the maximum possible level of **maximum metabolic rate** (MMR).

Measuring standard metabolism directly is usually impractical, as even in a darkened chamber most animals show some spontaneous activity. Therefore it is necessary to record a series of metabolic rates along with an estimate of the activity level during the recording, and extrapolate back from these to the standard metabolic rate at zero activity. For laboratory measurements of routine or sustained or active metabolism the actual level of activity during the recording has to be carefully specified, and interspecific comparisons are therefore particularly difficult. It is crucial to control for stress, feeding history, acclimation temperature, activity and body size (some of these aspects are considered in more detail below) and to compare like with like in terms of developmental stage or age, and lifestyle.

In mammals, the range of possible metabolic rates, from basal to maximum, may be very large (Table 6.5). The ratio between maximum and basal rates of oxygen consumption is termed the **factorial aerobic scope**; the difference between the two levels is the **absolute aerobic scope**. Factorial aerobic scopes are easier to use in comparisons between species, and are usually in the range 5–12 for vertebrates, though larger and more athletic mammals may have somewhat higher values. Invertebrates tend to have values in the range 2–10, the lower values occurring in smaller and less active species and in cool aquatic habitats where body temperatures during activity are lower; some terrestrial invertebrates (especially flying insects) may have much larger aerobic scopes. Insects are very different in design and in

Table 6.5 Resting and peak metabolic rates (MR) and factorial aerobic scopes.

Species	Resting /standard MR (ml O_2 g^{-1} h^{-1})	Active /peak MR (ml O_2 g^{-1} h^{-1})	Factorial aerobic scope
Invertebrates			
Barnacle	14	58	4
Periwinkle	0.2	0.4	2
Fruit fly	2.3	30	13
Butterfly	0.6	100	170
Fish, reptiles			
Sunfish	0.03	0.29	9
Salmon	0.08	0.60	8
Monitor lizard	0.08	0.38	5
Desert iguana	0.05	0.89	17
Turtle	0.03	0.64	20
Birds, mammals			
Budgerigar	1.7	18.9	11
Hummingbird	2.8	42.0	15
Mouse	2.5	20.0	8
Dog	0.33	4.0	12
Human	0.23	3.2	14
Elephant	0.11	1.2	11
Whale	0.04	0.6	16

metabolic responses from most other animals, and they may have very large factorial aerobic scopes during flight, with values commonly in the range 80–120 for large bees and moths, whose body temperature (T_b) is considerably elevated when flying; even when temperature-corrected these factorial scopes may be about 30–50, reflecting the huge metabolic rates of insect flight muscles.

Comparisons between sustained metabolic rates and basal metabolic rates are also of interest, and in mammals this value (the **sustained metabolic scope**) ranges between 1.3 and 7.0, with a value of 4.3 in the best highly trained human athlete. The apparent ceiling of about 7 on this parameter may be due to constraints in the energy-consuming machinery (mainly the muscles) or in the energy-supply machinery (gut, liver, circulation or lungs perhaps). Or it could be that animals have evolved with the capacities of some or all of these limiting factors being matched to each other. The fact that sustained metabolic scope may differ for different kinds of energy expenditure in the same animal (e.g. in mice it is 3.6 for physical exercise, 4.8 for heat production and 6.5 for lactation) suggests that more peripheral limitations on the energy-consuming systems may be most important. However, the observation that sustained metabolic rate tends to be coupled with routine metabolic rate (perhaps because increases in the former demand increased masses of energy supply organs, which in turn require high maintenance costs that disproportionately raise routine levels) complicates the whole issue.

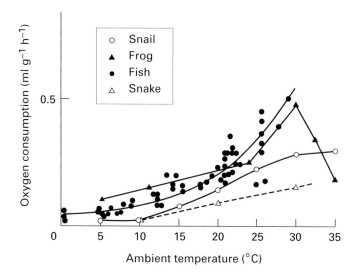

Fig. 6.17 Relationship between ambient temperature and metabolic rate (as oxygen consumption) for four ectotherms; the Q_{10} is commonly 2–3. (Data from Haugaard & Irving 1943; Aleksiuk 1971; Jusiak & Poczopko 1972; Santos *et al.* 1989.)

6.5.2 *Metabolic rate and temperature*

It is unusually difficult to measure the metabolic rate of an animal in relation to changing thermal conditions, not least because the animal itself continually alters the imposed experimental regime. Rates of oxygen consumption are the most commonly measured variable, and these will certainly show variation in response to a sudden shift of temperature. But the response will be erratic initially (as the animal's behaviour alters) and will often overshoot and then oscillate for a period of several hours or even days; even in quiescent stages such as eggs or pupae, or in anaesthetized animals, these overshoots occur, suggesting that they reflect perturbations to an underlying cellular steady state.

To be useful, then, experimental analyses involving an acute change of temperature must carefully control for handling and shock effects, and must last long enough to reach the new stabilized condition, which may be particularly difficult for sudden large temperature changes. This immediately introduces a further problem, though, as animals will commonly acclimate to a changed thermal regime by changing their own metabolic rate in compensation for the direct rate-changing effects of temperature; and the extent to which they can do this may be variable through a season, or may change in relation to the immediate history of the particular animals tested. As these acclimatory changes may take only a few days to implement, they can counteract the effect being studied, and reduce the slope of the rate–temperature curve that we are trying to determine. Any study is therefore a compromise between avoiding the influence of the initial shock reaction, and avoiding the onset of physiological adjustments by the experimental animal.

Metabolic rate commonly has a Q_{10} of about 2–3 for the majority of animals; i.e. a 10°C increase of ambient temperature

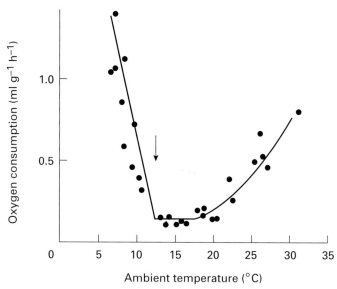

Fig. 6.18 Relationship between ambient temperature and metabolic rate (MR) (as oxygen consumption) for an endotherm that undergoes torpor. MR declines to a new level during torpidity, but increases sharply again if ambient temperature falls below a species-specific minimum (arrow).

produces a 2–3-fold increase in metabolic rate (Fig. 6.17; see also Chapter 8, p. 193). Values may vary depending on whether the 10°C range studied is distant from the normal environmental temperatures of the species studied; some fish and crustaceans from polar seas appear to have a relatively high Q_{10} for a temperature range around 0°C but a reduced Q_{10} as the test temperatures approach their lethal limits (which may be only 4–8°C).

For endotherms (primarily the mammals and birds) the normal Q_{10} value of 2–3 is obscured, because these animals control their body temperature so precisely. In such cases, only when the animals are in states of metabolic depression, either torpid, hibernating or aestivating, does the underlying effect of changing temperature on metabolism become apparent (Fig. 6.18).

6.5.3 *Metabolic rate and size*

Many of the component processes that contribute to metabolic demand in a resting animal (e.g. ventilation rate, heart rate) are related to body weight. Therefore, estimates of standard metabolic rate and of active metabolic rate always strictly have to be compared between individuals of the same size. In other words, as in so many areas of comparative biology, scaling effects must be very carefully considered.

The relationship between metabolic rate and body mass, both interspecifically and within a species, remains one of the most important and controversial issues in comparative animal physiology. If we look initially at the pattern just for the mammals, a group which has inevitably attracted most attention, it is obvious that a large elephant has a higher oxygen consumption than a

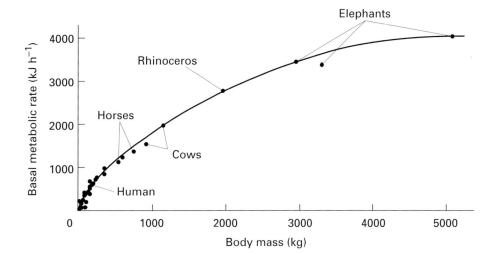

Fig. 6.19 Linear plot of metabolic rate against mass.

small mouse. The problem concerns the proportionality of this relationship, which takes the general form of a power curve:

metabolic rate $= a$ mass^b

On a simple plot, a curve results with a crowded left-hand side (Fig. 6.19) where b is the mass exponent of the curve. It is more useful in practice to take logarithms of the metabolic rate and body mass values, so that the relationship becomes:

log metabolic rate $=$ log $a + b$ log mass

The persistent controversy concerns whether or not this log plot is a straight line, and what the value of b, the gradient, might be. An example of the log plot is shown in Fig. 6.20. Most of the

analyses that have been done for mammals give a value for b of about 0.75 (see Table 6.6).

Whatever the true relation here, this kind of plot has its uses in revealing the 'aberrant' physiological performances of atypical mammals. Any species whose oxygen consumption is notably greater or less than that for other animals of the same size becomes of special interest to the comparative biologist, and reasons for the anomaly can be sought. The same is true intra-specifically; a value close to (often rather lower than) 0.75 is usually obtained within any one species (Fig. 6.21), and outlying individuals can again provide interesting insights.

Moving to a broader scale, we can do the same kind of comparison for the whole of the animal kingdom, and an example

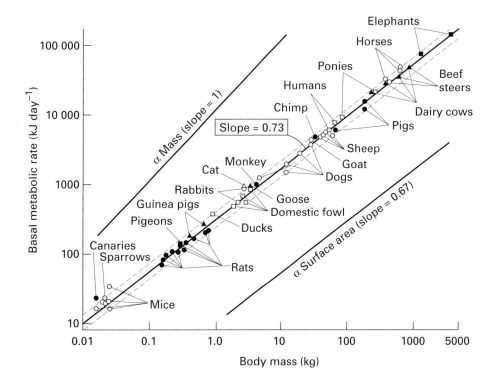

Fig. 6.20 Log plot of metabolic rate against mass for birds and mammals. (Data from Kleiber 1932; Heusner 1982.)

Table 6.6 Mass exponents for the relationship between mass and metabolic rate.

Animal group	Temperature (°C)	Mass exponent (*b*)
Ectotherms		
All combined	20	0.80–0.88
Nematodes		0.72
Annelids		0.61–0.82
Molluscs		0.75
Echinoderms		0.65–0.80
Crustaceans		0.81
Spiders		0.65
Insects		0.62–0.67
Reptiles		0.77–0.80
Snakes		0.81–0.98
Amphibians		0.75–0.86
Fish		0.70–0.88
Endotherms		
All birds	41	0.69–0.75
Passerines		0.72–0.73
Honey-eaters		0.68
Ratites		0.73
All mammals	39	0.73–0.76
Marsupials		0.75
Edentates		0.66
Bats		0.72
Rodents		0.72
Shrews		0.75

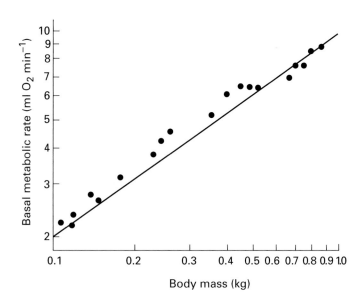

Fig. 6.21 Log plot of metabolic rate vs. mass for a single species, the guinea pig. The line drawn has a slope of 0.67. (From Wilkie 1977.)

of this kind of plot was used in Chapter 3 when we considered scaling problems in general. It turns out that all organisms have a similar relationship between body mass and metabolic rate, though the slope is set within somewhat narrower bounds for birds and mammals (0.70–0.75) than it is for other animals (0.6–0.9 range, with many invertebrate groups having a mass exponent of 0.66–0.72, see Table 6.6), and it is somewhat lower again for unicellular organisms than for the multicellular animals. The intercept values, *a*, are also of course lower for the majority of animals that cannot generate much heat internally (the ectotherms, see Chapter 8), with mean values about 3 J h⁻¹, than they are for endothermic mammals and birds (mean 116 and 139 J h⁻¹, respectively; see also Fig. 6.22).

The metabolic differences between ectotherms and endotherms seem to be explained by a number of small structural and physiological differences rather than by any one factor. For example, cells in endotherms have somewhat greater mitochondrial volumes, and slightly greater membrane areas per unit of mitochondrion, with raised enzyme activities, and rather more 'leaky' membranes around both cytoplasm and mitochondria. These animals also have proportionately larger organs (liver, kidneys and brain in particular).

As to why metabolic rates always show a mass exponent of 0.75 in relation to body mass, there are several hypotheses but

no clear-cut answers. Metabolic rate could theoretically be proportional to surface area (mass$^{0.67}$) because crucial metabolic processes and exchanges are occurring across surfaces, but it could also be proportional to volume (mass$^{1.0}$) since catabolic reactions occur and heat is generated within all the three-dimensional volume of the animal. Perhaps the exponent 0.75 represents a balance of these two effects. If so, it is a little unclear why the value of 0.75 should be found in ectotherms as well as endotherms, since the balances of body temperature, heat exchange and surface area differ substantially for the two strategies (see Chapter 8). However, support for this explanation based on a balance of surface-related and volume-related factors comes from considering the scaling of purely anaerobic metabolism; here processes occur entirely within cytoplasm, requiring no surface-related exchanges, and the mass exponent for a range of animals is close to 1.0, as would be expected for purely volume-related effects.

However, other explanations for the aerobic mass exponent have been offered, bringing in mechanical considerations, fractal scaling factors and the effects of four-dimensional rather than three-dimensional scaling (perhaps with time as the fourth dimension); these have not yielded any real new insight into the problem, but the debate continues. It may be that there are different explanations for the 0.75 figure applying between thermal strategies, between taxa and between individuals.

Given these controversies about metabolic rate, it is clearly often more useful to compare the metabolic rate of different animals per unit of body mass by determining the **mass-specific oxygen consumption** (mass-specific metabolic rate). This shows that a small mammal actually has tissues that are very much more metabolically active than a large mammal (Table 6.7); the rate of oxygen consumption per gram of tissue *decreases* consis-

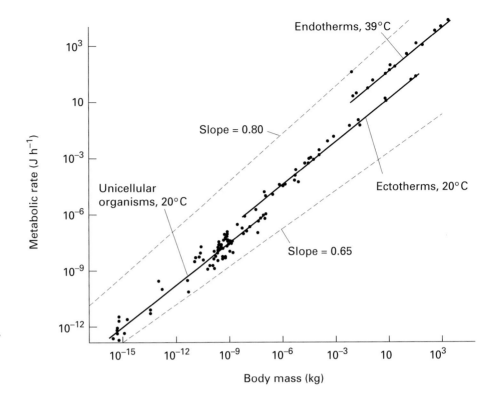

Fig. 6.22 Log plot of metabolic rate vs. body mass for unicells and for both ectothermic and endothermic animals. (From Phillipson 1981.)

Table 6.7 Mass-specific oxygen consumption in mammals.

Species	Mass (kg)	O_2 consumption (ml g^{-1} h^{-1})
Shrew	0.005	7.40
Harvest mouse	0.009	1.50
Kangaroo mouse	0.015	1.80
House mouse	0.025	1.65
Ground squirrel	0.095	1.03
Rat	0.290	0.87
Cat	2.5	0.68
Dog	11.7	0.33
Sheep	43	0.22
Lion	50	0.23
Human	70	0.22
Eland	240	0.17
Elephant	3 850	0.07
Blue whale	100 000	0.04 (est.)

tently as body weight increases (Fig. 6.23). This semi-logarithmic plot indicates that the mouse tissue has about 100 times the oxygen consumption of the elephant tissue (around 7 ml O_2 g^{-1} h^{-1} for mice and shrews, and only 0.07 ml O_2 g^{-1} h^{-1} for an adult elephant). Hence the blood flow supplying the oxygen must be about 100 times greater per unit of tissue in the mouse, and all related variables of the cardiovascular systems, such as heart function and respiratory rate, must be similarly scaled up for the mouse. Even more critically, the specific food intake of a mouse must also be 100 times greater, whence the familiar observation

that small mammals almost never stop eating during their waking hours.

The scaling of maximum metabolic rate in mammals suggests that mass-specific oxygen consumption scales to body mass to the power −0.25. Extrapolation of the data to blue whales weighing 100 000 kg would suggest a mass-specific oxygen consumption of only 0.02–0.04 ml O_2 g^{-1} h^{-1} (though such extrapolations are not strictly valid, as we discussed in Chapter 3). What are the biochemical consequences of these large differences? Are there upper and lower limits to energy metabolism that are related to body size?

It has been shown for fish, birds and mammals that mass-specific enzyme activities in heart and in skeletal muscle scale in the same fashion as the mass-specific oxygen consumption of the whole animal, varying with body mass to the power −0.21. Data for birds and mammals are shown in Fig. 6.24a. The roughly parallel scaling of maximal oxygen demands (Fig. 6.24b) and maximal enzyme activities ensures that the carbon flux through an animal can be increased during intense activity, in all species of all sizes. Substrate concentrations are thus essentially scale independent, and are kept within the range for maximum sensitivity of regulation at all body sizes.

The cellular units of aerobic metabolism are the mitochondria (see section 6.3). The fraction of muscle fibre volume (the volume density) that is occupied by mitochondria also scales with a slope of −0.21 on double-logarithmic plots. For most species it would appear that 1 ml of muscle mitochondria consumes about 200 μmol O_2 min^{-1} during maximal exercise, regardless of the oxygen source. Since specific oxygen consumption increases

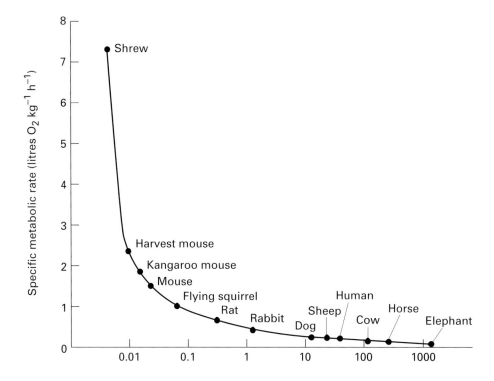

Fig. 6.23 Semi-logarithmic plot of specific oxygen consumption against body mass for mammals—the 'mouse to elephant curve'. (From Schmidt-Nielsen 1997; courtesy of Cambridge University Press.)

with decreasing size, this implies that the volume density of muscle mitochondria also increases systematically with decreasing body size. However, there is also an effect of lifestyle, in that for a given body size more athletic species have a higher volume of mitochondria per unit muscle mass; for example, the ratios of maximum oxygen consumption to body mass in dogs and horses are 2.5 times greater than in goats and cattle, respectively, matched by 2.5 times more mitochondria per unit mass in their muscles.

Clearly, in the tiniest mammals and birds there must be some limit to the number of mitochondria that can be packed into a muscle fibre without compromising its function for doing external mechanical work. The packing of mitochondria in flight muscles of hummingbirds (birds that can hover) appears to be close to the upper limit possible in vertebrates, at around 50–55%. Volume densities of mitochondria higher than this would probably be counterproductive, since this would decrease the space available for myofibrils and sarcoplasmic reticulum, the structural elements required for force generation and calcium cycling within the muscle fibres. Hovering flight is energetically expensive, requiring wing beat frequencies of 60–80 Hz, and metabolic rate is very high due to the small size of the birds (1–2 g), so it is logical that small hovering hummingbirds should be at the upper limit of mitochondrial packing.

Double-logarithmic plots of the surface density of the inner cristae membranes of mitochondria (which contain the electron transport chain assemblages, see section 6.3) against body mass also scale to about the power −0.21, but there are some very interesting exceptions, again including hummingbirds. Mitochondria from hummingbird flight muscles have around twice the surface density of cristae and hence twice the rate of oxygen

consumption of other birds of similar size. This can be viewed as an evolutionary design innovation to overcome the general constraints imposed by the scaling of the structural elements responsible for oxygen uptake. The increased density of cristae effectively doubles the maximal rate of oxygen consumption that can be achieved with a volume density of 50% mitochondria.

However, there is evidence that the real limit to aerobic capacity in hummingbirds is not at the enzyme level but is related to the cardiovascular system's ability to deliver oxygen-loaded blood to the capillary circulation. In mammals the rate of oxygen delivery per ml of capillary is practically invariant, but capillary density scales negatively with size (and thus positively with mass-specific metabolic rate) over the size range 20 g (mice) to 500 kg (horses). Insects, with their unique tracheal system for delivering oxygen, have overcome these constraints in an entirely novel way, and achieve much higher levels of mass-specific oxygen consumption than birds and mammals.

The limits set on muscle activities can be overcome in other ways, but only at some cost. Some animals do have muscles that contract much faster than hummingbird flight muscles, but at the expense of the work done per contraction cycle. For example, the tymbal (sound-producing) muscles of cicadas operate at a contraction frequency of 550 Hz. The volume densities of mitochondria, sarcoplasmic reticulum and myofibrils are in the ratio of approximately 1 : 1 : 1 (i.e. the mitochondrial volume density is only about 33%). A very high frequency, required to produce the necessary sound, has been achieved, but because there are more of the energy-supply systems there are fewer myofibrils, and the gain in frequency of contraction is at the expense of the muscles' ability to do mechanical work.

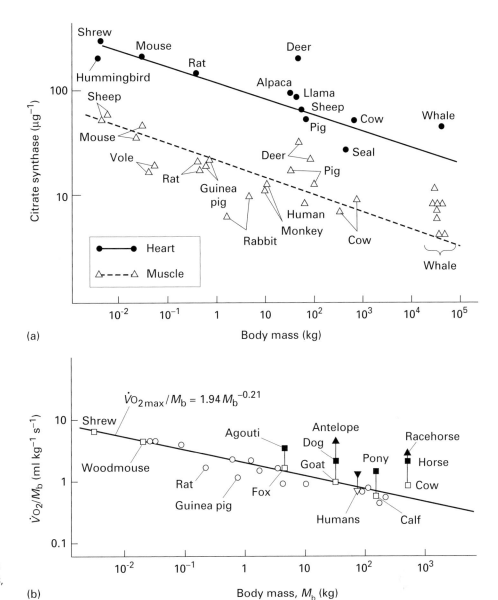

(a)

(b)

Fig. 6.24 Scaling of (a) mass-specific enzyme activities in bird and mammal skeletal muscles and heart muscles, and (b) maximal oxygen demands, with body mass. In (b) the more athletic species (closed symbols) always have values of $\dot{V}_{O_{2max}}$ well above those for sedentary species of similar body mass (for humans a sedentary adult and a highly trained athlete are compared). (a, From Hochachka *et al*. 1988; b, from Hoppeler & Weibel 1998, courtesy of Company of Biologists Ltd.)

6.5.4 *Metabolic rate and oxygen availability*

Given that metabolic rates are often assessed by measuring oxygen consumption, it is important to understand the relation between metabolism and oxygen supply (though here again we are really only considering aerobic animals). There are two main patterns to be found, corresponding to the distinctions between regulators and conformers made in Chapter 1. **Metabolic conformers** have standard (basal) metabolic rates that are directly proportional to ambient O_2. By contrast the **metabolic regulators** maintain their standard (basal) metabolic rates even when the partial pressure of oxygen (P_{O_2}, see Chapter 7) is reduced, at least down to some critical value below which their metabolism slows down markedly and they enter a hypometabolic state (as discussed in section 6.4.4). There is no obvious phylogenetic or environmental pattern to the occurrence of this strategy; we will

look at examples in more detail in Chapter 7 when we consider respiratory adaptations.

It is worth noting that metabolic rates also vary with other variables, including developmental stage and age (usually increasing through development, and declining in old age), food availability, diet and exercise regime, photoperiod (with a daily or 'diel' patterning), salinity, and hormonal balance. In each chapter in Section C of this book we will meet examples of these kinds of interactions. Indeed there is a common theme of **hypometabolism** associated with various physiological problems; animals may adopt a hypometabolic state ('turning down the metabolic burners') when they enter torpor daily or seasonally, when they acclimate to unusual drought or cold or heat, when they dive, when they burrow, or when they starve. The machinery underlying this response, at the biochemical level, is common to all these conditions.

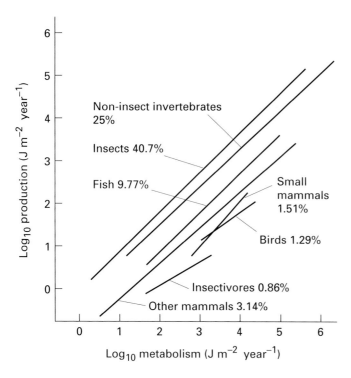

Fig. 6.25 Energy losses in metabolism: the relation between production and metabolic rate, with the mean for net production efficiency shown for each taxon. (Adapted from Humphreys 1979.)

6.6 Energy budgets

All the energy acquired by animals through the ingestion of food is ultimately either used in metabolic processes, or deposited as new body tissue (growth or reproduction), or lost as waste products. Energy budgets are used to describe the partitioning of the ingested energy into these various components. The simplest form of balanced energy budget can be represented as:

$$E_{in} = E_{out} + E_p$$

where E_{in} is the energy ingested as food, E_{out} is the sum of the energy losses, and E_p is the energy associated with production (Fig. 6.25). The equation is usually expanded to further subdivide each of the terms on each side:

$$E_{food} + E_{drink} = E_{faeces} + E_{urine} + E_m + E_p$$

where E_m represents the energy losses in metabolism. Metabolism can be further subdivided into basal metabolism (energy used to maintain bodily functions), active metabolism (energy used in locomotion and other activities), and energy used in the digestion, absorption and processing of food. The E_p term in the equation is also often partitioned into somatic growth and reproductive growth (energy expended on the production of gametes).

All components of energy budgets must be expressed in the same units, usually joules or kilojoules. Clearly ingestion rate

sets the upper limit to all other variables in the equation, so ingestion must be considered first.

6.6.1 *Foodstuffs and ingestion*

Ingested foodstuffs, whether as pure compounds or as the mixtures provided by natural foods, provide varying levels of energy content (Table 6.8). The biological materials that most animals eat have energy values in the range 15–30 kJ g^{-1}, with those of high lipid content being most valuable if seen solely as a source of energy. However, foodstuffs may not be utilized with complete efficiency by an animal, and their **apparent digestible energy** must also be considered: this is the difference between the energy that they could provide and the energy from them that is lost in the faeces. The percentage of the energy consumption that is actually assimilated is often termed the **assimilation efficiency**. This is usually much lower for herbivores than for carnivores due to the indigestibility of cellulose; by contrast it may be very high for fluid-feeders such as blood-suckers, phloem-feeders and suckling mammals, as their liquid diets are very readily assimilated (see Table 6.9).

A power function provides a good description of the variation in ingestion rate in relation to body mass. Figure 6.26 provides an example for ectothermic tetrapods, where the slope of the regression line is 0.82 and the ingestion rate of a 1 kg animal (obtained from the intercept) is 0.78 W. This implies that ingestion rate varies with body size but is relatively independent of diet. The same conclusion is reached from a comparison of the scaling of ingestion rates in endothermic herbivores and carnivores (Fig. 6.26). In this case ingestion rates scale to body mass$^{0.70}$, but the intercept is much higher at 10.7 W. The mass exponents obtained in this kind of analysis do not differ much from the 0.75 slope describing the effects of size on respiration rate (see above). The similarities of these scaling relationships suggest that species that deviate significantly from the general constraints of body size on respiration rate will show corresponding deviations in ingestion rate. For example, mammals with unusually low metabolic rates, such as sloths, should eat more slowly, and metabolically active animals, such as weasels, should eat more quickly than predicted on the basis of body mass alone.

Numerous factors influence ingestion rate, including seasonal cycles of growth, reproduction and the accumulation of fat stores, as well as many environmental variables such as daylength, oxygen availability and temperature. Acute changes in body temperature also have a major effect on the amount of food consumed in ectotherms. For example, increases in temperature initially lead to an increased rate of ingestion in fish (Fig. 6.27). Food intake peaks at some intermediate temperature, before declining markedly as the temperature approaches the upper thermal tolerance limit of the species. The energy available for growth, E_p, initially increases with temperature. However, at high temperatures the combination of a declining ingestion rate and an increasing metabolic rate markedly reduces the scope

Table 6.8 Energy contents and composition of different diets and foodstuffs.

Foodstuff	Energy		Protein (%)	Lipid (%)	Carbohydrate (%)
	KJ g^{-1} wet wt	KJ g^{-1} dry wt			
Pure carbohydrate		17	0	0	100
Pure protein		23	100	0	0
Pure lipid		38	0	100	0
Phytoplankton	~1	15–20	3	1	4
Macroalgae	~1		1	<1	4
Pulses	2–8	17–21	10	2	70
Plant sap	~0.1		0	0	2–5
Leaves	~1		2	<1	1
Fruits	1–2		~1	0	6–15
Earthworms	3		11	1	1
Crabs	4–5		13	3	2
Insects (larvae)	3–6		10–16	1–5	2
Fish (oily)	6–8	23–25	14–18	10	1
Fish (non-oily)	5–6		17	3	1
Meat (vertebrate muscle)	8–10	23–28	20	20–30	0
Milk (cow's)	2–3		3	4	5
Egg		25–28			

for growth. Seasonal temperature changes may serve to modify the responses shown in Fig. 6.27.

Terrestrial herbivores have a special problem in relation to ingestion and metabolism. Plant cell walls contain cellulose, which is not normally broken down by digestion. This material, referred to as roughage or fibre in the human diet, increases gut transit times and the volume of the faeces. The material in plant cell walls can be digested by symbiotic microbes, and animals that eat mostly plant material have evolved a variety of behaviours and structures to promote microbial fermentation; these are dealt with in Chapter 13 when we look at land animals in detail. The gastrointestinal tract (GIT) is an anoxic environment and may house symbiotic bacteria (in ruminants and other fermenters) that convert monosaccharides, mixed amino acids and fats into volatile fatty acids, methane and hydrogen. Volatile fatty acids represent a good source of energy (about 20% less than glucose on a molar basis).

The metabolic cost of processing ingested food is usually

Table 6.9 Assimilation efficiencies for different diets.

Feeding habit	Consumer	Food	Assimilation/consumption (% absorption efficiency)
Herbivory	Ectotherms, aquatic	Algae	30–70
	Ectotherms, aquatic	Macrophytes	30–60
	Ectotherms, land		40–50
	Endotherms, land		60–70
Granivory	Ectotherms, land		75–80
	Endotherms, land		70–77
Nectarivory	Ecto- or endotherms, land		95+
Carnivory	Ectotherms, aquatic	Invertebrates	65–85
	Ectotherms, aquatic	Fish	80–90
	Ectotherms, land	Flesh	85
	Endotherms, land	Flesh	85
	Endotherms, land	Milk	95
	Various	Blood	85+
Detritivory	Ectotherms, aquatic		40–45
	Ectotherms, land		10–20
Endoparasitism	Ectotherms		70–80

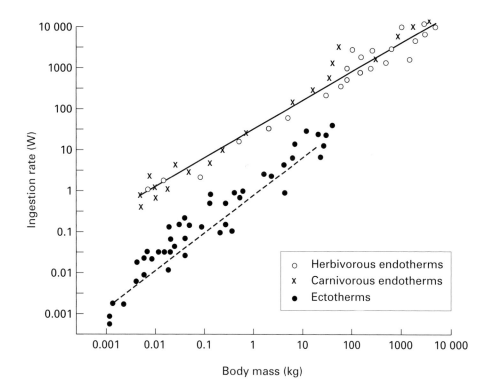

Fig. 6.26 Ingestion rate (IR) as a function of body mass for vertebrates. In ectotherms the regression line for the data is IR = 0.78 $M_b^{0.82}$, while for endotherms ingestion rate is substantially higher and IR = 10.7 $M_b^{0.70}$. (Data from Farlow 1976.)

referred to as the **specific dynamic action** (SDA) or specific dynamic effect (SDE); it leads to a raised metabolic rate after feeding, sometimes termed 'postprandial thermogenesis' (Fig. 6.28). In fasted animals, oxygen consumption rises markedly following a satiating meal; 30–40% is not uncommon in mammals, and in very infrequent feeders such as boas and pythons metabolic rate during meal processing may be as much as 20–40 times normal resting rate, more extreme than the change incurred in a mammalian sprinter and sustained for much longer. This partly reflects the low fasting metabolic rate in pythons, which is 13 times less than in a mammal of equivalent size (after correcting for differences in body temperature). The very large increase in metabolic rate arises from the costs of regulatory processes including the up-regulation of enzymes, transporters and gastrointestinal secretions, and the rapid growth of organs that have atrophied in fasting periods. For example, within 24 h of feeding the python's small intestine doubles in dry mass as a

result of a six-fold increase in microvilli length; and within 3 days there are also 50–100% increases in the masses of stomach, heart, liver, lung and kidney.

The duration of the SDA varies with the size and composition of the meal, but it appears that the metabolic costs of processing particular classes of foodstuffs are relatively independent of taxon. The ingestion of 1 kJ of protein requires approximately 310 J of additional metabolic power, regardless of whether the protein was consumed by an invertebrate, a fish, a bird or a mammal. The SDA associated with various diets varies considerably, and is generally lower for most plant foods, with the exception of nectar, pollen and seeds, which contain little or no cellulose.

In fasted individuals the relative increase in oxygen consumption after a single satiating meal is independent of body temperature. However, the duration of the SDA is markedly temperature dependent. For example, in teleost fish the dura-

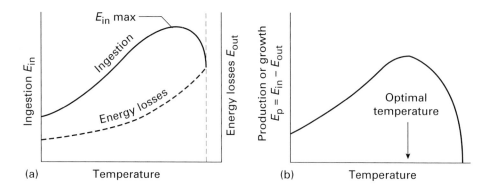

Fig. 6.27 The influence of temperature on (a) the rates of ingestion and metabolism (energy loss) in fish, where the vertical line shows the upper thermal tolerance for the species; and (b) the consequences for the available energy remaining for growth, which has a clear optimum. (From Jobling 1993, Fig. 1.2; with kind permission from Kluwer Academic Publishers.)

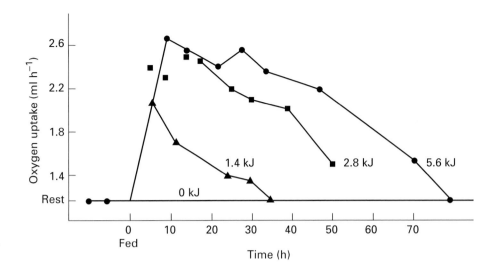

Fig. 6.28 The specific dynamic action (SDA) in terms of increased oxygen uptake after feeding at three different levels in the plaice. (From Jobling & Davis 1980.)

tion of the SDA is around four times shorter for tropical than Antarctic species (Fig. 6.29). Note that, after correcting for differences in body mass and meal size, the area under the curves in Fig. 6.29 is found to represent a similar proportion of the energy ingested for both the warm-water and cold-water fish; it simply reflects the composition of the meals, which was the same in both cases. The slow rate at which ingested food can be metabolized in Antarctic fish limits the rate at which energy can be made available for all other purposes, providing a general constraint to somatic and reproductive growth.

Fig. 6.29 The specific dynamic action (SDA) effect in fish given a satiating meal of shrimp; (a) is the polar fish *Notothenia* at –0.9°C, and (b) the tropical fish *Cirrichthys* at 25°C. The dashed line shows the fasting oxygen consumption in each case. (From Johnston & Battram 1993.)

The increase in metabolic rate with feeding is a reflection of numerous metabolic processes, including the digestion, absorption and storage of nutrients, the deamination of amino acids, the synthesis of excretory products and the increased synthesis of protein and other macromolecules associated with growth. The relative contributions of each of these processes is very difficult to quantify with any accuracy. Intravenous infusion of amino acids in vertebrates also results in an increase in oxygen consumption, which is a significant fraction of the maximum SDA observed with feeding. This suggests that the SDA largely reflects postabsorptive metabolism, and that costs associated with processing the food in the gut are relatively small. Tissue components are constantly being synthesized and degraded and it has been suggested that as much as 60–80% of the increase in metabolic expenditure associated with the ingestion of food may result from increased protein and lipid synthesis.

(a)

(b)

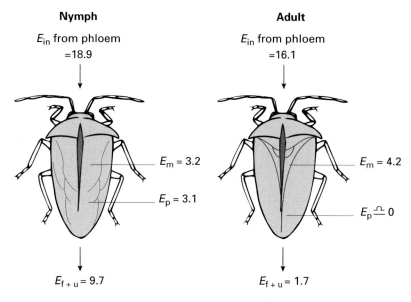

Nymph

E_{in} from phloem
=18.9

$E_m = 3.2$

$E_p = 3.1$

$E_{f+u} = 9.7$

Adult

E_{in} from phloem
=16.1

$E_m = 4.2$

$E_p \stackrel{\frown}{=} 0$

$E_{f+u} = 1.7$

Assimilation
efficiency (J day^{-1}):   48%   89%

Production
efficiency (J day^{-1}):   15% $\stackrel{\frown}{=} 0$

Fig. 6.30 Pictorial summaries of energy budgets for nymphs and adults of the common meadow spittlebug insect (*Philaenus*). (Data from Weigert 1964.)

6.6.2 *Energy losses in excreted products*

Some of the food eaten by animals is lost as a waste product, usually as a component of the faeces (though some simpler animals lack through-guts and lose the waste material in other ways). In addition to unabsorbed food the faeces may also contain mucus, cells sloughed off from the wall of the gut, bile salts, digestive enzymes and bacteria derived from the gut microflora. In some animals, particularly insects, reptiles and birds, the nitrogenous excretory products are also passed into the gut and exit the body together with the faeces; in these cases, E_{urine} and E_{faeces} cannot be distinguished.

There are a variety of other exit routes for excreted materials, including gills, skin and kidney tubules, and any or all of these may excrete forms of nitrogenous products, including amino acids, creatine, creatinine, choline, betaine and glycine. However, the most important nitrogen waste products are ammonia, urea and uric acid (see Chapter 5). Even fasted animals excrete nitrogenous end-products resulting from the transamination and deamination of amino acids that arise as a result of basal protein turnover (the endogenous nitrogen excretion). Exogenous excretion results from the direct deamination of amino acids ingested and absorbed in the food and is influenced by factors such as feeding rate, the protein content of the food, and the amino acid composition of the diet.

6.6.3 *Energy, growth and production*

Ingested energy that it is not lost as faecal or excretory products

or used for metabolism is available for growth and production. Biosynthesis is relatively cheap in energy terms, and may be around 95% efficient in principle, though this is somewhat reduced in practice by the losses due to ongoing metabolism and excretion. It can be directly assessed where embryos grow rapidly from the closed supply of stored yolk in the egg, and values for efficiency in such studies are around 50–70%.

The partitioning of ingested energy within an animal may vary with the composition and mineral content of the food. In insects, for example, the utilization of ingested energy has been shown to increase with the nitrogen content of the food. Energy partitioning changes during ontogeny as somatic growth slows, and the production term in the energy budget is then largely accounted for by the production of gametes and of secretions associated with reproduction. Figure 6.30 illustrates the energy budget derived for the meadow spittlebug (*Philaenus*). Rapidly growing nymphs have a production of about 3 J day^{-1}, whereas adult stages have a negligible production, mostly associated with the energy costs of eggs, sperm and reproductive secretions. In other terrestrial animals much greater amounts of energy are invested in reproduction (see Chapter 13).

The energy that females invest in reproduction (E_{rep}) is usually assessed indirectly by some index of **reproductive effort**. E_{rep} scales with a mass exponent, b, between 0.5 and 0.9, which means that larger species tend to invest relatively less in their offspring per unit time. But larger animals also tend to live longer: **lifespan** scales with a mass exponent of 0.15–0.29. Thus over a female's lifetime, these two factors combine, with the result that the energy invested in reproduction varies more or less isometrically with respect to body mass ($b = 1.0$). This is also a reflection of the fact that lifetime energy assimilation and energy expenditure also scale isometrically.

In species in which body mass does not change much during the reproductive phase of their lives, the energy invested in reproduction (E_{rep}) will equal the total energy assimilated minus the individuals' energy requirements excluding reproduction and growth (E_{req}):

$$E_{rep} = E_{in} - E_{req}$$

However, if females reproduce over a wide range of body mass then the mass exponent will be larger, as some of the available energy in smaller individuals will go to growth (E_{gro}):

$$E_{gro} + E_{rep} = E_{in} - E_{req}$$

It is often of great interest to measure the growth rate of individual species in ecology, agriculture and fisheries.

Age at maturity can be related to **growth rate** and **generation time**. Since growth scales to $M_b^{0.7}$ it follows that age at maturity would be expected to scale between species according to the relationship:

$$M_b/M_b^{0.7} = M_b^{0.3}$$

Studies of generation time and age at maturity in types of organism as diverse as viruses and giant sequoia trees, and including a wide range of metazoans, have found that, as expected, these variables scale with body mass to the power 0.21–0.33.

Mortality, **production** and **life history theory** are also interrelated because of links with energy budgets. The growth and reproductive characters of animals show genetic variation within populations and can be subject to selection. However, any gains in fitness achieved by increasing rates of growth and reproductive output may be at the expense of an increased risk of mortality. It would appear that there is some kind of trade-off between personal survival and production, such that it is not advantageous to select for maximum production. For example, it would be possible to maximize production by more intense feeding, but such a change in behaviour may bring increased risks of predation leading to increased mortality. Interspecific studies have shown positive correlations between growth rate and mortality for echinoderms and for fish. Intraspecific correlations for these variables have also been demonstrated for aphids and springtails.

It is also worth considering the influence of **genotype** on maintenance energy requirements and production. Many enzymes occur in a number of polymorphic forms within populations (see section 2.4). By scoring the number of individuals in a population that are heterozygous for polymorphic alleles, an index for heterozygosity can be calculated. Positive correlations between maintenance energy requirements and heterozygosity have been demonstrated in molluscs, salamanders and fish, particularly under conditions of energy limitation.

Marine bivalves show particularly high levels of genetic variation on both micro- and macro-geographical scales, with

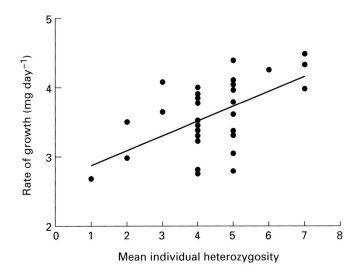

Fig. 6.31 The relationship between heterozygosity (mean number of heterozygotes per individual at seven sites) and growth rate for the mussel *Mytilus galloprovincialis*. (From Bayne & Hawkins 1997; courtesy of University of Chicago.)

between 12 and 35% of enzyme loci showing allelic variation. The common edible mussel, *Mytilus edulis*, has been the subject of many studies on the relationship between genotype and energy budgets. Changes in the growth rate and mortality of *Mytilus* following reciprocal transfer between populations have shown that genetic differences can account for up to 27% of the variation in production. Several studies have shown positive correlations between growth rate and heterozygosity at glycolytic and lysosomal enzyme loci. Typically, heterozygosity at the polymorphic enzyme loci can explain around 10–30% of the variation in individual growth rates (Fig. 6.31). Body proteins are continuously being synthesized and degraded, both energy-consuming processes. Protein turnover costs are thought to represent about 20–30% of the total maintenance energy requirements in ectotherms. There is a trend for the more heterozygous individuals to have lower rates of protein turnover and hence reduced energy requirements for maintenance, which means that more of the ingested energy is available for growth. In fact relatively small differences in maintenance energy requirements can have a major impact on energy balance. Simulations have shown that a 20% decrease in maintenance energy requirements relative to a 'normal' mussel could theoretically double the cumulative production over a 10-year period (Fig. 6.32). Mussels with the lowest rates of protein turnover also have the lowest rates of both protein synthesis and protein degradation, and show the best rates of survival under a range of stressful environmental conditions studied. It is possible that faster rates of protein turnover in more homozygous mussels may be advantageous under other conditions (e.g. by facilitating metabolic regulation, detoxification and/or replacement of damaged proteins), although this has yet to be established.

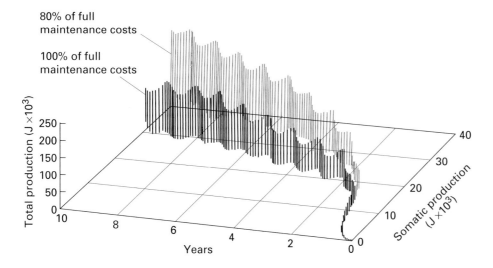

Fig. 6.32 Simulated effects of a 20% reduction in energy requirements for maintenance in *Mytilus edulis* (the common mussel) upon its performance over 10 years after first settlement. Energy is partitioned between growth, reproduction and storage depending on the amount already stored and current energy balance. After 10 years the mussels with only an 80% energy requirement have achieved a one-third increment in somatic production and almost double the total production. (From Hawkins *et al.* 1989.)

The genetic basis for these correlations between heterozygosity and individual differences in protein turnover, energy expenditure and growth are unknown, although there is evidence that it reflects functional differences in the studied genes rather than merely indicating linked loci. Here again we have an area where molecular and genetic analyses may soon give us a better handle on understanding the nature of environmental adaptations.

Further reading

Books

Alexander, R.McN. (1999) *Energy for Animal Life*. Oxford University Press, Oxford.

Blaxter, K. (1989) *Energy Metabolism in Animals and Man*. Cambridge University Press, Cambridge.

Hochachka, P.W. & Guppy, M. (1987) *Metabolic Arrest and the Control of Biological Time*. Harvard University Press, Cambridge, MA.

Hochachka, P.W. & Somero, G.N. (1984) *Biochemical Adaptation*. Princeton University Press, Princeton, NJ.

Jobling, M. (1993) Bioenergetics—food intake and energy partitioning. In: *Fish Ecophysiology* (eds J.C. Rankin & F.B. Jensen). Chapman & Hall, London.

Martin, B.R. (1987) *Metabolic Regulation: A Molecular Approach*. Blackwell Scientific Publications, Boston, MA.

Prosser, C.L. (1991) *Environmental & Metabolic Physiology*. Wiley-Liss, New York.

Salway, J.G. (1999) *Metabolism at a Glance*, 2nd edn. Blackwell Science, Oxford.

Stryer, L. (1988) *Biochemistry*. Freeman, New York.

Townsend, C.R. & Calow, P. (1981) *Physiological Ecology: An Evolutionary Approach to Resource Use*. Blackwell Scientific Publications, Oxford.

Reviews and scientific papers

Balban, R.S. (1990) Regulation of oxidative phosphorylation in the mammalian cell. *Journal of Applied Physiology* **73**, 737–743.

Elgar, M.A. & Harvey, P.H. (1987) Basal metabolic rates in mammals: allometry, phylogeny and ecology. *Functional Ecology* **1**, 25–36.

Fothergill-Gilmore, L.A. (1986) The evolution of the glycolytic pathway. *Trends in Biochemical Sciences* **11**, 47–51.

Guppy, M., Fuery, C.J. & Flanigan, J.E. (1994) Biochemical principles of metabolic suppression. *Comparative Biochemistry and Physiology B* **109**, 175–189.

Hawkins, A.J.S. (1995) Effects of temperature change on ectotherm metabolism and evolution: metabolic and physiological interrelations underlying the superiority of multi-locus heterozygotes in heterogeneous environments. *Journal of Thermal Biology* **20**, 23–33.

Hochachka, P.W. (1997) Oxygen—a key regulatory metabolite in metabolic defence against hypoxia. *American Zoologist* **37**, 595–603.

Hochachka, P.W., Buck, L.T., & Doll, C.J. (1996) Unifying theory of hypoxia tolerance: molecular/metabolic defense and rescue mechanisms for surviving O_2 lack. *Proceedings of the National Academy of Sciences of the United States of America* **93**, 9493–9498.

Poppitt, S.D., Speakman, J.R. & Racey, P.A. (1993) The energetics of reproduction in the common shrew (*Sorex araneus*): a comparison of indirect calorimetry and the doubly labeled water method. *Physiological Zoology* **66**, 964–982.

Secor, S.M. & Diamond, J. (1998) A vertebrate model of extreme physiological regulation. *Nature* **395**, 659–662.

Speakman, J.R., McDevitt, R.M. & Cole, K.R. (1993) Measurement of basal metabolic rates: don't lose sight of reality in the quest for comparability. *Physiological Zoology* **66**, 1045–1049.

Storey, K.B. & Storey, J.M. (1990) Metabolic rate depression and biochemical adaptation in anaerobiosis, hibernation and estivation. *Quarterly Review of Biology* **65**, 145–174.

van Waarde, A. (1991) Alcoholic fermentation in multicellular organisms. *Physiological Zoology* **64**, 895–920.

7 Respiration and Circulation

7.1 Introduction

Life is thought to have originated in an anoxic (reducing) environment, with simple anaerobic heterotrophs present in the oceans within 1 billion years of the formation of the Earth; anaerobic life has therefore been in existence for at least 3500 million years, utilizing the pathways outlined in the previous chapter. Even small amounts of oxygen would produce real problems for any existing life in such an environment. It produces some highly reactive reduction products, including the free radicals superoxide (OO^-) and hydrogen peroxide (H_2O_2). These are more reactive than 'normal' molecular oxygen (O_2) because their oxygen atoms have unshared electrons in their outer shells, and they are effectively poisonous to living cells because of their potential to damage macromolecules. It seems likely that early microorganisms had to develop methods of detoxification to deal with oxygen so that these free radicals were not generated. In modern organisms, enzymes such as superoxide dismutases, peroxidases and catalases break down free radicals to give harmless products, and antioxidant organic molecules such as vitamin E, vitamin C and β-carotene will scavenge any free radicals that have otherwise escaped attention. Any or all of these safeguards may have appeared in the early Precambrian prokaryotes. The complex cytochrome systems in modern mitochondria may have originally developed in their prokaryotic ancestors as oxygen detoxification mechanisms, working alongside simple anaerobic fermentation pathways; later they were selected for their more lucrative role in aerobic respiration, producing far more energy per unit of carbon substrate than the original fermentation pathways.

Once molecular O_2 became more readily available (see section 6.1) its potential advantages could begin to override the problems it caused, and organisms would have the chance to exploit these advantages. Rising atmospheric O_2 permitted the formation of an ozone layer, and ozone is the only atmospheric gas that is a good absorber of harmful UV-B radiation from the Sun, so its accumulation meant that nucleic acids would become less prone to radiation damage. Increased O_2 may also have allowed the synthesis of certain crucial biological macromolecules, such as collagen (and the plant equivalents including lignin), that permit the assembly and mechanical integration of multicellular bodies. Thus planetary O_2 levels came to permit the evolution of metazoan life—animals—that could make use of this gas in oxidative cellular metabolism (or 'internal respiration'). As the size of such early metazoans increased and their surface area to volume ratios decreased they came to require external respiration, the exchange of gases between the surrounding medium and the tissues of a living organism, with substantial O_2 uptake and approximately equivalent carbon dioxide (CO_2) excretion.

As plants continued to diversify and spread, giving net oxygenation, animals radiated to use the gas in their own metabolic processes. Atmospheric levels of O_2 gradually rose towards present-day levels, and although the pattern of this atmospheric accumulation remains controversial, a useful 'best guess' is shown in Fig. 7.1. Current evidence suggests that the level may have been about 10% of the present atmospheric level (10% PAL) at the Cambrian boundary, not settling at around 100% PAL (i.e. to 21% of the atmospheric gases) until after the end of the Mesozoic (>65 MYA). There may, however, have been a period of unusually high O_2 levels (perhaps 30–40%, approaching twice PAL) in the Carboniferous era (around 350–300 MYA).

Oxygen availability in the modern world varies in both time and space (Fig. 7.2). In the seas it is reasonably constant at 4–6 ml l^{-1} through much of the water, but decreases in the oceanic 'oxygen minimum layer' (see Chapter 9) and in deep water in the Pacific basin. It is also drastically reduced in the benthic sludges and in deep marine vents. In the terrestrial world the air contains a standard 21% or 210 ml l^{-1}, but microhabitats in burrows and nests, within the litter layer, or air pockets underground or underwater may be substantially less O_2-rich (as low as 10% O_2 and up to 5–10% CO_2), and any habitat at altitude also experiences lowered O_2 availability. Where animals live at the interface of water and air (e.g. intertidally) but can only breathe from one or other of these media, they may suffer periodic withdrawal of available O_2. Furthermore, in any of these habitats, an animal undergoing vigorous exercise may be unable to acquire O_2 quickly enough to offset its consumption, as we saw in Chapter 6. Many animals are therefore subject to temporary or

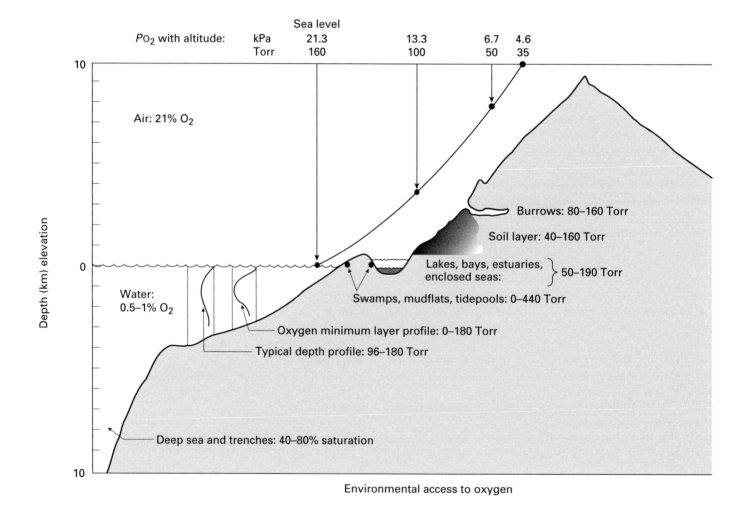

Fig. 7.1 Patterns of atmospheric O_2 from the late Precambrian, through the Palaeozoic and Mesozoic to the present; details are controversial, but there appears to have been a steady rise, with a 'burst' around the Carboniferous period.

Fig. 7.2 The general pattern of availability of O_2 in different habitats in the modern world. (Adapted from Dejours 1981.)

Table 7.1 Values for P_c, the critical partial presure at which regulation of O_2 uptake ceases and animals become O_2 conformers.

Organism	P_c (kPa)	Habitat
Aurelia (cnidarian)	16	SW
Mytilus (bivalve)	10	SW/shore
Loligo (squid)	6	SW
Gnathophausia (crustacean)	0.8	SW (O_2 minimum layer)
Uca (crab)	1	SW/land
Pelmatohydra (cnidarian)	8	FW
Tubifex (oligochaete)	3	FW
Cambarus (crustacean)	5	FW
Cloeon (mayfly)	4	FW
Lumbricus (oligochaete)	10	FW/terrestrial
Helix (pulmonate)	10	Terrestrial
Hyalophora (caterpillar)	3	Terrestrial

SW, sea water; FW, fresh water.

permanent low O_2 **(hypoxia)**, and some will experience a complete absence of O_2 **(anoxia)**. They may also experience accompanying rises in CO_2 levels, or **hypercapnia**.

The respiratory responses that animals show when subject to hypoxia can be placed in one of two categories, as with so many response systems. Some animals are **oxygen conformers**, where O_2 consumption decreases in proportion to decreasing ambient O_2 concentration. These include a few protozoans, most cnidarians, most marine worms and most parasitic worms, some molluscs and crustaceans, most echinoderms and a few aquatic insects. Generally, O_2 conformity is common in marine invertebrates, especially in large sedentary forms or smaller but less active forms that lack highly specialized respiratory organs and circulatory systems. In contrast, other animals will maintain their own O_2 consumption at ambient oxygen partial pressures (P_{O_2}) well below normal; these are termed **oxygen regulators**, and metabolism is maintained at normal levels down to some species-specific P_{O_2} level termed the **critical partial pressure**, P_c, values of which are shown in Table 7.1. Regulators include most protozoans, many freshwater and terrestrial annelids, many molluscs and crustaceans, some aquatic and probably all terrestrial insects, and almost all vertebrates. Some examples of O_2 usage in different animals are shown in Fig. 7.3.

In this chapter we look at the fundamental design of respiratory systems whereby aerobically respiring animals take up the O_2 they require (and may also release CO_2). These systems always involve an **exchange site**, with a large surface area where the epithelial layers in contact with the medium are of minimal thickness. For larger and/or more active animals, respiratory

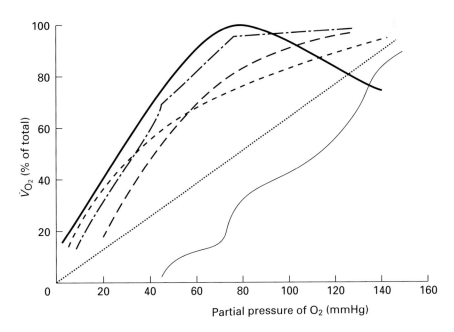

Fig. 7.3 Oxygen conformers and regulators: O_2 consumption expressed as percentage of total O_2 consumption is shown in relation to ambient P_{O_2} for six species of invertebrate. The pycnogonid, annelid worm (*Nereis*), crab (*Callinectes*) and horseshoe crab (*Limulus*) are conformers, while the earthworm *Lumbricus* and the crab *Libinia* are regulators but only above 70–80 mmHg. (Data from deFur & Mangum 1979; Davenport *et al.* 1987.)

efficiency also requires **ventilatory systems** to ensure that sufficient volumes of the external medium pass across the exchange site. Such animals also commonly have internal perfusion of the exchange site linked to internal **circulatory systems** to move the dissolved O_2 from the exchange site to the respiring tissues; this is essential because while diffusion is very effective at very short range it is exceedingly slow over even moderate distances (e.g. only about 30% of diffusing O_2 molecules arrive at a point 1 m from their source in 4 years!). These circulating fluids are known in more complex animals as cardiovascular systems, and in most cases serve many other functions such as moving nutrients and wastes around the body.

The presence of adequate ventilation and/or transport systems may be the most important determinant of whether an animal is a conformer or a regulator, since conformers seem to be limited mainly by an inability to transport O_2 quickly enough to the metabolizing tissues; evidence of this is seen with isolated tissue slices from such animals, which may have higher rates of O_2 consumption than the whole animals.

The distinction between O_2 conformity and O_2 regulation is therefore not a sharp one, particularly when the critical partial pressure (P_c) of a species is high. Moreover, both invertebrates and ectothermic vertebrates may be O_2 regulators at low temperatures when O_2 consumption is quite low, but conformers at high temperatures when their metabolic rates are much higher. The value of P_c may also change during acclimation; it decreases with previous exposure to hypoxia, reflecting acclimation in one or more of the supply components (respiratory system, blood properties or cardiovascular functioning). Thus the metabolic response to O_2 availability is not absolutely fixed, but can range from conformity to regulation depending on external conditions and individual experience.

7.2 Uptake and loss of gases across respiratory surfaces

7.2.1 Diffusion, partial pressures, gas concentrations and gas solubilities

Fundamentally, the uptake and loss of gas by animals is always achieved by simple diffusion. Diffusion does not rely on metabolic energy but is merely the result of the random thermal motion of molecules, at a velocity proportional to the absolute temperature. Although individual molecules move at random, where there is a concentration gradient there will be a net movement of the gas down the gradient; from an O_2-rich environment into the respiring (O_2-using) animal, and from the increasingly CO_2-rich animal back into the environment. At this level of analysis, respiratory exchange is very simple. However, the time taken for a randomly moving molecule to move in a particular direction increases with the square of the distance moved. Thus diffusional transport becomes extremely slow over large distances.

In a mixture of gases such as air, the total pressure exerted is the sum of the pressures exerted by each gas (Dalton's law). The **partial pressure** (P) exerted by a gas can be calculated from its fractional content of the total gas (F) and the total barometrically measured pressure (p_b):

$$P_{O_2} = F_{O_2}\, p_b$$

$$P_{CO_2} = F_{CO_2}\, p_b$$

For normal dry air at sea level, p_b is of course '1 atmosphere', or 760 mmHg (760 Torr), or 101 kilopascals (kPa). Under the same conditions, the partial pressure of O_2 is 21.2 kPa, and that of CO_2 dioxide is 0.03 kPa.

Uptake of respiratory gases always depends eventually on the gas (O_2) being dissolved in a liquid phase, though this may only occur deep within the tissues in some terrestrial animals. In an idealized system at equilibrium, the gas pressure in the two phases (gaseous and liquid) will be equal. This remains true when the gaseous phase is a mixture of several gases, such as air; each of the components will have equal partial pressures in the gaseous and liquid phases. However, to calculate the amount of gas dissolved in a volume of liquid we also need to take solubility into account, given by Henry's law as follows:

$$Q = \alpha\, P$$

where Q is the molar concentration of dissolved gas, P is the partial pressure of that gas and α is the solubility coefficient (in mol l^{-1} kPa^{-1}). Water and saturated air in equilibrium with each other will have the same gas partial pressures, but may have very different molar concentrations of the gases. In practice, gases can therefore diffuse 'up' a concentration gradient because they are diffusing down a partial pressure gradient. Most obviously this occurs when a gas dissolves into a liquid in which it is very soluble, even if already present there in high concentration, because this reduces the partial pressure difference between liquid and gas.

In general, of the three important atmospheric gases, O_2 and nitrogen (N_2) have fairly low solubility but CO_2 dioxide has high solubility; therefore O_2 and N_2 are present in water at much lower molar concentrations than in air, while CO_2 is present at about the same concentration in the two media. But solubility for any gas is affected by the nature of the solvent (water, in biological systems), and also by temperature and ionic strengths in the solution (Table 7.2). Thus at higher temperatures there is less CO_2 in water than in air, whereas at low temperatures there will be more CO_2 present in the water. Similarly at normal biological temperatures more O_2 will dissolve in solutions of low ionic concentration; at 10°C, the solubility coefficient for O_2 (in μmol l^{-1} kPa^{-1}) varies from 16.9 (fresh water) to 13.4 (sea water). However, within animal tissues the solubility of the important gases is fairly insensitive to the presence of other gases or to the nature of the electrolytes present (although as we will see gases and ions may critically affect the binding of O_2 to carrier pigments).

Table 7.2 Respiratory gas solubilities (in μmol l⁻¹ kPa⁻¹) and effects of temperature and salinity.

Temperature (°C)	Oxygen			Carbon dioxide	Nitrogen
	FW	50% SW	SW	FW	FW
0	21.7	18.9	16.6	767	–
10	16.9	–	–	531	–
20	13.7	12.2	10.8	386	6.8
30	11.6	–	–	294	–
40	10.2	9.3	8.3	235	5.5

FW, fresh water; SW, sea water.

7.2.2 *The major respiratory structures in animals*

In the simplest and smallest animals, adequate exchange of gases can take place over the general body surface, and this can be shown by simple consideration of diffusion rates (the relationship of time taken to the square of the distance moved, discussed above) to be adequate for animals up to about 1 mm in diameter, or 1 mm in thickness if they are of elongated flattened form as in the platyhelminths, or elongated tubes, as in nematodes. For a body of given shape, a two-fold increase in diameter increases the surface area by a factor of 4 and the volume by a factor of 8. Any tissue mass greater than a 1 mm sphere would have insufficient surface to supply the O_2 needed by the contained volume of metabolizing tissue, and would thus have an effectively anoxic zone at its core, in the absence of any extra specialization. Since metabolic rate, and therefore O_2 consumption, scale with body mass with an exponent of only 0.75, as discussed in Chapter 6, there is a small mitigating factor here in that larger bodies need proportionally less O_2; but this effect is far too small to compensate for the surface area to volume effect for O_2 uptake. In fact the limit may be set well below the theoretical 1 mm level because of the existence of unstirred O_2-depleted boundary layers around the surface of the living organism, delaying diffusion even further. Most single cells are considerably less than 1 mm in diameter, and where they do exceed this

they are usually of irregular shape to keep the maximum diffusion path small. Most animals bigger than 1 mm in thickness, and also many smaller animals that regularly experience lowered external O_2 concentrations or that are relatively active, have evolved specialized gas exchangers, at the tissue or organ level.

Commonly this occurs at the outside of the animal, and where the structure is a protruding evaginated surface it is usually called a **gill**; most aquatic animals have one or more pairs of gills. Three main designs of gill can be identified, with increasing three-dimensional complexity (Fig. 7.4). **Tuft gills** can be little more than a raised thinned area of the general skin, as seen in the papulae and modified podia (tube-feet) of some echinoderms (starfish and sea urchins), but range through to the segmental tufts and the elaborate terminal plume-like structures common in many aquatic worms. **Filament gills** are best exemplified by aquatic arthropods, where a series of feather-like structures occur, supported by a thin external cuticle, and with a more elaborate internal flow system than is seen in any of the tuft arrangements. Filament gills also occur in some vertebrates, for example in embryonic elasmobranch fish, in some salamanders and tadpoles, and on the pelvic fins of some adult lungfish (dipnoans) as a supplementary breathing mode. **Lamellar gills** are found as flat platelets extending out from a central strut or 'gill arch', with a specific orientation to water flow. The pleopods (modified abdominal limbs) of most of the isopod crustaceans take this form, and beat gently against the ventral surface of the body. Lamellar gills also occur in the larger and more active decapod crustaceans (crabs etc.), and littoral species often use such gills for air-breathing: the shore crab *Carcinus maenas* has gill lamellae that are rather more sturdy than in fully aquatic crabs, and can be ventilated in air as well as in water; other species will allow water from their branchial chambers to flow out over a rather hairy region of their carapace, where it becomes oxygenated, subsequently trickling back into the gill chamber through the inhalant openings at the leg-bases. Lamellar gills of particularly elaborate form also occur in both elasmobranch and teleost fish, where there may be 4–14 pairs of gills along the axis of the body, protected in teleosts by a flap-like covering (the operculum).

Tuft gills *or*   Filament gill   Lamellate gill

Fig. 7.4 A schematic view of the three main types of gill. Arrows show blood flow.

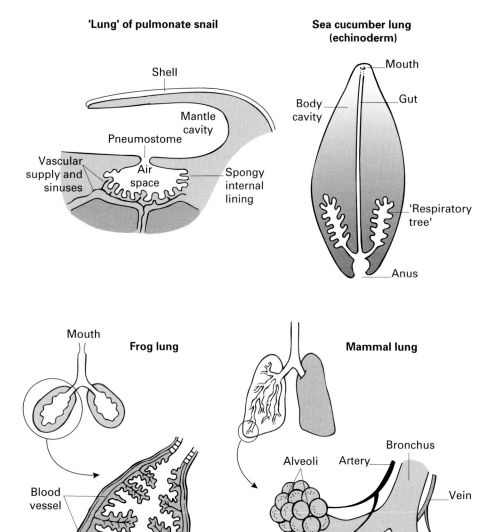

'Lung' of pulmonate snail

Shell

Mantle cavity

Pneumostome

Vascular supply and sinuses

Air space

Spongy internal lining

Sea cucumber lung (echinoderm)

Mouth

Gut

Body cavity

'Respiratory tree'

Anus

Mouth

Frog lung

Blood vessel

Mammal lung

Bronchus

Alveoli

Artery

Vein

Pulmonary capillaries

Fig. 7.5 Different types and complexities of lung.

Horseshoe crabs (not actually crabs, but related to the arachnids) and some isopods have respiratory structures that are gill-like, though somewhat enclosed and ventilated by being waved through water, and these tend to be termed **gill books** or book gills. In arachnids, such as spiders, a design of respiratory organ that appears rather gill-like also occurs, but it is well-protected within a chamber having only limited access, and functions in air rather than water. Here the lamellae may be fused more securely to the chamber wall for extra rigidity, held apart by cuticular struts to stop the lamellae collapsing on to each other, and supported by internal props to keep the blood space open. This arrangement is termed a lung book or **book lung**.

In relatively few animals the respiratory surface is completely internalized and forms by invagination rather than evagination of the body surface, and it is then usually termed a **lung**. This kind of structure occasionally occurs in marine animals, notably in the somewhat enigmatic 'respiratory tree' of holothurian echinoderms (sea cucumbers). But lungs are particularly associated with partially or fully terrestrial animals, where water loss from a superficial structure would be an overwhelming disadvantage. Lungs may vary from simple sacs to complicated subdivided structures with internal alveoli (Fig. 7.5).

Simple lungs occur in at least one earthworm, where the rear portion of the body is tucked back on itself to form a simple vascularized chamber; and in some land-dwelling molluscs and crustaceans. Sac lungs are achieved in some terrestrial crabs where the gill lamellae are lost and the branchial chamber becomes vascularized; ocypodids (ghost crabs) can pump air through the water in these chambers. Crabs that are more fully terrestrial have air-filled and fully vascularized branchial chambers, losing their gills either partially or fully; some soldier crabs are obligate air-breathers and drown if their mud-walled burrows do not keep out the incoming tide. At the extreme, the Trinidadian mountain crab *Pseudothelphusa* has a highly vascular branchial wall surrounding a network of interconnected airways, approaching the complexity of many vertebrate lungs. Molluscs show a similar trend towards elaboration of the lung. In intertidal prosobranch snails, such as *Pomacea*, we find the intermediate situation where half the branchial chamber is filled with gill filaments and the other half is a vascularized simple lung. Most pulmonate snails and slugs have quite simple lungs, albeit with highly complex vascularization of the mantle wall; but some slugs have finely divided cavities with air diverticula interspersed with blood sinuses.

Invaginated sacs for breathing also occur in a range of fish; in many relatively primitive fishes that live in swamps or ponds the swim-bladder is used as a lung, while in some of the perch family a 'supra-branchial pouch' is developed as a lung-like structure, and in a few species from various families parts of the gut are used for breathing. The lungs of modern air-breathing fish are relatively little subdivided, whether they occur in chondrosteans (e.g. *Polypterus*), holosteans (gars and bowfins) or dipnoans (true lungfishes).

'True' lungs, pouching off from the lower pharynx, are most obviously associated with the tetrapods: amphibians, reptiles, birds and mammals. Amphibian lungs exhibit the whole range from secondary absence (some salamanders), through simple sacs, to finely alveolar structures, though in all cases a substantial part of the oxygen requirement is supplied via the cutaneous route (see below). In reptiles lungs are of course always present but their complexity is very variable.

In the birds and mammals the demands of endothermic life always require a high O_2 consumption and thus a highly alveolar lung, giving a dense spongy tissue; 1 cm³ of mammalian lung has at least 600 cm² of alveolar surface, and the two lungs together may have 300 million alveoli, each only 150–300 μm in diameter, the overall internal surface area scaling in simple proportion with metabolic rate (Fig. 7.6). Air enters the paired bronchi from the trachea before branching into ever finer airways (Fig. 7.5); in mammals, ventilation largely ceases beyond the tertiary branches and air merely diffuses to the final alveolar exchange sites. However, in birds the lungs are additionally ventilated by air sacs with a two-stroke one-way through-flow system (see below and Fig. 7.11), and ventilation extends throughout the bronchial network down to parabronchi of only 500 μm diameter, the final passages before the alveoli (often termed 'air capillaries' in birds). Lung volume, pulmonary capillarity and surface area of the blood–gas barrier all scale linearly

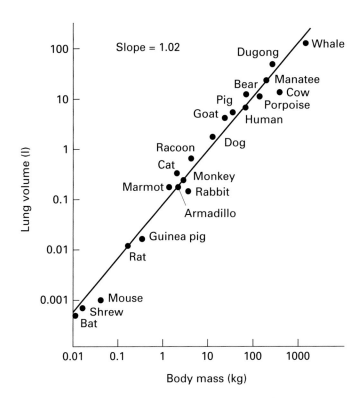

Fig. 7.6 Lung volume in mammals, scaling in simple proportion to body mass. (Data from Tenney & Temmers 1963.)

with body weight in birds, while the weight-specific blood–gas surface area scales negatively, indicating that the smallest birds (with very high metabolic rates) have the most specialized and highly adapted of all lungs.

One other respiratory exchange system is found in the animal kingdom, the **trachea** system of insects and some other terrestrial arthropods (myriapods, onychophorans, some spiders and many other arachnids). This again involves invagination into the body, but of many separate tubular arrays rather than a single pair of sacs from a single opening as with lungs. Spiders may have a combination of book lungs and simple 'sieve tracheae' (branching in bundles from a book-lung-like cavity); but in more advanced spiders and in other arachnids only true 'tube tracheae' are present. Some terrestrial isopod crustaceans have a combination of gills and tracheae, in that their abdominal pleopods contain internal air-passages called 'pseudotracheae'. The pterygote insects always have tube tracheae; the system starts with a series of spiracles (primitively arranged with one pair per segment but some pairs often being lost), each leading to an air-filled primary tracheal tube up to 1 mm in diameter, then branching into finer secondary and tertiary tracheae ramifying to all the tissues and down to about 2 μm in diameter. These retain a moderately thick and probably rather impermeable cuticular lining, and are not sites for respiratory exchange. They lead ultimately to the very finely divided **tracheoles**, blind endings that may be 400 μm long but only 0.6 μm across, penetrating next to and even within individual cells, with a per-

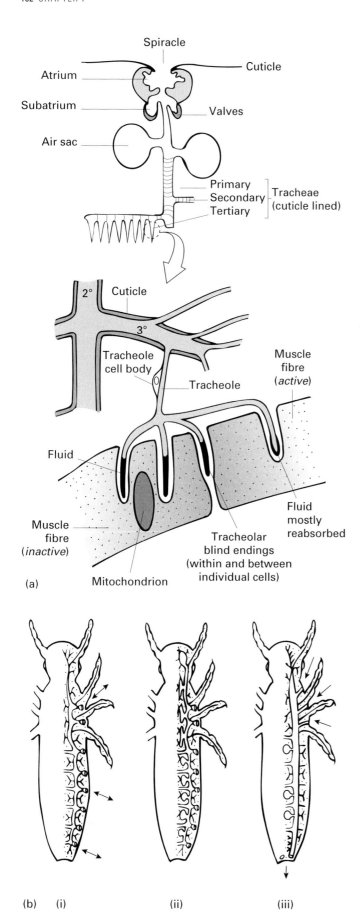

Fig. 7.7 (a) Basic tracheal respiratory systems in insects. The ultimate blind-ending tracheoles supply individual cells with inspired air. They are fluid-filled at rest but during activity fluid is withdrawn (see text) and there is a higher surface area for O_2 uptake. (b) Varying patterns of tracheal network: (i) with many spiracles and only partly interlinked tracheae; (ii) with air sacs added for ventilation; and (iii) with air sacs and reduced numbers of functional tracheae giving unidirectional flow from front to back.

meable cuticle only 0.01 µm thick. In a very large insect flight muscle fibre there may be 5–10 million effectively intracellular tracheolar indentations. Air is delivered directly to the tissues without ever being pumped around the body dissolved in liquid (blood); if a blood system is also present in the animal, it tends to have no respiratory role. In a simple tracheal system flow is essentially tidal; but in larger more active insects (and probably in all flying insects) the tracheae may fuse together into a complex network throughout the body (Fig. 7.7), with air sacs at intervals for storage purposes or to act as bellows.

A tracheal system may also be modified for aquatic respiration (Fig. 7.8, see also Chapter 11). Some aquatic insects merely restrict their spiracular openings to their posterior end, make the cuticle and hairs there hydrophobic, and poke the tip of the body through the surface film to breathe from air. But many freshwater insect adults and larvae have other solutions: there may be extensions of the body (commonly laterally or terminally on the abdomen) containing dense arrays of tracheae (tracheal gills); or 'plastrons' of air contained within a dense fur on the cuticular surface; or tufted extensions from the spiracular openings. In dragonflies water is drawn in and out of the rectum and O_2 diffuses into tracheoles there, giving the additional advantage of using the expelled water for jet propulsion! Yet other insects carry air bubbles under water with them when they dive (see Chapter 11).

No listing of respiratory structures would be complete without mentioning the **skin** itself, however. A surprising range of animals possess no special 'structures' for gas exchange, and get most of their O_2 supply through their skin (cutaneous exchange). Even quite large and complex animals use the cutaneous route for much of their O_2 supply, without external ventilation; they can do this because they get over the 1 mm maximum diffusion problem by having efficient internal perfusion systems, with a flow of deoxygenated blood or haemolymph passing close to the respiratory surface. This combination of direct cutaneous diffusion externally and efficient perfusion internally works perfectly well in a range of small aquatic invertebrates (including most larval forms and many adult aquatic worms), and in semiterrestrial earthworms and tropical slugs. It also occurs in some aquatic vertebrates, including the early larval stages of many fish, adult eels, catfish, salamanders and sea snakes, and most notably in the more terrestrial anuran amphibians (frogs and toads); in all these vertebrates it is probably a secondary development as the immediate ancestors were more highly dependent on gills or lungs. All these skin breathers, where the skin is

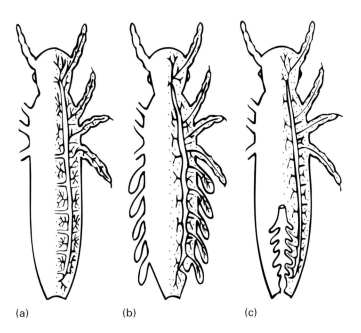

Fig. 7.8 Aquatic tracheal systems with (a) no spiracles, cuticular diffusion only; (b) abdominal gills; and (c) rectal gills.

Table 7.3 The thickness of respiratory gas-exchange barriers (ambient medium to capillary or cell) in air-breathing animals.

Animal group/species	Diffusion distance (µm)
Crab	0.2–0.4
Insect	0.1–0.3
Fish (air-breathing)	
Haplochromis	
Gill	0.3–2.0
Saccobronchus	
Gill	3.6
Air sac	1.6
Anabas	
Gill	10
Branchial wall	0.2
Amphibian	
Toad	1.3–3.0
Birds	
Pigeon	0.1–1.4
Swallow	0.09
Mammals	
Rat	0.13–0.26
Human	0.36–2.5

Table 7.4 Diffusion coefficients for gases in air, in water and in biological tissues.

Medium	Diffusion coefficient ($cm^2\ s^{-1}$)		
	Oxygen	Carbon dioxide	Water vapour
Air			
0°C	0.18	0.14	0.24
20°C	0.20		
Water			
20°C	20×10^{-6}	18×10^{-6}	
37°C	33×10^{-6}		
Tissues			
Human lung	23×10^{-6}		
Muscle	14×10^{-6}		
Amphibian skin	14×10^{-6}		
Amphibian egg	10×10^{-6}		
Fish egg capsule	6×10^{-6}		
Eel skin	2.4×10^{-6}		
Insect chitin	0.7×10^{-6}		

permeable to O_2, are inevitably at risk from desiccation and the phenomenon therefore only occurs in aquatic animals or those from rather moist terrestrial microhabitats; in fact in the few modern frogs that live permanently in water respiration is entirely cutaneous.

7.2.3 Diffusion barriers

All the respiratory exchangers discussed above have in common the presence of a very thin gas-exchange barrier (Table 7.3). In many animals this diffusional barrier will be no more than a single epithelial cell. In birds and mammals the diffusion barrier across the alveolar surface may be as thin as 0.1 µm, and in fish gills it may appear equally thin but is often in practice augmented by a protective mucus layer, which produces a somewhat thicker boundary layer.

In arthropods there will inevitably be an additional layer of chitinous cuticle at the gill surface. In crab gills this can vary from 2 µm in active shore-dwelling species to 15 µm in some deep-sea low-metabolic-rate species, though in land crabs the barrier within the lung tissues may be only 0.2–0.4 µm. Since chitin has a much lower diffusion coefficient than cytoplasm and vertebrate connective tissues (Table 7.4), the actual barrier this imposes will be substantial. In terminal insect tracheoles the whole diameter may be only 0.1–0.2 µm so the cuticular barrier is evidently very thin.

However, insects may be unusual in being able to alter the depth of their diffusion barrier according to need, by variation in the amounts of fluid within these tracheolar endings (see Fig. 7.7a). The system appears to operate automatically in that when

an insect is very active and its metabolic rate rises, the tissues around the tracheolar endings are breaking down stored sugars into smaller metabolites so that the osmotic concentration in the fibres rises. This draws water out of the tracheolar fluid, so that more of the space is air-filled rather than fluid-filled, with just a thin surface coating on the tracheolar wall in the aqueous phase, and gaseous exchange can therefore occur more quickly to supply the needs of the active animal.

7.3 Ventilation systems to improve exchange rates

Ventilation involves convective flow of the external medium, maintaining a movement of fluid over the respiratory exchange site, reducing unstirred boundary layers and replenishing the source of O_2 more or less continuously. The rate of convective transport of oxygen (Q) from the medium (high concentration, C_2) into the animal (low concentration, C_1) is then given by:

$$Q = V_w (C_2 - C_1)$$

where V_w is the rate of mass flow of the medium. This convective transport rate is always much higher than for diffusional transport alone, but there is the disadvantage that energy is expended in creating the ventilatory current, especially for aquatic animals where the fluid to be moved is relatively dense and viscous water.

7.3.1 *Mechanisms of ventilation*

In some animals ventilation can be achieved quite passively by exploiting the properties of moving fluids at interfaces. Where a fluid flows past a surface there is usually a velocity gradient

Fig. 7.9 Gill ventilation systems. (a) The parapodial 'fans' in the tube-dwelling annelid *Chaetopterus*. (b) The scaphognathite, or 'baler', in a decapod crab (longitudinal section).

(slower nearer the surface), and pressure is therefore automatically higher at the surface and lower further up (Bernoulli's theorem); thus for a perforated sessile animal attached to the surface fluid will inevitably flow in at the base and out at the top, explaining why a dead sponge still gets ventilated. The same principle applies to ventilation of flow-through burrows, if one end of the burrow is raised above the surface pressure is lower at that end and water or air will flow through the burrow automatically (as in prairie dogs' terrestrial burrows, and in termite nests).

Ventilatory mechanisms where some force is generated by the animal range from ciliary systems to complex muscular machinery; all these in effect function either as pumps or as paddles. Cilia are the most common way of achieving a 'paddle' effect, especially in non-arthropod aquatic invertebrates. The sponges (Porifera), which are amongst the simplest and probably most primitive of invertebrates, can attain very large body masses with a design based on water-filled channels, where sea water is propelled along with the aid of cilia borne on the specialized choanocyte cells. Here the cilia are effecting both ventilation and perfusion, and as in many aquatic animals the current they generate serves both respiratory and filter-feeding purposes. In more complex marine animals such as molluscs, coordinated cilia on the gills produce substantial bulk-flow water currents, which are particularly obvious when channelled through the siphons of larger bivalves, for example clams.

(a)

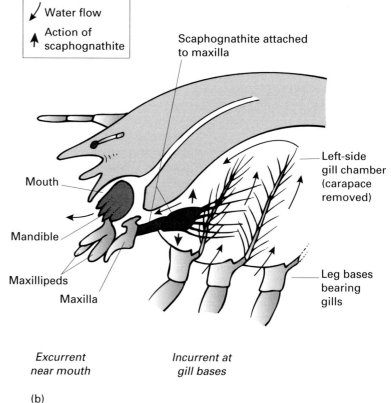

(b)

Muscle-related respiratory ventilation at its simplest merely involves moving the animal's body in a suitable manner. Swimming or crawling movements in a benthic animal may stir up the water enough to provide adequate oxygenation. Positioning the body across an existing water current is a more direct way of achieving ventilation, exhibited in several freshwater aquatic insects such as stoneflies and blackflies. However, muscles may be used much more specifically for ventilation where the whole gill structure is moved around, as with the pleopods of isopod crustaceans. Muscles more commonly produce a pumping effect; this may be seen at its simplest in tube-dwellers (marine annelids such as the lugworm *Arenicola*) where peristalsis of the body wall draws water in and out of the tube, and in a somewhat more complex form in the related annelid *Chaetopterus*, where the appendages of two segments are modified as fans to draw water through the tube (Fig. 7.9a). Pumps are more common where the respiratory surfaces are enclosed, however, as in the gills of crustaceans and prosobranch and cephalopod molluscs. Most larger crustaceans, and especially the decapods such as crabs and lobsters, have a paddle-shaped 'baler' (technically the scaphognathite) that acts as a pump to force water in and out of the branchial chamber (Fig. 7.9b), with a fairly fixed stroke volume but a variable frequency. The smooth-muscle layers of

the mantle in *Nautilus* and in octopuses contract rhythmically to provide a pulsatile flow over the gills within the mantle cavity (but note that here again the respiratory current can have an additional function, this time in locomotion). In most fish, the muscular pumping effect is more complex again, with the mouth and the gill cover (operculum) coordinated to produce a two-phase pumped one-way flow over the intervening gills (Fig. 7.10), and with both stroke volume and frequency capable of modification according to metabolic demand. However, in certain actively swimming fish, such as sharks and the tuna and mackerel family, the mouth is kept open and water is forced through the gills continuously by the passage of the animal through the water, a system known as 'ram ventilation'. Note that in the gills of an active animal, whether decapod or fish, the optimal ratio of ventilation to perfusion (water flow to blood flow) is rather high, often 10–20. This is because of the low O_2 content of water relative to the O_2-carrying capacity of blood, so that ventilation has to be highly organized and rather rapid to support the animal's O_2 demand. Ventilation becomes a significant cost in animals with high metabolic rates, and ram ventilation is a way of saving energy in an active fish, though it means the animal has to keep moving indefinitely.

Whereas gills are usually ventilated unidirectionally, most of the lungs and tracheae of terrestrial animals have to be tidally ventilated with alternate filling and emptying along common channels. Tracheal systems in small insects are entirely diffusional, but in many larger forms ventilation may be assisted

Fig. 7.10 Ventilation of teleost fish gills, also showing the patterns of buccal and opercular movements to produce a two-stroke pump action.

Fig. 7.11 Birds' lungs, air sacs and ventilation patterns. DB, dorsobronchus; VB, ventrobronchus; P, parabronchus running between these; AS, air sac.

Cervical sac

Lung

Interclavicular sac

Abdominal sac

Anterior thoracic sac

Posterior thoracic sac

Main lung

Neopulmonary network

DB
P
VB

AS

Trachea

AS

AS

AS
(air sac)

AS

Model

Parabronchi

Ventrobronchi

Dorsobronchi

Anterior air sacs

Posterior air sacs

Trachea

Mesobronchus

Inhalation
(fills posterior air sacs, air forward through 'lung')

Exhalation
(posterior air sacs empty, air forward through 'lung' to anterior air sac)

Detail of a parabronchus

Air

Dorsobronchus

Blood capillary

Parabronchus

Pulmonary artery
Pulmonary vein

Blood flow

Ventrobronchus

Air

Air

Parabronchus

Air capillaries (alveoli)

Blood capillaries

by muscular contractions (e.g. of the flight muscles) adjacent to air sacs, allowing them to act as positive-pressure bellows. Movements of the abdomen associated with air pumping are easily seen in larger bees and locusts, and the rate of pumping increases with metabolic rate, so that supply tracks demand. Air movements through the system may be organized into unidirectional pathways, usually with air entering at the thoracic spiracles and leaving through the abdominal ones; in fact in large flying insects air may be forced into the anterior spiracles in a manner analogous to the ram ventilation of swimming fish, and in some beetles the spiracles are shaped so that air is 'scooped in' at the thorax in flight and automatically flows through the animal from front to back. The tracheae may thus have unidirectional airflow, though the tracheoles are always blind-ending with simple steady-state diffusion.

Lung books in arachnids appear to be simple diffusion systems, with little or no ventilation. 'True' lungs may be ventilated by positive-pressure inflation as in the systems described above, using the cloacal muscles in holothurians, the scaphognathite in crabs, or the buccal muscles or opercular muscles (or both, in some lungfish) as a pump, or again by utilizing air sacs as bellows in birds. Alternatively, they can be ventilated by aspiration breathing, where negative pressure is created to suck air in. In *Pomacea*, the prosobranch snail with both gills and lungs in the branchial chamber, a flap of mantle is modified as a siphon which can pump air into the 'lung' section from above the water surface when the water becomes O_2-depleted. In the common snail, *Helix*, air is drawn into the lungs by lowering the mantle floor. Aspiration ventilation in vertebrates may be achieved by ribs alone, for example in most reptiles where the rib cage muscles (intercostals) cause the expansion of the thorax and hence lung inflation. Alternatively it can be achieved by a diaphragm alone, though this system is described in only one fish (the Amazonian *Arapaima*), and in effect in a few reptiles, including crocodilians, where the liver is moved back and forth as a functional diaphragm. In tortoises (where the ribs are fused to the carapace and cannot help to inflate the lungs) the limbs are extended to decrease the intracarapace pressure, so pulling down the diaphragm and inflating the lungs. In most cases, though, ribs are accompanied by a muscular diaphragm to isolate the thoracic space and enhance the negative pressure resulting from rib movements, as in birds and mammals.

In birds, the extra demands of flight are met by adding air sacs to the system; these work with the ventilatory musculature to achieve a one-way fully ventilated flow through much of the system (Fig. 7.11). The lungs remain at fairly constant volume but are ventilated by a bellows effect from the neighbouring non-alveolar air sacs which surround the viscera and even extend into some of the bird's bones. Expansion of the anterior sacs draws air in through the trachea and mesobronchi, and on to the posterior air sacs via the dorsal bronchi, parabronchi and ventral bronchi of the lung. Collapse of the air sacs in expiration forces air from the posterior sacs and back out by the same route of dorsal bronchi, parabronchi and ventral bronchi. Thus flow through the parabronchi adjacent to the alveoli is continuous and always in the same direction, quite unlike the tidal flow of the mammalian lung. Parabronchial flow is roughly countercurrent to the main blood flow (see below), and in the alveoli, which are radially arranged around the parabronchi, it is roughly crosscurrent or at best nearly countercurrent to the finest capillaries, allowing bird lungs to be proportionally smaller and thus lighter than in other tetrapods.

Since air is both far less dense than water and has a much higher O_2 content, ventilation with air is significantly cheaper than aquatic ventilation for an equivalent O_2 demand, though always with the drawback of inescapable evaporative water loss. On the plus side, air does not carry away solutes, so there is no respiratory exchange of ions in terrestrial animals. For land animals, some further savings of energy are made by ensuring that the ventilation rate is tied in with other patterns, particularly of locomotion; birds ventilate at the same rate as they flap their wings, or in a strict ratio to wingbeat, and many mammals also breathe in synchrony with their gait. This probably provides greater mechanical efficiency, and may even assist ventilation, as for example in jumping kangaroos where the rising and falling viscera act as an extra pump against the elastic diaphragm and so reduce the cost of lung deflation.

7.3.2 *Cocurrents, crosscurrents and countercurrents*

Ventilation systems can be usefully characterized by the relative orientation of the flow of the external medium and of the internal perfusing fluid (Fig. 7.12; see also Box 5.2 for a review of general principles). If the two flows are parallel to each other the flow is termed **cocurrent** and the gas content (e.g. the partial pressure of O_2) of the internal perfusate leaving the exchanger cannot exceed that of the external medium leaving the exchanger; at best the two fluids will equilibrate, with 50% exchange occurring. However, if the two flows are arranged as a **countercurrent** the internal perfusing fluid may reach almost the same level of Po_2 as is present in the bathing medium; here the effluent perfusate is in contact with the incoming and maximally oxygenated external medium, so that even if the perfusate has already picked up a good loading of O_2 within the exchanger the gradient is still such that it will continue to load up, leading to almost complete exchange between the two. The **crosscurrent** exchanger has properties lying between these two extremes, and the effluent perfusate will load to a Po_2 somewhere between those of the incoming and outgoing external media.

Many of the more complex gills in animals achieve countercurrent flow, and in filamentous and lamellar gills there are normally parallel vascular channels within the gill to ensure that the blood flows in the right directions. In fish and the larger more active invertebrates such as cephalopods, countercurrent flow is almost essential to ensure that O_2 uptake is adequate to compensate for the difficulty and energetic expense of ventilating with a viscous medium. In practice, the advantages of full countercurrent flow (and the generally lower metabolic rates of most

Amphibian skin	**Fish gills**	**Mammalian lung**	**Bird lung**

Fig. 7.12 Simple models for respiratory gas exchange. For the blood (shaded, *y*) V = venous, A = arterial; for the water or air flow (stippled, *x*), I = inhalant, E = exhalant. (Adapted from Piiper & Scheid 1982.)

aquatic animals) offset the difficulty of ventilation, so that aquatic animals generally require less respiratory surface area than lung-breathers (Fig. 7.13), which may be helpful in limiting their osmotic and ionic exchanges with the water.

Crosscurrent flow approximates to the situation found in many bird lungs, discussed above, and is adequate in a terrestrial animal even given the high metabolic demands of flight. But mammalian lungs, with a strictly tidal flow and effectively unventilated alveoli, cannot be characterized in these terms, and are sometimes described as 'uniform pool' gas exchangers (Fig. 7.12d), with the alveolar air reaching a steady state by diffusion. The same term could be applied to cutaneous respiration. It may also be applicable to many insects, though in the species that have unidirectional flow with air sacs and controlled ventilation there may be an approximation to countercurrent flow in parts of the system such as the flight muscle, where air flows primarily front to back and may be pulsed with the muscle contractions and resultant wingbeats.

7.4 Circulatory systems

Virtually all metazoan animals have some form of circulatory system, designed to move fluids in bulk within the body and thus reduce diffusion distances by fluid perfusion of all the exchange sites where gases, and also foods and wastes, must be taken up or disposed. In most larger animals this system serves to reduce all diffusion paths to 0.1 mm or less, ensuring very rapid diffusive transfer at the tissue level. In effect, a circulatory system becomes a system of forced convection of respiratory gases around the body, so that only the final exchanges at the point of uptake and at the point of delivery to the tissue mitochondria are reliant on the slow process of diffusion. Using such forced convection, a large mammalian body can move a molecule of O_2 from the uptake point in the lung over a distance of about 2 m to peripheral capillaries in a limb in about one minute.

Circulatory systems, once evolved, can serve a multiplicity of roles. They will transport O_2 and CO_2, water and salts, food components such as glucose, fatty acids and amino acids, vitamins and trace elements, and waste products such as ammonia, urea and creatine. Hormones may be distributed, so that the circulatory system becomes a key player in maintaining homeostasis; the cellular and chemical components of bodily defence systems

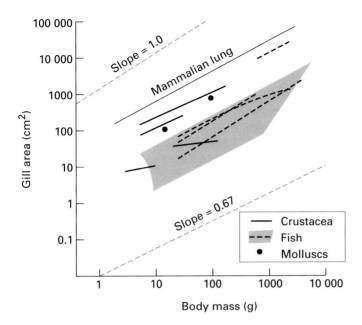

Fig. 7.13 Relationship between gill surface areas and body masses for various taxa, shown in relation to mammalian lung surface area.

are also distributed (see Chapter 15). The circulated fluid may also have a hydraulic role, either as a hydrostatic skeleton for the whole body (in many marine worms) or as a means of protruding tissues and organs under pressure in special areas of the body. Examples here include echinoderm tube-feet, legs in spiders (which have retractor muscles but no extensors), wings in newly moulted adult insects, gills ('branchial crowns') in many annelids, and erectile tissues such as throat display pouches and penile tissues in many animals including vertebrates. Thus the circulatory system can come to play a role in locomotion, in reproduction and in behaviours associated with aggression, dominance and mating.

7.4.1 Open-plan and closed plumbing

The circulation in an animal can range from a slightly controlled pumping of the external medium through the body, to continuous circulation of a contained regulated internal fluid endowed with specialized cellular components through highly organized closed conduits.

Use of the external medium is generally associated with smaller animals, but can be effective in larger species with relatively sedate or sessile lifestyles, such as sponges or cnidarians. It is mainly found in marine animals, where the sea water provides a suitable osmotically balanced and well-oxygenated medium to bathe the tissues, but also occurs in some freshwater species; in both cases, the circulating medium may also be used as a source of filterable food.

Circulation of internal fluid can be achieved by a movement of the general extracellular fluid, so that the extracellular spaces of all tissues are confluent with the circulating fluid, which moves

through loose sinuses within tissues and larger lacunae or cavities between the tissues. In this situation the circulatory fluid is usually termed **haemolymph** and the system is described as an **open circulation**. Open circulation occurs in most of the smaller animals that have bodies designed around the acoelomate or pseudocoelomate grades, where fluid in the blastocoel bathes the tissues directly (see Box 1.3). No specific pump is needed in these animals, as the movements of the body wall maintain an adequate circulation. Such a system is also found in a more sophisticated version in most of the arthropodan groups (Fig. 7.14d,e), in many molluscs (Fig. 7.14f), and in less well known invertebrates such as tunicates. Here the haemolymph is partially directed around the body through vessels and may flow through short sections of finely divided networks of thin-walled vessels in key sites such as excretory organs, gills or central ganglia, but it then passes into the general open haemocoel, which lacks any lining separating it from the muscles and other tissues, before returning to one or more simple hearts. In a few cases there are specific vessels leading from the respiratory system back to the heart; for example, some spiders have 'lung veins' draining the book lungs.

Alternatively, the circulating fluid may be highly specialized and largely separated from the rest of the extracellular fluid, and it is then usually termed **blood**. This separation results in a **closed circulation**, where the blood travels in vessels lined with endothelial cells, and only comes into contact with the tissues when it traverses the walls of the finest vessels, known as **capillaries**. Within such a system there may be areas of apparently 'open' circulation where larger blood sinuses occur, but these will always be endothelium-lined. Clearly in this situation, where the whole circuit is in narrow vessels, one or more powerful pumps will be needed to force the blood around the conduits of the body; in larger and more active animals the pump is a highly muscular heart (see below), which itself needs a substantial blood supply.

The distinction between open and closed circulatory systems is not absolute. The presence of capillaries is sometimes said to be the defining difference, but capillaries can occur in the most finely divided reticulate sections of the essentially 'open' systems in large crustaceans, and within the phylum Mollusca the complete range from open to closed systems occurs, with cephalopods having fully formed endothelium-lined capillaries. Open systems are also usually characterized by large blood volumes (at least 30% of body weight is commonplace, compared with about 8% in most mammals; see Fig. 4.1), by low pressures relative to the values of 100–200 mmHg found in many vertebrates, and by long circulation times. In terrestrial crabs the peak ventricular pressures can be as high as 50 mmHg, although once the blood has passed beyond the gills the pressure falls almost to zero, with virtually no pulse detectable. In terrestrial pulmonate molluscs, too, pressures of 20 mmHg have been recorded. Perhaps the best way to distinguish open and closed systems is by the presence or absence of an identifiable extracellular space separate from the blood; note that the distinction is not then

Fig. 7.14 Diagrammatic views of circulatory patterns in a range of animal taxa.

between invertebrates and vertebrates, since the cephalopods at least would qualify as having fully closed circulations. This allows them to generate pressures of over 100 mmHg during activity, comparable to vertebrate levels.

In both open and closed blood systems, the carriage and distribution of blood to tissues is carefully regulated, with fluid pressure and fluid volume monitored by central or peripheral receptors (see section 7.7). Closed systems are in no way inherently more complex or more highly regulated than open ones. However, in the higher vertebrates the fluid composition of the blood is more stable, and perhaps more rigorously monitored than in other animals, due to their high O_2 demand coupled with the susceptibility of their haemoglobin properties to environmental variability (see section 7.5.2).

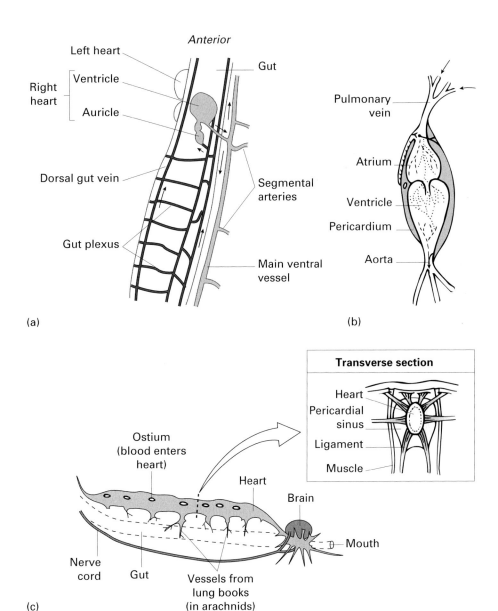

Fig. 7.15 Structure of the heart in (a) earthworms, (b) molluscs and (c) arthropods.

7.4.2 *Pumping stations*

In animals of moderate activity and/or large size, some form of pumping station is needed to keep the blood or haemolymph circulating adequately, and this pump is conventionally termed a **heart**. In essence the heart is a section of one of the principal blood vessels where the vessel wall is thickened and muscularized, and this ancestral design can be traced even in the complex multichambered hearts of tetrapods. To qualify as a fully functional heart, a pumping structure should also have a degree of nervous synchronization of its contractions, and be equipped with internal valves to ensure one-way flow.

Invertebrate hearts

Many invertebrate groups are rather flexible in their arrange-

ment of hearts, with one or many being present and sited according to need, so that there may be accessory 'branchial hearts' near the gills, for example. Hearts appear at their simplest in annelid worms, where sections of the main dorsal vessel, carrying blood forwards along the body, are thickened and show rhythmic contractility. In some of the oligochaete annelids (earthworms and their relatives) these dorsal hearts are supplemented or even supplanted by a series of paired hearts on the lateral vessels that link the dorsal and ventral longitudinal vessels (Fig. 7.15a). In the giant megascolecid earthworms of Australasia, which are up to 3 m long and weigh 0.5 kg, these lateral hearts can generate pressures of up to about 20 mmHg.

Molluscs show a greater diversity of hearts, with a two-chambered arrangement becoming common (Fig. 7.15b). In most classes of mollusc there is a thin-walled **auricle** (or atrium) collecting blood from venous vessels and sinuses, then a thick-walled muscular **ventricle** as the main pump, supplying several main arteries leading in parallel to the important regions of the body (the kidney, the visceral hump where digestion and absorption occur, and the head region). The heart is enclosed in a fairly rigid sac, the **pericardium**, so that the sequential contractions of the auricle and then the ventricle provide suction pressures to speed up the refilling process. In cephalopod molluscs this same structure exists but is proportionately larger and thicker-walled, and the pumping action is supplemented by a pair of branchial hearts at the base of the gills, providing a pressure boost in the venous system as it enters the complex gill vasculature where viscous resistance must be overcome.

Rather different kinds of hearts are found in arthropod groups, where the absence of a piped venous return makes the return of blood to the heart more problematic. Most crustaceans solve this with a series of valved holes or 'ostia' in the heart itself (Fig. 7.15c), taking in blood from the pericardial cavity during cardiac expansion; as the heart then contracts with the ostial valves closed, blood is expelled forwards into the arteries and the negative pressure created in the rigid pericardium draws further venous blood in. A similar system operates in insects, and in all these arthropod groups the degree of vascular development (particularly of arteries) is strongly positively correlated with size and activity.

Vertebrate hearts

Fish hearts represent only a slight increase in complexity on the molluscan and crustacean hearts already described. In addition to the **auricle** (or atrium) and **ventricle**, two further sections are found in series to the system. The **sinus venosus** precedes the auricle and is the principal convergence point of the great veins, which empty into it via valves; neither the sinus nor the auricle are particularly muscular, and their weak contractions generate only small pulse pressures. By contrast, the ventricle (lying ventrally to the auricle) is usually extremely muscular. In less active fish it may become reticulate and spongy with only a small open lumen, but in fast-swimming fish there is a good-sized lumen and an outer layer of circumferential muscle with its own coronary blood supply, producing a fast and forceful contraction. In both elasmobranch and teleost fish the ventricle leads into a fourth chamber, either a **bulbus arteriosus** (a thickened elastic structure in teleosts) or a **conus arteriosus** (both elastic and muscular, in elasmobranchs). The main function of this extra section to the heart is to smooth out the pressure changes created by the ventricle.

In a few fish where lungs have been added to the respiratory system, notably the dipnoan lungfish, we see the first indications of a separation of the oxygenated and deoxygenated bloods in vertebrates. The return vessel from the lungs, carrying oxygenated blood, enters the sinus venosus separately from the other veins, and an **intra-auricular septum** keeps this oxygenated blood largely separate from the deoxygenated systemic venous blood as it traverses the auricle. The ventricle is also partially divided by a plug of tissue, and the conus arteriosus is modified with a spiral twist and complex valves. The net effect is that the two bloods are only partially mixed and the lungs get a relatively deoxygenated supply compared to the O_2-rich blood entering the systemic arterial loop. This partial separation is continued in the amphibians, with an improved distinction of the pulmonary and systemic return circuits brought about by a full separation of the auricle into right and left chambers, and a more effective partial **ventricular septum**. Differential timings of the flows from each return system, and channelling of flows by mechanical obstacles (ridges within the heart), seem to ensure the supply of rather fully oxygenated blood to the systemic circuitry. However, the amphibian heart, at least when compared with those of many active fish, is relatively weak, with a spongy poorly muscularized ventricle and no independent coronary blood supply.

In the amniote vertebrates there is increasing separation of both the venous and the arterial circuits, with septation inside the heart increasing and separation of the auricles evident externally as the sinus venosus is progressively lost. In most reptiles the ventricle is not quite fully divided, but the sinus has been largely incorporated into the right auricle, while the bulbus has been divided into two separate channels opening via valves right back at the ventricular wall. It should be noted that partially separated circulations in the 'lower' vertebrates are not necessarily inferior, and allow the individual to choose the pattern of blood oxygenation in different parts of the body, particularly useful in diving animals.

In adult crocodilians, and in all birds and mammals, the ventricular septation is complete, with independent right and left ventricles opening separately to the pulmonary arteries and the aorta, respectively. The detailed anatomy of the system differs somewhat in each of these three groups, but all have achieved the pattern of two side-by-side pumps, with the right auricle and ventricle handling deoxygenated blood for transit on to the lung, and the left auricle and left ventricle handling the oxygenated blood from the lung for transit on to the rest of the body (Fig. 7.16). All these amniote vertebrates also have highly muscular ventricles and independent coronary arteries supplying the heart muscles. Birds have the most powerful of all hearts, coping with the 10–15-fold increase in O_2 demand during flight and giving a 5–7-fold higher rate of tissue perfusion compared with a similarly sized mammal.

In the avian and mammalian heart, the muscular ventricle fills with blood while it is relaxed (**diastole**), and then contracts to force blood out into the arteries (**systole**). The terms diastolic pressure and systolic pressure can also be applied to less sophisticated circulations, with highest pressures always recorded at peak systole.

7.4.3 *Circuitry*

With invertebrates generally, there is a single circuit of blood around the body, and the circuit runs from heart to systemic sys-

Fish heart

Mammal heart

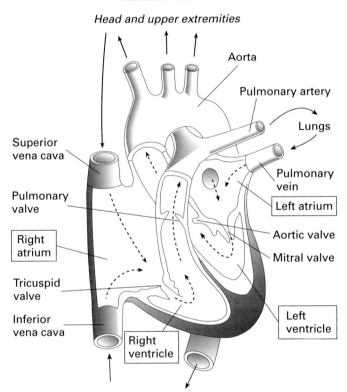

Fig. 7.16 Vertebrate heart complexity in fish and in mammals.

tem to gills or lungs and then back to the heart (Fig. 7.17a). In vertebrates, however, even where there is only a single-circuit system there are two differences in the pattern of the circuitry. Firstly, in vertebrates such as fish the blood from the heart goes first to the respiratory exchange site (gills or lungs) and thence to the main systemic organs before returning to the heart (Fig. 7.17b). Secondly, all invertebrates with vascular systems have a dorsal

vessel where the flow runs from posterior to anterior and a ventral vessel with flow in the opposite (backwards) direction, whereas in all vertebrates in the main ventral vessel blood flows anteriorly and in the dorsal vessel it flows posteriorly (Fig. 7.18).

Where dual circuitry is found, in birds and mammals (and crocodiles), the two halves of the circulation have very different physiological properties. The **pulmonary circuit** is low pressure, but fairly high volume (Table 7.5), driven by the right ventricle. In mammals the average pressure here is only about 13 mmHg, peaking at 25 mmHg. The system is very compliant to changes in output from the heart, and there is little capacity for control (except in crocodiles, where the unusual circuitry allows flow to be diverted away from the lungs by increases in vascular resistance). By contrast, the **systemic circuit** is highly variable in its needs, with the demands of muscles, kidney and skin varying with activity levels, and with water and temperature balance; thus the pressure is always relatively high, but pressures and volumes in different parts of the system may be varied and controlled on a minute-to-minute basis.

In all vertebrates, the blood supply to most organs arises directly from the dorsal aorta, so that each organ is supplied in parallel and at full pressure. However, a few organs are supplied in series with other organs, receiving the venous outflow of another organ, giving a **portal circulation**. This entails a delivery of partially deoxygenated blood, but it permits particular metabolites to be localized in the circulation, so that they do not reach the body as a whole. This is an obvious advantage in some circumstances, for example where the liver receives digestion products directly from the intestinal and splenic circulation; this is the hepatic portal system, responsible for 75% of the supply to the liver in mammals. The liver can then deal with any toxins present in the food before they reach other tissues. Portal systems also operate in the vertebrate kidney, where it is effluent blood from the glomeruli that supplies the nephron tubules (see Chapter 5), and in the brain where a portal vein carries blood leaving the hypothalamus (and thus carrying several key peptide hormones) directly to the anterior pituitary gland.

7.4.4 Blood flow and pressure

Systolic pressures close to the heart in any animal are always

Table 7.5 Typical systemic and pulmonary pressures in animals with closed circulatory systems.

Animal group	Systolic systemic pressure (kPa)	Systolic pulmonary pressure (kPa)
Cephalopod	6.0	–
Lungfish	3.5	2.8
Teleost	3.9	–
Amphibian	4.4	3.2
Reptile	5.5	3.5
Mammal	15.5	2.5
Bird	21.0	2.7

Invertebrate

Vertebrate

(a)

Invertebrate

Vertebrate

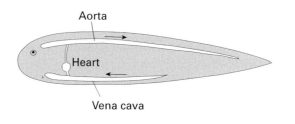

(b)

Fig. 7.17 (a) Differing patterns and (b) directionalities of flow in major vessels in invertebrates and vertebrates. In invertebrates the respiratory organs are downstream of the tissues, whereas in vertebrates (until two separate circuits are established in reptiles) the respiratory organs are served directly from the heart. Dorsal vessels also carry blood forwards in almost all invertebrate groups and backwards in vertebrates.

well above atmospheric pressure, whilst other parts of the circulatory system may be at almost zero pressure. It is this pressure gradient that drives the blood around the circuit and from arteries to veins via the capillaries. Because hearts beat intermittently, the systemic arterial pressure is not constant but decays between heartbeats; in a human it decays from around 120 mmHg to about 80 mmHg. This gives a pulsatile driving force and a pulsatile flow.

The relationship between pressure and flow is therefore rather complex. But we can model it for a more simple situation with flow along a rigid tube at constant pressure, where the flow, Q, is directly proportional to the difference between the inlet pressure (P_1) and the outlet pressure (P_2):

$$Q \propto P_1 - P_2$$

Note that flow is by definition a rate (mass or volume per unit time), so that it is unnecessary to refer to a 'flow rate'.

If a constant is inserted into the above equation, it is transformed into a simple expression describing flow:

$$Q = K (P_1 - P_2)$$

K is called the hydraulic conductance of the tube, and its reciprocal is called the resistance to flow or hydraulic resistance (R):

$$Q = (P_1 - P_2)/R$$

Thus the flow in the tube is proportional to the 'driving pressure', i.e. the pressure differential ($P_1 - P_2$), and is inversely related to the hydraulic resistance. This is a form of Darcy's law of flow (which is in turn analogous to Ohm's law relating resistance, voltage and electrical current). It tells us that in any animal's circulatory system, flow can be altered either by changing the driving pressure exerted by the pump, or by altering the vascular resistance.

In practice, blood pressure tends to be kept relatively constant and flow is regulated by resistance changes brought about by the contraction and relaxation of vessels—**vasoconstriction** and **vasodilation**.

7.4.5 *Blood vessels*

There are various kinds of blood vessel in any animal's circulatory system, but those in the vertebrates are best known and may be used as a general model. The pattern is certainly fairly constant: an aorta leading from the heart to one or several major arteries, progressively breaking into narrower high-resistance vessels (arterioles) then branching into a network of capillaries that converge again into venules that unite into the veins. The characteristics of the various types of vessel in a mammal are given in Table 7.6.

With the exception of the capillaries, all blood vessels are made up of three layers of tissue (Fig. 7.19). An inner (intimal) layer of flat endothelial cells rests on a middle (medial) layer of circular smooth muscle cells embedded in a matrix of elastin and collagen fibres. Outside these is a connective tissue layer (the

Fig. 7.18 (*Opposite*) An overview of circulation patterns in various vertebrate taxa.

Ancestral and embryonic pattern
(6 main gill arch arteries)

Carotid to head

1
2
3
4
5
6

H

Oxygenated
Portal
Deoxygenated

Fish

2
3
4
5
6

H

Fish

Dorsal aorta

Head

Gill

Heart

Ventral aorta

Anterior cardinal vein

Gut

Liver

Kidney

Renal portal

Posterior body

Hepatic portal

Posterior cardinal vein

Frog

3

4

6

H

Skin   Skin

Lung

6 = Pulmocutaneous
4 = Systemic
3 = Carotid

Frog

Aorta

Skin

Pulmocutaneous artery

Lung

Heart

Head

Gut

Liver

Kidney

Renal portal

Posterior body

Hepatic portal

External jugular

Posterior vena cava

Reptile

3

4

Variable

6

Lizard

Aorta

R. systemic arch

L. systemic arch

Pulmonary vein

Lung

Pulmonary artery

Heart

Head

Gut

Liver

Kidney

Renal portal

Posterior body

Hepatic portal

External jugular

Posterior vena cava

Mammal
(birds retain the *other* systemic arch)

3

4

6

Mammal and bird

Aorta

L. systemic arch

Lung

Heart

Head

Gut

Liver

Kidney

Posterior body

Hepatic portal

External jugular

Posterior vena cava

Table 7.6 Properties of different kinds of blood vessels and distribution of blood in mammals (total volume = 8–9% body volume).

	Typical diameter	Wall thickness	Composition of vessel wall (% diameter)				% of total blood volume contained	
			Endothelium	Muscle	Collagen	Elastin	Systemic	Pulmonary
Aorta	25 mm	2 mm	5	22	33	40	–	–
Artery	4 mm	1 mm	10	35	25	30 ⎫		
Arteriole	30 μm	20 μm	10	50	20	20 ⎬	12	7
Capillary	8 μm	<1 μm	100	0	0	0	11	1
Venule	20 μm	2 μm	40	10	40	10 ⎫		
Vein	20 mm	1 mm	8	32	40	20 ⎬	41	14
Heart								14

Smooth muscle + elastin + collagen

Artery — 10–25 mm

Endothelium

Arteriole — 30–50 μm

Capillary — 8 μm

Venule — 10–20 μm

Vein — 1–20 mm

Fig. 7.19 Transverse sections of vertebrate blood vessels, showing typical dimensions.

adventitial layer), containing nerves and often (in the large arteries in particular) some small blood vessels to provide circulation to the vessel wall.

Within this general structure, certain distinct types of vessel can be distinguished. **Elastic arteries** include the pulmonary artery, aorta and major limb arteries; these have very extensible walls due to the presence of large amounts of elastin fibres in the medial layer (elastin being six times more extensible than rubber). These major arteries therefore expand during ventricular emptying and recoil again later, thus having the effect of converting the intermittent pumping of the heart into a relatively continuous flow through the more distal vessels. They also have a good proportion of collagen fibres in the medial layer; these fibres are stiffer than elastin and probably ensure that the vessels do not get over-distended to cause aneurysms or bursts.

Muscular arteries are less important in smoothing out blood flow but have a rich autonomic nerve supply and can contract. Examples include many smaller arteries such as the cerebral and coronary arteries. They have thick walls to prevent arterial collapse at sharp bends such as the knee and elbow joints. In diving animals they contract sharply during a dive to limit the blood supply to the limbs and ensure that the brain and heart are well supplied.

Arterioles have the highest wall thickness to lumen ratio of any vessels. The larger arterioles are richly innervated by vasoconstrictor neurones, while the smaller ones are poorly innervated and have a particularly narrow lumen. The pressure drop here is greater than at any other point in the circulatory system, indicating that the systemic arterioles represent the chief resistance to flow (Fig. 7.20). The smaller arterioles are therefore referred to as **resistance vessels**, and by changes in their luminal diameter (vasoconstriction and vasodilation) they control the blood flow to organs and tissues, mediated by a variety of chemical and neural signals. Full contraction of the terminal arterioles can shut off the blood flow to a particular region of capillaries (a capillary 'bed') entirely.

Capillaries have the smallest diameter of any vessels (only 3–100 μm) and are generally very short (250–1000 μm). They are present in very large numbers, reducing the diffusion distances from circulation to cell. Their thin walls consist of a single layer of flattened endothelium, through which gases and metabolites can permeate. Blood passes along capillaries in short, smoothed pulses ('bolus flow'), and because of this and their very high density and thus large total cross-sectional area they offer rather little resistance to flow. The transit time of a red blood cell in mammalian muscle capillaries is around 1–2 s, allowing sufficient time for the unloading of O_2 and uptake of CO_2.

Certain tissues, such as the skin and the mucosa of some chemosensory organs, possess shunt vessels, 20–130 μm in diameter and called **arterio-venous anastomoses**. These enable

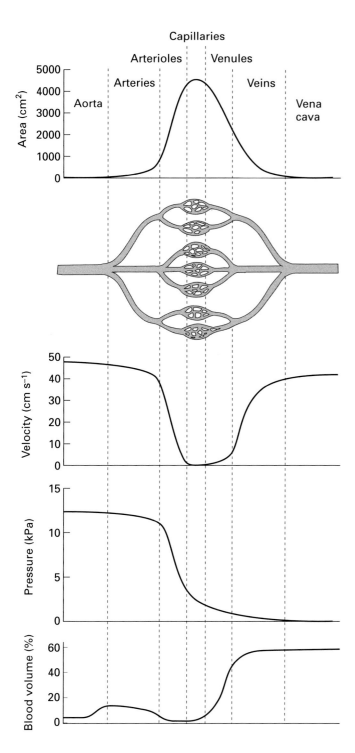

Fig. 7.20 Pressure variation, in relation to surface area and velocity, through the mammalian circulatory system. Area peaks in the capillary beds, where velocity is lowest; the major drop in pressure occurs through the arterioles.

blood to pass directly from arterioles to venules without going through the capillary bed. Such an anastomosis has a rich sympathetic innervation, and in many mammals the anastomosing vessels have an important role as bypasses in regulating blood flow to the surface of the body as part of the process of temperature regulation.

Venules and small **veins** are more numerous than arterioles and arteries (see Table 7.6), and hence they provide a low resistance to flow. For example, in humans a pressure differential of only 10–15 mmHg is sufficient to drive the cardiac output from the venules back to the vena cava (the great vein leading to the heart). In addition to returning blood to the heart, the venous system of vertebrates serves as a variable reservoir for blood; the vessels have thin walls that are readily distended or collapsed, having only low elastin content, enabling their blood content to vary enormously according to their degree of vasoconstriction. The venules and veins are therefore referred to as **capacitance vessels**. In the limbs of vertebrates, the veins have pairs of semilunar valves that prevent the backflow of blood even when operating at high capacity.

7.4.6 Blood and haemolymph

All animal circulatory fluids are primarily water, with low and variable concentrations of ionic solutes, waste products, dissolved gases, foodstuffs, some proteins and some cells. The solute composition is largely related to osmotic regulation and habitat, and was dealt with in Chapters 4 and 5, where waste products were also discussed. Concentrations of food breakdown products (especially glucose, amino acids and free fatty acids) are variable with an individual's state of activity and recency of feeding, so that systematic differences between taxa or between habitats are very hard to detect. However, the protein and cellular components are more relevant here in the context of respiration and gas transport. Many marine invertebrates have very low levels of protein in the blood or haemolymph or coelomic fluid (Table 7.7), though the level is higher in decapod crustaceans and very high in cephalopods. Most vertebrates have moderately high plasma protein levels.

These high blood protein levels are responsible for a raised **colloid osmotic pressure** within the blood vessels, since the proteins are too large to permeate the vessel walls. A few large proteins have less effect than many smaller protein molecules, and animals with high concentrations of small blood proteins have a substantial Donnan effect influencing the distribution of other charged solutes in the body fluids (see section 4.3.3). These proteins are predominantly the respiratory blood pigments, dealt with below.

Most animal bloods also contain cells, generically termed **haemocytes**. These have many functions (Table 7.8), particularly involving defensive systems (e.g. clotting, repair, phagocytosis and chemical immune responses; see Chapter 15) and storage (of foods, excretory products or blood pigments). In some animals a subpopulation of the blood cells are specialized as containers of the respiratory pigment, and these may lack cell nuclei; they are commonly given the special name of **erythrocytes**.

Blood cells make up a highly variable proportion of blood volume, which is usually expressed as **haematocrit** (packed cell

Table 7.7 Protein contents of plasma in different taxa.

Animal group	Plasma protein (g l^{-1})
Most marine invertebrates	
(e.g. cnidarians, sipunculans, echinoderms)	0.2–2.0
Marine crustaceans	10–90
Freshwater crustaceans and molluscs	10–50
Cephalopod molluscs	80–110
Insects (varies with moult cycle)	20–100
Vertebrates	30–80

volume as a percentage of total volume). This ranges from about 20–30% in most amphibians and reptiles to about 30–45% in birds and mammals (up to 50–55% in some very small mammals and in some diving marine mammals).

7.5 Delivering and transferring gases to the tissues

7.5.1 *Oxygen carriage: pigments and their properties*

We have seen that when animals become too large to rely on diffusion alone, the solutions they generally adopt include ventilation systems externally, and circulatory systems internally to carry respiratory gases to and from the metabolizing tissues. Further sophistication can be added to this. The simplest way to carry gases in the blood is by dissolution, since, for example, O_2 in the blood will move down the diffusion gradient into any tissue where O_2 is being consumed. However, the quantities of gas that can be moved around in this manner will be limited by the solubility coefficients of the gas concerned (see equation for Henry's law, p. 158). The blood of a marine animal, having roughly the ionic composition of sea water, will hold about 5.5 ml l^{-1} of O_2 at 20°C when exposed to normal air. Under the same conditions this blood will hold 159 ml l^{-1} of CO_2, a gas with much greater solubility (see Table 7.2), giving about a 29-fold

increase in capacity. Oxygen limitation will therefore become a serious problem for larger animals with a high metabolic rate, especially where raised body temperature further compromises the blood plasma's ability to hold O_2. Therefore in nearly all of the more complex animals there are special components of the blood that enhance its O_2-carrying capacity—the **blood pigments**, or **respiratory pigments**.

Pigment types

These compounds (Table 7.9) are highly pigmented because they are metalloproteins, containing metallic ions bound to one or more polypeptide groups, so that the metal as it undergoes partial changes in its valency state confers varying colour on the pigment as a whole.

The **haemoglobins** (Hb) are closely allied to the non-circulating tissue myoglobins found in many organisms (including fungi, some plants and many protists as well as all animals). They all have iron as their metallic component, in its ferrous state (Fe^{2+}), and bound in the centre of a porphyrin ring (Fig. 7.21) to give a functional haem group. The haem group is located within a protein (globin) molecule to give a functional pigment subunit. There are four such subunits in a typical mammalian haemoglobin molecule, with two α and two β globins. Oxygen is bound reversibly to the iron by a partial transfer of one electron to the O_2, producing a transient ferric state in the iron and a superoxide of oxygen (OO^-); the result is oxyhaemoglobin HbO_2. The iron-porphyrin functional group is loosely bound to proteins, which are of variable form. Haemoglobins are very diverse, with subunits ranging from monomers of low molecular weight to huge aggregates of up to 3 million daltons in some annelids (where they are often green and are termed erythrocruorins or chlorocruorins). Haemoglobins can be found in some nematodes and nemertines, many annelids, a few molluscs and insects, some crustaceans, and virtually all vertebrates (the exceptions being a few icefish and a few amphibians). They are normally extracellular, but occur within **red blood cells** (RBCs)

Table 7.8 Functions of blood cells in different animal groups.

Function	Annelids, sipunculans, echiurans	Molluscs	Crustaceans, myriapods, insects	Echinoderms	Vertebrates
Phagocytosis	✓	✓	✓	✓	✓
Encapsulation	✓	✓	✓	✓	
Wound plugging	✓	✓	✓	✓	✓
Clotting	✓		✓	✓	✓
Immune responses	✓	✓	✓	✓	✓
Glycogen storage	✓	✓		✓	
Lipid storage			Some	Some	
Hb storage (few)	Few	Some	Some	Few	✓
Storage excretion	✓		Some	✓	
Nacrezation (pearl formation)		Some			

Table 7.9 Respiratory pigments.

Pigment	Structure	Colour (+ change)	Oxygen capacity (ml g^{-1})	Molecular weight (kilodaltons)	Cells or solution
Haemocyanin	Protein + Cu^{2+}	Blue (colourless)	0.3–0.5	25–7000	Solution
Haemoglobin	Protein + haem + Fe^{2+}	Red (purple/blue)	1.2–1.4	16–2000	Either
Chlorocruorin	Protein + haem + Fe^{2+}	Green	0.6–0.9	3000	Solution
Haemerythrin	Protein + Fe^{2+}	Violet (colourless)	1.6–1.8	16–125	Either

Fig. 7.21 The structure of haem (porphyrin ring plus a ferrous ion), of myoglobin and of tetrameric haemoglobin.

in at least seven different animal phyla, mostly marine and intertidal in habitat. Vertebrate haemoglobins are always intracellular and of relatively low molecular weight (~17 kDa).

Haemerythrins (He) are less common, occurring in three phyla, which from recent molecular evidence may have a common phylogenetic origin (sipunculids, priapulids and bra-

chiopods), as well as in one family of marine annelids. Their most common forms are trimeric or octomeric, each polypeptide containing an iron-binding site within a fold of its tertiary structure where two Fe^{2+} ions are located. The probable binding system involves forming a peroxo compound (O_2^{2-}) of ferric iron. **Haemocyanins** (Hc) also involve direct reversible binding of a metal to protein structures, but in this case the metal is copper and a peroxo compound of cupric (Cu^{2+}) ions occurs, the active site involving six histidine residues. Haemocyanins

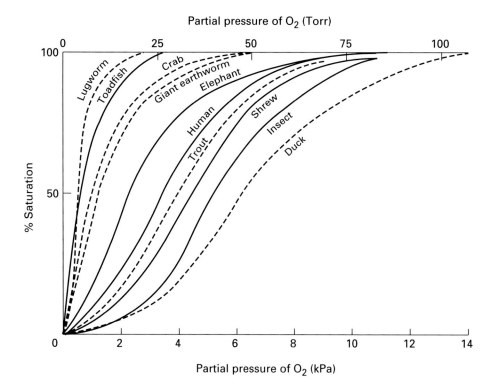

Partial pressure of O$_2$ (Torr)

% Saturation

Partial pressure of O$_2$ (kPa)

Fig. 7.22 Oxygen dissociation curves as percentage saturation.

are very large proteins, occurring in multimeric forms with up to 48 of the original polypeptide units. They occur in molluscs (some gastropods and all cephalopods, plus chitons and a few bivalves) and in some arthropods (crustaceans, the horseshoe crab, arachnids and centipedes), but the protein components in molluscs and arthropods are sufficiently distinct to suggest independent origins.

Most animals, whatever their pigment, use it in an extracellular form. However, haemoglobin as we saw above is sometimes packaged into blood cells instead, and haemerythrin is also intracellular in most animals where it occurs. Traditionally it was argued that intracellular pigments allowed higher pigment concentration without compromising the colloid osmotic pressure of the blood or making the blood too viscous, but neither of these arguments holds up well. Alternatively, it might be that an intracellular site allows closer control of O$_2$ binding by allosteric modulators. However, most of the invertebrates with pigments contained in cells do not have complex modulatory systems. It seems likely that the vertebrate RBC was simply an inherited phenomenon, to which allosteric modulators were added as respiratory demands became more sophisticated.

Oxygen binding

The degree of oxygenation of a pigment depends on the partial pressure of oxygen, Po_2; the pigment is fully saturated at some specific Po_2 where all its O$_2$-binding sites are occupied. The extent of saturation can be calculated as:

$$\% \text{ saturation} = \frac{100 \times [\text{HbO}_2]}{[\text{Hb}] + [\text{HbO}_2]}$$

The relation between this value and Po_2 is traditionally used to show pigment properties, and is termed the **oxygen equilibrium curve** (also known as the loading curve or dissociation curve).

To be physiologically useful, any respiratory pigment must bind O$_2$ at relatively high partial pressures, and release it at the kind of partial pressures found in metabolizing tissues. In most animals, the O$_2$ equilibrium curve is sigmoid (Fig. 7.22). Nonlinearity is to be expected; at low Po_2 and low pigment saturation, any O$_2$ molecule is likely to hit an unoccupied site on a deoxygenated Hb molecule, while at higher Po_2, its chances of hitting a free site are lower and increasingly more O$_2$ molecules are needed to increase the pigment loading by a particular increment. This might produce a hyperbolic curve. But in addition, the **oxygen affinity** of the pigment as a whole (at least of haemoglobins and haemocyanins, which have been best studied) is a function of its precise molecular shape (its tertiary and quarternary structure), and binding O$_2$ to any one haem group slightly affects the shape of the rest of the molecule. In fact all the four haem components in vertebrate haemoglobins interact, so that the binding of O$_2$ to the first haem increases the affinity of the second one, and so on, giving the sigmoid effect. A useful quick comparison between such curves is given by the P_{50}, the partial pressure at which 50% (half) saturation occurs; a high-affinity pigment has a low P_{50} and a curve well to the left, whereas

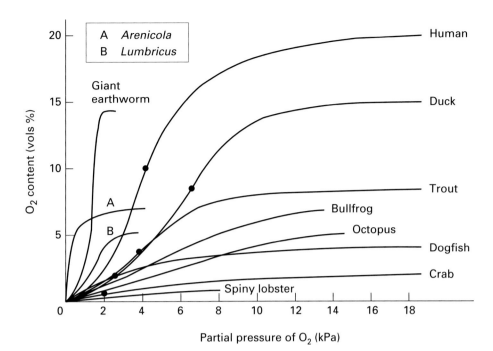

Fig. 7.23 Oxygen dissociation curves as actual O_2 content, stressing differences in O_2 capacity.

a low-affinity pigment has a high P_{50} and a curve shifted to the right.

When O_2-binding curves are expressed as percentage saturation the relative shapes of the curves in different species can be readily compared (as in Fig. 7.22), and this kind of comparison will often be used in later chapters; invertebrate Hbs commonly have lower P_{50} values than vertebrate Hbs, and Hbs from more active species amongst the vertebrates (carnivores and insectivores) have particularly low affinities. But note that where the vertical axis of a dissociation curve is actual O_2 content then differences not only in shape (affinity) but in overall capacity are more clearly revealed (Fig. 7.23). On any such plot, the concentrated pigments of mammals turn out to be very effective indeed. Arterial mammalian blood is usually at least 95% saturated, whilst venous blood may be less than 75% saturated (but it is by no means fully unloaded in the tissues and does not therefore normally turn blue in the veins).

A summary of O_2-carrying capacity for fully saturated bloods of a range of animals is shown in Table 7.10, again emphasizing the considerable augmentation produced by pigments, especially in the birds and mammals.

7.5.2 Modulating oxygen-carrying capacity

Oxygen-carrying capacity could be altered either by altering the amount of respiratory pigment or by altering its properties. As we saw in Chapter 2, the functional properties of a protein can be changed either by the action of extrinsic cofactors or by internal changes in molecular structure. Both mechanisms do occur, but while both are common for haemocyanins the latter is rather rare in haemoglobins, especially in endothermic vertebrates. Here

Table 7.10 Oxygen-carrying capacity of bloods with pigments, from a range of animals.

Animal group	Pigment	ml O_2 100 ml^{-1}
Nematode	Hb	1–3
Annelid	Hb	0.1–20
	He	3–6
Echiuran	Hb	4–5
Sipunculan	He	2–3
Cephalopod mollusc	Hc	3–4
Gastropod mollusc	Hc	1–3
Crustacean	Hb	2–3
	Hc	1–4
Insect	Hb	5–12
Elasmobranch	Hb	4–5
Teleost	Hb	4–20
Amphibian	Hb	6–10
Reptile	Hb	6–12
Bird	Hb	10–22
Mammal	Hb	14–32
Water	–	0.65

Hb, haemoglobin; He, haemerythrin; Hc, haemocyanin.

we briefly survey factors that control pigment function, and the control of pigment production.

Temperature

Nearly all haemoglobins and haemocyanins suffer decreasing O_2 affinity (raised P_{50} values) as temperature increases (Fig.

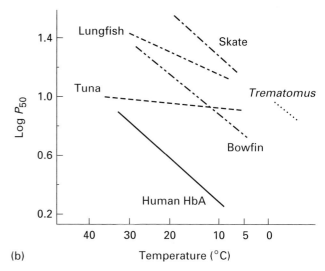

Fig. 7.24 The decreasing affinity of blood pigments with increasing temperature: (a) for haemocyanin in the spiny lobster; (b) as a semi-log plot (P_{50} against temperature) for various vertebrates (the thermal insensitivity in tuna is probably related to the rapid changes in T_b occurring in its swimming muscles—see Chapter 9). (a, Data from Redmond 1955; b, data from Hochachka & Somero 1973.)

7.24). Though irrelevant for bird and mammal haemoglobins (where T_b is constant), this is probably advantageous to most ectothermic animals, since raised temperature will be associated with higher tissue respiration rates and unloading will be assisted. In a range of crustaceans from different thermal niches, it seems that the haemocyanin P_{50} is fairly similar when measured at the normal ambient condition; the values for the Antarctic isopod *Glyptonotus* at 0°C, for the shore crab *Carcinus* at 10°C and for the terrestrial crab *Pseudothelphusa* at 25°C are all in the range 15–25 mmHg, with some seasonal variation in the latter two species. For species living at lower temperatures, a higher proportion of the O_2 transport can take place in simple solution in the blood, so that a lower haematocrit but higher blood volume is sometimes seen (in *Glyptonotus* and in some Antarctic fish).

pH levels and carbon dioxide levels

Haemoglobins and haemocyanins are almost invariably affected by altered pH levels, usually with a shift of the O_2 equilibrium curve to the right as the pH decreases (Fig. 7.25). This effect seems to be universal in vertebrates, and almost so in invertebrates. It is again due to subtle changes in the higher order structure of the protein, with the pH of the solution altering the charge characteristics of the amino acid residues and thus the precise shape of the polypeptide chains. Most animals therefore have mechanisms to buffer pH change; buffering involves weak acids and alkalis, and the most important components in animals are

bicarbonate, phosphate and weak protein anions, especially histidine residues intracellularly. Thus animal bloods show a surprising degree of uniformity in their pH and also in the response of pH to body temperature (Fig. 7.26), ensuring a constant 'relative alkalinity' (and hence protein charge) as temperature changes.

Since pH is decreased by increasing concentrations of CO_2, it follows that raised CO_2 levels will also shift the O_2 equilibrium curve to the right. This is the so-called **Bohr shift**, a term now used for both pH and P_{CO_2} effects (Fig. 7.25). This shift can be very pronounced with both haemoglobins and haemocyanins, and is useful in accentuating unloading of O_2 on the venous side of capillary networks. In the haemocyanins of some molluscs and of the horseshoe crab (*Limulus*), however, a reverse Bohr shift has been reported with both pH and CO_2. Carbon dioxide also seems to have a more direct effect on O_2 binding to blood pigments, independent of its pH effect, by forming carbamino compounds (NHCOO- groups) with the constituent proteins.

Carbon dioxide may affect the maximum O_2 binding, as well as the shape of the equilibrium curve, particularly in the haemoglobins of some fish species. High CO_2 levels decrease the total bound O_2, a phenomenon known as the **Root effect** (Fig. 7.27). This is used to good effect in controlling the activities of fish swim-bladders (see section 9.6.2). Root effects have not been observed with haemocyanins.

Ionic levels

Since haemoglobin has weakly dissociated groups in its own structure, it is affected by overall ionic strength in the blood, and in general any increase in inorganic ion concentrations will lead to decreased O_2 affinity (increased P_{50}). This is probably not very important in vertebrates, where the Hb is packaged inside tightly regulated cells, but it may be very significant in many euryhaline and osmoconforming invertebrates, where the blood concentration varies with that of the medium. More specific

(a) Partial pressure of O$_2$ (kPa)

(b) Partial pressure of O$_2$ (kPa)

Fig. 7.25 The Bohr shift due to the effects of pH or $P\text{co}_2$ on respiratory pigments: (a) in the haemocyanin of the mangrove crab with varying $P\text{co}_2$; (b) in the haemoglobin of the dog with varying pH. (Data from Roughton 1964.)

effects also occur, with calcium and magnesium both affecting Hb and Hc binding directly (Fig. 7.28).

Organic modulators

Various small organic compounds are known to be modulators of haemoglobin affinity in animals. In vertebrate water-breathers

Fig. 7.26 The relation between blood temperature and pH across a range of taxa, suggesting that animals regulate their blood at a fairly constant 'relative alkalinity' at which there is a constant protein net charge. (From Feder 1987; courtesy of Cambridge University Press.)

the main strategy during hypoxia is the release of an organic phosphatic modulator which alters the P_{50}. In some aquatic animals, including many fish, ATP itself seems to regulate oxygen affinity. For air-breathers one of the most important modulators is 2,3-diphosphoglycerate (2,3-DPG), the addition of which decreases the O$_2$ affinity of the blood of most mammals. This may be significant in high-altitude adaptation (Chapter 14), and it perhaps also serves to enhance fetal–maternal gas exchange. It is also used in developing bird and reptile eggs.

Pigment production

Many species show varying patterns of expression of O$_2$ transport proteins during ontogeny. For example, larval chironomid flies express a haemolymph Hb that has very different properties and Bohr effects from the adult form. More rapid and reversible changes in O$_2$-carrying capacity are also possible in relation to need, and are often brought about by increments of the pigment concentration, which in vertebrates means production of more RBCs (**erythropoiesis**). This is controlled by the hormone erythropoietin. In many vertebrates, especially strong runners such as dogs and horses and also aquatic species, some RBCs are stored in the spleen and can be quickly mobilized when needed.

7.5.3 *Oxygen storage*

Respiratory pigments can transfer O$_2$ between themselves, and this frequently occurs in invertebrates, where there may be separate pigments in cells, in plasma and in the coelomic fluids. In such cases the intracellular pigment may serve as a short-term store during transient anoxia. A similar phenomenon occurs in many animals with muscle **myoglobins**, where the relatively small and mobile myoglobin facilitates diffusion across the cell membrane.

Fig. 7.27 The Root effect, a right-shift in the O_2 dissociation curve due to increased P_{CO_2} or pH, with a decrease in the O_2-binding capacity. (a) For fish haemoglobin and (b) for octopus haemocyanin, showing the non-linearity of the effect.

Fig. 7.28 The effects of specific ions on O_2 dissociation in crustacean haemocyanin; the log P_{50} is shifted up as salt concentrations increase.

Pigments have a particularly important role as O_2 stores in animals that burrow, such as benthic annelid worms, many littoral species, and certain freshwater fly larvae. Here the pigment-bound O_2 is generally insufficient to support aerobic

respiration for more than about an hour, so that it could not be used for a full intertidal cycle in the littoral zone; but it is enough to support aerobic metabolism between periodic bouts of gill ventilation in a benthic burrower, or during an escape response.

Stores are also important in air-breathing animals that dive (see Chapters 9 and 11). In aquatic mammals the blood is the most important store of O_2 during the period of apnoea, but muscle myoglobin is a very important secondary source; the muscle myoglobin store in seals is particularly high relative to that in similar-sized non-diving mammals such as man.

7.5.4 Carbon dioxide transport

Carbon dioxide is a relatively soluble gas, and its transport in solution presents no particular problems for most animals; it will diffuse into the blood from active tissues, and be disposed of into the surrounding medium at the external exchange surfaces. Most of the CO_2 in an animal's fluids is actually chemically combined with water to form carbonic acid and thus bicarbonate ions, as follows:

$$CO_2 + H_2O \rightleftharpoons H_2CO_3 \rightleftharpoons H^+ + HCO_3^-$$

The first of these reactions is relatively slow but the second very fast, so that under normal conditions the formation of the carbonic acid would be the rate-limiting step. However, the enzyme **carbonic anhydrase** catalyses the reaction, and is present within vertebrate red cells. Thus CO_2 is very readily taken up into such cells at its sites of formation in the tissues (Fig. 7.29), and the resulting HCO_3^- then diffuses out into the plasma in exchange for chloride ions (the **chloride shift**, which slightly increases the osmotic concentration of the RBCs so that water enters and they swell a little; thus venous blood has a detectably higher haematocrit than arterial blood). The remaining H^+ is effectively buffered by the haemoglobin inside the blood cells, and the acid haemoglobin (HHb) has a lower affinity for O_2,

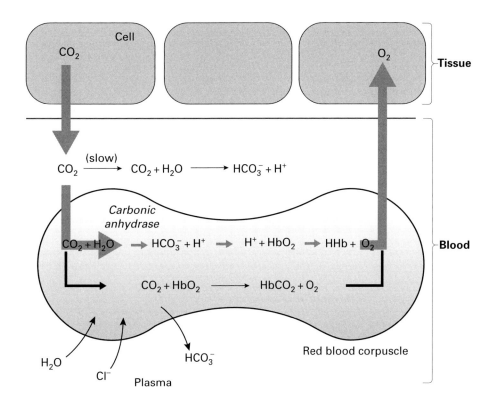

Fig. 7.29 Carbon dioxide carriage in red blood cells (details in text).

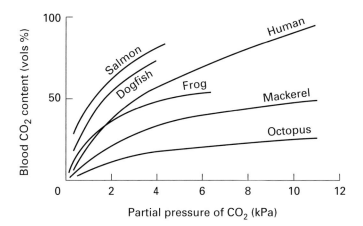

Fig. 7.30 Carbon dioxide dissociation curves in vertebrates.

allowing O_2 to be unloaded to the tissues. About 65% of the CO_2 ends up in the plasma in the bicarbonate form, and no carriers of any kind are involved. The carriage of CO_2 is therefore not saturable, showing a rapid increase at low P_{CO_2} and thereafter a slow steady increase (Fig. 7.30).

In invertebrates CO_2 transport has been less well studied but shows similar patterns. Carbonic anhydrase is sometimes involved, and carriage as bicarbonate is the norm, with Hb or Hc acting as efficient H^+ buffers.

However, the issue of CO_2 distribution does become somewhat complicated because of the effects of CO_2 on acid–base balance in the animal as a whole, so that CO_2 levels in the circulating fluid have knock-on effects on various other physiological systems.

Hypoxia *per se* has little effect on acid–base relations, but since in most animals it leads to increased ventilation more CO_2 is lost from the body and **respiratory alkalosis** can set in. However, where the hypoxia was itself a result of exercise, there will also be an increase in metabolism and of metabolic end-products which are acidic, so that respiratory alkalosis is countered by a degree of induced **acidosis**.

Hypercapnia (an increase in P_{CO_2} in the blood) may be caused independently of hypoxia by an inadequate excretion of CO_2 or an increase in the environmental P_{CO_2}. It has a rapid effect on animal body fluids, with pH decreasing sharply. In air-breathing animals this acidosis will normally be quickly rectified by increased ventilation, since blood P_{CO_2} directly triggers the ventilatory control centres (see section 7.7); and in the longer term the kidney will excrete an excess acid load. However, in aquatic animals, including fish and crabs, blood P_{CO_2} and hence HCO_3^- levels are routinely low, and regulation of pH cannot be achieved by varying P_{CO_2}; thus ventilation is less important as a response to CO_2 levels or acidosis. Here the gills are used as an ion regulatory exchange site, performing the classic Na^+/H^+ and Cl^-/HCO_3^- exchanges (Chapter 5), though this may leave a degree of alkalosis due to accumulated bicarbonate, so that complete compensation is rare.

7.5.5 Other functions of pigments

Oxygen-transporting proteins have taken on several additional roles in animals. In deep-sea vent animals they can transport and detoxify hydrogen sulphide, as we will see in Chapter 12. In vertebrates the RBCs are involved in the regulation of blood pressure, because the haemoglobin is able to bind nitric oxide, which is an important signal molecule (Chapter 2) acting as a relaxing factor in blood vessel endothelium.

7.6 Coping with hypoxia and anoxia

Animals in many different kinds of habitat may have to deal with periods of hypoxia or anoxia; examples include aquatic animals in benthic muds or on shores when the tide goes out or in a pond that dries up or freezes over, and terrestrial animals in underground burrows or at altitude. Many other situations lead animals to enter an induced **hypometabolic state**, or **metabolic depression**, as we discussed in Chapter 6. All such animals commonly show varying degrees of 'hypoxia adaptation', depending upon down-regulation of energy turnover, and up-regulation of energetic efficiency. Different combinations of these two strategies may be used, but it appears that suppression of energy turnover (often by an order of magnitude) supplies the best protection against severe hypoxia, and the process has been likened to a boiler that has its main burners switched off and is 'turned down to the pilot light'. The O_2 sensor at the cell level may well be a haem protein, and from this a transduction process ultimately leads to the activation of some genes and suppression of others, with enzyme phosphorylations as a major outcome so that metabolic depression results.

The changes to fundamental cellular metabolism that accompany hypoxia and anoxia, including switching to various anaerobic pathways, were covered in Chapter 6 (especially section 6.4.4). As a result of these changes, protein metabolism, glucose synthesis and maintenance of electrochemical gradients are all depressed. Cellular permeabilities are also reduced, so that the reduced sodium pumping rate can still keep excitable membranes functioning, albeit at a lower firing frequency. In the brain of vertebrates, hypoxia tolerance is geared to maintaining the crucial ionic gradients, so that increased blood supply, reduced electrical activity, increased release of inhibitory transmitters and neuromodulators and down-regulation of excitatory ion conductances are all coordinated. In hypoxia-intolerant species (e.g. most mammals) neurones lose ATP progressively, depolarizing and suffering toxic increases in intracellular calcium levels, with cell swelling and membrane damage. However, in hypoxia-tolerant vertebrates, such as diving turtles, the neurones are able to decrease their energy usage and ion channel activity in a coordinated fashion, regulating intracellular calcium at a new setpoint.

The best-known molecular mechanism of hypoxia-dependent regulation involves the action of **hypoxia-inducible factor 1** (Hif-1); this is a helix-loop-helix transcription factor, and binds to various sequences on the regulated genes whenever hypoxia occurs (e.g. in humans in response to high altitude, anaemia or severe wounding). Hif-1 is seen as the master regulator of oxygen homeostasis in mammals, with particular effects on erythropoietin (controlling RBC production) and on vascular endothelial growth factor (controlling capillary density). Long-term hypoxic adaptation also requires binding of regulatory proteins to other sections of mRNA to increase their stability. Coupling between the oxygen-sensors and these regulatory systems is still being explored; it excites interest in mammals because of the role of hypoxia in mortality from heart attacks and strokes.

The further problem posed by bouts of hypoxia or anoxia is that of the recovery phase when cells and organs become reoxygenated. This phase may be accompanied by high levels of particularly reactive oxygen species, notably the free radicals superoxide and hydrogen peroxide referred to in the introduction to this chapter. These radicals can damage macromolecules, especially lipids; they produce conjugated dienes and lipid hydroperoxides, which are useless and potentially damaging in themselves. Animals therefore need protection by the use of detoxifying **antioxidants** during this phase. Species that regularly experience bouts of anoxia therefore tend to have particularly high levels of the key enzymes such as superoxide dismutases, peroxidases and catalases. Levels of these are generally higher in ectothermic animals than in endotherms, and higher in animals from variable habitats, indicating a good match between antioxidant defences and the probability of experiencing temperature change and hypoxia. Non-enzymic antioxidants also occur: invertebrates often use the small antioxidant glutathione, and vertebrates sometimes show high levels of the antioxidants vitamin E and vitamin C. During anoxia the levels of these may decrease, but they rapidly increase (especially in tissues such as the brain, heart and kidney) during aerobic recovery.

7.7 Control of respiration

Since environmental conditions may vary so markedly, and with the additional problem of varying animal activity levels, all animals require some further system for controlling their gas exchange and respiratory rate. Metabolic demand for O_2, and (secondarily perhaps) the need for CO_2 excretion, are the major determinants of the rate needed; Po_2, Pco_2 and pH are likely to be the main feedback triggers. To regulate the needs of the organism as a whole, an animal is likely to need both internal and external receptors, and may also need specialized 'defence receptors' to ensure appropriate protective responses to unexpected noxious stimuli in the environment.

7.7.1 Aquatic animals

In most aquatic animals the respiratory surface is a rather delicate site of high surface area at the periphery of the body, even protruding into the medium, so that it is peculiarly vulnerable to physical or chemical damage. Receptors to detect damaging conditions—**nociceptors**—may therefore be sited close to the inhalant openings. Such receptors have been identified in the nares (nasal passages) of certain fish, and in the anterior segments of tube-dwelling annelids. They may elicit an inhibition of gill ventilation, or even a transient reverse ventilation; in fish this results in a 'cough' response.

Many aquatic animals must also have external receptors sensitive to Po_2, though these are rarely identified. Some teleost fish gill nerves clearly respond to O_2 levels, and in elasmobranchs O_2-sensitive receptors seem to be widely distributed over the branchial cavity, though their precise role remains controversial. Some crustaceans and the horseshoe crab *Limulus* also appear to have O_2-sensitive areas in the gill filaments. In molluscan gills the cilia have an intrinsic beat, though their frequency may be altered in relation to the external O_2 concentration.

However, **internal chemoreceptors** are probably more important in determining the overall respiratory response in most aquatic animals. Fish can detect blood Po_2 with chemoreceptors in regions just distal to the ventral aorta, where venous O_2 tension is readily monitored. In aquatic animals the primary response is nearly always to O_2 rather than to CO_2, as Pco_2 is low and changes only slightly with varying metabolic rate because of its high solubility. However, osmotic or mechanical stresses, and a range of external chemical stimuli, will also affect respiratory patterns.

All these inputs are integrated centrally, and in many animals the organizing centre then feeds its integrated output to a pacemaker system affecting both heart and breathing rates. The central integration of respiratory systems in invertebrate aquatic animals is almost completely unexplored. In the lugworm *Arenicola* and in some freshwater insects respiratory integration is known to involve a group of neurones in the ventral nerve cord. In decapod crustaceans the pacemaker neurones are in the suboesophageal ganglion; these are non-spiking oscillatory cells, providing the basic frequency signal, and are modified by neurones that are fed by the gill receptors. The oscillator cells link to motor neurones in two groups, stimulating muscles that raise and lower the balers (scaphognathites, Fig. 7.9) that ventilate the branchial chambers on each side of the body. The principal response shown is an increased ventilation rate when the animal is hypoxic; there is little response to environmental hypercapnia, and CO_2 regulation is simply dealt with by excretion at the body surface.

In fish the regulation of respiration occurs in the medulla of the hindbrain, the respiratory control input signals arriving via the vagus nerve (cranial nerve X) and to a lesser extent via nerves V, VII and IX. Efferent motor pathways also originate at the medullary level, and the fore- and midbrains do not appear to affect respiratory rhythms. The efferent signals control three principal factors in fish:

1 The **ventilation volume**, by adjustments to the mouthparts, operculi and gill lamellae.

2 The **cardiac output**, with the heartbeat pattern determined largely by the intrinsic rhythmic discharge of the sinoatrial node (SA node), inhibited to varying degrees by the vagus nerve.

3 The **perfusion pattern** within the gills.

In a fish faced with a hypoxic environment, the commonest responses (i.e. the result of integrating all these inputs and outputs) are increased locomotion to move it out of the hypoxic area, and increased ventilation. The magnitude of each response is species-specific, as fish vary greatly in their tolerance of reduced O_2. Many fish also show a slowing of the heart rate (bradycardia), but with an increase in stroke volume to maintain blood flow; and changes in blood distribution within the gill, so that transit time and gas exchange are increased. As with crustaceans and other aquatic animals, respiratory rhythms are not very responsive to hypercapnia.

7.7.2 *Amphibious animals*

Amphibious animals, in terms of respiration, come in two categories. Firstly, there are water-dwellers with additional air-breathing organs. These include a variety of fish, including not only the lungs of lungfish but also modifications of stomach areas in catfish, of buccal areas in the electric eel, of part of the gill chamber in gouramis, and even of parts of the skin or fins in a few species. This category also includes a range of invertebrates, such as land crabs and some littoral molluscs. Secondly, there are animals of terrestrial origin that live temporarily or permanently in water: here there are many insects and an occasional spider, plus a range of vertebrates including amphibians, some reptiles and some mammals.

Again we need to use the better-studied vertebrates as models for how control systems might be organized. Air-breathing fish do not generally monitor the gas concentrations in the air itself, as that is effectively invariant, but instead monitor gas using **venous chemoreceptors** in the veins draining the air-breathing organ. In addition they may have **gill arch chemoreceptors** to monitor hypoxia in the aquatic environment and initiate the switch to air-breathing. Most of the fish that breathe air lose CO_2 primarily through the gills and skin rather than from the lungs, so lack significant CO_2 sensitivity in the vicinity of the air-breathing organ. The other major control of aerial respiration is from **mechanoreceptors**, detecting lung distension with both tonic and phasic receptors. All these inputs are integrated in the medulla of the brain, but (unlike fish) there appears to be no central pattern generation for air breathing; instead a variety of reflex responses occur, giving an on-demand intermittent breathing pattern.

Within the amphibians and reptiles there is an almost complete spectrum from fully aquatic species (some frogs, toads such as *Xenopus*, many neotenous salamanders, sea snakes and a few turtles) to fully terrestrial forms (desert frogs and toads, most reptiles). In the middle of this spectrum are animals with genuinely bimodal breathing, including many frogs and toads and some turtles. Despite this diversity, the respiratory controls found are fairly similar, and indeed match those found in nearly all other tetrapods. Submergence usually invokes closure of the mouth and nostrils (nares), with receptors around these openings triggering the **apnoeic response** (breath-holding). A back-up is provided by water-sensitive taste receptors in the tongue. In animals that dive for significant periods these back-up circuits become the main ones, as the mouth is often left open; the receptors of the tongue, larynx and posterior nares then become the first part of a reflex arc that produces apnoea and may cause lung contraction and reduced pulmonary blood flow. The lung itself again has stretch receptors with tonic and phasic responses. To monitor O_2 levels, the internal chemoreceptors of the **carotid arch** blood vessel are particularly important, often elaborated into a carotid labyrinth. Denervation of this area produces reduced breathing and a lack of response to hypoxia or hyperoxia. Other areas of chemosensitivity also exist, in the vicinity of the pulmonary (pulmocutaneous) arch or the aorta, but these are more taxon-specific and probably affect blood flow distribution more than ventilation.

Most amphibians and reptiles breathe only intermittently, with bursts of breathing separated by apnoeic periods. This contrasts both with the majority of fish and with birds and mammals. The overall control of this pattern is complex and relatively poorly understood, not least because of the variety of breathing musculature found, varying from buccal breathing in frogs to aspiration breathing in turtles (by limb and carapace movements) and in crocodiles (having the liver-diaphragm action). With the added complication of extensive cutaneous respiration in many of these animals, itself subject to controls by modifications to cutaneous blood flow and also by ventilating the skin through bodily movements, it is clear that amphibious respiration in vertebrates is one of the more complicated breathing control systems found in animals.

The same may well be true of amphibious invertebrates, but too little is known to generalize here. Some land crabs have been investigated, and in parallel with the fish they do show a response to hypercapnia in addition to the normal aquatic response to hypoxia; but the receptors and the neural circuitry of this are unknown.

7.7.3 Terrestrial breathers

As a general principle, control of breathing in land animals is more commonly related to **perception of hypercapnia** than to hypoxia, continuing the trend seen in amphibious animals as compared to aquatic ones. This difference may be partly because of the increasing importance of acid–base balance; certainly blood P_{CO_2} levels are usually markedly higher in amphibious and land animals (Table 7.11), so that changes in this value should be more easily detected. The switch may also be partly because blood P_{O_2} will be an increasingly less reliable predictor of available O_2 as pigments that store O_2 become more important.

Table 7.11 Blood P_{CO_2} and HCO_3^- concentrations for a range of water-breathing (W) and air-breathing (A) animals.

Species	Breathing	P_{CO_2} (kPa)	HCO_3^- (mM)
Tadpole	W	0.3	4
Trout	W	0.35	5
Crab	W	0.4	7
Dogfish	W	1.0	6
Salamander	W/A	0.8	9
Garfish	W/A	1.2	8
Toad	W/A	1.6	22
Bullfrog	W/A	1.8	31
Lungfish	W/A	2.2	33
Electric eel	W/A	4.1	30
Lizard	A	2.3	13
Turtle	A	2.0	40
Human	A	5.4	24
Ground squirrel	A	5.7	23

Invertebrates on land breathe with either simple lungs, with book lungs or gill books, or most commonly with tracheae. Only for the last of these do we have significant understanding of the control mechanisms involved (though it is at least clear that land crabs have made the switch to relying on P_{CO_2} for ventilatory control). In insects there are very local controls at a cellular level, since epidermal cells that are short of O_2 are able to send out processes to 'capture' and drag back a nearby tracheole. There are also controls of ventilation pattern in larger insects where the tracheal system forms a single interconnected network, with air sacs acting as bellows. But the major controls for the whole animal lie at the level of the **spiracles**, which can be fully closed to prevent water loss and where the timings and durations of closure are significant controllers of respiratory rate. In most insects the spiracles open to a small atrium and have both opener and closer muscles (Fig. 7.31). Rhythmic bursts from the ventral nerve cord may stimulate these and give an intrinsic rhythm of opening and hence of ventilation. Each segment may have its own respiratory centre, and these may be differentially sensitive to P_{CO_2}, with trigger levels varying from 0.2% CO_2 to 15% CO_2. To a lesser extent the tonic activity in the nerves is also modulated by O_2. However, the spiracular muscles can also be directly affected by CO_2; in locusts the spiracle closer muscle is relaxed by a direct application of CO_2, so that the spiracle opens due to its own elasticity. In some insects local pH of the tissues (affected by activity levels and lactate build-up) may also have an effect. These controls in turn may interact with a direct effect of humidity in at least some insects, ensuring that respiratory needs do not excessively dominate the need for water conservation.

All land vertebrates breathe with fully alveolar lungs, and all have a common form of control (Fig. 7.32). The most important element on the receptor side is provided by chemoreceptors in the **carotid bodies**, adjacent to the carotid artery and innervated by the vagus nerve. These primarily transmit information

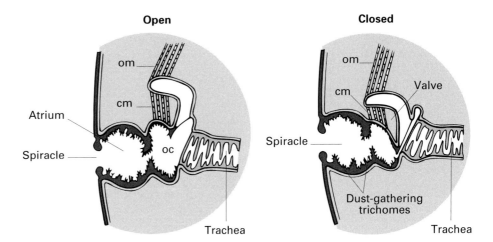

Fig. 7.31 Spiracle opening and closing mechanisms in the abdomen of an ant. The internal occluding chamber (oc) is constricted when the closer muscle (cm) contracts, and opened again by the opener muscle (om) contracting. (Adapted from Richards & Davies 1977; with kind permission from Kluwer Academic Publishers.)

about arterial Po_2, and will invoke responses to both hypoxia and hyperoxia. They may also respond to hypercapnia, but this role is less clear and varies between taxa. Further receptors occur in scattered locations along the aorta and major arteries, apparently mostly concerned with pressure regulation. Within the lungs, chemoreceptors are commonplace in birds, and also in some lizards, snakes and crocodilians, with a primary response to Pco_2. This is in contrast with the lungs of the water-dwelling amphibians and reptiles, which have only stretch receptors. However, this difference is not habitat related, as mammals too lack chemoreceptors and rely on stretch receptors. It appears to reflect phylogenetic history, as the mammals are descended from a stem reptilian lineage predating the divergence of most

modern reptile and bird groups; probably both stretch receptors and CO_2 receptors in the lungs are modifications of a common ancestral mechanism, and either will serve to monitor the efficacy of ventilation.

Central regulation in birds and mammals is primarily related to Pco_2 and pH levels, again with little O_2 sensitivity. The response to Pco_2 and/or pH (it is not clear which is the primary trigger) is centred in the **medulla**, and is much slower than the response to peripheral changes, probably giving a low-level background regulation feeding in to the central integration processes. This integration reverts to the use of a rhythmic pattern generator or pacemaker (as in aquatic fish), involving areas of the brain cortex, pons and medulla, and producing oscillatory

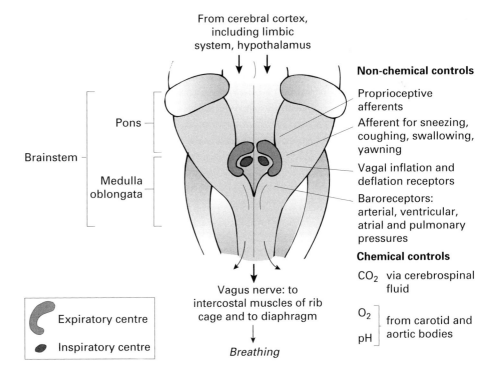

Fig. 7.32 Control of respiration in a terrestrial vertebrate.

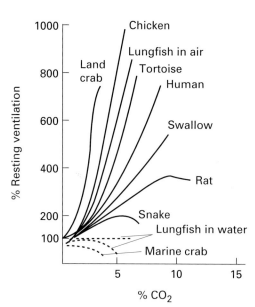

Fig. 7.33 The roles of O_2 and CO_2 in control of respiration. In aquatic animals, falling O_2 has a large effect and rising CO_2 is relatively unimportant. In land animals (air-breathing), decreasing Po_2 has relatively little effect but increasing Pco_2 causes a massive increase in ventilation.

sequences in the intercostal and diaphragm muscles controlling inhalation and exhalation. Both hypoxia and hypercapnia produce very rapid increases in the rate of this ventilatory pattern, the former via the carotid bodies and the latter probably triggered from the lungs, the carotid receptors and the brain medulla. Animals and birds that normally experience raised CO_2 in their environments—for example, species that burrow, and some that dive—show reduced CO_2 sensitivity.

Viewing the progression from aquatic to air breathing in the whole range of animals that have been studied, the overriding change is from reliance on control by Po_2 levels to control by hypercapnia via Pco_2 levels. Figure 7.33 shows some comparative examples of Po_2 and Pco_2 responses for a variety of species to underline this point.

Further reading

Books

Boutilier, R.G. (ed.) (1990) *Vertebrate Gas Exchange: from Environment to Cell.* Advances in Comparative and Environmental Physiology 6. Springer-Verlag, Berlin.

Bryant, C. (ed.) (1991) *Metazoan Life Without Oxygen.* Chapman & Hall, London.

Cameron, J.N. (1989) *The Respiratory Physiology of Animals.* Oxford University Press, New York.

Dejours, P. (1988) *Respiration in Water and Air.* Elsevier, Amsterdam.

Gilles, R. (1985) *Circulation, Respiration and Metabolism.* Springer-Verlag, Berlin.

Hochachka, P.W. (1980) *Living Without Oxygen: Closed and Open Systems in Hypoxia Tolerance.* Harvard University Press, Cambridge, MA.

Houlihan, D.F., Rankin, J.C. & Shuttleworth, T.J. (1982) *Gills.* Cambridge University Press, Cambridge.

Johansen, K. & Burggren, W.W. (1985) *Cardiovascular Shunts.* Munksgaard, Copenhagen.

Lamy, J., Truchot, J.-P. & Gilles, R. (1985) *Respiratory Pigments in Animals: Relations, Structure and Function.* Springer-Verlag, Berlin.

Prange, H.D. (1996) *Respiratory Physiology: Understanding Gas Exchange.* Chapman & Hall, New York.

Vogel, S. (1993) *Vital Circuits: on Pumps, Pipes and the Workings of Circulatory Systems.* Oxford University Press, New York.

Reviews and scientific papers

Bramble, D.M. & Jenkins, F.A. (1993) Mammalian locomotor-respiratory design; implications for diaphragmatic and pulmonary design. *Science* **262**, 235–240.

Dejours, P. (1994) Environmental factors as determinants in bimodal breathing: an introductory review. *American Zoologist* **34**, 178–183.

Feder, M.E. & Burggren, W.W. (1985) Cutaneous gas exchange in vertebrates: design, patterns, control and implications. *Biological Reviews* **60**, 1–45.

Hadley, N.F. (1994) Ventilatory patterns and respiratory transpiration in adult terrestrial insects. *Physiological Zoology* **67**, 175–189.

Hochachka, P.W. (1997) Oxygen—a key regulatory metabolite in metabolic defence against hypoxia. *American Zoologist* **37**, 595–603.

LaBarbera, M. (1990) Principles of design of fluid transport systems in zoology. *Science* **249**, 992–1000.

Lighton, J.R.B. (1994) Discontinuous ventilation in terrestrial insects. *Physiological Zoology* **67**, 142–162.

Mangum, C.P. (1994) Multiple sites of gas exchange. *American Zoologist* **34**, 184–193.

Mangum, C.P. (1998) Major events in the evolution of the oxygen carriers. *American Zoologist* **38**, 1–13.

Paul, R.J., Bihimayer, S., Colmorgen, M. & Zahler, S. (1994) The open circulatory system of spiders: a survey of functional morphology and physiology. *Physiological Zoology* **67**, 1360–1382.

Smatresk, N.J. (1994) Respiratory control in the transition from water to air breathing in vertebrates. *American Zoologist* **34**, 264–279.

Storey, K.B. & Storey, J.M. (1990) Metabolic rate depression and biochemical adaptation in anaerobiosis, hibernation and estivation. *Quarterly Review of Biology* **65**, 145–174.

Wasserthal, L.T. (1997) Interaction of circulation and tracheal ventilation in holometabolous insects. *Advances in Insect Physiology* **26**, 298–351.

Weber, R.E. & Jensen, F.B. (1988) Functional adaptations in hemoglobins from ectothermic vertebrates. *Annual Review of Physiology* **50**, 161–179.

Wenger, R.H. & Gassmann, M. (1997) Oxygen(es) and the hypoxia-inducible factor-1. *Biological Chemistry* **378**, 609–616.

8 *Temperature and its Effects*

8.1 Introduction

The lowest temperature yet recorded in the Earth's biosphere is −89.2°C, in Antarctica, whilst the highest temperatures range from 80°C in large deserts to 100°C in certain geothermal hot springs and to over 350°C at high hydrostatic pressures in deep-sea hydrothermal vents. Prokaryotic life can be found over much of this temperature range, but active animal life is restricted to a relatively narrow range of thermal conditions. For example, in hot springs there may be thermophilic bacteria living at temperatures in excess of 90°C, but for metazoans the upper limits are set well below this: for land animals the

upper limits are attained by certain desert insects and reptiles, in which body temperature may sometimes exceed 50°C; for aquatic invertebrates maximum heat tolerance occurs in some specialized ostracod crustaceans, which can tolerate 49°C for short periods; and among the vertebrates a few species of fish can thrive at 44°C. The bottom end of the range for aquatic animals is inevitably set by the freezing point of sea water at −1.86°C, but on land large polar mammals and birds can tolerate ambient temperatures of −60°C. Some indications of the broader range of ambient thermal conditions survived by active terrestrial animals are shown in Fig. 8.1. Some animals survive in inactive states somewhat outside these limits, most spectacularly the

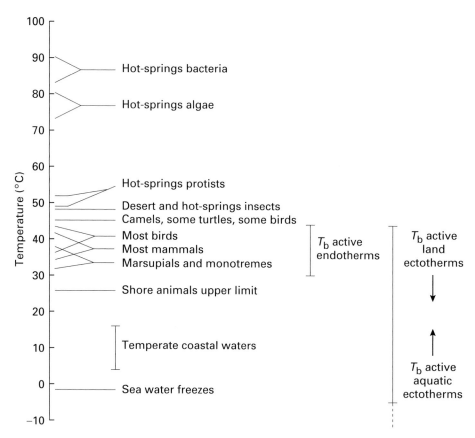

Fig. 8.1 The range of temperatures occurring naturally at the surface of the Earth, and thermal limits to animal life there. T_b, body temperature.

nematodes, tardigrades and eggs of insects that can withstand freezing in liquid helium at –269°C (see Chapter 12).

Despite the rather narrow range of temperatures most animals encounter, they are drastically affected at all levels of their biological organization by any change in their thermal surroundings. Thus temperature changes pose a real challenge to animals. Climate change over geological time periods is thought to have played a significant part in evolution, influencing adaptive radiations and in some cases leading to mass extinctions. Temperature may also change relatively dramatically over a few hundreds or thousands of years, in cyclical patterns related to astronomical anomalies and to continental (plate tectonic) movements. Moreover, temperatures may alter very widely even during the course of a year, especially at moderate and high latitudes. However, on annual and circadian timescales we cannot easily make broad generalizations, because of the major differences between the thermal characteristics of terrestrial and aquatic habitats. Water is a reasonably good conductor of heat and it has a much higher specific heat capacity than air. As a result, the temperature of relatively large bodies of water is highly buffered and a large heat input/output is required to modify the temperature. In contrast, air is an effective insulator and it has a low specific heat capacity, such that relatively minor inputs of solar radiation can produce dramatic rises in temperature. Thus the surface air temperature just north of the Arctic circle in Siberia can change from –70°C in winter to +36.7°C in summer (more than 100°C of temperature change); while close to the Antarctic continent the temperature of the ocean water varies by less than 0.2°C through the entire year. Similarly in low-latitude deserts there may be exceptionally high daylight temperatures but very cold nights, while just offshore the sea temperatures show no daily variation. Thus dramatic daily temperature changes are the norm for many terrestrial animals, but (with the exception of species living in small bodies of water such as puddles and rockpools) are relatively rare for aquatic organisms.

In this chapter we review the effects of temperature on animals, and the kinds of adaptation they show to withstand or to counter temperature change.

8.2 Biochemical effects of temperature

Temperature is a measure of the average molecular motion within a material; at higher temperatures, there is more molecular vibration. In fact the mean kinetic energy of molecules is proportional to the absolute temperature. However, particular molecules within a material have differing kinetic energies about this mean value, and at any particular temperature a certain proportion of the more active molecules in a system are moving fast enough to react with other molecules when they collide—i.e. their energy is above the required activation energy ($\Delta G°$) for a given reaction. As the temperature rises, the number of molecules whose activity exceeds this threshold also rises, so that any chemical reaction becomes faster at higher temperature.

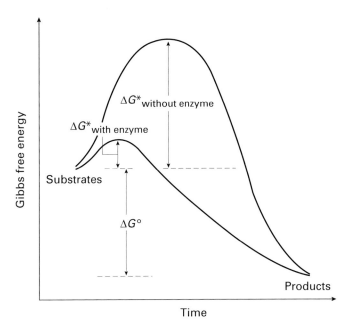

Fig. 8.2 Free energy changes during a chemical reaction, showing the decreased activation energy (ΔG^*) for an enzyme-catalysed reaction where the enzyme-substrate complex facilitates breakdown of substrate(s) to product(s).

In fact, while a rise of only 10°C may raise the kinetic energy by just a few per cent, it may increase the reaction velocity by several hundred per cent by pushing a large proportion of molecules across this activation energy barrier (Fig. 8.2). The same is true, within limits, of all biochemical reactions catalysed by enzymes; the reaction rate may rise very sharply with temperature due to thermal effects on the reaction of substrates with enzyme catalysts and the consequent faster release of products. The energy barrier of enzyme-catalysed reactions is itself subject to evolutionary modification. However, there is normally an upper thermal limit beyond which reactions are drastically reduced or cease altogether as the biochemical structures involved are damaged.

Many biological functions are made up of lots of steps, the individual components of which cannot easily be measured. In such cases it is not appropriate to derive an activation energy and instead Q_{10} values are used to describe the effects of temperature on the rate of the function being studied (Table 8.1).

Usually reaction rates are compared at two or more different temperatures to show how thermally dependent a particular function is. The Q_{10} value is then the ratio of the velocity constants k_1 (at $t°C$) and k_2 (at $t + 10°C$):

$$Q_{10} = k_{2(t + 10°C)}/k_{1(t°C)}$$

or more generally, the ratio between k_1 and k_2 corrected for a 10°C difference, i.e.:

$$Q_{10} = (k_2/k_1)(10/t_2 - t_1)$$

Table 8.1 Q_{10} values for a variety of reactions involved in biological processes.

	Temperature (°C)	Q_{10}
Physical reactions		
Diffusion	20	1.03
Biochemical reactions		
Cytochrome reductase (possum)	>20	1.5
	<20	2.5
Pyruvate kinase V_{max} (rat)	>25	1.7
	<25	3.2
Haemoglobin coagulation	60	13.8
Physiological reactions		
Anemone oxygen consumption	5	2.0
	15	2.3
Crayfish heart rate	5	2.4
	15	1.6
	25	0.8
Crustacean gill movements	5	3.8
	15	1.7
Beetle oxygen consumption	10	2.4
	20	2.1
Insect thermal induction of diapause	20	1.4
	10	3.7
Torpid mammal oxygen consumption	20	4.1

In practice the rate of most processes changes as a logarithmic function of temperature. This means that Q_{10} values are not con-stant for different 10°C increments (e.g. 10–20°C and 30–40°C), but instead normally decrease at higher temperatures; the value therefore has to be quoted for a *particular* 10°C interval. This may not be easy, as effects are rarely measured exactly 10°C apart, so the Q_{10} value is in practice extrapolated from observations at any two experimental temperatures.

The Q_{10} values for physical processes such as diffusion, and for many inorganic chemical reactions, are normally about 1. Values for photochemical reactions (as in photosynthetic systems) are often only a little above 1. However, Table 8.1 shows that for enzyme-mediated biochemical reactions, and for physiological processes, such as heart rate or metabolic rate, that rely on underlying enzymic processes, the Q_{10} is normally higher, usually in the range 2–3 (i.e. a doubling or tripling of reaction rate per 10 degree temperature change, so that the rate plot is an exponential upward curve). This could mean that for an animal able to survive over a 30°C temperature range, the metabolic rate might be raised at least 27-fold at the upper temperatures. In reality, since we know that the Q_{10} values tend to be variable in different parts of the temperature range, such a spectacular rise in metabolic rate or heart rate probably will not occur. For marine invertebrate heart rates, the Q_{10} values are commonly over 3 below 10°C, around 2 between 10 and 20°C, and less than 2 when the temperature exceeds 20°C; marine animals will not encounter such a range (except in the special case of hydrothermal vent animals considered in Chapter 12), but simple calcula-

tion shows that the increase in heart rate incurred might be up to 12-fold at the higher temperatures. In some cases, for processes such as ventilation rate, for example, there may be extra complications due to sharp transitions in Q_{10} values at particular ambient temperatures, suggesting rather fundamental underlying changes in the materials or tissues or mechanisms involved. Indeed the Q_{10} for oxygen consumption may show a reduced value in the region of maximum activity temperatures for some animals, which has been interpreted as an adaptive means of controlling energy expenditure ('metabolic homeostasis'). There is good reason to suppose that Q_{10} values are under strong selective pressure with regulatory processes in place, given that phenomena that need to be temperature independent (e.g. the functioning of biological rhythm controllers or 'clocks') are commonly found to have a Q_{10} of 1.0.

Thus although Q_{10} values are often helpful and illuminating, and are frequently quoted in relation to biochemical and physiological adaptations (see Table 8.1 for examples), they clearly have to be treated with care. We need a more detailed understanding of processes and effects at these levels to appreciate how temperature changes really influence animals.

8.2.1 *Enzymic adaptation to changing conditions*

Enzymes are proteins, and their performance is inevitably susceptible to thermal changes, but as temperatures rise or fall their performance can be modified in a variety of ways that may be adaptive, extending the thermal range tolerated by individuals of the species. The nature of these modifications varies with the timescale of the temperature change.

Short-term changes

The activity of many enzymes is controlled directly or indirectly via hormones and/or the nervous system. These provide mechanisms for modulating enzyme activity over the timescale of a few seconds to several hours, compensating for changes in temperature. There are at least four possible types of change that could be instituted quickly by hormonal or neurotransmitter intervention:

1 The enzyme's effective concentration or activity could be altered.
2 The concentration of the relevant substrate could be altered.
3 The energy supply for the reaction being catalysed could be increased (or decreased).
4 The intracellular environment could be altered to modify the enzyme's effects.

A great many animals have been said to use the first of these adaptive responses, by increasing effective enzyme concentrations in colder conditions. This could occur by direct modifications to the enzyme (such as phosphorylation to an active form), or by changed compartmentalization within the cell (such as mobilization from membrane-bound areas into the cytoplasm) or by altered activity due to allosteric modulation by activators or inhibitors.

In practice, though, it is usually impossible to disentangle mechanisms 1–3 given above, as all of them interact. A neat example of hormonal control of enzyme concentrations, substrate concentrations and energy supply is provided by cold-tolerant insects, where falling temperature triggers synthesis of cryoprotectant molecules (these are dealt with specifically in section 8.3.1). The first step is release of a hormone that binds to receptors on the fat body cell membranes, causing an increase in intracellular cAMP and Ca^{2+} levels via a classical protein kinase cascade (see Chapter 2). The result is that an inactive enzyme, phosphorylase b, becomes phosphorylated to the active form, phosphorylase a, giving a net increase in the cellular concentration of phosphorylase a. This in turn allows the breakdown of more glycogen, which is the 'fuel' to provide the extra substrates for the synthesis of the cryoprotectants.

Enzyme performance can be modulated in a less direct manner by changing the intracellular environment in which the enzyme works. Thus ionic concentration within the cell commonly changes slightly during short-term thermal exposures, and both pH and potassium concentration are known to alter enzyme activity. In particular, pH is usually observed to increase (i.e. become more alkaline) in animal cells as the temperature drops, and this provides a significant stabilization of protein structure and thus reduced enzyme temperature dependence (sometimes termed 'alphastat regulation'). Since this effect could influence the thermal responses of large numbers of proteins, it should represent a particularly efficient way of dealing with acute thermal stress. Certain low molecular weight solutes also affect protein stability; the sugar trehalose is a powerful stabilizer of both protein and membrane structures, and is synthesized in some yeasts after heat shock. It also appears in many metazoan organisms entering cryptobiotic states (see Chapter 12), where it protects against extreme dehydration as well as temperature changes.

Medium-term or acclimatory changes

Over the timescale of days to weeks, temperature change often influences the concentration of enzymes more fundamentally, via differential effects on protein synthesis and protein degradation. Normally this involves just a few of the enzymes that govern a complex reaction sequence, those that are rate-limiting, and it may be reflected in a rise in mitochondrial protein content as the temperature falls, by as much as 50% in some animal tissues. Over a similar timescale, new enzymes may also be synthesized that can regulate other enzymes' activities. Here the best known examples are enzymes that influence the fatty-acid composition of membrane phospholipids (see section 8.2.2), resulting in improved performance of other membrane-bound enzymes following a period of low-temperature acclimation.

However, since most enzymes exist as a variety of isozymes with subtle differences in kinetic properties, the best and the commonest solution for maintaining the catalytic efficiency of an enzyme may be to change the suite of isozymes present. The net effect is often to alter the fate of metabolites rather than overall reaction rates; for example, in frogs and fish low temperatures favour the synthesis of the 'heart type' H_4 lactate dehydrogenase (LDH) isozyme in skeletal muscles and thus the oxidation of pyruvate, at the expense of its reduction to lactate by the more normal 'muscle-type' M_4 isozyme.

In practice it seems to be quite rare for organisms to carry a range of isozymes that are specific to particular parts of their thermal range. Perhaps the best documented example is myosin ATPase. In goldfish and common carp subjected to several weeks of low temperatures, an altered pattern of expression of the genes for the heavy and light myosin chains is found, so that ATPase activity is increased and muscle contraction speed and force are improved. This contributes to an increased maximum swimming speed at low temperature, having obvious survival value in escaping from predators; but it occurs at the expense of a trade-off in performance at warm temperatures in the cold-acclimated fish.

Clearly there must be a reasonable time period for switching on the new isozyme locus and synthesizing the new protein; for example, trout living at 17°C apparently have an acetylcholinesterase enzyme quite distinct from the acetylcholinesterase found in trout living at 2°C, and a fish moved from one habitat to the other is not immediately able to mobilize the new form of its enzyme to cope. However, trout living at 12°C seem to have copies of both isozymes and need very little time to adapt to a change in thermal conditions. This is perhaps an example of 'on-off' gene expression, where one or other isozyme may be switched off according to thermal history. In other cases, for example the malate dehydrogenase (MDH) of the goby *Gillichthys*, regulation of expression is more subtle, with adjustments of the ratio of expression of two isozymes occurring.

Although this strategy of having a suite of isozymes with different temperature optima is actually not very common, nevertheless rather large numbers of animals do express different isozymes in different tissues (such as liver, heart and muscle). The ability to alter gene expressions and thus switch between protein components in response to seasonal (or even diurnal) change is a key factor in the adaptive repertoire of some animals with particularly wide temperature tolerances.

Long-term or evolutionary changes

Natural selection offers the possibility of evolving enzymes tailored to particular thermal environments both in terms of maximal activities and regulatory properties. In some cases the mechanisms are similar to those observed in acclimatory changes, with a quantitative increase in enzyme concentrations. For example, individual mitochondria in Antarctic fish have relatively low rates of oxygen uptake, but their concentration is greatly increased: they occupy 30–60% of the fibre volume of red muscles in Antarctic fish, compared with 15–25% in tropical fish and only 8–10% in terrestrial reptiles.

Other possibilities include genotypic changes in enzyme sequence and structure to give altered properties, resulting from

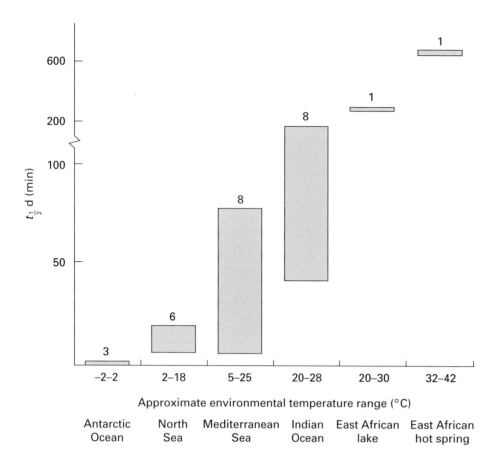

Fig. 8.3 Thermal stability of enzymes in marine animals is related to their thermal habitat. The thermal denaturation half-times at 37°C for myofibrillar ATPase from fish species taken from six habitats are compared; those from cold seas denature within a few minutes, while those from the Indian Ocean (normally 20–28°C) have half-times of up to 3 h. Note that enzymes in fish from warm freshwater lakes and springs are even more thermally stable. Numbers above bars indicate numbers of species tested. (Adapted from Johnston & Walesby 1977.)

the combined effects of mutation, recombination and selection to give more suitable protein variants for particular temperature regimes. The ligand-binding site is usually highly conserved, but adjacent areas are altered to change the thermal stability and/or flexibility of the enzyme. In general, early studies indicated that enzymes of animals in thermally stable environments have lower thermal stabilities and less resistance to thermal perturbation of substrate binding than the homologous enzymes of an animal from a thermally variable habitat; and enzymes of warm-adapted species tend to show more thermostability than the homologues from cold-adapted species.

Adaptive changes in the thermal properties of homologous enzymes from different populations of a species may reflect underlying allelic variation maintained by selection; the enzyme variants are then termed allozymes (or orthologous homologues), as discussed in Chapter 2. A classic example of these multiple allele effects is the muscle lactate dehydrogenase (M_4-LDH) enzyme that we met earlier. In southern populations of the fish *Fundulus* this is predominantly the LDH$_a$ type, which is progressively replaced by the LDH$_b$ type in more northern populations. Measured under certain conditions *in vitro* the kinetic properties of the enzymes appear to be optimal at 20°C for the cold-adapted northern fish, but at 30°C for the southerners living in warmer seas. It has been suggested that these differences are maintained by strong selection, although non-adaptive explanations (such as random genetic drift) are possible.

Analysis of enzymic function from animals that have evolved to cope with different thermal regimes has become more sophisticated in recent years. Firstly, thermal denaturation temperature has been studied, and this correlates closely with normal environmental temperature (Fig. 8.3). Another example is the M_4-LDH enzyme, where the homologue from polar fish species starts to unfold at a much lower temperature than the homologue from thermophilic species such as desert lizards. Work on this and other enzymes indicates that only tiny changes in protein structure—perhaps only a few residues—are enough to affect the enzymes' thermostability. It has been suggested that enzymes from animals with low body temperatures have a more 'open' structure (stabilized by fewer and weaker bonds) than those from animals with warmer body temperature (T_b) values, and that (while reducing thermostability at low temperatures, as described above) this permits more efficient conformational changes during catalysis; enzymes from animals with higher body temperatures need more compact and rigid enzyme structures because of the higher kinetic energy of the cell.

Secondly, enzymes from cold-bodied species show major differences in substrate turnover rates (K_{cat}, the moles of substrate converted to product per mole of enzyme per second). In general, enzymes from species with lower T_b values have higher K_{cat} values at a given temperature than the homologous enzymes from warmer-bodied species (Table 8.2). Thirdly, it has been observed that the activation energy of homologous enzymes

Table 8.2 Substrate turnover rates (K_{cat}) for lactate dehydrogenase (LDH) from barracuda (see also Fig. 9.9), showing the similar (conserved) values for different species when measured at normal operating temperatures.

Species (distribution)	Normal mean temperature of habitat (T_m) (°C)	K_{cat} at 25°C (s⁻¹)	K_{cat} at T_m (s⁻¹)
Sphyraena argentea (northern)	18	893	667
Sphyraena lucasana (intermediate)	23	730	682
Sphyraena ensis (southern)	26	658	700

across a range of species from different habitats is such as to counter the effect of temperature. That is, cold-adapted enzymes have a lower activation energy than warm-adapted enzymes. Thus they can avoid the reduction in reaction velocity that would otherwise take place, and in some cases even allow an increase in reaction rate in cold conditions.

Enzyme affinities present a somewhat more complicated picture. We now know that substrate concentrations in cells are generally rather similar at all body temperatures. The ability of an enzyme to bind its substrate (its 'affinity') can often be approximated by the apparent Michaelis–Menten constant (K_m) of the reaction (see Box 4.5 for a reminder). Cold-adapted isozymes also tend to show an increased affinity (lower K_m value) compared to warm-adapted ones when tested at the same temperature (Fig. 8.4). But this also shows that cold-adapted enzymes have rather similar K_m values to warm-adapted isozymes when each is determined at its relevant working temperature, i.e. under physiological assay conditions. These careful interspecific comparisons of enzyme activities measured *in vitro* indicate that ligand-binding characteristics are also highly conserved. Evolution seems to have favoured the conservation of the important kinetic properties of enzymes within fairly narrow ranges, so that enzymes that differ within species and even within individuals retain maximal responsiveness to changes in the normal cellular concentrations of substrates and allosteric modulators.

Differences in K_{cat} and K_m may occur in closely related species with only a small difference in mean habitat temperature, with enzymes differing by only a few amino acids, indicating the potent effects that can be achieved with tiny selective adjustments in tertiary protein structure. Indeed there are indications that virtually *no* changes in primary structure are needed to produce changes in thermal stability for LDH. The enzymes from different species of goby (*Gillichthys*) are indistinguishable in most respects with cDNAs differing only by four synonymous nucleotides, yet have different temperature sensitivities, suggesting that in some cases the same isozyme may exist in different folded conformations and have different functional properties.

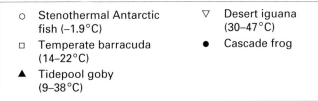

Fig. 8.4 The effects of temperature on the K_m values for pyruvate breakdown by lactate dehydrogenase (M_4-LDH) from vertebrates. For each species the normal body temperature (T_b) is shown. K_m values vary widely, but when comparisons are made at the animal's preferred T_b all values are in a similar range (around 0.1–0.3 mM). (From Somero *et al.* 1996.)

8.2.2 Thermal effects on membranes and cellular structures

The physical properties of biological membranes, and especially of their lipid components (which may be 25–50% of the membrane dry weight), are markedly influenced by temperature. The lipids normally exist in a 'liquid crystal' state that is intermediate between a highly fluid material and a rather rigid structure. This delicately balanced structural state can be disrupted rather easily, and this is reflected in substantial and abrupt changes in viscosity in preparations of membrane phospholipids exposed to changing temperature *in vitro*. Heat death may result from disruptions to membrane function, especially at very sensitive sites, such as synapses, where any malfunction would obviously produce immediate behavioural effects. In living cells, however, a phenomenon known as **homeoviscous adaptation** (HVA) occurs to keep the membrane in its normal moderately fluid state with its protein components properly functioning.

The most important contributor to this acclimatory or adaptive process appears to be variation in the fatty acid composition of the lipids. Shorter chain fatty acids, and those that are unsaturated (with some carbon–carbon double bonding) are inherently more fluid than longer chain or fully saturated fatty acids. This is because a long molecule with a saturated and therefore straight molecular backbone can achieve more stable bonding

All lipids saturated

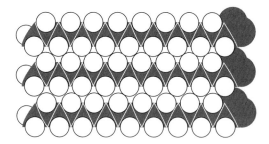

One unsaturated kinked molecule
with a double bond

Fig. 8.5 Double bonds in membrane lipids producing a less densely packed structure.

with adjacent molecules compared to the rather kinked unsaturated fatty acid molecules (Fig. 8.5). The introduction of the first unsaturated bond has the greatest effect on membrane physical properties. If this first unsaturation is at the central 9–10 position from the carboxyl terminus then the largest increases in fluidity are observed. Cholesterol also has some effect on HVA, increasing membrane order and affecting phase behaviour and permeability. In general more cholesterol is present in membranes from endotherms than from ectotherms, but changes in cholesterol content with temperature may be in either direction, probably reflecting its multiple roles.

Temperature-induced changes in membrane fluidity can be detected by relatively new techniques such as electron spin resonance spectroscopy or fluorescence polarization spectroscopy. In the light of the resulting new understanding of membrane structure and membrane domains (Chapter 5), the use of terms such as viscosity or fluidity in relation to membranes has been partially superseded by motional rate parameters and measures of static molecular ordering. The concepts of changing 'fluidity' or of changing states of 'order' and 'disorder' are both useful ways of envisaging the effect of lipid alterations on the properties of the membrane.

The membrane lipids of a wide range of animals show varying fatty acid compositions at varying ambient temperature that are just as we would expect for adaptive maintenance of the homeoviscous state (Table 8.3). The exact nature of the fatty acids that are varied differs according to taxon, and is complicated because membrane composition also depends upon diet. Indeed a diet artificially high in polyunsaturated lipid (resulting in membranes of lower viscosity at a given temperature) has been shown to lead to a lowered preferred T_b in lizards, with a reduction of as much as 5°C. This may indicate that seasonal changes in diet can automatically bring about adjustments of thermal physiology. Certainly both the type of fatty acids and the proportions of pre-existing saturated and unsaturated forms are varied naturally in most species that have been studied. This also means that direct inference of adaptation by changing fatty acid saturation is quite difficult.

Perhaps the most impressive evidence of thermal adaptation of biomembranes comes from studies of species that have evolved in particular thermal habitats or maintain elevated and constant body temperatures, as in birds and mammals. Figure 8.6 compares brain membranes isolated from a variety of teleost fish species and from a representative mammal and bird. In this study the measure of membrane physical condition was the fluorescence anisotropy of two standard fluorescence probes, a technique which provides information on the degree of hindrance to the free rotational motion of the rod-shaped probe. There is a clear relationship between the position of the curve and the temperature at which the species lives, so that at a given temperature membranes from the Antarctic species, *Notothenia*, were more fluid, and less ordered, than equivalent membranes from the perch, *Perca fluviatilis*, and so on. The mammalian and

Species	Body temperature	Phospholipid		
		Choline	Ethanolamine	Serine inositol
Arctic sculpin	0	0.59	0.95	0.81
Goldfish	5	0.66	0.34	0.46
	25	0.82	0.51	0.63
Desert pupfish	34	0.99	0.57	0.62
Rat	37	1.22	0.65	0.66

Table 8.3 The ratio of saturated to unsaturated fatty acid residues in three phospholipids from cell membranes in brain tissues of various vertebrates acclimated to different temperatures.

Fig. 8.6 Homeoviscous adaptation (HVA) seen in brain membranes, for various vertebrates. HVA is measured as the polarization of fluorescence (see text for details: decreased polarization indicates increased 'fluidity' and decreased 'order'). Values at the normal body temperature for each species (given in parentheses) are shown as large circles. (From Behan-Martin *et al.* 1993.)

Fig. 8.7 The effect of cold acclimation in carp on the levels of desaturase enzyme (measured as Δ^9-desaturase mRNA transcript from liver, relative to a standard 18S rRNA signal). Filled symbols represent data from individual animals, and open symbols represent mean values, showing the clear increase in the first few days of cooling. (From Gracey *et al.* 1996; courtesy of Cambridge University Press.)

avian species possess relatively ordered membranes, which matches their need to maintain stable membranes at high body temperatures of 37°C and 41°C, respectively. The figure also indicates the value of anisotropy at the respective body temperature for each species, showing that these were always very similar. Thus it appears that selection designs the brain membranes of vertebrates to give equivalent physical properties under *in vivo* conditions of environmental or body temperature, the implication being that the condition is in some way favourable for the functional properties of that membrane system. Evolutionary responses to cold regimes therefore combine a higher proportion of unsaturated fatty acids with a marked increase in the number of unsaturated bonds in each unsaturated fatty acid, to maintain overall membrane fluidity.

All these studies on HVA are correlative, and do not establish causal relationships between increased phospholipid unsaturation and decreased membrane order. Molecular genetics offers some alternative and less equivocal approaches for establishing a degree of experimental control over membrane adaptation, by intervening in the expression of enzymes involved in the response. The over-expression or abolition of a specific enzyme, or its induced expression in the absence of temperature change, allow its role in the adaptive regulation of membrane structure

to be demonstrated more confidently. The changes in fatty acid composition are now known to be brought about by a group of **desaturase enzymes**; for example the Δ^9-desaturase enzyme incorporates the first double bond into stearic or palmitic acid. The activity of desaturases is directly related to temperature, so that following cold transfer in fish this activity is induced, with enzyme activity increasing up to 30-fold. Diet also affects this, so that feeding on more saturated lipids causes an up-regulation of desaturase activity, even at high temperatures.

It is possible to make a chemical probe that binds to the mRNA for the carp liver Δ^9-desaturase, and then look for the induction of the desaturase transcript relative to that of some other gene not thought to be involved in temperature adaptation. Figure 8.7 shows the variation in Δ^9-desaturase transcript levels relative to 18S ribosomal RNA from carp acclimated to 30°C, and at different periods through a progressive 3-day cooling regime to 10°C. Transcript levels were low in warm-acclimated carp and after 24 h of cooling, but increased 8–10-fold after 48 h. Levels peaked at 2–3 days and then quickly subsided. The transcript levels after long periods in the cold remained about double those observed in warm-acclimated carp liver. Thus although cooling led to a dramatic increase in transcript levels, the effect was transient. Evidently the change in lipid saturation requires only a transient increase in desaturase expression, and low levels of desaturase activity are then sufficient to maintain the *status quo* on long-term exposure to cold.

In prokaryotic cells and in plants the genetic manipulation of unsaturation of fatty acids in membrane lipids is easier to perform and is known to influence growth and chilling toler-ance. For example, the introduction of cytoplasmic DNA for glycerol 3-phosphate acyltransferase from a squash plant (which is chill-sensitive) or from *Arabidopsis* (which is chill-resistant) into *Nicotiana* (tobacco, which has intermediate chill-sensitivity) results in altered chilling sensitivity and resistance, respectively. In the former case, the level of unsaturated fatty acids in glycerophosphatides falls significantly and chilling sensitivity is markedly increased, whilst in the latter there is an increase in both the unsaturation of fatty acids in glycerophosphatides and in chilling tolerance.

In terrestrial animals that experience rapid temperature fluctuations these changes can be extremely rapid, with a time course of just a few hours. Recent evidence also shows that the changes vary in different tissues within an individual; the degree of compensation of membrane order (the 'homeoviscous efficacy') varies from 75% in the carp intestinal mucosa mem-branes to 25% in brain myelin and almost zero in the sarco-plasmic reticulum of skeletal muscle.

At present we know relatively little about the nature of the temperature sensor in what may be a long and complex feedback pathway. Some transcriptional activation seems to be linked to changes in the fatty acid saturation and/or the physical properties of the plasma membrane, so that the ultimate sensor may be some aspect of membrane physical condition. In at least some cases the thermally induced change in enzyme activity is probably due to configurational changes within the membrane, such that the active site of the desaturase is masked in relatively fluid/disordered membranes but becomes exposed as the temperature falls and the membrane becomes less fluid and more ordered.

This neatly indicates the complex interrelation of membrane effects and enzymic effects in a thermally stressed animal. Indeed a similar process of membrane fluidity/enzyme masking may be part of the explanation for the widespread changing of activation energies of membrane-bound enzymes as tem-perature changes, discussed above.

In addition to membrane components, other subcellular com-ponents will be adversely affected by temperature changes, and the best-studied examples are microtubules. In mammals the constituent tubulins are cold-labile, depolymerizing at 0–4°C, but the tubulins from polar fish are much more stable with temperature and will self-assemble *in vitro* at –2°C. These dif-ferences seem to be intrinsic to the tubulin subunits them-selves, and arise from relatively minor variations in amino acid sequences, largely in the alpha chains of the tubulin structure at a domain responsible for contact between tubulin dimers. This produces more hydrophobic interactions in the microtubules of the polar fish.

8.2.3 *Thermal induction of stress proteins*

A third effect of temperature at the biochemical level is the induction of certain proteins that have a protective effect, commonly called **heat-shock proteins** (HSPs) or more generally **stress proteins**. The HSPs have been a focus of interest in the 1980s and 1990s, as they have proved to play a key role in regu-lating protein folding and in coping with proteins affected by heat and other stresses.

HSPs belong to several families, defined by their sizes (molecu-lar weights, in kilodaltons). In animals the relatively large pro-tein families dominate (HSP60, HSP70, HSP90, HSP100 and Lon), whereas in plants there are many more of the small HSPs (including HSP10, HSP27 and ubiquitin). The structures of these are highly conserved even across kingdoms, and their action also seems to be highly conservative. They can be induced by many stresses: natural changes such as hypoxia or hyperoxia, osmotic shock and pH change, and less natural stresses such as the presence of alcohols, ionizing radiation, heavy metals, toxins and free radicals. Hence they can bring about 'cross-tolerance', i.e. once induced in relation to a particular stress, they make the organism more tolerant of other stresses. Some of the smaller HSPs appear to be developmentally patterned in animals, with HSP27 regulated in part by ecdysone in at least some insects. The higher molecular weight HSPs are more commonly the forms that are inducible by temperature change, though the 70 kDa family includes both inducible (HSP70) and non-inducible or 'constitutively expressed' (HSC70) members.

Induction of HSP expression in relation to temperature is related once again to normal thermal regimes, so that species from warm environments undergo a stress response at a sub-stantially higher threshold temperature than those from cold environments. Induced HSP concentrations also differ substan-tially between tissues within an individual. It is evident that the ability to synthesize HSPs is strongly correlated with thermal tolerance, in all animal taxa.

Expression of the proteins and the resultant increased thermal tolerance occurs within minutes to hours, substantially faster than the acclimatory changes of lethal limits normally recorded in animals. Molecular studies using gene deletions or insertions of extra copies show that there is a direct match between protein expression and this inducible thermo-tolerance, both in the laboratory and recently in wild populations of fruit fly larvae (*Drosophila*) living in the stressful alcoholic environment of rot-ting fruits.

The mechanisms of HSP action are now understood in terms of 'molecular chaperonage'. Proteins are normally folded into specific tertiary configurations, requisite for their proper func-tioning; but they may be unfolded during the early stages of their synthesis, during intracellular transport across membranes, or as a response to various kinds of stress. Such unfolded proteins may refold wrongly, and may be susceptible to interac-tions with other cellular components, including other unfolded proteins, rendering them useless. Molecular chaperones are used to limit these interactions, by recognizing and binding to the exposed side groups of the unfolded proteins, stabilizing the unfolded state. When appropriate (and often in association with ATP breakdown) the chaperone molecule may release its bound protein to allow correct refolding and a resumption of normal

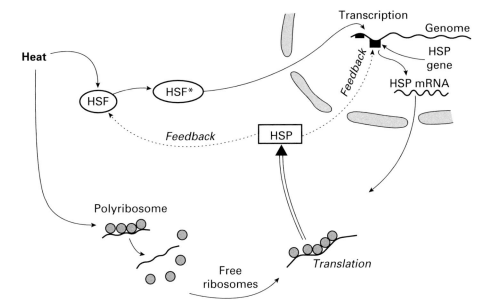

Fig. 8.8 A model for regulatory systems during heat shock. Changing temperature induces a cytoplasmic factor, HSF, which then changes from an inactive monomer to an active trimeric form, HSF*. This has a regulatory effect on the promoter region of the heat-shock protein (HSP) gene, initiating production of HSP mRNAs. Since existing ribosome complexes have also been disrupted by the thermal shock, RNAs that were bound to them are released or degraded, and the new mRNA is preferentially translated, the HSPs then being distributed to the most sensitive nuclear and cytoplasmic sites. There is also a feedback interaction between some HSPs and HSF (see text).

function; or alternatively the chaperone may release its bound protein for degradation or export from the cell. HSPs are seen as the primary molecular chaperones synthesized in response to any stress that causes protein unfolding; in some cases HSPs from different families work together as co-chaperones.

The regulation of the heat-shock response is now known to be mediated via a cytoplasmic detector, the heat-shock factor (HSF), which exists as a monomer until heat shock occurs, when it trimerizes and moves into the nucleus. There it binds to a particular site on the promoter region of HSP genes, initiating transcription. One of the main HSPs, HSP70, appears to have a feedback interaction with HSF, since it acts to inhibit trimeriza-

tion. Thus unfolded proteins and HSF monomers may compete for interactions with HSF—if there is an excess of unfolded protein, HSP70 will be mopped up and the HSF will be free to trimerize and bring about more HSP transcription (Fig. 8.8).

8.3 Physiological effects of temperature

Biological processes generally exhibit a biphasic effect of temperature (Fig. 8.9). The first phase is a consequence of the rate-enhancing influence of temperature on enzyme function discussed in section 8.1, so that rates of activity rise as temperature increases up to a certain point. But in a second phase beyond this (and normally beyond the natural range of temperatures experienced) the destructive effects of temperature supersede and rates of activity decline. The rate–temperature curve therefore reveals an 'optimum' temperature for any particular activity.

Determining the rate–temperature curve is relatively simple for systems of no greater complexity than single-enzyme reactions, but it becomes more difficult where tissues are used, and nearly impossible for whole animals whose own reactions continually affect the conditions of the experiment. It is therefore appropriate to move from the effects of temperature on cellular activities first, to the integrated effects on whole animals.

8.3.1 Low-temperature effects on cells and organisms

We have seen that enzymes are intrinsically less effective at low temperature, and that membranes might become more viscous, but that cells show a range of adaptive mechanisms to maintain reasonable activity. If enzymes and membranes are so adaptable, what then actually brings about cell death at low temperature? Some animal cells can tolerate extensive freezing, even down close to absolute zero in a few very special cases,

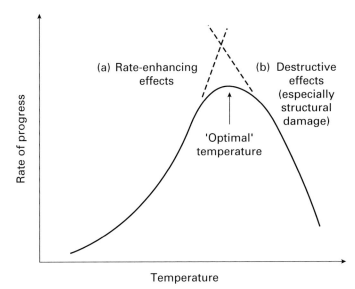

Fig. 8.9 Biphasic effects of temperature increase on the rate of biological processes.

yet others (e.g. cells from tropical fish) are so sensitive to cold that the cells are dead well before 0°C is reached on a cooling cycle.

Cells and animals that can survive low temperatures are exhibiting the general phenomenon of 'cold hardiness', and most of them have some ability to survive subzero temperatures, raising the general problem of the potential freezing of their contained water. Surveying the animal kingdom as a whole there appear to be two broad strategies for dealing with this problem. Some animals show **freeze-tolerance**, and can cope with extensive freezing with ice formation occurring within the body. Others show **freeze-intolerance**, surviving at –40 to –50°C without ice forming in their bodies, but usually dying very quickly if any ice crystals do begin to form; they survive by avoiding freezing, and are therefore also described as showing **freeze-avoidance**. However, the two different options are not entirely exclusive (e.g. a species can show tolerance as a juvenile and intolerance as an adult, or vice versa), and they show no clear phylogenetic distributions. Indeed, animals may be killed by **chill injury** before they show any signs of freezing, and within species there is often marked variation in the extent of chill injury and resultant mortality.

Any solutes present in water lower its freezing point in a predictable manner: remember that freezing point is one of the colligative properties, from which the concentration of dissolved solutes can be determined experimentally (Chapter 4). Most marine animals are isosmotic and therefore have blood that freezes at a similar temperature to sea water, –1.86°C, whilst vertebrate body fluids (around 300 mOsm) have a freezing point of about –0.6 to –0.7°C. Invertebrates living in sea water therefore have no particular likelihood of ice formation, as the high latent heat of fusion of water coupled with the thermal inertia of the marine system makes it most unlikely that the sea itself will freeze; but the hyposmotic vertebrates in sea water may be at risk. Freshwater animals have body fluids more concentrated than their surroundings, and hence a lower freezing point, so will not be at risk of tissue freezing unless their entire habitat is frozen (Chapter 11). But intertidal invertebrates may be exposed at high tide to very cold air and will rapidly freeze, many of them undergoing a regular twice daily freezing and thawing cycle in winter (Chapter 10). Small terrestrial animals such as insects may also experience exceptionally cold temperatures on a seasonal basis, while some desert forms such as scorpions only encounter cold temperatures at night; these too will undergo internal freezing.

Freeze-tolerant animals

These perhaps represent the lesser puzzle. When any animal does freeze, ice crystals can potentially grow very rapidly indeed if there are any nuclei present on which the water molecules can condense, and the crystals may severely damage cells in the process. In a biological system there are bound to be particles present in the extracellular and intracellular solutions

to act as nuclei. Ideally, therefore, ice crystals should be allowed to form rather slowly at relatively high subzero temperatures, perhaps –2 to –10°C, and they should be confined to the extracellular fluids, leaving the cellular machinery undamaged down to much lower lethal temperatures, even when as much as 65% of the body water is frozen. This can be achieved, somewhat unexpectedly, by having **ice-nucleating agents** (INAs) within the extracellular body fluids to encourage ice to form there; at the same time, much of the intracellular water is kept highly ordered by reactions with macromolecules and is relatively unfreezable.

Nucleating agents in freeze-tolerant invertebrates appear to be proteins (**ice-nucleating proteins** or INPs), with a very hydrophilic character. Their structure provides many hydrophilic sites that order the water molecules into small crystals, reducing the energy barrier for nucleation. They thus allow a slow controlled ice formation that involves only the actual water content of the fluids, and only the extracellular compartment normally freezes; the solutes (ions and metabolites) therefore concentrate in the cells and in the limited body fluids that remain unfrozen, giving membranes time to undergo controlled shrinkage. Once a high percentage of the body water is frozen the cells must be tolerating very high osmotic concentrations, which incidentally render them even less likely to freeze intracellularly. The tissues do become very distorted by the surrounding ice formation, with cells becoming shrunken and very changed in morphology.

Protecting the cells from injury during all these drastic changes is largely achieved by a second group of molecules, the **cryoprotectants**. These are of two kinds. **Colligative cryoprotectants** accumulate in high concentrations (0.2–2 M) and raise the osmotic concentration of the body fluids so that only a limited percentage of total body water can turn into extracellular ice. The commonest such molecules are polyhydric alcohols (polyols) or sugars, which are non-toxic even at high concentrations and are metabolically inert, being the end-products of biochemical pathways rather than intermediate stages. Glycerol (C_3) is the most commonly encountered (at up to 30% of body weight in some insects), but other polyols such as sorbitol (C_6) and ribitol (C_5) are also used. **Non-colligative cryoprotectants** occur at lower concentrations (usually <0.2 M) and are membrane protectants, binding in place of water and thus preserving the subcellular structure from long-term damage; trehalose and proline are common examples in insects. Many freeze-tolerant animals have multiple cryoprotectants. They are synthesized from glycogen reserves in the liver or fat body (Fig. 8.10), and must be in place before the first freezing exposure of winter.

The cells of freeze-tolerant animals must survive and maintain reasonable homeostasis in the face of long-term lack of blood flow (ischaemia) imposed by the surrounding ice. Their enzymes need to be able to function at high salt concentrations and very low temperatures, with a reduced metabolic rate and little or no respiration or circulation of body fluids. In the frozen state energy production often switches over to a phosphagen-hydrolysis system (using phosphoarginine or phosphocreatine),

Fig. 8.10 Synthesis of the major cryoprotectants in insects.

followed by anaerobic glycolysis, so that lactate and alanine accumulate as end-products.

Freeze-tolerance is quite common in many groups of invertebrates including some insects, bivalves, gastropods, annelids and nematodes. Freezing survival to –25 or –35°C for many weeks is common, with a number of Arctic species having lower lethal temperatures at –55 to –70°C. This strategy is rare in the vertebrates, where it apparently only occurs in a few specialized amphibians and reptiles that overwinter in cooler terrestrial habitats (see Chapter 14).

Freeze-intolerant animals

Animals whose strategy is freeze avoidance are decidedly puzzling. They live with their body fluids well below 0°C but with no ice formation. For example, many Antarctic mites can survive unfrozen at –10 to –20°C, and some species may remain active at these temperatures. This situation arises for two main reasons: because of the peculiar phenomenon of supercooling, and because of the additional presence of specific antifreezes.

Any liquid can undergo **supercooling**, i.e. being cooled below its freezing point without actually solidifying, but water is particularly good at supercooling (Fig. 8.11). The actual freezing process depends on three main variables: temperature, time and the presence of nuclei on which crystals of ice can begin to form. If foreign nucleating agents, such as dust particles, are carefully excluded, pure water can readily be supercooled to –20 or even –40°C without any ice appearing. If dissolved substances are present in the water we know that the freezing point is always

lowered anyway. But the presence of dissolved solutes also generally leads to the supercooling point (SCP) being lowered, so that body fluids have an inherently greater ability to supercool than pure water.

Animals that routinely encounter low temperatures (but are not continuously exposed to them) commonly enhance their own supercooling abilities by voiding their guts (where food particles could act as ice nucleators) and by adding extra solutes—known as **antifreeze** compounds—to their blood. Any solute, such as a sugar or salt, will have a limited antifreeze effect in accordance with its colligative properties (Fig. 8.12). But certain kinds of low molecular weight osmolytes are particularly effective antifreezes; glycerol and its relatives are, once again, the most famous examples in small terrestrial animals such as insects and arachnids. Glycerol occurs in high concentrations in many cold-hardy invertebrates that do not actually freeze, some of which can survive winter temperatures of –60°C. But these polyols still only have a simple colligative effect in lowering the freezing point. As we saw above they occur in some freeze-tolerant species too, probably to prevent the many small ice crystals recrystallizing into more stable larger crystals over time, which would be inherently dangerous. Thus animals with this strategy tend to show accumulation of polyols or sugars in autumn, lowering the supercooling point; then additional antifreezes are synthesized at the onset of winter to stabilize the supercooled state.

In polar marine fish there is a more severe problem to be overcome, because for them the threat of ice formation may be continuous; the animal's fluids are more dilute than the surrounding medium and thus can be below their own natural

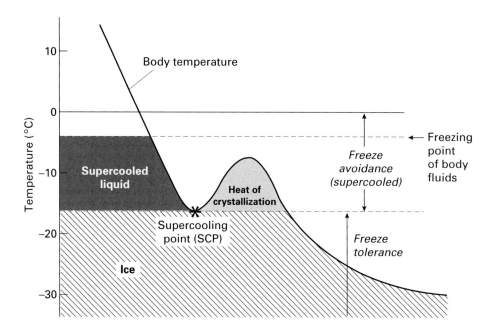

Fig. 8.11 Supercooling in aqueous fluids: freeze-avoiding animals may spend long periods in the supercooled state.

freezing point of about –0.7°C. They are in part protected by higher plasma NaCl levels than are normal for teleosts. But these animals also need compounds that will actually prevent existing ice crystals from growing. At least 11 different families of teleost fish have a range of **antifreeze protein** (AFP) or **antifreeze glycoprotein** (AFGP) molecules that can do exactly this (see Fig. 8.12 again). They bring about a protective phenomenon called 'thermal hysteresis', such that the freezing point of the solution is somewhat lower than its melting point (Table 8.4);

hence they are also sometimes called 'thermal hysteresis proteins', or THPs. On a molar basis, these antifreezes are far more effective than expected, reaching peak effectiveness at only 4% by mass, and thus being about 200 times as effective as the equivalent concentrations of NaCl. Again, it has recently been shown that they may also occur in freeze-tolerant species, and in both types of animal they confer protection against sudden unexpected freezing. Since their main effect is in marine fish, more details are given in Chapter 9. The AFGP and AFP molecules act by

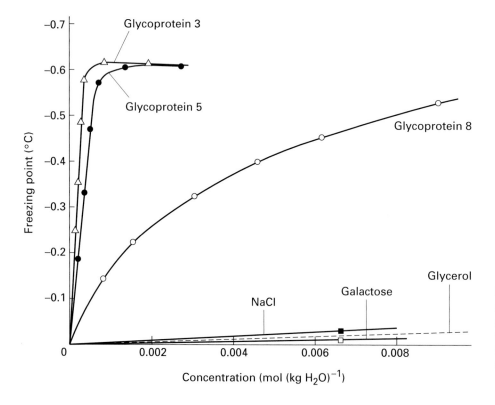

Fig. 8.12 Freezing points of aqueous solutions of important sugars, salts and glycoproteins used as antifreezes, the last of these being much more effective mole for mole and non-linear (suggesting a non-colligative action). (Data from de Vries 1980, 1988.)

Table 8.4 Freezing points and melting points of body and blood fluids from some cold-hardy animals showing thermal hysteresis (all values as °C).

Genus	Freezing point (blood)	Melting point (blood)	Supercooling temperature
Spiders			
Philodromus	−5.2	−2.7	−26.2
Clubiona	−4.8	−2.9	−15.4
Insects			
Meracantha	−5.0	−1.3	−10.3
Dendroides	−7.0	−3.4	−9.5
Fish			
Gadus (cod)	−1.1	−0.7	
Rhigophalia (eel pout)	−2.0	−0.9	
Myxocephalus (sculpin)	−2.0	−1.1	
Eleginus	−2.2	−1.2	
Pagothenia	−2.7	−1.1	

linking to the edges of ice crystal lattices and preventing new water molecules adding on there. They probably attach by their hydroxyl groups, and once in place they break up the ice front into small highly curved mini-fronts (Fig. 8.13), increasing the surface area in relation to the volume. Water molecules cannot add on to these curved ice fronts unless the liquid temperature is much lower; in other words, the freezing point of the solution is decreased.

A summary of the characteristics of cold survival shown in the two kinds of animals is given in Table 8.5; there are obvious overlaps, but the essential difference is the synthesis of ice-nucleating agents in freeze-tolerant species which override the supercooling effects of polyols. Freeze-tolerance and freeze-

Table 8.5 Freeze tolerance and freeze avoidance.

Characteristic	Freeze-avoiders	Freeze-tolerators
Ice formation	Lethal	Extracellular ice tolerated
LCT	−5 to −20°C (rarely −60°C)	−20 to −70°C
Supercooling capacity	High	Low
Supercooling point	Close to LCT	Well above LCT
Ice-nucleating agents	Absent or masked	Present and active (proteins)
Antifreezes	Polyols ± peptides in fish	–
Cryoprotectants	–	Polyols, ± trehalose, ± proline
Occurrence	Many invertebrates	Common in invertebrates
	Some vertebrates, notably polar fish	A few frogs

LCT, lower critical temperature.

intolerance have different advantages and disadvantages. Freeze-tolerant animals have a lower lethal temperature for particularly high levels of blood osmolality. Freeze-tolerance could be regarded as the 'cheaper' strategy, as with so many kinds of environmental tolerance; it ensures a sharp decline in metabolic rate in cold spells, and avoids the need to synthesize large quantities of polyols or other antifreeze molecules. It is perhaps most useful in areas where there are consistently severe winter climates. However, freeze-intolerance (freeze avoidance, with supercooling) may be the commonest overall strategy in animals, and may in particular be the best strategy in environments that are thermally quite variable. Here switching rapidly from inactive to active states during winter thaws or summer frosts is advantageous. Freeze-intolerance may also be linked with lower rates of cuticular permeability, and hence reduced problems with water loss. Some animals, notably larval beetles, can actually switch from freeze-tolerance to freeze-intolerance, giving yet greater flexibility. However, we should also remember that at an individual level many animals (particularly overwintering insects) are not really either tolerant or intolerant because they die from chill injury well above freezing point.

8.3.2 High-temperature effects on cells and organisms

In general terms, enzymes and membranes are active and functional over a broader range of temperature than is the case for individual cell organelles, which are active over a broader thermal range than whole cells, and organisms in their turn have a narrower thermal tolerance zone than do cells. There is an upper thermal limit for any given cell, beyond which cell death occurs; and a lower maximum for an animal as a whole. Clearly lethal limits are only partly determined by temperature effects on protein denaturation and coagulation so far described, though these are almost certainly part of the story. Protein (enzyme) inactivation is a less drastic but potentially equally

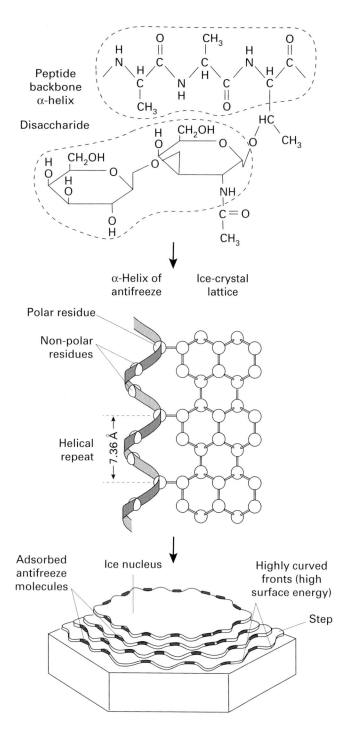

Fig. 8.13 The repeating structure of glycopeptide antifreezes; and a model of how this may allow adsorption to ice crystals, preventing growth at the sites of attachment and forcing intermediate sites to grow out into a series of curved ice fronts of high surface energy which require lower temperatures for further growth.

lethal effect, when inactivation rates exceed the rate at which new proteins can be formed via the ribosomes. However, both these effects really only apply in those cells whose upper lethal temperature is in the range 30–55°C, and cannot account for heat

death at 6°C in some polar fish. Some other mechanisms must also be involved, and three are particularly important:

1 There may be a disruption of the balance between reactions of differing thermal sensitivity, so that metabolic pathways are subverted. Several hundred enzymes contribute to cellular intermediary metabolism, and each has its own particular thermal properties. If one part of a pathway is greatly slowed the pathway it contributes to must also slow down for lack of appropriate substrates, affecting the balance in the system. Similarly where particular steps are accelerated the immediate end-product of that step will build up and may itself affect the operation of steps elsewhere in the system, again distorting the balance between pathways. Thus small changes of temperature can have drastic effects, and the cells within an animal may run out of energy supplies.

2 There may be effects on membrane structures (lipid–lipid and lipid–protein interactions), so that the various transport systems into and between cells (discussed in Chapter 4) become out of balance with each other (or even entirely inoperative), again upsetting the intracellular conditions on which enzyme function depends. In crayfish the first sign of heat death is a breakdown of the normal permeability of the gills, so that haemolymph Na^+ concentration decreases as gradients run down; abnormal neural activity soon occurs and the animal becomes severely uncoordinated just before death. Many other animals also show extracellular decrease of Na^+ levels and increments of K^+.

3 Animals may also be forced into death by the upper thermal limit for particular tissues, with nerves and heart muscle often showing a lower thermal tolerance than skeletal muscles, for example, so that uncoordinated movement may persist for a while after the heart and central nervous system (CNS) have succumbed.

For some or all of these reasons, there will be a distinct thermal maximum for a particular organism. This critical temperature may be 70–90°C for bacteria and other prokaryotes, but is rarely above 50°C for animals (where to avoid complications due to varying tissue sensitivities it is often measured as the temperature eliciting loss of neuromotor control). In fact this **upper critical temperature** (UCT) in animals commonly lies between 30°C and 45°C for all but deep-sea animals and polar aquatic species (Table 8.6). UCT is reasonably consistent within taxonomic groups, but with variation superimposed through ontogeny (caterpillars are more sensitive than butterflies, and tadpoles more sensitive than adult frogs and toads). Variation also occurs with habitat; for example, temperate lizards have lower UCTs than tropical and desert lizards, and on a smaller scale within a particular habitat lizard species from open grassy areas are more tolerant than nearby forest dwellers.

8.3.3 *Overview*

Biochemical and physiological effects clearly underlie the thermal tolerances, preferences, responses and susceptibilities of animals. The thermal phenotype of a metazoan comprises many

Table 8.6 Upper critical temperature (UCT) for a variety of animals from different habitats.

Group	Example (habitat)	UCT (°C)
Prokaryotes	Bacteria (aquatic)	70–75
	Bacteria (thermophilic)	90–91
	Cyanobacteria	75
Molluscs	*Modiolus* (SW bivalve)	38
	Nassa (SW gastropod)	42
	Clavarizona (SW gastropod)	43
Annelids	*Lumbricus* (land earthworm)	29
Echinoderms	*Asterias* (SW starfish)	32
	Ophioderma (SW brittlestar)	37
Crustaceans	*Palaeomonetes* (SW/littoral prawn)	34
	Porcellio (SW crab)	39–41
	Uca (littoral/land crab)	39–45
	Armadillidium (land woodlouse)	41–42
Insects	*Lepisma* (land springtail)	36
	Thermobia (land firebrat)	40+
	Sphingonotus (land moth)	41
	Bembex (land sandwasp)	42
	Onymacris (desert beetle)	49–51
	Dasymutilla (land sandwasp)	52
	Ocymyrmex (desert ant)	51.5
	Melophorus (desert ant)	54
Arachnids	*Buthotus* (land scorpion)	45
	Leiurus (land scorpion)	47
Vertebrates		
Fish	*Pagothenia* (polar SW)	6–10
	Fundulus (cold SW)	35
Amphibians	Salamanders (FW/land)	29–35
	Anurans (FW/land)	36–41
Reptiles	Alligators (land/FW)	38
	Turtles (SW/land)	41
	Lizards (land/desert)	40–47
	Snakes (land)	40–42
Birds	Passerines	46–47
	Non-passerines	44–46
Mammals	Monotremes	37
	Marsupials	40–41
	Placentals	42–44

FW, fresh water; SW, sea water.

hundreds of traits; it is important not merely to describe these traits but to understand their ensemble activities for the organism as a whole. Some traits will co-vary in evolutionary terms, while others will act almost independently but may on their own exert major effects on evolutionary fitness. Unambiguous attribution of a difference in thermal phenotype to any particular genetic change (or group of changes) is always going to be problematic.

It is also worth noting here that biological processes do not commonly show anything like complete **compensation** for the effects of temperature—they seem to be adapted to keep going,

but not to keep going at the same rate irrespective of thermal regime. Thus, for example, there appears to be little or no temperature compensation in the rate of development of embryos for a wide range of marine invertebrates and fish. Figure 8.14 shows the relationship between temperature and development time for fish and for sea urchins with planktonic larvae, and it is striking that polar, temperate and tropical species all fit to the same relationship. The same lack of compensation is true for the resting and maximum metabolic rates in fish discussed in Chapter 6, and it is probably a widespread phenomenon (perhaps not surprising, since maintaining high rates at low temperatures would be very costly). This underlines again the caution needed in evolutionary physiology; adaptation and acclimation do not produce what we may see as theoretically ideal solutions, merely solutions that are 'good enough' and reasonably cheap.

8.4 Terminology and strategies in thermal biology

Different animals have very different ranges of body temperatures in their natural surroundings, and differing degrees of thermal constancy. Unfortunately, though, this is an area of comparative biology where the terminology used has often tended to obscure rather than clarify the underlying patterns and processes; so this section needs to begin with a review of the various categories that have been used, and of their meanings.

8.4.1 Patterns of body temperature

Initially, the terms **warm-blooded** and **cold-blooded** were coined to describe animal temperature patterns, and they are still used today in general conversation. The terms stress the perceived difference between animals that are always warm to the touch—the birds and mammals—and those that commonly feel rather cool, a grouping that includes all other animals but which has particularly been applied to other vertebrates (the fish, amphibians and reptiles). However, these terms never enjoyed much scientific favour, since it is evident that many supposedly 'cold-blooded' animals can become distinctly warm, for example after a period of basking; reptiles and many insects regularly achieve an elevated T_b of the same order as that found in the birds and mammals. Moreover, a hibernating mammal may have a T_b as low as 5°C. Such expressions as warm- and cold-blooded are therefore best avoided.

The more favoured terms that were in general use for much of this century were **poikilothermic** and **homeothermic**, referring to the constancy of the body temperature rather than its actual setting. Poikilotherms (from the Greek *poikilos* meaning changeable) have a variable T_b, whereas in homeotherms it is rather constant. In principle this makes perfect sense, except that the terms have been regularly misused and misunderstood. In particular, 'homeothermic' has often been seen as a synonym of 'warm-blooded' and applied in an exclusive sense to the birds and mammals, where T_b is nearly always around 37–40°C. All other animals are thus seen as poikilotherms. Yet in this usage

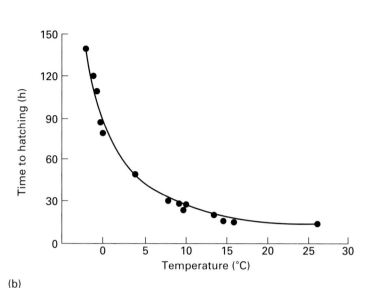

(a)

(b)

Fig. 8.14 Temperature and development time for (a) various marine fish and (b) sea urchin planktonic larvae. Note in (a) the similar form of relationship for very different latitudinal populations. NT, north temperate; Med, Mediterranean. (a, from Johnston 1990; b, from Bosch *et al.* 1987.)

the term is clearly misleading; a fish or a worm living in cold, deep water at 5°C is not a poikilotherm at all, but is every bit as homeothermic as a hamster or a human. An insect, or treefrog, in the litter layer of a tropical forest may also be close to homeothermy. In other words, homeothermy may be achieved physiologically in some specialist animals (the sense in which the term was primarily intended to be used); but it may also be brought about quite passively, by environmental constancy. To categorize such different conditions together seems unhelpful. The concept of poikilothermy can be similarly muddling—a hibernating mammal or a small tropical hummingbird with nightly torpor may have a body temperature almost as variable as that of a reptile living in similar conditions.

A more recent and more sensible terminology refers to **endothermic** and **ectothermic** animals. These terms emphasize the heat sources used, rather than the settings or constancies of achieved T_b; i.e. they refer to mechanisms in thermal biology rather than to conditions. As such, they are unambiguous:

• **Ectotherms** have a body temperature principally dependent on external heat sources, almost always ultimately the Sun, either directly ('heliothermy') or from a heated substrate ('thigmothermy'). Geothermal heat may also be available in some cases.

• **Endotherms** have a body temperature principally dependent on their internally generated metabolic heat.

Again the distinction is largely between birds and mammals, being endothermic, and the rest of the animal kingdom, using

ectothermic strategies. But on a broader view the distinction is actually independent of that between poikilotherms and homeotherms (Fig. 8.15). This arises because our new terminology is still not without its own problems. The main complication is that in practice *all* animals generate some internal heat, as a natural consequence of being alive and having metabolic conversions going on in their constituent cells (especially their muscles) that are not perfectly efficient. And nearly *all* animals also use external heat sources; most terrestrial birds or mammals will choose to bask, for example, maintaining their body temperatures as cheaply as possible, when solar radiation is available. However, the two terms remain readily distinct, for the principle *mechanisms* of heat gain do not merge as a continuum. While all animals may use external heat sources and will generate some internal heat, a few groups of animals are quite distinctly endothermic in that they have an additional special ability to raise their metabolic rate markedly (especially in the liver and viscera) in excess of all their normal needs, specifically benefiting from the resultant heat generation. In fact a useful complementary terminology exists, with the endotherms referred to as **tachymetabolic** (having a fast metabolism) and the ectotherms being **bradymetabolic** (slow metabolism); for a given body size, an endotherm will normally have at least five times the metabolic rate of an ectotherm.

Although ectothermy and endothermy are thus distinct and non-overlapping, there are some intermediate conditions, which are sometimes collectively called **heterothermy** although they occur for a variety of different reasons and would be better treated separately.

1 Partial endothermy occurs in some of the very smallest birds and mammals (hummingbirds, tiny rodents, etc.), where the normally high metabolic rate is turned down seasonally or even

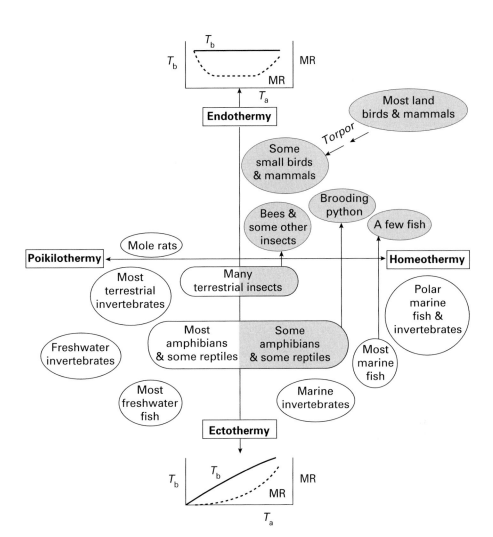

Fig. 8.15 Terminology in thermal biology—
how the terms used are distinctive or
overlapping. Shading shows
thermoregulators; white shows
thermoconformers; arrows indicate
temporal heterothermy. MR, metabolic rate;
T_a, ambient temperature; T_b, body
temperature.

each night, to reduce energy expenditure in conditions of cold
or of low food and water supply; a temporary torpor results,
with the body temperature falling substantially (see section 8.9
and Chapter 14). Newborn birds and mammals are also in effect
ectothermic conformers, relying on nest materials and parental
body heat to keep warm. At the extreme, the curious burrowing
naked mole rats of arid southern and eastern Africa, and also
the golden moles living in similar habitats, appear to be almost
entirely 'turned off', and operate more like non-mammalian
ectotherms, perhaps because they would lose too much heat by
conduction to the burrow walls if they attempted continuously
strongly elevated and regulated body temperatures. (Here is a
further reason why the term homeotherm should be avoided;
some traditionally homeothermic birds and mammals just aren't!)
2 Facultative endothermy is increasingly recognized in a variety
of animals that, while normally ectothermic, can 'turn on' an
endothermic heat generation system in some parts of their body
when they need to become active at low ambient temperatures.
Insects such as bumble-bees are a classic case, being essentially
ectotherms but with considerable scope for raised metabolic
rates leading to temporary endothermy. Most bees, and quite a

wide range of other insects, have this ability, and more detail of
the mechanisms involved will be found in Chapter 13.

Note that strategies 1 and 2 are sometimes collectively called
temporal endothermy or **temporal heterothermy**, as they involve periods of endothermy alternating with a slow metabolic
state closer to ectothermy. Figure 8.16 compares the relative
body size ranges in which these strategies occur in vertebrates
and in insects.

3 Regional endothermy occurs in a variety of fish and some
reptiles. In these cases, localized areas of the musculature
routinely operate at much higher temperatures than the rest of
the body, allowing faster or more sustained locomotory activity,
or maintained sensory abilities, in cold environments. Some of
these cases, involving specialized muscle cells that have lost
their contractility, will be described in Chapter 9.

Here we should also mention **regional heterothermy**, having
somewhat similar effects but for very different reasons. It occurs
in many birds and mammals in cold environments, where it is
common for the extremities to be much cooler than the body
core. This is usually the result of cardiovascular adjustments
serving to reduce heat loss (see section 8.8.2).

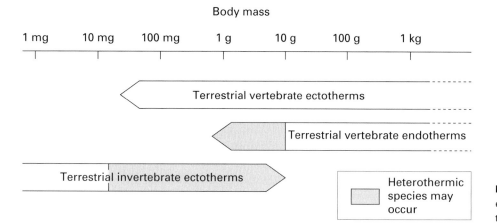

Fig. 8.16 Relative sizes of ectotherms and endotherms, showing the size ranges for the occurrence of temporal heterothermy.

4 Inertial homeothermy/endothermy should also be dealt with here. Animals that have no specific strategies for raised metabolic rate and are essentially ectothermic and bradymetabolic can still end up with a rather high and rather constant T_b if they are particularly large-bodied. They have a very high thermal inertia because of scaling effects, the animal's surface area being relatively too small to dissipate even the slowly generated internal metabolic heat from their comparatively enormous volumes, so that internally generated heat is the major contributor to their body temperature as with an endotherm. Inertial homeothermy is the main reason why the largest dinosaurs,

though reptilian and perhaps lacking any specific endothermic capacity or tachymetabolism, must have been effectively 'warm-blooded' all the time.

Despite these four kinds of equivocal case, there is no doubt that the ectotherm/endotherm terminology is the most satisfactory of those available to us. In this book we will therefore use only the terms ectotherm and endotherm to describe basic thermal adaptive strategy.

These terms separate essentially distinct thermal strategies, while all other systems, by concentrating on the result rather than the mechanism, involve the inappropriate bracketing together of quite different kinds of thermal strategy (see Fig. 8.15). This is because the *result* of all thermal strategies is rather often a relatively constant body temperature, achieved by whatever means are available. For most aquatic animals T_b is close to that of the medium, whilst for aquatic mammals and for most active land animals T_b during activity is commonly in the range 25–40°C (Table 8.7; cf. Fig. 8.1).

8.4.2 Patterns of temperature maintenance

So far we have dealt only with the terminology of overall thermal strategies. There is a further complication pertaining to the terminology of body temperature. It is very often implied that some animals have a special ability as **thermoregulators**, while most do not bother and are **thermoconformers**, at the mercy of the thermal variation of the environment. This is fundamentally untrue, for most animals do indeed thermoregulate to some extent. Probably the only 'conformers' are those animals that live in environments, such as the deep sea, that have almost no thermal variability, where body temperature is virtually identical to the invariant ambient temperature at all times (it will in practice be detectably elevated by internal metabolic processes, but normally by less than 1°C). The misconception reflects once again the long-standing over-emphasis on physiological mechanisms as the keystone of adaptation; it results from the view that only **physiological thermoregulation** is 'real' thermoregulation, whereas in practice **behavioural thermoregulation** is at least as crucial to many animals, and can be equally effective.

Table 8.7 Body temperatures (T_b) during normal activity in ectothermic and endothermic animals.

| Habit | Ectotherms | | Endotherms | |
	Group/species	T_b	Group/species	T_b
Aquatic	Marine deep-water fish	4–6	Fur seal	38
	Marine surface fish	5–10	Whale	36
	Temperate riverine fish	5–15		
	Reef fish	20–28		
	Tropical turtles	29–35		
	Warm temperate turtles	20–25		
Amphibious	Alligator in air	32–35		
	Alligator in water	20–35		
	Bullfrog tadpoles	22–30		
	Bullfrog adults in air	22–28		
Terrestrial	Blowfly	22–29	Man	37
	Flour beetle	25–30	Small rodents	35–37
	Bumble-bee (flying)	35–39	Bats	35–39
	Carpenter bee (flying)	38–42	Echidna	31
	Housefly (flying)	30–33	Armadillo	34–36
	Housefly larva	30–37	Chicken	40
	Chicken louse	38–42	Dove	39–42
	Temperate caterpillar	20–28	Owl	38
	Semi-tropical caterpillar	25–35	Zebra finch	41–42
	Tropical lizard	28–36		
	Desert iguana	36–41		

The muddled thinking about thermoregulation probably results from the association of the term with the 'homeothermic' endotherms. Physiological regulation is certainly used by most endotherms to help achieve their state of T_b constancy. They may have physiological mechanisms to ensure a high heat production or heat gain, and also mechanisms to reduce or to hasten heat loss. But they also use a great many behavioural tricks, such as basking, huddling and burrowing. Nor is physiological regulation their exclusive prerogative; for whilst ectotherms may have a very obvious repertoire of behavioural thermoregulatory options, many of them do also have a suite of physiological strategies, as we will see.

To summarize:

• ectotherms can be regulators or conformers, and use both physiological and behavioural mechanisms to regulate;
• endotherms are always regulators, but also use both kinds of regulatory mechanism to achieve constant T_b.

8.4.3 *Patterns of temperature tolerance*

There is a third and final problem with terminology that needs to be sorted out, but which is fortunately relatively straightforward. This concerns the familiar usage of the terms **eurythermal** and **stenothermal**. As with the equivalent terms euryhaline and stenohaline used in the discussion of water balance and osmotic problems (Chapter 1 and section 5.1.3), these concern the range of tolerance an animal will show. A eurythermal animal tolerates, and is active with, a rather wide range of T_b; for example, many temperate insects and reptiles will continue feeding and locomoting with body temperatures of perhaps 8–38°C, tolerating a range of 30°C, and similar ranges of tolerance exist in some intertidal rock pool fish. By contrast, a stenothermal animal only operates when its T_b is within a narrow range: most mammals and birds are stenothermal, but some lizards and even a few insects are too, and many aquatic animals are very strict stenotherms, with some polar fish tolerating only about 6°C in total range (from the freezing point of sea water to roughly 4°C).

Linking this back to our discussion of temperature effects at the biochemical level, the terminology can be applied at that level too. It is generally true that stenotherms have proteins and membranes that are themselves tolerant of only narrow temperature changes, i.e. are stenothermal, while eurytherms have biochemical components with wider (eurythermal) tolerances; the differences are in large part genetically fixed.

8.5 The thermal environment and thermal exchanges

The thermal exchange between any two objects is proportional to the difference in their temperatures, and the same will apply to exchange between an animal and the components of its environment. Temperature control in animals is largely concerned with manipulating the various avenues of heat exchange with

---

Box 8.1   Conduction, conductivity and insulation

Conduction involves direct transfer of heat between two solid materials, and rate of conductive heat transfer (Q_c) in Joules per second (or watts) can be calculated by a modified version of Fick's law as follows:

$$Q_c = -kA \, (T_2 - T_1)/d$$

where A is the area of contact and d the distance between the two measured temperatures T_1 and T_2. Here the constant k is the **thermal conductivity**, with units of J s^{-1} °C^{-1} cm^{-1}.

Since A and d are often difficult to measure, the equation is often rewritten as:

$$Q_c = C \, \Delta T$$

where ΔT is the thermal gradient; now C is the **thermal conductance** (or conductive heat transfer coefficient) with units of J s^{-1} °C^{-1}.

Conductivity (k) is a fixed property of materials, and is directly related to **insulation**, which is the reciprocal of the total heat flux per unit area per unit of temperature difference, and therefore with units of °C cm^{-2} s J^{-1}. In practice it is usually measured per metre squared, and a common unit given is the 'clo' (clothing unit), where 1 clo = 0.16 °C m^2 W^{-1}.

For biological use, an insulative layer is often described by a **resistance** value (r, measured in s cm^{-1}), where the resistance takes into account both the thermal conductivity and the thickness (l) of the layer:

$$r = \rho C_p l/k$$

Thus there is a reciprocal relation between resistance and thermal conductivity.

Here $\rho \, C_p$ is the specific heat capacity of the medium (air or water).

---

that environment in the animal's favour, depending on its immediate needs and circumstances. Since the thermal environment of many animals, especially those that are terrestrial, is very complex, these patterns of exchange control can be remarkably elaborate.

Before considering details of these control systems, we need to deal with the basic principles of heat exchange, which as in any physical system will involve **conduction**, **convection** and **radiation**. In addition, since animals are made up of watery tissues those that live on land may sometimes have the option of heat control by **evaporation** or **condensation**.

8.5.1 *Conduction*

Conduction is the direct flow of heat (kinetic energy) between two materials in contact, whether those materials are solid or fluid; heat flows from the warmer to the cooler material, by a direct transfer of the kinetic energy as molecules collide. The rate of heat transfer therefore depends upon the temperature difference (or temperature gradient), the area of contact, and the conductive properties of the two materials (see Box 8.1). Clearly an animal in direct contact with its environment has control over

Table 8.8 Insulation values of biological and other materials.

Material	Insulation (°C m² W⁻¹)
Silver	0.015
Steel	0.14
Ice	2.9
Water	11
Human tissue	14
Dry soil	20
Natural rubber	38
Fat	38
Cattle fur	50
Pigeon feathers (flat)	99
Sheep wool	102
Goose-down feathers	122
Husky dog fur	157
Lynx fur	170
Still air	270

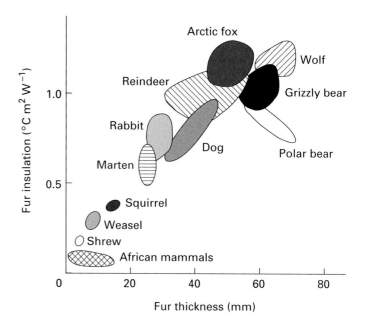

Fig. 8.17 Fur thickness and fur insulation values; small mammals of necessity have short fur, of relatively poor insulation value. Note that these examples are cool temperate and polar species; the fur in African mammals rarely exceeds 15–20 mm in length, and this decreases rather than increases with body size, being only 0.5 mm in the eland (the largest antelope) (see Chapter 14). (Data from Scholander *et al.* 1950; Hofmeyr & Louw 1987.)

some of these parameters. Any motile animal can exercise choice over the nature of the substrate on which it rests, and thus choose an appropriate temperature difference. It can also vary the area of contact rather easily by changing its posture, from minimal contact at the tips of appendages to a full exposure of up to half of all its surface area to the substrate (or 100% of its surface if it burrows).

Since these parts of the heat conduction relationship are variable in the short term by behavioural means, thermal conductivity is often the most important long-term adaptive element in determining conductive loss and gain of heat. A variety of ways of expressing this and related concepts are used (Box 8.1), of which **insulation** value is perhaps the most familiar. Insulation may be provided by internal air sacs or fat layers, or externally by cuticular bristles of chitin or by various hairs and other structures made of keratin (Table 8.8). Internal insulators are much less effective: a layer of blubber 60–70 cm thick has about the same insulating properties as 2 cm of good mammalian fur. In practice, most external biological insulating materials, including these familiar furs and feathers, have an insulation value rather similar to that of still air (Table 8.8), which is not surprising as they nearly all rely on trapping a layer of still air close to the animal's surface. Of course the insulation provided does depend rather directly on the thickness of this trapped layer (Fig. 8.17), so that a large mammal able to carry a thick fur layer (e.g. about 6 cm thick in a polar bear) has much better insulation than a tiny mouse with only a 0.2 cm fur layer, on which any thicker fur would severely restrict movement. Nevertheless, per unit thickness the mouse fur turns out to be a somewhat better insulator, as might be expected.

Mammals are also adept at varying fur thickness and therefore conductivity over the body, with denser fur layers commonly around the trunk and rather scanty fur on distal parts of the limbs (this may sound counter-adaptive at first, allowing huge heat loss from large surface areas, but linked to blood

and heat distribution patterns, discussed in section 8.7, it becomes more intelligible). Both birds and mammals naturally also make use of extrinsic insulators, in their choice of nesting materials.

For large terrestrial mammals there may be the opposite problem of eliminating excess metabolic heat, especially after periods of activity or burrowing in warmer climates. This problem is compounded because total heat production increases with $M_b^{0.75}$ (see Chapters 3 and 6), whilst surface area for heat dissipation only increases with $M_b^{0.67}$. In the tropics very large mammals such as the elephant and rhinoceros are therefore largely hairless, to aid heat loss and keep them in thermal balance.

8.5.2 *Convection*

Convection is the flow of heat between two bodies by the mass movement of an intervening fluid, whether gas or liquid. Convection currents may be 'free' or 'natural', caused by inherent temperature differences (since warm air or warm water will rise), or may be imposed by outside mechanical factors such as wind, water currents or movements by the animal itself ('forced convection').

Convection is naturally much faster than conduction. For animals, convective heat movements often occur continuously within the body, when hot blood is moved from the core to the periphery; in endothermic vertebrates with complex circulations, any disorder that reduces this convective flow may cause

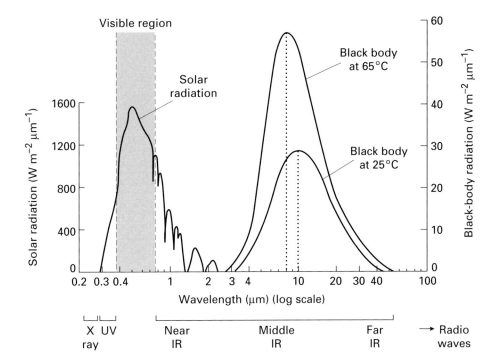

Fig. 8.18 Thermal radiation from bodies of different temperatures: the higher the surface temperature, the shorter the wavelength and the higher the intensity. Note especially the spectral distribution from the Sun (6000+°C), from a Sun-heated rock or animal surface at 65°C, and from a mammalian body surface at about 25°C. (Dotted lines show peak wavelengths.) (Adapted from Hardy 1949; Bond *et al.* 1967.)

an explosive rise in body temperature, since purely conductive heat transfer to the periphery is much too slow to cope.

However, the convective heat exchange of animals with their environment is the key issue here, and is essentially the process of conductive heat exchange through the boundary layer of fluid that separates that animal from (and differs in temperature from) the free mass of fluid constituting the environment, whether air or water. This boundary layer is extremely complex, modified on a very small spatial and temporal scale by water or air flow, and assessment of convective temperature exchange across animal skins is notoriously difficult. In both air and water, the temperature difference between the animal and its environment is an important determinant of heat exchange, as are the animal's dimensions, the curvature of its surfaces, and the fluid's thermal expansion coefficient, density, viscosity and velocity. But even when all these are known, the exact pattern of convective exchange across and just above the skin may defy accurate description or prediction. It is therefore common, and not too inaccurate, to treat conduction and boundary-layer convection together when dealing with animal heat balance.

In water, convective heat transfer is very rapid and it is very costly for all but the largest aquatic animals to maintain temperatures above that of their surroundings; few do so. In air, convection is slower and large temperature gradients between the body and its surroundings become feasible, again relying on the effects of insulating layers that effectively expand the boundary layer.

8.5.3 Radiation

Radiation travels with the speed of light and is emitted by all bodies whose temperature is above absolute zero (–273°C). The wavelength of this radiation depends on the surface temperature (T_s) of the emitting body, the predominant wavelength (l_{max}) being inversely proportional to T_s (Wien's law). For a living organism, which in absolute terms is relatively cool, l_{max} is around 9–10 μm, which is rather longwave infrared radiation (Fig. 8.18). The total rate at which such radiation is emitted from any object is, according to Stefan's law, proportional to the fourth power of T_s, so it rises very rapidly with small increases in surface temperature. It is also dependent on the radiative surface area and its **emissivity**.

This latter parameter is a dimensionless constant for any material and is related to the familiar concepts of **colour** and shininess of the surface. Emissivity for a particular surface actually varies for different wavelengths, which is what normally gives rise to perceived colour, a red object having a high reflectivity of (and low emissivity for) incident red light. For most biological materials, the emissivity at 'biological' wavelengths in the medium to long infrared is between 0.90 and 0.99, where a perfect 'black body' has an emissivity of 1.00. Some external skeletal materials, such as insect cuticle or snail shell, can be white and shiny, though, and are of very much lower emissivity (Table 8.9); in cuticles this is often due to tiny air bubbles, while in shells the properties of calcium carbonate are often exploited.

However, colour as perceived in relation to visible light wavelengths is not necessarily a good guide to emissivity for the important biological infrared radiation; black and white skins in humans, though they appear very different, have very similar properties in relation to emission of thermal radiation, and also differ little in their absorption of longwave radiation that is re-emitted from the ground and other solids. The difference, as with most differences in skin or fur colours, lies in their absorption of

Table 8.9 Emissivities and reflectances of various materials.

Material	Emissivity	Reflectance (%)
Water	0.95	
Cork	0.95	
Glass	0.94	
Wood	0.90	
Ice	0.92	
Steel	0.07	
Aluminium	0.20	
Copper	0.03	
Insect cuticle		
Black matt		2–5
Black shiny		4–23
Pale colours		5–25
White		10–35
Snail shell		
Dark		2–30
White		95
Human skin		
'Black'		5–9
'White'		5–10

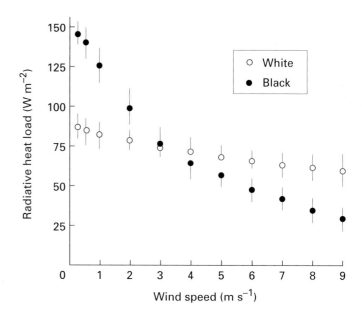

Fig. 8.19 Colour effects in the pigeon. Radiant heat reaching the skin under a high insolation regime is greater for black plumage at low wind speeds but greater for white plumage at high wind speeds. (Adapted from Walsberg & King 1978.)

the visible components (almost 50%, see Fig. 8.18 again) of direct solar radiation. Dark-coloured skins and furs do indeed absorb more of this incident light than pale skins and furs. The situation is further complicated by the 'penetrance' of the radiation. For example, black pigeons absorb a greater radiative heat load than white pigeons in still air, but this difference becomes reversed at high wind speeds because radiation penetrates more deeply into white plumage and therefore the white birds' heat gain is less affected by ruffling of the feathers in windy conditions (Fig. 8.19). Note that confusion can also arise due to structural colours (often seen as iridescence, e.g. in butterfly wings or metallic feathers), which can be generated by the scattering of small particles, by multilayered reflectors, or by diffraction gratings, and which do not relate closely to thermal properties.

In general, though, colour bears some relation to habits and habitats, especially on land; animals have pale surfaces where they need to stay cool in hot climates, and dark surfaces if they need to warm up by basking, etc. However, there are exceptions in every environment (black desert goats, white polar bears) which will be dealt with in the relevant chapters of Section C. For now, we should note that it is always important to avoid a simple-minded view of colour in relation to temperature.

It must also be stressed that emissivity, colour or reflectance, or any other similar parameter, affects only the rate of change of temperature and cannot alter the equilibrium temperature that an object will finally reach; for example, a white ball and a black ball will both eventually reach the same temperature if left in the sun, but the highly reflective white ball will take substantially longer to reach this equilibrium.

Objects at temperatures in the biological range can be calcu-

lated to emit radiant energy in the range 300–500 W m^{-2}. This is very low emission compared to hot bodies such as a light bulb or an electric heating element, but represents a substantial loss relative to energy production from metabolism; for a resting 10 kg mammal with about 1 m^2 of radiating surface the emissive loss would be 300–500 W whilst metabolic production would be only 20 W. In practice, of course, such an animal is also gaining radiative heat all the time from components of its environment. Solar irradiance perpendicular to the Earth's surface may exceed 1000 W m^{-2}, which is much greater than the metabolic heat production of a typical mammal, while surrounding surfaces (soil, rocks, vegetation) of similar temperature (T_{sur}) will also be radiating heat to the animal at roughly the same rate as the animal is emitting it (Fig. 8.20).

8.5.4 Evaporation

Evaporative loss of water is an excellent way of dissipating heat; the latent heat of vaporization of water is 2500 J g^{-1} at 0°C, and is still 2400 J g^{-1} at 40°C (or about 580 cal g^{-1}). Rate of evaporation depends not only on the surface temperature, but also on the difference in water vapour density between the animal's surface and the environment, and on the resistance to water loss from that surface (measured as permeability or some related parameter).

Air at the immediate surface of the animal is usually assumed to be saturated, at 100% relative humidity (RH), and to decrease in saturation in a regular manner away from that surface through the boundary layer. As with convection though, the actual situation in this layer is complex and difficult to model

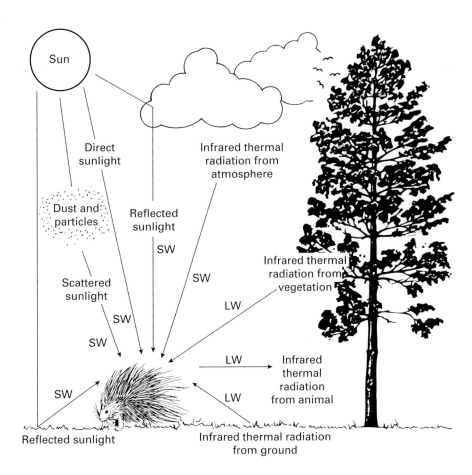

Fig. 8.20 Radiation exchanges for a land animal, emphasizing direct shorter wavelength inputs and reflected longer (infrared) wavelength inputs. LW, longwave; SW, shortwave.

accurately, being modified by very small-scale fluid movements. This makes rates of water loss and thus evaporative heat loss difficult to measure (and note that it also complicates the issue of determining permeabilities of biological surfaces, as Chapter 5 showed that these are generally estimated indirectly from rates of water loss!). Evaporation in relation to the water content of air also has to be very carefully expressed, in view of the interactive effects of temperature and water vapour concentration discussed in Chapter 5.

Evaporation as a means of temperature regulation is unique amongst the methods discussed here, as it can *only* be used by animals living in air. It is thus a fairly common technique for animals on seashores and in humid terrestrial zones, where heat stress can be a problem but water is reliably available. But evaporative cooling declines in importance in most terrestrial zones, and it is rarely used in deserts, even though these may be very hot and will put animals under considerable thermal stress, since the water loss that it entails cannot then be tolerated; only a handful of moderately large species, away from the depths of deserts, use it extensively (Chapter 14). Figure 8.21 summarizes the extent of evaporative cooling in a variety of land animals.

8.5.5 Overall thermal balance

The thermal balance of any animal is determined by the net exchanges of heat through all the avenues outlined above, together with metabolic heat production (*M*). The general equation describing heat balance is therefore:

$$M = h_{cond}\,(T_b - T_a) + h_{conv}\,(T_s - T_a) + h_{rad}\,(T_s - T_{sur}) + E + S$$

where *h*-values are the various heat transfer coefficients, which are in practice rather similar for the three main temperature exchange routes. *E* is any evaporative heat loss, while *S* is any heat storage which occurs, dependent on the change of temperature, the mass of the body, and the specific heat capacity of body tissues. A pictorial summary of these complex inputs and outputs for a terrestrial animal such as a tortoise is shown in Fig. 8.22.

This equation can be greatly simplified for a steady state, with no change of temperature in the animal, no heat storage, and no significant difference between T_s and T_{sur}, as:

$$M = h\,(T_b - T_a) + E$$

where *h* is the total heat transfer coefficient.

For a thermoconforming ectotherm in a thermally stable environment, *E* and *M* are negligible and the equation simplifies readily to $T_b = T_a$. However, for more complex thermal strategies, particularly in land animals, the balancing of this equation is clearly going to be a more difficult procedure, and a wide range of adaptations to control different elements of the animal's heat-exchange systems have evolved.

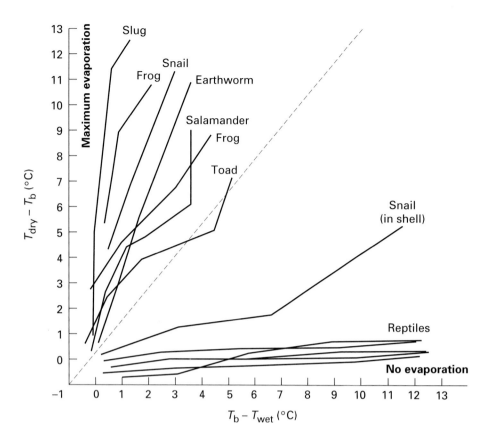

Fig. 8.21 Body temperatures (T_b) of animals as a function of the differences between T_b and the wet and dry bulb temperatures. The wet bulb temperature (T_{wet}) is that of a thermometer with a moist covering, and many soft-bodied animals act as wet bulb thermometers, so that T_b is close to T_{wet}. Hence the y-axis indicates the body temperature depression due to evaporation, and thus the extent of evaporative cooling, which is high for slugs and frogs, especially at high T_b. Note that some of these animals (toad, salamander) show a sharp increase in evaporation as T_b rises. In contrast, reptiles act more like dry bulb thermometers, with T_b close to T_{dry} (T_{dry} is the normal air temperature without any evaporative cooling effects). (Adapted from Withers 1992.)

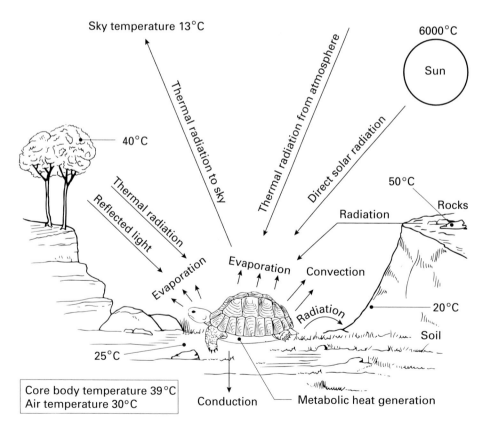

Fig. 8.22 An overview of the thermal exchanges between a terrestrial animal and its environment.

8.6 Avoidance, tolerance and acclimation in thermal biology

Now that the avenues of heat exchange have been considered, we need to examine the specific strategies that animals use to reduce, accentuate or control these exchanges. As with so many other aspects of comparative physiology, these can be considered under the general categories of avoidance strategies, tolerance strategies and regulatory mechanisms, the latter including both acclimatory responses and a variety of strictly regulatory systems. Thermal regulation is the most complex of these topics because so many different possible techniques exist in animals. It is therefore covered in depth in sections 8.7 and 8.8, dealing with regulation of heat gain and of heat loss separately.

8.6.1 Avoidance

Avoidance of environmental temperature extremes is an important part of the behavioural repertoire of a huge range of animals; virtually all land animals use avoidance as a first line of defence and thus as a key part of their thermal biology, as do many littoral, estuarine and freshwater animals. The strategy is less obvious in truly marine animals as the temperature variation in their environment is rather limited, but thermal avoidance behaviours certainly do operate in marine animals that encounter permanent or seasonal thermoclines (steep environmental temperature gradients—see Chapter 9).

Avoidance in partly or fully terrestrial animals may take many forms, but can be generally viewed as the location and use of appropriate **climatic conditions** in time and space. In this context an animal's size becomes a crucial variable influencing its thermal balance, in two key respects. Firstly, cooling and heating rates will be faster for a small ectothermic animal than for a large one given the same thermal gradient; the larger animal has a larger **thermal inertia**. The time for thermal equilibrium to be reached increases with body size and with the initial temperature differential (Fig. 8.23). Within each taxon the time to equilibration increases around 10-fold for a 100-fold increase in body mass. Large insects and reptiles can make use of their relatively large thermal inertia, for example, by heating their bodies in a warm microhabitat and then foraging in a previously inhospitable cooler one. Conversely species that forage in warm environments may gain too much heat, which then has to be lost by returning to shade. In general the activity bouts and cooling-off bouts will have to be shorter for smaller species.

Size is obviously crucial in a second respect in that the concepts of microclimates and microhabitats are very labile depending on the size of the animal experiencing the environment. For organisms with dimensions in millimetres the environment is experienced as a very 'fine-grained' one, with conditions varying widely in time and space. It is vital to appreciate the huge range of microclimates available to a small terrestrial invertebrate; conditions vary enormously on a scale of millimetres or centimetres, and change from second to second when there is air

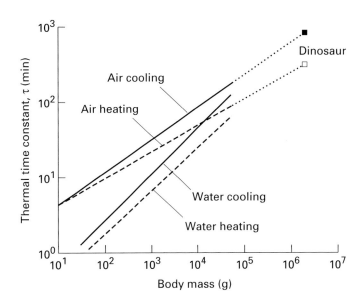

Fig. 8.23 The relationships between the thermal time constant (τ) and size in reptiles. The time constant scales with mass over four orders of magnitude for extant species, varying according to whether the animal is heating or cooling and in water or in air. Warming occurs more quickly (i.e. lower time constant) than cooling, especially in water (having a higher conductance). Extrapolation to reptiles the size of a large dinosaur indicates the very substantial thermal stability that these animals must have had, with a time constant for cooling of up to 16 h.

movement. Familiar examples of microclimate exploitation by smaller animals include the occupation of thermally equitable refugia such as burrows or nests, which on a seasonal or diurnal or moment-to-moment basis can make the difference between survival and death for animals in very hot or very cold climates (Chapter 14). Less familiar, but equally dramatic in effect, are the subtle movements of insects or spiders up and down a grass blade, around a plant stem, or in and out of the boundary layer above hot sand by standing up on the tips of long limbs (stilting), movements which can allow the animal to avoid highly adverse conditions and achieve an entirely equitable microclimate within seconds. Many other kinds of thermal avoidance strategy will be met in Chapters 13 and 14, dealing with terrestrial adaptations; their importance for small animals can hardly be overstated.

But large animals do not have this option of avoidance to anything like the same extent; they physically do not fit within the varied microhabitats open to a small creature, and their thermal inertia would be too large to allow them to take advantage of these temporally variable microhabitats anyway.

8.6.2 Tolerance

Animals vary substantially in their tolerance of varying body temperatures, and thus in the temperatures that elicit acclimatory or regulatory responses. We saw in Table 8.6 that some polar marine animals have an upper lethal temperature as low

Fig. 8.24 The match between preferred body temperatures as measured in the laboratory and observed body temperatures in the field, for reptile species ranging from monitor lizards (A) through small lizards (B), marine iguanas (C) and snakes (D). (From Cossins & Bowler 1987, Fig. 3.4; with kind permission from Kluwer Academic Publishers.)

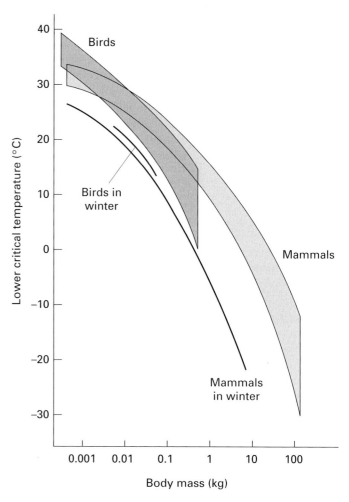

Fig. 8.25 The relation between lower critical temperature (LCT) and body mass for mammals and birds, showing the lower temperatures survived by larger species, and the seasonal effects particularly in mammals (lowered LCT in winter). (From Peters 1983; courtesy of Cambridge University Press.)

as 6°C, while for desert terrestrial animals this value may be close to 50°C. Looked at more closely, these kinds of data show clear-cut phylogenetic and geographical patterns. Certain kinds of animal tend to have broader tolerance ranges than others; for example, insects on land, and crustaceans and molluscs in aquatic habitats, are generally rather catholic in their thermal preferences, whilst vertebrates in all habitats commonly have more restricted tolerance zones. Within a taxon, temperate animals will usually show broader tolerances than either polar or tropical forms.

Most animals have a **preferred body temperature** (T_{pref}) at which they operate, and which is usually rather close to the 'optimal' temperature measured for many of their physiological processes. They are often very good at maintaining their T_b very close to their measured preferred temperature during activity (Fig. 8.24). In many terrestrial ectotherms, though, the T_{pref} for different activities varies; for example, in snakes the digestive processes are more sensitive to temperature than are the locomotory speeds, so there can be no single optimal T_b for prey capture. It is often more useful therefore to consider a **thermal performance breadth**, a range of T_b values for which performance is still high (a value of 80% of maximum performance value is often used to define the limits of this term). Beyond this again is the **thermal tolerance range**, the full range of body temperatures over which an animal can survive indefinitely. Finally there is the even broader **thermal survival zone**, be-

tween the UCT and the lower critical temperature (LCT). As a broad generalization it would be fair to say that these limits are roughly matched to the environmental temperatures experienced, thus varying with latitude. Tolerance ranges are also generally broader, and critical temperatures spread further apart, for larger animals, so that the LCTs of birds and of mammals are clearly size dependent (Fig. 8.25).

However, measuring any of these parameters for a particular species is fraught with difficulty. There is inherent variability between individuals, as with so many biological issues. Therefore critical temperatures are often defined in terms of 50% effects; for example, the lethal temperature for a species is that temperature at which 50% of the animals die (T_{L50}). This would be tedious and dreadfully wasteful to determine exactly in practice, but if desired it can be established graphically by plotting percentage survival against temperature for a range

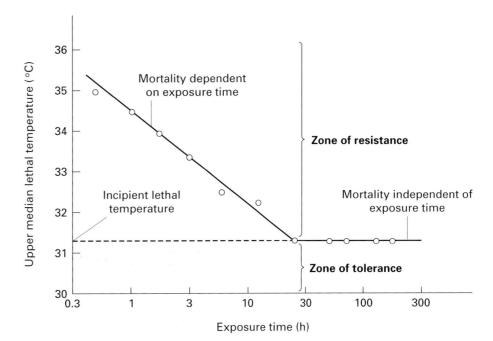

Fig. 8.26 Interactions of exposure time and temperature on mortality. The median lethal temperature (T_{L50}) can be established from standard mortality curves; it decreases as exposure time is increased, and reaches a plateau where mortality is independent of time, known as the 'incipient lethal temperature', separating the zone of resistance and the zone of tolerance. (From Cossins & Bowler 1987, Fig. 6.4; with kind permission from Kluwer Academic Publishers.)

of experimental conditions and reading off the predicted value for 50% survival. However, we have already noted that the destructive effects of temperature extremes are time-dependent, so that a temperature that can be tolerated for a few minutes may not be survivable over a longer period; the point at which exposure time begins to have an effect is sometimes termed the **incipient lethal temperature** (Fig. 8.26), and can be used to distinguish the 'zone of resistance' from the 'zone of tolerance'. A better analysis of thermal survival is therefore given by temperature against time to 50% mortality (Fig. 8.27a), which commonly gives a straight line on a semi-logarithmic plot (Fig. 8.27b).

An even more complete picture of a species' thermal tolerance is illustrated by the **tolerance polygon** (Fig. 8.28). The animals are given sufficient time to acclimate to the maximum possible range of temperatures, and the lethal temperature at each acclimation temperature is then determined as before. The area between the two curves represents the combination of acclimation and lethal temperatures that 50% of the population can withstand for an indefinite period. The greater the enclosed area (measured as °C²) the greater that species' thermal tolerance range. Table 8.10 shows this kind of analysis for a range of aquatic organisms, revealing the generally higher values for freshwater forms over their seawater relatives.

In practice, it is still difficult to quote a single figure for thermal tolerance for a species, even using the tolerance polygon, for a number of reasons:
1 Plots of percentage mortality vs. temperature are usually sigmoidal, i.e. most individuals in a population die over a restricted range of temperatures but some are more and some are less susceptible.
2 Tolerance of high and low temperatures is rarely fixed in a species and may be modified by conditioning. For example, many

perennial animals are more cold-tolerant and heat-susceptible in the winter. Birds that are artificially exposed to hot environments can increase their T_b by as much as 3–4°C.
3 Evolution may produce marked differences in thermal tolerance throughout a species' geographical range.
4 Any or all of the commonly quoted thermal parameters may be variable through the life of an animal (e.g. Fig. 8.29). In particular, thermal performance breadth and thermal tolerance polygons tend to be reduced in young animals, often substantially so in metamorphic invertebrate animals where larval stages are especially vulnerable to thermal stress.
5 Temperature tolerance may be modified by interactions with other factors, such as salinity in aquatic organisms and humidity in terrestrial organisms.
6 In mammals and birds, higher core temperatures may be tolerated with internal heating (exercise) than with external environmental heating.
7 Finally, of course, thermal tolerance may be modified within a rather short timescale for a particular individual due to its own adaptive responses, discussed below.

8.6.3 *Thermal acclimation*

Thermal tolerance as discussed above is partly determined by an animal's ability to acclimate to temperature change, and in this sense acclimation is a response intermediate between tolerance and thermoregulation. Although most biochemical and physiological processes have a Q_{10} value between 2 and 3 (section 8.2), the rates of many such processes can be gradually adjusted in living animals to compensate for variations in the thermal environment, whether on daily, monthly or annual timescales. As we have seen before, this adjustment tends to be called

(a) Time to death (min)

(b) Time to death (min)

Fig. 8.27 (a) Determinations of the lethal times for 50% mortality at raised temperatures in the freshwater crayfish *Austropotamobius pallipes*; in (b) the same data are plotted logarithmically. (From Cossins & Bowler 1987, Fig. 6.2; with kind permission from Kluwer Academic Publishers.)

Table 8.10 Thermal tolerances of fish and crustaceans, as given by the area of tolerance polygons (units of °C²). Compare Fig. 8.28.

Species	Tolerance	Habitat
Goldfish	1220	Freshwater, widespread
Bullhead	1162	Freshwater, widespread
Lobster	830	Marine, widespread
Greenfish	800	
Silverside	715	Mostly marine, widespread
Flounder	685	Marine, temperate
Trout	625	Freshwater/marine, temperate
Puffer fish	550	Marine, north temperate
Chum salmon	468	Marine/freshwater, north temperate
Rock perch	47	Antarctic

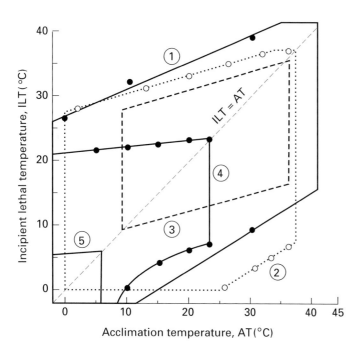

Fig. 8.28 Tolerance polygons for five species of fish. At any particular acclimation temperature, the upper critical temperature (UCT) and lower critical temperature (LCT) can be read off. (1 = goldfish, eurytherm; 2 = bullhead, eurytherm; 3 = puffer fish, temperate stenotherm; 4 = chum salmon, cold stenotherm; 5 = Antarctic rock perch, extreme cold stenotherm).

acclimatization when it occurs in nature, and acclimation when it is induced in laboratory studies; but since all reliable studies are based on laboratory analyses the effects are generally called acclimatory responses. Thus many animals have lower T_{pref} in winter than in summer—they have acclimatized, to the extent that in winter they may even be killed by raised temperatures at which they would have been normally active in summer. The measured underlying acclimatory responses may involve heart rate, ventilation rate, metabolic rate, neural conduction persistence, and enzyme reaction rates. And the effect may be expressed not just in T_{pref} but also in thermal tolerance range and in LCT and UCT.

As a general rule, acclimatory responses act to keep biological processes operating at roughly the same rate whatever the temperature; the changes in enzymes and membranes discussed in section 8.2 are clear examples of acclimatory responses. Perfect acclimation is rare, however, and numerous attempts have been made to classify the exact nature of acclimatory response. One useful scheme (based on the Prosser scheme) is shown in Fig. 8.30. Type 1 involves no acclimation of rate with temperature; Type 2 moves the rate response up or down in a translational movement;

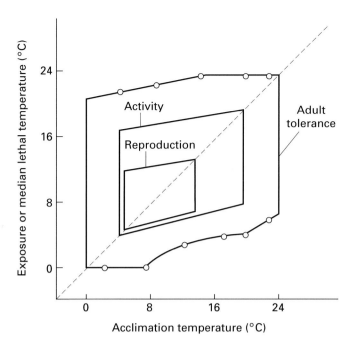

Fig. 8.29 The varying polygons for particular activities in the sockeye salmon. (Adapted from Fry & Hochachka 1970.)

while Type 3 involves a rotational movement of the rate curve (implying a change in Q_{10} value as well as in rate). The more complex Type 4 includes both translational and rotational effects.

Another widely used classification uses the terms **capacity acclimation** (or capacity adaptation) for a change in the rate of some function during long-term exposure, and **resistance acclimation** (adaptation) for a change in upper or lower tolerance limits. Both kinds of response are widespread, and they

Fig. 8.30 Four types of acclimation as defined by Prosser. The solid line indicates the initial rate–temperature curve, and the dashed line indicates this curve after thermal acclimation (to higher temperatures). C, cold acclimatized; W, warm acclimatized.

very often occur simultaneously. The usage of this terminology can therefore be confusing, and has probably been overworked in the literature.

Examples of these varying kinds of thermal acclimation occur in all kinds of animal and in virtually all habitats. In invertebrates and ectotherm vertebrates acclimation occurs in response to varying T_b, and examples of acclimation of LCT, UCT, T_{pref} and metabolic rate can all be found for creatures ranging from sea anemones to desert lizards. Some of the 'lower' organisms perhaps show relatively little compensation; some littoral invertebrates, for example, really do become quiescent in winter with little or no acclimation of their thermal properties, and are sluggish unless artificially warmed (Type 1 response); at best they may show resistance acclimation of their lower thermal limits. Ectothermic vertebrates and terrestrial invertebrates are more likely to show the more complex acclimatory responses, Types 3 and 4, with an upward translation to higher rate and a clockwise rotation to lower Q_{10} value, i.e. with both capacity and resistance acclimation occurring. But there are certainly no clear-cut rules here. It is also particularly important to recall the points made in Chapter 1 about adaptation and acclimation, and realize that a changed response is not necessarily evidence for a beneficial adaptive response. It is possible to think up 'reasons' and 'advantages' for each of the types of acclimatory response described above, but this does not prove that they are specific adaptations of benefit to the animals exhibiting them.

From ectothermic vertebrates, from littoral invertebrates and from the insects, particularly from the well-studied *Drosophila* fruit flies, come some of the best known cases of naturally occurring acclimatory gradients, along latitudinal or altitudinal clines. For example, amphibians show clear clines of LCT and UCT with latitude; southern populations of oysters (*Crassostrea*) spawn at higher water temperatures than more northern populations; and southern populations of *Drosophila funebris* are more resistant to high temperatures than are northern (cooler climate) populations. These effects could be due to acclimation, or could represent genotypic physiological races of the animals. Both

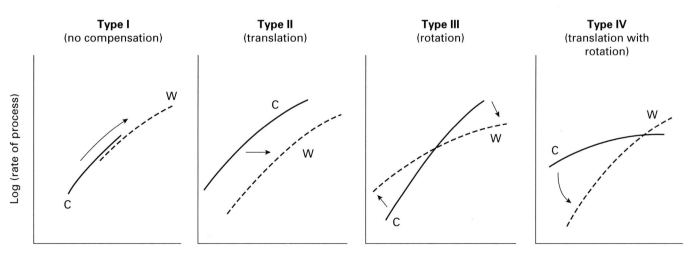

situations do occur, but in *Drosophila* at least some of these effects can be directly correlated with chromosomal differences or even with allelic variation at specific genetic loci. This is an area where real and significant advances are being made in our understanding of comparative adaptation.

Endotherms show acclimation in a similar fashion to the ectotherms thus far considered, but here the effects are of course in response to variation in T_a rather than T_b. The acclimatory effects in endothermic birds and mammals may be very elaborate, ranging from behavioural patterns (e.g. reduced activity outside a burrow, or gathering of extra nest insulation) through morphological features (e.g. fur or feather characteristics, or patterns of fat deposition) and physiological features (e.g. rates of thermogenesis, or nerve conduction velocity), to biochemical features such as the nature and properties of membrane constituents, or the concentrations of enzymes.

8.6.4 *Rates of acclimation and effects of history*

Rates of acclimation generally follow a hyperbolic curve, with complete acclimation in a laboratory setting taking from a few hours to several days depending on the kind of animal involved. This is most easily expressed as the time for 50% acclimation. In small terrestrial animals, such as insects and woodlice, acclimation is often particularly fast. Acclimation times again have to be expressed rather carefully, as for any given species they tend to vary with temperature and can be drastically affected if oxygen

supplies are less than optimal. Furthermore, acclimation to a particular increment of temperature change tends to be quicker from low to high than from high to low.

Acclimation rates also depend on age, size and previous thermal history. Even for animals kept in constant conditions for very long periods, and thus lacking any apparent thermal cues, acclimation can occur, albeit slowly. For example, thermal preferences may change to a colder range in winter for a laboratory goldfish, in the absence of any water temperature change; the fish is apparently responding to a photoperiod cue.

8.7 Regulating heat gain and keeping warm

8.7.1 *Metabolic heat and body temperature*

Heat is produced by all metabolic activity, and in a tachymetabolic (endothermic) animal it is produced in substantial quantities in the core of the body. The tissues of endotherms have substantially higher mitochondrial densities and higher activities of mitochondrial enzymes than do those of similarly sized ectotherms (Fig. 8.31), reflecting this ability to produce metabolic heat. For birds and mammals at rest, the thoracic and abdominal organs (gut, liver, kidney, heart, lungs, etc.) produce up to three-quarters of the metabolic heat, whereas when active, with the metabolic rate perhaps elevated 10-fold (Chapter 6), it is principally the muscles that produce extra heat. The core temperature of the animal therefore varies according to activity,

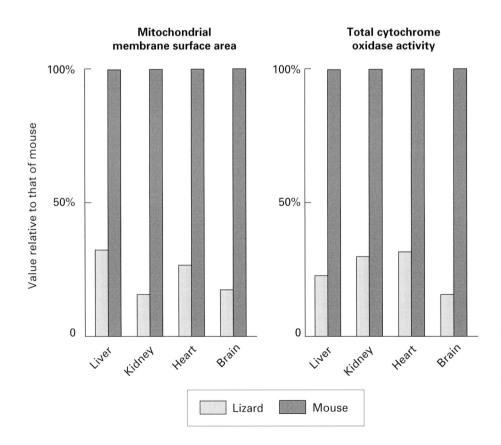

Fig. 8.31 A comparison of mitochondrial density and cytochrome oxidase activity in the tissues of an ectotherm (lizard) and endotherm (mouse) of similar body mass. Values are given relative to the mouse as 100%, and are roughly $1/3$–$1/4$ as high for the lizard. (From Else & Hulbert 1985.)

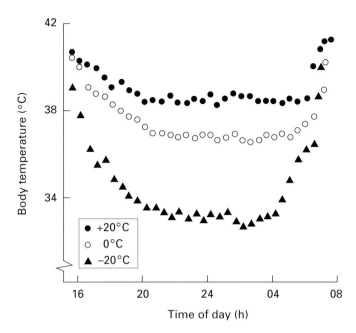

Fig. 8.32 Diurnal variation in body temperature for a small bird (the willow tit) during the night at three different ambient temperatures. (From Reinertsen & Haftorn 1986.)

Table 8.11 Core body temperatures (T_b) characteristic of endothermic vertebrates.

Taxon	Common names	T_b (°C)
Mammals		
Monotremes	Echidna, platypus	30
Edentates	Anteaters, etc.	33–34
Marsupials	Possums, kangaroos, etc.	36
Insectivores	Hedgehogs, moles, etc.	36
	Shrews	37–38
Chiropterans	Bats	37
Cetaceans	Whales, etc.	
Pinnipeds	Seals, etc.	
Rodents	Mice, rats, etc.	37–38
Perissodactyls	Tapir, rhinoceros, horse	
Primates	Monkeys, humans, etc.	
Carnivora	Dogs, cats, etc.	
Artiodactyls	Cow, camel, pig, etc.	38–39
Lagomorphs	Rabbits	
Birds		
	Penguins	38
	Ostrich, petrels, etc.	39–40
	Pelicans, parrots, ducks, gamebirds	41–42
	Passerine songbirds	42

so that quoting a standard core temperature for a species is difficult. This is particularly so for some large desert mammals, such as the camel, that have spectacular daily variations in body temperature as part of their thermal strategy (Chapter 14). But core temperature of most birds and mammals also varies a little with a diurnal periodicity (Fig. 8.32) according to the light cycle, even when an individual is kept under constant thermal conditions. Diurnal species are warmer by day, and nocturnal species warmer by night, with a mean difference of up to 2°C, and in extreme cases (such as the tiny tree shrews) of about 5°C. Even when light cues are removed the diurnal effect persists, with a slightly different periodicity, indicating an endogenous rhythm of heat production. Further longer-term complications arise in relation to oestrous cycles in mammals, where females may be 0.5–1°C warmer at ovulation.

Table 8.11 shows the core body temperatures (allowing for the problems outlined above in determining this mean value) for a range of birds and mammals, whether measured in laboratory studies or by implanted thermistors transmitting telemetric records of T_b in natural surroundings. Birds generally maintain about 40°C, with smaller species (especially passerines) somewhat higher than larger species, and ratites (flightless birds) particularly low. In mammals, clear phylogenetic effects can be seen, with the monotremes at 31°C, marsupials at 36°C, and eutherians at about 38°C but with 'primitive' groups somewhat lower than more 'advanced' groups (e.g. insectivores, such as shrews, have T_b values of 34–37°C, slightly varying with ambient temperature). The reasons for lower body temperature in some groups are not at all clear, and the apparent evolutionary

trends are puzzling. Certainly the earlier groups in evolutionary terms (e.g. monotremes) are every bit as good at thermoregulating at their specific preferred T_b as the evolutionarily later groups that have 'chosen' a higher T_b, so that the differences are not due to any primitive failing in the ability to control thermal balance.

To maintain a constant body temperature, at whatever level, there must be a steady-state condition in which the rate of heat production (M) equals the overall rate of heat loss (Q, where $Q = h (T_b – T_a) + E$; see section 8.5). It follows that if T_b is to be constant then the only factors that can be regulated are M (the metabolic heat production, and the distribution of this heat to the surface), or the conductance factor (represented by h), or the evaporative loss E.

8.7.2 Heat production: shivering and non-shivering thermogenesis

Heat production in animals is continuous and inevitable; the important distinction is that in endotherms it occurs at a higher rate (4–8 times that of ectotherms at the same T_b) because of a greater concentration of mitochondria and perhaps also a different kind of mitochondria in which oxidative phosphorylation is always partially uncoupled so that more heat is released for a given level of ATP generation. Thermoregulatory heat production is extremely expensive, and makes up the single largest

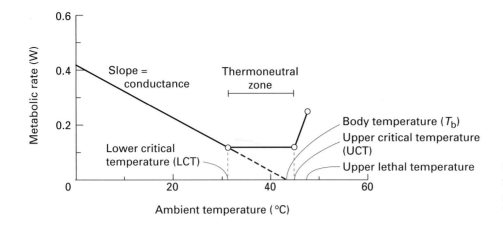

Fig. 8.33 The influence of ambient temperature on the metabolic rate (MR) of a small temperate endotherm, showing the 'thermoneutral zone' where MR is constant.

component of the energy budget in endotherms, reducing energy available for growth and reproduction.

For such endothermic animals the **thermoneutral zone** (TNZ) is a useful concept, being the range of ambient temperatures over which the animal's resting metabolic rate and thus heat production is not varied. A plot of ambient temperature against resting metabolic rate for a mammal or bird is shown in Fig. 8.33, and within its thermoneutral zone any particular endotherm is keeping its energy metabolism fairly constant. It is in this range that the resting metabolic rate (see Chapter 6) should always be measured. Below some specific lower temperature, a mammal will show a rise in metabolic rate with decreasing temperature in a linear fashion. Above some specific higher temperature metabolic rate may again rise due to the direct effects of temperature on metabolic processes and because of additional metabolic costs associated with thermoregulation (such as increasing the blood flow to the skin—see below); eventually these processes run out of control (before declining finally as the

cell machinery is irreversibly damaged and death ensues). The upper lethal temperature is often only 3–6°C above the normal body temperature.

Comparing thermoneutral zones between animals is difficult unless the resting metabolic rate of each is normalized as 100% (Fig. 8.34). A comparative plot then reveals the much greater width of thermoneutral zones in cold-adapted animals. An arctic mammal does not need to increase its metabolic rate to keep warm unless the external temperature is well below 0°C (e.g. −40°C for an Arctic fox), and even then the metabolic rate/ambient temperature plot is very shallow. In contrast, a tropical mammal may show an increased metabolism if the ambient temperature only drops to 25°C, and the rate of increase will be much sharper with declining temperature.

These differences in thermoneutral zone are largely due to reductions in the thermal conductance of the cold-adapted endotherms—they have thick insulating pelts (see sections 8.5 and 8.7.4) that greatly increase their thermal tolerance. Larger

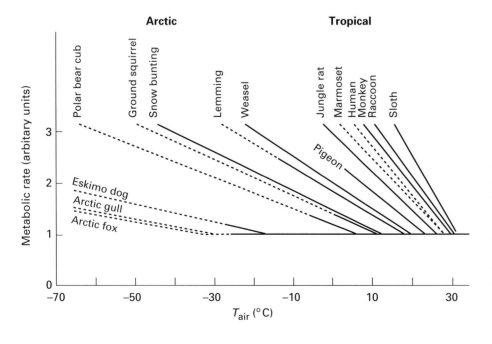

Fig. 8.34 Plots of metabolic rate (MR) against ambient temperature for a variety of tropical and polar mammals and birds. The plots are equivalent to the left-hand side of Fig. 8.33, with the MR in the thermoneutral zone (TNZ) taken as unity to allow direct comparisons. Polar animals have a much wider TNZ, surviving to lower ambient temperatures before MR is elevated: this is mainly due to their much better insulation. (Adapted from Scholander *et al.* 1950.)

mammals and birds have thicker fur or feather layers, and/or subcutaneous fat layers, and therefore are better protected against heat loss; specific heat conductance (C^*, heat flow per unit body mass) varies with $M_b^{-0.50}$ for birds, marsupials and eutherian mammals. Specific heat conductance decreases with body size more rapidly than does specific heat production, so that the decrease in ambient temperature an animal can tolerate without increasing its heat production also increases with body mass. This mismatch between heat production and conductance as size decreases means that a large endotherm can tolerate a much greater degree of cooling without increasing its heat production; it will inevitably have an extended thermoneutral zone in the lower temperature range.

Nevertheless, once conductive heat loss has been decreased to an irreducible minimum, a further fall in ambient temperature will necessitate an increase in heat production. Maximum heat production is generally around five times the basal metabolic rate. Heat production can be increased according to need in three ways. Muscular activity by physical exercise greatly raises the metabolic rate as we have seen, and could be used to warm the body. Involuntary muscular activity could also be initiated, this being what we perceive as shivering. As a third possibility, thermogenesis could be achieved without muscular contractions, by a variety of chemical means, so-called non-shivering thermogenesis.

Voluntary muscular activity to achieve warm-up is not uncommon in animals, and is of course used by humans. Some insects may use short periods of flight for no other purpose than to warm their bodies up, and recent evidence suggests that some can alter their heat production in flight by alterations of wing-beat frequencies and wing kinematics to help maintain a constant T_b. However, activity is not a simple solution to being cold for most animals, as exercise may in itself accelerate some avenues of heat loss; it tends to involve postures with large surface areas exposed, reduces the thickness of boundary layers,

and most seriously may reduce the value of insulation layers by wetting them (in sweating mammals) or making them dishevelled (as with bird plumage).

Shivering is used by nearly all adult endotherms, and by many ectotherms as well. It is a high-frequency reflex producing oscillatory contraction of mutually antagonistic skeletal muscles, giving little net movement. Thus the ATP hydrolysed to provide energy for contraction is producing minimal physical work but substantial heat output. The intensity of shivering is linearly related to oxygen consumption. Shivering is mediated via the normal neuromuscular junctions in the nervous system (largely controlled by the cerebellum), and it is therefore susceptible to neuromuscular blocking drugs such as curare.

The use of such drugs gives a convincing demonstration that other forms of thermogenesis are commonly present; curarized laboratory animals can still exhibit an increase in metabolic rate when cooled. This results from **non-shivering thermogenesis** (NST), a contentious but exciting phenomenon in the physiological repertoire of many classic endotherms, including placental mammals (especially juveniles and small species), some marsupials and rather few birds. It certainly occurs during arousal from torpor in specific mammalian tissues such as brown fat (brown adipose tissue, or BAT), which consists of fat cells with a very high mitochondrial content, and which is highly vascularized with a direct venous return to the heart, and with sympathetic nervous innervation. It probably also happens in liver, kidneys and elsewhere in chilled mammals. Some semi-aquatic mammals also use it to warm up after diving or prolonged swimming.

Non-shivering thermogenesis equally clearly occurs in at least some ectotherms, including leatherback turtles and a few insects, where for example warm-up can occur in immature forms (social wasps and others) lacking any proper skeletal muscle to shiver with. Its occurrence in adult insects such as bees has been disputed, but now seems incontrovertible, and will be discussed in Chapter 13.

Actual mechanisms and fuels are somewhat variable (Fig. 8.35), but the primary heat allowing warm-up before muscle activity begins is generated from futile cycling of substrates through a

Fig. 8.35 Possible heat-generating 'futile' biochemical cycles: (a) is used in some insects with endothermy, (b) in some fish heater cells, and (d) is probably the route in tetrapod brown adipose tissue (BAT).

series of linked anabolic and catabolic reactions simply to generate the by-product of heat. The best-understood mechanism is that in the BAT of mammals. Under resting conditions the BAT mitochondria function in the normal way, protons flowing back into the mitochondrial matrix through the ATP synthase enzyme, generating ATP and some heat (see Chapter 6 and Fig. 6.10). But when stimulated by norepinephrine (noradrenaline) from the sympathetic nerve endings the BAT mitochondria switch to a different proton route (Fig. 8.35d), normally inhibited in the resting state by purine nucleotides. Once this inhibition is released, the protons flow back into the matrix via proton translocase channels ('thermogenin' sites), and since there is no associated ATP synthetase all the energy is liberated as heat. When this proton shunt pathway is operating, respiration no longer depends on ADP supplies, and the release of heat is only limited by carbohydrate and lipid substrates. Heat production can be as high as 500 W kg^{-1}, almost 10 times higher than in normal mammalian skeletal muscle.

Other heat sources may play a minor role in heat generation in animals. Digestion certainly produces some heat, and in ruminants the additional heat from fermentation can substitute for shivering in the overall thermal repertoire. We can assume that large herbivorous dinosaurs gained significant heat in this way, adding to the effect of their large thermal inertia to ensure that they were effectively 'warm-blooded'.

8.7.3 Heat distribution: exchangers and countercurrents

Heat is produced in all body tissues, but particularly in the core organs of a resting animal, and in muscles during activity. This must be distributed around the body, principally in the blood or other body fluids, and dissipated suitably at the surface according to need. Clearly this need will vary; sometimes an animal will have to offload the excessive heat produced in a bout of activity, at other times it will be vital to conserve as much of the metabolic heat as possible. Most terrestrial animals, and many in aquatic environments, therefore have adaptations to ensure appropriate blood flow patterns in relation to thermal needs.

The simplest form of control is to manage the rate and volume of flow of blood to the surface relative to the core. In a cool and resting animal where most of the heat is generated in the core it can be effectively retained there with minimal peripheral vasodilation, keeping most of the circulatory flow in arterioles and venules and away from superficial capillary beds (Fig. 8.36). There may be arteriole-venule anastomoses in vertebrates, as we saw in Chapter 7, allowing this bypassing of capillaries and keeping heat in the core. In an active or hot animal peripheral arterioles dilate, sometimes through action of specific valves, and the capillary beds are incorporated into the circulation.

This technique in effect involves regulation of surface temperature, T_s, and it might be thought that it would inevitably become less effective for larger mammals and birds (with relatively less surface area) and especially for those with a substantial external insulating layer. However, the ability to control T_s can be measured in mammals as the **vasomotor index** (VMI), and this has been shown to scale positively with body weight with an exponent of 0.28. This means that large mammals use vasomotor control and regulation of T_s very intensively as a thermoregulatory strategy, whereas smaller mammals (with an

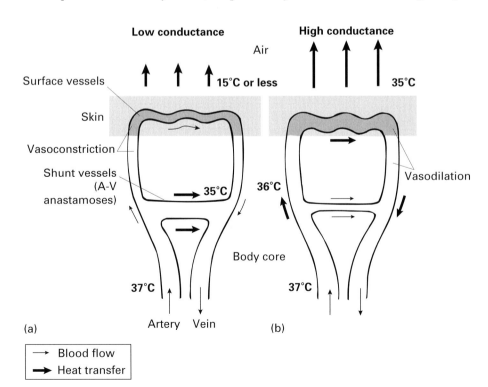

Fig. 8.36 The role of blood flow to the skin in regulating the heat conductance of the body surface and hence heat loss. Vasomotor control of peripheral arterioles and arteriovenous (A-V) anastomoses is mainly responsible for shunting blood either (a) to or (b) away from the skin.

intrinsically high level of thermogenesis and limited absolute surface area) instead have to rely more on reducing their metabolic rate when heat stressed. Indeed, conspicuous vasodilation is most frequently described in large and endothermic animals.

Nevertheless, physiological thermoregulation by peripheral blood flow control is certainly also present in many smaller and ectothermic animals, such as lizards and even insects. It is a major part of the explanation for the common observation that heating rates in animals can be substantially faster than cooling rates (see Fig. 8.23). During solar heating in a reptile, for example, the heart rate is kept high and blood flow to the skin is high, whereas during cooling the heart rate drops very quickly and little of the hot blood flows peripherally, so controlling and reducing heat loss (see Chapter 13, Fig. 13.25).

More subtle control of heat distribution comes from the addition of **countercurrent heat-exchangers** to the blood circuitry. These involve the principles of countercurrent flow discussed in Chapter 5 (see Box 5.2) and Chapter 7, and a model of their function in regulating thermal balance (together with examples) is shown in Fig. 8.37. Countercurrent heat-exchangers acting to conserve core heat are commonly found in the extremities of animals living in cold habitats, particularly in the legs or flippers of polar and cool temperate birds and mammals (e.g. whales and porpoises, seals, mink, platypus, gulls), the horns of some ungulates, the tails of beavers and other rodents, and the ears of Arctic hares. In all these cases, warm blood in the arterioles runs parallel and very close to the venous return, so that heat can be exchanged effectively and returned to the body, leaving the relatively poorly insulated appendage perpetually cool. This may work in conjunction with peripheral vasodilation and constriction patterns (Fig. 8.38). The result is a magnified thermal gradient along the limb or other extremity, giving rise to **regional heterothermy** (Fig. 8.39).

This kind of heat-exchanger also occurs in some tropical and temperate animals. To a limited degree the legs of a human have an artery–deep vein exchange system, while some sloths and armadillos and some of the more primitive tropical forest primates (lemurs and lorises) exhibit limb exchangers too, probably to counter the cooling that would occur for an inactive creature on a wet and windy night.

Heat-exchangers to conserve warmth certainly achieve significant thermal benefits for the animal, but they produce very cold extremities that could run the risk of freezing and tissue damage. Just enough heat must be added to the peripheral tissues to prevent this, and many birds and mammals will begin to pass occasional pulses of warm blood to their feet as the ambient temperature is lowered beyond 0°C. This is reflected in a slight increment in the metabolic rate curve at this same temperature (Fig. 8.40).

Heat-exchangers can also be used to conserve heat within specific parts of the body, rather than just in the trunk as a whole relative to extremities. Examples occur in some of the fish that show regional endothermy, maintaining the central parts of the

(a)

1 **Vena comitantes:** 2–3 veins surrounding central artery

2 **Central rete:** central artery and many small veins

3 **Artery–vein rete:** many interdigitating vessels

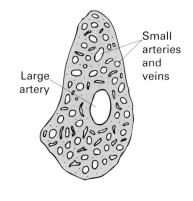

(b)

Fig. 8.37 Thermal countercurrent systems. (a) A model of flow and heat exchange. (b) Examples of three kinds of vascular arrangement.

Countercurrent in operation

Artery
Heat exchange
Dilated A-V anastamosis
Skin capillaries
Vein
Constricted superficial vein
Little heat loss

Countercurrent bypassed

Constricted core vein
Constricted A-V anastamosis
Vasodilated superficial vein
Large heat loss

Fig. 8.38 How a countercurrent system can supplement the peripheral dilation and constriction shown in Fig. 8.36.

Temperate mammal: dog

10 cm

39°C
37°C
24°C
15°C
$T_a = 16°C$
14°C
14°C
14°C
17°C
20°C
$T_{floor} = 20°C$

23°C
15°C
10°C
37°C
$T_a = 20°C$

Fig. 8.39 Regional heterothermy resulting from peripheral blood flow management and countercurrents, in limbs and noses.

Metabolic heat production

Heat loss from feet

Heat loss (W)

Heat loss (kcal h^{-1})

Air or water temperature (°C)

Fig. 8.40 At and below 0°C, intermittent pulses of increased blood flow to extremities are used to prevent freezing and tissue damage, reflected in increased heat loss from the feet of a duck and increased metabolic rate. (From Kilgore & Schmidt-Nielsen 1975.)

swimming musculature at an elevated temperature in an otherwise essentially ectothermic animal (see Chapter 9). In such fish there are parallel arrangements of arterial and venous pathways at the outer edge of the trunk muscle myotomes, so that heat is effectively retained within the musculature.

In certain circumstances heat-exchangers can be used to keep cool as well, as section 8.8 will show.

8.7.4 *Insulation effects*

Changing insulative properties is an obvious way of regulating heat gain, for example by an ability to vary the thickness of an insulating layer whether this be fur, feathers or a layer of fatty tissue.

For most mammals insulation change can be achieved in the short term by pilo-erection, which can decrease thermal conductance very substantially, retaining more heat in the body. In effect, the angle of each hair to the skin surface is changed by contraction of muscles attached to the hair root, maximal contraction bringing the hair perpendicular to the skin. However, while partial erection thickens the insulating hair and helps to conserve heat, full erection on the dorsal surfaces of the body may increase heat loss by opening up routes for convective transfer of the warm air rising away from the skin surface; some animals may pilo-erect more fully on their ventral surface when overheated, while keeping the dorsal fur flattened.

Birds achieve the same kind of heat loss effect as mammals by the equivalent process of ptilo-erection ('fluffing up' of their feathers), but since feathers are structurally more complex, and their position can be controlled and changed more radically, birds often have better insulation control than mammals. The soft down feathers closest to the skin provide the best insulation, but interlocking layers of contour and flight feathers above this add extra dimensions of insulating variability. All birds also possess a uropygial gland ('preen gland') near the base of the tail that provides an oily liquid to clean and lubricate the feathers. This may have little insulation value in itself, but by protecting the feathers from becoming waterlogged it ensures that air remains trapped within the feathers so they retain their insulating effect even in heavy rain or after immersion; hence the gland's output is particularly copious in aquatic birds.

For aquatic mammals using the third main insulation system, i.e. subcutaneous fat deposits, changed insulation value may be effectively achieved by shunting the blood away from the surface, below the adipose layer (Fig. 8.41). This adds sophistication to the general strategy of shunting blood away from radiating surfaces, and it is a major advantage of having the main insulative layer internal to the heat-dissipating surface, rather than externally as with the fur and feathers of terrestrial birds and mammals, where no bypass is possible.

The strategies described so far may reduce heat loss in the short term. In the longer term, many birds and mammals grow much denser winter coats, often with a change of hair structure and colour, while aquatic mammals may deposit thicker adipose layers as winter approaches (though of course this also serves the additional major function of providing a food store during migration between feeding grounds in whales or during beached periods for mating and giving birth in many seals). These winter strategies for retaining heat in endotherms are dealt with in Chapter 14.

Invertebrates and ectotherms in general lack any ability to change their heat conduction by varying the effective depth of insulative layers. However, some insects do have quite sophisticated insulation from hairs or from internal air sacs. In bumblebees there are outer guard hairs and underlying more feathery down hairs, a striking parallel with the situation in birds and mammals. These hairs may differ in density over the body, and the ventral part of the abdomen often serves as a 'thermal window', lacking the fur covering of the rest of the body. If the bee becomes overheated pulses of hot blood flow close to this

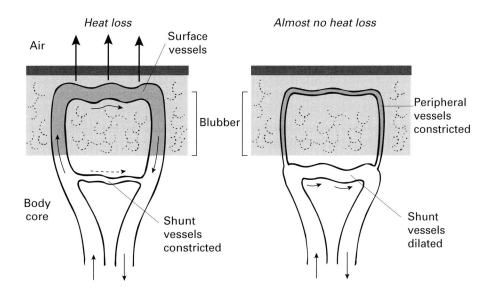

Fig. 8.41 The use of shunt vessels in aquatic endotherms to send blood either above the blubber layer for heat loss, or to keep it below the blubber where it is insulated.

uninsulated ventral abdominal cuticle to enhance heat loss (see Chapter 13).

8.7.5 Behavioural aspects

There are a great many ways of regulating heat gain (and heat loss) by behavioural means, but most of them can be viewed in general terms as ways of locating an appropriate microclimate, and ways of changing effective surface area. Techniques mainly operate for an individual, but in some cases group behaviours can help in thermal regulation.

Choice of microclimate is clearly going to be most important for smaller terrestrial animals, where the possibilities are so much greater. The issue of microclimates on land is dealt with more thoroughly in Chapter 13, but clearly intermittent basking in sunlit spots underlies many strategies; sun/shade shuttling behaviours, rotational movements around an area as the Sun moves across the sky, vertical movements up and down surfaces projecting through the boundary layer, and the use of hide-outs, nests and burrows are all of relevance here.

Individual behaviours often involve general posture, and thus control the amount of surface exposed to the Sun in basking. Basking is particularly useful for radiative warm-up in winged insects, with large exposable surfaces. Many of the crests, sails and ridges on early reptiles, such as *Dimetrodon* and the stegosaurs, were probably used as radiant heat gatherers in basking. In mammals, changing posture substantially affects the thermal balance by altering the insulation of the receiving surfaces—densely furred backs absorb very differently from nearly naked ventral and limb surfaces. Using posture to vary the surface exposed to conduction from heated substrata is another possibility; for example, many animals warm up by crouching against hot rocks, or in the heated sand at a burrow entrance.

Group behaviours to regulate warming are perhaps most obvious in the social insects, particularly those with colonies that overwinter such as hive-dwelling honey-bees in the genus *Apis*. Here a tight cluster of bees share their slowly generated body heat to raise the whole cluster to a temperature far above that attainable by any one individual. This can even be used as a form of aggression, since *Apis* will huddle around an invading hornet to overheat it. Huddling also occurs in many other terrestrial animals, and is therefore described more fully in Chapters 13 and 14.

8.8 Regulating heat loss and keeping cool

Keeping cool is not usually an enormous problem for the majority of animals. Aquatic animals rarely meet high temperatures, and cooling is only likely to be critical for some freshwater pond-dwellers, and for some rock-pool animals, in high summer. Most terrestrial animals behave in such a way as to avoid high ambient temperatures, and smaller ones can use the microclimates around plants, or underground, to great effect. Only the larger land animals, unable to evade or avoid the midday heat of the Sun, are obviously going to have a problem with cooling; though for them the high radiation loads around midday may pose an overheating problem even in polar latitudes.

However, all thermal issues have to be seen in relation to a particular animal's habitat and experience, and many other animals may in practice have to regulate the heat loss side of their thermal balance to maintain their own T_{pref}. They also have to achieve appropriate activity levels at appropriate times of day and in the right places. After all, avoidance may be the easiest strategy, but at some point every individual has to feed and reproduce, and during these periods heat may inevitably be gained from the environment or from the metabolic effort supporting the activity, so that regulating heat loss may need attention.

8.8.1 Heat storage

Many animals use transient hyperthermia (where T_b exceeds the normal regulated value) as a way of managing exposure to hot conditions or heat load due to exercise; they thereby retain a gradient for shedding heat slowly to the environment without having to employ specific heat-loss mechanisms that may have other associated costs. The larger the animal, the longer it can tolerate a degree of hyperthermia. For example, small desert rodents (100–250 g) may tolerate brief overheating to a T_b of 43°C during foraging bouts lasting 15–30 min, shedding this heat load on return to inactivity in the cool burrow; whereas a 250 kg camel may take all day to warm from its overnight T_b of 35°C to an evening temperature of 40°C, shedding the excess heat overnight again.

Having a high body temperature has two advantages. Firstly, it increases the thermal gradient for heat loss from the body, or at least (if T_a exceeds T_b) it reduces the heat gain. Secondly, for each 1°C increase in T_b around 3.4 kJ of heat is retained per kg body mass, and this can be dissipated when heat production returns to the resting level or ambient temperatures drop (and without wasting water in evaporative heat loss).

Heat storage is clearly a useful strategy for small animals with cool refugia, or for large animals in climates where night temperatures are much lower, but it is only a stop-gap solution for most animals; at some point, actual heat-loss regulatory systems must be engaged.

8.8.2 Heat-exchangers and blood distribution

One way for an animal to lose excess heat is to shunt more of the hot blood within the core to the surface, effectively increasing the thermal conductance of the outer layers; this is essentially the inverse of blood shunting techniques mentioned in section 8.7. Peripheral vasodilation is an essential component of homeostasis during exercise, and occurs in many situations of temporary overheating. It forms part of a general suite of blood shunting strategies, with flow increased to muscles, heart and brain during moderate exercise, and then with blood shunted

Table 8.12 Cardiovascular changes during exercise in vertebrates. Numerical values shown are for a maximally exercising adult.

Factor	Human male Resting	Human male Exercise	Pigeon Resting	Pigeon Exercise	Salmon Resting	Salmon Exercise
Heart rate (min^{-1})	75	180	100	600	38	51
Stroke volume (ml)	77	130	1.0	1.0	0.5	1.1
Cardiac output (ml min^{-1})	5700	25 000	–	–	18	53
Cardiac output per kg	85	360	–	–	18	53

Blood shunting during exercise

Blood flow increments:	Skeletal muscle, brain, heart, skin
Blood flow reductions:	Non-skeletal muscle, gut, kidney, liver, spleen

even more peripherally to the skin, with concomitant reductions of flow to the gut and viscera, as the body warms further (see Table 8.12; note that in mammals, birds and reptiles the increased flow is largely due to increased heart rate, whilst in fish it is mainly brought about by increased stroke volume of the heart).

More specific regulation of heat loss can again be achieved with heat-exchangers. The single most vital area that cannot be allowed to overheat in active animals is the brain, as neurones are extremely sensitive to thermal disturbance. Yet at the same time the brain requires a continuing high blood flow. The commonly adopted mechanism to keep the brain cool is to install a heat-exchanger on the supply side to the brain, so that the arterial input (the carotid artery in tetrapods) breaks up into a capillary network and runs parallel to a cooled blood supply just before it enters the brain itself. Such a structure is known as a carotid rete (Fig. 8.42), and may occur in mammals and birds from hot environments (see Chapter 14) and in those that use intense bursts of activity for prey capture, as well as in some lizards. It has the additional advantage of keeping the thermal receptors in the brain below the level at which they might trigger additional cooling by sweating or panting, which could allow excessive evaporative water loss in a desert species. A few insects have an alternative approach to brain cooling; nectar-feeders may achieve cooling of the head by regurgitating nectar onto the tongue and allowing this to evaporate.

It is also worth noting that when animals from cool environments do have overheating problems, for example after prolonged activity, they may use most of the peripheral heat-exchange systems mentioned previously as heat dissipators rather than as conservation systems—ears, flippers, horns and the like may become temporary thermal windows through which the excess heat is off-loaded, by sending arterial blood peripherally instead of through the deep arteries adjacent to the venous return path.

8.8.3 Changing surface properties

Another possible way to regulate heat loss is to modify the actual properties of the heat-dissipating surfaces of the body. Changing the insulative value of a furred or feathered surface is

an obvious example, and this is taken to extremes in humans, who can take off layers of clothing. The commoner response is to reverse any pilo-erection, flattening the fur or plumage and reducing the trapped air layer. Wetting the insulation by a brief immersion is even more effective, and is often employed by birds and mammals in hot weather, largely for thermal purposes. Two effects are achieved: there is heat loss by evaporation, and this is supplemented by extra heat loss through the greatly enhanced conductance of the wet surface layer. For example, the insulation of eider duck feathers is about 1.6°C m^2 W^{-1} in air, and only 0.8 in water. This is principally because the trapped water is a better conductor than air, but also because the feather layer is compressed into a reduced thickness by the water anyway.

Occasionally an apparently unlikely response occurs, when animals in hot environments partially raise their dorsal fur or plumage. This is observed when the air temperature exceeds the animal's T_b, such that a thick insulation layer will reduce conduction, convection and radiation into the body, thus reducing heat uptake. This also explains why some desert-dwellers, such as ostriches, goats, sheep and even camels (at least dorsally), have surprisingly dense pelage.

Coloration also affects radiative exchanges as we have seen, and some animals can change skin or cuticle colour in ways that are adaptive for keeping cool. Some lizards and a few frogs can do it; in lizards this may be due to 'iridophore' cells, where quite rapid thermally induced colour change arises from alterations in the packing between layers of reflective crystalline platelets (which are organelles containing crystals) inside the cells. Insect and crustacean examples are perhaps best studied, though. In insects colour change generally involves phase changes of lipid or other material within the cuticle, and tends to produce a change from a nearly black surface appearance in an insect that is warming up to a shiny almost metallic blue or green in the same insect with raised T_b, with a time course of minutes or hours. This sort of adaptive response is known in some grasshoppers, beetles and dragonflies. In littoral and terrestrial crustaceans, particularly some isopods and smaller crabs, colour change is brought about instead by dispersion or contraction of pigment-containing cells (chromatophores) in the skin, and may be a balanced response reflecting both camouflage and thermal needs.

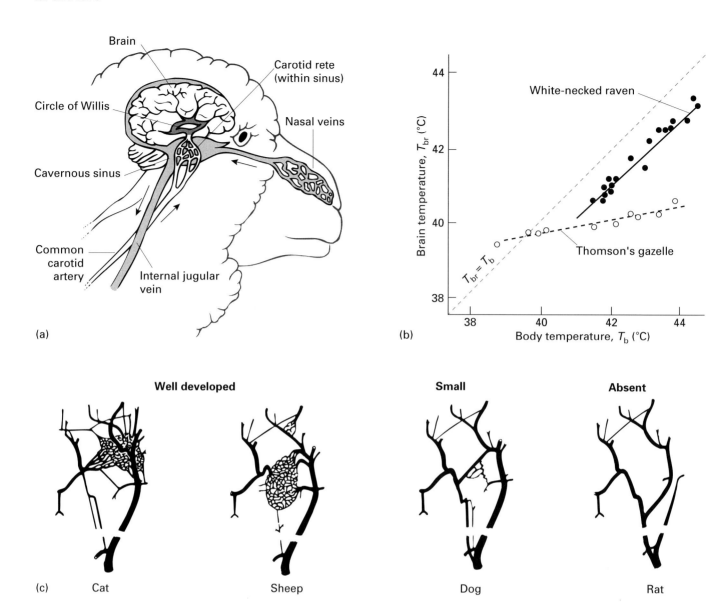

Fig. 8.42 The carotid rete to give brain cooling, as found in many mammals and birds. (a) Basic design; (b) effect on brain temperature in two species; and (c) patterns of rete blood flow in various mammals. (Adapted from Schmidt-Nielsen 1972; Withers 1992.)

8.8.4 *Evaporation*

Beyond the rather limited regulative possibilities already discussed, there is no doubt that evaporative heat loss is the key to keeping cool for most larger terrestrial animals, and for small animals in humid semi-terrestrial niches such as the littoral zone and tropical forest litter.

There are various ways of achieving evaporative cooling, as different kinds of body fluid can be used. Some insects and reptiles indulge in urinating or in regurgitating gut fluids over their own surfaces, and we have just seen that some nectar-feeding animals, like bees and butterflies, will regurgitate nectar and saliva onto their elongated tongues and wave the droplet about to achieve some cooling ('tongue-lashing'). Many animals will lick themselves to spread saliva over the chest, limbs or flanks to achieve a cooling effect. But most obvious are the two techniques used by birds and mammals: **sweating** from special glands at various points over the body surfaces; and **panting**, a rapid breathing out through the mouth (rather than the nose, which is commonly designed to conserve water—see Chapter 5) to lose water from the respiratory surfaces. In effect these two mechanism involve supplementary cutaneous evaporative water loss (CEWL) and supplementary respiratory evaporative water loss (REWL), respectively (see Chapter 5). Most species use only one of these possible avenues of evaporative heat loss, though some can switch between alternative mechanisms.

The amount of heat needed to evaporate a fixed amount of water is the same for all these sources, so it is not immediately clear as to why one technique might be preferred over another.

Sweating is virtually exclusive to mammals, though something similar happens in a few frogs and in one or two insects such as cicadas (Chapter 14). Since heat load is proportional to body surface area, and water available for evaporative cooling is proportional to body mass or volume, small mammals with a high SA/V ratio are at a disadvantage and do not normally sweat because of the danger of dehydration. Mammalian panting is therefore commoner in smaller herbivores and in nearly all carnivores; and panting is ubiquitous in birds (which lack sweat glands). Licking is frequently observed in Australian marsupials, and in many rodents, but is relatively ineffective and laborious as a means of evaporating water. Some mammals choose different methods in different circumstances; for example, kangaroos will sweat during exercise, but pant when stationary while subjected to a radiative heat load.

Sweating is clearly most effective when the skin is not hairy or feathered, and in relatively hairless mammals it can bring about a 100-fold increase in CEWL; perhaps not surprisingly it is more effective in humans than in almost any other mammal, with sweating rate being highly attuned to small changes in internal temperature (Fig. 8.43). Yet many larger furred mammals, such as antelopes and camels, also rely on it to lose heat, perhaps because they live in such dry atmospheres that water evaporates readily even through their fur.

An obvious advantage of panting over all the other systems mentioned is that it involves the loss of only water, in the vapour form from the lungs, whereas other mechanisms involve at least some loss of solutes as well (heavily sweating mammals notoriously run the risk of salt depletion). A second advantage is that panting allows the body surface to remain hot, whereas sweating cools the skin so that heat gain may increase and loss by radiation is reduced. Finally, panting creates its own airflow and can thus speed up the evaporative process.

However, evaporation by panting does require increased ventilation above that which the animal would otherwise need. This can lead to two problems. Firstly, it requires muscular work and so increases the heat production that it is designed to control. Secondly, it can lead to an unusually rapid loss of carbon dioxide and hence a degree of alkalosis in the tissues; humans can easily become dizzy and nauseous if they deliberately hyperventilate, because of this raised tissue pH. Both of these effects can be ameliorated by a commonly observed adaptation during panting—a switch to a much shallower and higher frequency respiration (Fig. 8.44a), often up to 10 times the normal respiratory rate. This reduces the muscular work substantially by utilizing the intrinsic elastic properties of the lungs, inflating them at about their own natural oscillation frequency (resonant frequency). Inhalation stretches the elastin elements in the alveolar walls, which return quickly to their unstretched state when exhalation begins, with little additional muscular effort. Shifting to a shallower respiration with a small tidal volume (inevitable when the respiratory rate is so much increased) also aids the alkalosis problem as much of the ventilation then involves only the upper dead space of the lungs, perhaps only 15% of the normal breath

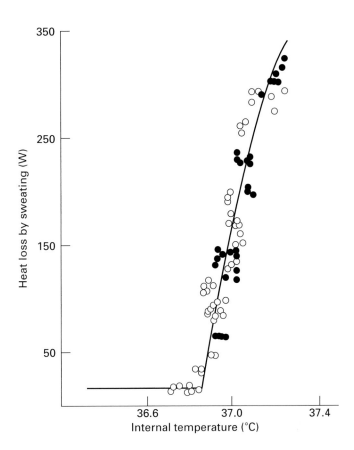

Fig. 8.43 Heat loss by sweating in a human, showing the small changes in body temperature (T_b) that elicit very substantial sweating rates. Core T_b was elevated either by exercise (closed circles) or by increased ambient temperature (open circles). (From Benzinger 1961.)

volume, and so does not remove carbon dioxide from the deeper alveoli. In birds this shallow, fast respiration is even more specialized as a 'gular flutter' (Fig. 8.44b), which largely involves rapid (>600 min⁻¹) vibration of the floor of the mouth by muscles pulling on the hyoid apparatus, with little lung ventilation; this can be used with or without simultaneous panting.

Above and beyond all these advantages and disadvantages of varying methods of evaporative cooling, it must be stressed that though the techniques are highly efficient for regulating heat loss they are also potentially highly dangerous. This is because of the close interrelation of thermal balance and water balance; very few terrestrial animals can afford too much use of evaporative cooling, as they lack the necessary water reserves. For many animals it is a 'last resort' strategy when the risk of overheating overrides all other considerations.

8.8.5 *Behavioural aspects of heat-loss regulation*

As with heat gain, the behavioural regulation of heat loss may operate on an individual or a group basis, and many of the techniques employed are the inverse of those dealt with in section 8.7

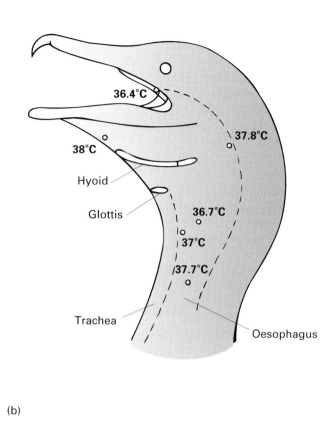

Fig. 8.44 (a) Respiratory patterns during panting in a dog: inhalation is predominantly through the nose and exhalation through the mouth, so avoiding heat retention in the nasal turbinals. Note that the respiratory rate is elevated to about 5 s⁻¹. (b) Temperatures in parts of the head of a bird during a period of gular fluttering, when the floor of the mouth and throat area is vibrated rapidly by movements of the hyoid apparatus, giving rapid evaporative heat loss and up to 2°C cooling in the mouth ($T_b = 38°C$). (a, From Schmidt-Nielsen *et al.* 1970a; b, from Lasiewski & Snyder 1969.)

above. Many small-scale movements of invertebrates, ectothermic vertebrates, and even of smaller endotherms are primarily thermally related behaviours. Postures where minimal surfaces are exposed to the Sun, and/or maximum surfaces to a breeze, are common in overheated insects. The opposite response to crouching for maximal heat gain is that of stilting for heat loss (or at least for minimal heat gain), most effective in long-legged insects and lizards.

On an individual basis there may be extra possibilities that involve exposing or erecting parts of the body that can act as radiators. These will be areas, such as ears, that are thin, vascularized and of large surface area, so that heat can be lost quickly from the peripheral blood supply. 'Pointing' this hot part of the body at a cool part of the sky may give particularly effective cooling, so that small mammals may sit in shade and point their noses or ears at the sky away from the Sun. Examples of appendages as radiators are most obvious in large animals, particularly endotherms, and are dealt with in Chapter 14.

Group behaviours pertinent to heat loss most obviously involve huddling, a phenomenon widespread in terrestrial animals from many habitats but perhaps most spectacularly effective in the tundra and polar animals considered in Chapter 14. Other huddling animals include many young birds and mammals within their nests, where they can achieve an energy saving of up to 30% at low temperatures. Aggregation behaviour akin to huddling is also used by many ectothermic animals, including some insects and littoral invertebrates such as isopods, to reduce heat loss.

8.9 Opting out: evasion systems in space or time

In seasonal climates at moderate and high latitudes, many animals adopt the simplest strategy of all when thermal problems become too severe, and effectively opt out of attempting to maintain a suitable body temperature. For a great many animals, particularly for small invertebrates, this involves a seasonal life cycle, where the periods of the year that are too hot or too cold are spent as a resistant stage, i.e. egg, cyst or (in holometabolous insects) pupa. For many other animals there is a period of juvenile or adult quiescence or torpor. In insects this is determined intrinsically by endocrine changes and is termed diapause. For adult endotherms there may be periods of torpor, either on a diurnal cycle or through the winter ('hibernation') or through the summer ('aestivation'). All of these options represent a way of evading thermal problems in time. There is a further alternative of evasion in space, which can be seen most spectacularly in the massive seasonal migrations undergone by some species, at least partly as a way of escaping adverse thermal conditions.

8.9.1 *Torpor and diapause in ectotherms*

Many aquatic animals show marked seasonal changes in activity that are related to water temperature. Some fish become relatively inactive and cease feeding, entering a state described as torpor or dormancy. This strategy spares their food reserves until prey become more abundant again, and it is more common in freshwater species than in marine taxa. On land, amphibians and reptiles that show winter torpor are described as hibernating, whilst insects go into diapause in the egg, pupal or adult stages. In all these cases, metabolic rates are lowered, and there is little or no activity or feeding. Fat development in subcutaneous layers of tetrapods, or in the insect fat body, is usually pronounced. Blood composition and muscle tone may be highly modified.

Ectotherms such as lungfish, frogs and desert snails may show the 'opposite' response of aestivation in summer, with a substantially reduced metabolic rate despite the high temperatures. These animals often maintain their tissues in an unusually dehydrated state, and have various ways of resisting the arid conditions they would normally encounter by burrowing, sealing the skin and/or orifices with relatively waterproof membranes, or using air layers to buffer the tissues from excess heat; details are considered in Chapter 14. Aestivation also occurs in a range of insects, where it is again termed diapause.

Diapause in insects is a rather special case of arrested development, and it may occur in any part of the life cycle according to species (Fig. 8.45b). It usually involves an 'anticipation' of forthcoming adverse conditions, and is generally triggered by changing day length; there may be a specific sensitive period, with the photoperiodic signals acting via the neuroendocrine system in several different fashions, suggesting several independent evolutionary origins. Before its onset, the larval or adult animal may accumulate extra nutrients and seek a sheltered microhabitat, often spinning a cocoon or constructing a shelter from leaves or litter or other extrinsic materials. Insect diapause can normally only be broken after a suitable period of cooling, again by an endocrine signal, this time from the brain to the prothoracic gland, normally serving to reinstate a moulting programme (see Fig. 8.45). In some species it has a strong genetic component in that certain populations are much more prone to diapause than others (with evidence of maternally controlled inheritance via the egg cytoplasm). Diapause is commonest in eggs and pupae, which are inherently inactive, and it is often

triggered in the previous phase of the life cycle, so that larval instars respond to the triggers promoting pupal diapause and the diapause of eggs may be predetermined in the mother (e.g. in crickets and in the bog-dwelling dragonfly *Somatochlora alpestris*) thus providing an example of a cross-generational effect of temperature on phenotype. Diapause serves to synchronize the whole life cycle with the environment, ensuring that further growth and development, and eventual reproduction, happen at a time when they will have the best possible outcome. Insects in the diapausing state have extremely reduced respiratory rates, with very viscous body fluids and (for winter diapause) a high level of antifreezes such as glycerol. Some herbivorous species may spend as much as 10 months of the year in diapause, only emerging to be active at the time when their particular host plant is flourishing.

8.9.2 *Hypothermia, torpor and hibernation in endotherms*

The terminology of hypothermia is very confused, and some clarification is needed at the outset. Many mammals, and relatively few birds, undergo regular winter periods of reduced metabolic activity, when T_b drops substantially. This **torpor** occurs primarily in small endotherms such as rodents, insectivores and bats, the small marsupial possums and the monotreme echidnas, and in some hummingbirds and swifts. In advance of seasonal torpor substantial deposits of fat are usually laid down, giving a fuel reserve. Then metabolic rate, heart and ventilation rates, and most other physiological processes are greatly reduced, and the animal becomes quiescent and unresponsive to stimulation. Some examples of animals that use torpor, and the percentage savings of energy over maintained basal metabolism, are shown in Table 8.13. Most of the animals that undergo true winter torpor weigh less than 200 g, and the largest animals in this category are marmots, weighing up to 8 kg. For these small animals, the energy savings of torpor are clearly potentially huge.

True torpor is *not* a state in which endothermy and thermoregulation are abandoned; instead, it involves regulation at a new level, with a new critical minimum temperature being maintained. If ambient temperature falls below this, metabolic rate will be increased to maintain the critical T_b. In many very small endothermic animals from temperate climates (such as hummingbirds and shrews), using torpor as an escape from transient seasonal or even nightly cold, the critical T_b is around 15°C, but in somewhat larger species from similar habitats (e.g. poorwills and honeypossums) it may be as low as 5°C. The terms 'shallow' and 'deep' torpor are sometimes used to distinguish these, though in practice there is a continuum of states. Many animals that undergo torpor on a daily basis use a shallow type, and those with longer-term and seasonal torpor normally use a deeper type, but there are many exceptions to this (see Chapter 14, p. 553).

Torpor is not easy to distinguish with clarity from 'winter

Two main types of photoperiod control

Induction and termination

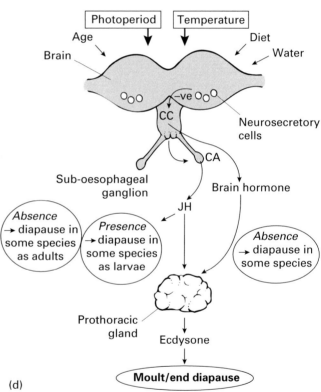

Sensitive periods for induction of (grey) and entry into (black) diapausing states

(a)

(b)

(c)

(d)

Fig. 8.45 Control of diapause in insects. CA, corpus allatum; CC, corpus cardiacum; JH, juvenile hormone. (c, Reprinted from Kerkut, G.A. & Gilbert, L.I., *Comprehensive Insect Physiology, Biochemistry and*

Pharmacology, Vol. 8. *Endocrinology II*, copyright 1985, with permission from Elsevier Science.)

sleep', often occurring in much larger animals such as skunks, beavers or bears, where there is a relatively mild degree of torpidity, with metabolic rate dropping a little and T_b only partially reduced, yet with many physiological functions continuing fairly

normally. Torpor and winter sleep together are often loosely termed 'hibernation'. Nor is torpor entirely distinct from the kind of physiological shut-down and inactivity that may occur in many invertebrates in winter. Furthermore, many animals

Table 8.13 Body temperatures (T_b) and metabolic rate (MR) changes in mammals and birds showing torpor. Basal MRs are similar to non-torpid mammals; here the ratio of torpid MR to normal basal MR is given. (Data from Geiser 1988.)

Species	Normal T_b (°C)	Torpid T_b (°C)	Ratio of torpid MR to normal MR
Monotremes			
Echidna	32.2	5.7	0.44
Marsupials			
Pygmy possum	33.7	10.1	0.22
Insectivores			
Shrew	34.7	14.0	0.10
Rodents			
Mouse	37.4	19.0	0.44
Dormouse	37.7	7.0	0.35
Ground squirrel	37.1	5.0	0.15
Carnivores			
Badger	37.0	28.0	0.50
Birds			
Poorwill	37.0	10.0	0.17
Hummingbird	40.0	21.0	0.13

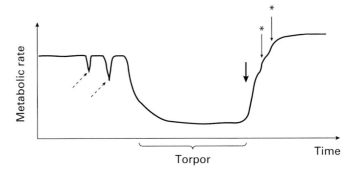

Fig. 8.46 The pattern of changes in metabolic rate during entry into and exit from torpor in a small mammal. Dashed arrows show preliminary 'trial runs' when metabolism transiently slows; then entry into torpor occurs smoothly but relatively slowly. Torpor ceases abruptly (solid arrow) when the brown adipose tissue (BAT) is activated, and metabolic rate may subsequently show further brief periods of increment as shivering occurs (*).

tissue. Later on, shivering also occurs as the muscles become activated (see Fig. 8.46).

Since these phenomena of torpor and arousal are particularly associated with land animals in cold environments they are dealt with in some detail in Chapter 14.

8.9.3 *Migration*

Migration is an alternative way of escaping seasonally adverse environmental conditions, though it is often triggered by dwindling food supplies, and/or by photoperiod, rather than being a direct response to temperature. It can occur on various spatial scales. Many animals show fairly local movements to more equable winter refugia. Littoral and terrestrial invertebrates commonly retreat deep into crevices in rocks or tree-barks; freshwater invertebrates, and many frogs and turtles, seek refugia in the deeper waters of ponds and streams, where the water will remain at 4°C below ice cover; a whole variety of land animals use caves (ranging from spiders, woodlice and hoverflies to snakes, bats and bears), or dig down 1 m or more to ensure that they stay below the frostline, often combining this with some form of controlled hypothermia as discussed above. Summer migrations may also occur to seek out damper and cooler refugia, often to the deeper soils in deserts.

On a grander scale, migration involves a translocation to entirely new biomes, by long-distance migrants. We discussed briefly in Chapter 3 (see section 3.5.2) the effects of scaling on migration, and while net costs of transport are lowest for swimming, especially in ectotherms (see Fig. 3.14), so that ectothermic fish can undertake prolonged seasonal migrations due to their low feeding requirement, it is flight that permits much longer-range movement for most animals over a wide range of sizes (Fig. 8.47). Migration by swimming or walking is only possible for endotherms if they are relatively large, and takes many weeks or months to complete. Amongst endotherms only the

that show a deep winter torpor also show a degree of much shorter-term nightly torpor, again with decreased metabolic rate and body temperature (e.g. hummingbirds and some bats, including many tropical species).

However, in the laboratory it transpires that the same rather common physiological principles probably underlie all these phenomena and they can best be treated together as variations on the theme of controlled hypothermia.

There is no single trigger for going into a torpid hypothermic state; entry may vary with food supply and with the imposed ambient temperature regime, though where there is a yearly cycle the onset of torpor is generally regulated through photoperiod cues. Torpor onset largely results from failure to increase the metabolic rate as T_a declines, because the setpoint for regulation has been changed. This leads to an inevitable decrease in T_b and a further decrease in metabolism, so that the animal slides rather gently into a torpid state (Fig. 8.46). In the fully torpid state, the metabolic rate may be only 1–20% of its normal resting value.

Arousal from torpor is a specific and very energy-demanding process, which may occur quite rapidly (1–6 h) relative to the process of entering torpor. Each species has a minimum temperature from which it can arouse, the **critical arousal temperature**. Arousal in most mammals is a two-stage process. At first it is not accompanied by obvious shivering or muscular activity, because a large part of the thermogenesis is achieved by non-shivering thermogenesis (NST) in the brown adipose

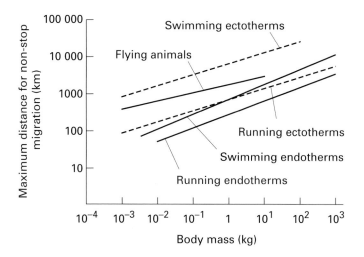

Fig. 8.47 Maximum possible migration distance in relation to size; fish and flying animals can achieve long distances relatively cheaply, and larger animals can achieve much longer distances than small ones (compare Figs 3.14 & 3.16). (From Peters 1983; courtesy of Cambridge University Press.)

birds are conspicuous and common migrators, together with some marine mammals; a few large land mammals do use this strategy (perhaps rather fewer now than in the recent past, due

Fig. 8.48 Major marine and terrestrial migration routes.

to man's impact on traditional migration routes) but most high latitude mammals use an *in situ* torpor as the escape strategy of choice.

All long-range movements require specialized navigational skills, with a need to detect the spatial relationships of the starting point and the expected endpoint. Many animals use the stars as navigational aids, and this may be backed up by a magnetic sense (at least in bees, and some birds); landmark learning and olfactory cues are also important parts of the navigational repertoires. Figure 8.48 shows major routes taken over land and through the oceans by migrating species.

Aquatic migrations by fish are particularly well-studied where they involve a movement between marine and fresh waters. Eels, for example, move out to deep oceanic waters to breed, returning to fresh water for most of their life cycle (a **katadromous** migration). A number of tropical fish species show similar behaviour. In contrast, many high-latitude salmonids show **anadromous** migration, where breeding occurs in the headwaters of rivers and the young then migrate out to sea to feed, often for several years. Olfactory cues seem to be the major basis for 'homing' behaviour, allowing the returning fish to locate the river system they originated from.

Cetaceans also show marine migrations, with various species of whales particularly well studied. Some of these baleen whales spend the months from April to September feeding on krill (*Euphausia*) in the Arctic and move to tropical waters around November; others spend November to April feeding similarly in the Antarctic, and move to the tropics around May. During breeding in the tropical waters there is little food intake and body weights may fall by 25–40%. Few baleen whales cross between hemispheres, and the populations rarely intermingle. Amongst the toothed whales (such as the sperm whale *Physeter*) migrations are less extensive, with populations shifting between subtropical and tropical waters of either hemisphere; feeding is more or less continuous and changes in blubber thickness are much less severe than in the baleen whales.

Migration by flight can be equally spectacular. Amongst insects long-range movement is relatively uncommon, but the most familiar examples are the monarch butterflies (*Danaus plexippus*) that migrate from northern USA to southern California and to Mexico each autumn (up to 3000 km), spending the winter in enormous aggregations in the warmer forests they encounter. Some bats also migrate over perhaps 100–1000 km, usually moving from northern coniferous forests to more southerly Mediterranean or chaparral habitats. Bird migrations are far more spectacular and can encompass almost the entire globe, with many species moving annually between hemispheres. Several hundred Palaearctic-breeding species (perhaps 5–6 billion individual birds) move southwards to feeding grounds in Africa each year, many of them using the narrowest water-gap routes to get maximum uplift from thermals over land, and thus concentrating over southern Spain, through Italy, and across the Middle East and the Sinai Peninsula; similar numbers from the eastern Palaearctic move into southeast Asia and Australasia. At the same time a rather high percentage of birds breeding in northern North America are migrating to the forested areas of Florida, Mexico and northern parts of South America for the winter.

On land, two groups of ungulates provide the major examples of long-distance migration. In the far north, elk and caribou (reindeer) migrate northwards into the tundra in spring, returning to the coniferous forests of the taiga belt in the autumn. In the savanna biomes of the world a variety of ungulates, including zebra, wildebeest and several gazelle species, move regularly between grazing areas as the pattern of precipitation changes with the season, covering perhaps 700–1000 km in the course of a whole year. Both these migratory systems are governed largely by food availability, though linked, respectively, with snowfall and rainfall patterns; neither involves a great alleviation of thermal problems.

8.10 Regulating thermal biology: nerves and hormones

Animals must have a range of thermal effectors used to control thermal exchanges, and hence T_b, by modifying a great variety of processes, from cellular enzyme concentrations to the behaviour of the whole body. There must also be sensors within the body, and control centres to integrate the various thermal responses of each individual. This has been studied most intensively in mammals and birds, where central sensing and pro-

cessing occur predominantly in interneurones of the preoptic centre of the hypothalamus, with additional input from sensors in the skin, spinal cord and abdomen. Similar broad principles of thermal coordination are likely to apply in other animals.

Current evidence indicates that there are dual setpoint regulatory systems in mammalian and reptilian nervous systems, with separate neural thermostats for controlling heat gain and heat loss. Some neurones ('warm receptors') increase firing when the hypothalamic temperature rises, and probably link to neurones that activate heat dissipation systems such as shade-seeking, sweating, or vasodilation. Other neurones ('cold receptors') decrease in firing under the same stimulus, and may initiate vasoconstriction, basking and other behavioural effects. Still others only increase in firing when the hypothalamic temperature falls below a specific setpoint, and these are likely to trigger heat production mechanisms (shivering etc.) and conservation mechanisms (pilo-erection, huddling behaviours, etc.). A highly schematic view of the possible control systems is shown in Fig. 8.49.

In many animals, including mammals and lizards from a range of aquatic and terrestrial habitats, the resultant T_b shows a negatively skewed distribution around T_{pref}. Normally this involves substantially more regulatory precision for the high setpoint. The desert iguana has a trigger point for mechanisms to increase T_b, such as shuttling into sun, at 36.4°C (also known as the **low body temperature setpoint**, LBTS); and a trigger point for cooling mechanisms (**high body temperature setpoint**, HBTS), at 41.7°C. Thus an iguana may sit in the sun until T_b reaches 41.7°C and then shuttle into shade until it cools to 36.4°C. In practice as these are averages each behavioural thermal shuttling cycle varies a little; and LBTS is generally more variable than HBTS. This is likely to be adaptive, as T_b rising towards the critical thermal maximum is inherently more dangerous to the animal than a T_b falling moderately below the preferred levels.

In all animals studied to date, the role of peripheral thermoreceptors is very much secondary to that of the core receptors and brain regulation centre (Fig. 8.50). Clearly reflex adjustments of local vasodilation or pilo-erection, when an extremity is close to a hot rock or dipped into cold water, cannot be allowed to override central needs determining the whole body's core temperature.

However, in some of the smaller endotherms the peripheral sensors are used in a key role to reset the setpoints of the central thermoregulating system, so that as the weather cools the central setpoint rises to allow for greater heat production in the difficult time ahead. For example, in lizards the reception of light of different wavelengths by the parietal eye (a photoreceptive structure on the dorsal surface of the head) affects the thermoregulatory setpoints. Similar but more drastic adjustment of the neural setpoints must underlie the torpor and hibernation responses discussed above.

Hormones seem only rarely to have direct controlling effects on animal temperature balance. Some of the vertebrate

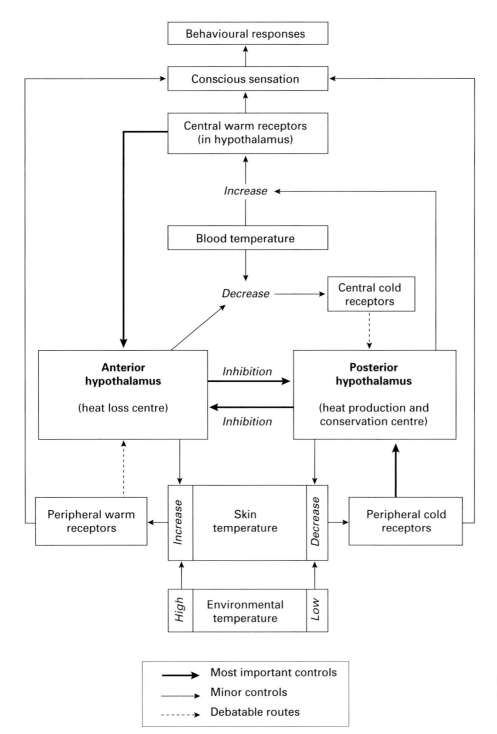

Fig. 8.49 Control systems regulating body temperature in mammals. (From Davenport 1992, Fig. 4.6; with kind permission from Kluwer Academic Publishers.)

hormones (some steroids, and adrenaline) do have effects on glycogen breakdown as we saw earlier, and hence may raise metabolic rate; the thyroid hormones (thyroxine) also have direct effects in most endotherms in causing an increase in metabolic rate, and may act similarly in some lizards, with thyroxine injections elevating daily average T_b and activity levels. Prostaglandins also have some regulatory effects, and are particularly involved in the fever response. However, other ver-

tebrate hormones that influence metabolic effects in the liver and kidney, or uptake processes in the gut, or even skeletal muscle activities, must also have indirect influences on T_b, which must be integrated into the overall body temperature control system.

Effects similar to those of thyroxine appear to operate at the fat body in insects and crustaceans, though details of the hormone action are unclear. Prostaglandins may be the main controlling factors, triggering behavioural hyperthermia in some arthropods.

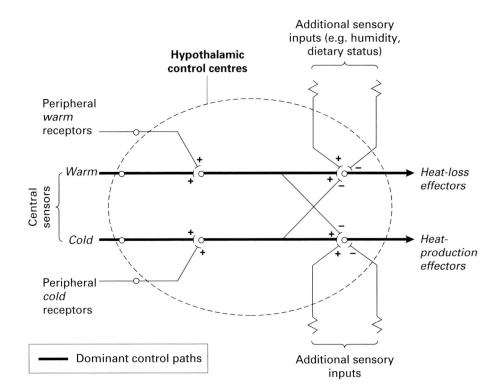

Fig. 8.50 The role of peripheral thermoreceptors.

8.11 Evolution and advantages of varying thermal strategies

8.11.1 *The relative merits of ectothermy and endothermy*

Ectotherms have an approach to living that essentially involves low energy flows, and this can work in a very wide range of body sizes. Terrestrial ectotherms also usually have less need for water, therefore a whole range of arid microniches are available to them that endotherms cannot use (though ectotherms cannot normally maintain activity in a nocturnal niche, except in tropical and summer temperate climates). Ectotherms can also devote a larger proportion of their energy budget to reproduction, with less investment in day-to-day maintenance, so that they are more likely to be good colonizers of poor-quality environments.

However, they are less able to sustain high activity bursts, running a risk of oxygen debt. They tend to use periods of anaerobic metabolism for activities more energetic than slow movement; therefore, although briefly active, they fatigue rapidly due to lactate accumulation. Many of them are therefore ambush predators themselves, but susceptible to intense sustained predation by endotherms.

In contrast endotherms are high-energy-flow systems, and they require larger body sizes, with relatively low surface area to volume ratios, to avoid excessive rates of heat loss and to enable them to eat enough to keep up with their metabolic needs. They have a continuous need for high-quality and/or high-quantity food, often coupled with a high water need, so that their num-bers are severely limited in very poor quality habitats. The extra respiration they require also gives them an increased risk of disrupted acid–base balance.

They can undergo much longer periods of intense aerobic activity than ectotherms, including prolonged nocturnal activity in all habitats and an ability to exploit colder environments at extreme latitudes or altitudes. Because they can maintain high-level activity for long periods, they are able to forage widely and migrate over long distances. Thus they have become the dominant terrestrial animals at all higher latitudes.

It may also be the case that a high and constant body temperature at all stages of the life cycle allows the development of more sophisticated neural processing in endotherms, and greater complexity of learned behaviours; but this is allied with, and difficult to disentangle from, a longer contact period with the mother necessitated by the reproductive constraints on endotherms.

8.11.2 *The evolution of endothermy*

Endothermy has clearly evolved a number of times in animals, and some of the points made above probably indicate why this should be so in general, given its advantages in particular kinds of habitat. But much speculation has also centred on the particular issue of why and when endothermy evolved in the birds and the mammals, two different offshoots from a reptilian stock. There are two possible theories currently in play.

1 Thermoregulatory advantage, with maintenance of a steady T_b being primarily selected for. Such maintenance would be facilitated when the T_b level was rather high, leading to selection for endothermic heat generation and a high resting metabolic rate.

A major difference between ectotherms and endotherms resides in their cellular machinery, with endotherms having increased net mitochondrial membrane density and rather 'leaky' membranes surrounding both mitochondria and cytoplasm. The leaky plasma membranes (to Na^+ and K^+ especially) require extra metabolic energy to be generated to maintain normal solute concentration gradients, as discussed in Chapter 4. This is associated with increments in tissue protein levels and phospholipid concentrations in endotherms, and perhaps with increased unsaturation of the plasma membranes. Similarly the mitochondrial 'leak' is reflected in the increased rate of futile proton cycling in endotherm mitochondria. It has been proposed that these differences are linked to a need for increased thermogenesis in endotherm cells; i.e. a need for increased heat and heat regulation *per se*, supporting the thermoregulatory theory of endothermy origins.

There is an important argument against this theory, however. The evolution of increased metabolism on thermoregulatory grounds alone is difficult to explain in view of its high energetic cost. Small increments in maintenance metabolism would not have been sufficient to establish a steady T_b, but would have added to energetic demands with little apparent benefit. Thus intermediate states may have had little benefit in terms of fitness, if thermoregulation alone was the selective driving force. It is also worth noting that endothermy has evolved in plants, insects, fish and possibly dinosaurs *without* being linked to routine thermoregulation; only in birds and mammals do the two go together.

2 Aerobic capacity advantage, with selection for high rates of aerobiosis allowing sustainable physical activity—effectively, selection for high maximum metabolic rates (MMR). The net costs of transport for terrestrial walkers and runners are substantially greater than those for swimmers of equal body size (see Chapter 4). However, the aerobic capacities of amphibians and reptiles are not greatly expanded relative to those of fish of the same body size and temperature. Thus the ectothermic land tetrapods reach the limits of their endurance at much lower speeds and levels of exertion than their aquatic ancestors, and there might well have been strong selection to improve their endurance. Increased levels of maximum oxygen consumption would result in enhanced performance in a variety of activities that would influence fitness, such as prey pursuit, predator avoidance, territorial defence and courtship. Ancestral animals in both the mammalian and avian lineages do seem to have been progressively more capable of vigorous activity, as shown by general skeletal arrangements such as changes in limb suspension and reduction of skeletal mass.

On this theory, elevated maximum metabolic rates led secondarily to elevated resting metabolic rates; there may be some fixed relation between them, and with carefully controlled comparisons the correlation between RMR (resting or routine metabolic rate) and MMR is certainly rather tight. It may be that

intense aerobic activity leads to increased permeability of membranes to supply oxygen, thereby requiring extra sodium pump activity all the time and linking to a permanently increased T_b. This might lead ultimately to metabolic rates high enough to achieve continuous endothermy and homeostasis. Resting metabolic rates in birds and mammals are both higher (by a factor of 5–15) and more constant than those in most ectothermic vertebrates. However, this maintenance heat is not myogenically derived in endothermic vertebrates but is mainly generated by visceral organs and the CNS (e.g. in man 70% of resting metabolic heat is generated by the metabolism of the heart, kidneys, brain, liver and intestines, even though they only constitute 7% of the body mass); it is not easy to see how this could have arisen as a way of supporting sustained muscular activities, although it is possible that once the advantages of 'leaky' membranes had been acquired they could be spread through the body. Supporters of the aerobic capacity theory claim that the leakiness of membranes in endotherms might also be linked to increased rates of protein and phospholipid turnover. Increased levels of sustainable activity, on this model, would require extra provision of high endurance oxidative-type muscle fibres, with increased demand for support and re-supply to these from the digestive and excretory systems, linking in turn to enhanced resting as well as maximum metabolic rates.

Perhaps the 'price' of increased maximal oxygen consumption was the disadvantage of inevitably but expensively increased resting metabolic rates. Thus it may be that the early stages in the selection for endothermy were indeed dominated by selection for increased endurance, and the later stages by selection for thermoregulatory homeostasis to ensure maximum benefit from the inevitably raised resting metabolism.

Most of the palaeontological evidence supports the aerobic capacity theory, particularly since incipient endothermy probably appeared in moderately sized predatory therapsids (mammal-like reptiles) at a time when they lived in warm equable climates and were likely to be moderate 'inertial homeotherms' anyway, and capable like modern reptiles of highly efficient behavioural thermoregulation. However, it is difficult to test these theories directly by looking for the real origins of endothermy in the fossil record. We can rarely be sure whether particular fossil groups (such as dinosaurs, therapsids, pterosaurs or *Archaeopteryx*) were endothermic or not, since endothermy is largely manifested in aspects of the soft tissues, of which no trace remains. Several possible lines of evidence for endothermy from palaeontological studies have been proposed, and most remain controversial. These include the gait and posture; the nature of the bones; the presence of fur or feathers on some dinosaur fossils; the proportions of predators and prey in particular populations; the size of the braincase; the apparent growth rates; and the oxygen isotope ratios in bones. More recently the development of the nasal respiratory turbinals (especially the maxilloturbinate, bearing respiratory epithelium) has been advanced as evidence of endothermy, this being a more direct correlate of endothermic ability as it reflects a high ventilation rate and the capacity for conservation of water (Chapter 5) and heat (this chapter) in the expired air. Complex turbinates occur in xeric and mesic mammals, and (convergently) in all birds, but are absent in reptiles. These structures, or at least the ridges to which they attach, are readily preserved in fossils, and thus may give a clue to the presence of endothermic respiratory rates. Such ridges are found throughout many of the mammal-like reptiles, even including pelycosaurs; endothermy seems to have accompanied or even predated the other aspects of 'mammalness'. However, in birds there is only a rather weak crest for turbinate attachment, and evidence from fossils is therefore less compelling, though it would indicate an absence of turbinates in *Archaeopteryx* and its immediate dinosaur ancestors. Thus it is inferred that endothermy postdated the origin of birds and avian flight by some 50 million years. Hence the selective regime favouring endothermy in birds was perhaps rather different from that in mammals, though in both cases a selection for increased aerobic scope was probably a major component.

8.11.3 *The evolution of high body temperatures*

Irrespective of the question of why endothermy might have evolved, there is a separate question as to why endotherms almost invariably settle for a body temperature of 35–40°C (see Table 8.11). These temperatures are also similar to those maintained by quite a range of active terrestrial ectotherms, notably reptiles and insects. This might suggest separate convergent evolution on this thermal range. Why should this be so?

One explanation would be the need for body temperature to be set rather high because biochemical reactions and particularly enzyme functions are then faster. Organisms can therefore respond more rapidly to any changes in cell conditions and thereby keep their own cellular environment more stable, allowing selection for enzymes with narrow thermal preferenda and for maximal efficiency within that narrow temperature range. Faster and more constant catalytic rates should also permit faster and more forceful muscle action.

Relatively high body temperatures will also favour faster neurological and hormonal function, hence allowing selection for quicker integration and behavioural sophistication.

Another point commonly made is that T_b needs to be above the highest T_a likely to be experienced, because then the animal can lose heat in a regulated manner by conduction, convection and radiation. If T_b is below ambient, then the body can only cool by evaporation, and that is usually a dangerous strategy in terms of water balance. On this argument, a routine T_b approaching 40°C would be adequate in many terrestrial habitats.

All these theories would predict a selection for higher temperatures, but may not explain why the selected T_b is not higher still (remember that proteins can certainly be constructed to function at much higher temperatures if necessary, e.g. as in hot-springs prokaryotes). An alternative explanation for the choice of 35–40°C comes from recent evidence showing that the physical properties of water are such that the ideal balance of

viscosity, specific heat, and state of ionization in living tissues is achieved at around 35–37°C; this input from physical chemistry may be all the explanation we need, though bird T_b values of 42°C remain somewhat problematic.

We may be able to see reasons for selection for higher body temperature in those cases where animals exhibit unusually raised T_b values. An important example of this is the fever response in endotherms; controversy continues as to whether this is an adaptive response of benefit to the host (helping to disable the pathogen), or an effect induced by the pathogen and deleterious to the host. Table 8.14 shows examples of fever responses. There is evidence that in some cases the host actually regulates T_b slightly higher than normal, and that the survival of a host with a fever reaction is enhanced, the higher temperature helping to overcome the pathogen. This view would be supported by observations that an equivalent effect also occurs in ectotherms; leeches, grasshoppers and some reptiles and amphibians will choose warmer habitats during an infection. *Anolis* lizards appear to show adaptive responses in that some species exhibit this form of 'fever', whilst others with lower preferred T_b values do not. (However, the issue is complicated

Table 8.14 Body temperatures, T_b, (°C) during fever in invertebrate and vertebrate taxa.

Species	Normal (or preferred) T_b	Fever T_b	Difference
Invertebrates			
Crayfish	22.1	23.9	1.8
Shrimp	31.0	35.5	4.5
Lobster	16.0	20.7	4.7
Cockroach	32.3	35.9	3.6
Desert beetle	33.0	34.5	1.5
Horseshoe crab	27.0	33.0	6.0
Leech	20.5	30.0	9.5
Scorpion	25.0	40.0	15.0
Fish			
Sunfish	30.1	32.3	2.1
Bass	29.6	31.9	2.3
Amphibians			
Frog	25.5	27.9	2.4
Reptiles			
Desert iguana	41.0	42.7	1.7
Birds			
Pigeon	39.7	41.5	1.8
Mammals			
Dog	38.2	39.4	1.2
Rabbit	39.5	40.8	1.3
Monkey	38.9	40.1	1.2
Man	37.4	41.3	3.9

because some parasites can apparently also 'control' whether they cause fever; parasites with vectors are especially likely to cause fever, perhaps because it is then in the parasite's interest to have the host prostrate and more likely to be bitten by further vectors, whereas parasites with direct transmission are favoured by keeping the host active, so fever is then less prevalent.)

There are also a few documented cases of animals selecting lower temperatures for particular reasons relating to interactions with parasites or prey items. Certain bumble-bees 'choose' lower temperatures to reduce the effects that parasitic nematodes have upon them, and will stay out of the nest at night to this end. Spider-hunting wasps have also been shown to store their paralysed prey items in chambers at lower temperatures than their normal living quarters, to slow deterioration and deter fungal growths; and some spiders will also move prey around their nests or temporarily shift them out of the nest to keep the prey storage temperature within a narrow optimal range. Thus the selection of particular temperatures by animals is not a simple matter, but reflects a balance of biochemical, physiological, behavioural and coevolutionary needs.

Further reading

Books

Bennett, A.F. (1987) Evolution of the control of body temperature: is warmer better? In: *Comparative Physiology: Life in Water and on Land* (eds P. Dejours, L. Bolis, C.R. Taylor & E.R. Weibel). Liviana Press, Padova.

Cannon, B. & Nedergaard, J. (1985) Biochemical mechanisms of thermogenesis. In: *Circulation, Respiration and Metabolism* (ed. R. Gilles), pp. 502–518. Springer-Verlag, Berlin.

Cossins, A.R. & Bowler, K. (1987) *Temperature Biology of Animals.* Chapman & Hall, London.

Davenport, J. (1992) *Animal Life at Low Temperature.* Chapman & Hall, London.

Heinrich, B. (1993) *The Hot-Blooded Insects.* Springer-Verlag, Berlin.

Johnston, I.A. & Bennett, A.F. (eds) (1996) *Animals and Temperature: Phenotypic and Evolutionary Adaptation.* Cambridge University Press, Cambridge.

Lyman, C.P., Willis, J.S., Malan, A. & Wang, L.C. (1982) *Hibernation and Torpor in Mammals and Birds.* Academic Press, New York.

Morimoto, R., Tissieres, A. & Georgopoulos, C. (1994) *The Biology of Heat Shock Proteins and Molecular Chaperones.* Cold Spring Harbor Laboratory Press, Plainview, NY.

Reviews and scientific papers

Atkinson, D. (1995) Effects of temperature on the size of aquatic ectotherms: exceptions to the general rule. *Journal of Thermal Biology* **20**, 61–74.

Atkinson, D. & Sibly, R.M. (1997) Why are organisms usually bigger in colder environments? Making sense of a life history puzzle. *Trends in Ecology and Evolution* **12**, 235–239.

Baker, M.A. (1982) Brain cooling in endotherms in heat and exercise. *Annual Review of Physiology* **44**, 85–96.

Bakken, G.S. (1992) Measurement and application of operative and standard operative temperatures in ecology. *American Zoologist* **32**, 194–216.

Bennett, A.F. & Ruben, J.A. (1979) Endothermy and activity in vertebrates. *Science* **206**, 649–654.

Block, B.A. (1994) Thermogenesis in muscle. *Annual Review of Physiology* **56**, 535–577.

Bozinovic, F. (1992) Scaling of basal and maximum metabolic rate in rodents and the aerobic capacity model for the evolution of endothermy. *Physiological Zoology* **65**, 921–932.

Brock, T.D. (1985) Life at high temperatures. *Science* **230**, 132–138.

Caceres, C.E. (1997) Dormancy in invertebrates. *Invertebrate Biology* **116**, 371–383.

Crockett, E.L. (1998) Cholesterol function in plasma membranes from ectotherms: membrane-specific roles in adaptation to temperature. *American Zoologist* **38**, 291–304.

Detrich, H.W. (1997) Microtubule assembly in cold-adapted organisms: functional properties and structural adaptations of tubulins from Antarctic fishes. *Comparative Biochemistry and Physiology A* **118**, 501–513.

Else, P.L. & Hulbert, A.J. (1985) An allometric comparison of the mitochondria of mammalian and reptilian tissues: implications for the evolution of endothermy. *Journal of Comparative Physiology* **156**, 3–11.

Else, P.L. & Hulbert, A.J. (1987) Evolution of mammalian endothermic metabolism: 'leaky' membranes as a source of heat. *American Journal of Physiology* **253**, R1–R7.

Franks, F., Mathias, S.F. & Hartley, R.M.H. (1990) Water, temperature and life. *Philosophical Transactions of the Royal Society of London* **326**, 517–534.

Gibbs, A. & Mousseau, T.A. (1994) Thermal acclimation and genetic variation in cuticular lipids of the lesser migratory grasshopper: effects of lipid composition on biophysical properties. *Physiological Zoology* **67**, 1523–1543.

Hazel, J.R. (1984) Effects of temperature on the structure and metabolism of cell membranes in fish. *American Journal of Physiology* **246**, R460–R470.

Himms-Hagen, J. (1985) Brown adipose tissue metabolism and thermogenesis. *Annual Review of Nutrition* **5**, 69–94.

Huey, R.B. & Kingsolver, J.G. (1993) Evolution of resistance to high temperature in ectotherms. *American Naturalist* **142**, S21–S46.

Johnston, I.A. & Walesby, N.J. (1977) Molecular mechanisms of temperature adaptation in fish myofibrillar adenosine triphosphatases. *Journal of Comparative Physiology* **119**, 195–206.

Kuhnen, G. (1997) Selective brain cooling reduces respiratory water loss during heat stress. *Comparative Biochemistry and Physiology A* **118**, 891–895.

Laszlo, A. (1988) The relation between heat shock proteins, thermotolerance and protein synthesis. *Experimental Cell Research* **178**, 401–408.

Leroi, A.M., Bennett, A.F. & Lenski, R.E. (1994) Temperature acclimation and competitive fitness; a test of the beneficial acclimation assumption. *Proceedings of the National Academy of Sciences of the United States of America* **91**, 1917–21.

Lovegrove, B.G., Heldmaier, G. & Ruf, T. (1991) Perspectives of endothermy revisited: the endothermic temperature range. *Journal of Thermal Biology* **16**, 185–197.

McNab, B.K. (1983) Energetics, body size and the limits to endothermy. *Journal of Zoology* **199**, 1–29.

Morrison, R.L., Sherbrooke, W.C. & Frost-Mason, S.K. (1996) Temperature-sensitive, physiologically active iridophores in the lizard *Urosaurus ornatus*: an ultrastructural analysis of color change. *Copeia* **1996**, 804–812.

Parker, A.R. (1998) The diversity and implications of animal structural colours. *Journal of Experimental Biology* **201**, 2343–2347.

Partridge, L. & Coyne, J.A. (1997) Bergmann's rule in ectotherms: is it adaptive? *Evolution* **51**, 632–635.

Pough, F.H. (1980) The advantages of ectothermy for tetrapods. *American Naturalist* **115**, 92–112.

Ruben, J. (1995) The evolution of endothermy in mammals and birds: from physiology to fossils. *Annual Review of Physiology* **57**, 69–95.

Somero, G.N. (1995) Proteins and temperature. *Annual Review of Physiology* **57**, 43–68.

Storey, K.B. (1997) Organic solutes in freezing tolerance. *Comparative Biochemistry and Physiology A* **117**, 319–326.

Storey, K.B. & Storey, J.M. (1996) Natural freezing survival in animals. *Annual Review of Ecology and Systematics* **27**, 365–386.

Turner, J.S. (1987) The cardiovascular control of heat exchange: consequences of body size. *American Zoologist* **27**, 69–79.

Turner, J.S. (1988) Body size and thermal energetics: how should thermal conductance scale? *Journal of Thermal Biology* **13**, 103–117.

Van Voorhies, W.A. (1997) On the adaptive nature of Bergmann size clines: a reply. *Evolution* **51**, 635–640.

Whiteley, N.M., Taylor, E.W. & El Haj, A.J. (1997) Seasonal and latitudinal adaptation to temperature in crustaceans. *Journal of Thermal Biology* **22**, 419–427.

Willmer, P.G. & Unwin, D.M. (1981) Field analysis of insect heat budgets: reflectance, size and heating rates. *Oecologia* **50**, 250–255.

Section C:
Coping with the
Environment

Introduction

Our planet is very unusual. It is so, not just in the trivial sense that all planets are unique, but in several distinct and important ways. It is at just the right distance from its star (the Sun), and of just the right size, to experience a crucially significant range of temperatures. These are the temperatures that permit a molten core on which a thin crust of solid rock can float, giving us the 'plates' of land making up the continents. These temperatures are also in the range that allow the existence of both liquid water and water vapour, giving oceans between the land masses and water vapour above them. Furthermore the planet is just large enough to have a sufficient gravitational pull to allow it to retain any atmosphere that forms. Our planet also has the only oxygen-rich atmosphere in the solar system (though this is primarily a consequence of the presence of life rather than a prerequisite for life to have evolved). Thus we have a lithosphere, a hydrosphere and an atmosphere, as a result of which an evolving biosphere becomes possible.

All these components interact and change relative to each other over time; in particular, the continents move around continuously over their molten foundations, creating changing climates and biogeographic patterns, and precipitating major geological events such as earthquakes and volcanoes. All these effects have major impacts on the rates and patterns of evolutionary change and adaptation in the biota.

Within this biosphere, a multitude of habitats are thus made available, determined by the range of interactions of the solid, watery and gaseous phases. Much of the living space on the planet is aquatic, and most of the aquatic phase is rather salty; this marine habitat is certainly where life first evolved, and where it is 'easiest'. Where evaporated water vapour precipitates back onto the land as rain it may collect to make freshwater habitats (always geologically short-lived, but giving rise to their own specialist fauna), and a range of 'brackish' zones where sea water and fresh water meet at the seashores and estuaries. Finally there are the more strictly terrestrial animals, always exposed to the atmosphere, and only transiently in contact with external liquid water via their food and drink. Most of these live on the surface of the land, while a few are closer to being genuinely 'atmospheric' inhabitants.

Cyclical effects on the planet's environments

Many cycles and rhythms are imposed upon planet Earth and its habitats by the physical behaviour of the solar system, organizing the dimension of time as perceived by the biota, and affecting patterns of energy flow on Earth. Most of these rhythms originate with the orbital motions of the planets and planetary satellites, particularly of Sun, Earth and Moon.

The most important and immediate cycle is the circadian rhythm of sunlight and dark, resulting from the rotation of the Earth around the Sun once every 23.93 h. Secondary to this is the lunar cycle, with the cycle of moonlight being 29.5 days, or 1 lunar month; the Earth is unusual among small planets in having its own companion satellite, and the gravitational effects of the Moon are a major influence on the movements of the Earth's seas, giving rise to regular lunar tides with a period of 24.8 h (imposed on smaller solar tides due to the gravitational pull of the Sun). Thirdly, there is a seasonal cycle, mainly resulting from the fact that the Earth is tilted on its axis, with the northern hemisphere tilted towards the Sun in June (northern summer) and away from it in December (northern winter), the southern hemisphere having the opposite experience. These daily, monthly and annual cycles have huge effects on virtually all the animals that inhabit the planet, principally through their effect on the pattern of solar energy received at any particular place; in addition, the life cycles of many coastal marine organisms must be organized to be in harmony with the complex tidal cycles.

There are also several longer-term elements of periodicity in the Earth's heat budget that result from astronomical factors. Firstly the Earth has a slight wobble or 'precession' in its spinning, so that we are not always in the same position relative to the Sun at a particular time of year. This has a periodicity of about 21 000–23 000 years. Secondly, the angle of tilt of the Earth's axis varies over time, between 22.1° and 24.5°, and with a periodicity of 41 000 years; at present we are in a period of decreasing tilt, which should give a slight cooling. Thirdly, we suffer from 'orbit eccentricity', having an elliptical path around the Sun but slightly off centre so that the ellipse varies, with a periodicity of 93 000 years. The elliptical variation makes a difference of up to 4.8 million kilometres (3 million miles) in our distance from the Sun, and hence of up to 7% in the solar radiation we receive. All these long-term cycles certainly have effects on evolutionary patterns on Earth; we can use carbon isotopes in ancient deposits that reflect the temperature of the seas in which they were laid down, and so track how much of the world's water was locked up as ice in different periods. It is clear that large cyclical natural fluctuations in the Earth's temperature have been happening throughout the planet's history, and have probably been one of the major influences on glaciations, sea levels, plate tectonic movements and hence the patterns of extinctions, radiations and physiological biodiversity on Earth.

In this section, a series of habitats offering different challenges to their occupants are considered in turn. We look at the nature of each habitat and the particular problems it poses, and the types of animals that live there; then we deal with the major themes of adaptation in a standard format to allow easier cross-reference:

- ionic and osmotic adaptation and water balance;
- thermal adaptation;
- respiratory adaptation;
- reproductive and life-cycle adaptations;
- mechanical and sensory adaptations;
- feeding and being fed on;
- anthropogenic problems.

9 *Marine Life*

9.1 Introduction: marine habitats and biota

The seas occupy about 70% of the surface of the Earth, and because of their great depth (a mean of about 3800 m, with trenches to 11 000 m) they form at least 99.9% of the available living space on the planet. Marine animals are found throughout the sea, from benthic organisms in the soft oozes of the cold and profoundly dark ocean floors to drifting or swimming pelagic organisms in the photic zone near the surface.

The marine world is the obvious place to start a habitat-based review of environmental adaptation, because existence in the sea should be the 'easiest' of all modes of life for animals on this planet. This simple truth arises because life originated in the sea, probably in shallow ocean fringes and pools, so that cells were originally adapted to function optimally in an ionic and osmotic environment probably not very different from that of modern oceans. The vast majority of contemporary inhabitants of the seas are therefore isotonic and isosmotic with their surroundings, and in most respects have rather limited need for complex physiological strategies. Only at the fringes of the seas—shorelines and estuaries—do problems become more severe, and these habitats are dealt with separately in the next chapter.

However, as vertebrates we tend to lose sight of this situation, given our own difficulties in coping with the salty, undrinkable and tissue-shrinking seas around us. The reasons for this, and the special adaptations needed by marine vertebrates and other 'invaders' of the marine environment, will be dealt with as a special issue at the end of this chapter, although some of the basic biology of marine vertebrates is included in earlier sections where it corresponds closely with that of invertebrates.

9.1.1 Chemistry of the marine environment

The oceans of this planet are all continuous with each other, and are relatively consistent in chemical composition throughout, being well mixed by currents and tides. This constancy applies on daily and seasonal and even geological timescales, and on coastal or global spatial scales. Why should the seas have been stable through such long periods, and why is the stable point the apparently strange mixture we find today, dominated by sodium and chloride ions? After all, the oceans are continuously fed by rivers, which consist of fresh water plus dissolved crustal rocks and some nutrients. Therefore we might expect to see gradually increasing levels of several of the typical terrestrial mineral elements in the seas, but this does not happen.

Ocean chemistry is actually created and maintained by complex geochemical cycling processes, which have apparently always gone on and hence have kept the seas in a fairly steady state. Several different components are involved in maintaining this stability. The rocks that are eroded into fresh waters from the land are mainly feldspars: these are the potassium, sodium and calcium salts of aluminium silicates. Thus rivers should gradually become rich in K^+, Na^+, Ca^{2+} and silicates, and be relatively poor in Cl^-. However, much of the K^+ gets removed; in particular, particles of kaolin (china clay silicates) in the sediments of rivers take it up, and the result is the mineral 'illite', which is deposited and held in muds. As the rivers disgorge into the seas, Ca^{2+} is progressively removed by the biota, and deposited after death as skeletal elements made largely of calcium carbonate and calcium phosphate. At the same time chloride ions are added to sea water, both via the land and rivers from terrestrial volcanoes and directly to the sea from mid-ocean ridge volcanic activity. Sodium is regulated as part of a complex cycle, being added to the sea via rivers, but replaced onto land via wind-borne sprays in the short term and via erosion of emergent sedimentary rocks on a geological timescale. Thus the sea does become, roughly speaking, dissolved NaCl, and it is maintained in that state indefinitely. Other minor elements get 'swept up' within the seas such that their levels remain constant: some accumulate into clays, others into algae (this is especially true for phosphates and for copper ions) and thus are ultimately deposited as sediment when the algae die.

The longer-term stability of composition of sea waters is still a very contentious subject, though. Certainly on a geological timescale volcanic action, rock erosion, sedimentation rates and the balance of precipitation and evaporation would all affect the marine environment, whilst biological processes themselves would change the oxygen and carbon dioxide levels and alter

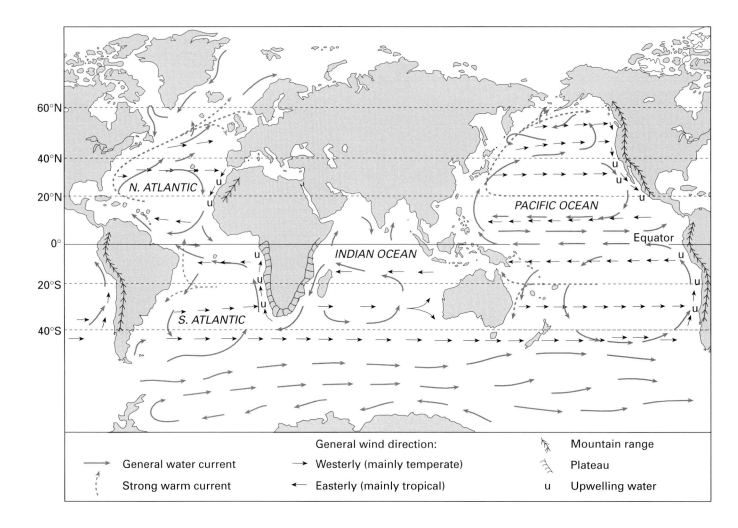

Fig. 9.1 Patterns of ocean currents and winds on the Earth. The winds tend to produce strong warm currents off eastern coasts and cold upwellings off western coasts. (Adapted from Crowley & North 1991.)

the distribution of ions such as calcium and carbonate. However, evidence as to what chemical features actually varied in the past, and to what extent, is incomplete and controversial; a gradual rise of dissolved oxygen through the Precambrian and early Palaeozoic is almost certain, and a gradual loss of potassium into clays and other deposits seems very likely, but all else remains contentious. A significantly lower overall concentration, perhaps about 60–70% of present levels, during early geological eras (Precambrian) has been inferred from data on the regulatory ranges, and points of greatest stability, of modern animals; most modern animals are better able to cope with moderate dilution to around 60–70% than with even slight concentration above 110–120%. It is also likely that an increase in salinity and in carbonate levels may have been accelerated from the Silurian period onwards, once life on land induced greater terrestrial rock weathering. However, the most severe changes were probably completed quite early in the story of the evolution of life,

and modern sea water may therefore be rather a good model of what animal groups have had to contend with through most of their evolutionary history.

9.1.2 *Modern-day oceans*

The distributions and physical characteristics of the seas on Earth have varied continuously through geological time as the continental plates coalesced, drifted apart and re-formed in new configurations, giving rise to changing patterns of a few or many separate circulating oceanic systems. Today we have two vast ocean basins, the Atlantic and the Pacific, with their own circulating currents, together with a fairly small relatively enclosed Arctic ocean and a number of smaller seas linked to the two main Oceans. But this pattern is of rather recent origin, dating from the time about 4 MYA when the two American continents joined at the Isthmus of Panama and for the first time in hundreds of millions of years separated the Atlantic waters from the larger Pacific area, so that there was no single equatorial ocean with a single circulation.

Each basin therefore now has its own circulation of major

Fig. 9.2 Distribution of primary productivity in the world's oceans.

Productivity
(g C m^{-2} year^{-1})

<35
35–55
55–90
>90

currents, shown in Fig. 9.1. These currents perturb the waters, promoting mixing and constant salinity but also bringing together cold and warmer waters. In some areas, large loops of water break off from the currents and form massive eddies or 'rings', moving across the open ocean, with a core of either cold or warm water; in the North Atlantic there are commonly about 10 cold-core rings and three warm-core rings at any one time, broken off from the Gulf Stream. Marine waters of different temperature and salinity also mix and mingle to create 'salt fingers' of interdigitating rising warm and falling cold waters. Thus animals adapted to different thermal regimes may be translocated *en masse* across apparently inhospitable marine zones.

9.1.3 Marine flora and productivity

Marine phytomass is dominated by thallophyte plants, especially the microalgae (phytoplankton) and macroalgae (seaweeds). There are only about 30 species of marine seed plants, and these are restricted to estuaries and coasts. Thus primary productivity derives mainly from algal activities, and for most of the oceans this means floating phytoplanktonic algae (in addition to bacteria and other prokaryotic organisms).

Currents and eddies have major effects on marine productivity in every ocean. Productivity is rather low in vast areas, especially of the Pacific, with low iron levels often implicated as the limiting factor; an illustration of this causal link is provided by the leaching of iron from sub-Antarctic islands, resulting in locally enhanced productivity. By contrast there are areas of very high primary productivity (Fig. 9.2) usually also associated with abundant animal life. Such areas traditionally occur in the turbulent regions where warm and cold waters meet, a classic example being the (now almost depleted) fishing grounds off Newfoundland known as the Grand Banks, where the Gulf Stream and the Labrador Cold Current collide. Further areas of high production occur where cold nutrient-rich waters are forced into upwellings (see Fig. 9.1) as they meet the west coasts of continents, best exemplified by the coasts of Ecuador and Peru, where a large proportion of the world sardine catch originates. The polar seas represent another area of good productivity, at least in the polar spring and summer when glacial meltwater introduces nitrates and silicates, and promotes rapid growth in the algal communities formerly held in the ice-pockets. The

Table 9.1 Distribution of major phyla in marine habitats.

Phylum/group	Marine benthic	Marine pelagic	Freshwater	Terrestrial	Parasitic
Porifera	++	l			
Cnidaria	++	++	+		
Ctenophora		++			
Platyhelminthes	+	+	++	+	++
Nemertea	++	l	+	+	
Nematoda	++		++	++	++
Acanthocephala					++
Rotifera	+	+	++		r
Mollusca	++	+	++	++	r
Annelida	++	+	++	+	+
Pogonophora	+	l			
Sipuncula	+	l			
Priapulida	+	l			
Arthropods					
Crustacea	++	++	++	++	r
Arachnida	r		+	++	++
Insecta			++	++	++
Onychophora				+	
Bryozoa	++	l	+		
Brachiopoda	+	l			
Echinodermata	++	lr			
Urochordata	++	+			
Chordates					
'Pisces'	++	++	++		
Amphibia	r		+	+	
Reptilia		+	+	++	
Aves		(+)	(+)	++	
Mammalia		+	+	++	

++, Common; +, present; r, rare; l, larvae only.

final major areas of production are the coastal and estuarine waters, especially in temperate latitudes where seasonal floodings bring in vast amounts of nutrients; these habitats come into the category of brackish waters, covered in Chapter 10.

9.1.4 *Extant marine animals*

Most animal phyla are represented in the marine habitat (Table 9.1); among the few absentees are the taxa that have evolved as terrestrial specialists (insects, myriapods and arachnids) or as parasites in or on terrestrial organisms. Some of the deepest sea areas are only just being explored, and are proving to have much greater biodiversity than previously thought. However, large parts of the pelagic marine environment are relatively low in species numbers, perhaps largely because the habitat has no real structural complexity to allow niche subdivision and speciation. Only in relatively shallow seas where there are areas of reef or kelp forest (Fig. 9.3) do we find any real architectural complexity to promote very high diversity. Coral reefs (see Plate 1d, facing p. 310) are largely restricted to areas where the surface temperatures are on average above 20°C even in the coldest month, whereas kelp forests thrive in much cooler waters. In both areas,

and in some deep-water benthic communities, representatives of the majority of the phyla listed in Table 9.1 can be found.

Marine animals are probably best subdivided according to where they live in relation to depth. Fundamentally they may be either bottom-dwelling (**benthic**) or living within the water column (**pelagic**). The benthos may be further subdivided into **epifauna** (or epibenthos), living (either attached, sessile and non-motile, or as errant crawling animals) on top of the substratum (Plate 1a); and **infauna** living within the substratum, the latter including both sedentary macroinvertebrates that burrow and a host of microinvertebrates living between the substrate particles and often referred to as the **interstitial fauna** or **meiofauna**.

Within the pelagic zone animals are again divided largely by size linked to locomotory abilities. Pelagic animals that are relatively active swimmers (and usually fairly large) are referred to as **nekton**, a category currently dominated by fish and cephalopods (Plate 1b), but with spectacular representation from the marine mammals (cetaceans, sirenians and members of the Carnivora, the last of these being amphibiously marine and including pinnipeds, sea otters and polar bears). In the Mesozoic these marine mammals were absent and various radiations of marine reptiles occupied the large carnivore niches.

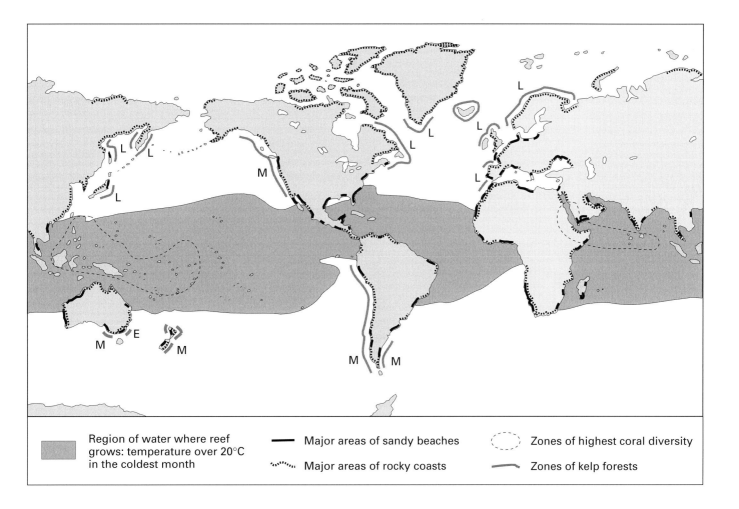

Fig. 9.3 Major regions of reef and of kelp forest in the world's oceans. For the kelps, L = *Laminaria*, M = *Macrocystis* and E = *Ecklonia*.

The smaller animals that essentially float and drift with the currents (though they may also be good swimmers) are termed **plankton**, or more strictly zooplankton (Plate 1c). The plankton of most oceanic areas is often extremely diverse, with many jellyfish (cnidarians), comb jellies (ctenophores) and arrow worms (chaetognaths), though usually numerically dominated by a variety of classes of Crustacea. It also includes a vast range of larval forms (including juvenile fish), varying in composition with the season, since many marine animals have a pelago-benthic life history with a relatively sessile adult form and a more or less prolonged planktonic larval stage.

9.2 Ionic and osmotic adaptation

9.2.1 *Sea water*

The chemical composition of 'normal' open sea water is shown in Table 9.2; this gives the major components, but virtually every other element is also present in trace amounts. This mix of ele-

ments results in an osmotic concentration of 1000–1150 mOsm, a salinity of around 35–36 parts per thousand (ppt or ‰), and a freezing point depression of around 1.86°C. Below a depth of about 1000 m, sea water is practically invariant, and always has a salinity of 34.5–35.0 ppt. Near the surface the two opposing effects of higher rates of loss of water by evaporation, and gain by precipitation, produce greater local variation in salinity. This property can therefore change somewhat with latitude (Fig. 9.4a), reaching 37.5 ppt in parts of the tropical seas. Only in a few unusual situations do the composition parameters change more substantially. Some very large enclosed 'bays' such

Table 9.2 Composition of normal sea water.

Component	mmol per litre sea water	g per litre sea water
Na	470	10.8
K	10	0.39
Mg	54	1.30
Ca	10	0.41
Cl	548	19.4
SO_4	28	2.7
HCO_3	2	0.14

Great Ocean Conveyor Belt

Fig. 9.4 Variation in salinity (a) in surface waters through the major oceans, and (b) in deep-water circulations, showing the deep salty water flowing out of the north Atlantic, around Africa and past Antarctica, then northwards into the Pacific. Here some of the water wells up and warms and there is then a shallower warm current of less salty water back through the Indian Ocean. This salt conveyor system compensates for water vapour transport through the atmosphere. (a, Adapted from Ingmanson & Wallace 1989; b, adapted from Kerr 1988.)

as the Hudson Bay and the Baltic Sea, with substantial fresh-water inputs from surrounding mountains, may be moderately diluted, and polar seas may also be somewhat dilute for parts of the year due to melting ice. By contrast, enclosed areas such as the tideless Mediterranean Sea and the Red Sea, in high temperature zones and hence with particularly high evaporation, may become more concentrated. Variation of salinity with depth also occurs, with a cool and salty current running at depth southwards through the Atlantic and north again through the Pacific, as shown in Fig. 9.4b.

9.2.2 Composition of fluids in marine animals

Virtually all truly marine invertebrates are stenohaline, and their body fluids are nearly isosmotic with their surroundings; a range of examples is given in Table 9.3. However, it is evident that there is commonly a slight osmotic excess in the tissues, and this is attributable to the Donnan effect due to the proteins and other metabolites contained in all living organisms (see section

Table 9.3 Osmotic concentration of marine animals' body fluids.

Phylum/genus	Osmotic concentration (mOsm)
Cnidaria	
Aurelia	1050
Mollusca	
Sepia	1160
Eledone	1061
Mytilus	1148
Annelida	
Arenicola	1120
Neanthes	1108
Crustacea	
Carcinus	1100
Ligia	1220
Nephrops	1108
Chelicerata	
Limulus	1042
Echinodermata	
Echinus	1065
Chordata	
Elasmobranch fish	
Scyllium	1075
Squalus	1096
Teleost fish	
Paralichthys	337
Gadus	308
Sea water	1000–1050

5.3.3). Since metabolic activity within the cell creates more of the relatively non-diffusible small molecules, any cell would tend to increase slowly in osmotic concentration, thus drawing in water and so gradually swelling up. Hence all cells, even in nearly isosmotic sea water, have a constant need for a minimum degree of volume regulation. A membrane-bound sodium-activated ATPase (the Na^+/K^+-ATPase or 'sodium pump'—see section 4.4.2) actively contributes to this by extruding cations. At the same time, amino-acid regulation systems also help to control the internal osmotic concentration and thus the cell volume (see section 4.6). In most stenohaline marine invertebrates volume control is limited and cellular swelling does occur in moderately dilute media; but some compensatory reduction in intracellular osmotic concentration is brought about by losses of intracellular constituents. The principal osmotic effectors that move out of the cell during hyposmotic stress are the common neutral amino acids (especially alanine and glycine, with proline also important in many crustaceans). However, a group of amino-acid derivatives are also expelled from the cells according to need, and in molluscs and crustaceans, which have been best studied, these include betaine, taurine and trimethylamine oxide (TMAO). Long-term isosmotic balance is thus maintained.

With respect to ionic balance and particular ionic concentrations in cells the situation is rather different (see Table 4.4). The osmotic concentration in marine animals' cells may be very like that of sea water, but ions in fact make up only about half of this total intracellular concentration in most animals, the remainder being due to the cell's amino acids and other small organic molecules already mentioned. The ion fraction in the cell is dominated by potassium (150 mM, or rarely up to 400 mM) rather than sodium (25–80 mM), with chloride (25–100 mM) making up only a part of the anionic total as with any intracellular situation (see Chapter 4). A single cell in a marine organism therefore corresponds to the simple model of osmotic and ionic regulation shown in Fig. 9.5. Marine cells are indeed the archetype for all others, and they differ little in proportional composition from the cells of the freshwater or terrestrial animals that have evolved from them.

Moving outside the cell, examples of the ionic compositions of the extracellular body fluids in marine organisms were given in Table 5.1, and are shown relative to normal sea water in Table 9.4; the values there make it clear that many organisms, despite being nearly isosmotic, are specifically regulating at least some of their ionic levels at the 'skin' interface between the sea and the body fluids. Sodium and chloride rarely deviate much from being isionic in marine bloods, but potassium is commonly hyper-regulated, and calcium may also be high at times during the moult cycle of many arthropods. Sulphate ions may be roughly as in sea water or very low (in crustaceans and in pelagic animals), and magnesium may also be kept low in crustaceans. Some groups, notably echinoderms, show virtually no ionic regulation.

It should be noted that there are just a few marine invertebrates that are not isosmotic with their surroundings. Some

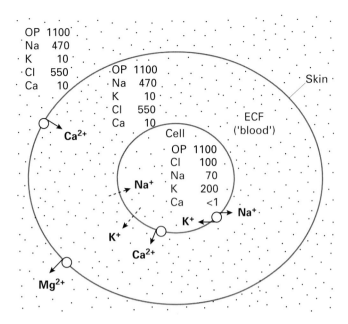

Fig. 9.5 A model of salt movements between compartments (cells and extracellular fluid (ECF)) in a typical marine animal. Circles with solid arrows indicate active transport; OP, osmotic pressure.

Table 9.4 Ionic composition of marine animals' body fluids given as percentages relative to sea water.

Phylum/ species	Ion					
	Na^+	K^+	Ca^{2+}	Mg^{2+}	Cl^-	SO_4^{2-}
Cnidaria						
*Aurelia**	99	106	96	97	104	47
Mollusca						
Pecten	100	130	103	97	100	97
*Sepia**	93	205	91	98	105	22
*Loligo**	95	219	101	102	103	36
Annelida						
Arenicola	100	104	100	100	100	92
Aphrodite	99	103	100	101	99	92
Sipuncula						
Phascolosoma	104	110	104	69	99	91
Crustacea						
Maia	100	125	122	81	102	66
Carcinus	110	118	108	34	104	61
Nephrops	113	77	124	17	99	69
Echinodermata						
Marthasterias	100	111	101	98	101	100
Echinus	99	100	101	99	99	99
Urochordata						
*Salpa**	100	113	96	95	102	65

* Indicates a pelagic species.

grapsid crabs and some palaeomonid shrimps are always hyposmotic to sea water (often around 80–90% concentration), and these same animals are commonly fairly euryhaline, able to survive in both brackish and hypersaline habitats when necessary. Another oddity is the common octopus, which is always somewhat hyperosmotic to sea water at about 1300 mOsm, with a continuous need to produce a slightly dilute urine and to absorb salts from imbibed sea water.

9.2.3 *Adaptive mechanisms*

Marine invertebrates clearly require some ion-pumping capacity, and must retain some degree of control over the ion fluxes across their surfaces. Hence, although relatively permeable compared to close relatives living in other habitats (see Chapter 5), few seawater animals act as perfect osmometers, but instead have a somewhat lower water permeability at their outer surfaces than elsewhere in their bodies. Some examples of marine skin permeabilities were given in Table 5.3; the general range for water flux rate in invertebrates is around 5–20 mg cm^{-1} h^{-1} (1–10 mg cm^{-1} h^{-1} for vertebrates) corresponding to a skin resistance (see Chapter 5) of 6–15 s cm^{-1} (up to 35 s cm^{-1} for vertebrates). Marine animals also require impermeability to their own organic constituents, and restriction on the movements of divalent cations and anions in particular. In fact calcium permeability must be especially low in all the animals' membranes, to keep internal cellular levels correct, as this ion in turn affects all other membrane permeabilities.

But how and why are the blood ions regulated in marine animals? A basic mechanism for ion regulation was discussed in Chapter 4, but that largely accounted for regulation of Na^+ and Cl^-, the two ions that are least varied in the standard marine animal's blood. The standard Na^+/K^+-ATPase enzyme that effects such regulation has a relatively high K_m value in marine animals (see Table 5.4), and achieves quite rapid sodium transport rates (cf. Table 10.5) out of the cells to the already high levels of sodium concentration in their surroundings. Beyond this, though, special mechanisms are clearly needed to raise calcium or potassium levels, or to extrude magnesium and sulphate, and though little information is available on these topics, ionic 'pumps' requiring the splitting of ATP are inevitably involved, though probably indirectly via symport and antiport mechanisms (Chapter 4). In some special cases, such as the floating organisms that increase their buoyancy by eliminating heavy sulphate ions (see below), we know that exchange mechanisms (extruding sulphate in exchange for lighter chloride ions) are involved. In most marine mammals, ion-regulating mechanisms operate at the general body surface rather than in the excretory system; in a range of stenohaline invertebrates, the body surface is responsible for 75–95% of salt loss and uptake.

Thus the marine animal's body is not just a passive equilibrium system, but is involved continuously in minor regulatory adjustments to keep its cell volumes (and hence overall volume) constant, and to maintain the ionic gradients between blood and medium, and between cells and blood. This degree of regulation must have been instituted very early in evolution, probably as soon as the cell membrane was 'invented' by the first marine cells in the Precambrian era.

Since most of the work of osmoregulation is performed at the level of individual cells, marine invertebrates require only very simple osmoregulatory/excretory organs, generating a gentle flow of fluid from which any nitrogenous wastes (those that have not diffused away from the gills or general body surface) can be lost. The marine mussel *Mytilus* and a marine ostracod crustacean provide simple examples in Fig. 9.6, with collecting areas leading via short unspecialized ducts to the discharge point; the octopus is somewhat unusual in having nitrogenous wastes added to the urine from a separate area distinct from the filtration site. Marine animals commonly produce a daily urine volume equivalent to 3–15% of their body volume, the volume being particularly low in stenohaline crustaceans (3–5%); this is in marked contrast to daily production of 15–50% in many euryhaline intertidal animals (Chapter 10).

9.3 Thermal adaptation

9.3.1 *The thermal environment of the seas*

The seas are remarkably stable in thermal terms, due to their enormous volume, the high heat capacity of water, and the currents that tend to disperse any local cool or hot spots. Therefore they again provide the easiest of all environments for their inhabitants. Heat can only be gained geothermally (important only in very localized areas near hot 'vents') or by surface absorption from the sun, particularly at longer wavelengths. Sea temperatures therefore have a much narrower range than air temperatures. Salt water would freeze at –1.86°C, but formation of a surface ice layer in very cold habitats buffers the deeper waters and prevents substantial deep freezing. The deep waters where the sun's radiation cannot penetrate, below the levels of continental shelves and making up 90% or more of the total marine living space, are at a constant 1–4°C all year round. They are kept very cold because of inputs from oceanic bottom water. This cold water (always just above –1.86°C) is formed near the polar ice-sheets, where it sinks due to its high density and flows towards lower latitudes beneath the surface currents of warmer, less dense water that flow towards the poles.

Surface waters, especially over the shelves, may therefore be both warmer and thermally more variable. Since warmer water

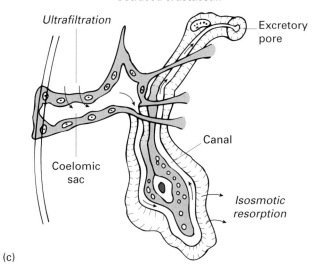

Fig. 9.6 A variety of kidneys from marine invertebrates. In molluscs (a,b) and crustaceans (c) there is a filtration system (from heart to pericardium in molluscs) followed by a short tubule where volume may be reduced by isosmotic resorption, but little or no change of concentration occurs. A terminal bladder or sac may be present. In the cephalopod (b) the renal appendages are a site for extra diffusion of NH_3 into the sac, where it is held by conversion to NH_4^+ by the pH conditions.

Fig. 9.7 (a) Patterns of temperature gradient in the upper ocean showing seasonal and permanent thermoclines. (b) The associated patterns of mixing and stratification for the open ocean. The strongly mixed layer extends only to the top of the diurnal thermocline, though there may be some deeper mixing where night-time temperatures are very low and at the onset of winter (the 'autumn overturn').

is less dense than cold water and tends to float on top of it, mixing between different thermal masses of water is reduced, which may give rise to stratification of the water and the existence of a **thermocline** (Fig. 9.7a). However, cycles of mixing and stratification can occur on a daily or seasonal basis to modify the thermocline. These cycles result from the interactions between heat input, night-time convection and day-time wind-driven turbulence (Fig. 9.7b), the last of these being augmented in shelf areas by the extra turbulence generated by tides interacting with the seabed.

Thus in regions where radiation is strong all year round the thermocline is permanent and lies quite deep down, giving a large living space of fairly warm water in tropical seas; here the surface water is usually 24–28°C throughout the year. With increasing latitude the magnitude and depth of the thermocline decrease, although the density gradients associated with a stable thermocline do help to resist penetration by turbulent eddies of water. Nevertheless in temperate zones the seasonal permanence of a thermocline is usually partly lost, and there may be an 'autumn overturn' with fairly sudden mixing as the thermal and density gradients break down. In such zones we see a temporary 'summer' thermocline superimposed on the permanent one (Fig. 9.7a). In polar seas there is usually no thermocline at all.

However, even with a thermocline the temperature range for many sea areas may be only about 5–8°C through a year, and it is normally only 2–5°C in the tropics and 4–6°C in Arctic waters or

less than 1°C in the Antarctic. Only in some temperate zones with temporary thermoclines over continental shelves is there a really marked seasonality; the western shores of the UK are a good example of this, with sea temperatures of 2°C in January rising to as high as 18°C at the end of a good summer. In such zones the mixing in the warmed surface layers may be very limited.

A global survey of the mean annual sea temperatures for surface waters (conveniently measured using cameras from satellites these days) is shown in Fig. 9.8; values can vary from about 30°C in the tropics to −2°C around Antarctica. Note also that in enclosed areas of sea the recorded means and extremes can be exceptional: temperatures of up to 40°C are found in shallow seas such as the Red Sea and Arabian Gulf, whilst enclosed high-latitude seas such as the Baltic and Hudson Bay may freeze over substantially in winter.

Superimposed on fundamental latitudinal effects the world's oceans also show pronounced local variations in temperature in both surface and deep waters due to particular currents determined by geomorphology and crustal plate positions. The North Atlantic Drift/Gulf Stream is a classic case, where the warm surface current from the southwest keeps western shores of Europe milder than expected. In terms of marine animals' adaptations it is therefore important to know the local thermal situation and water current patterns through the year rather precisely.

9.3.2 *Temperature in marine animals*

The vast majority of marine animals live in water that is below 5°C all the time. Equally the great majority of them are ectotherms and have body temperatures that are within a degree or so of their surroundings at all times and thus essentially invariant. Most of these animals are also stenothermic, and cannot tolerate artificially imposed changes in temperature.

Indeed, virtually *all* the primary inhabitants of sea water are

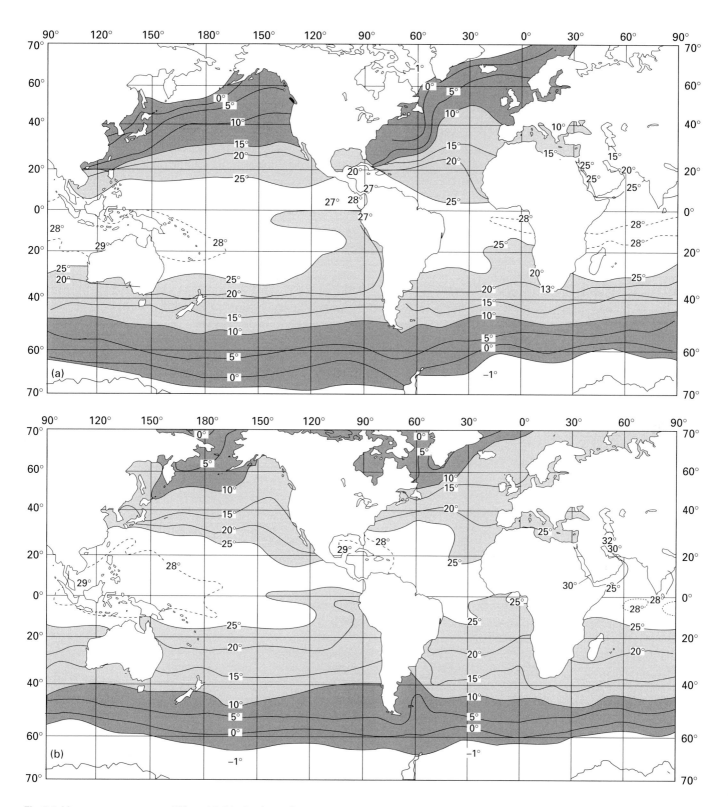

Fig. 9.8 Mean sea temperatures (°C) worldwide, for the surface waters in (a) February and (b) August. (Adapted from Ingmanson & Wallace 1989.)

essentially ectothermic, partly because water is a sufficiently good conductor of heat that marine endothermy is particularly hard to maintain, with any internally generated heat being rapidly dissipated at the body surface, especially at the large-surface-area gills that are associated with active marine life. It is normally only some secondary invaders, with their previous history of life on land, that show true endothermy. Even then only relatively large endotherms such as the cetaceans (whales and their kin), with highly efficient insulation and no gills, can achieve permanent residency in the sea with a raised body temperature (T_b) (see section 9.10); though it seems likely that large marine dinosaurs and other Mesozoic marine reptiles may also have achieved this.

Thus, for one reason or another, the great majority of marine animals have a marked constancy of T_b and little need for acclimatory or regulatory strategies. However, some pelagic ectotherms may experience variation in T_b in association with changing environmental temperature, especially in migrating species. Diurnal migrations from depths of 400–800 m by day to the top 100 m by night are rather common for planktonic species, and in the tropics this may mean a temperature change of 15°C or more in less than 2 h. For larger nektonic crustaceans and fish there may be latitudinal migration, or movement from mid-ocean to offshore regions following food supplies, or seasonal migration through a thermocline; for example, some sculpin species and other fish migrate to deep water in winter to avoid freezing surface waters.

9.3.3 Adaptive mechanisms

For all those animals dwelling at a particular latitude and below the surface waters, thermal acclimation is not a problem, and cell machinery and enzyme biochemistry can be adapted to work optimally over a very narrow and rather cold temperature range, with little need for isozymic variation or for allosteric modulators (see section 8.2.1). Many cases of evolutionary adaptation of the cellular machinery to particular environmental temperatures are well documented. The myosins or 'molecular motors' from a range of fish are perhaps the best known. The myosins from typical fast muscles can all generate similar stresses when operating at the normal habitat temperature for each species, although function is lost at unphysiological high or low temperatures. Some interspecific adaptations in the structure of the molecule are evident in relation to the ease at which enzymes can be thermally denatured; Fig. 8.3 showed that the myosins from Antarctic fish have denaturation half-times about 100 times less than those recorded in fish from the Indian Ocean. This reflects the more 'open' character of cold-adapted marine enzymes compared with the rigid, highly bonded structures of enzymes from animals adapted to other aquatic (and terrestrial) biomes. However, the maximal contraction speed and power output of the intact fast muscle fibres from these species is approximately a function of the water temperature at which they are tested, providing little evidence for interspecific adjustment in performance at different latitudes.

One clear example of adjustment of enzyme function across a latitudinal range is afforded by the lactate dehydrogenases in the eastern Pacific barracudas (genus *Sphyraena*). There are three species having similar morphologies and ecologies but different and overlapping temperature ranges. The turnover numbers (K_{cat}) of their muscle lactate dehydrogenase (M_4-LDH) enzymes at 25°C are inversely related to average habitat temperature (T_m), such that K_{cat} values are similar at the physiological temperature of each species (Fig. 9.9a). The figure shows that

(a)

(b)

Fig. 9.9 Enzyme property variation with temperature in marine fish. (a) Latitude effects, shown here for three species of barracuda (*Sphyraena*) with different distributions in the eastern Pacific. Values of K_{cat} and K_m for M_4 lactate dehydrogenase are given, both at 25°C for all species (circles) showing latitudinal variation, and also at the normal environmental thermal midpoint for each (squares) showing much more consistent properties. (b) Seasonal effects. K_m values for malate dehydrogenase from the goby, *Gillichthys*, in winter and in summer. Dotted lines show the normal body temperatures of the fish in that season, indicating conservation of K_m at different seasons. In fact the fish contains two isozymes of this enzyme in both seasons, but the ratio of the thermolabile form to the thermostable form varies to produce the differences shown. (b, From Somero *et al.* 1996.)

under physiological assay conditions, K_m values for pyruvate (the substrate for LDH) in these muscles are also highly conserved. DNA sequencing studies have shown that interspecific differences in the properties of the barracuda LDHs involve only a few base substitutions, all coding for amino acids remote from the substrate- and ligand-binding sequences, but nevertheless affecting the conformation of the ligand-binding pockets.

Membrane adjustments are also necessary to achieve efficient functioning at particular environmental temperatures, as we saw in Chapter 8. Studies of a range of crustaceans from polar waters indicate that both branchial and neural membranes are more fluid (less 'ordered', with more unsaturated lipids) than those from temperate relatives, with the neuronal membranes being even more fluid than the gill membranes.

Acclimating to different temperatures

Relatively few cases of thermal acclimation in marine animals are described as yet. Most aquatic ectotherms prefer a thermal niche close to their normal ambient water temperatures. Large inshore crustaceans, such as lobsters, do show a degree of behavioural thermoregulation, selecting water temperatures of 13–19°C that are a little above their mean ambient temperature, with the exact preferred temperature varying by 1–2°C according to acclimation history. They may move into warmer waters in spring to enhance their growth and reproduction rates. Some fish are also known to exhibit seasonal acclimation of their general metabolic machinery, by increasing the concentration of many enzymes and of metabolites in cold periods; in at least some species the mitochondrial volume in cells also rises to ensure an adequate supply of ATP. Such fish also show specific patterns of heat-shock protein (HSP) synthesis, with induction temperature varying between species, and with concentration varying between species, individuals and, indeed, between tissues within an individual.

Ciliary function should also be under strong selective pressure, since it underlies respiration, feeding and larval dispersal in many invertebrates. Thermal latitudinal clines in ciliary function have been reported in some bivalves. In the common oyster, strains from colder temperature regimes have more active cilia at lower temperatures than do strains from warmer temperature regimes.

The clearest examples of the physiology of acclimation again come from work on muscle enzyme function in fish, operating at different acclimated T_b values; for example, Fig. 9.9b shows seasonal acclimation of K_m values in malate dehydrogenase from a goby. With any fish, initial exposure to reduced temperatures gives deleterious effects on swimming speed and endurance, but with prolonged exposure there is an improvement in swimming performance. This reflects numerous adaptations in nerve and muscle properties, including an increase in the amount of red muscle and in the density of mitochondria within muscle fibres. For example, the maximum speed attained by the long spine sea scorpion, a bottom-living ambush predator from coastal areas of northern Europe, increases at both low and high temperatures following a period of acclimation lasting several weeks. At 15°C (this fish species' maximum summer temperature), the fast muscle fibres of a 15°C-acclimated fish are capable of generating more than three times the power of muscles from a 5°C-acclimated fish. This is sufficient to increase the rate of successful predation strikes on prey (shrimps) from 23% to 73%. The mechanism involves altered expression of myosin light-chain subunits, which modulate the maximum contraction speed of muscle fibres. The composition of myosin heavy chains, and the ATPase activity, are unchanged by temperature acclimation in this species. This contrasts with the situation in freshwater cyprinid fish, where thermal acclimation modulates both the maximum tension generated and the contraction speed of muscle fibres, via altered expression of both heavy- and light-chain myosin genes.

Clines of genetic loci also occur, within populations where little individual acclimation can be observed, so that the proportions of different polymorphic forms at a given enzyme locus appear to vary with temperature and, consequently, often with latitude. An oft-quoted example (see Chapter 8) occurs in the common killifish, *Fundulus heteroclitus*; the occurrence of different M_4-LDH enzyme morphs is apparently correlated with latitude along the eastern seaboard of the USA, with the allozyme that is commonest at northern latitudes being the best catalyst at low temperatures. However, more recent work on mitochondrial DNA in this species has revealed that there are in fact two rather distinct races, northern and southern, with a boundary between the two lying off the coast of New Jersey; it seems likely that here the earlier allozyme work has been over-interpreted in terms of a neatly selected cline.

Regional endothermy in large ectothermic vertebrates

The major exception to ectothermy in primarily marine animals occurs in a few very fast active fish that come into the category of 'regional endotherms' as described in section 8.4.1. The ability to maintain elevated body temperatures is known in just 27 species of large oceanic fishes, from the Scombroidei (mackerels, tunas, bonitos and billfishes) among the teleosts and from the Lamnidae (mako, porbeagle and white sharks) among the elasmobranchs. Conditions for regional endothermy are a large body size, a heat source, and heat-exchangers to conserve the heat. In all cases muscles or modified muscles provide the source of heat.

The tuna and some of its relatives can retain endogenous heat within the core of the swimming musculature, where the temperature may be more than 10°C above the surrounding water temperature (Fig. 9.10a). In contrast with most terrestrial endotherms, they do this by regulation of heat retention and not of heat production, the latter being independent of water temperature. The red muscle used for cruising in tunas is abundant and is internalized close to the body core. In most teleosts the blood supply to the red muscle comes from a central dorsal aorta and the venous blood is returned to a central postcardinal vein. In contrast, the blood supply to red muscle in most tuna is

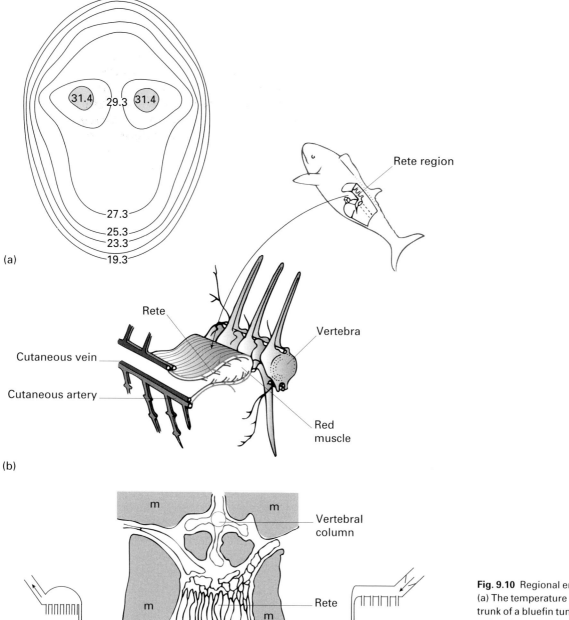

(a)

(b)

(c)

Rete region

Rete

Cutaneous vein

Cutaneous artery

Vertebra

Red muscle

Vertebral column

Rete

Blood flow H E A T Arterioles of rete

da

m m

m m

da

pcv

Kidney Kidney

Blood flow H E A T Venules of rete

pcv

Fig. 9.10 Regional endothermy in tuna. (a) The temperature profiles across the trunk of a bluefin tuna, showing the 'hot' swimming muscle at the core. (From Bone *et al.* 1982.) (b) The paired lateral 'cutaneous retes' of the big-eye tuna, where the cutaneous artery and vein run in parallel so that heat generated centrally is exchanged between the two vessels and retained at the core. (c) The central rete of the skipjack tuna, where a capillary bed beneath the vertebral column contains arterioles and venules running in opposite directions (up from the dorsal aorta (da), and down from the postcardinal vein (pcv)—see insets) between the main muscle blocks (m).

delivered via subcutaneous arteries, and the venous blood is also returned to subcutaneous vessels. Heat (resulting from the inefficiency of chemical energy conversion to muscle contractions) is therefore retained by a classic countercurrent heat exchanger system (Fig. 9.10b); warm blood leaving the muscle is used to heat the incoming cold blood. There is evidence that the internalization of red muscle preceded the evolution of vascular countercurrent heat-exchangers and predisposed this taxon for conserving metabolic heat from muscle. The bonitos are the ectothermic sister group to the endothermic tunas, and some species have a degree of red muscle internalization intermediate between tunas and other scombroids.

The countercurrent rete design varies somewhat in different species; larger forms, like the bluefin tuna, have a cutaneous rete as described above, where the peripheral artery and vein enter and leave the muscle mass as dense interdigitating sheets of arterioles and venules (Fig. 9.10b), whilst smaller species, like the yellowfin tuna, have multiple central retia where cool arterial blood from the gills meets warm venous blood draining inwards from the surrounding muscle (Fig. 9.10c). The exchanger is always arranged to keep heat in the core of the trunk muscles, ensuring that up to 90% of the heat from the venous blood flowing out from the muscle is returned to the arterial blood (note that since heat-exchange is markedly faster than oxygen-exchange the arterial system can still maintain its normal oxygen delivery function). Thus although this regional endothermy keeps these core locomotory muscles above ambient, the surface temperature of the tuna is close to that of its surroundings, with a sharp temperature gradient across the body. The raised temperature of the swimming muscles gives the tuna a higher sustainable speed than a non-endothermic fish of equivalent size, though maximum speeds remain similar and neither is there much improvement in time to exhaustion during a burst of activity.

There is little evidence that most tuna species can regulate their body temperature by physiological means; the temperature difference between the tuna and its surroundings does not change much as ambient temperature changes (Fig. 9.11). However, the large bluefin tunas do have considerable thermal inertia, which helps to stabilize T_b so that rapid changes in water temperature are not reflected in changes in body temperature. They may therefore thermoregulate behaviourally by operating in sea temperatures (T_a) that result in an acceptable excess temperature at routine cruising speeds. Tunas caught by anglers and played on fishing lines so that they cannot select a suitable T_a may actually overheat, producing what is known as 'burnt tuna', of much reduced commercial value.

A variety of other marine fish from the mackerel and shark groups also use regional endothermy. Certain sharks have elevated visceral temperatures, routing the arterial blood via a rete close to the liver; indeed, recent evidence indicates that the white shark, *Carcharodon carcharias*, is a remarkable regional endotherm with a stomach temperature up to 14°C above ambient and strongly regulated, allowing this and related species to

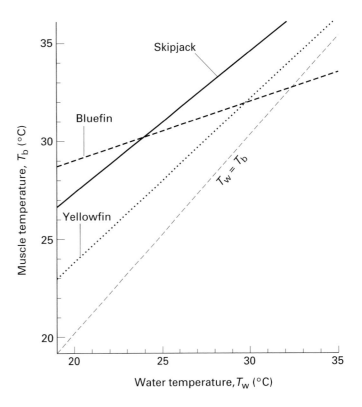

Fig. 9.11 The relation between water temperature and muscle temperature for some tuna species, showing their ability to maintain a high core T_b. Note that there is often little ability to thermoregulate since the lines are parallel to $T_w = T_b$. However, the bluefin tuna (large and with an efficient cutaneous rete arrangement) does regulate efficiently (see text). (From Carey *et al.* 1971.)

continue as active predators in very cold waters. Brain heaters occur in the 'billfishes', comprising the families Xiphiidae (swordfish) and Istiophoridae (marlins, sailfish, spearfish), and independently in a primitive member of the tuna family, the butterfly mackerel, *Gasterochisma melampus*. Endothermy in billfish is restricted to the warming of the brain and eyes. Swordfish spend most of the night at the surface, but during the day undergo vertical excursions as deep as 600 m, thus encountering water temperature changes of up to 19°C for short periods (Fig. 9.12). Telemetry of free-swimming fish reveals that cranial temperatures remain above, and buffered from, short-term decreases in water temperature. In the swordfish the brain temperature elevation is partly brought about by a carotid rete arrangement, but it also involves a so-called 'heater organ'. This is derived from a group of small muscles used to control eye movement in other fish (the superior rectus muscles). These heater muscles from each side of the head converge at the base of the braincase; the basisphenoid bone, which forms the lower braincase, is reduced to a membranous sheath, so that the brain is embedded on to the dorsal surface of the heater organ. Large amounts of adipose tissue surround the brain and insulate the brain, eyes and heater organ. The amount of heater tissue varies between species and correlates with the ability to maintain

Fig. 9.12 Telemetric recordings of depth in the swordfish *Xiphias gladius*, showing its diurnal movements, upward at sunset and downward at sunrise, exposing it to substantial water temperature change.

elevated brain temperatures; a 40 kg swordfish has around 60 g of heater tissue. At the base of the heater a countercurrent heat-exchanger, formed from the carotid artery and the venous return from the heater, prevents dissipation of the heat from the head region.

Heater cells in these fish are packed with mitochondria (63% by volume), contain high concentrations of myoglobin, and have an extensive internal smooth membrane system (tubules, cisternae and membranous stacks, equivalent to the sarcoplasmic reticulum (SR) of normal muscle) making up 30% of the cell volume. Thermogenesis appears to involve the uptake and release of calcium from these SR membrane systems, providing a version of futile cycling and thus an energy 'short circuit'. Release

of calcium through the SR Ca^{2+} channel promotes uptake back into the SR via the Ca^{2+}-ATPase, thus using ATP generated by the mitochondria and producing heat as a by-product. Calcium may also stimulate the calcium–proton transporter of the mitochondria (Fig. 9.13). The calcium coupling ratio of the SR (i.e. the calcium ions pumped per ATP) is low compared with normal muscle, so that more heat is generated per calcium ion transported. Thus heater organ thermogenesis is rather different from other systems; in particular, the electron transport chain in the billfish heater organ mitochondria is tightly coupled to oxidative phosphorylation, contrasting with the uncoupled mitochondria of mammalian brown adipose tissue (see section 8.7.2).

Endothermy must have evolved at least three times within the

Fig. 9.13 A model for calcium-related thermogenesis in the heater cells of certain fish. Calcium is released from the sarcoplasmic reticulum (SR) (possibly in response to excitation from the axon, as in normal muscle) then recycled back into the SR by the calcium ATPase; thus ATP is split and heat released. Some calcium may also be taken into mitochondria, where it stimulates the calcium–proton transporter and generates heat via the electron transfer system. (Adapted from O'Brien & Block 1996.)

Scombroidei, and cranial endothermy involving a thermogenic organ must have evolved twice (once in the billfish and once in the butterfly mackerel). The repeated evolution of endothermy suggests that there is a strong selection for this energetically expensive metabolic strategy. However, the advantages of regional endothermy in all these fish are not entirely obvious. It is not clear that muscle power or sustained swimming speeds are any greater in fish such as tuna than in any other tropical fish of the same size and habit. However, they can penetrate below the thermocline into cold water without suffering the same drop in temperature as other fish. Thus pursuit speed and predator–prey interactions may be rather independent of rapid changes in water temperature, allowing these fish to be versatile predators at the top of the marine food chain. Recovery from exercise may also be quicker than in other fish, which may be an advantage for foraging in an environment where prey are patchily distributed. Tunas have relatively small guts, and a high visceral temperature may be advantageous in speeding the rate of food digestion and assimilation. The bluefin tuna and salmon sharks, the warmest of the endothermic fish, have ranges which extend far to the north, enabling them to exploit the food-rich seas of temperate and subpolar latitudes and to achieve rapid somatic and gonadal growth. Buffering of brain temperature in the billfish may also allow them to forage in relatively cold water, permitting extended periods of foraging beneath the thermocline, opening up the resources of the mesopelagic zone to exploitation. Thus in most cases the selective pressure driving the evolution of endothermy may well have been niche expansion.

Finally, some large aquatic turtles also have marked regional heterothermy. Metabolic heat production from the muscles is again conserved by countercurrent heat-exchangers. Adult turtles have significant thermal inertia due to their large size, and the insulative properties of the enclosing carapace also aid heat retention. Leatherback turtles can generate local excess temperatures of up to 18°C. Again the advantages are not entirely clear, but the locally raised T_b may increase swimming speed or endurance, helping them to complete their long migrations.

9.4 Respiratory adaptation

Oxygen occurs at relatively constant concentrations in sea water; levels of 4–6 parts per million (equivalent to < 6 ml l^{-1}) are usually recorded, although the deep northern Pacific Ocean yields only 2–3 ml l^{-1}. Only in a few relatively enclosed seas does substantial deoxygenation of the bulk of the water ever occur. However, because oxygen has a somewhat lower solubility coefficient in warm than in cold waters, there may be rather less oxygen available for pelagic animals at the surface of the sea than there is at depth, especially at low latitudes. Furthermore, an **oxygen minimum layer** (OML) often occurs, somewhere between 400 m and 1000 m, where biological oxygen usage is greatest in relation to the replenishment of oxygen by convection. Such layers are normal in areas of highly productive tropical oceans (especially the eastern Pacific and northern Indian oceans, where the oxygen concentration may fall as low as 0.1–0.5 ml l^{-1}), and in some temperate seas. At points where the OML abuts on continental margins, stable hypoxic conditions also exist for the benthic communities.

By contrast, carbon dioxide is exceedingly abundant in sea water, and forms about three-quarters of all the gas dissolved in the sea (compared to only 0.04% of the gases in air). Most of the carbon dioxide is dissolved as bicarbonate ion (HCO_3^-), which has an important role as a major buffer of sea water. Expired carbon dioxide can readily be added to this pool without creating any difficulties for the respiring organism.

Respiration is therefore not too much of a problem for most marine animals, in terms of acclimation to changing availability of gas. The exceptions are:

1 pelagic (and some benthic) animals exposed to the oxygen minimum layers;
2 benthic animals where decomposing sludge and unstirred waters can produce an almost anaerobic environment;
3 animals that migrate, whether vertically or latitudinally, diurnally or seasonally, between these problematic environmental zones;
4 some fish and most endothermic vertebrates, where metabolic rate tends to be high, and for whom bursts of swimming can lead to temporarily anaerobic conditions at the muscle level.

The major physiological problem in all these cases is to avoid hypoxia, by establishing a respiratory system with enough exposed and/or ventilated surface area to extract the requisite oxygen by diffusion; the relation between this necessary surface area and body size was considered in general terms in Chapter 7 (see section 7.3.2). The respiratory surface area needs to be quite high for marine animals because water is viscous and thus quite difficult to move compared to air, tending to form a thick boundary layer next to the respiring surfaces. Large surfaces are also needed because sea water has a rather low oxygen content. These problems are partly compensated for by the possibility of having unidirectional flow of water over the exchange surface, rather than the bidirectional tidal system used in lungs. Nevertheless a marine animal must move about one litre of water to get 1 ml of oxygen, compared to only 25 ml air per 1 ml oxygen for an air-breather.

The general body surface is certainly the primitive site of gas exchange, and may still be the only surface used in smaller animals. In benthic and moderately active marine invertebrates the larger requisite diffusive surface can be achieved by a highly perforated body structure (e.g. sponges, tunicates); or by frilled tentacles or gills (many worms and molluscs, but also many 'simpler' animals such as sea anemones); or by dermal papillae (some anemones, many starfish). Within just one group, the echinoderms, different species can be found using papillae, tube-feet, invaginated respiratory trees, and even the gonad cavities as sites of oxygen uptake.

Examples of some of the more elaborate respiratory structures used in a range of marine invertebrates are shown in Fig. 9.14. In

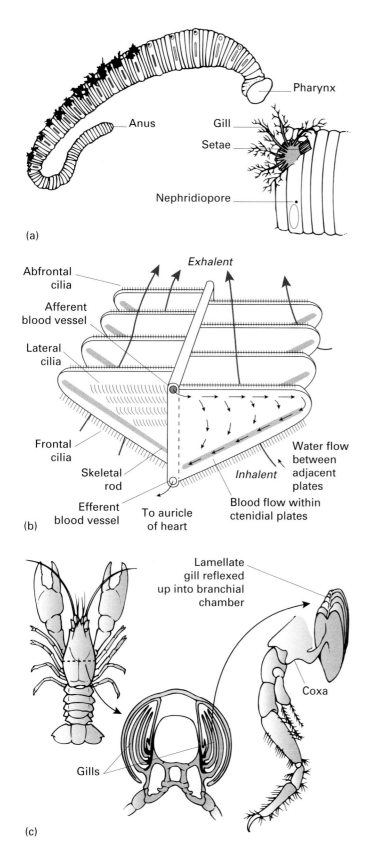

(a)

(b)

(c)

Fig. 9.14 A variety of respiratory structures in marine invertebrates: (a) polychaete; (b) gastropod mollusc; and (c) crustacean.

many molluscs and crustaceans very elaborate filamentous or lamellar gills occur. In crustaceans these are borne on the limbs, where the covering of chitinous cuticle somewhat increases the diffusion path-length (to 10–15 µm) and thus reduces rates of diffusion. However, the more active malacostracan crustaceans, such as prawns and crabs, have these gills tucked under a protective carapace, and ventilate the resulting branchial chamber either by gill movements or by using the scaphognathite baler on an anterior limb to give a one-way flow (see Fig. 7.9); the frequency of beating by this structure is the main determinant of water flow. In relatively active invertebrates, such as swimming crabs and cephalopods, the gills and their vascular supply are arranged to give countercurrent flow (e.g. Fig. 9.14b). Gill areas in most of these reasonably active animals generally scale with body mass to the power 0.7–0.8 (Fig. 9.15), giving a figure of about 5–10 cm² g⁻¹.

Marine fish may achieve some gas exchange through a cutaneous route (usually only about 5% of the total uptake), but invariably breathe principally with gills, sometimes external in embryos but always more or less internal in adults. Marine fish

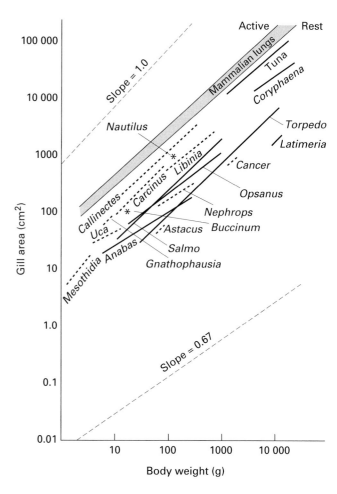

Fig. 9.15 The relationship between gill area and body mass for a range of crustaceans (dashed lines), molluscs (*) and fish (solid lines). (From Withers 1992.)

Fig. 9.16 The structure of fish gill lamellae showing the countercurrent flow of water and of blood.

gills have short diffusion distances and therefore need a relatively lower surface area than many similarly sized crustaceans (Fig. 9.15), with a range of about 1–15 cm^2 g^{-1}, the higher values occurring in active forms such as tuna. These gills are ventilated with an organized one-way flow of sea water in most species, over a complex arrangement of filaments and lamellae (Fig. 9.16). Blood circulation is normally such as to give countercurrent flow, with water flow rates being substantially faster than blood flow rates to achieve maximum difference in oxygen partial pressure (Po_2) across the diffusing surface. Path lengths from water to blood may be only 2–5 μm in such gills. Ventilation may be through a spiracular opening, by active pumping from buccal and opercular muscles (see Fig. 7.10), or by 'ram ventilation' achieved by fast forward swimming with the mouth open. Some fish use opercular pumping at slow speeds and ram ventilate when swimming faster and consuming more oxygen.

But achieving adequate oxygen uptake is still going to be particularly problematic for those species that are active swimmers with high demand, or those that inhabit the OML or experience marked temperature change. Animals from the OML may show three key approaches to cope with the reduced oxygen gradient into their tissues: increased oxygen extraction efficiency, reduced metabolic rate, or the use of anaerobic pathways. These may be used singly or in concert. For example, ctenophores in the OML probably cope because most of their oxygen demand is at the body surface in the comb-rows of cilia, and inward diffusion to the mitochondria is fast enough to supply them. Pelagic crustaceans from the OML off California show a more sub-

stantial suite of features that improve oxygen removal from the water: enhanced ventilation, larger gill surface areas, reduced diffusion distances across gill surfaces, and higher-affinity haemocyanins. These crustaceans therefore rarely have to use sustained anaerobiosis. Pelagic crustaceans, such as euphausid shrimps ('krill'), that may experience moderate seasonal temperature change may have haemolymph with oxygen-binding properties suited to take up oxygen at lower Po_2 in summer than in winter (Fig. 9.17). For many marine fish modification occurs at a structural level, so that there are clear correlations between normal activity levels and various aspects of gill surface areas (Table 9.5). In relation to the fast-swimming fish referred to in the preceding section though, it is worth noting that periods of anaerobic activity are rather common. The marlin, for example, which often pursues and preys on tuna, has a 9 : 1 ratio of white to red muscle and can tolerate very high tissue acidity after a long chase and prolonged anaerobic work; lactate is well buffered in the blood, and there is a remarkably pronounced Bohr effect (see section 7.5.2) to overcome this buffering and permit oxygen delivery when aerobic respiration is again possible. The tuna itself uses anaerobic activity effectively, and can recover from the associated acidosis in only 1–2 h, where other fish lacking warm musculature may take 8–24 h to clear lactate build-ups.

When structural sophistication is not enough and marine animals are faced with a degree of hypoxia, most (whether invertebrate or fish) will show two physiological responses: an increase in the ventilation rate, and/or a slowing of gill perfusion, usually achieved by slowing the heartbeat (bradycardia). In the molluscan sea slug *Aplysia*, for example, mild hypoxia produces an increase in respiratory pumping as the animal attempts to maintain its oxygen uptake (behaving as an 'oxygen

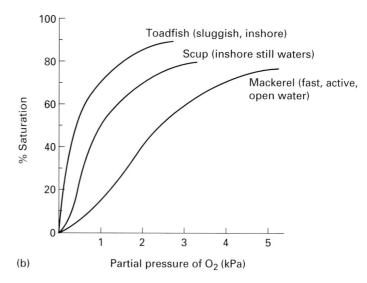

Fig. 9.17 Oxygen binding curves for the haemolymph of (a) euphausid shrimps (krill) at 5°C and at 10°C, and (b) various fish in relation to depth and activity. (a, Adapted from Brix *et al.* 1989; b, from Stewart 1991.)

Genus	Body mass (g)	Activity	No. of filaments	No. of lamellae (mm⁻¹)	Gill area (cm² g⁻¹)
Crustaceans					
Callinectes		High			14
*Uca**		High			6
*Ocypode**		High			3
Panopeus		Moderate			9
Libinia		Low			6
Fish					
Lucioperca	70	High	1 811	15	18
Salmo (salmon)	390	High	1 606	19	2
Katsuwonus (tuna)	3 258	High	6 066	32	14
Thunnus (tuna)	26 660	High	6 480	24	9
Pleuronectes (plaice)		Moderate			4
Callionymus (dragonet)	3	Low	478	16	2
*Saccobranchus**	42	Low	658	23	1
*Anabas** (climbing perch)	54	Low	567	21	1
Ictalurus (catfish)	239	Low	–	10	1
Opsanus (toadfish)	251	Low	660	11	2
Tinca (tench)	268	Low	1 764	22	2

Table 9.5 Gill structure and activity levels in marine animals.

*Indicates fully or partially air-breathing species.

regulator', see Chapter 7); but if hypoxia persists or worsens its pumping rate declines and it becomes an oxygen conformer, entering a hypometabolic state. The capacity to control ventilatory or circulatory changes is greatest in species that routinely migrate through the regions of lower oxygen tension or that use bursts of sustained activity for predation or escape. One specialist at dealing with relative hypoxia is a continuous inhabitant of the OML, the crustacean *Gnathophausia ingens*, which has unusually large gill surfaces with very thin walls (hence low diffusion distances), an unusually fast circulation for a crustacean, and a

rather low routine metabolic rate. It also possesses a haemocyanin with high affinity and a large Bohr shift. This and other crustaceans from the same habitat can take up oxygen from water with a Po_2 of only 0.4–0.8 kPa. Other inhabitants of this layer, including some fish, are probably largely anaerobic for much of the day, and migrate to the surface at night. In common with other fish they may have a store of 'extra' red blood cells in the spleen which can be mobilized if hypoxia is a problem.

While respiration in the marine environment may be fairly straightforward, and relatively unaffected by depth in a direct

Table 9.6 A comparison of respiratory performance in marine animals.

Genus	Mass (g)	O$_2$ consumption rate (ml O$_2$ min^{-1})	Gill water flow (ml min^{-1})	P_{O_2} excurrent (kPa)	O$_2$ extraction efficiency (%)
Molluscs					
Octopus	10 000	2.8	2 050	12.5	27
Crustaceans					
Callinectes (crab)	200	0.206	111	9.7	53
Homarus (lobster)	460	0.225	224	15.4	23
Cancer (crab)	900	0.504	288	13.6	34
Fish					
Callionymus (dragonet)	100	0.108	30	10.6	55
Scyliorhinus (dogfish)	200	0.134	24		40–70
Salmo (trout)	210	0.193	31	11.1	46
Acipenser (sturgeon)	900	0.86	380	14.0	30
Katsuwonas (tuna)	2 250	13.3	4 000	14.5	56

sense, it should be noted that there may be specific physiological problems associated with anaerobic zones and with the general sensitivity of respiratory pigments to high pressure and to low temperature. Animals living in low-oxygen benthic zones usually have oxygen dissociation curves well to the left on a standard plot (see section 7.5.1). Respiratory pigments in such animals can also show highly specific (adaptive) responses to temperature, pressure, carbon dioxide and pH levels, according to the precise nature of their habitat. Deep-sea animals therefore often have haemoglobins or haemocyanins that are unusually temperature-independent (e.g. the tuna, and a deep-sea vent crab, *Bythograea*), while others use oxygen-affinity modifiers such as organic phosphates or lactate ions to bind to the pigments and modify their properties as needed. The deep-sea shrimp *Glyphocrangon* has a haemocyanin that is virtually insensitive to temperature but markedly influenced by pH, probably to allow a substantial Bohr shift for unloading oxygen to transiently active tissues.

In summary, the respiratory performance of a range of marine animals is compared in Table 9.6. The most telling information perhaps comes from the values of efficiency of oxygen extraction shown there, this factor varying from around 20% in some relatively inactive invertebrates and benthic fish to well over 50% in some crabs and in fast-swimming fish with good countercurrent organization.

9.5 Reproductive and life-cycle adaptation

Marine animals use both asexual and sexual modes of reproduction, with asexual processes being more abundant than in any other habitat, perhaps favoured by the stability and predictability of the habitat, where the genetic variation arising from recombination is less vital. A vast range of 'lower' invertebrates use systems such as budding to reproduce, either budding off free-living asexual larvae (e.g. in many anthozoans and scyphozoans) or budding incompletely to produce attached clonal organisms, as in so many colonial animals (e.g. hydroid corals, bryozoan sea mats and tunicate sea squirts).

Sexual reproduction is present at some point in the life cycle of virtually all animals, however, and the process must include gametogenesis, gamete release and/or sperm transfer, and fertilization with a gamete from another individual. The majority of marine animals have simple uniflagellate sperm (presumed to be the 'primitive' form) and relatively small buoyant eggs (oocytes). Rather large numbers of marine invertebrates (and a few fish, such as wrasse) are hermaphrodite. Some have an 'ovotestis' that produces both male and female gametes at the same time or sequentially, although they tend to exchange gametes with another individual rather than self-fertilize. Other animals have a progression within their own lifetime (see Box 9.1) from young male to old female (protandrous hermaphroditism) or less commonly young female to old male (protogynous hermaphroditism). The switch between sexual phases may be environmentally (seasonally) determined, or may depend on received signals (pheromones) from nearby potential mates.

9.5.1 Pelago-benthic alternation

There are a number of different reproductive options available to marine animals, ranging from broadcast spawning in massed populations to careful one-to-one copulations and viviparity, each of these having its own advantages and problems. Most marine animals have a larva (Fig. 9.18) as the pelagic stage of the life cycle, with relatively few hatching directly as juveniles (small adults) without metamorphosis. An alternating pelago-benthic lifestyle therefore predominates. This may be because it allows small larvae to feed on seasonal surface plankton

Box 9.1 Hermaphrodites and sex change in marine animals

Explanations of sex change in animals are based on the fact that the reproductive success achievable as a male or a female changes with body size.

Male success via sperm can begin in very small individuals, because tiny sperm are cheap to produce and easy to store in large numbers. Reproductive success for males increases steeply with reproductive investment, as sperm released into the sea can potentially reach many females. Beyond a certain level, however, further investment fails to produce more success, perhaps because the limited longevity of sperm puts an upper limit on the distance over which they can disperse and still retain fertility.

Females produce eggs which are much more expensive than sperm, so that a larger minimum body size is required for female reproductive function. Once this minimum size is reached, female function increases more or less linearly with body mass, as more eggs can be made and/or stored, and there is no clear upper limit to this.

The relationships of reproductive success to size for males and for females are therefore as shown below:

The lines for males and females intersect at some particular body size/investment value. For maximum reproductive success through a single lifetime it therefore pays off for an individual to follow the arrowed track in the diagram, changing from male to female at the body size where the intersection occurs. This is the most common direction of sex change, and is seen in animals such as pandalid shrimps, and clown or anemone fish on tropical reefs.

Cases of female-to-male reversal tend to occur when the relation between size and reproductive success is distorted. In certain wrasse, males are territorial and big males achieve many matings, while small males cannot maintain a big territory and get very little reproductive success. Hence the preferred track for sex change is modified, and individuals become males only when they are big enough to do well:

The size at which sex reversal occurs is known to depend on the make-up of the population. For example, pandalid shrimps usually show male-to-female switches, and the size at which the switch occurs changes in at least two situations.

1 Invasion of lots of small shrimps during a good year for larvae, producing lots of small competing males; the value of being male at a given size falls, so it pays to become female at a smaller body size:

2 Enhanced mortality of large females (produced by selective fishing by man), where the success of smaller females is enhanced, again favouring the sexual switch at a smaller size:

Given the advantages of hermaphroditism, why is it not more common? It may be that the competing requirements of male and female functioning in a single body are too expensive, especially where there is marked sexual dimorphism requiring substantial morphological change at the time of the sexual switch.

blooms and grow rapidly, before becoming large and benthic. It also reduces competition between the generations, as well as allowing larval or juvenile dispersal. In addition, there are no particular 'extra' physiological burdens on small immature stages as compared to larger adults, for this is the one habitat whose equable ionic, osmotic and thermal characteristics make the surface area/volume ratios of most animals relatively unimportant. Osmotic and ionic control, respiration, and temperature relations are fairly straightforward whatever the animal's size, so that small larvae are perfectly feasible and may even have advantages in relation to buoyancy and dispersal.

Evolutionary choices as to exact reproductive mode are therefore likely to be determined more by adult habit and habitat (benthic/interstitial vs. pelagic, sessile vs. active, and so on), and by phylogenetic constraints, than by physiological considerations.

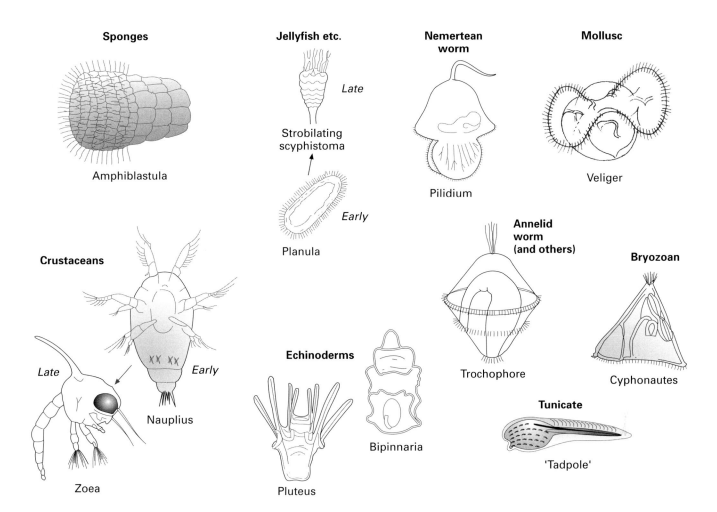

Fig. 9.18 Marine larval forms.

9.5.2 *Planktotrophic and lecithotrophic larvae*

A major distinction is usually made between these two modes of larval development. A planktotrophic larva feeds for itself on the abundant plankton, whereas lecithotrophic larvae feed off yolk reserves supplied by the mother.

This distinction has a limited range of physiological correlates. Most marine invertebrates adopt the pelagic planktotrophic option, producing very large numbers of simple gametes that are fertilized externally to give small larvae which then fend for themselves; this is the 'easiest' reproductive/life-history option, and is not open to animals in most other habitats. The major control system needed, whether a species is hermaphrodite or with separate sexes (dioecious), is an ability to accumulate and then to release gametes *en masse*, preferably in synchrony with many others of the same species ('epidemic spawning') to ensure good cross-fertilization, and if possible also in synchrony with planktonic blooms. This is commonly achieved by an integration of hormonal control systems (ultimately the major gonadotropic hormones) with environmental cues. Such cues may be limited in deep marine environments, where there is little temperature variation to give periodicity; however, recent evidence shows that deep-sea clams (*Calyptogena*) can spawn in response to very small thermal changes of just 0.1–0.2°C that occur with hydrographic water movements. For shallower waters more consistent cues are likely to be available relating to tide and to season, and it is no surprise that many marine animals use photoperiod and/or lunar cycles to trigger spawning. The best known example here is the South Seas palolo worm (*Palola viridis*), a polychaete where gametes accumulate and mature in the coelom and are then shed by rupture of the parental body wall. At a precise time of year (a night of full moon, usually September or October), the gravid individuals migrate upwards through the water column and form vast surface swarms where the gametes are shed and fertilized as milky clouds in the sea water. Mortality may be huge, as is usually the case with planktotrophic life cycles; survival rates of well under 1% are commonplace.

Lecithotrophic larvae have some inbuilt food supply, and tend to spend a shorter time in the planktonic phase, often settling out again within a few hours or days of being shed. They serve primarily to disperse the young away from the parents, without too much risk of long-range dispersal to entirely unsuitable habitats. The strategy is quite common in deep-water animals from cold and rather unproductive seas, where plank-tonic food organisms are relatively scarce; fewer gametes are released, and each fertilized egg has a better chance of survival, rarely reaching the hazardous surface waters but settling again among the deep benthos. Species with lecithotrophic larvae require a period of vitellogenesis in the mother, to add yolk (lipid and protein) to the developing eggs; the female reproductive tract needs to be more complex, with accessory yolk glands, and more sophisticated physiological control systems involving hormonal signals.

Larvae that have matured enough to undergo metamorphosis, whatever the food supply used, are termed 'competent', and may then respond to environmental cues that trigger settling and metamorphosis responses. Negative phototaxy and/or positive geotaxy bring the larva to a substratum, which is then assessed for suitability on the basis of physical, chemical and biological information.

9.5.3 Direct development

Relatively few truly marine invertebrate animals use a degree of direct development. Those that do produce a fairly small number of large and yolky oocytes that are usually fertilized by direct insemination, the female often storing the male sperm or sperm packages (spermatophores). Vitellogenesis (yolk production) is essential, with follicle cells or nurse cells in the ovary to provide nutrient transfer to the oocytes; and embryos are invested with some degree of parental care, either by encapsulation or by brooding in or on the parent's body. Most marine invertebrates that have direct development are either oviparous (laying the fertilized eggs externally, albeit in some sort of capsule) or occasionally ovoviviparous (retaining the eggs internally, but relying on the yolk to feed the young, with no additional feeding from the mother's body).

Direct development occurs in species where dispersal is not a problem, for example where the adult is pelagic, as in the squids and the arrow worms (chaetognaths), or where the environment is relatively harsh and/or unpredictable, which in the context of the seas often means in high-latitude zones. Examples include *Octopus*, where spermatophores are transferred to the female cloaca by a specialized arm of the mature male. Some of the opisthobranch molluscs known as sea slugs also copulate in elaborate fashions, though being hermaphrodite sperm transfer occurs in both directions. In both these cases the fertilized eggs are then packaged internally into fairly elaborate jelly-like capsules and laid on the substrate as discrete egg masses. The female octopus then takes the extra step of guarding the eggs and turning them repeatedly to ensure adequate oxygenation; however, her maternal care does not extend beyond hatching, and in some species she normally dies soon after the eggs mature. Rather less elaborate direct development occurs in some echinoderms, where the fertilized eggs are brooded on the upper surface of the body, in the gonads or gonadal bursae, or even in the stomach.

In some cases direct development is followed by release of a larval form that must still metamorphose to reach its adult morphology. This kind of 'mixed life history' is particularly common in marine arthropodan groups, where both yolk production and a protective cuticular eggshell are characteristic features of the taxa, with elaborate mating behaviours and specially modified limbs (gonopods) for sperm transfer, but where larval dispersal and metamorphosis are also desirable for the predominantly benthic species. Mixed life histories also occur in many of the colonial tunicates, with eggs brooded in a common cloacal chamber until the relatively large 'tadpole larvae' are ready to emerge.

9.5.4 Semelparous and iteroparous life histories

Semelparous species are those in which the adults breed once and then die, as in the *Octopus* example above, whereas iteroparous animals breed continuously through a season, or repeatedly over more than one season. This distinction does affect physiological sophistication in the adult quite profoundly. Iteroparous animals must control rates of germ-cell maturation more carefully, and thus need to organize the partitioning of their own trophic resources in time and space; their endocrine systems, and the receptors that feed into them, have to be both more complex and more plastic. Semelparity is, perhaps not surprisingly, quite common among marine invertebrates of all taxa and size ranges; iteroparity occurs more commonly in species that have direct development of very few eggs in each batch.

9.5.5 Vertebrate reproduction

Fish can adopt life-history strategies that are similar to those of invertebrates; they often produce many thousands or even millions of tiny floating eggs for external fertilization, with a semelparous life cycle. They tend to endow their eggs with a small amount of food reserve, to allow the young to hatch slightly later and thus slightly larger, limiting the problems of hyposmotic regulation at extremely small sizes. The egg and larva may also be endowed with a small oil droplet to confer buoyancy (see below). At certain times of the year, fish larvae make up a high proportion of the zooplankton in temperate seas and provide a particularly rich food resource for the nekton.

The marine tetrapods are faced with very different problems, and cannot discharge large numbers of eggs and sperm into the sea as most invertebrates and fish do. Reptiles and birds are derived from essentially terrestrial egg-laying ancestors, and most of the species associated with the sea have no alternative but to return to land to breed, as the amniotic egg is designed for aerial rather than aquatic breathing. Many reptiles then run into the problems of finding thermally suitable nesting sites, given their temperature-sensitive sex-determining systems (TSD—see Chapter 13). Today only sea snakes are viviparous and can bear live young while still submerged (though there is evidence that Mesozoic ichthyosaurs also achieved this). In this they parallel the mammals, some of which (cetaceans, including whales and dolphins) can achieve live birth at sea. The young of all the permanently marine tetrapods are inevitably relatively precocious, already able to swim, breathe air and achieve homeostasis; the only essential difference from the adults' biology arises from the provision of an unusually concentrated and lipid-rich milk (see Chapter 13) from the parent mammals (matched by a milk-like uterine secretion in sea snakes).

Other than cetaceans, most marine mammals routinely come ashore and raise the pups terrestrially at first. This can put substantial strain on the maternal physiology, particularly with respect to temperature balance, as the adult is adapted for the lower temperatures associated with submersion and, encased in insulating blubber, can overheat rather quickly when hauled out on a sunny beach. Seals and walruses seek (and often compete fiercely for) shady patches and cool substrata, and those that achieve the most favourable microhabitats may rear young most successfully. These animals may also take rapid cooling dips in the sea whenever possible, but thereby often risk predation from sharks and other large carnivores waiting just offshore from the breeding sites.

9.6 Depth problems, buoyancy and locomotion

The sea has an average density (g cm^{-3}) of 1.025 (tropical surface waters) to 1.028 (colder and deeper polar seas). Cytoplasm has a density of 1.02–1.10 g cm^{-3}, and most organisms also have some heavy skeletal materials in their make-up giving an overall density markedly greater than their surroundings. All organisms therefore tend to sink, so must be able to cope with the adverse effects of depth and/or must have adaptations to maintain themselves at a particular depth, leading us on to the issue of buoyancy.

9.6.1 The effects of depth and pressure

The major problems of depth lie not in salinity, temperature, or gas availability, as we saw in preceding sections, but in the effects of pressure. For every 10 m below the surface the pressure increases by 1 atmosphere (101 kPa). Since life is found at all depths of the sea, right down into the deepest trenches where depths can reach 11 000 m, some animals are clearly adapted to coping with pressures in excess of 1000 atmospheres (101 MPa), and may be termed **barophilic**. Many more animals migrate vertically and must be able to cope with a wide range of pressures; for example, the pygmy shark moves up and down daily by about 1500 m to feed on bony fish and squid, while diving mammals and birds can reach depths of 500–1500 m (see section 9.10.3).

A pictorial summary of the kinds of life found living permanently at depth is shown in Fig. 9.19. Some rather 'simple' animals, such as sponges, cnidarians (anemones and sea pens) and echinoderms (sea stars, sea cucumbers and the like), survive very well at 5000–6000 m, together with more complex forms such as crustaceans and squids. Highly specialized teleost fish

are particularly prominent, reaching depths of at least 7000 m (4.5 miles down), although those at greater depths lack the gas-filled swim-bladders described later.

Adaptations to high pressure occur at all the levels discussed in the introductory chapters of this book, but are most evident in the realms of physiology and biochemistry. All effects of pressure are in effect a consequence of volume change (ΔV). Pressure primarily affects molecular structure, altering the volumetric packing of enzymes and other proteins, and of lipid-based membranes. At a higher level of organization structures such as cilia and other microtubular structures may be disrupted, leading to problems with spindle formation in mitotic (and meiotic) cell division. Pressure can also be particularly disruptive for solute/solvent interactions; the hydration shells around individual ions change their effective size when pressurized, so that diffusion rates are affected and events dependent on permeation of cell surfaces—such as excitability and impulse transmission—can be grossly modified. Locomotor activity and behaviour are therefore particularly sensitive to pressure in all kinds of animals, and the most critical adaptations often occur in nervous systems. When shallow-water organisms are exposed to gradually increasing pressures their motor activity tends to increase and they exhibit enhanced excitability, culminating in convulsions and spasms at about 100 atmospheres. Thereafter immobilization occurs, which soon becomes irreversible. In contrast, animals collected from moderate depth in isobaric traps show no such responses to increasing pressures, but some are reversibly immobilized by normal shallow-water pressures ('decompression paralysis').

Only recently have the adaptations at the level of reaction rates, enzymes and membranes attracted much attention. Hydrostatic pressure may perturb protein and membrane function wherever volume change occurs at specific molecular sites, and hence phenomena such as structural stability, ligand-binding affinity and bilayer fluidity are all affected. Metabolic reactions can be accelerated, retarded or remain unaffected by increasing pressure, according to the molecular geometry of enzymic reaction sites; consequently, complex multistep enzyme pathways can be completely disrupted at depth as constituent reactions become unbalanced. Ion transport systems, and hormone or neurotransmitter binding, can also be disrupted.

Enzymes must be modified to operate at high pressure by having reduced pressure sensitivity, probably achieved by increasing the hydrophobic (or in some cases electrostatic) bonds within the molecular structure, and by increasing bond strength between protein subunits. Where necessary, enzymes should ideally retain a fairly constant K_m value irrespective of pressure, and this has been shown to be true for LDH binding to its NADH cofactor in several deep-sea fish, whereas the same enzyme in shallow fish has its K_m value roughly doubled for a pressure equivalent to 700 m depth. In diving mammals and birds, however, the LDH is apparently similar to that in non-diving species and is inherently pressure-insensitive; this is perhaps because the LDH of endotherms is much less fluid

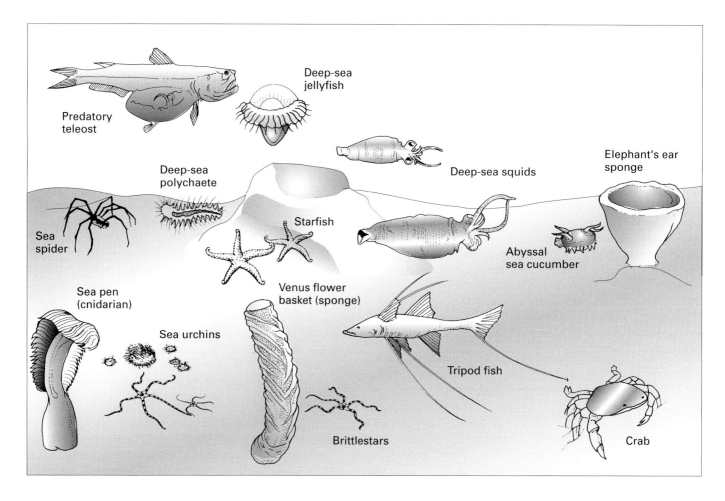

Fig. 9.19 A pictorial view of deep-sea fauna.

in structure anyway, to allow it to function at a T_b of 37°C. Pressure-insensitive enzymes are also particularly useful for non-endotherms that migrate vertically. For example, the fish *Sebastolobus* has larval and young stages that are pelagic but the adults are deep-sea specialists; its muscle enzymes can function without modification over a range of 130 atmospheres, with no switching to alternative suites of isozymes during development.

For membranes, a classic strategy is the modification of the lipid composition, replacing standard phospholipids with less saturated forms to counteract the effects of pressure (and of the routinely low temperatures also experienced at depth) in making normal membranes more viscous. This is sometimes called 'homeoviscous adaptation' (see Chapter 8). In fish, we know that mono-unsaturated lipids, such as phosphatidylcholine and phosphatidylinositol, become much more important in the cell membranes of deep-sea forms, to retain membrane fluidity. In the sunfish and other well-studied cases these effects are largely due to the desaturase enzyme systems discussed in Chapter 8, rather than to any change in dietary lipids or transfer enzyme activities.

For whole animals, physiological performance can perhaps best be compared between species in relation to their **minimum**

depth of occurrence (MDOC), so that permanent deep residents and vertical migrators can be considered together. For pelagic fish, cephalopods and crustaceans, rates of oxygen consumption (metabolic rates) decrease rapidly with increasing MDOC (Fig. 9.20), and to an extent much greater than can be explained by the decreasing temperature. Most of the decrease occurs in the first 200–400 m, so that at even moderate depth most animals have very low metabolic rates for their size. But the decrease with depth may be more to do with ecological factors favouring limited locomotion (low prey density, absence of light and therefore of plant-feeding opportunities, etc.), rather than with physiological constraint. This would accord with the particularly low activity of white muscle enzymes at depth.

9.6.2 *Buoyancy mechanisms*

Buoyancy could be seen as a means of avoiding the effects of variable pressure, by a combination of behaviour and physiology. There may be three very general ways of achieving neutral buoyancy (see Fig. 9.21): removing heavy material, adding very light material, or improving flotation by greatly increasing the surface area/volume ratio. For very small animals this last strategy is commonly found, as organisms of only a few milli-metres in size are dominated by viscous drag forces, and having

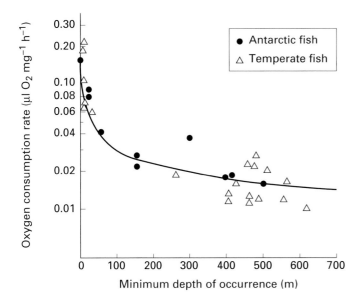

Fig. 9.20 Effects of depth (pressure) on oxygen consumption rates for a range of Antarctic and temperate fish. (Reprinted from *Comparative Biochemistry and Physiology B* **90**, Torres, J.J. & Somero, G.N., Vertical distribution and metabolism in Antarctic mesopelagic fish, pp. 521–528, copyright 1988, with permission from Elsevier Science.)

an elaborately frilled or spiky outline can in itself increase the drag sufficiently to prevent sinking. A great many marine larvae and small planktonic creatures exploit this property (Plate 1c), and are strikingly different in form from the relatively stream-lined larger animals in the same habitat. For these larger animals, there are various different mechanisms for achieving neutral buoyancy.

Reduction and substitution of heavier ions

This technique involves replacement of heavier (higher molecular weight) ions in the body fluids and tissues (such as magnesium and sulphate) with lighter ions, using some kind of exchange pump system, leading to some of the oddities of body fluid composition discussed above. The jellyfish *Aurelia* and the cuttlefish *Sepia* both show unusually low sulphate levels, with chloride making up virtually the entire anionic component of their body fluids. A range of comb jellies (ctenophores), jelly-fish (cnidarians), salps (tunicates) and sea slugs (molluscs) are known to extrude sulphate in similar fashion, and their measured lift correlates with the degree of this extrusion (Fig. 9.22).

Even more extreme is the deep-sea squid *Heliocranchia*, which has a large fluid-filled pericardial cavity (up to 65% of the body volume). This contains very little magnesium or even sodium, the cationic component being dominated instead by light ammonium ions, produced from protein metabolism and trapped in the pericardium by the low pH of that cavity, probably in turn resulting from a Na^+/H^+-exchange pumping system. Again, the anionic component is almost entirely chloride, with virtually no

sulphate; the resultant fluid has a specific gravity of only 1.01, and is isosmotic with sea water. Other squid in the same family have similarly buoyant fluid vesicles containing NH_4Cl in their arms or mantle.

Many elasmobranchs use urea and TMAO as important balancing osmolytes in relation to maintaining their body fluid composition (see section 9.10.1), and recent calculations indicate that in species such as the Port Jackson shark these solutes also have a significant positive buoyancy effect. TMAO is also useful in this sense in some decapod crustaceans, and the effect may be widespread.

Losing ions without substitution

If the heavier ions are simply extruded actively from the body a degree of buoyancy can be achieved, though there will be inevitable osmotic costs as the body fluids will be relatively dilute, the animal becoming hyposmotic to the sea. This strategy is uncommon in invertebrates, but serves as an important buoyancy device in teleost eggs. It may in fact be a major advantage to the vertebrates of having retained hypotonic body fluids, giving a degree of buoyancy to all marine teleosts. It is taken to a greater extreme in some deep-sea teleosts, where skeletons and muscle mass are also reduced (see below).

Reduction of heavy minerals

In particular, buoyancy can be increased by the loss of calcium carbonate or phosphate in the skeleton. A number of pelagic gastropod molluscs (especially the heteropods and pteropods) that live at depth have substantially reduced their shells, to the extent that some have become the delicately buoyant animals known as 'sea butterflies'. Vertebrates also commonly exploit this strategy, especially in the elasmobranch fish. The same effect is found in marine mammals, which include the largest animals on Earth (the blue whale, at up to 4×10^5 kg); many of these cetaceans have limb bones and ribs with depressed skeletal maturation and no marrow cavity, the bone being very porous and containing significant quantities of lipid (see below). Similar phenomena are found in marine reptiles and in penguins. Alternatively, skeletal materials in some parts of the body may be much reduced whilst others that are crucial to feeding are strengthened. Here, some deep-sea pelagic teleosts are the best examples, often having shrunken trunk musculature and reduced backbones with cartilage (density 1.1 g cm^{-3}) replacing bone (density 2.0 g cm^{-3}), the vertebral processes being reduced leaving just a narrow bony collar around a persistent gelatinous notochord. A vivid example is provided by a few Antarctic notothenioid fish which (following glaciations that reduced the pre-existing pelagic fauna) have become secondarily adapted to a pelagic mode of life from a benthic ancestor. They lack the normal teleost swim-bladder, and achieve buoyancy in the water column by reducing their mineralization and accumulating lipids.

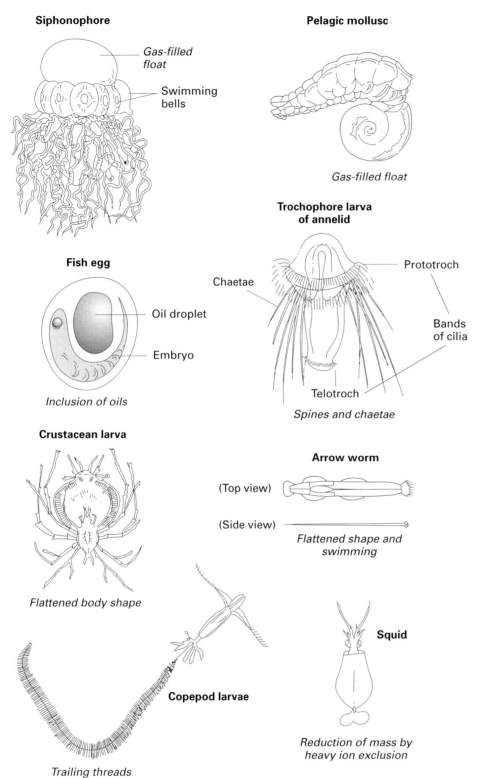

Siphonophore

Gas-filled float

Swimming bells

Pelagic mollusc

Gas-filled float

Fish egg

Oil droplet

Embryo

Inclusion of oils

Trochophore larva of annelid

Chaetae

Prototroch

Bands of cilia

Telotroch

Spines and chaetae

Crustacean larva

Flattened body shape

Arrow worm

(Top view)

(Side view)

Flattened shape and swimming

Copepod larvae

Trailing threads

Squid

Reduction of mass by heavy ion exclusion

Fig. 9.21 Examples of methods to increase buoyancy in marine animals.

Fig. 9.22 The effects of extrusion of sulphate on buoyancy in a range of pelagic cnidarians, ctenophores and tunicates. (From Denton & Shaw 1961.)

Increasing the amounts of light fats and oils

Fats and oils commonly float on water, having densities less than 1 g cm^{-3}, so that if present in an animal's body they can produce an overall density less than that of sea water. This strategy is very common in planktonic crustaceans, where droplets of oily material may be visible within the body. It is also evident in many fish eggs, where oils may account for 50% of the organic matter present. More spectacular examples occur in some sharks in the family Squalidae (including the massive basking shark), which never have swim-bladders but may have greatly enlarged livers (up to 20% of the body weight) in which up to 75% by weight is oily material, particularly the unsaturated hydrocarbon squalene ($C_{30}H_{70}$: specific gravity 0.86). In other fish, and in marine mammals such as whales, a similar function is performed by a range of other lipid materials (Table 9.7). These lipid compounds are all rather inert, with very low metabolic turnovers.

Increasing the amount of gas in the body

Using gas 'floats' by enclosing gas within the body is the 'best' solution to the buoyancy issue, since compressed air has a

Table 9.7 Lipids providing upthrust in marine animals.

Lipid	Specific gravity	Upthrust (g upthrust per g lipid)
Triacylglycerol	0.93	0.103
Alkyldiacylglycerol	0.89	0.127
Wax ester	0.86	0.193
Squalene	0.986	0.193

density of only 0.0012 g cm^{-3} at sea level and even a small float will provide substantial lift (the swim-bladder of a marine fish is normally only 5% of its body volume). Many organisms exploit gases, including marine plants such as the seaweeds. In marine animals there are two major kinds of gas float:

1 the rigid incompressible float characteristic of various cephalopod molluscs such as *Nautilus* and the cuttlefish (and even more common in various extinct cephalopod groups);

2 the soft-walled chamber found in *Physalia* (the Portuguese man o' war) and other cnidarians, and also in a more complex form as the swim-bladder of fish. The disadvantage here is that the float is compressible so that its volume changes with depth, and lift reduces during a descent. Species that migrate vertically may therefore have problems with this system.

In cephalopods gas is retained within a shell—either in a complex internal shell, which may be coiled (*Spirula*) or straight (*Sepia*), or in an external coiled shell where the most distal chambers are given over to gas storage (*Nautilus*). A solid float like this has the merit of being constant-volume and thus relatively independent of depth. These animals do not have specific gas-generating tissues, but rather the gas (principally nitrogen) just accumulates as a consequence of active salt transport. Initially the chambers of *Nautilus* shells are filled with a NaCl-rich isosmotic fluid, from which Na$^+$ ions are actively pumped by the mitochondria-rich cells of the central strand of living siphuncle tissue that extends up the middle of the shell. Chloride follows passively, and water then follows down the osmotic gradient, leaving a gas space. This equilibrates at about 0.8 atmospheres N$_2$ (as in air), though the chambers maintain a rather lower than ambient oxygen pressure. The system in *Sepia* (cuttlefish) is rather similar (Fig. 9.23), except that the shell is tightly laminated with no clear chambers, and is internalized, making up about 10% of the body volume. The gas composition within the laminae is as in *Nautilus*, but the relative proportions of gas and of NaCl-rich fluid are varied in different parts of the cuttlebone and at different times; when the animal descends to greater depth, salts are extracted from the shell fluids by surrounding epithelial cells, so that the fluids become hyposmotic and the tendency of the increased hydrostatic pressure to force more water into the cuttlebone is counteracted. A freshly extracted cuttlebone therefore has a specific gravity varying between 0.55 and 0.65. In this way the cuttlefish varies its own density, and can live with neutral buoyancy at various depths from the surface to about 200 m (whereas complete withdrawal of fluid from the cuttlebone would theoretically permit penetration to about 240 m).

Other cephalopods using this kind of buoyancy system can descend to about 450 m (even down to 800 m in *Spirula*), suggesting that additional mechanisms operate, possibly to enhance blood concentration in the siphuncle with a countercurrent multiplier. But as the rigid flotation device has to be relatively rather massive, and strongly calcified to withstand the imposed pressures, it is also heavy, and at these depths around 80% of the gas is needed just to support the float itself.

Clearly a non-rigid float, although compressible as depth

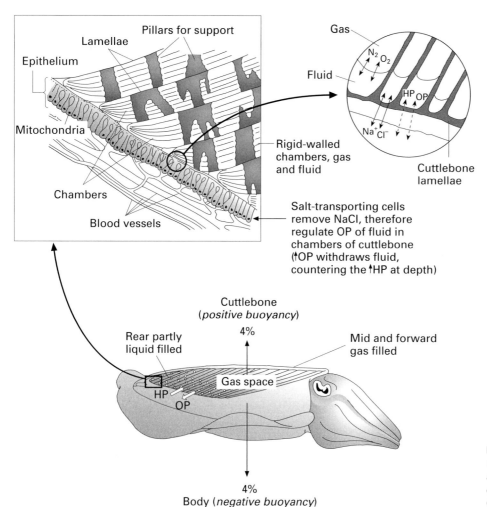

Fig. 9.23 Form and functioning of the *Sepia* cuttlebone; salt transport and gas accumulation. HP, hydrostatic pressure; OP, osmotic pressure. (Adapted from Denton & Gilpin-Brown 1961.)

increases, would have the advantage of being much lighter to carry around. It would be particularly suitable over a range of depths if the quantity of enclosed gas could be regulated to counter the hydrostatic effects of pressure (otherwise the float volume would be halved for every 10 m of descent). Thus in floating cnidarians and in fish there are distinct gas-secreting tissues to give rapid control. In the surface-dwelling colonial siphonophore *Physalia* (Portuguese man o' war) the chamber is up to one litre in volume (Fig. 9.24a), filled with the normal components of air but also with 12–15% carbon monoxide, CO, derived from the amino-acid serine. Other siphonophores live at variable depth and produce rather more CO, which is present at up to 90% in the tiny floats of *Nanomia*. In most of these animals there are specific gas-generating cells ('gas gland') at one side of the float (with cells rich in folic acid, which appears to be the storage precursor for the gas), and an exit pore lying at the other end (Fig. 9.24b). The exit pore is important in any soft-walled float operating below the surface because a rising animal would have an expanding float and positive buoyancy, causing increased upward acceleration; gas must be released quickly to avoid unduly fast ascent and possible rupture.

In teleost fish swim-bladders are the common, though by no means ubiquitous, buoyancy structures. These soft-walled sacs lie above the gut and just below the spinal column, taking up about 4–5% of the body volume, and there are two basic arrangements. In the **physostomes**, such as the salmonid fish, the bladder retains a connection with the gut, so that gas can pass in either direction, and this may be the sole means of buoyancy control; the system can be filled by gulping at the surface, and the intervening duct is controlled by the autonomic nervous system. However, some physostome fish also have a simple version of the rete and gas gland system described below.

The more 'advanced' fish (**physoclists**) have a separated swim-bladder, which is filled with gas (oxygen and nitrogen principally) from the blood supply, via a **gas gland**. No gas is in itself actively secreted of course; gaseous movements are purely diffusive, even at great depth and apparently against enormous gradients. The gas gland has an associated rete of exceptionally long and unusually highly ordered blood capillaries (Fig. 9.25), the length of these capillaries being correlated with the depth to which the species can descend (they are up to 25 mm long in the

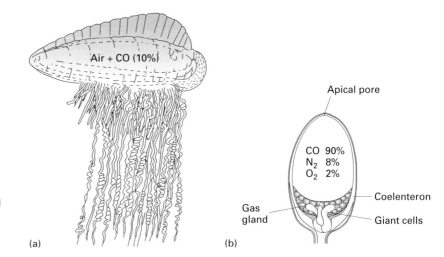

Fig. 9.24 (a) *Physalia* (the Portuguese man o' war) and its gas float. (b) Cross-section of the small gas float of *Nanomia*, with gas gland and apical pore for gas release.

Fig. 9.25 The gas gland rete and its function in the swim-bladder of fish. (Adapted from Wittenberg 1958; Denton 1961.)

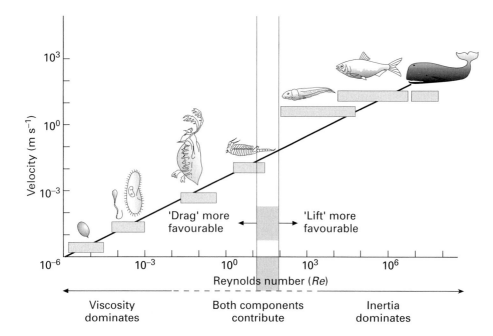

Fig. 9.26 Effects of size and Reynolds number (*Re*) on swimming patterns and the relative importance of viscous forces and frictional forces. *Re* is given by $\rho VL/\mu$, where *V* is velocity, *L* is dimension of moving surface, ρ is the density of the medium and μ its viscosity, so giving a dimensionless ratio related to size. A high Reynolds number applies to a large or a very fast animal, where flow patterns are dominated by inertia and streamlining is useful. At low Reynolds numbers (small or very slow swimmers, with high surface areas relative to volume) viscosity is much more important and streamlining is irrelevant. (Adapted from Nachtigall 1983.)

deep-sea genus *Bassozetus*). In effect the gas gland is the distal end of a **countercurrent multiplier** (see Box 5.2) formed by this rete, so that gas from the blood accumulates there at high concentration and diffuses into the swim-bladder. The cells of the gas gland mainly function by undergoing intense glycolysis (an unusual occurrence in the presence of high O_2 levels), with lactic acid and CO_2 as the principal end-products. This creates a very low pH so that in the efferent venous blood the Root effect operates very effectively and there is also an enhanced Bohr effect (see Chapter 7). Hence oxygen is very rapidly unloaded from the venous haemoglobin and can diffuse across the rather permeable rete into the afferent arterial blood. In the gland this enriched arterial blood gets even more acidified and gives up its oxygen to the chamber. In the eel (*Anguilla*) the swim-bladder receives 83% of the oxygen removed from the blood, with just 17% being metabolized in the gas-gland tissues. This countercurrent multiplier may be aided by a 'salting out' effect, the gland adding ions (lactate and bicarbonate, and possibly NaCl) to the blood to free extra oxygen; this would also explain how nitrogen and other inert gases are accumulated in the swim-bladder.

The chamber of the swim-bladder retains this gas very efficiently, having highly impermeable walls with a thick layer of connective tissue, in some species impregnated with flat plate-like structures of crystalline guanine; the permeability of such a layer to oxygen is only about 10% that of normal connective tissues. However, in one particular area of the swim-bladder there is a window of vascularized tissue capable of resorbing gases from the chamber, often with a muscular valve so that it can be exposed to the gas or shut off from it. By balancing the activity of the gas gland and the resorptive window the volume of gas in the chamber can be regulated.

Pelagic fish with gas-filled swim-bladders normally only function at depths less than 1000 m, and very-deep-sea fish in

mid-ocean trenches have oil-filled bladders instead. However, a few species of near-bottom teleosts have been caught at depths of up to 4500 m with normal gas-filled bladders, and it may be that these have extra physiological specializations to maintain gas secretion beyond those already described.

9.6.3 *Locomotion*

It is clear from our consideration of buoyancy that some part of many marine animals' locomotory need is linked to station-keeping in relation to depth, and many achieve this mainly or partly by swimming mechanisms. Size plays an important role here (Fig. 9.26), since for very small animals (including most larvae) locomotion in water is dominated by viscous forces and a highly elaborated non-streamlined profile is possible (and, in terms of reducing sinking rates, appropriate). For larger animals frictional drag becomes increasingly important instead, and swimmers tend to converge on a relatively streamlined shape, culminating in the essentially torpedo-like shape of fast-swimming cephalopods, elasmobranchs, teleosts and marine mammals, as well as the extinct reptilian ichthyosaurs. Here the power requirement to overcome drag increases very dramatically with speed, so that fast swimming is extremely expensive and streamlining is crucial.

Swimming mechanisms to propel animals through the water come in many varieties. Many of the smallest animals rarely swim actively but instead drift passively with the currents, relying on viscous forces to maintain stability and height in the water column. Small invertebrates, and most larvae, tend to rely on ciliary locomotory systems, though in ctenophores (sea gooseberries) there are complex cilia arranged as comb-plates and individuals can be up to a foot long. Normally, for animals above a few millimetres in diameter, cilia are inadequate and muscles are universally employed. The muscles may be used to operate a water jet system, giving jet propulsion—as in many rapidly darting cephalopod squids, cuttlefish and octopuses, in many planktonic jellyfish, and in some benthic bivalves that can clap the two shell valves together forcefully to give a jet and thus 'jump' away when threatened by predators. Alternatively, and more commonly, the muscles operate appendages or true limbs functioning as 'paddles'—in crustaceans (including their vastly diverse larval forms), in tetrapods, such as turtles, and in some birds swimming underwater; here propulsion is drag-based. In the third option, representing the commonest swimming mode, some part of the body is modified as a muscular undulating flap—as in the fins or tails of many fish and marine mammals (and the tails of some crustaceans), the lateral mantle edge of cuttlefish (in their 'slow' gait), or the edges of the body in large flatworms. Waves of muscle contraction occur in this structure, usually alternating contractions of longitudinal and circular muscles; the undulating part of the body usually has an aerofoil profile, and is moved with a gentle vertical flapping motion, with a combination of lateral and backward forces. The lateral forces cancel out on either side, and the backward force against the water propels the body forward. The extended mantle flaps of sea butterflies (pteropod molluscs) operate in a similar but more spectacular fashion, looking rather like a pair of wings. The extreme (and very common) version of this swimming motion comes when the whole body experiences the muscular wave, usually in a lateral plane, giving the typical anguilliform ('eel-like') swimming of certain fish and the fast swim/crawl modes in many worms. All these modes based on waves of contractions and oscillating hydrofoils are forms of lift-based swimming, and are more energetically efficient than paddling systems. Most swimming aquatic animals have a considerable capacity to vary the speed and energy of their swimming, with a factorial aerobic scope of 5–7 (though varying with temperature); species with a capacity for long-distance endurance swimming (notably some crabs, lobsters and fish) may have somewhat lower values.

Crawling mechanisms in benthic animals are often a variation on the undulatory swimming systems described above, aided by some contact with the substratum. Molluscs crawl using waves of contraction in their mucus coated muscular foot, but with a substantial contribution from the cilia on its surface up to surprisingly large sizes. Crustaceans crawl by a slower and more forceful use of the legs pushing against their surroundings. Many polychaetes reduce the lateral undulation of their body and increase the paddle effect from their parapodia when in a slower crawling gait. Some fish also use their pectoral and/or pelvic fins to push themselves along the sea bottom or over rocks.

Burrowing is the remaining form of locomotion open to marine animals. Marine burrowers exploit the physical security and often also the food reserves of the benthic sludge. However, active burrowing is rather uncommon in fully marine animals; they may form a shallow burrow and live in it with one end intermittently protruding into the water, but they do not often burrow continuously for a living. By contrast, for animals in the littoral and estuarine zones, this is the main strategy of a high proportion of the fauna, affording protection, physiological stability, and major food-gathering opportunities. Burrowing mechanisms are therefore dealt with in Chapter 10.

9.7 Sensory issues: marine signalling

Life in the deeper waters of the seas creates a number of sensory problems. Animals have no light to provide for normal vision, and high pressures create problems for sound and vibrational signalling systems. Modifications in the senses and in effectors are therefore major features of physiological adaptation in these habitats.

9.7.1 *Light and vision*

Water represents a far richer and more varied photic environment than that encountered by terrestrial animals, because of complex interactions between light and suspended or dissolved materials, varying with depth. Light in the sea varies in intensity, frequency and polarization as it progressively penetrates to greater depths. The water molecules themselves scatter and attenuate light, particularly at the blue (short wavelength) end of the visible spectrum; and polarization of the light is also affected by scattering from the water. Dissolved salts have little effect on intensity, but suspended particulate matter attenuates all components of the transmitted light. Thus for typical sea water 50% of the light is absorbed within 2 m of the surface, and 90% within 8 m. At the same time, the frequency of transmitted light varies with depth in a predictable manner, with red light absorbed the most readily to give mostly blue wavelengths below 50 m (Fig. 9.27), though in coastal waters the greater amounts of dissolved matter shift this relation somewhat. The net result is a relatively predictable vertical gradient of light intensity, dividing the open ocean into the **euphotic zone** (epipelagic, where down to 200 m there may be enough light for photosynthesis), the **dysphotic zone** (mesopelagic, 200–1000 m, where blue light is still present but at insufficient intensities for photosynthesis to balance respiration), and the **aphotic zone** (bathypelagic, >1000 m, where all solar light has been absorbed; here most species in certain taxa are eyeless). Away from the open ocean the photic zones may be much attenuated, and also shifted towards greener wavelengths, by plankton blooms, sediment-laden run-off from land, and tidal turbulence.

The presence of so many distinct light regimes is reflected in the evolution of a great diversity of visual systems. For animals that live in the euphotic and dysphotic zones, photoreception is always a major feature of the sensory repertoire, and 'eyes' of some sort occur in virtually every phylum of animals (with clearly convergent designs recurring across the animal kingdom). Most marine animals with eyes make use of simple spherical lenses, with refractive index decreasing from the centre to the periphery, and this type of eye must have arisen on at least six separate occasions. Multilayered lenses also occur (e.g. in

Fig. 9.27 Light attenuation and selective absorption in the ocean in relation to depth; and the associated trend towards blue-shifted rhodopsins in the eyes of fish from increasing depth.

some copepods), as do concave mirrors to channel incident light (e.g. some bivalves and ostracods). In a range of arthropodan animals compound eyes are also found, with at least three different designs, based on lens-cylinders, mirrors, or a combination of lens and parabolic mirror. For many of these animals, light (photoperiod) is a major influence on behavioural cycles, particularly for reproduction in polychaetes, starfish and decapod crustaceans.

Visual pigments are the essential components of visual systems and the source of much habitat-related variation. Their role is the transduction of light energy into a nervous impulse, and they are usually housed in a layer of cells forming a retina, the cells being termed rods in vertebrates (or cones where they give colour vision). The visual pigments are membrane-bound proteins, composed of a **chromophore** group covalently bound to a lysine residue on the protein **opsin**. Opsin is a member of the same protein family as the adrenergic and muscarinic membrane receptor proteins, with seven transmembrane α-helix segments (see Chapter 2). When a photon is absorbed, the chromophore changes from one isomeric form to another and is expelled from the opsin protein, presumably revealing an enzymic binding site. This site then binds G protein, leading to a typical enzymic cascade eventually resulting in a neural response via cGMP (see Chapter 2). The cascade is shut off again by a light-dependent phosphorylation at the opsin C-terminal, and the chromophore is converted back to the original isomer, which reassociates with the opsin.

The maximum sensitivities of visual pigments are determined by the character of the chromophore. The commonest chromophore is an aldehyde derivative of vitamin A, particularly 11-*cis*-retinal, which results in the classic visual pigment **rhodopsin**, with maximal absorption at 500 nm. The exact peak varies according to the interaction between the chromophore and the opsin. In invertebrates and fish, absorption peaks are somewhat variable according to habitat (Fig. 9.27), many species showing two or more spectral peaks, and with a trend to shorter wavelengths and broader sensitivity curves as one moves offshore. In general the peaks for invertebrate eyes occur at about 450–500 nm, and are never above 560 nm. This contrasts with the situation in some marine fish where 11-*cis*-3-dehydroretinal is used instead, and peak sensitivity may sometimes be towards 600 nm, perhaps giving better contrast vision. Fish in the euphotic zone have both rods and cones ('duplex retinas') with several different peak sensitivities; sensitivity and resolution are adapted to different parts of the day–night cycle, with lower wavelength (green to blue) maxima as depth increases. It has been suggested that the fish's visual pigment absorbency peaks at the dominant wavelength occurring at the typical depth and the time of day most important to the active fish; for example, in reef fishes the rhodopsins match the spectrum of twilight underwater irradiance, since twilight is when visual sensitivity is most crucial for survival.

Some of the eyes found in cephalopods and crustaceans (and perhaps some fish) are also known to be sensitive to variations in polarization, though this has been little studied.

Marine animals may have particular problems with camouflage, given that threats may come either from above or below them. Many marine fish and squid exemplify the conflicting needs. To achieve camouflage from visual predators from above in daytime, the back of the animal needs to be very dark, so that the reflected light equals the light passing upwards on either side of its body. The sides of the body need to be near-perfect mirrors so that the light reflected back to a lateral observer is almost the same as if the animal were not there. And the ventral

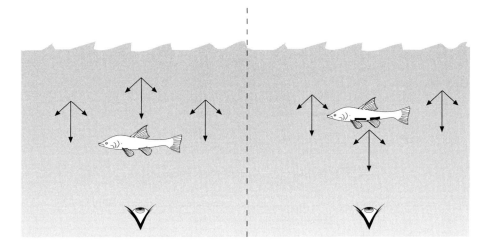

Fig. 9.28 The use of ventral camouflage light emission in marine animals. On the left, an observer from deeper water sees the animal because it blocks the limited daylight coming downwards. On the right, the animal itself emits light from its lower surface to replace the daylight and thus becomes camouflaged.

surface can only be rendered invisible if the intensity and spectral distribution of light from it matches that of the surrounding sea. Below 250 m the spectral distribution of light is fairly constant, so that in practice this can be achieved by the animal sensing the intensity of light falling on its dorsal surface and matching it with emitted (bioluminescent) light from its ventral surface (Fig. 9.28).

9.7.2 *Bioluminescence*

The abyssal plains of the oceans average 3800 m in depth, well into the aphotic zone; so the majority of the Earth's surface receives no sunlight. Although vision would seem to be unimportant at aphotic depths where solar light cannot penetrate, in practice many animals there retain eyes and can see rather well. This is because **bioluminescence**—light (photons) produced by the biota itself—is a prominent feature of a wide range of deep-sea animals. At least 700 genera from 16 phyla are known to use bioluminescence.

Bioluminescence may occur as a continuous glow or in a great variety of flash patterns, and it is used for many different functions. Single cells (photocytes) or complex glands (photophores) can be used, and these may be sited deep within the animal tissues or superficially in the epidermis. Some bioluminescence involves intrinsic photocytes with intracellular reactions, while some originates from an extracellular process. Extracellular bioluminescence is particularly common in marine invertebrates, often involving two separate groups of cells within a gland. When both cellular products are released under the influence of muscular contraction, the reaction between them occurs in the gland lumen and light is emitted, with the other end-products excreted from the gland. A great deal of both intracellular and extracellular bioluminescence involves symbiotic bioluminescent bacteria. In any of these cases, accessory structures such as lenses, reflectors, light guides and pigment 'shutters' are commonly found (Fig. 9.29a). This diversity clearly indicates multiple convergent origins for the phenomenon, probably as a serendipitous acquisition of symbionts and/or as by-products

of other metabolic pathways. Many different kinds of cell or tissue may be modified to emit bioluminscence, the most unlikely recent example being the suckers of some deep-sea octopuses.

There is considerable biochemical diversity in bioluminescent systems, though all are commonly referred to as using a **luciferin-luciferase** system, as in bioluminescent bacteria. The process involves the oxidative breakdown of a high-energy, complex organic molecule ('luciferin') by a specific enzyme ('luciferase') in a process that releases photons of light. The 'true' luciferin of bacteria is a modified form of a reduced flavoprotein, complexed with an aldehyde; under the influence of oxygen and the luciferase enzyme this complex breaks down to the flavoprotein and a carboxylic acid. Marine animals may also have a variety of more specific luciferins, of which the most famous are the calcium-activated molecule, aequorin, found in certain jellyfish and other cnidarians (but also in a squid, several crustaceans and some fish), and the *Vargula*-type luciferin originally isolated from the ostracod *Vargula*, but also found in other crustaceans and some fish.

In the fish taxa, bioluminescence is a purely marine phenomenon, and can be taken to extravagant lengths. Again, it may arise from either the inherent tissue biochemistry or from symbiotic bacteria. Bacterial organs are commonest in fish from the aphotic zone, occurring in 20 families, nine of which are in the order of deep-sea anglerfishes (Ceratioidea). Each species has just one species of bacterium, though the same bacterium may occur in more than one fish species or even in more than one family. Inoculation of the light organs by the bacteria probably occurs from the surrounding sea water at the larval stage, since at least some species of the bacteria (e.g. *Photobacterium*, *Vibrio*) can be readily cultured outside the fish.

Fish from the dysphotic zone commonly use non-bacterial luminescence, where the luciferins are chemically variable but are never flavoproteins. Such fish may use either the cnidarian/ aequorin-type luciferin or the *Vargula*-type. In some species, bioluminescence is supported by luciferin ingested in the diet, by swallowing either the cnidarian or the ostracod from which the luciferin originates. For example, the midshipman fish, *Porichthys*

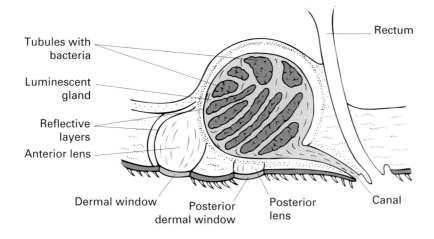

Tubules with bacteria
Luminescent gland
Reflective layers
Anterior lens
Rectum
Dermal window
Posterior dermal window
Posterior lens
Canal

Photogenic cells
Reflective layer
Blood vessels
Epidermis
Lens
Nerve

(a)

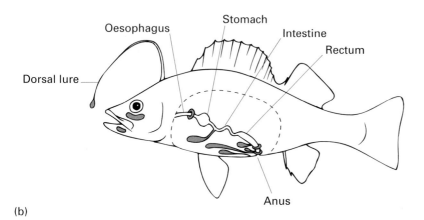

Oesophagus
Stomach
Intestine
Rectum
Dorsal lure
Anus

(b)

Fig. 9.29 Bioluminescence in fish. (a) Structures that emit and control light, including shutters, lenses and light guides; and (b) the distribution of bacterial light organs in fish (shaded areas). (a, Adapted from Herring & Martin 1978.)

notatus, has several hundred superficial ventral photophores that contain *Vargula* luciferin only when the fish is in *Vargula*-containing habitats. A single dietary intake of *Vargula* confers a lifetime of luminescence due to an efficient recycling system.

Whatever the source of the luciferin, fish luminescent organs are widely distributed (Fig. 9.29b) and are often very similar to the eyes in construction, probably having coevolved with the fish eye. For example, in both eyes and luminescent organs reflectors occur and are composed of purine platelets, while the pigmented layers contain high concentrations of melanin. The retinas of bioluminescent fish from aphotic depths have only rods, which achieve better photon capture than cones.

The light-emitting organs often contain bacteria that emit luminescence continuously, but there may be a covering mechanism or shutter (rich in melanin to give a dark absorptive cover), so that flashes can be produced. Light may be reflected from the internal swim-bladder, and may be modulated by superficial chromatophores or by filters of varying colour. In fish with non-bacterial light organs the nervous and endocrine systems may also regulate light intensity.

Functions of bioluminescent organs in all taxa appear to be very varied. In pelagic marine animals they are most frequent on the ventral surfaces, and towards the periphery, thus presumably playing some role in ventral camouflage as described above. Emission is often in the blue-green range (450–490 nm), where transmission in sea water and visual sensitivity of most marine animals are both greatest. Benthic and deep-water animals may have slightly longer wavelength emissions (500–520 nm), suited to the more turbid waters they encounter (Fig. 9.30). For many animals the bioluminescence may serve to aid predation (e.g. as a 'lure' for food-gathering, as in the anglerfish), or as a deterrent to predation (e.g. as a 'startle' effect in various small fish that are heavily preyed upon and may hope to escape while a predator is confused by their display, or as a directly off-putting signal especially to negatively phototactic predators). Intraspecific luminescent signalling is also occasionally found, in relation to recognition systems and prior to mating, and in such cases the signal may involve unusual and highly tuned wavelengths. For example, the rhodopsins of the deep-sea *Bathylagus* have peaks at 466 nm and 500 nm, detecting and discriminating between inter- and intraspecific bioluminescent red emissions.

9.7.3 *Chemical signalling and chemoreception*

Chemoreception plays a major part in the lives of marine organisms. Chemicals can diffuse over relatively long distances in aquatic systems, retaining directional information where currents are of moderate and consistent velocity and with minimal turbulence. In inshore waters they may be more problematic, however, and are normally only useful over a short range or as chemical deposits on a substrate.

Detection of food by chemical cues is an obvious case, and the signal is frequently an amino acid, or a mixture of several of these. This has been particularly well-studied in fish, where 'taste buds' occur on various parts of the body, and in marine cnidarians where chemical feeding signals (sometimes in concert with a mechanical trigger) elicit the firing of specific nematocysts (stinging organelles). Thresholds for such chemicals can be extraordinarily low, perhaps only 10^{-9}–10^{-11} M.

More sophisticated cases of chemical signalling in the form of pheromones are also becoming increasingly well documented

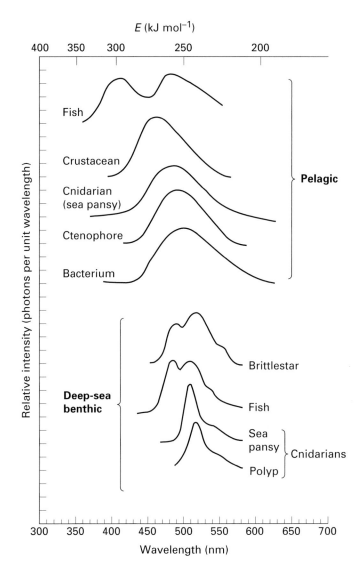

Fig. 9.30 Emission wavelengths of various bioluminescent marine animals. (Adapted from Wampler 1978.)

in the marine environment. Pheromones are used by animals in the sea for alarm purposes, as settling or metamorphosis cues, and perhaps most obviously as mating signals. Orientation to such chemicals is primarily by movement upstream ('up-current') when they are detected, but at least in crabs also involves sampling of attractant distribution across the stream to take account of turbulence. In this respect aquatic pheromone systems are unlike the better-studied system of airborne detection in terrestrial insects, which appears to be entirely by upwind flight. This difference is presumably because structures of chemical plumes arising from a point source are more stable in liquids than in air, and eddies are at least 100 times less disruptive to the flow, so that concentration-based information is more reliable.

The sea anemone *Anthopleura elegantissima* has been especially well studied as an example of a marine alarm pheromone; when wounded it releases 'anthopleurine', which causes rapid contraction and convulsion in neighbouring anemones within a few seconds. This pheromone is particularly intriguing because it can be carried by the anemone's chef predator, a sea slug; when the slug eats some of the anemones it receives a dose of anthopleurine, which is stored in its tissues for several days and gradually diffuses out; thus when the slug approaches further anemones they are forewarned of its presence.

A few marine animals from the benthos of shallow waters also release trail pheromones, the most obvious examples here being molluscs that deposit a chemical signal with the mucus trail. The sea slug *Navanax inermis* uses intraspecific trails to locate mates and also to locate other slugs as (cannibalistic) food. However, when attacked by other predators the slug adds another component ('navenone') to its mucus, which serves as an alarm pheromone to conspecifics, and they then avoid rather than follow its trail.

Perhaps the most ancient function of pheromones is as a gamete attractant, released from female cells and bringing about the release and approach of male gametes; this phenomenon occurs quite commonly even in relatively simple marine organisms such as the brown algae, and has been analysed in great detail in certain mass-spawning invertebrates such as sea urchins, ascidians, hydrozoans and corals. Chemical regulatory systems are particularly important for such animals, since mass-spawning of many species may occur simultaneously and each sperm must find a conspecific egg or be wasted; the chemicals used are rather varied, including unsaturated fatty alcohols and modified terpenoids. In planktonic organisms the problems may be rather different, requiring two mature adults to come together and then release gametes. Some sea urchins, some crustaceans and one marine rotifer have recently been shown to use surface glycoproteins as contact-mating pheromones. These glycoproteins are used in the crustacean *Tigriopus* as a cue for mate-guarding behaviour; they arise mainly from the female's terminal appendages, the male having appropriate protein receptors on his antennules. However, there are good theoretical reasons why small pelagic animals should not use long-distance pheromones; depending on the pheromone diffusibility and threshold concentrations, calculations show that efficiency of detection would be inadequate below a minimum size of between 0.2 and 5 mm. Microorganisms and small planktonic animals, such as rotifers and ostracods, are therefore unlikely to use sex pheromones. The more spectacular cases of reproductive pheromones can thus be found in larger and more complex invertebrates. For example, in certain gnathiid isopods males release an attractant from their burrow that spreads over a long range as a pheromone 'plume' in the water and attracts a number of females to the burrow, where they are captured and kept as a harem. Another particularly well-studied marine pheromone working at close range occurs in barnacles (*Semibalanus balanoides*), which are inseminated by a highly extensible 'pseudo penis' from a neighbour and then brood the fertilized eggs over winter. In spring the onset of algal blooms

permits filter feeding, which in turn promotes the production of a 'hatching pheromone', a modified fatty acid derived from the algal/invertebrate diet. This is secreted into the egg cavity and causes the dormant eggs to hatch as free-swimming nauplius larvae. (It is noteworthy that this same barnacle species also has a well-known 'settlement pheromone', in this case a protein which gradually leaches out from its cuticle and serves as an attractant to other barnacles to settle nearby, so allowing the close-range fertilization to operate as the barnacles mature.) The crab *Callinectes sapidus* uses a sex pheromone to elicit more specific and complex behaviours, with a compound from the almost-mature female's urine exciting the male into his full courtship display, after which he grabs the female and carries her around until she moults to her final mature state, when copulation occurs. Sex pheromones in marine animals may also have even longer term effects; the sex and gross morphology of a marine echiuran worm such as *Bonellia* are determined by pheromones, the larvae developing into small and effectively parasitic males if they encounter a signal ('bonellinin') from a female, but growing into large females themselves if they settle on a substrate where no females and thus no pheromone are present.

Olfactory organs in marine fish are large with a high density of receptors, linking to further chemical sensors on gills, skin and fins, and suggesting a more sophisticated chemical environment and signal detection system than we have previously supposed. Fish certainly use pheromones as sexual attractants, and in some species (e.g. the lamprey, *Petromyzon*) the females can be attracted in their hundreds by fishermen who trap a single mature male as lure. Fish have also been shown to use pheromones in regulating pair interactions, in parent–young interactions, and in spawning. The fairly widespread phenomenon of *schreckstoff*, or alarm substance, in the skin of fish, such as minnows, that shoal (where a damaged fish becomes a source of alarm to its conspecifics), and the extraordinary 'homing' ability of salmonids that detect their own particular spawning stream with extremely low 'taste' thresholds, certainly also attest to the sophistication of aquatic chemoreception. Indeed, the limited evidence available as yet suggests that the timing, pulsing and detection systems of pheromones in marine habitats are of similar complexity to the far better studied terrestrial systems of insects and of mammals, discussed in Chapter 13.

9.7.4 *Mechanical senses*

In the sea mechanoreceptors can be used to detect gravity, pressure, water currents and oscillatory particle movements (vibration), as well as the more obvious 'sound' or far-field vibrations. All are essentially based on one of two fundamental designs. **Statocysts** (Fig. 9.31a) are composed of one or more statoliths, which are accretions of calcium carbonate held in a chamber where movement of the statolith alters the position of cilia on surrounding sensory epithelial cells. They are common in invertebrates, from radial arrangements in the umbrella margins of

Crustacean statocyst **Vertebrate hair cell and its directional sensitivity**

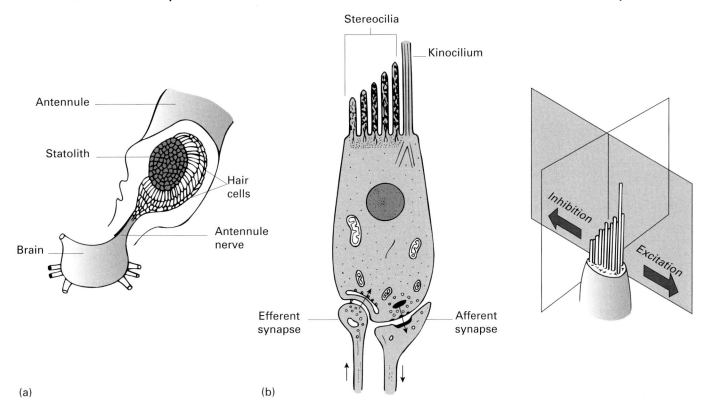

Fig. 9.31 (a) Statocysts (with calcareous statolith resting on an array of receptors) used in marine invertebrates for mechanoreception, and (b) hair cells as used in the lateral line of a marine fish. (Adapted from Flock 1965.)

jellyfish to multiple arrays in the heads of crustaceans and cephalopods. **Hair cells** (Fig. 9.31b) are more complex receptors, usually with one main 'kinocilium', often in a particular spatial relationship to multiple 'stereocilia'; distorting the kinocilium in some planes disturbs and triggers the stereocilia, whereas other distortions avoid the stereocilia, giving overall directional sensitivity. Fish use ranks of such hair cells in their lateral-line systems, either as small free-standing groups, called neuromasts, or embedded within a complex subepidermal canal system. This system permits even the monitoring of many neighbours' movements in a shoal, and precise coordination of manoeuvres by all the constituent fish of that shoal. At least some fish show the beginnings of an inner ear, derived from this lateral-line system, and in some species its sensitivity and frequency range is enhanced by an association with the swim-bladder, whose walls magnify the oscillations via modified vertebrae known as Weberian ossicles. Curiously, coleoid cephalopods (squid, etc.) with a similar lifestyle have no real mechanosensory system and are effectively 'deaf'.

Acoustic energy is transmitted more efficiently through water than almost any other form of energy, and sound therefore persists without serious attenuation or interference to the signal over a long range, allowing submerged animals to use it as a primary method of communication. Sounds that seem rather diffuse and unfamiliar to us can give vital information to fish and mammal ears—the noise of surf, or of ice creaking, or of the faint disturbance created by the passage of a potential predator such as a shark. Many marine mammals go further than this and capitalize on the long-range potential of underwater sound by using it to communicate, with calls related to the presence of danger, of food, and of conspecifics, as well as giving information about their own identity, position and territory, or their reproductive state. The toothed whales use calls of fairly high frequency (1–20 kHz), while baleen whales (when not echolocating) mainly use rather low-frequency calling noises (20–500 Hz, the wavelength being hundreds of metres at the lower frequencies). Behavioural audiograms for a range of species are shown in Fig. 9.32. Calls are variously described as grunts, moans, whistles and barks, but there is no agreement on their classification or even on their function. For the toothed whales, which are often very gregarious, certain sounds do appear to serve as individual recognition markers ('signature calls'); these are well-documented in dolphins and their relatives. Pinnipeds (seals, etc.), which are able to hear both in water and in air, often lack very long-range underwater communication systems, though they do call using barking noises within a group and between mother and pups.

Underwater calls are particularly effective at moderate depth, where sound can be channelled by the properties of water

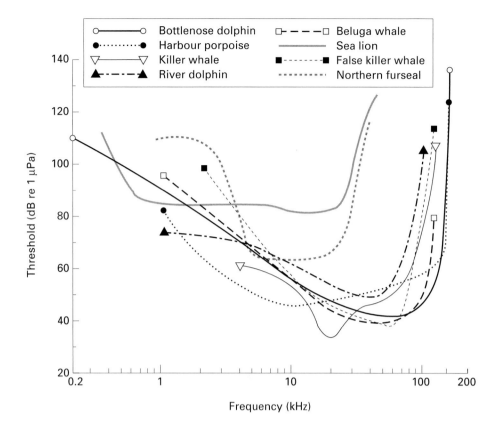

Fig. 9.32 Long-range sound and hearing in the sea: audiograms showing the frequency spectra of various cetaceans and pinnipeds using long-range sound communication under water.

layering in the oceans. For example, sounds are propagated halfway round the globe through the 'deep sound channel', typically occurring at 600–1200 m depth in low and middle latitudes; this is the depth of minimum sound speed, where sounds are focused and propagated without loss from bottom and surface reflections. A similar phenomenon occurs at much shallower depths in polar waters.

9.7.5 *Other senses*

Marine organisms may also use more unusual senses to navigate and locate objects in the very dark or turbid parts of the oceans, particularly echolocation and electrical sensitivity. Marine mammals, such as dolphins, some whales and a few seals, are well-studied exponents of aquatic **echolocation,** and demonstrate how useful a sense this can be in the seas where sound attenuation is much less than in air. Usually the system involves high-frequency sound (12–150 kHz), above the frequencies used for intraspecific calls, and commonly described as 'clicks'; each click may be only 50–200 microseconds (µs) in duration, and can be useful over a range exceeding 100 m, possibly further for very large targets. Weddell seals can echolocate prey and also (crucially) find their breathing holes in the ice. Toothed whales echolocate at very high frequencies (20–150 kHz), and dolphins in particular have very complex clicking emissions at about 30–150 kHz (peaking in amplitude at around 50 kHz), using nasal and laryngeal structures to emit a highly directional 'cone'

of sound with a beam width of only 5–12° according to species. Unusually, they appear to receive most of the returning sound through their bones, especially the lower jaw, which is filled with fatty material; from here it is transmitted to the middle ear and inner ear, which are embedded in unusually dense bone. Dolphins can discriminate a difference of just 1 decibel (dB) in the amplitude of echoes, allowing them to detect very subtle differences in targets.

Electroreception is one sensory modality that is confined to aquatic habitats and is reasonably common in marine fish. There are several alternative levels of sophistication. Sharks and rays can detect very small electric emissions (only a few microamps) from the hearts or gill muscles of other active animals that form potential prey, but they do not produce any external electric discharge of their own. However, a number of fish take matters further and make their own electric field; the so-called 'weakly electric' fish achieve an endogenous electric field of AC discharge at some species-specific (but often sexually dimorphic) frequency between 50 and 200 Hz, usually from an electric organ of modified muscle, sited near the tail (Fig. 9.33). They can electrolocate intruders by sensing the disturbance to this field, or can signal to conspecifics. In all these cases, the actual electroreceptors are small ampullary organs closely associated with the lateral-line systems, and with rather long entrance canals that help to maximize the voltage difference between different receptors and improve discrimination. In elasmobranchs these structures are known as the ampullae of Lorenzini. In teleosts rather

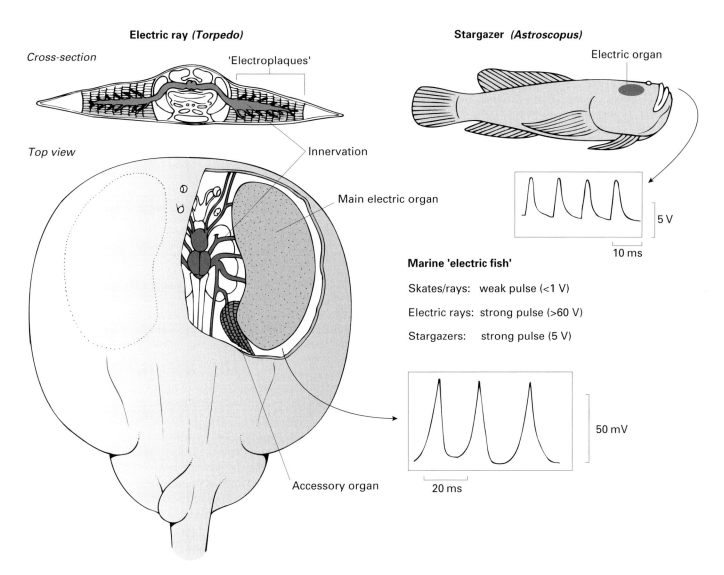

Fig. 9.33 The electric organs of marine fish: occurrence, structure and discharge patterns.

similar electroreceptors may be scattered over much of the body surface. These electroreceptors may be sufficiently sensitive to detect the Earth's magnetic field, because current flow is induced in any conductor moving through a magnetic field, and sea water (unlike fresh water or air) provides a sufficiently good conducting medium to allow completion of an electric circuit. This opens up the possibility of **magnetoreception** for marine fish, and sharks and rays with their ampullae of Lorenzini have indeed been shown to use this detection mode.

However, representing the final level of development of exploitation of electricity, there are a few species of 'strongly electric' marine fish, such as electric rays (*Torpedo*) and stargazer fish, which have very powerful discharges of up to 600 V, originating from 'batteries' of modified muscle cells. Such discharges are used for actually stunning the prey, and in these animals there are usually no special electroreceptors.

9.8 Feeding and being fed on

In the pelagic marine environment, many feeding options are open to animals. They may filter-feed on phytoplankton (though it is normally impractical to separate out phytoplankton and zooplankton, so there are no real 'herbivores' at this level); or prey on other nektonic or planktonic animals, which are often present in huge numbers (but, as we saw earlier, not necessarily with great diversity of species); or in a few special areas, such as the Sargasso Sea, they may graze on floating seaweeds (in similar fashion to coastal algal grazers dealt with in the next chapter).

Filters may be provided by cilia and flagella, by mucus sheets (on their own or combined with cilia), or by cirri (cuticular bristles) in invertebrates; vertebrates use various modifications of the jaw structures, such as gill-rakers, culminating in the impressive sieving baleen plates of various whales. Pelagic feeders vary from small passive floaters, doing little more than unselectively filtering out the surrounding life, to highly active swimmers, up to and including sharks and whales, which may be very selective and often fiercely predatory in their feeding. Taking the middle ground are many pelagic feeders, such as crustaceans, that are selecting their food on size and palatability, and migrating to keep up with the best food supply. Planktonic species commonly show diurnal vertical migration, while many of the nekton have prominent seasonal latitudinal migrations (e.g. the whales that migrate from poles to equator and back, covered in section 8.9.3).

Nutrition in the benthic environment, at all depths, is rather different. Some animals in the shallower seas again operate as filter-feeders, exploiting all the floating life forms; these benthic filterers are often colonial (tunicates, bryozoans and many hydroid cnidarians), or in the case of molluscs may be individually massive (giant clams). Others feed on dead remains, as deposit-feeders and scavengers; these more commonly occur at depth, and are usually solitary (many worms, crustaceans and brittlestar echinoderms). There are also options for more active feeding, by preying on other benthic animals: but at depth there is only a very low-density biota, and food is scarce (e.g. one fish per 1000 m³). Therefore deep-water marine predators have to be very specialized structurally, while being catholic in their tastes: many have very extensible mouths and guts, and a reduction of less vital skeletal components, with a very low metabolic rate. Fish that make their living in this way often also have reduced trunk muscles and skeletons as we saw earlier; but some of these species also have strengthened and highly ossified jaws, so that they become rather static feeding traps for smaller organisms. To escape moderately active predators, many of the smaller deep-sea inhabitants that might serve as prey tend to be highly transparent.

Many benthic marine animals have been found in recent years to rely at least partly on direct uptake of simple dissolved chemicals from the surrounding sea water. Some of the marine annelids are particularly good at this (Fig. 9.34). Better again are the pogonophorans, which have no gut at all and absorb some of their nutrients directly through their own skins; they often live in areas unusually rich in nutrients. Probably living in a tube helps them, by creating a local concentration of nutrient; furthermore, being immensely long and thin, they have a remarkably high surface area to volume ratio, which also helps.

We should not forget that some benthic marine animals live in and around deep-sea ridge crests and at subduction zones where there is geothermal activity, or around oil seeps, or in highly reducing sediments. These animals have very 'odd' metabolisms where sulphur bacteria living as symbionts within their tissues or body cavities provide food from the unusual sulphurous cocktail of chemicals that occur. Here again the best examples are the pogonophorans; we will deal with the physiology of these hydrothermal vent organisms as a special case in Chapter 12.

9.9 Anthropogenic problems

There is no longer any part of the ocean system that does not contain pollutants, though in the deep oceans, beyond contin-

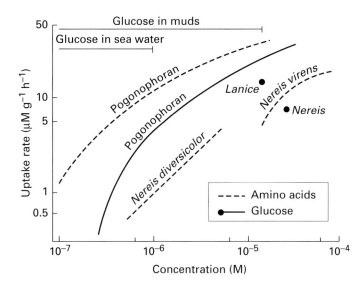

Fig. 9.34 Amino acid and glucose uptake rates through the cutaneous surface in various polychaete annelids and (at even lower concentrations) in the gutless pogonophoran worms.

ental shelves, it remains true that most pollutants get diluted in such a huge volume of sea that they probably represent no real threat. Random dumping of 'loose' chemicals at sea is undoubtedly unwise, although far too little is known of deep currents and sediment movements to predict how such things might get concentrated in food chains and where they might surface again. But waste disposal that is genuinely deep-sea, and in properly sealed well-designed weighted containers, may indeed be a relatively cost-free solution to some of the disposal problems that can be so difficult on land.

However, there are other threats to the marine biosphere from man's activities. Some of these come from the various atmospheric changes that man precipitates: global warming caused by increased CO_2 and CH_4 levels, acidification of rainfall, and atmospheric heavy metals such as lead and copper. There is mounting evidence that warming sea temperatures are already having effects on coral reefs, which lose their photosynthetic symbiotic algae and become bleached as a result, with knock-on effects on all the associated biota. Even small changes in temperature can clearly be hazardous to marine animals that are almost inevitably stenothermic. Particular attention has been drawn to sea turtles, already under threat when they come ashore on tourist beaches to lay eggs, and for whom small temperature changes may have chaotic effects on sex ratios (due to their temperature-dependent sex determination; see Chapter 13), and hence on population dynamics.

Marine transport also causes problems, chiefly arising from oil lost from tankers, which can be especially dangerous in relatively enclosed areas such as the Red Sea, the Caribbean, areas around the Philippines, and the Mediterranean. The major causes are uncontrolled discharge, leakage and tank cleaning; all this has declined somewhat since oil prices rose in the late

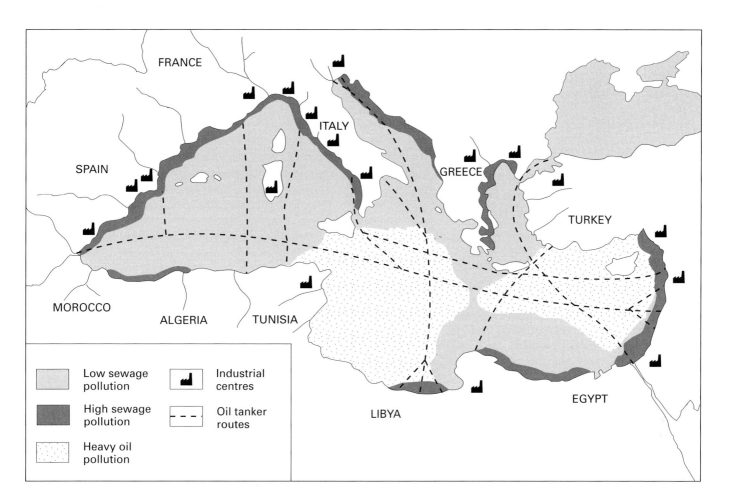

Fig. 9.35 Oil and sewage pollution zones in the Mediterranean.

Table 9.8 Effects of hydrocarbons on marine birds.

Site of oil	Effects
Plumage	Reduced buoyancy
	Increased swimming activity and metabolic rate
	Blockage of nasal gland
Gut (ingestion)	Reduced gut absorption
Blood (after ingestion)	Haemolytic anaemia (damaged red blood cells); hypoxia
	Disruption of adrenal cortex function; reproduction disrupted
	Malfunction of nasal glands; salt accumulation, kidney failure
Eggs	Disrupted embryonic development; death or teratogenic effects

1970s, and with better policing. But still about 3 million tonnes of oil get into the seas each year. Figure 9.35 shows areas of the Mediterranean commonly found to be polluted with oil, corresponding closely to the oil-refining areas and shipping lanes. Then there are also periodic oil transport accidents, the most publicized recently being the grounding of the *Exxon Valdez* in Alaska in 1989 (this episode released around 36 000 tonnes, killing at least 35 000 birds and most of the sessile organisms in Prince William Sound), and of the *Sea Empress* in February 1996, when about 72 000 tonnes of light crude oil were released into the seas around the coast of southwest Wales, a region renowned for the diversity of its coastline, of which more than 100 km became seriously polluted.

The high mortalities accompanying such disasters result because oils are potentially directly lethal to all filter-feeding organisms and indirectly to the organisms higher in the food chains that feed on them. Oils are also commonly fatal to sea birds and some sea mammals that rely on their surface properties for buoyancy or for insulation, and that will therefore groom off and swallow surface deposits from fur or feathers. A summary of the effects of hydrocarbons on birds is shown in Table 9.8.

Nevertheless, natural degradation and wave action may often be the best option for cleaning up beaches, because many of the oil dispersants used are detergents with effects worse than the oil, disrupting the biological lipid layers on which so many animals depend for their intrinsic impermeability, and especially destroying the integrity of their respiratory surfaces. Bacterial remediation may be the long-term solution, whereby nitrogen and phosphorus are added to oil slicks to encourage bacteria to degrade the oil rapidly, though this has not really been shown to work well *in situ* on a large scale as yet.

Apart from oil, marine cargoes of toxic wastes and chemicals also get into the seas; usually this does not matter too much because of the vast volumes and diluting effect of the oceans, but very persistent toxins, such as dioxin and polychlorinated biphenyls (PCBs), can become locally concentrated, and PCBs in particular have been linked with loss of disease resistance in marine mammals, birds and invertebrates.

The North Sea is undoubtedly badly polluted; volumes of waste deposited have all decreased (by about 40% overall) in the last decade but are still high. Fish stocks have seriously declined, though there are additional reasons for this (climate, silting and over-fishing). Sewage sludges and oils have been the worst pollutants; heavy metals and radionuclides are also a problem though. Conditions in the Mediterranean are even worse, since it is a shallow enclosed sea surrounded by industrial areas to the north and oil-producing areas to the south. Any foreign material dumped into this sea tends to stay in it; this also applies to all the waste coming down the rivers Rhône, Po, Danube, Don and Nile, as well as dozens of lesser rivers. Each year the Mediterranean gets 1.5 million tonnes of industrial and agricultural effluent, and 30–50 million tonnes of untreated sewage. Figure 9.35 also demonstrates the huge areas of this sea that are affected by severe or moderate sewage pollution. Coastal resorts that are major tourist centres are affected, some almost terminally; marine life and fish stocks are reduced, with detectable contamination by mercury and other metals, and some key species are now endangered. An action plan was put together by the United Nations in 1982 involving 18 countries bordering this sea, and there have been resultant improvements in some respects in the last 15 years.

Both the Mediterranean and North Seas, and also the Baltic and localized sea areas around Japan and Indonesia, have recently become infamously prone to red algal blooms that can release toxins and deplete fish stocks. There have been incidents in Japan of multiple severe poisoning from eating affected fish; and the tourist industry along parts of the Italian Adriatic coast has been seriously affected in the last few years after great blankets of seaweeds grew offshore. This was probably a consequence of unbalancing the nutrients in the ecosystem due to pollutants combining with climatic changes and CO_2 levels; a salutary lesson in the way that differing environmental problems actually interact to make things even worse.

9.10 Secondary invasion of the seas: marine vertebrates

Much of the discussion so far in this chapter has dealt with invertebrate animals in the sea, for whom life is relatively straightforward in physiological terms. But marine vertebrates (and also

the very rare marine insects) have a special set of problems when living in the sea. Both groups have arrived in the sea secondarily, after a long history of life in fresh water or on land. (Note that while there is an argument about the original habitat of the first vertebrates, it is clear that a large part of teleost history, and radiation from crossopterygian fish into the tetrapods, occurred in fresh water.) The modern animals in these taxa all have low blood concentrations, usually in the range 250–450 mOsm, and are effectively in a state of water shortage ('physiological drought') when they live in the seas, as their body water is continuously susceptible to leaking out down an osmotic gradient, causing tissue shrinkage.

Three particular problems therefore need special consideration. Firstly, we need to know how these animals deal with the osmotic and ionic stresses imposed upon them. Secondly, there will be problems of temperature regulation for many of the vertebrates, some having an endothermic inheritance. Thirdly, there will be problems with breathing in the sea, in cases where the respiratory systems were evolved for use on land—a situation that applies to reptiles, birds and mammals, all having air-breathing lungs, and to insects with their tracheal systems. But it is also worth noting in passing that all the other problems of life in the sea—pressure and buoyancy, life-history strategies, and sensory systems—will be affected in differing ways by the altered 'starting conditions' for these secondary invaders, and in particular any marine adaptation that involves ion pumping may be rendered more difficult for a hyposmotic animal. It may be no coincidence, for example, that vertebrates have adopted a buoyancy system that does not involve ion-pumping, and that the cephalopods, endowed with such a system, have never been able to leave the sea and colonize freshwater habitats.

9.10.1 *Ionic and osmotic regulation in marine vertebrates*

The body fluid composition of a range of marine vertebrates is shown in Table 9.9. The most obvious feature is the marked hyposmotic condition of many of the animals listed. In fact, most marine vertebrates retain osmolarity around one-third that of sea water (about 300–400 mOsm) that their ancestry gave them. They have blood dominated by Na^+ and Cl^-, with low potassium levels and relatively little organic content.

Ureo-osmoconforming

There are some clear 'anomalies' in Table 9.9, notably the cyclostome hagfish and the various cartilaginous fish (Chondrichthyes). Figure 9.36 reveals that only the first of these isosmotic curiosities has body fluids that are also roughly isionic with its surroundings, at least in respect to the major sodium and chloride ions; in other words, the hagfish is unique amongst vertebrates in having blood very like sea water, being an osmoconformer and an ionic conformer. This may reflect a marine ancestry, though the levels of ions in the animal's cells are surpris-

Table 9.9 Body fluid composition in marine vertebrates.

Species	Osmotic concentration (mOsm)	Na	K	Ca	Mg	Cl	Urea	TMAO
Agnatha								
Myxine (hagfish)	1 062	558	10	6	18	546	–	–
Petromyzon (lamprey)	317							
Chondrichthyes								
(elasmobranchs)								
Mustelus (dogfish)	1 011	288	8	5	3	270	342	~100
Raja (ray)	1 050	289	4			——444——		
Osteichthyes								
Latimeria (coelacanth)	1 181	181	51	4	14	200	355	~100
Teleosts								
Lophius (anglerfish)	452	180	5	3	3	196	<1	–
Opsanus (toadfish)	392	160	5				<1	–
Anguilla (eel, in sea water)	371	177	3				<1	–
Gadus (cod)	308	174	6	7	3	150	<1	–
Amphibia								
Rana cancrivora (crab-eating frog)	830	252	14			——350——		

TMAO, trimethylamine oxide.

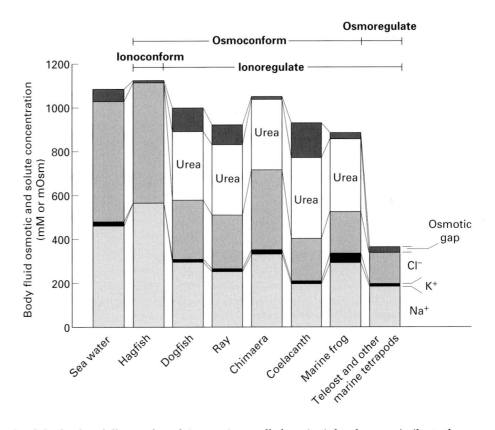

Fig. 9.36 Body fluid composition in marine fish and a marine frog, showing the ionic and osmotic conforming or regulating patterns. 'Osmotic gap' is the blood solute concentration not accounted for by the listed components. (Adapted from Withers 1992.)

ingly like those of all other fish, and the hagfish also has fully glomerular kidneys, neither of these features being particularly helpful to a primitively marine animal.

In contrast to the hagfish, the cartilaginous elasmobranch fish though isosmotic actually have ionic levels more similar to those of other vertebrates (see Fig. 9.36, the dogfish and ray), and less than half of seawater levels. In these animals the osmotic deficit is restored in a very different fashion, by incorporating high

Component	Osmotic concentration (mOsm)	mM					
		Na	K	Ca	Mg	Cl	Urea + TMAO
Blood/ECF	1096	296	7	3	4	276	504
Intracellular (muscle)	1189	18	130	2	14	13	905
Rectal gland fluid	1050	512	3	1	3	520	–

Table 9.10 Body fluids, intracellular concentrations, and rectal gland fluid concentrations in the elasmobranch fish, *Squalus*.

ECF, extracellular fluid; TMAO, trimethylamine oxide.

levels of organic material into the blood, in particular the waste product urea, at about 300 mOsm, and its derivative TMAO, at about 100 mOsm (with small amounts of other osmolytes such as betaine). The methylamine derivative TMAO acts as a counterbalance to the perturbing effects of high urea concentrations on protein structure. TMAO levels are down to 40 mOsm in shallow water species, and up to 400 mOsm in some very deepwater species, suggesting that it may also help to counteract the destabilizing effects of pressure on proteins (and remember that it can also help with positive buoyancy; see section 9.6.2). These animals are therefore osmoconformers but have to be strong ionic regulators, and the inward ion diffusion is principally countered by secretion of NaCl by a rectal gland, which can secrete a fluid containing about 500 mM Na$^+$. The composition of different body fluids in the elasmobranch *Squalus*, including those of its rectal gland, are shown in Table 9.10.

Only three other marine animals are known to use this particular technique of 'ureo-osmoconforming' (Fig. 9.36 & Table 9.9). Two are fish—the chimaera (cartilaginous rat-fish) and the coelacanth *Latimeria* (a 'living fossil', close to the root of the tetrapods and lungfish, and only distantly related to other 'fish'). In addition there is the single species of modern amphibian that can live in sea water, the crab-eating frog, *Rana cancrivora*, of southeast Asian mangrove swamps (oddly enough, the tadpoles of this frog can also live in sea water but do so without urea accretion, adopting instead the ionic and osmotic regulation patterns of teleost fish). In all these cases the level of urea + TMAO in the blood may be around 350–400 mOsm, and the presence of these materials substantially limits the osmotic stress for the animal by preventing or greatly reducing the continuous outward leakage of water.

Coping with hyposmotic body fluids in teleosts

All the rest of the marine vertebrates, including the teleosts and the tetrapods, are substantially hyposmotic to their surroundings and will therefore tend to lose water and gain salts. They must regulate continuously, both osmotically and ionically. The water and ion leakages will occur especially through those parts of the body surface that are most permeable, such as the gill epithelia of teleosts; other parts of the outer surface generally have a rather reduced water permeability.

In teleosts the inward passage of salts and outward passage of water are countered by four main systems (Fig. 9.37).

1 Drinking the surrounding salty water to get water into the body.
2 Uptake (70–80%) of this water and of the univalent ions (Na$^+$, K$^+$, Cl$^-$) from the imbibed sea water through the gut wall, particularly the oesophagus; but minimal gut uptake of divalent ions (Ca^{2+}, Mg^{2+}, SO$_4^{2-}$), these being passed with the faeces.
3 Elimination of most of the excess salt thus gained by active transport of sodium and of chloride across the gill surfaces, with varying levels of cutaneous transport also occurring. The gills contain 'chloride cells' (see Chapter 5), responsible for much of the Cl$^-$ secretion, via sodium pumps, Na$^+$/K$^+$/2Cl$^-$ cotransporters and chloride channels.
4 Excretion of excess divalent salts through the kidney, via an isotonic urine in which Mg^{2+} and SO$_4^{2-}$ concentrations are nevertheless unusually high. Urine volume is generally rather low, and glomeruli are reduced in number, sometimes even lacking (e.g. in the toadfish, *Opsanus*, and some goosefish and pipefish, where urine is produced by secretion rather than filtration). Figure 9.37(a) shows the major ion and water movements in the nephron of a marine teleost.

The net result is retention of a fluid that (if it could be isolated) would be hypotonic both to the ingested sea water and to the urine, and which therefore restores the balance of the blood; in effect, these fish are distilling sea water in their bodies. The marine teleosts also use the gill epithelia as a major route for excretion of nitrogenous wastes, with NH$_3$ and NH$_4^+$ leaking passively outwards (perhaps with some Na$^+$/NH$_4^+$ exchange). This may be aided by NH$_4^+$ substituting for the similarly sized K$^+$ in baso-lateral sodium pumps.

The tissues are regulated relative to the blood in a rather similar fashion to the invertebrates considered earlier, i.e. by the manipulation of amino acid levels in the cells. Some teleosts use low levels of TMAO as an osmotic effector in addition, indicating that the strategy of urea/TMAO retention in elasmobranchs discussed above is perhaps just an extension of the general principle of using small organic osmotic effectors to maintain isosmotic conditions.

Using these kinds of mechanism, marine teleosts can maintain their blood and cells at remarkably constant concentrations. Indeed some can migrate into freshwater environments for part of their life cycle without substantial osmotic problems, by modifying the relative rates of these various processes through careful hormonal control. These anadromous fish (breeding stage in

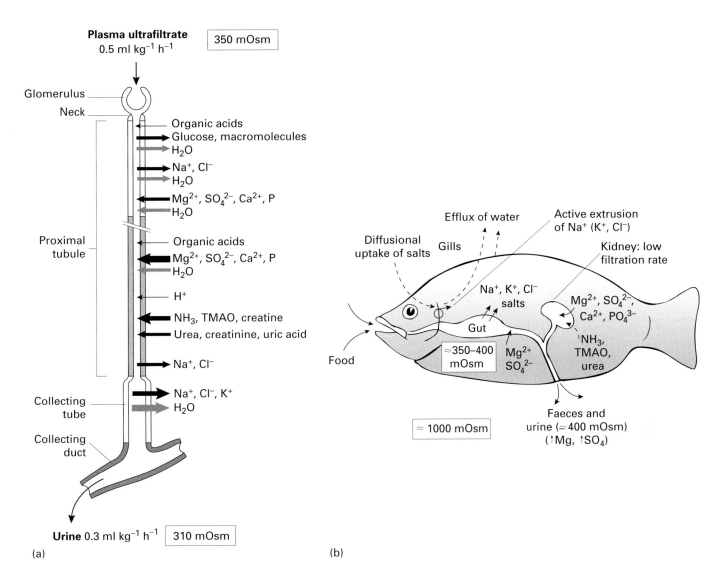

Fig. 9.37 Movements of salts and water: (a) in the nephron of a marine teleost, and (b) in the whole organism. TMAO, trimethylamine oxide. (a, From Hickman & Trump 1969.)

fresh water—lampreys and salmon) and catadromous fish (breeding stage in sea water—eels) have the regulatory patterns characteristic of the inhabitants of whichever environment they are in, thus being able to pump sodium ions outwards in sea water, and inwards when in fresh water; although their body fluids are fairly constant throughout (see Chapter 10).

Hyposmotic body fluids in marine tetrapods

Marine tetrapods have rather different strategies from those so far described, and may conveniently be considered in two groups: firstly, the reptiles and birds, and second, the mammals.

There are a number of reptiles and birds that live in or around sea water for most of their lives (Table 9.11); in the case of sea snakes this residency is permanent, whereas turtles must breed

on land, while marine iguanas and all marine birds are essentially terrestrial animals that submerge themselves for varying periods. Like the teleosts they are unable to produce a significantly hyperosmotic urine to get rid of excess salts, yet they have no gill surfaces at which salts could be excreted. Instead they have a very low P_{Na} in their skin to slow inward salt leakage as far as possible, and they have specific **salt glands** (Fig. 9.38) where salt from drinking and from the slow leak into the body can be actively extruded. These glands can produce NaCl at very high concentrations (Table 9.11); for example, 1170 mM Na^+ and 1330 mM Cl^- has been recorded in the salty extrusions of the marine iguana, which can be sprayed out some distance from the head, whilst 1100 mM NaCl in Leach's petrel is the record for any fluid concentration achieved by a bird. They operate using standard sodium pumps across a rather impermeable epithelium so that there is little entrained osmotic water flow; in the avian salt gland this is an ouabain-sensitive Na^+-transporter, apparently sited on the basal membrane of these epithelial cells. The glands may be variously situated in reptiles:

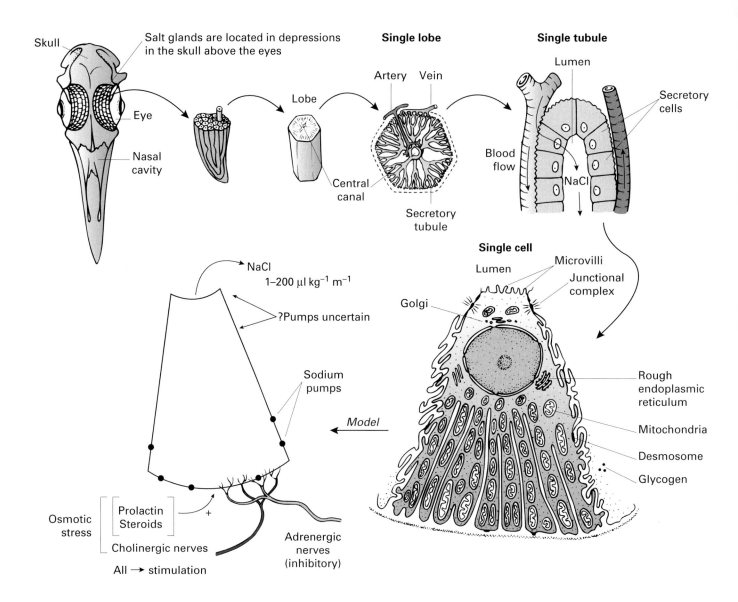

Fig. 9.38 Salt gland positions and functions in marine tetrapods. (Adapted from Peaker & Linzell 1975.)

Table 9.11 Salt gland performance in marine tetrapods.

Species	Na+ concentration (mM)	Flow rate (μl kg⁻¹ h⁻¹)
Sea snake	620	35
Terrapin	680	4
Turtle	690	19
Marine iguana	1000–1400	8
Cormorant	500–600	70
Pelican	600–750	100
Gull	600–900	30
Penguin	720–850	40–70
Petrel	900–1100	–

the nasal cavity (iguanas), eye sockets (turtles), mouth (sea snakes) or tongue (estuarine crocodiles). In birds they are usually situated on top of the skull, and invariably open to either side of the 'nose', near the top of the bill. They seem to be modified mucoid or lacrimal glands, and probably have multiple convergent origins. These various salt glands are remarkably effective in removing salt from the body, and they allow their possessors to drink only salt water. They can be switched on and off, according to osmotic need, by cholinergic nerves, and their activity level is also affected by prolactin and steroid hormones. In birds they can become enlarged or shrunken within an individual's lifetime according to the extent of its encounters with a salty environment and diet. However, many partially marine birds and reptiles ameliorate their problems by taking as much water as possible from their food instead; the marine iguana gets about 58% of its water from food, only 38% from drinking sea water, and 4% from metabolism.

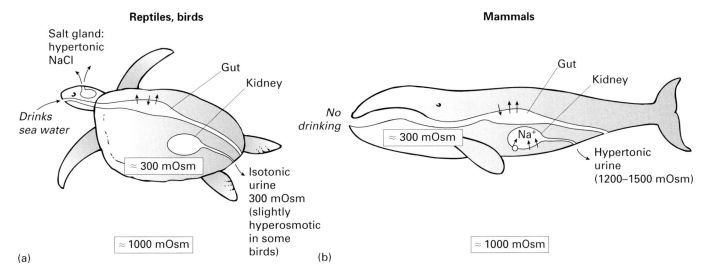

Fig. 9.39 Overview of salt and water balance in marine tetrapods.

Secondly, there are several groups of marine mammals, in different families. Most obvious are the permanently resident cetaceans (whales, porpoises) and sirenians (dugongs), and the amphibious pinnipeds (seals and their kin). These have no salt-secreting glands, because they have inherited the benefits of the mammalian kidney (see Chapter 5), superior to that of other tetrapods in that by using the elongate loop of Henle it can produce a very markedly hyperosmotic urine, more concentrated than the animal's own blood and more concentrated even than the sea water in which the animal lives; thus these animals get rid of excess salt directly from the blood. A whale drinking one litre of sea water gets a net gain of about one-third of a litre of pure water once the kidneys have done their work. (A human can produce urine more concentrated than human blood, but less concentrated than sea water, so that drinking sea water is of no benefit to a thirsty human, producing a net water loss just to get rid of the salt load.) However, marine mammals do not normally drink much sea water and so do not get as much excess salt as the reptiles and birds anyway. This is at least partly because all marine mammals are predominantly carnivorous, and take in dietary water only as a component of the flesh they get as food, having relatively dry stomach contents. In a sense then, the marine mammals are 'parasitizing' the osmoregulatory abilities of their various prey, by taking in only these preregulated fluids.

A summary of the fluid compositions, and of the variety of osmoregulatory mechanisms operating, in a range of marine vertebrates is shown in Fig. 9.39.

The more complex ionic and osmotic regulatory strategies of marine vertebrates necessitate rather more complex control systems than are found in invertebrates. Prolactin from the anterior pituitary (in the brain) affects chloride uptake in general, and also more specifically controls mucus production by skin and by fish gills, and ion uptake in salt glands. Vasopressin and other posterior pituitary peptides act on kidney glomeruli in most marine vertebrates (though on tubules in marine mammals), and the size of these peptides appears to correlate with the need for water conservation, being especially large in the marine mammals. In addition, a range of steroid hormones, including aldosterone, affect sodium secretion and resorption, in kidney, gut, salt glands and gills, thus regulating Na^+ and K^+ levels in plasma. These steroids are in turn affected by the renin/angiotensin hormone system, which is activated by changes in fluid volume and electrolyte levels.

9.10.2 Temperature in marine vertebrates

Surviving in cold seas with dilute body fluids

Marine fish, amphibians and reptiles, like most of the invertebrates, are essentially ectotherms with T_b levels very close to the temperature of their surroundings. This can pose severe problems for those living in cold polar seas, as their body fluids (being more dilute) may freeze even while the surrounding sea water remains liquid. Furthermore, their surroundings are likely to contain small ice crystals that would precipitate rapid freezing in any unprotected animal. Hence polar fish, in particular, are very well endowed with the 'standard' antifreeze mechanisms discussed in Chapter 8.

The body fluids of most marine fish are hyposmotic to sea water, freezing at −0.7°C. Since the temperature in shallow ice-laden water is close to −1.9°C, most Antarctic fish are supercooled by about 1°C. They may have at least three further adaptations that help them to avoid freezing. Firstly, the freezing point of sea water is depressed by 0.0075°C for each 10 m increase in depth, so that many polar fish avoid freezing by living at or migrating to sufficiently deep water. Deep-living Antarctic fish caught on hook and line through holes in the ice therefore have no real defences against ice formation, and commonly do freeze solid on being brought to the surface. Secondly, Antarctic fish have a higher plasma NaCl concentration than

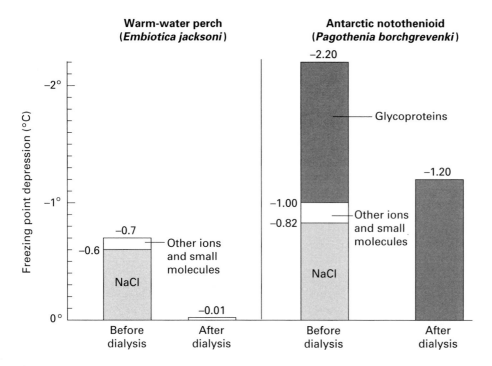

Fig. 9.40 Levels of NaCl in the blood of temperate and polar teleosts. In the warm-water perch, the freezing point of the blood before dialysis was −0.7°C, and after dialysis (removing small solutes) it rose to −0.01°C, indicating that NaCl and other small solutes are the main cause of freezing point depression (FPD). In the Antarctic notothenioid the NaCl levels are elevated, and their removal raises the freezing point by a full degree; however, the non-dialysable glycoproteins account for more than half the total FPD. (From Eastman 1993.)

typical marine teleosts (Fig. 9.40), but this can only account for 40–50% of their recorded plasma freezing point depressions. The remainder is due to their third adaptation, the presence of particularly effective antifreeze proteins (AFPs) or glycoproteins (AFGPs), whose mode of action in preventing any existing ice crystals from growing was described in Chapter 8. These molecules are found in at least 11 different families of teleost fish (Table 9.12).

The structure of the antifreeze glycoprotein in an Antarctic fish (e.g. the notothenioid *Pagothenia borchgrevenki*, which lives in crevices in sea ice) is shown in Fig. 9.41(a). This AFGP consists of many repeating units of a disaccharide molecule linked to the third amino acid of a tripeptide (alanine–alanine–threonine). In fact in this species a range of eight antifreeze glycoproteins have been identified, ranging in relative molecular mass from 2600 to 33 700. In a related species, *Notothenia coriiceps*, one of the AFGP structural genes has been shown to consist of 46 tandemly

repeated segments encoding 44 copies of AFGP8 and two copies of AFGP7. This is one of the highest gene copy numbers presently known for any organism, and may reflect the need for high and seasonally constant levels of the AFGP.

Unrelated species of fish living at opposite extremes of the Earth (Antarctic notothenioids and Arctic cod) have almost identical AFGPs. However, while the original AFGP gene is thought to have evolved from the trypsinogen gene, it seems that the Ala–Ala–Thr repeats in the cod antifreeze are derived from duplications in an unrelated ancestral coding element; thus the similarity of structure may be a rare example of protein sequence convergence.

When Antarctic notothenioid fish are placed in ice-free water, they do not freeze until their temperature drops to −6°C. Apparently they do not usually develop ice crystals unless ice first penetrates from the outside. The main roles of their AFGPs may be to keep ice from propagating across their integument,

Genus	Environment (temperature, °C)	FP organism (°C)	FP blood (°C)	MP blood (°C)	Antifreeze molecules
Gadus	Deep, ice free	−1.0	−1.1	−0.7	AFGPs
Chaenocephalus	Shallow, ice free (−1.0)	−1.5	−1.5	−0.9	
Rhigophalia	Deep, ice-free (−1.9)	−1.9	−2.0	−0.9	AFP 1, 2, 3
Notothenia	Shallow, icy (−1.9)	−2.0	−2.1	−1.1	AFGPs
Pagothenia	Shallow, icy (−1.9)	−2.4	−2.7	−1.1	AFGPs 1–8

Table 9.12 Freezing points (FP) and melting points (MP) due to antifreeze glycoproteins in polar marine fish.

AFGP, antifreeze glycoprotein; AFP, antifreeze protein.

(b)

Fig. 9.41 The antifreezes of Antarctic fish. (a) The structure of the antifreeze glycoprotein (AFGP) from an Antarctic fish, with the typical Ala–Ala–Thr repeat and attached disaccharides. (b) The freezing point depressions produced by antifreeze proteins (AFPs) and AFGPs (N.B. 3, 4 and 5 are larger than 7 and 8), in comparison with the much lesser effects of NaCl and of galactose; but note also that the AFPs of all kinds have little or no effect on melting points.

and to coat the edges of ice crystals that happen to get into their body along with the sea water they drink. They have a very substantial effect on freezing point but little or no effect on melting point (Fig. 9.41b). The AFGPs are synthesized in the liver, and secreted into the blood. They appear in most body fluids except urine, ocular fluid and the cytoplasm of cells (with the exception of the hepatocytes where they are made). They are excluded from the urine because most Antarctic fish have aglomerular kidneys: the urine is formed by secretion, so that the cells lining the kidney tubules leave the antifreeze in circulation. However, the tissues surrounding the urine and ocular fluids are fortified with antifreeze, constituting a barrier to ice propagation. It has recently been reported that antifreeze glycoproteins are found in the urine of winter flounder, sea raven, ocean pout and Atlantic cod, at a concentration that approaches that of plasma; thus in some species antifreeze glycoproteins probably also increase the freezing resistance of urine. The other site at risk of ice formation is the gut, since its contents are often hyposmotic to plasma, and in most polar fish antifreezes (AFGPs 6, 7 and 8) are secreted into the gut lumen via the bile (translocated into the gall bladder

from blood). Salt, water and digested foodstuffs are absorbed during passage down the intestine, concentrating the AFGPs; they are eventually lost with the faeces, representing an energetic cost of freezing avoidance. The loose anal sphincter is a potential danger-spot for ice entry, but at this point the intestinal tract is quite concentrated and only freezes at –2.2°C.

At least five different classes of AFPs have also been identified in polar fishes, most being either alanine- or cysteine-rich. Since these types all have very different structures and amino acid sequences it seems likely that their mechanism of action is different in each case. In Arctic fishes production of AFP is seasonal (Fig. 9.42a). For example, the appearance of AFP in the plasma of winter flounder is suppressed when day-length exceeds 15 h, and it appears to be primarily under the control of photoperiod rather than low temperature (Fig. 9.42b). Exposure of winter flounder to temperatures of 10–12°C and long day-lengths results in a loss of blood antifreeze within 3 weeks. Synthesis is controlled at the level of DNA transcription and translation, as AFP mRNA appears about 4 weeks before the peptide is detectable. Control of transcription by photoperiod is hormonally mediated. Growth hormone represses transcription of the AFP gene in the summer, but as day-length decreases this inhibition is reduced, resulting in higher levels of liver AFP mRNA. Control of translation then comes into operation, and is related to water temperature; low temperature reduces the rate of AFP mRNA degradation, which together with high levels of tRNA[Ala] and of alanyl-tRNA synthetase favours the translation of the AFP mRNA and the appearance of AFP in the blood.

Marine endotherms

Birds and mammals that have adopted a marine habitat have very different thermal problems to those of marine ectothermic vertebrates. They are permanently endothermic and tachymetabolic, and must maintain constant T_b even when resting, despite the high thermal conductance of their surroundings. Permanent or semi-permanent residents of the sea therefore tend to be rather large animals (cetaceans, pinnipeds, penguins, etc.), with reduced relative surface area. They have exceptionally effective insulating layers, whether as internal fat or external pelage (fur or feathers), or both. Where fatty 'blubber' is used, the overlying skin is normally at roughly the same temperature as the sea, with a sharp temperature gradient over the outer 5–6 cm of fat and almost constant temperature inwards of this. Most marine mammals and birds also use sophisticated countercurrent heat-exchangers in their feet or flippers, with a central artery surrounded by 15–20 small veins (Fig. 9.43), to avoid heat loss at the extremities. The thermal consequences of these features when marine mammals are beached for pupping were considered earlier (section 9.5). In the permanently aquatic whales, not only are the thermal losses through the fins and flukes controlled by countercurrent systems, but so too are the temperatures of the reproductive organs (which must not be allowed to overheat), and the mouth area, for it has been shown that the tongues of

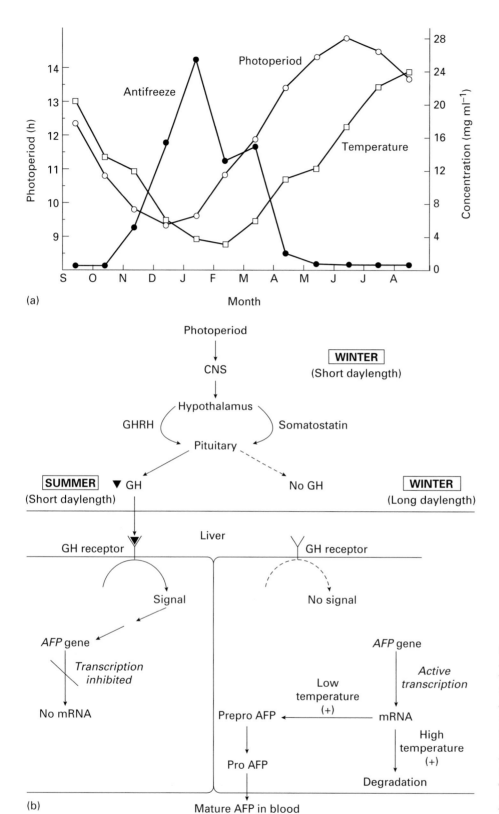

(a)

Month

(b)

Fig. 9.42 (a) Seasonal production of antifreeze proteins (AFPs) in the winter flounder. (Adapted from Cossins & Sheterline 1983.) (b) Regulation of this AFP production by photoperiod and the intermediary effects of growth hormone (GH) initiating transcription. Note that post-transcriptional control by temperature also operates, to ensure that AFP is only present if temperatures are low. CNS, central nervous system; GHRH, growth hormone-releasing hormone. (Adapted from Chan *et al.* 1993.)

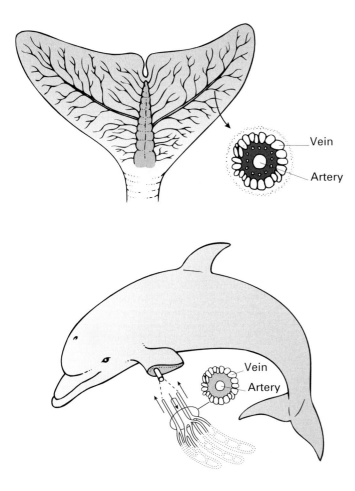

Fig. 9.43 Countercurrents in the tails and flippers of marine mammals.

baleen whales, feeding fairly continuously with open mouths and in cold high-latitude waters, are well endowed with heat-exchangers to reduce oral heat loss.

9.10.3 *Respiration in marine vertebrates*

In marine fish, breathing is a relatively straightforward issue comparable to that in other marine animals, and thus has already been dealt with in section 9.4. However, in the context of fish as secondary invaders it is worth mentioning that the very short diffusion distances and hence decreased areas and perfusion rates possible in many teleost gills have the additional advantage that the gains and losses of ions at the gill surfaces are reduced.

In tetrapods breathing is very different, as their terrestrial origins dictate a reliance on intermittent supplies drawn into the lungs from the air above, so that all marine tetrapods are intimately associated with the sea surface. Their 'normal' respiratory modes and adaptations are therefore similar to those for lung-based respiration, discussed in Chapter 13.

Most of them do, however, submerge for long periods, deriving their food from below, and they show complete cessation of breathing (apnoea) during the dive, posing more interesting physiological problems. Even when using an artificial air supply, fully terrestrial tetrapods, such as man, experience blackouts when submerged for even short periods in shallow water; they exhibit severe nervous disruption in deeper waters, and suffer nitrogen narcosis. Yet emperor penguins can dive to at least 500 m, sperm whales have been recorded below 1000 m, and elephant seals penetrate down to 1500 m. Many sea mammals spend only about 10% of their time at the surface, and may be at 250–500 m the rest of the time; Weddell seals stay submerged for at least 60 min at a time, and sometimes up to 2 h. Green turtles actually seem to 'hibernate' at 10–15 m depth for several months. During submersion, though, there must always be sufficient oxygen to maintain the animal, as the mammalian central nervous system in particular cannot survive even brief anoxia (more than 1 min can be fatal in humans). Dive durations in many animals appear to exceed the theoretical 'aerobic dive limit' (the time they could stay submerged and still have enough oxygen to retain aerobic metabolism), so they were assumed to switch to anaerobic respiration. But in fact most tetrapod divers do not return to the surface for long enough to repay the kind of oxygen debt that this would imply. Instead of prolonged anaerobiosis, they use several linked adaptations to achieve maximum dive duration, collectively often termed the 'diving response':

1 Apnoea, with suppression of the normal respiratory muscle action and its control systems.

2 Some system of oxygen storage, primarily the blood and its haemoglobin, and to a lesser extent muscle myoglobins. Blood volume may be enlarged in divers (in birds and mammals, though not apparently in marine reptiles); haemoglobin also occurs at a raised concentration in diving mammals (30–40 ml per 100 ml blood, compared to 20 ml per 100 ml in most mammals); and the haemoglobin may have reduced temperature sensitivity. In mammals blood cells (RBCs) are also stored in the spleen, which undergoes regulated emptying in a dive.

3 Reduced blood flow (hypoperfusion) to all visceral organs other than the brain and heart, with carefully controlled patterns of vasoconstriction (Fig. 9.44).

4 Reduced metabolic rate (hypometabolism), partly because organs are oxygen-starved. This metabolic depression is the main factor reducing the need for anaerobic metabolism. Some seals show an associated reduction in body temperature, particularly in smaller species, and the abdominal temperatures of penguins have been recorded as low as 11°C, possibly due to extra heat loss from the exposed flippers and brood patch.

5 Anaerobic metabolic pathways in most visceral tissues and in muscles, with some lactic acid formation, but much less than expected, due to the reduced metabolic rate. In mammals lactate is not released to the blood until the dive ends, when there is a phase of 'lactate washout'. By contrast, lactate concentrations can reach 200 mmol kg^{-1} in the muscles of diving turtles, five times normal levels in active vertebrates.

6 Reduced heart rate (bradycardia) and reduced cardiac output (Fig. 9.45). Bradycardia is triggered directly by chemosensitivity to oxygen levels.

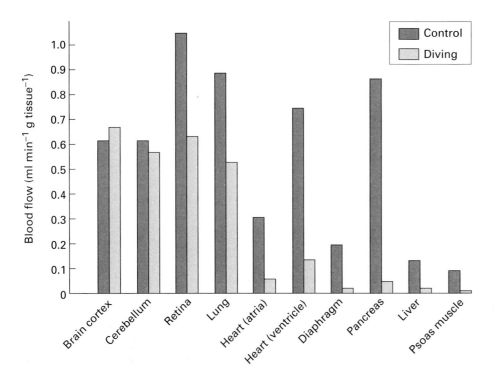

Fig. 9.44 Selective blood distribution in a diving seal. Blood flow to the brain cortex and cerebellum is maintained, and that to the lungs and eyes remains moderately high, but all other organs experience a very marked reduction in circulation. (Data from Zapol *et al.* 1979.)

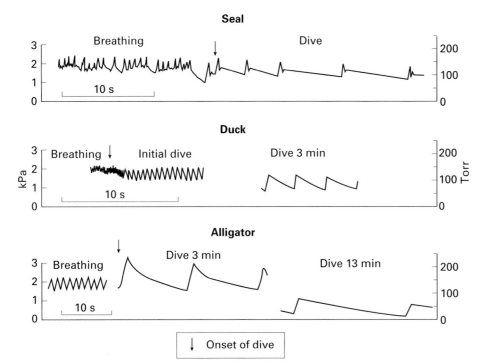

Fig. 9.45 Patterns of blood pressure recorded during diving in seals, ducks and alligators, showing substantially reduced heart rate and in some cases also a decrease in overall blood pressure. (From Scholander 1964.)

7 Neuronal and hormonal control of cardiac and spleen action, mainly by catecholamines.

8 In marine reptiles, some cutaneous respiration—up to 30% of the oxygen requirement comes through the general skin in sea snakes, and a small proportion through the oesophagus and rectum in some turtles. The incomplete separation of venous and arterial blood in reptiles also means that increasing proportions of the blood can bypass the lungs entirely during diving.

Of all these strategies, bradycardia was traditionally regarded as the centrepiece. But recent evidence from phylogenetically controlled comparisons in pinnipeds shows that bradycardia and hypoperfusion are ancestral characteristics, and it is the spleen and blood volumes, and the RBC changes, that are specific to divers.

Many diving animals also have surprisingly small lungs, especially in relation to the volume of the upper airways; and (counterintuitively) they exhale forcibly, rather than inhaling a good lungful of air, just before submerging. This is because a full lung would be a liability rather than a useful oxygen store. As the animal penetrates to greater depth, the increasing hydrostatic pressure causes lung compression, and air is forced back out of the alveoli into the (large) tracheoles and bronchi. This avoids diffusion of unwanted nitrogen through the alveolar surfaces into the blood supply, which would otherwise cause bubble formation and 'the bends' (decompression sickness) when the animal resurfaced. Reptiles are again unusual in that they can avoid this problem by allowing some cutaneous escape of nitrogen.

Even oxygen diffusing from the lung to the blood under the high pressures during diving would be undesirable, as high Po_2 can cause convulsions in mammals. During the dive, the inspiratory muscles are inhibited via receptors that sense the presence of water near mouth and nose; and the signals from the carotid O_2 and CO_2 receptors (see Chapter 7) are also overridden, to prevent the normal stimulus to ventilate as blood gas composition changes through the dive.

9.10.4 *Vertebrate overview*

Marine vertebrates, and air-breathing tetrapods in particular, clearly have a range of special features, relating to ion and water balance, temperature balance and respiratory supply, that permit their survival in what is for them a rather hostile environment. Since the extinction of the large marine reptiles such as ichthyosaurs and plesiosaurs at the end of the Cretaceous (65 million years ago), the marine mammals in particular have proved adept at exploiting the marine world and are the only really large predators there apart from elasmobranch sharks. Most other tetrapods, those not primarily adapted for life in the sea, cannot survive the osmotic problems posed by marine conditions. This is clearly illustrated by the fact that humans are unable to live afloat without access to fresh drinking-water; if we drink only sea water our bodies accumulate too much salt and lethal dehydration effects set in.

9.10.5 *Marine insects*

Insects have an entirely terrestrial evolutionary history, and though many are secondarily aquatic very few have ever invaded the sea. This seems very surprising given their overwhelming diversity in all other habitats. Those few species that do occur are essentially intertidal or surface dwellers, including some collembolans, heteropteran water striders (especially the genus *Halobates*), and beetles. They therefore avoid respiratory problems, and are unaffected by pressure or temperature regimes in the sea. Indeed, they can only really be considered 'marine' in the sense that they face the osmotic and ionic prob-lems of having access purely to salt waters; but the insect osmoregulatory and excretory system (based on Malpighian tubules and rectum working together) can readily cope with many worse problems than this, as we will see in Chapter 13. The major point of interest is perhaps *why* insects do not live in the sea. The likeliest answer is that the very features that make them particularly successful on land (a hydrophobic cuticle with very low permeability; reduced numbers of specialized legs; wings and flight; waterproof eggs; tracheal respiratory systems) are precisely those that will be of neutral value or even disadvantageous in the sea. Insect tracheal systems may be of much reduced efficiency at high hydrostatic pressures in the sea, yet insects would need to be able to dive deeper to escape fish predators. The insect cuticle may also make it very difficult to invade the sea, since penetrating the surface tension film is hard; amphibious freshwater insects largely make the transition by dragging themselves down stout vegetation stalks, but this is particularly difficult in disturbed waters such as the littoral and estuarine fringes of the sea. Indeed the prevalence of insect interactions with angiosperm plants may be important in a broader sense too; angiosperms are absent from open oceans, and insects therefore cannot benefit from their many coevolutionary links to such plants. It has also been argued that while insects are very good at osmoregulation they are in many ways poor competitors. The oceans already had a vast diversity of crustaceans, and these were endowed with most of the advantages of insects, with none of the potential disadvantages, and perhaps with a particular advantage of their own in the form of numerous complex cirri-bearing legs that could make excellent filter-feeding apparatus. These crustaceans would compete in the same size range for the same niches and resources, so it is perhaps not surprising that insects have not made the transition to the seas.

9.11 Conclusions

Most phyla of animals live in the sea and have done so throughout their history; indeed, most of the evolutionary history of the planet has been marine, and most of the really important evolutionary innovations in design occurred there. At the cellular level, physiological systems also evolved against a marine background, with fairly constant conditions of salinity, temperature and gas availability. Because of this history marine environments do not impose a great physiological challenge to most components of an animal's living machinery. It is primarily those animals that have invaded the seas secondarily that have real difficulties with osmotic and ionic control, with respiration, and sometimes with their thermal biology, and that therefore require complex regulatory systems at the tissue and organ levels.

With an understanding of the essentially marine legacy of structure and functioning that all living cells have inherited, it should be easier to follow the progressively more difficult physiological challenges, and the appropriate adaptive changes, encountered in more dilute, more desiccating, and above all more variable environments, discussed in the following chapters.

Further reading

Books

Atema, J., Fay, R.R., Popper, A.N. & Tavolga, W.N. (1988) *Sensory Biology of Aquatic Animals.* Springer-Verlag, New York.

Clark, R.B. (1992) *Marine Pollution.* Oxford University Press, Oxford.

Gage, J.D. & Tyler, P.A. (1991) *Deep-Sea Biology: A Natural History of Organisms at the Deep-Sea Floor.* Cambridge University Press, Cambridge.

Herring, P.J., Campbell, A.K., Whitfield, M. & Maddock, L. (1990) *Light and Life in the Sea.* Cambridge University Press, Cambridge.

Jobling, M. (1995) *Environmental Biology of Fishes.* Chapman & Hall, London.

Kooyman, G.L. (1989) *Diverse Divers: Physiology and Behaviour.* Springer-Verlag, Berlin.

Laverack, M.S. (1985) *Physiological Adaptations of Marine Animals.* Society for Experimental Biology, Cambridge.

Rankin, J.C. & Jensen, F.B. (eds) (1993) *Fish Ecophysiology.* Chapman & Hall, London.

Vogel, S. (1994) *Life in Moving Fluids.* Princeton University Press, Princeton, NJ.

Reviews and scientific papers

Block, B.A. (1986) Structure of the brain and eye heater tissue in marlins, sailfish and spearfishes. *Journal of Morphology* **190**, 169–189.

Boyd, I.L. (1997) The behavioural and physiological ecology of diving. *Trends in Ecology and Evolution* **12**, 213–217.

Childress, J.J. (1995) Are there physiological and biochemical adaptations of metabolism in deep-sea animals? *Trends in Ecology and Evolution* **10**, 30–36.

Childress, J.J. & Seibel, B.A. (1998) Life at stable low oxygen levels: adaptations of animals to oceanic oxygen minimum layers. *Journal of Experimental Biology* **201**, 1223–1232.

Croll, D.A., Nishigushi, M.K. & Kaupp, S. (1993) Pressure and lactate dehydrogenase function in diving mammals and birds. *Physiological Zoology* **65**, 1022–1027.

Crossin, G.T., Al-Ayoub, S.A., Jury, S.H., Howell, W.H. & Watson, W.H. (1998) Behavioural thermoregulation in the American lobster, *Homarus americanus. Journal of Experimental Biology* **201**, 365–374.

Davenport, J. (1997) Temperature and the life history strategies of sea turtles. *Journal of Thermal Biology* **22**, 479–488.

de Vries, A.L. (1982) Biological antifreeze agents in coldwater fishes. *Comparative Biochemistry and Physiology A* **73**, 627–640.

Dietz, T.J. & Somero, G.N. (1993) Species- and tissue-specific synthesis patterns for heat shock proteins HSP70 and HSP90 in several marine teleost fish. *Physiological Zoology* **66**, 863–880.

Dusenbery, D.B. & Snell, T.W. (1995) A critical body size for use of pheromones in mate location. *Journal of Chemical Ecology* **21**, 427–438.

Fedak, M.A. & Thompson, D. (1993) Behavioural and physiological options in diving seals. *Symposia of the Zoological Society of London* **66**, 333–348.

Fish, F.E. (1996) Transitions from drag-based to lift-based propulsion in mammalian swimming. *American Zoologist* **36**, 628–641.

Fujiwara, T., Tsukahara, J., Hashimoto, J. & Fujikura, K. (1998) In situ spawning of a deep sea vesicomyid clam: evidence for an environmental cue. *Deep Sea Research* **45**, 1881–1889.

Gillett, M.B., Suko, J.R., Santoso, F.O. & Yancey, P.H. (1997) Elevated levels of trimethylamine oxide in deep sea gadiform teleosts: a high pressure adaptation? *Journal of Experimental Zoology* **279**, 386–391.

Goldman, K.J. (1997) Regulation of body temperature in the white shark, *Carcharodon carcharias. Journal of Comparative Physiology* **167**, 423–429.

Handrich, Y., Bevan, R.M., Charassin, J.-B. *et al.* (1997) Hypothermia in foraging king penguins. *Nature* **388**, 64–67.

Heyning, J.E. & Mead, J.G. (1997) Thermoregulation in the mouths of feeding gray whales. *Science* **278**, 1138–1139.

Johnston, I.A. (1990) Cold adaptation in marine organisms. *Philosophical Transactions of the Royal Society, Series B* **326**, 655–667.

Kirschner, L.B. (1993) The energetics of osmotic regulation in ureotelic and hyposmotic fishes. *Journal of Experimental Zoology* **267**, 19–26.

LaBarbera, M. (1984) Feeding currents and particle capture mechanisms in suspension feeding animals. *American Zoologist* **24**, 71–84.

MacDonald, A.G. (1997) Hydrostatic pressure as an environmental factor in life processes. *Comparative Biochemistry and Physiology A* **116**, 291–297.

Maddrell, S.H.P. (1998) Why are there no insects in the open sea? *Journal of Experimental Biology* **201**, 2461–2464.

Newton, C. & Potts, W.T.W. (1993) Ionic regulation and buoyancy in some planktonic organisms. *Journal of the Marine Biological Association of the UK* **73**, 15–23.

Pelster, B. & Scheid, P. (1992) Countercurrent concentration and gas secretion in the fish swim bladder. *Physiological Zoology* **65**, 1–16.

Rasmussen, A.D. & Anderson, O. (1996) Apparent water permeability as a physiological parameter in crustaceans. *Journal of Experimental Biology* **199**, 2555–2564.

Sanders, N.K. & Childress, J.J. (1988) Ion replacement as a buoyancy mechanism in a pelagic deep sea crustacean. *Journal of Experimental Biology* **138**, 333–343.

Sebert, P., Simon, B. & Pequeux, A. (1997) Effects of hydrostatic pressure on energy metabolism and osmoregulation in crab and fish. *Comparative Biochemistry and Physiology A* **116**, 281–290.

Siebenaller, J.F. & Somero, G.N. (1989) Biochemical adaptation to the deep sea. *Aquatic Sciences* **1**, 1–25.

Snyder, G.K. (1983) Respiratory adaptations in diving mammals. *Respiratory Physiology* **54**, 269–294.

Somero, G.N. (1990) Life at low volume change: hydrostatic pressure as a selective factor in the aquatic environment. *American Zoologist* **30**, 123–135.

Somero, G.N. (1992) Adaptation to high hydrostatic pressure. *Annual Review of Physiology* **54**, 557–577.

Suchanek, T.H. (1993) Oil impacts on marine invertebrate populations and communities. *American Zoologist* **33**, 510–523.

Toulmond, A. (1985) Circulating respiratory pigments in marine animals. *Symposia of the Society for Experimental Biology* **39**, 163–206.

Van der Hage, J.C.H. (1996) Why are there no insects and so few higher plants in the sea? New thoughts on an old problem. *Functional Ecology* **10**, 546–547.

Wilkie, M.P. (1997) Mechanisms of ammonia excretion across fish gills. *Comparative Biochemistry and Physiology A* **118**, 39–50.

Withers, P.C., Morrison, G. & Guppy, M. (1994) Buoyancy role of urea and TMAO in an elasmobranch fish, the Port Jackson shark. *Physiological Zoology* **67**, 693–705.

Wright, S.H. & Manahan, D.T. (1989) Integumental nutrient uptake by aquatic organisms. *Annual Review of Physiology* **51**, 585–600.

Zimmerfaust, R.K., Finelli, C.M., Pentcheff, N.D. & Wethey, D.S. (1995) Odor plumes and animal navigation in turbulent water flow: a field study. *Biological Bulletin* **188**, 111–116.

Shorelines and Estuaries

10

10.1 Introduction: brackish habitats and biota

Life in the sea is, as we have seen, a relatively easy affair for most animals. But problems start increasing as organisms move towards the edges of the sea, for both benthic and pelagic fauna. There are particular difficulties (and advantages) for life at the shifting interfaces of water, land and air; animals here are in a cyclically changing habitat, with new challenges but also with new kinds of strategies opening up to them.

Seashores and estuaries therefore constitute the subject matter for this chapter. Shores occur where the sea meets up with the land/air interface, and estuaries are where the sea meets a freshwater interface as rivers flow from the land masses. These habitats collectively involve what are termed **brackish waters**. They share certain key characters:

1 they are not very common, in terms of surface or volume;

2 they are not at all easy to live in because cyclical variation is a continuous problem;

3 they are variable over quite short periods of evolutionary time, as the interface between media advances and recedes due to erosion, deposition, or changes in sea level.

These habitats are nevertheless zones of quite high diversity and of characteristic patterning among the biota. This arises from the multiplicity of problems of life in these habitats (Fig. 10.1). Most obviously, a shore is not just at the interface of air, land and water, but has complex spatial and temporal variations in the balance of these three media, with environmental gradients much in evidence. The rise and fall of water level due to tides brings problems particularly affecting breathing patterns, osmotic balance, temperature balance and predation pressure. Wave action may lead to crushing or dislodgement, and may increase water turbidity with sand and mud becoming suspended in the

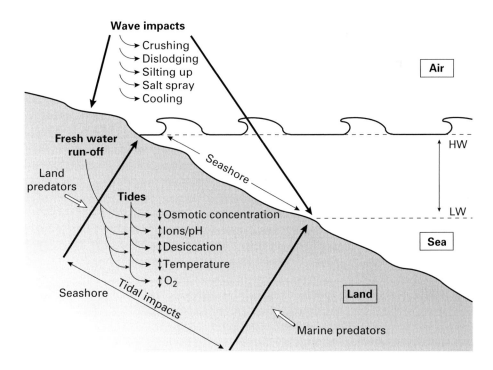

Fig. 10.1 A summary of the problems experienced by animals living on seashores. HW, high water; LW, low water.

Table 10.1 Productivity of brackish ecosystems (in comparison with marine and freshwater systems).

Ecosystem	Productivity (g C m^{-2} year^{-1})
Open ocean water	5–50 (cf. Fig. 9.2)
Coastal waters (temperate)	
Shallow continental shelf	30–150
Zones of upwelling	50–220
Coastal bays	50–120
Subtidal	
Seaweeds	800–1500
Coral reefs	1700–2500
Sea grass	120–350
Intertidal	
Sandy beaches	10–30
Rocky beach seaweeds	100–250
Estuarine mudflats	500–750
Supratidal	
Salt marsh	700–1300
Mangrove swamp	350–1200
Sand-dunes	150–400
Eutrophic freshwater lake	400–600

water column. Freshwater inputs lead to large variations in water and ion balance. Coastal boundaries also exert strong effects on the air and water movements of the planet as a whole, and oceanic air masses often move laterally along shores rather than inland over rough or steep terrain. Heat exchange between air and land is also complex and may affect the movements of maritime air masses. Thus shores have their own 'weather', and storm fronts often track along them (e.g. as in the English Channel or the Gulf of St Lawrence) rather than over land. Oceanic water masses impinging on land also have effects on water mixing, often with downwelling close in and upwelling further offshore, which may help to promote and localize inshore productivity.

But there are also some key advantages of life on seashores. Above all these relate to food supply. The sea water is penetrated by light and there is a substratum for marine plants to attach to, so that a wide range of multicellular algae (seaweeds) can thrive and produce a strong base to the food web. These seaweeds also stabilize the substratum in shallow waters, allowing invasion by some specialist angiosperms, notably the sea grass *Zostera*. Coastal benthic habitats are therefore zones of high primary productivity (Table 10.1). With abundant algae to eat, most herbivorous marine animal taxa do extremely well either just offshore or on the shore itself. There will also be plenty of suspended matter, primarily marine plankton but also larger organic flotsam, which can be filtered or gathered by sessile animals and which is continually renewed by tide and wave action. The herbivores and filterers in turn provide opportunities for

predators, both marine animals (especially crustaceans and fish) coming in to feed, and terrestrial animals (especially birds) combing the shore at low tide. Finally, with damage and death resulting from waves, dislodgement and other hazards, there is a ready supply of food for a host of scavengers, either on the surface or amongst the sandy or muddy substrata. The abundance of resources leads to periods of rapid depletion and strong interspecific competition for food, so that population densities of many animals are cyclical.

All the organisms on a seashore are faced with the same problems and are exploiting the same resources within a narrow strand. Inevitably therefore there is a great deal of competition for space as well as for food. This is a major reason why there is also very precise patterning on most seashores. They provide the best possible example of where severe physiological constraints and severe ecological competition come together to give observable and definable structure to a biological community.

10.1.1 *The littoral habitat*

Seashores are in many ways extremely diverse in character. Coastlines may be classified very broadly according to their geological nature (Box 10.1, Geological tectonic categories), imparting a fundamental character to long stretches of shore. Within this, matters of scale impose a more varied outcome and a classification into six main categories (Box 10.1, Geological/biological

---

Box 10.1 Classification of coastlines

Geological tectonic categories

1 Collision coasts (leading-edge coasts, active coasts): abrupt and steep coasts, eroding cliffs, narrow continental shelves.
  (a) Continental collision coasts, e.g. Pacific coasts of North and South America.
  (b) Island arc coasts, e.g. Philippine islands and Caribbean islands.
2 Trailing-edge coasts (passive coasts): flat shores, with lagoons and wide continental shelves.
  (a) Recent rifting coasts, e.g. Red Sea.
  (b) Spreading basin coasts, e.g. southeast Africa, eastern USA.
  (c) Basins with one trailing and one leading edge, e.g. North Pacific.
3 Marginal coasts.
  (a) Back-arc basins, e.g. China Sea.

Geological/biological categories

1 Mountain coast.
2 Narrow shelf coast.
  (a) Headlands and bays.
  (b) Coastal plains.
3 Wide shelf coast.
  (a) Headlands and bays.
  (b) Coastal plains.
4 Deltaic coast.
5 Reef coast.
6 Glaciated coast.

Fig. 10.2 Major types of coastlines around the world's continents. (From Carter 1988.)

categories) is more useful. Worldwide distributions of these major categories are shown in Fig. 10.2. At a local level coasts may include not only the traditional sandy, muddy and rocky shores that surround all continents and islands, but also the sea cliffs and sand-dunes where wave splash may impinge on more terrestrial zones (see Plate 2a, facing p. 310). In some respects even the coral reefs (Plate 1d) are 'shores', where an essentially marine zone providing dwelling space on living coral is periodically exposed around the littoral margins of tropical land masses (see Fig. 9.3 for reef zones).

However, all these littoral fringes also have some key similarities, making the littoral zone as a whole a particularly fascinating place to study environmental adaptation. First of all, and most obviously, all three of the major habitat zones—air, land and water—come together, each imposing its own stresses. Secondly, the balance between the three habitats changes cyclically and with several interacting rhythms relating to the Sun and Moon, giving daily tides and predictable variations in emersion (exposure to the air) and immersion (return of a water covering) for the biota. Thirdly, and as a consequence of this, the organisms on a seashore are all laid out in a neat two-dimensional array, with distinct 'layers' of animals and plants, and with the precise nature of the layering varying according to the characteristics of the particular shore.

Tides

Tides are primarily the results of astronomical phenomena, with the gravitational forces of the Moon and Sun pulling on the Earth's surface water. Since the Moon is much closer to us, it has a much greater effect, and in effect it pulls an ellipse of water around the Earth as it moves on its own rotational path, creating a 'bulge' of water pointing towards wherever the Moon is, and a reciprocal bulge on the other side of the Earth as the planet's rotation adds a centrifugal effect. This means that each site on the planetary surface gets a surge of water—a high tide—every half lunar day, where the lunar period is 24 h and 50 min, thus cycling through low–high–low every 12 h and 25 min. Hence shores normally experience two high tides a day, each delayed by a fixed period of around 25 min on each successive day.

The effect of the Sun modifies this basic pattern by superimposing a lesser gravitational pull with a different periodicity. At new and full moons, the Sun and Moon are in line and the overall pull is therefore a bit greater, so that we get particularly extreme high and low tides known as **spring tides**. At half moons the Sun and Moon are working against each other so the tidal ranges are reduced, producing **neap tides**. Thus springs and neaps occur twice in each lunar cycle, or roughly twice a month (Fig. 10.3). Spring-tide low waters are particularly good times for ecological work on beaches, as a long stretch of the shore will be exposed and low-tide animals will be accessible.

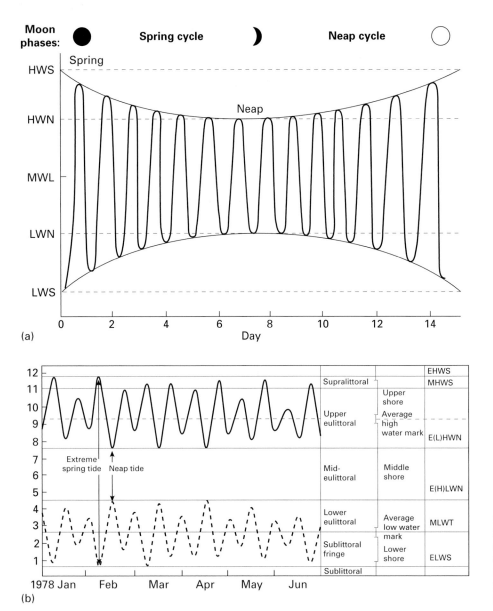

Fig. 10.3 Basic patterns of tidal cycles (a) with twice-daily and twice-monthly cycling related to the phase of the Moon, and (b) records over 6 months and the relation of tide to terminology used on beaches. (See List of Abbreviations, p. IX, for definition of terms.)

A third periodicity affects tides, since they also vary with the declination of the Sun, which runs on a yearly cycle. Declination is greatest at the equinoxes, so the most extreme spring tides usually come in March and September.

Thus far we have only been generalizing on a global scale about tidal patterns. The actual degree to which tides rise and fall on any shore is highly variable and must be computed for each location; likewise the precise time and daily progression of high and low tide varies with the constraints of geography. Tides are greatly modified by the shapes of the land masses around which the seas are pulled by gravitational forces, by the shape of local coastlines, and by the effects of prevailing winds. For example, average tide ranges for Great Britain are about 3.5 m (neaps) and 4.7 m (springs), but the east coast has smaller tides as tidal motion is constrained by the relatively small basin of the North Sea, while in enclosed seas, such as the

Mediterranean, there is effectively no tide at all. On a smaller scale, long, thin inlets pointing upstream from prevailing winds can have massive tidal ranges, often experiencing a 'tidal bore' when the fast incoming tide is increasingly channelled into a narrowing profile; the Bristol Channel in England is famous for a massive tidal range (12–13 m) and periodic large bores. Tides are also modified from day to day by weather, particularly by strong winds and the after-effects of storms out at sea.

Waves

Waves form whenever the water surface is disturbed, whether by winds, earth tremors or underlying planetary rotation. Most wave action is due to the effects of wind, and it is therefore restricted to surface water, having little effect on most marine organisms. It is also variable seasonally in mid-latitudes, with

Plate 1 Marine environments

(a) Marine benthos. Shallow soft mud with sea pen *Virgularia mirabilis* and the anemone *Sagartiogeton laceratus*, Loch Laxford, Scotland, UK. (Courtesy of R. Holt/Joint Nature Conservation Council.) (b) Marine nekton. A shoal of fish (hussars *Lutjanus amabilis*), Heron Island, southern Great Barrier Reef. (Courtesy of L. Newman & A. Flowers.) (c) Marine plankton, including invertebrate larvae, fish eggs and phytoplankton. (Courtesy of P. Parks, Imagequest.) (d) Coral reef. Underwater scene with soft corals, off Madang, Papua New Guinea. (Courtesy of L. Newman & A. Flowers.)

Plate 2 Littoral environments
(a) Sea cliffs, Lundy Island, UK. (Courtesy of A.S. Edwards.) (b) Vertical zonation on a moderately exposed temperate shore, Devon, UK. (Courtesy of P.G. Willmer.) (c) Mixed sandy/rocky shore with sealions, Galapagos Islands. (Courtesy of L. Crooks.) (d) Rock pools and rocky shore, Jervis Bay, southern Australia. (Courtesy of P.J. Sunnucks.) (e) Sand-dunes with marram grass, South Wales, UK. (Courtesy of P.G. Willmer.) (f) Volcanic lava shore, with basking iguanas, Fernadina, Galapagos. (Courtesy of L. Crooks.)

(a)

(b)

(c)

(d)

(e)

(f)

Plate 3 Littoral and estuarine environments
(a) Boulder beach, Cape Nat, Australia. (Courtesy of P.J. Sunnucks.)
(b) Muddy shore, Jervis Bay, Australia. (Courtesy of P.J. Sunnucks.)
(c) 'Ria' type estuary (drowned river valley), South Devon, UK. (Courtesy
of P.G. Willmer.) (d) Aerial view, River Eden estuary, Scotland, UK, with
sand spit at mouth and enclosed mudflats and salt marsh. (Courtesy
of D.M. Paterson.) (e) Mudflats, intertidal creeks and salt-marsh
formation, Skeffling, Humber estuary, UK. (Courtesy of D.M. Paterson.)
(f) Mangrove swamp with emergent roots (pneumatophores), northeast
Brazil. (Courtesy of I.A. Johnston.)

Plate 4 Freshwater lotic environments
(a) Lowland tropical river, Trauna River, Mt Hagen, Papua New Guinea. (Courtesy of G.N. Stone.) (b) Dry river bed in summer in 'Mediterranean' type climate, Mootwingee, Australia. (Courtesy of P.J. Sunnucks.) (c) Upland stream in arid zone, southern Chile. (Courtesy of J.M. Lambert.) (d) Upland stream, cool temperate Kyrgyzstan. (Courtesy of J.M. Lambert.) (e) Tropical eutrophic stream, Java, Indonesia. (Courtesy of G.N. Stone.) (f) Humic river in rain forest, northern Queensland, Australia. (Courtesy of I.A. Johnston.)

(a)

(b)

(c)

(d)

Plate 5 Freshwater lentic environments
(a) Upland oligotrophic lake, Mt Rugby, New Zealand. (Courtesy of P.J. Sunnucks.) (b) Tropical lowland seasonal eutrophic lake, Dindira, Tanzania. (Courtesy of G.N. Stone.) (c) Small mesic lake, Bosherston Ponds, Pembroke, UK. (Courtesy of D.M. Paterson.) (d) Crater lake, Rinjani, Lombok, Indonesia. (Courtesy of G.N. Stone.)

(a)

(b)

(d)

(c)

(e)

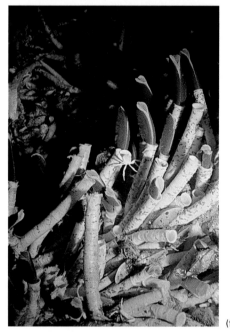

(f)

Plate 6 Unusual aquatic environments

(a) Pitcher plant, Mt Kinabalu, Borneo. (Courtesy of J.M. Lambert.)
(b) Soda lake, hypersaline and sulphurous, northern Chile. (Courtesy
of J.M. Lambert.) (c) Salt flat (resulting from completely dried out salt
lake), northern Argentina. (Courtesy of J.M. Lambert.) (d) Rift Valley
soda lake, Lake Magadi, Kenya, dominated by flamingos. (Courtesy of

I.A. Johnston.) (e) Hot-water springs and small geysers feeding
soda lake, with thermophilic algae, El Tatio, Chile. (Courtesy of J.M.
Lambert.) (f) Deep-sea vent fauna. Giant, tubicolous, pogonophoran
worms, *Riftia pachyptila*, found abundantly around deep-sea fumaroles.
(Courtesy of R.R. Hessler.)

Plate 7 Semi-terrestrial and terrestrial environments
(a) Tropical flooded forest, Amazonas, Brazil. (Courtesy of I.A. Johnston.) (b) *Melaleuca* swamp (swamp cypress), southern Australia. (Courtesy of P.J. Sunnucks.) (c) Temperate southern beech (*Nothofagus*) forest, Tierra del Fuego. (Courtesy of J.M. Lambert.)

(d) Tropical grassland, Mt Hagen, Papua New Guinea. (Courtesy of G.N. Stone.) (e) Temperate moorland, Isle of Hoy, Orkney, UK. (Courtesy of A.S. Edwards.) (f) Mediterranean 'garigue', Mt Carmel, Israel. (Courtesy of B. Vulliamy.)

(a)

(b)

(c)

(d)

(e)

(f)

Plate 8 Terrestrial environments
(a) Tropical rain forest, New World, Ecuador. (Courtesy of J.M. Lambert.) (b) Monsoon forest, Kakadu, northern Australia. (Courtesy of P.J. Sunnucks.) (c) Temperate upland grassland ('paramo'), with llama, Ecuador. (Courtesy of J.M. Lambert.) (d) Sub-tropical savanna thorn scrub dominated by *Acacia*, Kenya. (Courtesy of G.N. Stone.) (e) Sand desert, windblown dunes, Namibia. (Courtesy of N.P. Ashmole.) (f) Semi-desert scrub with goats, Turkana, northern Kenya. (Courtesy of G.N. Stone.)

Plate 9 Extreme terrestrial environments
(a) Cold mountain desert, Andes, northern Chile. (Courtesy of J.M. Lambert.) (b) Rock desert, with dry river bed (wadi), Sinai, Egypt. (Courtesy of S.A. Whiten.) (c) Arctic ice floes with seals, Spitsbergen. (d) Tundra with reindeer, Siberia. (e) Northern coniferous woodland (taiga), Canada. (Courtesy of A.S. Edwards.) (f) High-altitude mountain peaks, Mt Kinabalu, Borneo. (Courtesy of J.M. Lambert.)

(a)

(b)

(c)

(d)

(e)

Plate 10 Man-made environments
(a) Rice paddy, eastern Java, Indonesia. (Courtesy of G.N. Stone.)
(b) Industrialized river with transport and heavy chemical influences,
River Severn, UK. (Courtesy of D.M. Paterson.) (c) Intensive agriculture,
ploughed field, southern Australia. (Courtesy of P.J. Sunnucks.)
(d) Extreme urbanization, Athens. (Courtesy of B. Vulliamy.) (e) Arpley
landfill site, Warrington, Cheshire, UK. (Courtesy of David Drain/Still
Pictures.)

wave effects being much greater in winter when storms occur. In the littoral zone, and especially on beaches, waves are often the most important component affecting the interplay of marine and terrestrial conditions and impacting upon shoreline organisms.

Waves appear to move smoothly for long distances across the ocean, often for hundreds of miles (but remember that the water mass itself is *not* moving laterally, it is only the wave form that moves). In light winds the waves are seen only as slight ripples, but in very strong winds waves may reach over 50 m in height and 1000 m in length. When they reach shallow water, the waves become even higher and shorter and eventually unstable, 'breaking' onto the shore itself either by a gradual 'spilling' or by more dramatic 'plunging' especially on steeper beaches. Wave height is often critical to shore biology, affecting the energy that has to be absorbed by the shore. Where wave action is strong the effects of saltwater splash are also considerable and may occur for quite some distance inland. Wave height depends on three main factors: the wind velocity, the wind duration, and the 'fetch' of open water seawards of a particular coastline (effectively the distance over which the waves have travelled uninterrupted before reaching the land). On west coasts of Britain (Plate 2a) the fetch can be the whole of the Atlantic, and for the surf beaches of California it is most of the Pacific. But in the English Channel (Plate 2b) it may be only a few miles.

Waves are largely responsible for the shape and the character of beaches. As they hit a coastline they undergo complex processes of reflection and refraction, so that their direction is re-oriented. Although resultant water movements are extremely complex, and interact with local currents, a net longshore drift is commonly created and material is normally moved laterally. The waves therefore cause continuous erosion and deposition, leading to all the characteristic geological and geographical formations of coastlines (cliffs and caves, blowholes, stacks and coves, wave-cut platforms, dunes, spits, etc.).

Sediments

Coastal sediments, whether rocky, sandy or mixed (Plate 2c), represent a relatively short-term 'staging post', where material accumulates briefly in transit from mountain tops to ocean benthos. Some sediments may persist for thousands or even millions of years as part of the coastline, while others last for only a few seconds or minutes before being moved on by currents, tides, storms or waves.

The coastal sediments may be divided into primary and secondary sources. Primary sediment comes from cliffs, platforms or offshore submerged rock outcrops. Where gravel and sand predominate, much of the material comes from erosion of relatively loose glacial deposits. Additional primary material is from biogenic deposits, i.e. from the shelly remains of marine organisms. Secondary sediment arrives from rivers or glaciers and as wind-blown matter. Here the sediment is usually 'sorted' in some fashion (commonly by particle size or density) before arriving at the coast.

Sediment type has an enormous influence on the biota of shorelines, so we will often deal with two major types (rocky; sandy/muddy) separately in this chapter.

Patterns of biota on rocky shores

Vertical **zonation**, where dominant species occur in distinct horizontal bands, can be readily observed on shores, often from a long way away (Plate 2b), and it is surprisingly similar in its basic elements for all shores worldwide. It is often described mainly in terms of upper and lower limits for particular organisms (Fig. 10.4), with an upper 'black lichen zone', then a 'periwinkle zone', then a 'barnacle zone' and lowest of all a fourth band where the dominant species are more variable. The width and number of distinct zones, and the diversity within these zones, then depends on local effects: geography, climate, the magnitude of tides, the timing of low and high tides, the substratum, and above all **exposure**. This is a useful composite term involving beach aspect, fetch, wave force and winds as influenced by all the other astronomical and geographical factors; it is the most widely used term when comparing shores and their fauna and flora.

An exposed shore has a long fetch, and heavy wave action, with wave splash occurring to a considerable height and with all the zones fairly broad and spread out. By contrast a sheltered shore, such as occurs in an enclosed bay with reduced wind action, gentle waves and a short fetch, may have only limited wave splash upwards, and the zones are compressed together in terms of height. So a simple representation has to include the element of exposure, related to degree of wave action, as in Fig. 10.5; high-exposure shores are those on the left, with the indicator species' zones all wide, while low-exposure sheltered shores are on the right, their zones compressed together. On a very exposed shore the lichens and the small winkles (gastropod snails in the genus *Littorina*) may extend a long way up cliffs

Fig. 10.4 Fundamental patterns of zonation described by the limits of key types of biota.

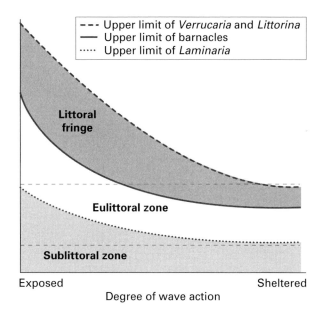

Fig. 10.5 A more realistic description of beach zonation where the degree of wave action modifies the height of the beach over which the patterns in the biota are spread: highly compressed on a sheltered beach, greatly spread (often well up the cliff face) on an exposed shore.

behind the seashore. This figure also shows some of the terminology used in studying seashores: the maritime zone, the littoral fringe, the eulittoral 'shore', and the sublittoral zone.

In terms of actual species composition, we can identify particular zones quite clearly wherever shores of relatively rocky substratum occur. There is an upper zone of grey-green lichens, maritime dicotyledonous plants and grasses, seen at the tops of cliffs worldwide. Below this is the black lichen zone, where the animals include isopods such as *Ligia*, collembolan springtails such as *Anurida*, and a range of small to medium-sized periwinkles (gastropod molluscs) that can fit into cracks and crevices. Further down the winkle species become somewhat larger, more diverse and more dominant, and merge into a zone dominated by barnacles (sessile crustaceans), usually with a high percentage covering of one or a few species of barnacle (*Balanus, Semibalanus, Chthamalus*). Seaweeds include the tougher fucoid species. Amongst the weeds and barnacles are gastropod limpets (*Patella*) and whelks (*Nucella*), some sea anemones (*Actinia*) and often very dense colonies of mussels (*Mytilus* or *Modiolus*). The common brown seaweeds, such as bladderwrack, abound. Finally we meet the red algal and kelp zone, with dense beds of long strappy seaweeds such as *Laminaria*. Here animals are often very diverse, including crabs, top-shells and cone-shells, anemones, tunicates, sponges, echinoderms, bryozoans and polychaete annelids—note that the list includes representatives of many of the soft-bodied phyla of animals. These will be particularly abundant and diverse on the more sheltered shores with a long shallow slope and plenty of rocky crevices (Plate 2d).

It is thus possible to build up a more complex view of a zonation scheme for a typical rocky shore (Fig. 10.6). A scheme such

as this can be used to assess exposure on shores and to compare them with all other shores worldwide; the patterns of occurrence are therefore giving us a way of *quantifying* exposure, usually given on a scale of 1–8. In other words, if we get information on species' abundances and their heights on shore, we can easily describe that shore and can make directly useful comparisons between shores. The exact species may differ on shores in Europe, Australia or in North America, but the principles will be the same. The major exceptions are rocky shores in Antarctica where ice-scouring almost entirely eliminates intertidal macrofauna.

Patterning of biota on sandy shores

Sandy shores (here taken broadly to include muddy beaches and shingle shores) have very different characteristics from rocky shores. Sand is predominantly silica (SiO_2), together with $CaCO_3$, both deriving from rock and from shelled animals, and in varying proportions; beaches in areas where deposition is predominantly from coralline material are dominated by calcium carbonate and tend to be very white (Plate 2e), while those on volcanic islands (Plate 2f) may be of very dark sand with rock erosion products dominating. There is also usually some complex chemical clay material present, and a small component of organic matter. Sands, of course, also contain water, because the 'packing factor' between roughly spherical grains means that there is normally a minimum volume of fluid (calculated as 26% for perfect spheres) that is hard to remove.

The main variable for sands is the **grain size**, which in turn affects a number of other features. Drainage is the most important of these, being the outcome of gravity working against capillarity, so that fine grains are more water-retentive when wetted. Oxygenation is also affected, with well-drained large-grain sands being more highly oxygenated. Small-grain fine sand may retain considerable volumes of water for long periods and so readily become anoxic; this condition merges into even finer-grained muds, which can get unpleasantly smelly as anaerobic decomposition products build up. Grain size is a parameter commonly measured and frequently quoted for beaches; one of the schemes most widely used is Wentworth's classification, shown in Table 10.2, listing types from boulders (see Plate 3a, facing p. 310) to fine clay and mud (Plate 3b). Grain size correlates rather well with the exposure factor discussed above. In exposed sites the fine-grained material gets removed, leaving rather coarse or gritty sands or shingles ('clastic' beaches). The smaller grains are deposited again in areas of gentle wave action, producing shallow sheltered beaches of soft sands, often with some intrinsic patterns of low ridges, cusps and bars resulting from waves interacting with local topography. Grain size therefore directly affects animal distributions in sandy shores, as shown for the crab species in Fig. 10.7(a).

Another factor differing between beaches is that of **sorting**. Some beaches have a very uniform particle size and are said to be well sorted, while others are highly variable in different areas, with a poor sorting factor. Thus two beaches can clearly have the

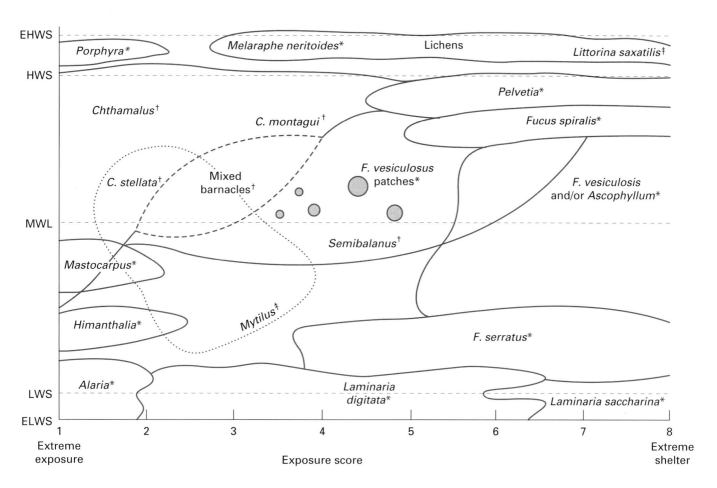

Fig. 10.6 A biotic analysis of zonation in terms of key genera and species encountered at different heights in relation to exposure (using a scale of 1–8 from exposed to sheltered). *, Seaweeds; †, barnacles; ‡, molluscs. (Adapted from Ballantine 1961.)

Table 10.2 Wentworth classification of surface deposits.

Grade	Description	Particle size (mm)
1	Boulder	>256
2	Cobble	64–256
3	Pebble	4–64
4	Granule	2–4
5	Very coarse sand	1–2
6	Coarse sand	0.5–1
7	Medium sand	0.25–0.5
8	Fine sand	0.125–0.25
9	Very fine sand	0.062–0.125
10	Silt	0.031–0.062
11	Clay/mud	<0.031

same mean grain size but very different sorting factors, in turn affecting their drainage and oxygenation. Note also that the presence of animals in itself can affect sorting; the processes involved are **bioturbation**, **biodeposition** and **biosecretion**.

These are brought about by burrowing and feeding animals that pass sediments through their guts, such as polychaete worms, thalassid shrimps, sea cucumbers and many bivalves. These animals may move finer egested particles up to the surface, giving a layer of fine material on top of rather coarser sands; but where the sand particles are bound together with faecal pellets the effective grain size may increase. Such processing of the sediments often shows distinct seasonal patterns (Fig. 10.7b); for the lugworm *Arenicola*, high summer rates of bioturbation are associated with the lowest sediment shear strength, the surface mud having a higher water content and low bulk density. The presence of tube structures made by animals can also have a marked effect on the stability of sediments. High tube densities can result in a reduced flow regime between the tubes, often leading to sediment accumulation. Figure 10.7(c) shows that high densities of the worm *Pygospio* (tens of thousands of tubes per square metre) on an estuarine sandflat can greatly increase the shear strength of the sediment, and this may over time raise the worm bed well above the surrounding sand.

The small-grain beaches provide the kind of material that animals can dig in, where they will be able to keep wet fairly reliably, so they can make a living without too much physiological stress. Foods will be available, because organic material and bacteria are usually associated with the surfaces of sand grains, and fine particulate sediments have relatively much greater surfaces available. Inevitably it is sheltered beaches that we are mainly

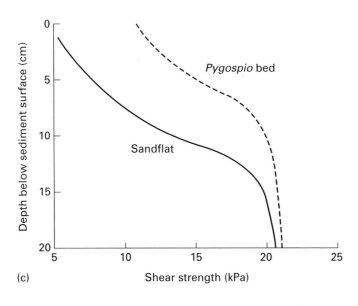

concerned with here; the exposed sandy beaches do not have a great deal of organic material or metazoan life associated with them. But note that in sheltered beaches only the upper sandy part is oxygenated, and there is usually a sharp transition vertically to deoxygenated mud. This transition layer is termed the **redox potential discontinuity** (RPD), where the sand changes from an oxygenating to a reducing environment. This is often visible as the interface between sand and black mud and it may be at various depths; in fact its depth affects how much mixing of oxygen occurs across it, and therefore how sharp it is. The lower layer below the RPD is a highly deoxygenating (reducing) environment; ferric ions are converted to ferrous, nitrates turn to nitrites and ultimately ammonia, and sulphates end up as hydrogen sulphide, producing the characteristic 'bad egg' smell. Some anaerobic fermentation by microbes also occurs, yielding alcohols and fatty acids. Animals and aerobic bacteria therefore occur only in the upper part of sheltered beaches and are always at some risk of anoxia. However, the activities of animals themselves affect the mixing a little, and in particular the tube-dwelling animals may penetrate into the anoxic mud by keeping a circulation of reasonably oxygenated fluid through their burrows or tubes; when excavated, the edges of such tubes often appear as a zone of altered coloration because the substratum is locally oxygenated.

In effect there are three main habitat options for animals in sandy shores: living amongst the sand grains, living on top of the sandy layer, or projecting downwards within tubes or burrows. For practical purposes the biota of sandy beaches is divided into a number of categories largely related to size and how the organisms can be collected, rather than to taxonomic categories.

1 Microfauna are organisms with dimensions less than about 0.1 mm, living amongst sand grains. The category includes protists and some very tiny invertebrates; these can only be collected by fine sieving of the sand.

2 Meiofauna are around 0.1–0.5 mm, and are also called the **interstitial fauna**, again living within the sand. Included here are some hydroids, small platyhelminths, a range of pseudo-coelomates (especially nematodes), some of the curious rather simplified groups of annelid worms termed 'archiannelids', and a few small crustaceans. These animals tend to be rather convergent in body form, all being vermiform (worm-like) with bristles or stubby 'legs' to help push through the sand.

Fig. 10.7 Interactions of animals and particle size on beaches. (a) Effects on the distribution of crabs; *Helice* is found higher on the beach and in finer sediments, while *Macrophthalmus* occurs lower down and in a more restricted habitat of rather coarse sand grains (particle diameter is given as $\log_2$ of the particles in mm). (From Little 1990; courtesy of Cambridge University Press.) (b) Seasonal bioturbation changes brought about by burrowing lugworms (*Arenicola marina*), mirrored by changes in the sediment shear strength. (c) Increased shear strength due to sediment accumulation between tubes of *Pygospio* worms in a sandflat. (b & c, From Lockwood *et al.* 1996.)

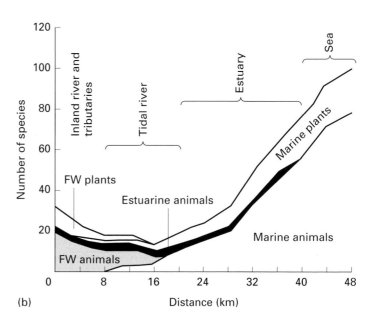

Fig. 10.8 The species diversity of estuaries. (a) Species number is plotted arithmetically against number of individuals for different marine and estuarine habitats. (Adapted from Sanders 1969.) (b) Number of species is plotted against distance along a typical estuary showing the seaward decline in freshwater (FW) species and increase in marine species, with truly estuarine species always in a small minority. (Adapted from Alexander *et al.* 1935.)

3 **Macrofauna**, which is divided into two types:

(a) **Infauna**, which are the visible large creatures living in burrows and tubes within the sediment, often leaving characteristic traces on the surface. Annelids and molluscs are most dominant, especially the filtering bivalve molluscs like cockles and clams (*Cardium, Mya, Mercenaria, Donax*), and polychaetes such as the lugworm *Arenicola* and the sand mason *Lanice*. The carnivore niche is filled by a few specialist polychaete worms, such as *Glycera*, a few nemertean worms, and boring gastropods such as *Natica*. There are also normally some echinoid and asteroid echinoderms (urchins and starfish), a number of other worm groups, and a profusion of small burrowing crabs, amphipods and isopod crustaceans scuttling or hopping about amongst and over the sand grains. Towards the top of the beach more terrestrial amphipod sandhoppers such as *Talitrus* may be dominant. At the lowest levels, especially where the beach merges into mud, more fragile filterers such as the peacock worms (*Sabella*) and burrowing shrimps (e.g. *Callianassa*), and rarer deposit-feeders, such as peanut worms (sipunculans) and sea cucumbers (holothurian echinoderms), occur patchily.

(b) **Epifauna**, which are the visible surface dwellers, including everything from seagulls to bivalves to crabs, with some essentially marine animals in the shallow water (including stingrays and small sharks, various flatfish and gobies) and

some essentially terrestrial animals on the upper shore such as insects (beetles and flies in particular). There is actually rather little epifauna (or epiflora) on most sandy beaches.

10.1.2 *The estuarine habitat*

Estuaries are semi-enclosed water bodies linked to the sea within which the water is measurably diluted by freshwater inputs. The category of estuaries could be said to include all zones where we get sea water and fresh water coming together cyclically, such that as water depths go up and down (varying the interaction with the aerial environment) so too does the salinity vary. This would therefore include salt marshes, mudflats, coastal swamps, and mangrove forests, and the upper reaches of sea lochs and fjords. The three-way interaction of seawater, freshwater and terrestrial habitats (dictated by tides and waves as above) means that life in estuarine conditions is particularly difficult. Plant diversity (in terms of number of species) is usually low. There are also far fewer animal species in such habitats than in either sea water or fresh water (Fig. 10.8); the critical salinity range is around 10–25% sea water, where few essentially marine animals survive and equally few freshwater species can remain active, in both cases mainly due to a breakdown of the normal cell volume regulatory systems (see below). Whether this is the primary reason for the low species number of brackish-water animals remains unclear; it may also partly be brought about by the transience of estuaries through evolutionary time, so that there is never a long enough period for a large number of species to establish themselves (except perhaps in the very largest estuaries, such as the Amazon). A combination of physiological stress and evolutionary ephemerality is likely to explain the paucity of estuarine species; it is certainly reflected in a rather similar fashion in every taxon, as shown in Table 10.3.

Region	Approximate salinity (ppt)	Scaphopods	Bivalves	Gastropods	Cephalopods
North Sea	35	11	189	351	32
Kattegat	26	1	92	162	14
Danish Belts	20	1	42	68	5
Outer Baltic	8	0	11	19	0
Gulf of Bothnia	4	0	4	4	0

Table 10.3 The numbers of species in various taxa of molluscs as salinity decreases from the North Sea into the Baltic Sea.

Habitat characteristics and biota

Estuaries are inevitably very variable in type and in terms of key characteristics such as salinity regime, depth and width, water flow patterns, and substratum. Most of the 'true' estuaries result from drowned river valleys (rias, Plate 3c), and they are therefore also geologically impermanent, greatly altered in conformation by small changes in sea level. Their average lifespan may be less than 10 000 years, and many of those observed today resulted from marine flooding immediately after the last Ice Age. Others arise from the gradual occlusion of a river mouth by spits and bars (Plate 3d).

But despite their ephemeral nature they are biologically very productive, with a high biomass. Estuaries can act as substantial nutrient traps, with nutrient run-off from the land and nutrient cycling occurring just within the bay, being retained by flow characteristics from both ends. Thus they can be areas of very high primary productivity (cf. Table 10.1), especially in tropical latitudes, limited usually by nitrogen availability. This productivity is reflected in high faunal density, despite the low faunal diversity; some estuaries have become major sites of fisheries (notably the large Chesapeake Bay area and the huge Hudson Bay), supporting large populations of molluscan and crustacean 'shellfish' as well as fish. The resident animals present tend to be specialists, particularly drawn from these three groups (molluscs, crustaceans, and fish such as mullet, flounder and bass). There is also a great range of 'amphibious' animals, dominated by wading birds, for whom coastal mudflats and other brackish wetlands are crucial stopovers during migrations, for winter feeding, or for summer breeding; also in this category would be tetrapod vertebrates such as otters, living on rocky shores of inlets and sea lochs. In the more freshwater parts of estuaries many of the crustaceans and annelids are replaced by insect larvae and adults, especially midges, caddis flies and beetles, often able to survive a few hours of submersion in brackish water. Few animals are able to tolerate the full range of estuarine conditions; only those relatively rare fish and crustaceans that can regulate across the entire salinity range are able to migrate right through estuaries.

Mudflats and salt marshes (Plate 3e) commonly form on the sheltered sides of estuaries in temperate zones, where wave energy is low and there is a surplus of fine sediment, and these areas may retain water through almost all the monthly and annual tide cycles. Mudflats may be bare of vegetation or show a patchy covering of sea grass. In the tidal saline creeks some green algae survive. But the tops of mudflats may get desiccated and very warm; thus there are often dense populations of filmy green algal species through winter, which get burnt off in summer and replaced by tougher brown algae with denser growth forms. In these creeks, certain mussels, crabs and gastropods thrive, and the habitat may be an important nursery ground for fish. Waters in the creeks may become unusually rich in sulphides and nitrites leaching out of the adjacent anaerobic mud, so that resident animals require tolerance or regulation of these compounds.

Mudflats usually occur in various successional stages that lead eventually to marsh and then solid ground. The original bare mud gradually forms into a more solid marsh as plants colonize and begin to reduce the flow rates, their roots trapping and holding the finer sediments. Salt marshes are initially dominated by relatively few specialist salt-tolerant plants, notably *Spartina* and other cord-like grasses with extensive rhizomes, gripping into the mud and spreading asexually over vast areas. Succulent species, such as glassworts, also invade and act as pioneer stabilizers. When the sediment surface rises above the normal neap high tide level, more terrestrial species such as reeds, rushes, sea blites and sea lavenders may colonize. Mud and peat continues to accrete amongst the marsh plants, giving a steady rise in height, until eventually the marsh is normally above the high tides and it becomes an essentially terrestrial habitat, colonized by insects and birds. This whole process may take perhaps 150–350 years if the system is left undisturbed, again stressing the ephemeral character of estuarine habitats.

In tropical areas, salt marshes are usually replaced by mangrove swamps (Plate 3f), intertidal zones where emergent mangrove plants are rooted in an essentially marine anoxic sediment. There are just a few mangrove genera (e.g. red mangrove, *Rhizophoa*, with surface prop roots, and black mangrove, *Avicennia*, with protruding pneumatophore roots for aeration), with a succession of species from seaward to landward. Small offshore islands may be formed from the root mats of mangroves. Each mangrove swamp has its own associated submerged fauna, usually again dominated by small crustaceans, annelids and gastropods that are essentially detritivores; many mangroves are important fish nurseries and may provide daytime resting places for oceanic species such as sharks. But unlike the temperate salt marshes, these mangrove swamps also have a strong above-ground fully terrestrial community, which may

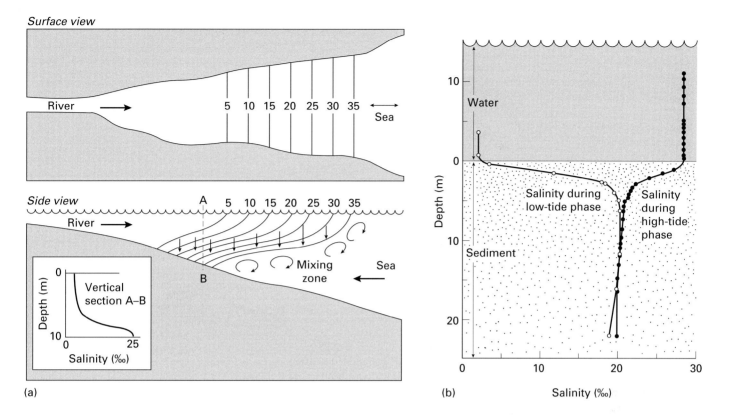

Fig. 10.9 Zonation in estuaries. (a) The distribution of salty and fresh water in top and side view, with a deep 'salt wedge'. (b) The variation in salinity in the water column and in the sediment at low tide and high tide, with the deep sediment salinity constant at all phases of the tide cycle.

include birds, bats and monkeys as well as the inevitable insects. Tropical mangroves are exposed to normal-salinity sea water at around 28°C, but with variation due to intense evaporation, torrential tropical storms, and run-off from shallow freshwater pools inland.

Patterning of biota in estuaries

Estuaries are most obviously characterized by a gradient of decreasing salinity as we move upstream from the mouth, giving a zonation of less tolerant to more tolerant species. But the pattern is complicated by two other factors.

Firstly, many estuaries also have a vertical gradient (stratification), resulting from the different densities of fresh and salt waters. Salinity will be greater at the bottom than in the surface water, with the fresh water moving over the top of a 'wedge' of saline water, while the underlying sediment may contain water of relatively stable intermediate salinity giving a **halocline** (Fig. 10.9). Thus burrowing and benthic animals can often penetrate further into an estuary than planktonic animals can (and adults may survive better than their planktonic larval forms). Flow is usually seaward in the surface water, except briefly at

the peak of an incoming tide; animals can exploit this to achieve station-keeping in an estuary, adjusting their depth within the water column to come inwards near the surface on the landward flood tide and then becoming benthic and drifting only a little way out again on the seaward ebb.

Secondly, the flow patterns of the river may override the expected mixing pattern within the apparent estuarine zone. For large tropical rivers, like the Amazon and the Congo, the flow coming out from the river is usually so forceful that the 'zone of mixing' is actually well out to sea. For moderately large rivers, with limited tides, flow progressively forces silts and mud out into a fanning 'birdfoot type' estuary (Fig. 10.10) where mixing occurs. Only with smaller rivers does the mixing happen strictly within the river mouth. Thus true estuarine conditions do not occur exclusively within the apparent geographical confines of an estuary.

There may also be strong seasonal patterning in brackish waters. Temperate estuaries in particular commonly show seasonal changes in biota, with a spring bloom of phytoplankton when inputs from the land and river run-off peak. This bloom of primary productivity is followed by successive peaks of animal species that feed on the phytoplankton and on each other (Fig. 10.11).

10.1.3 Fauna of brackish waters

Brackish-water faunas, as we have seen, are not normally very diverse or speciose, but can be of very high biomass and density.

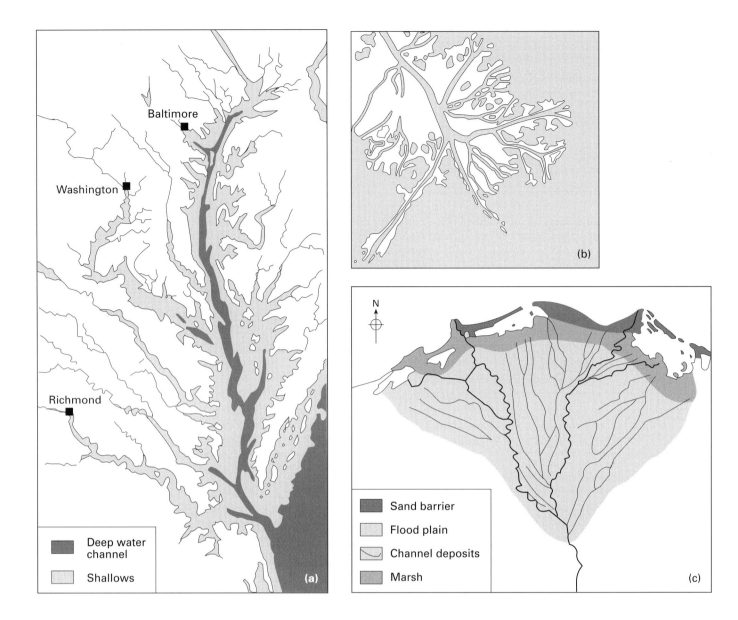

Fig. 10.10 Different types of estuary and mixing patterns. (a) An example of a traditional drowned valley or ria (Chesapeake Bay). (b) A birdfoot estuary where high flow rates push sediment out to sea (the Mississippi delta). (c) A classical triangular river delta, again mixing well out to sea (Nile).

Most marine invertebrate groups achieve some representation, but some groups are too stenohaline to cope with even moderate dilution of the medium and are conspicuously absent, found only below the littoral zone and so encountered by beach-combers only on extremely low tides; these include cephalopod molluscs, tunicates, sipunculans, most echinoderms and scyphozoan cnidarians (jellyfish). There are occasional examples of sponges, anthozoan cnidarians (anemones), bryozoans ('sea mats'), and minor groups such as echiurans and hemichordates.

A few amphibians tolerate brackish conditions (*Rana cancrivora*, already mentioned in Chapter 9 as being able to tolerate full sea water, but also a few species of *Bufo*), and one crocodile species lives in estuaries in Australasia. But brackish waters are clearly dominated by four major groups:

1 Annelid worms, including deposit-feeding sand masons and terebellids, filter-feeding sabellids, and omnivorous mobile ragworms.

2 Gastropod molluscs (including the grazing winkles, limpets, etc.) and filter- or deposit-feeding bivalve molluscs (including mussels, clams and razor shells).

3 Crustaceans, both omnivorous and predatory, including amphipods, shrimps, crabs, etc.

4 Fish, especially the gobies, blennies, killifish, etc. on shores and in pools, and larger fish migrating in and out of estuaries (including sawfish, stingrays, bull sharks, etc.).

Fig. 10.11 Seasonal patterns of abundance in a temperate estuary; spring peaks of phytoplankton occur and are followed by successive peaks of animals that eat the plankton and are themselves eaten by other animals such as the predatory planktonic arrow worm, *Sagitta*. (From Lockwood *et al.* 1996.)

10.1.4 *Central problems and strategies*

The problems of living on the fringes of seas largely arise due to ongoing change through time, mainly due to tidal patterns and wave actions. Most obviously there are the effects of being at the habitat interface, affecting osmotic and thermal balances, respiratory exchange, and feeding strategies for both predators or herbivores. Tidal variation adds another layer of hazards, affecting all those same parameters on a complex rhythmic time-base, and also altering ions, pH and light levels. Waves produce the further problem of crushing (with a pressure of up to 22 tonnes m^{-2}), and a major risk of dislodging and displacement. They also bring in suspended material and cause turbidity, which may silt over and bury animals clogging up gills and filtering surfaces, and reduce the rate of photosynthesis for algae. Freshwater run-off down a beach may add further variability in osmotic concentrations, in ionic levels and in pH. In estuaries there is the extra variation of cyclical water concentration changes as the sea retreats and the river adds fresh water.

In the face of all these problems, the simple strategy of avoidance is extremely important, and we will meet many examples of this. Where change is cyclical, and conditions can be guaranteed to reverse and ameliorate within an established time-frame, it makes very good adaptive sense to avoid the predictably short period of adverse conditions by simply shutting down activity, hiding, closing a shell, or burrowing deeper. This requires a detection system, and many littoral animals have chemoreceptors (osmoreceptors) in peripheral sites suited to sampling the changing medium; crabs use their antennules, other crustaceans their feet, molluscs the osphradium (near the gills) or the mantle edge generally, and polychaetes the cirri or tentacles on their heads. Avoidance can then take many forms. For example, some lower-shore organisms may only get exposed to air briefly twice a day at low tide, and closing up valves or lids, with some sea water within to provide an osmotic buffer and an oxygen store, may be the simplest and almost cost-free strategy. However, this may not be enough at times of the year where avoidance would need to work on quite a long timescale. Some animals at the extremes of a beach or estuary may experience cycles not just involving the 12 hours or so between tides, but of monthly recurrence, where tides only cover them (or uncover them) for a few hours on a couple of days each month. For these animals, as well as for many unprotected softer-bodied organisms that cannot readily avoid conditions, some degree of tolerance and regulation is essential, involving physiology and biochemistry at the cell or whole organism levels. These organisms need to be **euryhaline**, **eurythermal** and **euryoxic** to have any real chance of long-term survival in an estuary or seashore.

10.2 **Ionic and osmotic adaptation and water balance**

We know that all animals regulate their ionic balance to some extent, and that the necessary pumps and channels are a basic property of cells. However, on shores and in estuaries all animals need to cope with more external variation. Shores experience seasonal and diurnal/tidal variations; there will be periods of hyperosmotic conditions when the tide is out, due to evaporation of water from rock pools and water in crevices, etc., and periods of hyposmotic stress due to rain and freshwater run-off. Seasonally the interstitial water in a beach may vary markedly in salinity; in the example shown in Fig. 10.12(a), this water is hyposmotic in most winter months but roughly as concentrated

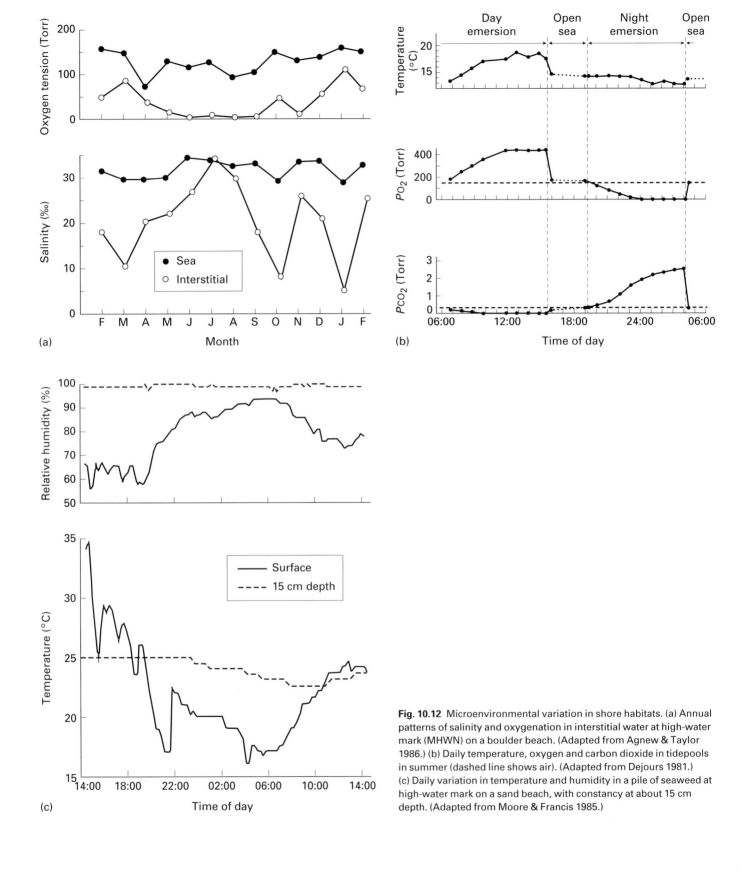

(a)

(b)

(c)

Fig. 10.12 Microenvironmental variation in shore habitats. (a) Annual patterns of salinity and oxygenation in interstitial water at high-water mark (MHWN) on a boulder beach. (Adapted from Agnew & Taylor 1986.) (b) Daily temperature, oxygen and carbon dioxide in tidepools in summer (dashed line shows air). (Adapted from Dejours 1981.) (c) Daily variation in temperature and humidity in a pile of seaweed at high-water mark on a sand beach, with constancy at about 15 cm depth. (Adapted from Moore & Francis 1985.)

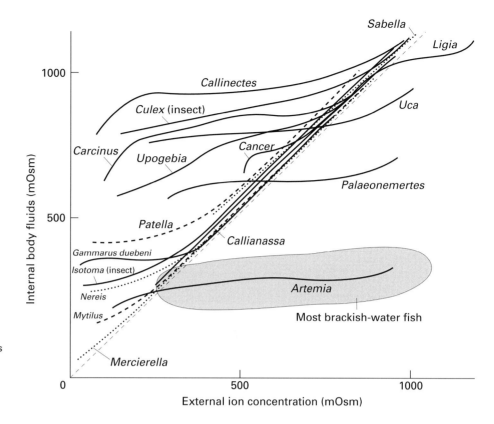

Fig. 10.13 Variation of internal body fluid concentrations with external salinity for a range of brackish-water species. Arthropods are shown by solid lines, molluscs by dashed lines and worms by dotted lines, with teleost fish represented by the tinted area.

as the sea water in the summer, when it also becomes much more anoxic. Superimposed on this, of course, is a substantial tidal variation. Animals that live in such beaches and cannot reach protected microhabitats may also become exposed to air and thus to extreme desiccation stress.

Estuarine waters range from almost fully marine to substantially hyposmotic, but alter to a varying extent through time and with distance upstream, to be nearly fresh water at low tide close to the river outlet or after prolonged rainfall.

10.2.1 Strategies

Avoiders

It is possible to avoid osmotic stress and/or actual drying out of tissues to a large extent, and a great many animals of brackish waters do this. They can easily stay covered in water by retreating into crevices, remaining in rock pools, or burrowing deeper in sand or mud. However, these options may give extra problems of oxygen depletion or of exposure to unusual pH levels or (in pools) to raised water temperatures, as shown in Fig. 10.12(b). For shoreline animals that are more terrestrial, excessively dry air may be avoided by burrowing into the piles of seaweed and debris that accumulate at the strandline, where temperature and humidity stay relatively constant (Fig. 10.12c); this niche always has a thriving community of small invertebrates associated with it.

Another solution is to carry a water store ('exosomatic water'),

perhaps within a gill chamber as in many crustaceans, or in the mantle cavity for molluscs. This provides exposure of the soft tissues to water whose salinity differs from that in the external environment, and may allow the animal to 'smooth out' the variations it actually experiences. Many isopods and some amphipods also carry an external water source in their ventral groove, a fluid-conducting system using capillarity to move water over the lower surfaces and allowing 'aquatic' methods of gas, water and ion exchange in otherwise semi-terrestrial beach dwellers.

A third possibility is to have an intrinsic resistant covering. This may be mucus, of limited value but present in certain animals such as the common beadlet anemone, *Actinia*; or more specifically it may involve tubes or boxes, manufactured by the animal, and usually capable of being sealed off from the outside world temporarily with a lid or sphincter. Obvious examples include calcareous shells or cuticles in molluscs, brachiopods and crustaceans, chitinous boxes in bryozoans, and tubes of sand, chitin, consolidated mucus or calcium carbonate in a range of annelid worms. It is no coincidence that so many shores are dominated by animals with intrinsic covers; most spectacularly by barnacles, limpets and mussels.

Conformers and regulators

The basic pattern of conformers and regulators within some common euryhaline organisms is shown in Fig. 10.13 (cf. Fig. 5.2). In essence, following the concept of four different 'spatial levels' of control that we considered in Chapter 1, animals that are conforming have no control at the skin/blood interface, but are to some extent regulating (in terms of maintaining reasonable cell volume) at the blood/cell interface; in contrast, regulating animals concentrate their controls at the skin/blood interface and are thus maintaining a relatively constant body fluid concentration.

Animals like *Carcinus maenas* (the common shore crab) are very good ('strong') hyperosmotic regulators, keeping their blood osmotic concentrations quite high and stable over a wide range of external concentrations. But there is intraspecific variation here: shore crabs from the North Sea are less tolerant of dilution than those from the more brackish waters of the Baltic Sea, and also regulate rather less strongly at any given dilution. *Cancer* (the edible crabs) and *Nereis* (the polychaete ragworms) are moderate ('weak') regulators, allowing their body fluids to fall in conformity with external dilution (while using moderate volume regulatory measures) down to some critical point, and then instituting a degree of skin- or gill-based regulation to cope with any further dilution. Animals such as *Actinia* and *Anemonia* (the commonest shoreline anemones) and *Mytilus edulis* (the common mussel) are essentially conformers, with only a little regulation when in moderately or (for *Mytilus*) very dilute habitats, to give an internal osmotic concentration slightly above that of the medium. Over most of the concentration range they do not attempt to control their water balance but act as simple osmometers, incurring substantial tissue volume changes. (However, remember that *Mytilus* is a bivalve mollusc, so does have the option of temporary avoidance by shutting the shell, and will probably only need to conform on the higher shore levels during periods of spring and neap tides.) At the extreme, there are a few invertebrates that come close to being perfect conformers, with body fluid concentrations that even in tapwater are scarcely elevated; the tiny reef-forming polychaete worm *Mercierella* (also known as *Ficopomatus*) is a well-known example, surviving both in tap-water and in 200% sea water.

The kind of plot shown in Fig. 10.13 clearly reveals that there is a continuum of strategies, not a simple dichotomy. Several points emerge when we look at specific examples. Firstly, there are no perfect regulators. Organisms exhibit varying degrees of maintaining their blood fluid concentration, but most brackish-water animals that survive dilution seem to settle at around 30–70% of original seawater concentration when in dilute media. Secondly, note that crustaceans usually regulate more than worms and molluscs, and this is probably due to their exoskeleton. Since they have a very hard covering they are unable to swell significantly and therefore cannot help but regulate somewhat; their cuticle gives an inbuilt physical aid to being a regulator, which also makes the process cheaper. Thirdly, conforming *can* be almost perfect, and is a very cheap solution as it avoids the need for using up substantial energy in active transport, so reducing the metabolic rate and food requirement. Fourthly, not many animals hypo-regulate (i.e. keep their blood concentration lowered when the medium gets more concen-

trated); amongst the invertebrates there are just a few crustaceans able to do this (some crabs and prawns, branchiopods and isopods). Most upper-shore animals instead just 'conform upwards' when their pools or crevices get hyperosmotic on warm days. Finally, as we should by now expect, vertebrates do not fit the general pattern, as they have dilute blood anyway and they regulate it fairly constantly at a moderately dilute level, whether the medium is above or below normal sea water. It is amongst the crustaceans (such as the blue crab, *Callinectes*, and the mitten crab, *Eriocheir*) and the vertebrates (such as salmon, *Salmo* spp., and eels, *Anguilla*) that we find the animals with the greatest osmoregulatory powers, capable of migrating right across the estuarine zones as they move from fresh water to sea water and back cyclically (the anadromous and catadromous species, respectively, spawning in fresh water and in sea water but moving regularly between the two media). However, their osmotic ability is matched or exceeded by some of the 'amphibious' shore animals that are essentially terrestrial rather than aquatic; for example, the sea slaters in the genus *Ligia* are good hypo- and hyper-regulators when immersed but achieve even better osmotic constancy when in air, as do supra-littoral beach-hoppers, such as *Orchestia*.

10.2.2 *Mechanisms*

For all brackish-water animals, there are three main mechanisms that may be involved in the adaptation to osmotic changes: alterations of external permeability, varying salt uptake, and cellular osmoregulation.

Permeability

For any animal living in an uncertain environment, it makes obvious sense to reduce surface permeability as far as possible, since this will reduce the rate of change of concentration, and the rate of change in blood volume, to levels where they can be manageable. This is potentially even more useful in brackish forms than in true freshwater organisms. Notice how many of the common shore and estuary animals are preadapted in this regard, especially the molluscs and crustaceans, by having 'built-in' low-permeability external surfaces as part of their body plan.

As we saw in Chapter 4, there are various separate components when considering permeabilities, and values for both water permeability and salt permeability for some molluscs, crabs and fish are summarized in Table 10.4. The permeability to water (P_w) is clearly reduced in most brackish-water groups. For example, non-marine crabs show lower fluxes for water than marine crabs, and they may also have measurably thicker cuticles. The value of P_w can sometimes be at its lowest in the brackish species, below that of freshwater congeners, underlining the point that this zone is most difficult and it is really worthwhile to invest in low permeability thus 'smoothing out' the environmental changes. Examples for the polychaete worm genus *Nereis* and the amphipod crustaceans allied to *Gammarus* are shown

Table 10.4 Values of water permeability (P_w) and sodium permeability (P_{Na}) in brackish-water animals in comparison with marine and freshwater species and races.

Genus and habitat	P_w (10⁻⁴ cm s⁻¹)	P_{Na} (10⁻⁶ cm s⁻¹)
Bivalves		
Brackish water		
Mya	3.8	5.7
Geukensia	1.3	4.0
Crustaceans		
Sea water		
Libinia	12.8	13.2
Porcellana	4.7	1.0
Eupagurus	12.4	1.6
Maia	3.4	12.9
Carcinus	9.0	5.0
50% sea water		
Carcinus	2.8	5.0
Cancer	1.3	2.7
Fresh water		
Eriocheir	0.22	0.87
Astacus	0.88	0.10
Vertebrates		
Fresh water		
Carassius	3.9	0.4
Salmo	1.6	0.3
Anguilla	1.2	0.6

Animals in these zones may also be able to modify P_w within their own lifetime by acclimatory change, as shown in Fig. 10.14 for *Gammarus duebeni*; here the rate of exchange of labelled water depends on the salinity, changing at best from about 8.3 h⁻¹ (at 35 ppt) to 1.7 h⁻¹ (at 1 ppt). It should be noted that some animals which also undergo periods of aerial exposure, such as the bivalves *Mya* and *Geukensia*, do *not* change P_w with decreasing salinity, having, respectively, a deep burrow or a routinely low P_w value to reduce water loss during such exposure. Where it does occur, permeability acclimation has a very rapid time course, probably due to hormonal action. Certainly in vertebrates there are a series of hormones, such as antidiuretic hormone (ADH), that control P_w variations in the skin and/or the kidney tubule cells.

Salt loss through the skin is also very important for most invertebrates; it usually makes up 80–95% of the total salt loss, with the 'kidneys' not very important. A few shoreline animals, such as sea anemones, are surprisingly permeable to ions at their epithelial surfaces, which allow rapid diffusion in either direction; but in general the sodium permeability (P_{Na}) of skin is as low in brackish-water animals as in freshwater species (Fig. 10.15a). There is little evidence that this can be acclimated in invertebrates, except in a few hyposmotic regulators such as the prawn *Palaeomonetes serratus*, which has decreased P_{Na} in 120% sea water. However, in vertebrates aldosterone and prolactin hormones control P_{Na}; this occurs in fish gills and frog skins, for example, as well as in the kidney (see Chapter 5). P_{Na} also varies genotypically, where populations are exposed to different media concentrations, as with the isopod *Mesidotea* shown in Fig. 10.15(b).

Changes in permeability may be accompanied by changes in cellular morphology in some brackish-water animals. In the hydroid *Cordylophora*, and in some crab gills, cells in the exposed

in Table 10.5. However, the trend is not universal; in some brackish-water bivalves that burrow deeply (e.g. the clam *Mya*) P_w is quite high, whilst in others that are fairly exposed (e.g. the mussel *Geukensia*) it is lowered.

Table 10.5 Rates of sodium loss in annelids and crustaceans from aquatic habitats.

Taxon	Habitat	ECF* [Na⁺] (mM)	Sodium loss rate†
Annelida—Polychaeta			
Nereis succinea	SW/BW	200–440	54
N. diversicolor	BW	150–350	13
N. limnicola	FW		8
Crustacea—Amphipoda			
Marinogammarus finmarchicus	SW	300	80
M. obtusus	SW	–	57
Gammarus tigrinus	BW	220	20
G. duebeni	BW	273	11
G. zaddachi	BW	252	14
G. lacustris	FW	135	5
G. pulex	FW	135	4

BW, brackish water; FW, fresh water; SW, sea water.

* Sodium concentration in the extracellular fluid (ECF).

† Sodium losses are expressed as μmol g⁻¹ h⁻¹ for annelids, and as μmol ml blood⁻¹ h⁻¹ for crustaceans.

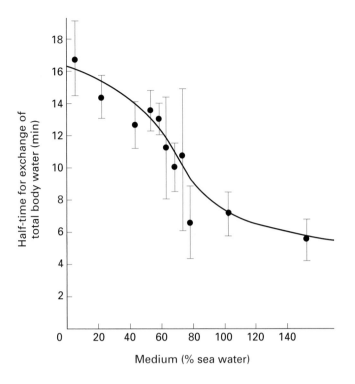

Fig. 10.14 Changes of water permeability (P_w) in *Gammarus duebeni* acclimated to different salinities. Vertical axis shows half-time for exchange of tritiated water, so that high values indicate reduced permeability. (From Lockwood 1961.)

epithelia become more rounded and have a reduced total surface area in estuarine conditions, presumably helping to reduce net fluxes.

Reduced surface permeability is the first line of defence, and it can be varied and controlled. But it is only a delaying tactic. Living organisms can never be totally impermeable and completely isolated from the variations in the world outside. Thus the next part of the total strategy must be a mechanism to keep the internal fluids at acceptable levels.

Ion transport mechanisms

For an animal with reduced ion and water permeability, but in water more dilute than sea water, ions and water will still move passively down their gradients—water moving in, and ions leaking out. This will happen especially at parts of the surface that have to be permeable for gas exchange, which for aquatic animals normally means the gills. Thus it is frequently the gills that are also used as ion-exchange sites, allowing control to be localized. In crustaceans the gills commonly have two cell types: 'thin' cells important in gas exchange; and 'thick' cells assumed to be the main site of ion exchange. In many developing embryos the ability to osmoregulate independently coincides with gill development, although in some crustaceans (e.g. *Gammarus*) regulation can be detected before the adult gills grow (on the coxal leg-bases) and seems to be a function of the vitelline membrane of the developing egg and a special 'dorsal organ' in the

Fig. 10.15 (a) Ionic permeability (P_{Na}, expressed as salt loss per hour into 100% sea water (SW), measured with sections of isolated cuticle) for a range of crustaceans from different habitats, showing the reduced values in freshwater and brackish-water species. (b) Adaptive changes in sodium permeability; a comparison of P_{Na} in the Baltic race and freshwater race of the isopod *Mesidotea entomon*. (Adapted from Croghan & Lockwood 1968.)

(a)

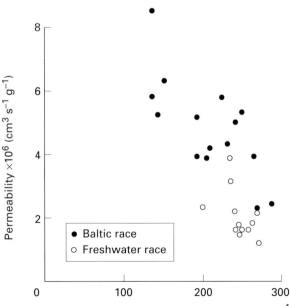

(b) Concentration of sodium in haemolymph (mM l^{-1})

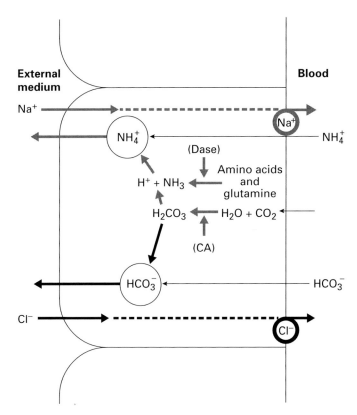

Fig. 10.16 Ion uptake in brackish-water species; the basic mechanism for Na$^+$/NH$_4^+$ exchange and Cl$^-$/HCO$_3^-$ exchange in uptake epithelia. CA, carbonic anhydrase; Dase, deamination enzymes. (Adapted from Maetz 1973.)

larva, which later regresses. Salt uptake in the midgut area is also likely in juvenile and adult crustaceans. In some adult aquatic animals the kidneys may also be used for some salt uptake, tubule cells resorbing ions from the filtrate into the body, producing slightly hyposmotic urine (5–15% body volume per day). This is fairly uncommon in brackish-water invertebrates, reported only for a few annelids (e.g. *Nereis diversicolor*) and some gammarid crustaceans. Salt uptake at the kidneys does occur in most brackish vertebrates, and is well-studied in the estuarine crocodile, where urine is routinely about 80% of the concentration of the plasma with around 95% of all the salts being resorbed in the tubules after glomerular filtration. Some further processing may occur in the cloaca, particularly of potassium levels.

The major contributor to blood osmolarity in all animals is sodium chloride, so the regulation of the fluxes for Na$^+$ and Cl$^-$ ions, whether at gills or kidneys, is bound to be at the centre of any osmoregulatory response. The basic principle for salt uptake by sodium transporters was described in Chapter 4 (Fig. 4.10), and the details of the system found in most aquatic animals are shown in Fig. 10.16. It is based on a classic sodium ion pump, located on the inner (basal) epithelial membrane. This involves an exchange system, which though based on the transporter (ATPase) that normally exchanges Na$^+$ for K$^+$ in practice often appears to utilize a Na$^+$ for NH$_4^+$ exchange. (Direct evidence for such active exchange is controversial, and it may be that NH$_3$ diffusion accounts for most of the ammonia excretion; a dependence of NH$_4^+$ loss on Na$^+$ uptake is perhaps caused secondarily, by boundary-layer effects on acidification.) There is commonly also a separate basal Cl$^-$/HCO$_3^-$ exchange in salt-uptaking epithelia. This gives the animals the two ions they most want in exchange for two waste products (ammonia and dissolved carbon dioxide). Details of what happens at the outer or apical membrane to move sodium into the cell are more controversial, as Fig. 10.16 shows: Na$^+$/H$^+$ antiporters, Na$^+$/K$^+$/2Cl$^-$ cotransporters and simple Na$^+$ channels have all been proposed. The first of these is strongly supported by recent molecular evidence showing the presence of an appropriate mRNA sequence in the gill tissues.

Cells involved in these transport processes always have a characteristic structure (see Fig. 4.12) with highly infolded apical membranes to give a large diffusive surface, a somewhat indented basal membrane on which the main cell/blood Na$^+$ transporters are found, and very large mitochondrial volumes. This architecture allows the pumps to work fast, with a high rate of fluid flow.

Ion transport normally increases with environmental concentration up to a limiting value (J_{max}), determined at least in part by the number of pump proteins, and it shows saturation kinetics, with the K_m value (the 'affinity') being the external concentration for half-maximal transport (see Box 4.5). Thus the lower the value of K_m the lower the range of concentration over which the transport system can achieve its maximum rate. It turns out that K_m (ultimately a function of pump protein structure) is well correlated with habitat, as Table 10.6 shows. Rates of salt uptake can thus be varied in the short term by acclimation, and also by evolutionary changes at the genotypic level, although the ability to survive low salinity is in fact a complex function of changes in J_{max}, K_m and permeability parameters. Many species are known to increase the number of ATPase pump proteins in their membranes during hyposmotic acclimation (e.g. in *Mytilus* nerve membranes, and in crab gills), and these changes can often be seen as an increase in membrane-bound particles in freeze–

Table 10.6 The K_m and J_{max} values for sodium influx in brackish-water osmoregulating animals.

Species	Habitat	K_m (mM)	J_{max} (mM kg^{-1})
Carcinus maenas (crab)	Brackish	20	
Mesidotea entomon (isopod)	Brackish	9	8.5
Gammarus duebeni (amphipod)	Brackish	1.5	15
Rana cancrivora (frog)	Brackish	0.4	5
Freshwater species (general range: see Table 11.9)		0.1–2 (lower)	0.1–20 (similar range)

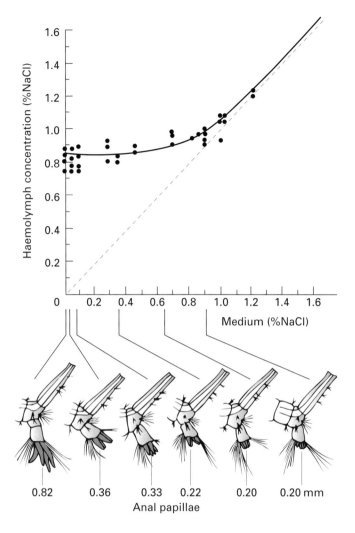

Fig. 10.17 The relationship between haemolymph and medium NaCl concentrations for the brackish-water mosquito *Culex*, also showing the associated growth of the anal papillae (where uptake occurs) in dilute media. (Adapted from Wigglesworth 1938.)

of other fish, such as tilapia and guppies, that tolerate a wide range of conditions. When adapting to fresh water, these fish show increases in gill Na^+/K^+-ATPase, and also increases in a bicarbonate-dependent ATPase and a Ca^{2+}-ATPase. An additional and unusual mechanism operates in the gills of these fish (and many other freshwater fish, considered in Chapter 11), whereby the chloride cells (see Chapter 5) can be physically covered or uncovered by volume changes in the nearby pavement cells, effectively turning the chloride cells on or off. In fish that can live in fresh or salt water, the gill chloride and pavement cell systems can take up NaCl in fresh water and secrete it in salt water; they also vary in function according to the pH of the medium in freshwater systems. But the change from one functional mode to another takes some time, with morphological change in the chloride cells and a change in the molecular components of the pumps, under the control of the hormone prolactin. Indeed many of the adaptive changes in moderately euryhaline fish are under tight hormonal control, with calcitonin and calcitonin-gene-related-peptide also implicated in gill membrane changes, and with arginine vasotocin and isotocin levels operating in longer-term adjustments. In salmonids growth hormone and cortisol also act to promote long-term seawater adaptation via actions on the differentiation of the chloride cells towards their marine condition. One signal for this change of function appears to be the shrinkage of the chloride cell itself during high salinity exposure. Recent evidence indicates that there may be two kinds of chloride cells in these fish, termed alpha and beta; the alpha cells play a key role in chloride-dependent sodium absorption in fresh water and sodium excretion in sea water, while the beta cells have some other unknown function in fresh water and are transformed into additional alpha cells on exposure to higher salinities. In conditions where ion regulation is particularly difficult, chloride cell populations proliferate, but this also tends to thicken the gill tissue and reduce the respiratory gas transfer.

A brackish-water animal thus has a means of taking up the extra salts it needs to counter passive outflow, and the blood concentration is taken care of by a regulatory system. Osmoregulation is not perfect as we have seen; the extracellular fluid (ECF, usually blood) is regulated but in a state that is still hyperosmotic to the medium, reducing the gradients for further exchange to a level where they can be balanced without excess expenditure. In crustaceans that are hyper-regulating, the ECF concentration may be reduced to roughly half that of sea water and the P_w is perhaps three-fold reduced, the net effect being a reduction in osmotic water loading of about one order of magnitude.

This provides the first level of regulation as we discussed in Chapter 1 (between outside world and circulating fluid). But there remains the problem of keeping the cell concentrations under control—how do these animals maintain the primary level of regulation, between relatively diluted circulating fluids and more concentrated individual cells?

fracture studies. Adaptation may also be exhibited at a higher morphological level by adaptive growth of the regulatory organs or surfaces, giving a greater membrane area for active pumping of ions; for example, in the brackish mosquito larva *Culex*, the anal papillae where salt uptake processes are concentrated grow very markedly in more dilute media (Fig. 10.17).

These changes must be triggered by sensors that detect changing external salinity. This aspect of adaptation is relatively little studied, but in some crabs it is known that the main sensors are specific 'hair-peg' organs on the legs, which may set in motion a hormonal cascade that involves a small peptide hormone, and/or dopamine.

Ion uptake processes and their regulation have been particularly studied in anadromous and catadromous fish, such as salmon and trout, which migrate through estuaries, and a range

Cellular osmotic adaptation

Cells need to hold on to ions, even when the world outside them, and their own immediate surroundings (i.e. the blood), are becoming dilute. This is because ions affect so many internal cellular processes dramatically, especially via allosteric effects on enzyme activities and DNA-histone interactions (see Chapter 2). Thus when faced with dilute surroundings, cells tend to keep their ions but lose instead some other small osmotically active solutes, known as **osmotic effectors**, or **compensatory osmolytes** (Chapter 4). These are a fundamental characteristic of cells, and are predominantly amino acids, which make up a substantial proportion of the normal cellular osmotic concentration (see Table 4.4).

Amino acid reduction during acclimation to dilute media could theoretically be achieved equally effectively in two ways: by increasing protein synthesis to turn the small osmotically effective amino acids into a few large, osmotically less effective, proteins; or by having a net export of the amino acids from the cell into the blood. Both systems do occur, but the latter is more important, i.e. when blood ion concentrations fall, many cell types in brackish water animals respond by exporting amino acids. The control of this process is complex, involving transaminations and deaminations, *de novo* synthesis controlled by protein kinases, and variations of efflux and influx rates. Some of the enzymes that control the synthesis of key amino acids (see Fig. 4.20) turn out to be very sensitive to Na^+ concentration, and cease to function at low Na^+ levels. Transmembrane movement of amino acids is also Na^+-dependent, so that at low external Na^+ concentration the net outward movement of amino acids is facilitated. Therefore, as Na^+ in the blood varies, the amino acid levels are automatically adjusted. Cell osmotic concentration gradually gets re-equilibriated downwards, and blood osmotic concentration is also corrected upwards slightly by receiving the amino acids. The system also works in reverse, whereby in hyperosmotic stress amino acids are synthesized and retained in the cells; this system has been particularly well studied in mammalian cells, where it appears that genes coding for enzymes and transporters that produce accumulation of organic osmolytes contain 'osmotic response elements' that are directly implicated in the activation of the transcription process. Over-all, the system of osmolyte regulation clearly forms a complex and neatly controlled automatic regulatory feedback mechanism.

The key osmotic effectors differ in different animals. In crabs, the amino acid loss is particularly of alanine, proline and glycine, in the lugworm it involves alanine and glycine, and in molluscs the modified amino acids taurine and betaine are particularly important. Inevitably the key enzyme in the mobilization of amino acids as osmotic effectors therefore also varies from the scheme shown in Fig. 4.20. In the copepod *Tigriopus*, inhabiting splash pools on Pacific coasts of North America, it is glutamate-pyruvate transaminase, which leads to alanine synthesis. This enzyme shows a two-allele polymorphism, where fast (F/F) homozygotes have around 50% higher specific activity for alanine synthesis than slow (S/S) homozygotes. In laboratory tests, hyperosmotic stress therefore induces faster alanine accumulation in F/F and F/S individuals, and the S/S individuals die more quickly and are less successful in more exposed sites. In the common mussel, *Mytilus edulis*, the aminopeptidase-I enzyme (LAP) is crucial and is polymorphic, with the LAP[94] allele, having higher kinetic capacity, showing a cline between open ocean and estuarine conditions. However, each year some carriers of this allele invade very low-salinity habitats as larvae and suffer weight loss and eventually high mortality; their disadvantage seems to be an excessively rapid use of their nitrogenous reserves, leaking away as excess amino acids, in these low salinities.

The regulatory system is even neater than we have so far described. Remember that the salt uptake systems that are working to restore blood concentration are using a Na^+/NH_4^+ exchange system. Therefore the very amino acids that have been shed by the cells into the blood can now be used (usually at the gills) to make the NH_4^+ by deamination; commonly some alanine is converted to proline, and proline oxidase in the gills then completes the deamination and so can 'fuel' the regulation of Na^+ uptake. Low Na^+ levels have thus stimulated high amino acid levels in blood, and this in turn stimulates uptake of Na^+ from the medium into the blood, giving a classic feedback regulation circuit (Fig. 10.18).

Therefore, all shore and estuarine animals have built-in biochemical and physiological circuitry allowing them to maintain blood Na^+ and keep their cells at a reasonable osmotic concentration and volume, despite regular and drastic changes in their environment.

Volume regulation

The net effect of exposure to lowered salinity for all these animals, and for tissues within animals (muscles having been particularly well studied), is the process of volume regulation, characterized by an initial swelling, and a subsequent gradual restoration of a less swollen (though rarely completely unswollen) state, known as the **regulatory volume decrease** (RVD). Figure 10.19 shows examples of the magnitudes and timescales of this response; it varies with taxon, but also with the pattern of exposure to osmotic stress (being largely eliminated if the change in concentration is introduced gradually enough).

We can now understand what underlies the phenomenon of RVD. During initial exposure the animal is gaining water into (and losing salt from) its blood through its finitely permeable surfaces, and the speed of volume increase will largely reflect the effectiveness of permeability control (it may be exceedingly slow in bivalve molluscs where the closed valves reduce rates of exchange even further, as seen in Fig. 10.19). Volume control of the whole animal may be aided by reduced drinking, especially

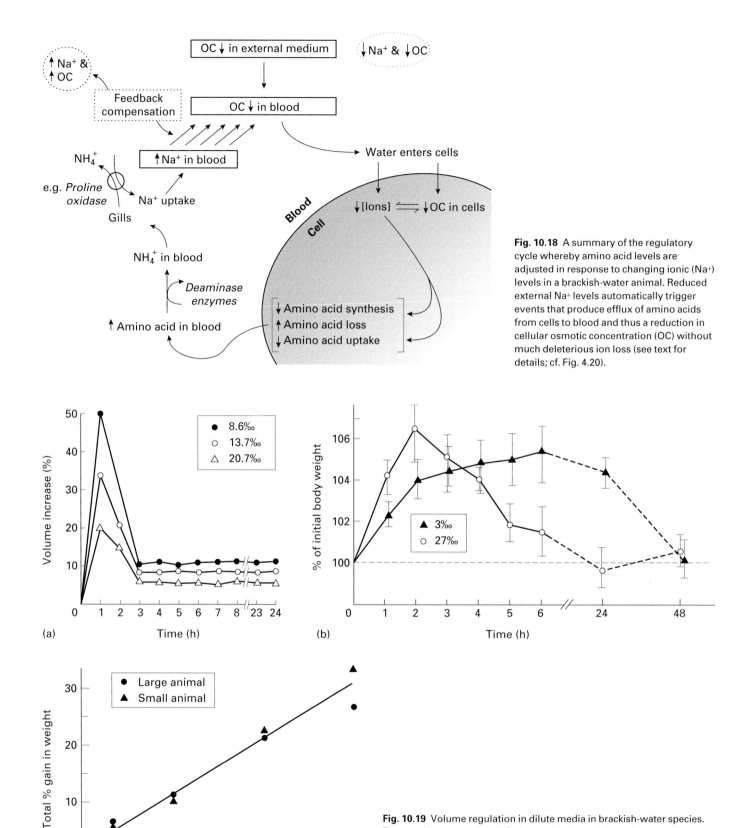

Fig. 10.18 A summary of the regulatory cycle whereby amino acid levels are adjusted in response to changing ionic (Na+) levels in a brackish-water animal. Reduced external Na+ levels automatically trigger events that produce efflux of amino acids from cells to blood and thus a reduction in cellular osmotic concentration (OC) without much deleterious ion loss (see text for details; cf. Fig. 4.20).

Fig. 10.19 Volume regulation in dilute media in brackish-water species. The regulatory response is fast in a polychaete (a), moderate in a mussel (b), where the gain of volume is also much less, and almost non-existent in a starfish (c), where water gain and swelling are very marked and the animal hardly deviates from the behaviour of a perfect osmometer. (Adapted from Binyon 1961; Pierce 1971.)

in fish, and by increased urine production, but these options cannot entirely eliminate tissue swelling, and the resulting osmotic gradient between blood and cells causes the cells in turn to take up water, a process which is often very fast but which is also corrected very quickly. This correction can only be due to solute loss, since we know that there can be no active water extrusion. The amino acid content of the cells drops very sharply in this RVD phase, until blood (or other extracellular fluid) and cells are approaching equilibrium again at some new level. While this is going on the active uptake of extra solute is also instituted (from the medium or sometimes from the urine) and the blood levels are corrected on a somewhat slower timescale by this uptake from outside the animal, until a final equilibrium volume level is attained, which may be little different from the starting condition. Fluid-filled body cavities such as the coelom may be used to buffer volume change, so that cell volume returns to normal even when total body volume remains elevated.

In some mid-shore bivalves, such as mussels, where the external concentration rises and falls predictably with a relatively short time course, there is little or no RVD in many of the cells, which are allowed to swell and shrink cyclically, presumably thereby saving energy that might be used in trying to regulate their volume. However, heart cells do show an RVD response, as do nerves.

Volume regulation may be aided by circulatory adjustments in crustaceans and other animals with arterial blood systems. In the crab *Cancer magister*, which suffers relatively large volume changes in dilute media, exposure to low salinity produces an increased heart rate, with a reduced stroke volume. Blood flow is redistributed, with anterior flow (mainly to the gills) reducing and with an increase through the posterior aorta. This may help to reduce the average exchange gradient for inward movement of water at the gill surfaces.

Maintaining ionic levels for nerve and muscle functioning

The mechanisms so far discussed allow cells to continue to function in dilute media, with relatively little loss of internal ionic content. However, certain kinds of cell are especially reliant on ionic gradients across membranes for their functioning; excitable cells depend upon axonal, synaptic and neuromuscular transmission systems that require inwardly directed concentration gradients of sodium, calcium or chloride ions and outwardly directed potassium ion gradients. Thus it might be expected that muscles and nerves in particular would need special adaptations in euryhaline brackish animals, above all in osmoconformers where concentrations of these ions in the body fluids vary widely. In muscles the primary adaptations relate to maintaining cellular volume by regulation of osmotic effectors, thus preventing damage to membranes caused by swelling; in both muscles and nerves the main osmotic effectors are aspartate and glutamate, since their net negative charge helps to maintain the normal cell resting potential (see Chapter 4). Maintaining cell volume using amino acids also ensures ade-

quate ionic gradients for conduction of sufficient electrical potential changes to trigger the contractile or conductive systems.

The problems of nerve functioning in estuarine animals are even more complex since the size and frequency of the action potential, which depend upon the magnitude of ion gradients, also carry information. Regular changes of ion gradient due to changing body fluid concentrations risk subverting the supply of information. The strategies adopted have been studied in relatively few cases. In the more stenohaline animals nerves show a rapid hyperpolarization and reduced action potential amplitude when in hyposmotic media, apparently due to increased potassium leakage as P_K increases to assist volume regulation. The hyperpolarization is reversible on return to normal sea water if exposure has not been too long or too severe. However, conduction block (no further propagation of action potentials) often sets in irreversibly at 60–70% sea water.

In the moderately euryhaline polychaete worm *Sabella*, which copes with a 50% dilution of its medium, changes in neuronal responses occur much more slowly, primarily because hyposmotic conditions produce a swelling of the protective glial cells that surround the nerve cords, thus reducing the access of ions to the extracellular spaces around the axons and helping to maintain their function. This purely physical response may be enough to sustain the nerves through many tidal cycles, ensuring that in the natural habitat hyposmotic conditions are experienced only gradually at the axon membrane surface. Meanwhile, a slow decline in internal K^+ concentration is matched to the declining external K^+ levels, so that the resting potential is relatively unaltered and conduction can continue. Axons also show an increased sodium selectivity in hyposmotic media, compensating for the reduced sodium gradient; hence the overshoot of the action potential can be reasonably well maintained.

The common mussel, *Mytilus edulis*, can osmoconform down to about 25% sea water, and shows another suite of responses that allow it to maintain nearly full-sized action potentials. During acute exposure to low salinity the axons, all of which are of small diameter, swell up and lose both sodium and potassium, together with non-ionic solutes such as amino acids, and they therefore depolarize and lose their ability to conduct. However, this effect is reversible and the axons rapidly recover in normal sea water. In longer-term and more gradual exposures (more typical of these animals' normal experience) axonal swelling is resisted by an onset of thickening (with increased collagen production) in the acellular neural sheath that surrounds the bundles of axons, allowing the nerve cells to maintain a relatively normal volume. Over a similar time course the axon membranes also synthesize extra sodium pump proteins (Na^+/K^+-ATPases), with an increment of about 70% for a dilution to 25% sea water, which will help to maintain an adequate Na^+ gradient and also sustain internal potassium levels.

Nerve function has been examined in the even more spectacular conformer *Mercierella enigmatica* (*Ficopomatus*), an annelid worm which can function even in tapwater. Here there is no obvious neural sheath around axons, yet they again seem to have a mechanism to counter swelling in the form of extensive hemi-desmosome systems in the axon membranes, which are the anchor points for an elaborate internal network of neuro-filaments, strapping the axons together from within and pre-venting much change of volume. The giant axons in this species also show reduced intracellular electrolyte concentrations in hyposmotic media, but there is a disproportionate retention of potassium so that the nerve hyperpolarizes markedly, main-taining a large overall action potential amplitude despite the reduced overshoot resulting from sodium depletion.

Overall, the nerves of osmoconformers thus seem to have various options to cope with osmotic dilution:

1 They have systems for limiting damage due to axonal swelling; volume increase of surrounding glial tissues, or intra-cellular strapping within particularly large diameter cells, or synthesis of extra neural sheath as a physical protection.

2 The kinetics of responses to hyposmotic stress are approxim-ately matched to ecological requirements; the polychaetes cope with very rapid change, whereas *Mytilus* nerves can only main-tain excitability if dilution is gradual, normally being protected from more rapid change by the simple trick of closing their shell valves.

3 All the invertebrate osmoconformers tested show a reduction in intracellular potassium, but the effect of this on the cell resting potential varies. In *Mercierella* intracellular loss is not propor-tionate to extracellular dilution, so there is a hyperpolarization, whilst in *Sabella* it is in proportion to reducing external K+ so that the resting potential is maintained, and in *Mytilus* the accom-panying losses of other solutes lead to a net depolarization.

4 The resultant effects on the crucial action potential spike amplitude are equally varied: in *Sabella* an increased sodium selectivity allows the action potential size to be maintained, whilst in other species the internal sodium level is reduced to help maintain a Na+ gradient, and hyperpolarization may ensure that the total spike amplitude is largely unaltered. Over a longer timescale extra synthesis of sodium pumps may be a common strategy to assist sodium gradient maintenance.

By some combination of all these techniques, operating over appropriate timescales, osmoconforming animals avoid losing electrical conduction and can adapt their nerves to varying osmotic regimes.

Salinity tolerance and water loss tolerance

The ability to tolerate a range of salinities, resulting from some mix of all the above strategies, is often seen as the key adaptation for brackish-water animals; their tolerances may be listed by species, as in Table 10.7, reflecting the lower point of all the lines in plots such as Fig. 10.13. It is worth remembering three key points here, however. Firstly, this lower limit is not absolute but is exposure-time-dependent—most species will survive consid-erable periods below this 'threshold', though they may be slug-

Table 10.7 Lower salinity tolerance of brackish-water species.

Species	Salinity tolerance (% sea water)	Strategy
Annelids		
Sabella	~50	Conformer
Nereis diversicolor	15–20	Moderate regulator
Nereis limnicola	~5	Moderate regulator
Mercierella enigmatica	<1	Conformer
Molluscs		
Nassarius	25–30	Weak regulator
Mytilus edulis	15–20	Conformer
Patella vulgata	10–15	Moderate regulator
Crustaceans		
Cancer	~50	Weak regulator
Gammarus obtusatus	~30	Moderate regulator
Callianassa	25–30	Conformer
Ligia oceanica	40–50	Hypo-hyper-regulator
Palaeomonetes varians	20–30	Hypo-hyper-regulator
Upogebia	15–20	Moderate regulator
Callinectes	10–20	Strong regulator
Carcinus maenas	10–15	Strong regulator
Artemia salina	~10	Hypo-hyper-regulator
Gammarus duebeni	<1	Strong regulator

gish and unable to maintain a positive energy balance. Secondly, there will be considerable individual variation in this tolerance limit within many species, affected in part by size but also by the variability in all the other factors that interact with salinity toler-ance (oxygen uptake and varied acid–base balance, nitrogenous balance, thermal tolerance, continued ability to feed or to loco-mote, etc.). For example, the whelk *Nassarius*, listed as having a species lower limit of 11.5 ppt (about 30% sea water), may have individual limits varying from 7 ppt to as high as 25 ppt, and the mean tolerance will also vary somewhat between populations. Thirdly, and perhaps most crucially, tolerance may be very dif-ferent in juvenile stages. For example, in supratidal grapsid crabs larval stages have some euryhaline tolerance from the moment of hatching (15–35 ppt), but tolerance of low salinities gradually increases through development to 5–45 ppt in the third (zoea) larva and 5–55 ppt in the final (megalopa) stage and the adult. This may be mainly a reflection of the decreasing surface area to volume ratio of the growing individual.

Tolerance of water loss is also highly developed in many estu-arine and littoral animals (see Table 5.2): the loss of body water without fatality can be 35–60% in limpets, and up to 75% in chi-tons. As desiccation sets in, many of these intertidal animals can use metabolic depression (see Chapters 6 & 7) to reduce their oxygen need and thus limit the need for any shell-gape for re-spiratory exchange, thereby reducing their water loss rates further. In a few species, such as the copepod *Tigriopus*, from supra-littoral splash zones where pools of water may be very

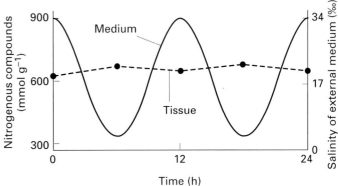

Fig. 10.20 The effects of 'smoothing out' environmental change in a bivalve mollusc (here the mussel, *Mytilus edulis*), where mantle fluid osmolarity changes much less than the external salinity and tissue amino acid concentrations can remain almost unchanged through a double tidal cycle.

ephemeral, there is a capacity to be almost completely dried out and yet to recover when wetted, an example of the phenomenon of 'cryptobiosis', covered in Chapter 12.

Finally we should return to the key point, which is that many littoral and estuarine animals eliminate the potential problems of osmotic change and dehydration not by complex physiological or biochemical means but by simple evasion. Thus even animals that have the machinery to adapt may rarely need to use it. The prawn *Palaeomonetes*, though a good regulator, moves around an estuary to minimize its need to regulate, following the salinity displacement pattern and being commoner higher up river in summer and autumn when inwards salinity encroachment is greater. Figure 10.20 shows that for *Mytilus*, which we have seen also has a suite of possible mechanisms to protect its tissues, the strategy of shell closure at low tide is commonly enough to make these other mechanisms unnecessary; the mantle fluid remains relatively concentrated even in dilute media, and the tissues show hardly any change in amino acid level through a normal 24 h tide cycle. Similarly for many crabs with good osmoregulatory physiology, such as *Uca*, much of their normal 'regulation' is in practice achieved by the use of the burrow during periods of environmental stress, conditions inside the burrow being thermally and osmotically much more stable. Physiological change is a last line of defence here (perhaps rarely called upon except when very low tides coincide with warm dry weather); it is not the mainstay of the adaptive response.

10.3 Thermal adaptation

Essentially, life is thermally very much more variable away from the buffering of constant cover by the seas, so that every animal living in the seashore zone and in estuaries needs to be eurythermal to cope with increased thermal variation. This variation can be manifested in several ways and over differing timescales.

Tidal effects mean that on shores there are marked differences over the 12 h cycle as the water advances and retreats (e.g. Fig. 10.12b). Seasonally the temperature of both water and air will vary, especially in temperate and high-latitude estuaries; Fig. 9.7 showed the thermocline effects in surface sea waters, so that the waters lapping temperate shores and estuaries may be 7–10°C higher than most of the sea. Some kinds of shoreline animals are therefore required to be more tolerant of thermal extremes than almost any other animals. For example, the rock pool goby, *Gillichthys seta*, is one of the most eurythermal of animals, tolerating body temperatures (T_b) from 8 to 40°C. But in addition to the tidal and seasonal variations there are also globally varying interactions of warm surface currents and cold deeper currents, causing substantial intermittent thermal overturns on some shores; of these, the El Niño events in South America are best known (Box 10.2). Finally there are long-term global changes in climates that may need to be accommodated by genotypic change within a species; and on a similar magnitude and timescale there may be effects arising from animals invading along shores and colonizing new thermal habitats.

Thermal changes on the scale of tidal and seasonal effects require some level of phenotypic adaptation. Usually we find behavioural change first, and then an intervention of physiology or biochemistry; remember the mechanisms for thermal acclimation and adaptation of enzymes and membranes in animals discussed in Chapter 8. But what further choices does a shoreline organism have when it comes to temperature regulation? There are various sources of heat loss and gain, and ways of using and controlling these; for essentially aquatic organisms the main problem is often to avoid getting excessively hot when the tide is out.

10.3.1 *Problems of overheating*

Heat stress is likely to be a cyclical problem, overlaid with a seasonal component on all but tropical shores. Summertime low tides will bring the greatest risk of overheating for most littoral and estuarine animals; large proportions of intertidal mussels can be killed or suffer severe tissue damage in unusually hot dry summers.

Box 10.2 The El Niño Southern Oscillation (ENSO) and its effects on coasts

The phenomenon of El Niño is a periodic reversal of the main currents across the Pacific Ocean, occurring every few years and having substantial consequences for world weather patterns.

It arises in an oscillatory (periodic) manner when tropical Pacific water warms up (by up to 2°C in places) and instead of remaining local (and causing intense tropical rains) the surface waters from the western Pacific, around the Indonesian area, migrate eastwards, accompanied by storm systems. Recent ENSO events have occurred in 1982 and 1987 (for 1 year), 1991 (with effects lasting for up to 3 years), and 1997 (even more prolonged). There is evidence that the effects are becoming more frequent, more intensive and more prolonged, probably because of an interaction with global atmospheric warming.

On land, the meteorological consequences commonly include risk of drought and wildfire in Australia, Brazil, East Africa and parts of India, and risk of torrential rain and flooding in normally desert regions of North America, the Peruvian coast of South America, and the western side of the Indian subcontinent.

Such effects clearly also impact upon coastal ecosystems, and in the eastern Pacific these changes have been well documented:

1 There may be a substantial deepening of the thermocline, with the phytoplankton redistributing downwards.

2 Californian fish have occurred well north of their normal latitudes, as have crabs and squid; commercial catches in southern California are often greatly reduced.

3 More northerly species have been displaced, with salmon fisheries in Oregon much depleted and knock-on effects on sea birds and on sea lions and fur seals.

4 Coastal planktonic communities have changed, with correlated decreases in filterers, predators and top predators; Peruvian anchovy stocks have therefore crashed, again affecting sea lions and pelicans, with many sea birds abandoning attempts to breed.

5 Relatively stenothermal low-shore animals have declined on sensitive beaches.

Being the right size, shape and colour is part of the solution to thermal stress, so that we often find intraspecific clines of size or shell height or shell colour on beaches, particularly within the molluscs. Small size gives a rapid heat gain or loss, but also a rapid desiccation rate; larger size helps to reduce heat exchange rates in very variable conditions such as the middle shore, but may lead to overheating through prolonged exposure to the Sun on the upper shore. Thicker shells in dog whelks (*Nucella*) appear to help reduce thermal stress (though they were probably selected principally to reduce predation). Shells with elaborate ridges appear to help increase convective heat loss in some beach gastropods, and such shells are much commoner in warmer latitudes. Paler colours help to reflect some of the radiation so that heating rates are reduced. For example, dark shells of winkles have a reflectivity of 2–10%, whereas bright-yellow or white shells may give a value of 30–50%; and even in mussels the difference between black and brown morphs is enough to produce a significant thermal difference, so that populations are nearly 100% black in eastern Canada and only about 40% black in the southeastern USA (Fig. 10.21). An ability to change colour may also be particularly useful to seashore animals; it has been demonstrated in a wide range of isopod and amphipod crustaceans, as well as in certain fish. The isopod *Ligia* changes colour using cells (melanophores) in which dark pigments can be either widely dispersed or tightly condensed, producing a reflectance varying from 2% (almost black) to 10% (very pale beige). Here the cuticle colour is determined both by body temperature needs and by the requirements of background matching for camouflage; in addition there are underlying circadian and circannual rhythms of colour change. As these animals are dawn and dusk scavengers the main priority is to ensure crypsis at such times, but this can be adjusted according to hygrothermal need (paler to stay cool or darker to warm up).

In addition to morphological adaptation, increased tolerance

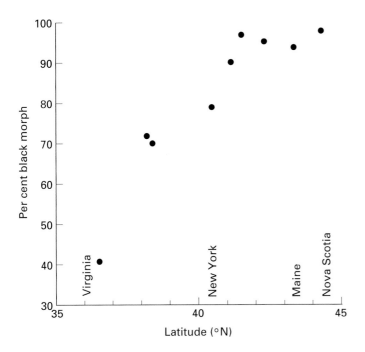

Fig. 10.21 A cline in coloration in the mussel *Mytilus edulis* along the eastern shores of North America, with progressively more of the black morph (better at absorbing solar radiation) at higher latitudes. (From Levinton 1995.)

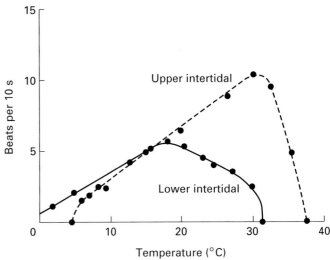

Fig. 10.22 Effects of height on the beach on thermal responses. The high-tide barnacle species show increasing cirri beat frequency with rising temperature up to and beyond 30°C, whereas low-tide species progressively lose cirri activity above just 18°C. (From Levinton 1995.)

in the animal's cells and tissues is also required, exploiting the mechanisms for protecting membrane function and enzyme activity that were discussed in Chapter 8. The mussel *Mytilus californianus* experiences 20°C cyclic change in body temperature on a daily basis, as well as marked seasonal effects; its gill membrane phospholipids show seasonal homeoviscous adaptation, optimizing membrane fluidity (order) for a winter mean T_b and for a summer upper extreme T_b despite the fact that they are likely to experience water submersion at about 10°C on a daily basis in both seasons. At least in high-shore populations they also show an ability to alter their membrane order within hours, to counter the thermal effects of the tidal cycle. Mussels also show local and seasonal variation in their levels of heat-shock proteins (HSPs, see Chapter 8), with ubiquitin and HSP70 increasing in gill tissue in summer months, but more so in intertidal than subtidal populations.

These phenomena of increased membrane and tissue tolerance are manifested in maintained or increased activity levels in many animals. For example, high-intertidal barnacles are able to keep their filter-feeding cirri beating at markedly higher temperatures than individuals from further down the beach (Fig. 10.22). Mean upper critical temperatures (UCT), and thermal niche width—the difference between UCT and LCT (lower critical temperature)—are generally positively correlated with maximum height of distribution for a range of species on any particular shore. However, some common shore animals such as the winkles (*Littorina*) show a lack of the normal temperature-related

changes in oxygen consumption expected in most ectotherms. This may reflect their experience of marked diurnal temperature change, where any benefits from longer-term temperature compensation would be negated by extreme daily fluctuations. Instead, the littorinids have an unusual ability for nearly instantaneous suppression of metabolic rate and entry into short-term metabolic 'diapause' at high temperature (20–35°C, depending on species and population). This kind of strategy may be present in other littoral invertebrates.

Behaviour is also a major part of the adaptive strategy to cope with heat stress, especially for invertebrates; examples include retreating to a burrow, seeking out cooler shaded crevices, hiding beneath dense clumps of seaweed, or excluding the warming air by clamping tightly to a rock. The pattern of burrowing may itself be highly variable within a species, as shown in the supralittoral amphipod beach-hoppers, which may dig deep high-shore burrows in winter to avoid freezing, shallower downshore burrows nearer the tide line in spring, and deeper burrows again in late summer to avoid overheating. Aggregation may also be used to increase the effective thermal mass (whether to keep warmer in winter or cooler in summer), and/or produce a more humid microclimate; mussels in aggregations have T_b values 4–5°C lower than similarly sized loners in short hot spells of weather, whilst winkles often aggregate more when on dry and cool substrata and then achieve significantly higher body temperatures over a longer term. Behavioural thermal regulation is also in evidence in seashore vertebrates. Blennies, for example, seek out deeper colder water in rock pools in the summer months. But probably the best-known example is on a larger scale, with the Galapagos iguana, *Amblyrhynchus cristatus*, living intertidally and subtidally on cool rocky shores. The Galapagos

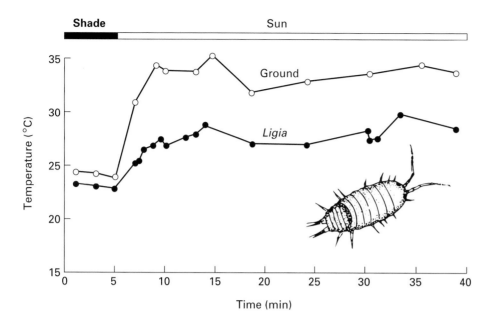

Fig. 10.23 The body temperatures (T_b) of the isopod *Ligia oceanica* (the sea slater), measured before and during exposure to the Sun, in comparison with nearby ground, at a humidity (RH) of 45–52%. The insolated animals are cooled by as much as 7°C relative to their surroundings by the effects of evaporative water loss, keeping their T_b below 30°C. (Adapted from Edney 1953.)

islands are equatorial, therefore receiving strong solar inputs, but they are bathed by cold currents, so that diving for their seaweed food cools the iguanas substantially. After a dive, when T_b falls despite peripheral vasoconstriction (see Chapter 8), they use a complex basking regime: they initially orientate sideways on to the Sun on fully exposed rocks to regain their preferred T_b of 36°C; or they may warm even faster by crouching down on to the black lava rocks in the middle of day, when these surfaces may reach 50°C. Once T_b reaches about 40°C, they need to avoid overheating, so turn head on to the Sun, with the head and neck raised so that the rest of the body is in shade, and with the thoracic region raised off the substratum, so that breezes coming off the sea will cool the chest region.

For seashores, though, there is a whole new area open for possible thermal regulation, and that is **evaporative cooling**. This is not available to fully aquatic animals (except in very rare cases of floaters that emerge above the surface, such as the Portuguese man o' war jellyfish), and it is usually too risky to be used except in emergencies for fully terrestrial animals without reliable water supplies. But it is the perfect solution on shores; animals can use up some of their water to keep cool, with the security that water will come back again to replenish the stores with the next tide. Remember that evaporative cooling is also very efficient, using up 2500 J g⁻¹ (580 cal g⁻¹) of water lost. For example, a littoral isopod scavenging in the sun at low tide may be able to keep its body temperature 5–8°C below the air temperature by use of evaporative cooling (Fig. 10.23), and a crab such as *Uca* achieves a similar reduction, with its burrow providing a source of renewable standing water when the tide is out.

Seashore and estuarine animals show a new range of strategies to take advantage of this opportunity. They may take up an extracorporeal water store (in the mantle cavity of molluscs, or the gill chambers of crabs, etc.); and they may control access of

this water to air, for example by a judicious gaping of the shell or dribbling of water from the gill space, to get their temperature balance right. Gaping and the use of evaporative water loss (EWL) is clearly adaptively exploited according to the environment experienced, and is not just a reaction to stress. For example, tropical oysters growing high up on the shore have a UCT of 40–42°C, roughly corresponding to the T_a during midday exposure, and they remain tightly closed during such exposure. In contrast, mangrove oysters in similar latitudes, which are exposed to less extreme ambient temperatures and remain in more humid air when open, have a lower UCT and do use gaping to increase their EWL and stay cool. Similarly, tropical limpets are more prone to use 'mushrooming' behaviour, lifting the shell off the substrate, than their temperate relatives, and this would appear to be a strategy for increasing EWL.

Shoreline animals able to gape and regulate EWL may also withstand temporary severe desiccation if necessary, and desiccation resistance tends to increase up a beach. The values of 60–75% water loss quoted in section 10.2.2 are far greater than would normally be needed for the 12 h tidal cycle, i.e. these animals are built to withstand the worst possibility of being exposed for a full 13–14 days between extreme high tides once or twice a year. They are also greater than the desiccation tolerances of most terrestrial animals (see Table 5.2), and are really only bettered by truly cryptobiotic creatures, dealt with in Chapter 12. Hence it is fair to say that water loss is a fundamental strategy for regulating T_b in shoreline animals, unlike almost any other habitat.

Because of this role for evaporative cooling on seashores, clearly we cannot isolate temperature studies from water balance studies for littoral species. Surviving high temperature depends on an ability to be euryhaline and cope with varying

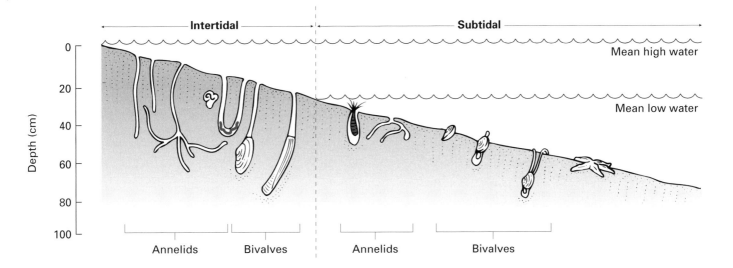

Fig. 10.24 Burrow depth vs. shore height in a soft sediment beach. Depth is greater in the intertidal zone, where the burrow provides microclimatic amelioration at low tide. (From Levinton 1995.)

blood concentration. For example, in various shore isopods the ability to withstand lower salinity decreases as the temperature rises, and the two systems are intimately linked. But again the settings for temperature tolerance and desiccation tolerance vary with species, and with the position on shore that each species occupies, as well as varying intraspecifically.

Thus far we have dealt mainly with animals on rocky shores and in estuaries, but of course temperature is also a problem for animals living in and on sandy shores. Once it is exposed and dried, sand can get very hot (often too hot for humans to walk on). It is usually not protected by weeds or other surface cover as on rocky shores. Problems of overheating are particularly bad on dark-substrate sands such as on volcanic islands, where the absorptivity is high, so that these kinds of beaches tend to have a different and rather specialist fauna, deeply buried during low tide. Deep burrowing can help solve most of the ecophysiological problems on shores for those animals that can actively move downwards reasonably fast. Intertidal animals do tend to burrow deeper than subtidal ones (Fig. 10.24); they need to get down to about 250 mm in temperate climates to avoid heat and drought stress effectively. Some (e.g. razor shells) can migrate vertically through the sand with the tide, which helps to maintain a fairly constant T_b. Animals with shells are also in a sense preadapted to reduce heating stress in sandy beaches, though they may lose out on the option of keeping cool by evaporation through the gape if they also become too water-stressed.

We should also note here the 'special' shoreline cases of coral reefs and mangrove swamps. Coral reefs constitute a 'manufactured' sublittoral and littoral zone, and are at their maximum development in up to 10 m of water at unusually high sea temperatures of 25–29°C. In some areas the reef may be almost exposed at low tide, giving even higher temperatures due to direct solar heating. Hence coral reef organisms (especially the coral cnidarians themselves) are all adapted to high thermal regimes, and have very high growth rates and hence enormous productivity. The same is true of many mangrove specialists, including the snapping shrimp, *Alpheus viridari*, which can survive 25–45 ppt salinity (regulating its body fluids fairly constant) and up to 35°C water temperature. However, animals within the mangrove forest will be more protected from the Sun because of shading by foliage. Nevertheless, all these warm-water animals may be at risk when temperatures rise beyond their limit. Many are at special risk since this may also be linked with increased surface evaporation, causing local sea-level lowering and an increased exposure of the coral or mangrove root habitats to direct unfiltered radiation from the Sun and thus to UV damage.

10.3.2 *Coping with cold*

Rock pools and interstitial waters in temperate and cool biomes may freeze for some part of the tidal cycle during winter months. Shore and estuarine animals in higher latitudes are therefore mostly freeze tolerators, with their body fluids able to tolerate freezing and thawing on a regular basis; this is true of most of the barnacles, bivalves and gastropods tested. The minimum temperatures survived by most intertidal invertebrates with freeze tolerance are not particularly low, often around –10°C; with re-immersion in sea water above –2°C always imminent with the returning tide, body tissues will rarely cool below this level. Indeed, in deeper rock pools animals may not freeze at all, since the surface ice both protects the water beneath and withdraws fresh water from it, leaving the pool water strongly hypersaline and therefore unfrozen as low as –8°C (Fig. 10.25). Hence there is often no clear relationship between freezing resistance and position on the shore.

Cold acclimation for shore animals involves the classic strategies of membrane and enzyme adaptation discussed in Chapter 8. For example, crabs undergo membrane lipid changes allowing increased membrane fluidity, specifically decreasing the

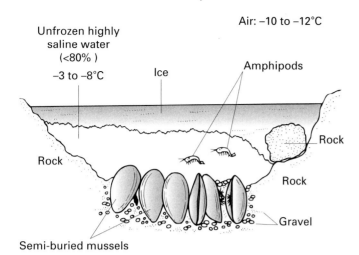

Fig. 10.25 The microenvironment of a high-shore rock pool in winter on a cold Norwegian shore; the upper pool is frozen, withdrawing fresh water from the water below, which becomes substantially hypersaline and therefore does not freeze even at −3 to −8°C, allowing amphipods also to remain unfrozen. (From Davenport 1992.)

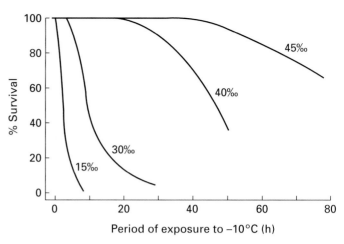

Fig. 10.26 Freezing resistance in the common cockle (*Cerastoderma edule*), showing acclimatory effects such that pre-exposure to high salinity substantially improves survival of freezing, presumably due to intracellular accumulation of amino acids. (From Davenport 1992; adapted from Theede 1972.)

cholesterol to phospholipid ratio but leaving the saturated to unsaturated fatty acid ratio largely unchanged. The stenothermal crab, *Cancer*, has slightly greater membrane fluidity and a lower cholesterol/phospholipid ratio than the eurythermal *Carcinus*. Enzyme adaptation has been particularly studied in the mussel *Geukensia*, in which freezing causes coordinated suppression of multiple cytosolic and mitochondrial enzymes involved in carbohydrate metabolism, primarily by reversible protein phosphorylations, so that biochemical pathways remain in balance.

For more extreme high-latitude seashores, air temperatures of −30°C are not uncommon, and resident animals on the beach may have 90% of their body water frozen. The cells must then be tolerating remarkably high osmotic concentrations (incidentally rendering them even less likely to freeze intracellularly). The tissues and cells do become very distorted by the surrounding ice formation, cells becoming shrunken and morphologically altered. Nevertheless, many of these animals, including the common upper- and mid-shore winkles and limpets, can survive repeated freeze–thaw cycles and have fully restored tissue function shortly after each thaw. Note too that salinity and freezing resistance are interactive; animals that are already acclimated to hypersaline conditions are substantially more freezing resistant (Fig. 10.26), since both phenomena involve increased concentrations intracellularly.

Polar inshore and shoreline invertebrates that are essentially marine but may experience transient aerial exposure are exceptions to the general rule of low UCTs in animals from cold seas. For example, Antarctic copepods may have UCT values of 14–18°C compared with only 4–6°C in fish that come from similar habitats but are mobile enough to guarantee staying submerged.

10.4 Respiratory adaptation

Oxygen is usually fairly freely available in littoral waters, but estuaries can become both significantly hypoxic (especially so where man's activities supervene) and significantly hyperoxic on sunny days when primary productivity is at its peak. In these habitats hypoxia tends to be accompanied by hypercapnia (elevated CO_2), and vice versa. Additionally, many of the animals on shores have respiratory problems because they are endowed with a system that only works well either during submersion or during exposure. The gills of molluscs, crabs, annelids and fish will commonly collapse and adhere together when unsupported by water, reducing the exchange surface area and producing hypoxia even through oxygen is freely available in the air. At the other extreme, lungs and tracheae are unable to function if water-filled, and their possessors rapidly drown. Relatively few animals are truly amphibious in this respect, and therefore many shore dwellers are facultative anaerobes. In particular, primarily aquatic forms may get stranded in pools or in mud as the tide goes out: in either site they can get short of oxygen quite quickly, because of rising temperatures and evaporation, leaving little or no O_2 in solution. Marine muds are particularly impermeable to gases, so rapidly go black and anoxic below the surface, with the resident bacteria converting nitrates and sulphates to nitrites and sulphides. We therefore need to consider both aerobic and anaerobic respiration for most shore and estuarine animals.

10.4.1 Aerobic respiration

Whenever oxygen is fairly freely available we find both regulators (i.e. animals that can maintain their consumption rate regardless of conditions) and conformers (i.e. those whose consumption varies in line with oxygen availability). Animals from more consistently highly oxygenated environments are more likely to be conformers (e.g. the fully marine swimming crabs, and the coastal polychaete worm *Marezelleria viridis*), and those from variable sites not surprisingly tend to be regulators (e.g. spider crabs from tidepools, shore crabs from estuaries, and the littoral polychaete worm *Nereis diversicolor*). The latter can genuinely be described as 'euryoxic' animals, although their respiratory responses may be modified in some conditions by other interacting stresses such as lowered salinity or raised temperature.

Conformers

Conforming strategies generally involve a reduction of metabolic rate (adaptive hypometabolism) and/or a gradual inclusion of the anaerobic pathways dealt with in the next section. The high-shore limpets, *Siphonaria*, can reduce their metabolic rate to just 18% and incur little or no oxygen debt, while in the lower-shore limpet, *Patella*, anaerobic respiration is commoner and tolerance much lower. Even within a genus, both regulators and conformers in this respect may occur according to height on the shore; the porcelain crab genus (*Petrolisthes*) has high-shore species that retain a reduced level of aerobic metabolism at low tide, and low-shore species that shift to anaerobic metabolism whenever they are in air. Oxygen usage limitation by slowing metabolism has also been demonstrated in shore crabs and in oyster larvae, amongst others, and in many intertidal animals it is greatly aided by reducing activity when exposed to air. Many species lower their heart rates and ventilatory rates, and bivalves reduce their heart rate and stop all filter-feeding activity when the tide goes out.

Regulators

By contrast, oxygen regulators can achieve a steady oxygen uptake by a greater variety of means, different species adopting a different combination of these.

Changes to the ventilation rate or depth

This may be the simplest response, with increased rate or depth in low-oxygen conditions. For example, many crabs can alter their ventilation rate by varying both the frequency and the 'stroke volume' of the scaphognathite (the flattened plate whose beating forces water through the gill chamber; see Fig. 7.9b).

Changes in heart function

In animals with closed or partly closed circulations, including the littoral crabs, there is the option of quite subtle changes in heart rate, or in the stroke volume of the heart; the crab *Callinectes* elevates its heart rate by up to 200% during aerial exposure.

Use of pigments

A further adaptation that assists maintenance of aerobic respiration is the predominance of respiratory pigments in littoral and estuarine animals. This is particularly common for annelids in sandy and muddy shores, where the lower layers are substantially deoxygenated; for example, some worms, such as *Arenicola* and *Sipunculus*, have enough haemoglobin to keep them oxygenated through the whole of the normal low-tide phase, and estuarine fish can mobilize extra blood cells (RBCs) from the spleen during periods of hypoxia. In the longer term, regulation may require changes to the oxygen carriers, switching gradually to higher-affinity pigments; for example, the annelid tube-dwelling worm *Pista pacifica*, from anaerobic mudflats, has three different haemoglobins with different oxygen affinities. Acclimation may occur within an individual's lifetime; for example, in the blue crab, *Callinectes*, a higher-affinity haemocyanin is produced during prolonged hypoxia. Evolutionary adaptation tends to produce higher-affinity pigments (ones having a lower P_{50}, the partial pressure for half-maximal loading—see Chapter 7) in the fully water-breathing species from shores and salt marshes. For example, aquatic littoral crabs with gills have much lower P_{50} values for their haemocyanin pigment than upper littoral and fully terrestrial crabs, where P_{50} is higher and arterial blood readily becomes fully saturated in air (Fig. 10.27).

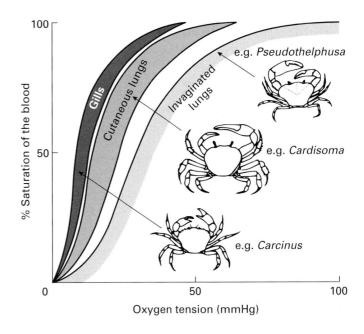

Fig. 10.27 Differences in oxygen-affinity curves and types of respiratory surface for air-breathing littoral crabs. *Carcinus* is fully aquatic with gills and has high-affinity haemocyanin (low P_{50}) whereas *Cardisoma* is upper-shore dwelling with vascularized gill chambers, while *Pseudothelphusa* is a tropical crab living inland with well-ventilated lungs, both these latter species having higher P_{50}s. (From Little 1990; adapted from Innes & Taylor 1986; courtesy of Cambridge University Press.)

Pigments in some animals from muddy shores and salt marshes, such as the salt marsh killifish (*Fundulus parvipinnis*), may also need to be unusually tolerant of sulphides, which can interact with haemoglobin by binding with the central porphyrin groups, thus making the pigment non-functional. The sulphide may be detoxified by reactions that also yield useful sources of energy, the S^{2-} entering the electron transport chain to be oxidized. The unusual echiuran worm (*Urechis*), found on sulphide-rich mudflats, appears to oxidize sulphides in its coelomic fluid, and can maintain aerobic respiration even when exposed between tides in mud with up to 30 µM sulphide. We will meet more elaborate versions of this strategy in deep-sea thermal vent animals (Chapter 12).

Use of a temporary oxygen store

This is an alternative (or supplementary) strategy for many regulators; water stores are used by primarily aquatic animals (which as we have seen may use the stores as a source for maintaining water balance as well), and air stores are used by primarily terrestrial air-breathers. Many intertidal molluscs retain a store of water in the mantle cavity when the tide is out, and crustaceans may use the branchial chamber in similar fashion. Many burrowing animals on sandy shores can draw in fresh oxygenated water when the tide is in, and store it in the burrow (perhaps aided by the mantle cavity or gill chambers). Most spectacularly, the mudskipper fish of tidal mudflats can accumulate air in the horizontal side-shafts of their burrows, apparently by air-gulping at the surface followed by bubble-release underground, providing safer conditions for their eggs in the roofs of the air-filled spawning chambers during potentially hypoxic periods. Littoral mites and some of the rare littoral insects use air films or trapped air bubbles (discussed in more detail in Chapter 11).

Switching between different sites of oxygen uptake

Maintaining aerobic respiration in regulating littoral animals may involve a more elaborate strategy that entails genuinely 'bimodal' respiratory systems that can function, at least to some degree, in both air and water, usually by altering the site of uptake.

Water-breathers in air

On shores, animals that normally breathe from water have to cope with periods of air-breathing when the tide is out, despite their collapsed gills. Thus many prosobranch molluscs have parts of the mantle edge and mantle floor vascularized to take up oxygen once the gills become useless. These molluscs, and many amphibious crustaceans, have reduced gill size and/or number, in parallel with the development of the accessory respiratory surfaces. One of the best-adapted examples is the high-tide and supra-littoral crab *Leptograpsus*, which shows fully bimodal breathing, maintaining oxygen delivery in both air and water equally, and regularly remaining emersed for 8 h. Switching respiratory mode raises the associated problem of varying acid–base balance; in *Leptograpsus* $P\text{CO}_2$ varies markedly during submersion–emersion cycling, producing alkalosis, which is apparently controlled largely by the behavioural trick of air–water shuttling, with a transient period of rapid ventilation and tachycardia on return to water.

Some quite unexpected animals can switch to taking oxygen directly from air to allow continued aerobiosis. For example, some North American mussels will gape to achieve gaseous exchange across the mantle tissue (though they avoid doing this in very dry air), while grass shrimps (*Palaeomonetes pugio*) can be observed jumping up into the air when the sea water becomes hypoxic. Some littoral crabs have invaginated 'lungs' within the branchial chamber; while others have vascularized membranous areas on the outer surfaces of their upper limb segments where they take up oxygen from air. Even within a genus the respiratory surface may vary; for example, the high-shore porcelain crab, *Petrolisthes cinctipes*, has leg membranes, and suffers high lactate accumulation when these are experimentally obscured, whilst the lower shore *P. eriomerus* has no leg membranes and breathes solely with gills.

One estuarine dweller that appears to 'prefer' air-breathing despite its aquatic ancestry is the salt marsh mudskipper fish, *Periophthalmus*. This feeds just above the tideline at low tide, and then clings to the creek walls just above the rising high tide, breathing in air for much of each tidal cycle, especially in periods of warmer weather.

Air-breathers in water

Animals that have air-breathing apparatus (lung books, lungs or tracheae) may be forced to shut these down during periods of submersion, most of them then getting their supplies through the skin instead. Small mites, and strandline insect larvae, probably achieve this readily for short periods, but it is uncommon in other arthropods. Frogs exposed to brackish waters in the supra-littoral zone certainly use cutaneous respiration.

10.4.2 Anaerobic metabolism

Intertidal habitats, including mudflats and salt marshes, can be extremely reducing in character, and have some of the highest known rates of sulphide production; although we have seen that a few animals can remain aerobic, anaerobic metabolism is much the commoner response. In a typical salt marsh up to 12 times as much of the primary productivity is degraded via anaerobic sulphate reducers as by aerobic processes. The hydrogen sulphide concentrations vary with water flow, organic content and season but, in general, sulphide is near zero at the sediment surface and increases with depth to the low micromolar range. The most outstanding anaerobic abilities are found in certain burrowing annelids living in reducing sediments where there is no possibility of pigment storage of oxygen being adequate to sustain aerobicity, and in some intertidal bivalves during aerial exposure. These abilities are accentuated in temperate and boreal habitats, where extreme temperatures also force animals into anaerobiosis.

When an organism becomes cut off from its normal source of oxygen, three critical needs must be satisfied (Chapter 6):

1 provision of suitable substrate(s) for fermentation;

2 provision to store, recycle, excrete or minimize the production of noxious waste products;

3 provision to re-establish metabolic homeostasis in the recovery phase when oxygen is again available, without undergoing damage from rapid increases in oxygen radicals.

Glycolysis is a highly conserved pathway, as we saw in Chapter 6. Reducing equivalents (NADH) formed by glyceraldehyde phosphate dehydrogenase are oxidized in the terminal step of the pathway, catalysed by another dehydrogenase. This enzyme is most commonly lactate dehydrogenase (LDH), because for most animals pyruvate is the terminal electron acceptor, producing lactate as the end-product; arthropods, echinoderms and vertebrates all rely on classical glycolysis for anaerobic energy production, resulting in the accumulation of large amounts of lactate. However, the anaerobic pathways incorporating amino acids (Chapter 6) are particularly common in littoral and estuarine species. Such pathways result in the accumulation of a different 'imino' end-product: arginine yields octopine, glycine yields strombine, and alanine yields alanopine. LDH is therefore replaced by an analogous 'imino dehydrogenase' enzyme. All of these imino dehydrogenases are functionally equivalent to LDH, and thus the ATP output per glycosyl unit is the same. The ATP yields of the different substrates of fermentation were shown in Table 6.2.

There is a distinct preference for octopine formation in many of the brackish-water species studied, and this appears to be linked with high resting levels of phosphoarginine (PAr) (see Chapter 6); the pathway is especially common in swimming bivalves and cephalopods, where PAr levels range from 20 to 30 μmol g^{-1} wet weight. The advantage of octopine production over lactate production may lie with the fact that arginine has serious effects on the catalytic properties of enzymes, whereas octopine is fairly benign; hence, it is preferable to maintain arginine at a low level and convert it to octopine. However, this advantage does not apply to alanopine or strombine, as the amino acid precursors alanine and glycine seem to have no deleterious effects on enzymes. In annelids, octopine dehydrogenase is usually absent, and alanopine dehydrogenase is more common; thus prolonged digging activity in the lugworm *Arenicola* results in the formation of relatively large amounts of alanopine. Aspartate and certain branched-chain amino acids can also be fermented in some invertebrates; in intertidal molluscs, aspartate fermentation is stoichiometrically coupled to glucose fermentation, resulting in the production of succinate and/or propionate (see Fig. 6.6).

Metabolic rate is greatly depressed during anoxia. This can be calculated from the rate of ATP utilization and rate of end-product formation. The ATP turnover rates for long-term environmental anaerobiosis vary. In molluscs the rates are rather low, at 5–20 mmol g wet wt^{-1} min^{-1}; in annelids and crustaceans they are higher, around 20–60 mmol g wet wt^{-1} min^{-1}. When compared with the corresponding aerobic resting rates it appears that molluscs can reduce their metabolic rate by as much as 75-fold, and both annelids and crustaceans by about 10-fold. However, since crustaceans seem to use only the lactate pathway they are probably less well adapted than annelids to long-term anaerobic survival. A case study of anaerobic patterns in the common lugworm is shown in Box 10.3.

Bivalves and other molluscs turn out to be peculiarly resistant to anoxia, and in part this is because they have good antioxidant defences for use in the recovery phase; enzymes such as superoxide dismutase (see Chapter 7) and high levels of the antioxidant glutathione have been recorded in mussels and periwinkles. Their survival of anoxia is also due to their inactive lifestyles, linked to this capacity for much reduced metabolic rates, and is probably further enhanced because the carbonates in their shells act as buffers for the acidic end-products of anaerobiosis. This anoxia tolerance may be one more reason for the preponderance of molluscs in sand and mud beaches.

Restoring metabolic homeostasis after a period of anoxia involves the clearance of anaerobic end-products and the recharging of glycogen stores. Intertidal species that regularly have to sustain low oxygen partial pressures need a quick re-establishment of normoxic cellular conditions, so that ATP and phosphagen levels are rapidly recharged. This recharging, as well as the resynthesis of aspartate and glycogen, the clearance of acidic end-products (by excretion or metabolism) and the resumption of normal activities such as feeding, all increase the energy demand. This is met by a temporarily increased oxygen consumption rate, but also by a small ongoing anaerobic component in the first 2–4 h of recovery, resulting in the accumulation of various pyruvate derivatives.

10.5 Reproductive and life-cycle adaptation

All the problems that are faced by adults, already discussed in the preceding sections, are much worse for the young (and thus almost invariably smaller) stages of animals. Physiological problems are accentuated by the higher surface area to volume ratios, and some studies indicate that desiccation is the greatest physiological threat to juveniles, particularly juvenile molluscs, on shores. Mechanical problems of holding on or station-keeping are also problematic for larval morphs lacking heavy protective shells or fully developed locomotory systems. Nevertheless it is worth noting that some otherwise oceanic vertebrates use inshore waters as breeding grounds, while many semi-terrestrial crabs come back to the shore to breed, so that shorelines can be seen as reproductive refugia for various species that have perhaps not fully adapted to a secondarily adopted adult habitat that is in some sense even more 'difficult'.

Box 10.3 Anaerobic lifestyles: a case study

Arenicola marina (the lugworm) lives in burrows up to 300 mm deep in sand and mud sediments of tidal flats. At low tide it may be exposed to several hours of anoxia. The switch from aerobic to fully anaerobic metabolism occurs in two phases.

During the first phase, or transition period, various metabolic pathways that are immediately available are used; the energy produced comes from utilization of the unusual phosphagen phosphotaurocyamine, and from glycogen. The glycogen is metabolized to pyruvate and then further converted to strombine or alanine. At the same time aspartate is catabolized, yielding oxaloacetate and then malate. This reaction re-oxidizes the NADH originating from the earlier stage of glycolysis. Malate is further metabolized to succinate, acetate and propionate in the mitochondria.

However, through the transition phase the energy content of the body wall musculature falls, indicating that energy production is failing to keep pace with requirements.

After about 3 h a second phase of prolonged anaerobiosis is initiated. The affinity of pyruvate kinase for the substrate phosphoenolpyruvate (PEP) is reduced, involving a phosphorylation of the enzyme. PEP is therefore converted to oxaloacetate instead of pyruvate, and feeds into the mitochondria via malate. The energy yield of glycogen catabolism by this direct route to propionate production is more than double that of conventional anaerobic glycolysis, though only about 15% of that gained by aerobic glycogen catabolism. At this stage phosphagen reserves are depleted but the ATP supply and demand pathways have come into closer balance, with the energy content of the cells falling only very slowly.

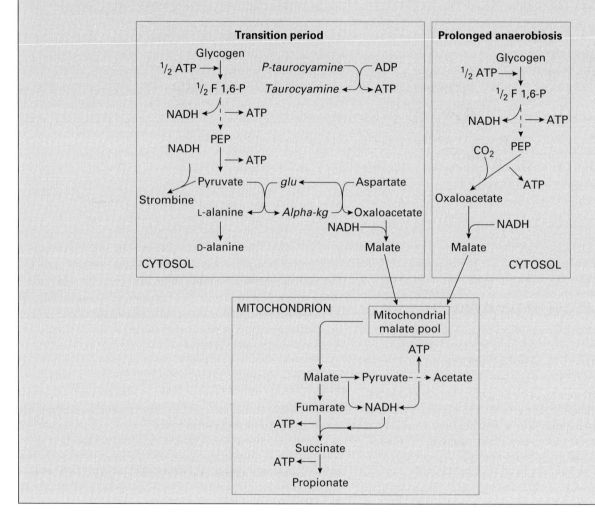

Many littoral and estuarine animals avoid some of the problems for their young by adopting more complex reproductive behaviours (Fig. 10.28) compared with marine species. Broadcast spawning is either carefully synchronized with tide patterns to achieve high fertilization rates, or is replaced with direct copulation; a startling example of the latter is found in the sessile barnacles, where a greatly elongated penis is used to transfer sperm to neighbours. Most of the invertebrates lose or greatly attenuate their pelagic larval stages; any larvae that are produced tend to be already rather large and advanced, with a reasonable ability to swim, and they will settle quickly and relatively locally. Larval production is timed to coincide with periods of reasonably benign weather conditions. Thus in high-shore grapsid crabs the eggs are yolky and substantially enlarged, and

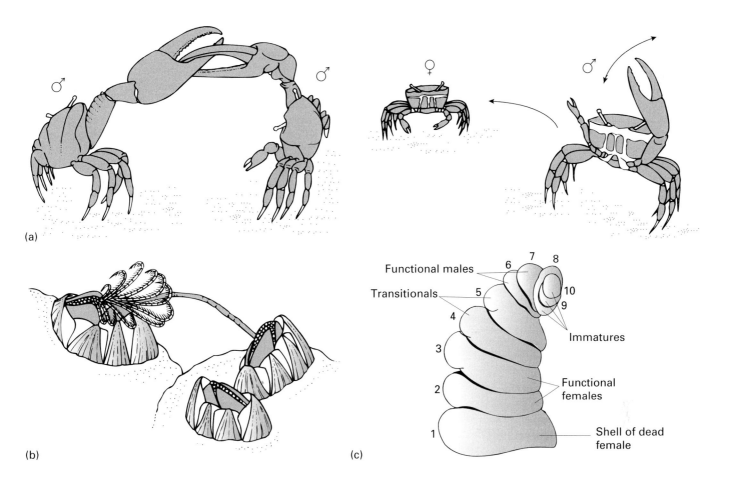

Fig. 10.28 Reproductive strategies in seashore and estuarine animals. (a) Courtship competitions and dancing in fiddler crabs. (b) Extended penis for mating in barnacles. (c) Piled-up slipper limpets (*Crepidula fornicata*) where sex of individuals changes as the pile builds up.

the larvae are relatively large and highly salinity tolerant, with their stages reduced from four to three; they are able to survive for at least two stages just by lecithotrophy. Many other invertebrates similarly switch to laying much yolkier eggs (Fig. 10.29), sometimes protected in toughened cases as in whelks, or laid in protected sites on seaweeds or in rock crevices and overhangs. In the gravel-beach dwelling gammarids, where no such protection is offered, egg size varies seasonally, being greater in winter when disruptive storms are likely. Other invertebrates use a brood pouch to protect the eggs, retaining the embryos until they are relatively large and potentially independent, as in many littoral crustaceans. An alternative strategy is to lay and then guard eggs carefully, a phenomenon found in many shoreline fish such as gobies.

Some species of littoral molluscs and some fishes can change sex within the lifetime of an individual (see Box 9.1). This may help in relatively dispersed mollusc populations by allowing any encounter between conspecifics to become a mating opportunity. In the slipper limpets (*Crepidula*) early arrivals in a location are young males, and they in turn become a settlement site for later arrivals; those at the bottom of the heap then gradually switch from being male to being intersexual and then female as numbers build up. The sex of each individual is controlled by the sex ratio of the whole aggregation, probably by the action of pheromones.

Life-cycle patterns vary with the exposure and extremity of shore conditions. On very 'difficult' shores such as some mangrove systems, where salinity change and heat stress are unusually marked and currents may be fierce, some specialist invertebrates adopt a life-history pattern dominated by extremely slow growth. For example, the mangrove gastropod *Thais kiasquiformis* grows at only 1 mm per year. Such species also show very high plasticity for shell shape and thickness, depending on their exact position on the shore, and during their lifetime may migrate according to the levels of desiccation and predation they experience.

The age of a particular shore also affects life cycles. Shores that are of relatively recent origin (where there have been marked coastal level changes, or on shores at high latitude covered by ice until the end of the last Ice Age around 10 000 years ago) tend to be dominated by brooders or direct developers that have arrived by rafting. This is presumed to be due to isolation from feasible

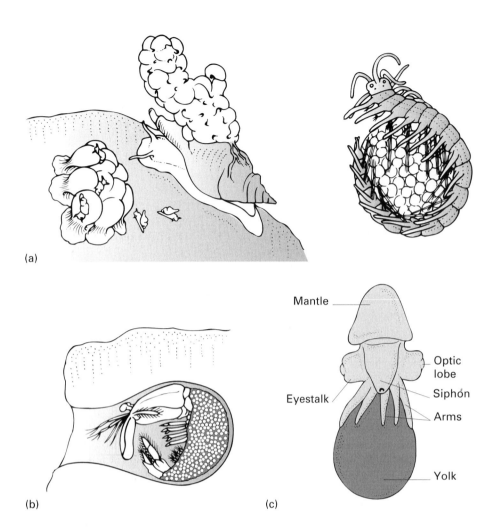

Mantle

Optic
lobe

Eyestalk

Siphon

Arms

Yolk

Fig. 10.29 Provision for care of eggs and juveniles in littoral animals. (a) Protected egg masses in whelks and polychaetes. (b) Mantis shrimp with egg masses in burrow. (c) Embryo with massive yolk sac in sublittoral octopus.

(a)

(b)   (c)

sources of larval colonization. The south Atlantic island of South Georgia, for example, lacks both mussels and barnacles, though the environment could certainly support them.

10.6 Mechanical, locomotory and sensory systems

10.6.1 Physical damage

Waves produce very considerable crushing and tearing forces. Not surprisingly, littoral zones require substantial investment in systems for protecting soft tissues against abrasion and pressure damage, and against drag forces as waves recede. Behavioural options may suffice (deep burrowing, sheltering under stable large rocks or in crevices); but a very high proportion of littoral invertebrates use the option of protective shells.

The design and mechanical properties of molluscan shells have been fairly extensively studied, and there is a relationship between microstructure and lifestyle. Mollusc shell is formed from small crystalline blocks of calcium carbonate each wrapped in a protein matrix ('conchiolin') and assembled together in different geometric arrays. Where the crystals are in parallel rows like flat bricks in a wall, in the plane of the shell, the material is termed nacre or mother-of-pearl; this gives maximum mechanical strength, at 35–115 N m^{-2}. The crossed lamellar structure is a much lower-strength material (9–60 N m^{-2}) where the layers of crystals alternate in direction rather like a plywood; however, this structure appears to be the most resistant to chemical attack from boring animals. A third alternative is the prismatic structure, with stacks of prisms perpendicular to the shell surface; this gives reasonable mechanical strength against crushing (around 60 N m^{-2}). Rarely, molluscs use a fourth option of a disorganized chalky structure. Many mollusc shells have a multilayered composition with more than one of these microstructures incorporated, presumably achieving the preferred balance of properties. Thus on very exposed sites where strong resistance to waves is crucial, this may mean an inner nacreous layer and a thick band of prismatic or crossed-lamellar shell distal to this, as in many mussel species; whilst in more sheltered habitats where predation is strong the crossed-lamellar format dominates, and in sheltered sublittoral zones a thin inner layer of nacre may be covered only with a thick chalky layer, as seen in oysters.

The macrostructures of shells are also potentially adaptive. Animals from outer coastal regions commonly have thicker coverings than conspecifics from inner, more sheltered waters; this applies both to shelled animals, such as mussels, and to endoskeletal animals, such as starfish, where the dorsal surface can be greatly strengthened by stronger calcareous ossicles. Ornamentation of shells is also much reduced in individuals from very exposed sites, though it may be useful in areas of moderate wave action where spines and ridges can help to anchor an animal on the beach.

Wave shock is also a major problem on sandy shores, where the impact of the water itself, and the debris the waves may carry, can crush animals. Because wave action is somewhat seasonal, usually fiercer in winter, the shape and slope of beaches change through the year and they normally become steeper in winter. The constant seasonal and daily shifting of the substratum is the main reason for the lack of macroalgae and of surface invertebrates on sand. The eastern surf clam, *Spisula solidissima*, is a rare example of a bivalve strong enough, and able to burrow fast enough, to live on exposed western US shores, while the rapidly burrowing streamlined crustacean *Haustorius* survives on many exposed beaches of western Britain. It should be noted in passing that tidepools on sandy beaches provide something of a refuge from damage for macroalgae and for many animals. They are often more common on 'groyned' beaches (where man-made breaks have been installed in an attempt to retain sand and stabilize a beach for human purposes). Some kinds of sheltered mudflats also provide refugia, where the substratum is not battered by strong waves and rarely dries out completely.

10.6.2 Dislodgement

Currents, tides and waves all combine to produce powerful forces tending to dislodge shoreline animals and so displace individuals from their adaptive niche. It should be evident by now that small displacements up or down or along a beach may produce substantial change in the conditions experienced, especially the regime of emersion and immersion, circumventing all the carefully balanced adaptations we have already discussed relating to osmotic and thermal survival and tolerance of anoxia.

Thus virtually all animals in these brackish-water ecosystems need adaptations for holding on and avoiding displacement from the shore. Algae have elaborate and very tough holdfasts, which help to create local zones of stability that a range of animals can live on or in. Anemones have adhesive pedal discs, and some worms, such as the sand mason *Lanice*, with filter-feeding crowns emerging above the surface, have radiating anchors of hardened mucus and sand grains. Gastropod molluscs use strong suction of the foot to the substratum, and in the case of limpets have the additional safeguard of a homebase where the rock is abraded to fit exactly the contours of an individual shell and improve the efficiency of the suction-pad effect; an absent limpet leaves a clear abrasion scar, and a limpet caught by waves when away from 'home' is much more likely to be dislodged.

Mussels and some other bivalves have strong byssus threads cementing them to a rock or to the other mussels in the bed. Many intertidal and rock pool fish have suckers anteriorly or ventrally. Burrowing again provides a solution for worms.

In the absence of systems for attachment, the alternative is a system of careful station-keeping to ensure that an individual stays at about the right level on a shore. The well-known limpet 'homing' responses are a way of achieving this; the animal leaves its circular rock homebase to graze over the rocks but returns to its own site reliably, mainly using chemical cues. Fixed behavioural responses may also help to keep an animal in the right site. The winkle *Littorina neritoides* uses negative geotaxis when submerged but switches to negative phototaxis when exposed and drying out, lacking any clear directional preferences when exposed but still moist. This suite of responses automatically maintains it in its high intertidal zone, after local feeding bouts or even after dislodgement by waves. Some sandy beach dwellers may stay in the 'right' zone in a rather different sense, leaving the sediment as they detect an incoming wave and thus riding the waves up the beach on an incoming tide then burrowing rapidly at the higher tide mark; this keeps them in the ideal balance of a moist but not extensively waveswept habitat for a greater proportion of each tide cycle. This 'swash-riding' technique is found in certain crabs and in some of the faster-burrowing clams.

10.6.3 Locomotion

Faced with problems of damage and dislodgement, littoral animals have both to move around rather carefully, usually just above or within the substratum, and to time their locomotory activities appropriately.

Rhythmic locomotory activities

Rhythmic locomotory activities are extremely common on seashores, regulated by **circatidal** and **circadian rhythms**. Those animals that are essentially aquatic and that swim and crawl in search of food or mates tend to do so as close as possible to the substratum, in the hour or two either side of high tide. Grazing molluscs and a host of creeping detritivores move slowly over and around rocks or across the sand surface throughout the period of emersion, but in many cases time their return to a homebase to occur just before the retreat of the tide. Some of the more motile animals, particularly predators such as crabs, have regular patterns of moving up and down the beach, and the crab *Carcinus maenas* appears to have at least three circatidal neural oscillators setting its behavioural rhythms, cued by salinity, temperature and pressure. Regular movements through beach levels also occur in some of the essentially terrestrial animals that inhabit sandy beaches and mudflats (Fig. 10.30a,b). Predatory birds invade shores and estuaries and move down with the retreat of the water. Insects and mites move up and down the vegetation on salt marshes, or up and down the strandline zone on beaches, to stay in proximity to the water surface (and thus

Fig. 10.30 Rhythmic behavioural strategies in brackish-water animals. (a) Patterns of climbing in the estuarine snail *Hydrobia*; given sticks above mud in a laboratory simulation, snails climbed at each tide, especially in early tides (1 and 3) when they were starved but less so at later tides once they had restored their nutritional balance. (From Barnes 1981.) (b) Respiratory rhythms in the isopod *Tylos*, showing an oxygen consumption peak just after low tide at night, with inactivity from dawn onwards (the line shows one individual, and the tint shows the range for six animals). (From Little 1990; adapted from Marsh & Branch 1979; courtesy of Cambridge University Press.)

potential food sources) without drowning. Certain shore insects are known to use upwind orientation, invoked by humidity changes, to stay in the 'safe' zone as the tide level changes.

Burrowing

Burrowing is probably the single most important locomotory mode for animals in the littoral and estuarine zones. As we have seen, it helps to solve nearly all their problems: osmotic, thermal and respiratory hazards may be avoided when the tide goes out, and predation is also reduced. It also opens up new food sources, either by poking one end out above the substratum to filter-feed or deposit-feed, or by eating the substratum itself to extract organic debris.

Burrowing animals need both to penetrate into the substrate and then to retain an ability to move about within it. The prin-

ciple for the main burrowing action is always the same, shown in Fig. 10.31. The idea is to have an alternation of two different kinds of anchorage. First the animal makes one part of the body expand widthways to serve as the **penetration anchor**, from which the front end can push away into the substrate. Then it expands the advancing tip into a **terminal anchor** to which the rest of the body can then be drawn. This in effect creates a single monophasic wave along the body; it is a modification of the basic alternation of circular muscle contraction waves and longitudinal muscle contraction waves that underlies most other forms of soft-bodied locomotions (see Chapter 9).

Figure 10.32 shows the principle in action in a bivalve, many of which burrow in sand and mud. The shell itself is allowed to gape open to give the penetration anchor, and the foot pushes out from within this. Then the end of the foot is expanded by fluid pressure in the haemocoel as the retractors and other body

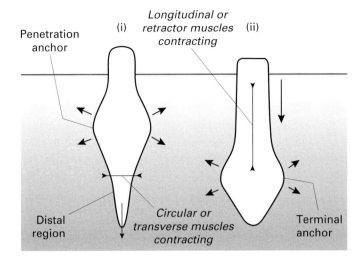

Fig. 10.31 The principle of alternating anchors used in burrowing animals: the penetration anchor holds the animal while the distal region is extended into the substratum, then the terminal anchor is expanded and holds the animal while the rest of the body is pulled downwards towards it. (From Trueman 1975.)

muscles contract, to give a terminal anchor; and the shell is closed as far as possible and drawn in after the foot. Thus shell and foot work together, each giving anchorage at different phases. Water expelled from the mantle cavity also helps to soften the substrate and ease penetration. In unsegmented worms the proboscis may be dilated to form the terminal anchor,

while in burrowing anemones, such as *Peachia*, the whole column of the body becomes the anchor. These unsegmented forms work better in softer sediments such as sand and mud because the lateral expansion of the body has to be substantial to allow any kind of purchase on the deformable substratum, and a segmented body with regular 'bulkheads' cannot expand outwards very much. Segmented worms do better at burrowing through quite compacted hard sediments (the earthworms being an obvious example).

Boring animals do something rather similar to burrowing, but in very hard substrates such as rock and wood on beaches. *Teredo*, a bivalve that bores into wood (and often known as the 'shipworm'), uses its foot and shell as anchors as before but with the extra trick of rasping with the serrated edge of the shell to soften the substrate. The bivalve piddock *Pholas* works its way into rocks, adding chemical secretions to soften and dissolve the substratum.

Crawling and swimming

Most worms and many crustaceans use a crawling mode for foraging and dispersing amongst the rocks and weeds of the littoral zone, modified in some of the more amphibious crustaceans into a scuttling or hopping mode when moving over dry sand or rock between tides. The sideways walk of crabs is a special case, where leading legs pull and trailing legs push; the adaptive value of this is not entirely clear, though it may reduce tripping and allow a longer stride length.

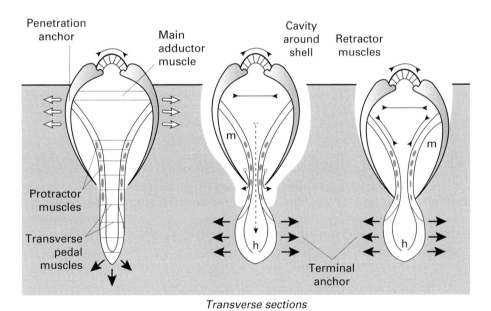

Transverse sections

Fig. 10.32 Successive stages in the burrowing of a bivalve mollusc in sand or mud, with the shell allowed to gape laterally to form the penetration anchor as the foot probes downwards, then the foot expanded to form a terminal anchor as the shell is drawn down (h, haemocoel; m, mantle cavity). (From Trueman 1975.)

⇐ Movement of shell	Hydrostatic pressure produced by adduction of valves
Water ejection from mantle cavity	
◀ Probing and extension of foot	Tension of muscles or ligament

Swimming is only appropriate in rock pools, or when the tide is well in, whether in littoral or estuarine zones. Animals swimming in the shallow rapidly moving surf zones have little control of their own direction, and in estuaries they may be swept too far inland to areas of low salinity if they enter the main water column on an incoming tide. It is essentially the marine animals of the sublittoral zone that retain a free-swimming mode, together with the specialist rock pool dwellers (including fish) that can enter a crevice or hold on to a rock with suckers at times when swimming is too dangerous. However, some of the relatively specialized invertebrate larval forms do become free swimming in the main water body, allowing currents to disperse them more widely.

10.6.4 Sensory systems

The littoral zone and estuarine habitats are both rather disturbed habitats with many physical and chemical factors undergoing cyclical change, so that sensory detection systems face particular problems. Perhaps the most fundamental of these is that all sensory neurones operate on the basis of ionic gradients, so that as we saw earlier (with nerves in general) there will be problems in maintaining a constant level of sensitivity in environments of changing salinity. This will apply to all internal sensors for all but the most precise of osmoregulators; and it will also apply to external chemoreceptors for all animals.

Visual senses

Visual sensitivity and responses to light will be more important for littoral animals than for most species dwelling in the ocean, where 70% of the living space receives no light at all. Most of the living space in brackish water is shallow enough to receive light, though this may be reduced by vegetation growth in summer. Animals that remain water-covered in the sublittoral zone and around coral reefs are perhaps particularly able to benefit from good vision and in particular from colour (polychromatic) vision; these zones are home to fish with up to five types of photoreceptors giving pentachromatic colour vision, and to some mantis shrimps with 10 or more types of photoreceptor. In contrast, animals from muddy shores and estuaries rarely have good visual acuity or colour discrimination.

Animals in the littoral zone are effectively amphibious and thus have the particular problem of focusing in two different media. Any lens system that focuses light accurately on a retina under water (refractive index (RI) = 1.33) will be severely myopic in air (RI = 1.00), focusing light in front of the retina. Similarly, eyes that work well on land lack sufficient focusing power to form an image on the retina when under water. Thus many littoral animals retain 'primitive' eyes that give information on light and dark, with little acuity. The problem can be overcome though by combining the properties of the lens and the cornea; a flat cornea and a very spherical lens (having high focusing power) is a common solution. The mudskipper represents a

good example of this. Another alternative is to have two pupils with different properties, and a few of the shoreline blennies do have paired side-by-side pupils. Amphibious shore birds that hunt under water tend to vary their focus using very strong accommodation instead, with powerful ciliary muscles that squeeze the lens so that it is more convex and bulges forwards almost through the iris.

Bioluminescence is relatively uncommon in littoral animals relative to the marine fauna, perhaps because it risks attracting attention from two lots of predators (from land and from the sea).

Chemical senses

Species that can survive and feed over a wide range of salinities must have olfactory receptors capable of working at highly variable Na^+ levels. There is little information on how this is achieved, though in salmonid fish the olfactory systems are similar in both pond water and sea water, the receptors themselves somehow providing inbuilt tolerance. Permitted reductions in internal K^+ levels at low salinity are likely to be part of the response, as with nerves, so that a reasonable resting potential is maintained, but little is known of the details of sensor adaptation. However, proper functioning is lost in artificial salinity media lacking adequate calcium, so that Ca^{2+} ions also appear to be essential for maintained sensor responsivity.

Chemoreception in littoral and estuarine animals serves similar functions and faces similar problems to those described for marine animals in Chapter 9. Many species detect their food chemically, or detect and respond to incoming predators, sometimes with defensive chemicals. Commensal and parasitic species locate their hosts chemically, and many species locate their mates with sex pheromones. However, the disturbed pattern of flow inevitably disrupts any pheromone plume and thus reduces the range over which chemical cues can act, and on exposed beaches it is unlikely that any chemical signalling system can be of use in the surf zone between tides. Nevertheless, the fact that salmon can home in on their spawning grounds through estuaries should caution us that very precise chemoreception remains a potent source of information.

Mechanical senses

We have already stressed that brackish habits pose a number of mechanical problems for animals, so it is no surprise that mechanoreception is important to many of them. Detection of flow in the external medium may be a crucial component of the response to tide and to waves, and can be readily signalled by stretch receptors in skin, by deformable projections such as the microvilli on hair cells (see Fig. 9.31), or by sensory bristles in arthropods and many fish. Most phyla of soft-bodied invertebrates have ciliated neuroepithelial cells that act as direct 'rheoreceptors' for flow and pressure. Any animal dislodged by wave action is also greatly aided by statocysts (Fig. 9.31) that act as gravity receptors, allowing it to right itself and find the substratum again. In fish the lateral line system containing neuromast cells (modified hair cells) provides the main sensor for flow and pressure, with otoliths (statoliths made of bone) giving orientation and gravity information.

Other senses

Electroreception in littoral fish is rare, since stable electric fields are unlikely and varying salinity also alters the field parameters. However, electric senses are particularly useful in turbid but otherwise more stable waters in the river mouth, and are important in some teleosts and sharks that penetrate large tropical rivers. Echolocation also occurs, with both dolphins and seals regularly fishing inshore and moving up estuaries. The fish and mammals using these sensory modes often occur very far up river in the case of turbid large estuaries such as the Ganges, Amazon and Yellow River; details were considered in Chapter 9.

10.7 Feeding and being fed on

All animals in the littoral and estuarine habitats have the potential benefit of two sources of food (from the sea and from the land) and the potential hazard of two parallel sets of predators. Predation may be the single most important population determinant on many shores, especially in taking out large proportions of the juvenile prey items.

Rocky shores contain an abundance of food: plankton for filterers, and detritus for scavengers, both renewed on each tide; macroalgae (seaweeds) providing a dense growth for herbivores at various depths; and a wide range of invertebrates for predators to pick off. Estuaries too contain an abundance of food; all the sources of the seashores, but also the local growths of phytoplankton supported by nutrient from the river. By contrast, at first glance there seems to be little food around on a sandy beach. However, animals can feed on the sand and mud itself, thereby extracting mostly bacteria, as with lugworms, fiddler crabs and mole crabs, and many flatworms and nematodes. They can also feed at the mud surface on the detritus deposited there; here examples include *Hydrobia* snails (especially on salt marshes), *Lanice* the sand mason (an annelid) and many bivalves, like the clam *Scrobicularia*, whose feeding tube (the inhalant siphon) comes up to lie on the surface. Other animals may live in the sand but feed from the water: some acheive this using structures projecting upwards, such as the tentacles of epifaunal hydroids or of tube-dwelling annelids like *Sabella* and other fanworms; others draw water though a burrow and filter from it with a mucous net (the annelid worm *Chaetopterus* being the classic example here; see Fig. 7.9a). Obviously these options depend on the duration of water immersion, which is often the limiting factor in feeding and growth rates; high-tide individuals feed less and grow more slowly than low-tide individuals.

It is worth dealing specifically with the problems of herbivory in these tidally influenced habitats. Marine plants are almost exclusively seaweeds, and varying proportions of their productivity are diverted into grazing animals—rather more in sheltered and low-latitude communities, and from the mid-shore downwards, but less towards the high-water mark and on exposed or high-latitude beaches, where decomposition is more important. Where herbivory is dominant, the key grazers are arthropods, molluscs and some polychaetes and echinoderms (sea urchins), together with specialist fish. In general, small urchins and gastropod molluscs are more important in temperate systems, with larger urchins and fish dominating on tropical shores, and with tropical reefs showing the highest herbivory rates of any known communities. Microalgal grazers, such as limpets and winkles, scrape or sweep over the substratum, eating diatoms and juvenile recently settled algae, whereas macroalgal grazers, such as sea urchins, scrape or bite off pieces of seaweed. All face substantial problems: how to find appropriate food (which may be widely dispersed or extremely patchy), when and how often to feed on it, and crucially when to stop feeding to avoid the onset of unfavourable or dangerous conditions as the tides or currents alter—inherent rhythmicity of behaviour is again a useful common feature. In addition they may face problems of predation while feeding, forcing them to select more protected or sheltered sites of poorer food quality. But above all they may have to overcome problems posed by the food itself, since many littoral weeds are well-defended against herbivory. Like terrestrial plants (see Chapter 13), seaweeds possess a range of chemicals that may serve to deter grazers. Polyphenolics and terpenes are commonplace, and certain aromatic or amino-acid derived compounds are also used, often with chloride, bromide or iodide incorporated; only the alkaloids are seemingly absent from the defensive repertoire compared with land plants. Most of the chemicals are broad-spectrum defences active against a wide range of herbivores; but herbivore size and mobility are often correlated with resistance to seaweed chemicals, and small sedentary grazers may eat weeds that are extremely deterrent to more active fish, for example. Just as with insect–plant interactions in the terrestrial world, some littoral grazers may use the compounds from their seaweed foodstuffs to their own advantage: as cues to locate and feed on these weeds, or as chemicals in their own bodies to act as defensive/deterrent antipredator agents. Coevolution between specific weeds and specific herbivores is relatively limited compared to terrestrial systems, since most littoral herbivores are fairly unspecialized. However, macroalgae do invest in chemical defences, and there may be morphological and/or chemical 'mimicry rings' among seaweeds, some harmless mimic species resembling toxic model species (Batesian mimicry) or various toxic species converging on a common phenotype (Müllerian mimicry). Seaweeds may also improve their chances of surviving herbivory by having the ability to pass through guts without being completely destroyed; pieces of vegetative tissue may regenerate after passing out with faecal pellets, and in some species algal propagules may also germinate after passage through a fish gut (a direct parallel to some kinds of seed dispersal on land).

Box 10.4 Corals and their microalgal symbionts

Many corals (including all the reef-building types), and some forms of jellyfish, harbour small algal cells (from two groups of protistans, the cryptomonads and the dinoflagellates) which are collectively termed zooxanthellae. In corals these occur in dense populations that may exceed 30 000 algal cells mm^{-3} of host tissue. Some freshwater cnidarians, including the familiar *Hydra*, also house green algal cells, termed zoochlorellae. All animals that house such symbionts must live in the photic zone where the algae can receive some light; thus corals are only present to a maximum depth of about 90 m.

The benefits to the coral are probably variable in different habitats. Some may derive relatively limited nutrition, or help in eliminating metabolic wastes (nitrogen and some phosphorus, which get used and recycled by the algal cells). Others appear to get very substantial nutritional gain, with a large part of the algal photosynthetic production being passed on to the host as glucose, glycerol or amino acids.

An additional advantage to the host comes in the process of depositing the calcareous coralline exoskeleton. By using up CO_2 in photosynthesis the zooxanthellae help to drive equilibria in favour of $CaCO_3$ formation as shown below.

One other form of nutrition on shores should be mentioned, as quite a range of animals in shallow waters have microalgae within their bodies, in a symbiotic relationship. Corals are the most obvious example (Box 10.4); these get the benefits of some carbon fixed by the algal cells' photosynthesis, as well as indirect assistance with deposition of their calcium carbonate skeletons. Certain clams and anemones in littoral waters also house microalgae; and perhaps the most spectacular example is the green flatworm, *Convoluta*, found on beaches around the English Channel, where it migrates in large numbers to the surface at low tide on sunny days to give the appearance of green sheets across the sand.

A large proportion of littoral animals run the risk of becoming food in their turn, and must possess strategies to avoid potential predators. Some invertebrates are assisted in this by chemical cues; the littorinids (winkles) appear to respond to 'alarm substances' in the presence of nearby crabs, migrating up shore or hiding in crevices—chemicals leaching from the crushed remains of conspecifics are the probable cue. Not surprisingly, many other littoral animals avoid predation primarily by hiding or by burrowing. This is not an absolute guarantee of safety, however, as some common beach predators have adaptations to reach down into muds and sands, or into crevices, or amongst small stones, to extract prey items. Figure 10.33 shows the adaptations of a variety of wading birds' beaks. Curlews and godwits have long bills able to reach deep-burrowing and softer-shelled clams; oystercatchers delve for burrowing worms and can open cockles; redshank, sandpipers and similar birds consume more shallow burrowers, with plovers and turnstones taking the superficial amphipods, etc. At high tide when the terrestrial predators have departed, and all these potential prey animals are active, they face the additional hazard of predators such as crabs and fish, especially flatfish and gobies, nipping off their extended siphons and tentacles. A study of one estuarine fish called the spot showed that the two sets of predators have effects that are not simply additive; spot avoid the aquatic predators (mainly other fish) by moving into shallower water, where they might be expected to be more vulnerable to birds, but in fact since they also then tend to aggregate more they reduce their susceptibility to bird predation, their behaviour ensuring that they suffer lower mortality in the presence of both predators than would be predicted by adding mortalities due to each predator acting alone.

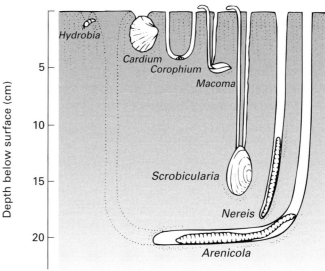

Fig. 10.33 Depths of burrows for common sandy and muddy beach animals, and the corresponding lengths of beaks of common waders that feed upon them. (Adapted from Green 1968.)

10.8 Anthropogenic problems

Man has always tended to settle along coastlines and thus to have an unusually forceful impact on shore environments, through food-related, industrial and leisure activities. Most of the current problems of coastline management bring up conflicts of interest, between developers and ecologists, landowners and economists; in many of these conflicts, the outcome for the biota is particularly hard to predict due to the very complexity of the brackish environment in time and space.

10.8.1 Pollution

All the habitats covered in this chapter are very susceptible to pollution. Most of the oils and chemical spills from seas accumulate on shores. Because humans are apt to live on rivers and estuaries their sewage is often also deliberately fed out there, at or below low-tide level, leading to increased coliform bacteria counts and zones of oxygen depletion. Deposition of solid wastes from pipes or from barges may also lead to choking of kelp forests and their holdfast communities; this occurred in Santa Monica Bay off Los Angeles, where the outcome was decreased diversity and a population explosion in sea urchin communities. The accumulating deposits of agrochemicals (such as nitrates, phosphates and DDT) and industrial wastes (including hydrocarbons and heavy metals) coming down from

rivers to the estuaries and nearby shores also add to the general problems of eutrophication and possible toxicity. Tributyltin (TBT) is used in antifouling paints and is a specific problem in some estuaries used as marinas and boat-moorings; it causes sexual dysfunctions in dog whelks and perhaps in fish. Some coastal species detoxify heavy metals by incorporating them into non-reactive complexes such as metallothioneins, which can reach quite high levels in some of their tissues; robust species such as the shore crab, *Carcinus*, can acclimate to some extent, gradually reducing their levels of copper uptake and more efficiently transferring it from their haemolymph to their tissues during prolonged exposures.

Apart from visible filth, the evidence of plankton blooms ('red tides') reveals an increasing problem, and excess macroalgal growth is also now regularly appearing on the shores of the Mediterranean, the Black Sea, and other areas of relatively closed-off inshore sea. Remember also that most of the oil pollution problems mentioned in Chapter 9 inevitably impinge mainly on the marine shorelines, especially where oil spillage results from the accidental grounding of tankers.

The worst problems at present and for the immediate future arise from polluted river outflows, inadequately treated sewage, non-degradable plastic waste, and insensitive use of the near-shore seabed for disposal. Inland deforestation is also becoming a problem as it often allows erosion and increases river sediment loads, leading to increased offshore deposition. Important areas of shore are being lost completely, especially coral reefs, wetlands and mangrove forests. In some parts of the world long sections of beaches are so polluted they become devoid of life and no longer usable for human leisure interests. Prime indicators of coastline pollution are the shellfish, particularly filter-feeding bivalves, and these often become too contaminated to be safely consumed by humans (or by the birds, crabs and fish that normally prey upon them).

Many coastal communities also rely on pumped groundwater for their fresh drinking-water supplies, and shallow coastal freshwater aquifers are easily damaged by abstraction for domestic and industrial use (often with induced seepage of salt water inwards), as well as by polluting influences.

10.8.2 *Coastline alterations and tourism*

Another problem impinging on many littoral zones and their fauna is man's tendency to alter and 'manage' coastlines, to reduce erosion of cliffs, stabilize dunes, drain wetlands or alter the movement of deposits along the coastline. These management practices can affect the balance of recycling processes whereby organisms normally recycle nutrients back to the land —especially from mudflats and wetlands. Most of the traditional methods of 'holding' the shoreline with groynes, seawalls, etc. are now being questioned due to their down-coast repercussions. Protecting shorelines against erosion or flooding is commonly more environmentally sensitive now, and may involve manipulation of the natural agents that give stability, with lower cost and better long-term outcomes.

Drainage schemes to produce tourist resources such as marinas may also destroy some rather important chemical cycling, and rebound on the land ecosystems. They may also destroy the breeding areas of fish and the important relatively sheltered zones where young fry mature, so damaging exploitable fish stocks. (Similar effects occur in certain areas from the use of shark nets, however, desirable these may otherwise be!) The loss of estuarine land may also lead to unforeseen modification of tidal and river channels and thus have effects far inland.

At the extreme, use of coastlines for tourism in areas such as the Mediterranean, with managed 'cleansed' beaches and offshore sporting activities, can produce complete destruction of the fragile ecosystem and also effectively destroy local human communities formerly dependent on fishing and local agriculture.

10.8.3 *Invasions*

There have been some spectacular cases of coastlines being affected by invasions of **non-indigenous species** (NIS) from other biogeographic areas, brought deliberately or accidentally (often as larval or adult settlers on ship hulls, or perhaps most commonly in ballast water) by man, or given access to new areas by the building of canals between seas (notably at Panama and Suez).

In Britain new barnacle species have been introduced, and the invader *Elminius modestus* is still spreading around the coastline from its first foothold in Southampton Water, outcompeting some native barnacles. The Chinese mitten crab, *Eriocheir*

sinensis, is also invading up many European rivers. In North America at least 400 NIS have been recorded along the Pacific and Atlantic coasts, of which perhaps 15% are classed as nuisance species. The non-native periwinkle, *Littorina littorea*, is now the single commonest winkle in most of New England, after an introduction to Nova Scotia about 100 years ago; the European shore crab, *Carcinus maenas*, is also now a common predator on these shores. Some such introductions are relatively benign, but others can seriously damage local communities. In San Francisco Bay the introduced snail *Ilyanassa obsoleta* has almost completely out-competed the local mud-snails, and the clam *Potamocorbula amurensis* has displaced a wide range of other endemic invertebrates. In reciprocal fashion, the introduction of the North American ctenophore *Mnemiopsis ledyi* into the Black Sea and Azov Sea has been implicated in a massive loss of fisheries revenue through predation of fish fry.

10.8.4 *Coastal energy*

Not only are relatively remote coasts commonly regarded as suitable sites for oil refineries and nuclear power plants, but there are significant plans for the use of wave power and tidal power as renewable energy sources for the future. Such ideas are superficially attractive but coastal energy is so far proving very difficult to harvest, not least because of the extraordinary variability of waves. Wave-energy devices are mainly at the testing stage, with limited success as yet. Tidal power is perhaps easier to harness, and small-scale tide mills have been in operation for centuries. Some small tidal power stations do now exist, and the impact of large tidal barrages in suitable rocky channels has been intensively studied in France, Canada, China and the UK, but with no very large projects in place as yet. However, these developments are bound to impact on coastal environments and biota in the future, perhaps especially on sessile species close to their environmental limits, and possibly on those more conspicuous fish and birds feeding (and breeding) at higher trophic levels.

10.9 **Conclusions**

Given all the problems covered here, it is clear that littoral communities are amongst the most finely patterned and perturbation-sensitive to be found anywhere. Every organism has its own very tightly defined niche, often within a vertically zoned habitat. Several of the figures in this chapter have shown the *interactions* of different environmental variables—notably salinity, desiccation, temperature and oxygen—stressing the point that these factors are strongly interdependent and that simplistic studies looking at just one aspect of the littoral environment rarely work. Figure 10.34 provides a particularly clear demonstration of this, with the mortality due to salinity and temperature change being interrelated and in turn modified by changes to oxygen availability. Reduced salinity exposure in any one species may lead to changes in oxygen consumption, altered

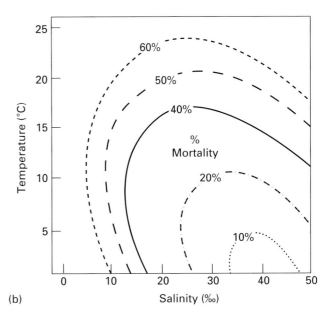

(a)   Salinity (‰)

(b)   Salinity (‰)

Fig. 10.34 Interactions between temperature, salinity and oxygenation in determining mortality of the common shrimp *Crangon*. In (a) where the water is fully oxygenated mortality is minimal at moderate temperatures and with salinities close to and somewhat below full sea water. In (b), with poorly oxygenated water, mortality is much higher but animals survive somewhat better with colder water and at a higher salinity range. (Modified from Haefner 1969, 1970.)

nitrogenous excretion, and altered acid–base balance, all by relatively simple mechanisms. But it may also produce reduced feeding, reduced activity levels and reduced reproductive output. Thus individuals of one species in different parts of a highly zoned shore may be operating very different physiological and life-history strategies; and apparently straightforward interindividual differences in salinity tolerance produce very subtle effects on the ability to penetrate estuarine conditions.

Zonation between species is in part *caused* by the highly vari-

able interacting physiological tolerances of desiccation, heat stress, freezing, oxygen deprivation and reduced feeding time. It is also inevitably influenced by larval preferences and tolerances and larval gregariousness. But above all it is also induced by competition, whether direct (mussels may overgrow other species, while limpets may undercut others) or indirect by more effective resource use; and it may be supplemented by differential predation, especially on the lower shore where fish and crustacean predators are most common.

The limits to distribution and success in brackish-water systems are therefore set by a mixture of intrinsic factors and biotic interactions. A few examples should underline this point. Firstly, Fig. 10.35 shows a general picture of the interacting factors in estuaries; each species is limited at the upper shore by temperature and desiccation, and especially at the upstream end by hyposmotic tolerance; while its lower limit is set by competition with other species. Secondly, a specific case from a shore is

Fig. 10.35 Limiting factors on animal distribution in shores and estuaries, for three representative species. Species A has high osmotic and thermal tolerance and could survive physiologically in the whole rectangle specified by A–A$_1$–A$_2$, but is in practice limited on the lower shore by competition with species B. B has a potential range B–B$_1$–B$_2$ but is only found well within its physiological limits, constrained upwards by this competitive interaction with A and downwards with C. C is able to live downshore to the limits of its physiological tolerance, but cannot extend up the shore where B out-competes it. All species have ranges into estuarine conditions limited by their osmotic tolerance in line with their potential upshore ranges. (From Newell 1979.)

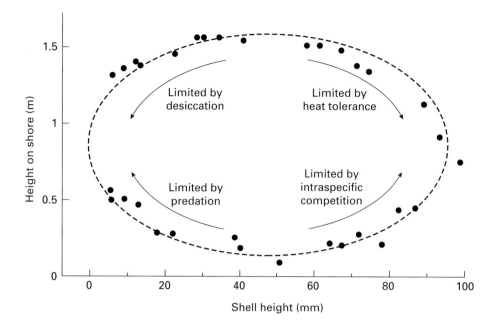

Fig. 10.36 Limiting factors for *Mytilus edulis* on seashores in relation to height of shore and size of shell (approximating age of individual). The dashed line represents the overall distribution limits, and the main limiting factor in each part of the range is shown; thus small mussels are limited on the lower shore mainly by predation, and so on. More details in text. (From Newell 1979.)

Box 10.5 Biotic and abiotic interactions: patterns of occurrence of barnacles

Two barnacle genera (each with several species) occur in the littoral zone around much of Europe and exemplify the interactions of physiological tolerances and biotic factors in determining shoreline zonation (Connell 1961).

Chthamalus is more heat tolerant with a higher upper critical temperature (UCT), and occurs alone on southern European shores, whereas *Balanus* or *Semibalanus* are relatively intolerant of heat and do better in higher latitudes, occurring on their own on the beaches of northern European countries such as Sweden.

At intermediate latitudes the two species co-occur, but with *Chthamalus* on the upper beach and *Balanus* lower down. The tidal height where one takes over from the other shows a latitudinal cline; even within Great Britain we find southern and western beaches that are dominated by *Chthamalus* with a narrow belt of *Semibalanus* low down, while colder beaches in the north and east have a broad zone of *Semibalanus* and a narrow high-tide strip of *Chthamalus*.

---

illustrated in Fig. 10.36 with an analysis of the factors controlling mussel distribution. The upper shore limits are again determined primarily by physiology, and the lower shore limits by competition and predation, but with the exact limits and constraints depending on individual size. Small/young individuals are limited at the top by water loss and at the bottom by predation (especially by starfish on many beaches); larger/older ones are limited at the top by their risk of overheating out of water, and at the bottom by competition with other mussels. The situation on some shores may also change seasonally with a strong impact from transitory populations of migrating predatory shore birds. A third example is the well-known interaction in

Europe of two barnacle genera, *Chthamalus* and *Semibalanus*, shown in Box 10.5.

As a final comment it is worth noting in particular the extraordinary number of cases where quite different littoral and estuarine animals arrive at the same designs and solutions to their problems, a point amply made by a simple comparison of barnacles, mussels, winkles and limpets, the most dominant taxa on so many shorelines. Convergent evolution is spectacularly apparent in these fringe environments; this is an issue that should become even more apparent as we go on to look at more terrestrial environments.

Further reading

Books

Baker, J.M. & Wolff, W.J. (eds) (1987) *Biological Surveys of Estuaries and Coasts.* Cambridge University Press, Cambridge.

Boaden, P.J.S. & Seed, R. (1985) *An Introduction to Coastal Ecology.* Blackie, Glasgow.

Carter, R.W.G. (1988) *Coastal Environments.* Academic Press, London.

Denny, M.W. (1988) *Biology and the Mechanics of the Wave-swept Environment.* Princeton University Press, Princeton, NJ.

Hawkins, S.J. & Jones, H.D. (1992) *Rocky Shores.* Immel Publishing, London.

John, D.M., Hawkins, S.J. & Price, J.H. (eds) (1992) *Plant–Animal Interactions in the Marine Benthos.* Clarendon Press, Oxford.

Little, C. & Kitching, J.A. (1996) *The Biology of Rocky Shores.* Oxford University Press, Oxford.

Moore, P.G. & Seed, R. (eds) (1985) *The Ecology of Rocky Coasts.* Hodder & Stoughton, London.

Newell, R.C. (1979) *Biology of Intertidal Animals.* Marine Ecological Surveys Ltd, Faversham.

Pequeux, A., Gilles, R. & Bolis, L. (eds) (1984) *Osmoregulation in Estuarine and Marine Animals.* Springer-Verlag, Berlin.

Reviews and scientific papers

Borgatti, A.R., Pagliarini, A. & Ventrella, V. (1992) Gill (Na+ + K+)-ATPase involvement and regulation during salmonid adaptation to salt water. *Comparative Biochemistry and Physiology A* **102**, 637–643.

Bridges, C.R. (1993) Ecophysiology of intertidal fish. In: *Fish Ecophysiology* (eds J.C. Rankin & F.B. Jensen). Chapman & Hall, London.

Brinkhoff, W., Stockmann, K. & Grieshaber, M. (1983) Natural occurrence of anaerobiosis in molluscs from intertidal habitats. *Oecologia* **57**, 151–155.

Burnett, L.E. (1997) The challenges of living in hypoxic and hypercapnic aquatic environments. *American Zoologist* **37**, 633–640.

Connell, J.H. (1961) The influence of interspecific competition and other factors on the distribution of the barnacle *Chthamalus stellatus*. *Ecology* **42**, 710–723.

Cronkite, D.L. & Pierce, S.J. (1989) Free amino acids and cell volume regulation. *Journal of Experimental Zoology* **251**, 275–284.

Crowder, L.B., Squires, D.D. & Rice, J.A. (1997) Non-additive effects of terrestrial and aquatic predators on juvenile estuarine fish. *Ecology* **78**, 1796–1804.

Deaton, L.E. (1992) Osmoregulation and epithelial permeability in two euryhaline bivalve molluscs: *Mya arenaria* and *Geukensia demissa*. *Journal of Experimental Marine Biology and Ecology* **158**, 167–177.

Deaton, L.E. & Greenberg, M.J. (1991) The adaptation of bivalve molluscs to oligohaline and fresh waters: phylogenetic and physiological aspects. *Malacological Review* **24**, 1–18.

Denny, M.W. (1995) Predicting physical disturbance: mechanistic approaches to the study of wave-swept shores. *Ecological Monographs* **65**, 371–418.

Gilles, R. (1987) Volume regulation in cells of euryhaline invertebrates. *Current Topics in Membranes and Transport* **30**, 205–247.

Gosselin, L.A. & Chia, F.S. (1995) Characterising temperate rocky shores from the perspective of an early juvenile snail—the main threats to survival of newly hatched *Nucella emarginata*. *Journal of Marine Biology* **122**, 625–635.

Graham, J.B. (1990) Ecological, evolutionary and physical factors influencing aquatic animal respiration. *American Zoologist* **30**, 137–146.

Heble, D.K., Jones, M.B. & Depledge, M.H. (1997) Responses of crustaceans to contaminant exposure: a holistic approach. *Estuarine, Coastal and Shelf Science* **44**, 177–184.

Helmuth, B.S. (1998) Intertidal mussel microclimates: predicting the body temperature of a sessile invertebrate. *Ecological Monographs* **68**, 51–74.

Hofmann, G.E. & Somero, G.N. (1995) Evidence for protein damage at environmental temperatures: seasonal changes in levels of ubiquitin conjugates and hsp70 in the intertidal mussel *Mytilus trossulus*. *Journal of Experimental Biology* **198**, 1509–1518.

Ishimatsu, A., Hishida, Y., Takita, T. *et al.* (1998) Mudskippers store air in their burrows. *Nature* **391**, 237–238.

Lockwood, A.P.M., Sheader, M. & Williams, J.A. (1996) Life in estuaries, salt marshes, lagoons and coastal waters. In: *Oceanography* (eds C.P. Summerhayes & S.A. Thorpe). Manson, London.

McGaw, I.J. & McMahon, B.R. (1996) Cardiovascular responses resulting from variation in external salinity in the Dungeness crab *Cancer magister*. *Physiological Zoology* **69**, 1384–1401.

McMahon, R.F. (1988) Respiratory response to periodic emergence in intertidal molluscs. *American Zoologist* **28**, 97–114.

McMahon, R.F. (1990) Thermal tolerance, evaporative water loss, air-water oxygen consumption and zonation of intertidal prosobranchs: a new synthesis. *Hydrobiologia* **193**, 241–260.

Morritt, D. & Spicer, J.L. (1995) Changes in the pattern of osmoregulation in the brackish water amphipod *Gammarus duebeni* during embryonic development. *Journal of Experimental Zoology* **273**, 271–281.

Murphy, D.J. (1983) Freezing resistance in intertidal invertebrates. *Annual Review of Physiology* **45**, 289–299.

Paine, R.T. & Levin, S.A. (1981) Intertidal landscapes: disturbance and the dynamics of pattern. *Ecological Monographs* **51**, 145–178.

Pannunzio, T.M. & Storey, K.B. (1998) Antioxidant defenses and lipid peroxidation during anoxia stress and aerobic recovery in the marine gastropod *Littorina littorea*. *Journal of Experimental Marine Biology and Ecology* **221**, 277–292.

Ruiz, G.M., Carlton, J.T., Grosholz, E.D. & Hines, A.H. (1997) Global invasions of marine and estuarine habitats by non-indigenous species: mechanisms, extent and consequences. *American Zoologist* **37**, 621–632.

Somero, G.N. (1986) From dogfish to dogs: trimethylamines protect proteins from urea. *News in Physiological Sciences* **1**, 9–12.

Towle, D.W. (1997) Molecular approaches to understanding salinity adaptation of estuarine animals. *American Zoologist* **37**, 575–384.

Treherne, J.E. (1980) Neuronal adaptations to osmotic and ionic stress. *Comparative Biochemistry and Physiology B* **67**, 455–463.

Warman, C.G. & Naylor, E. (1995) Evidence for multiple, cue-specific circatidal clocks in the shore crab *Carcinus maenas*. *Journal of Experimental Marine Biology and Ecology* **189**, 93–101.

Willmer, P.G. (1978) Electrophysiological correlates of ionic and osmotic stress in an osmoconforming bivalve (*Mytilus edulis*). *Journal of Experimental Biology* **77**, 181–205.

Willmer, P.G., Bayliss, M. & Simpson, C.L. (1989) The roles of colour change and behaviour in the hygrothermal balance of a littoral isopod *Ligia oceanica*. *Oecologia* **78**, 349–356.

Zanders, I.P. & Rojas, W.E. (1996) Osmotic and ionic regulation in the fiddler crab *Uca rapax* acclimated to dilute and hypersaline seawater. *Marine Biology* **125**, 315–320.

11 Fresh Water

11.1 Introduction: freshwater habitats and biota

11.1.1 The nature and occurrence of fresh water

'Fresh water' is not really a strictly defined concept, but is usually taken to be any water body of very low salt content, such that it is not detectably brackish. Commonly this corresponds to salt concentrations of between 0.01 and 0.5 ppt, i.e. usually less than 1% of sea water. It makes up only a tiny proportion of the water on the Earth (Table 11.1); about 3% of all water is fresh, and the habitable volume of liquid fresh water is less than 1%, because two-thirds of the total 3% is permanently frozen in polar ice caps and glaciers. Of that 1% free fresh water, much is bound up in underground aquifers or within soils, so that only about 0.1% of all the Earth's water is 'visible' liquid fresh water, as lakes, ponds and rivers. An even smaller percentage of the planetary water is within the biosphere at any one time.

Yet as a habitat fresh water is of great biological interest, because:
1 the high overall water availability, and the continual input of geochemically freshwater run-off from the surrounding land, makes floodplains and river deltas exceptionally productive areas; though they make up only 3% of the terrestrial surface of the planet, they may account for as much as 12% of the 'land-based' (non-oceanic) productivity;

2 the habitats concerned are highly variable, encompassing a great range of types and of tremendous chemical variability, more so than any other type of habitat, so that no two freshwater bodies are ever quite the same;
3 fresh water is an important driving force, cycling minerals and nutrients around the terrestrial environment, via the familiar hydrological cycle (evaporation and precipitation);
4 the habitats impinge on human activity very considerably, since our settlements have always been concentrated alongside rivers and lakes;
5 humans also impinge on the freshwater zones very considerably, so that they are a central cause for concern in environmental study and conservation.

Natural waters come in many forms with very varied physical and chemical characteristics; Fig. 11.1 shows patterns of temperature and pressure and of ionic strength and pH, for atmospheric, surface and subterranean (groundwater and hydrothermal) sources. In essence the surface fresh waters can be divided into moving, or **lotic** waters (rivers, streams and temporary trickles), and still, or **lentic** waters, such as lakes, pools, puddles and rain drops, but also including bogs, fens, marshes, swamps ('wetlands'), and even damp soils, moss cushions, etc. Note that these habitats also differ greatly in permanence (Table 11.2), and that they form a continuum in time and in space

Table 11.1 Distribution of water sources on Earth.

Source	% of total water	Volume (km$^3 \times 10^6$)	Renewal time (vol/vol added time^{-1})
Oceans	97.2	1350	300–11 000 years
Saline lakes	0.008	0.1	1–4 years
Fresh water (total)	2.8	37	
Glaciers and ice caps	2.15	29	12 000 years
Aquifers/ground water	0.62	8.2	60–300 years
Soil water	0.005	0.07	–
Lakes	0.009	0.12	1–100 years
Rivers	0.0001	0.001	2–10 days
Atmospheric water	0.001	0.013	~7 days
Water in biota	0.0001	0.001	–

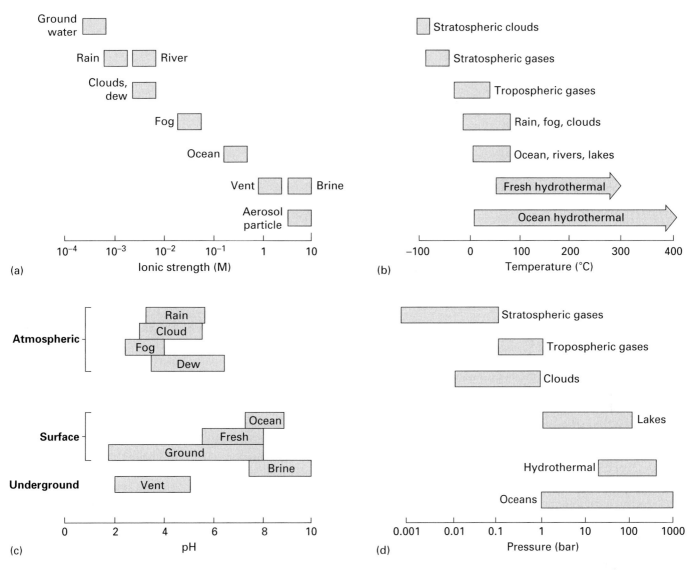

Fig. 11.1 A survey of the chemical characteristics of natural waters (atmospheric, surface and underground), showing (a) ionic strength, (b) temperature, (c) pH and (d) pressure ranges. (From *Atmospheric Change: An Earth System Perspective* by Graedel and Crutzen © 1993 by AT&T. Used with permission by W.H. Freeman and Company.)

Table 11.2 Types and permanence of freshwater habitats.

Moving (lotic) waters	
Rivers Streams Trickles	→ *Decreasing permanence*
Still (lentic) waters	
Lakes Ponds Puddles Rain drops	→
Wetlands: bogs, fens, swamps	*Increasing*
Interstitial zones: soils, moss cushions, etc.	*terrestriality* ↓

through to what we normally think of as 'terrestrial' habitats. Indeed, all surface freshwater habitats have a littoral zone where terrestrial influences increase, and in many cases this undergoes a periodicity of emersion and immersion as water levels change.

11.1.2 Flowing (lotic) waters

Rivers and streams vary considerably along their length, with the discharge (volume per unit time) and current (distance per unit time) interacting with the slope and local geology to determine the water and bed characteristics. It is helpful to divide lotic waters into two categories:

1 permanent (or 'reservoir') rivers, normally with permanent channels and without large flooding events outside the main channel or large seasonal reductions in water level (except in some very large tropical rivers);

2 more temporary 'streams'.

Reservoir rivers show a fairly clear pattern of alternating deep pools and shallow riffles and bars (see Plate 4a, facing p. 310),

Table 11.3 Major river basins of the world and river flow rates. Note that rivers in tropical and subtropical forest areas have much higher discharges than rivers draining similar areas of temperate or dry habitat.

River	Drainage area (km^2 × 10^3)	Discharge (km^3 year^{-1})
Amazon (Brazil)	7000	5500
Mississippi (USA)	4800	560
Congo (Africa)	4000	1800
Parana (Argentina)	3200	730
Nile (Egypt)	3000	90
Mackenzie (Canada)	1800	333
Volga (Russia)	1300	238
Niger (W. Africa)	1100	220
Murray–Darling (Australia)	1100	22
Mekong (SE Asia)	780	4800
Orange–Vaal (S. Africa)	650	12
Colorado (USA)	600	18
Rhine (Europe)	220	70

Table 11.4 Nutrient levels of important rivers. Values are high relative to most lakes, and European and North American rivers are generally higher in nitrate levels than African and South American rivers.

River	Nitrate (µg l^{-1})	Phosphate (µg l^{-1})
Amazon		
Whitewater	4–15	10–15
Clearwater	<1	<1
Blackwater	35	6
Mississippi	700–3000	40–440
Parana	500+	50–100
Nile	10–1000	1–40
Mackenzie	600	16
Volga	50–4000	1–250
Niger	1100–6300	500–3100
Orange–Vaal	300–1400	30–100

giving a range of benthic habitats which are consistently present even though their distribution may be altered by storms and varying erosion patterns. Streams lack this predictability of habitat and may lack any deep pools for most of their lifetime, even drying up completely (Plate 4b).

These flowing waters also vary greatly according to their climatic zone and topography, from torrential cold upland streams (Plate 4c & d) (high current, low discharge) through to warm and sluggish lowland tropical rivers (low current, high discharge). They also differ enormously in the area that they drain. Some of the largest rivers are compared in Table 11.3 where it is evident that large rivers in the tropics with massive basins achieve vastly

greater discharge volumes than any others. Table 11.4 shows the variation in nutrient status that is possible, from rivers rich in nitrate and phosphate and often choked with weeds (Plate 4e) to others that have almost undetectable nutrient levels or are dominated by humic acids, in either case being largely lifeless; indeed, the same river basin may have areas of entirely different nutrient status, as with the whitewaters, clearwaters and blackwaters of the Amazon.

This underlines the fact that in nearly all rivers we tend to find considerable variation along the length of the river bed, a **zonation** phenomenon (Fig. 11.2). This is traditionally classified by the commonest conspicuous animals, inevitably types of fish. In Europe the classic four zones are usually listed as follows:

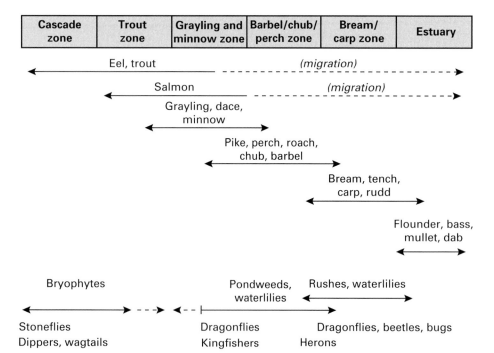

Fig. 11.2 A traditional view of river zonation by major fish zones. Species apply for palaearctic rivers.

Fig. 11.3 The river continuum concept, where zones are seen in terms of the relative abundance of various types of benthic invertebrates that collect, shred or graze on vegetation or act as predators. In (a) the patterns for each category are plotted against river order from origin to mouth (scale of 1–12); in (b) the resulting pattern of particulate matter (from coarse particulate organic matter (CPOM) to fine (FPOM)) is shown, reflecting twigs and leaves being shredded and recycled by invertebrates and microorganisms. (From Horne & Goldman 1994.)

1 The 'trout zone' in the upper reaches, with steep gradients, a rapid current, rocky bed, and well-aerated cool water, where the main inputs of organic matter are external (allochthonous), from leaves falling in from terrestrial vegetation.

2 The 'grayling and minnow zone', with moderate gradients, some pools, and gravelly surfaces.

3 The 'barbel and chub zone', with shallow gradients and plenty of quiet water, and with increasing plant growth giving internal (autochthonous) organic sources.

4 The 'bream zone' in the lower reaches, with slow currents, high summer temperatures when oxygen levels may be low, turbid flow and deeper water with substantial plant growth.

Other fish associated with each zone are shown in Fig. 11.2. Along with this there are associated zonations of other common animals, notably of flatworms, of caddis and of mayfly.

However, this simplistic view has been largely superseded by the concept of a **river continuum** (Fig. 11.3), where zones are classified by the relative abundance of different types of benthic invertebrate, from shredders through grazers to collectors, linked with the coarseness of particulate organic matter present in the water. Associated with this is a 12-point scheme of river 'order'. Upper reaches (low order) are dominated by inputs of entire leaves (plus twigs, fruits, etc.); further down these are gradually shredded into fragments and recycled into smaller

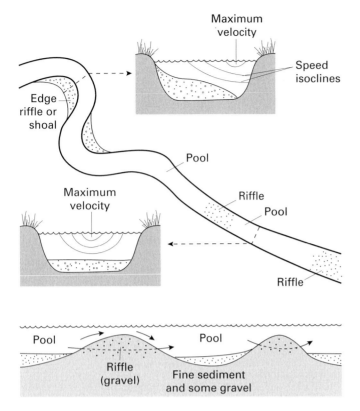

Fig. 11.4 Water flow and the structural patterns of small rivers and streams, with alternating pools and riffles where water passes through and over gravel bars. (From Horne & Goldman 1994.)

particles (FPOM—fine particulate organic matter—in the figure), to be collected by filter- and deposit-feeders in the lower reaches, where such feeders are often sheltered in the gravels of riffles (Fig. 11.4) away from the predators that are important in all parts of the river system. Plant zonation runs from predominantly bryophytes at the upper end, through the first angiosperm pondweeds to appear downstream, then abundant water lilies, rushes and green algae. In broad, slow rivers, vegetation may decrease again at high orders as light is blocked out by dense growths of riverside trees.

The primary cause of this vertical zonation is of course that lotic waters are fundamentally running downhill at all times, through a succession of altitudinal, climatic and vegetational zones, usually with a decreasing gradient, decreasing flow rate and increasing volume from source to sea, as hills flatten out into plains and small stream tributaries coalesce into larger rivers. Linked to this, the primary problem for life in all lotic waters is that they ultimately lead down to the sea; any animal living there has a continuous need to work against the current and maintain its position to achieve a degree of stability of conditions.

Nevertheless, animals in lotic fresh water are going to experience variation in many characters of their environment. Temperature will change, both seasonally (especially at higher latitudes) and daily if the water volume is small or the flow rate

slow. Ion and pH levels will change according to the rate of input from rainfall and patterns of run-off over surrounding soils and rocks, with pH values of 9–10 in chalk uplands and 4–5 in lowland humic forests. Oxygen levels will vary with temperature, with mixing patterns and flow rate, and with the amount of submerged plant growth and of decay. Food availability is likely to be strongly seasonal and floating or swimming food may be difficult to trap. The physical structure of rivers and streams in itself ensures a wide range of differing niches; there are areas of fast current contrasting with slow eddies behind sheltering rocks or trunks, well-lit sunny pools and cool, deep gravel beds, warm transparent upper surface waters and dark protected muddy corners, even completely underground (hypogean) streams. The animals that exploit different parts of lotic water systems therefore tend to be rather specialist in character.

11.1.3 *Still (lentic) waters*

In lentic waters there may no longer be the problems resulting from continual flow, ultimately to the sea, but there are new sets of problems posed by being in a closed volume of water, each pond or lake effectively being an ecological 'island' surrounded by inhospitable land. Problems also arise from the habitat being relatively ephemeral. Table 11.5 shows the area and volume of the world's biggest lakes, but even these are relatively transient on a geological timescale. Again there is also great chemical variability; most lakes have a pH between 6 and 9, but there are also small calcareous upland ponds and highly acid peat bogs well outside these limits, and some naturally acid and alkaline salt lakes (dealt with in Chapter 12).

No particular 'zonation' occurs lengthwise in lakes, but instead there may be other kinds of patterning known as **stratification** in the communities, largely depending on the size of the water body. It is therefore easier to deal with these habitats in an orderly size-related sequence.

Lakes

Lakes are tremendously variable in character, and three main factors determine this, though they interact heavily; these are depth, climate and nutrient status.

Firstly, the depth (or shape, or morphometry) of the lake basin most obviously affects how far from the shore rooted angiosperms can develop. Shallow lakes may have plant growth throughout, with floating and emergent leaves giving total cover; deep lakes have only a tiny marginal fringe of visible plants (macrophytes). Lake depth also affects water mixing patterns and water retention times (the time for all the water in a lake to be replaced); this is commonly 5–10 years for a moderately sized lake, but up to 100 years for Lake Victoria in the East African Rift Valley and much higher still for very deep lakes, such as Tanganyika and Baikal.

The second important factor is climate, especially temperature. In some areas lakes may be well mixed vertically by winds,

Table 11.5 The world's largest lakes. Note that Tanganyika and Baikal have very large volumes due to great depth rather than area.

Lake (location)	Area (km² × 10³)	Volume (km³ × 10³)	Salinity
Caspian Sea (Russia/Iran)	374	78.2	Marine/brackish
Lake Superior (USA/Canada)	82.1	12.2	Fresh
Lake Victoria (E. Africa)	68.5	2.7	Fresh
Aral Sea (Russia/Kazakhstan)	64.1*	1.0	Brackish
Lake Huron (USA/Canada)	59.5	3.5	Fresh
Lake Michigan (USA)	57.7	4.9	Fresh
Lake Tanganyika (E. Africa)	32.9	18.9	Fresh
Lake Baikal (Russia)	31.5	23.0	Fresh
Great Bear Lake (Canada)	31.3	3.4	Fresh
Great Slave Lake (Canada)	28.6	2.1	Fresh

* Aral Sea now much reduced from this area (see text).

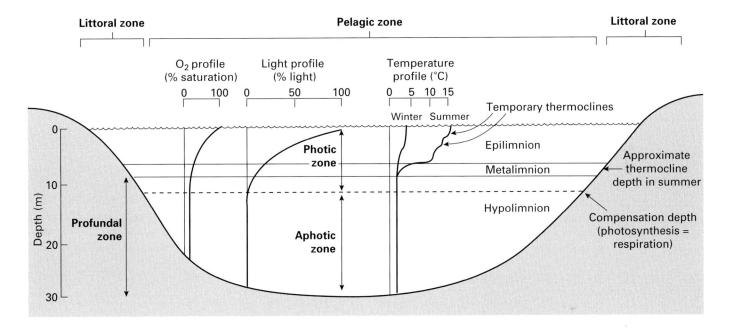

Fig. 11.5 Patterns of stratification in lakes, showing the deep hypolimnion and superficial epilimnion in relation to temperature profiles and oxygenation.

but particularly in seasonal (mid-latitude) climates the lake separates into two zones demarcated by a sharp thermocline (thermal stratification), with the **epilimnion** above and **hypolimnion** below (Fig. 11.5). The **thermocline** normally forms in spring and disappears again in autumn (the 'autumn overturn'); the level at which it forms is sometimes termed the **metalimnion**. There may also be an associated **oxycline**, with the deep hypolimnion being unmixed and gradually losing all its oxygen. This is also linked to a distinction between the photic zone (where light penetrates and photosynthesis is possible) and the deeper aphotic zone (where only respiration can occur); the compensation depth, at which the boundary between these two zones occurs, varies with the strength of the Sun (hence with both latitude and season) and with the water transparency,

determined largely by nutrient inputs. Because of all these climatic factors, many lakes can be clearly separated into a pelagic zone where variation is substantial and a profundal zone where conditions are physically and chemically uniform, each zone having its own characteristic biota.

Figure 11.6 shows the resultant different patterns of mixing. In seasonal climates where stratification patterns change predictably once a year, lakes are described as **monomictic** (mixing once). In seasonal continental biomes, there may be inversions of stratification in winter, with the warmest and densest water at 4°C at the bottom, covered by colder water grading up to the surface ice layer. Such lakes have both spring and autumn mixing periods, in opposite directions, and are termed **dimictic**. Very high-latitude lakes may be permanently ice-covered and therefore **amictic**, whilst tropical forest lakes mix rarely but rather unpredictably and are termed **oligomictic** or **polymictic**.

Climatic region, and hence the prevalence of winds, is also the main factor producing variation between lakes by the effects of

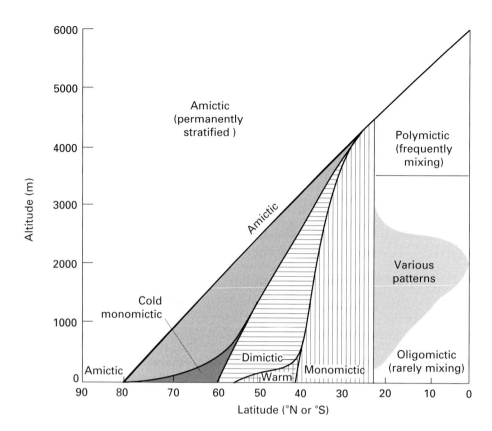

Fig. 11.6 Types of stratification and mixing in lakes in relation to latitude; high-latitude lakes are usually amictic, with progressively more mixing events occurring annually at lower latitudes. Note that increasing altitude usually also reduces mixing except in the tropics where montane lakes are often polymictic. (From Moss 1980.)

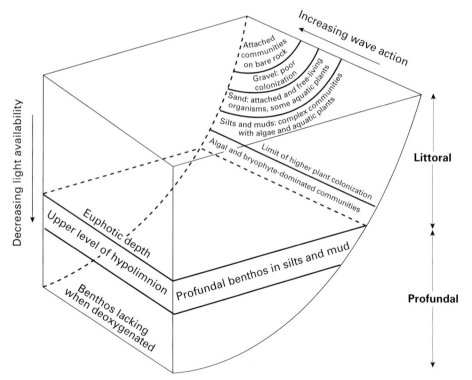

Fig. 11.7 Littoral and profundal habitats in lakes, and the effect of increasing wave action on the littoral characteristics. (From Moss 1980.)

wave action. The extent of wave scouring affects the horizontal patterning of the littoral communities particularly (Fig. 11.7), with the variation from muds and silts on sheltered shores to gravel and bare rock on windswept shores paralleling that found on marine shores (Chapter 10).

The third key factor determining lake biology is that of nutrient level. Two main categories are recognized, termed oligotrophic (very low in nutrients) and eutrophic (very high nutrient levels), with mesotrophic as intermediate.

Oligotrophic lakes are usually deep and clear, have low

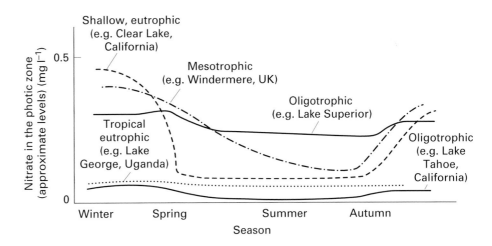

Fig. 11.8 Typical patterns of nutrient status (as nitrate availability) in different kinds of lakes through the annual cycle. (From Horne & Goldman 1994.)

nutrient levels, low ionic conductivity, and few algae. Characteristic examples include most of the upland waters in northern Europe and North America: Wast Water and Buttermere in the English Lake District are well studied cases, as are Lake Tahoe and Lake Superior in North America, and Lake Baikal (the oldest and deepest freshwater lake) in Siberia. Such lakes often arise due to glacial action, forming a long deep bed between hard rocks such as granites (see Plate 5a, facing p. 310). Their plant life is primarily floating plankton; but where even this is very sparse there may be some rooted vegetation at surprising depth, because of good light penetration through the clear water, giving an extended 'photic zone'. Biological production in an oligotrophic lake is usually mainly in the top 5–10 m. The hypolimnion usually stays oxygenated, so benthic mud-dwellers are characterized by *Orthocladius* chironomid midges (these are green rather than red, lacking the haemoglobin pigment found in midges from hypoxic habitats), and the characteristic fish are salmonids. In such lakes, as in no other freshwater systems, the physiological problems of depth (see Chapter 9) may become relevant; for example, in the extreme case of Lake Baikal some animals such as gammarid crustaceans occur at 1300 m. There may be a high diversity of shallow-water animals, but these are restricted to the lake fringes. The lake water is cold in winter, but then as the daylength and warmth increase there may be a spring burst of phytoplankton, before the lakes stratify in late spring. These changes will occur later if the lake surface has frozen (which is rare in Britain, but common in Scandinavia, Russia and Canada). In even more Arctic regions, there is a sufficiently long freeze to reduce the winter oxygen level substantially, so that such lakes cannot support any fish stocks at all.

Eutrophic lakes are often shallow, have high nutrient levels (especially high phosphate and nitrate concentrations), high ionic conductivity, and abundant phytoplankton and/or macrophytes (Plate 5b). This condition is often associated with a large drainage area. In Britain many of the Norfolk Broads are classic examples, as is Derwentwater in the Lake District; in North America, California's Clear Lake and Cayuga Lake in New York State are well known. These lakes are often not deep enough to show much stratification, but deeper examples do get a

deoxygenated hypolimnion in the summer; the classic indicator organism is therefore *Chironomus*, a red midge larva containing haemoglobin and thus surviving hypoxia. Often there is a high biomass of animals, but with low diversity. Such lakes may tend to be rather transient unless they are managed, gradually giving way either to more terrestrial wetland habitats or (especially when they have become eutrophic due to artificial addition of nutrients over a relatively short timescale) to unpleasantly smelly anoxic and effectively dead wastelands.

Obviously there are many intermediate types of lakes (Plate 5c), and an overview of the seasonal changes in nutrient status for classic lakes in each major category is shown in Fig. 11.8. There are also a few special types of lakes that are worth considering.

Large lowland tropical lakes

Many of these occur in the Rift Valley of Africa, and were formerly known as 'Rudolf', 'Albert', etc. but they now have more appropriate African names. These show permanent stratification into upper warm and lower cool regions, because of high temperatures and low wind so that mixing is minimal. Anything that dies falls through the thermocline and never gets recycled, so there is a huge nutrient-rich sediment. Thus wherever the water is moderately shallow the lakes may be dominated by flamingos, filtering the rich mud. There is some recent and current debate about forcibly mixing these lakes to release this nutrient and improve fish stocks. But the lakes do already have highly speciose fish communities; they are high-stability environments, so that groups such as cichlid fish (*Tilapia*, etc.) have had time to partition their niches very finely. Lake Malawi has 200+ species, most of them endemic. Mixing the lakes up could give short-term benefits, but would risk destroying all this diversity in the longer term.

Tropical mountain lakes

These are especially found in the Andes (such as Lake Titicaca), and in the high plateau of Ethiopia. They show a continuous cycle of stratification, layers forming by day when the Sun heats the surface water and mixing at night when it is cold and windy; they are truly polymictic (cf. Fig. 11.6). Often therefore these lakes are very productive indeed, and the fish are an important food source for many surrounding communities.

Crater lakes

These occur throughout the volcano belts of the world, when water fills the often doughnut-shaped cavities in the top of extinct eruption sites (Plate 5d). The volcano walls ensure that the water is very protected, therefore the waters do not mix much. The water itself is nominally 'fresh' but often with high levels of sulphur and other unusual chemicals, commonly within a layer of denser quite salty water at the bottom, so that there may be intriguing and unique communities of animals present. Crater lakes have varying patterns of stratification and may exhibit seasonal or sporadic overturns. These can occasionally be disastrous, releasing sulphurous fumes to the surroundings, as in the case of Lake Nyos (in Cameroon, West Africa) in 1986, when fumes from the overturning lake killed hundreds of people and thousands of cattle locally, as well as most life in the lake itself, largely due to CO_2 and SO_2 release.

Underground lakes and aquifers

Hypogean lakes occur in many parts of the world where caves are common, especially in limestone regions, and aquifers accumulate above the relatively impervious rock strata. The waters here may be strongly mineralized, and of course are usually lightless with no real plant growth; they may undergo marked seasonal drops of level and become hypoxic, stranding some animals on underground shores and in channels that become aerial. Inhabitants of these waters are invariably rather specialist, often eyeless and completely unpigmented, often with elongated sensory appendages, and sometimes with a trend to persistent larval morphology (paedogenesis).

Ponds

Ponds are defined loosely, only by their smaller size than lakes; in general, they are subject primarily to convective mixing rather than wind-stirring. They are invariably rather shallow, but may have highly stratified waters. Biological patterns therefore tend to be similar in kind to lakes, but much more rapidly changing—small ponds are notoriously prone to become eutrophic and anoxic very quickly. In temperate zones such small water bodies usually need management (especially periodic weed removal) to stay in a constant and productive state.

Ponds do tend to have a much more obvious surface fauna than lakes, because they are usually more sheltered from wind. Animals such as pond-skaters, water bugs and beetles occur, as do floating duckweeds and larger macrophytes.

There is a much greater likelihood of a pond totally drying up

in summer droughts, so the life cycles of resident species tend to be shorter, for both plants and animals, with each individual only experiencing a part of the total climatic cycle. There is nearly always a resistant stage in the life cycle, encysted and able to withstand extreme conditions, and commonly also suited for dispersal to another pond.

Wetlands

Wetlands lack the structured character of most lentic waters, being characterized by shallow fresh water with little thermal or density stratification and little current. The water and underlying soils are often anoxic, permitting the growth of only specialist plants. The wetlands therefore tend to be classified by the main vegetation present, though this may be spatially heterogeneous. Temperate examples include habitats such as: **fens**, with reed and sedge and willow floras; **swamps**, with emergent trees; and **bogs**, with *Sphagnum* moss and cottongrass often dominant. Strictly, fens are dependent on underground water and tend to be alkaline, whilst mires or bogs depend only on rainfall and are acidic, though the terminology is often confused in the vernacular. Productivity (Table 11.6) is usually in the order marshes >

Table 11.6 Productivity of freshwater habitats and freshwater plant communities. (Compare Table 10.1 for brackish and marine examples.)

Plant or habitat type	Primary production (g C m^{-2} year^{-1})
Plant types	
Open-water phytoplankton	10–3000
Submerged macrophytes	
Tropical	500–1700
Temperate	300–1300
Floating macrophytes	
Tropical	4000–6000
Temperate	100–1500
Lakes	
Oligotrophic	1–50
Eutrophic	400–600
Wetlands	
Marsh	
Oligotrophic	1000
Eutrophic	6000
Polluted eutrophicated	22000
Swamp	
Alder/ash	570–650
Reed	1500–3700
Papyrus	6000–9000
Bog	
Oligotrophic	100
Eutrophic	1800
Fen	
Eutrophic	350

swamps > fens > bogs; in all cases most of the primary production is not directly eaten by herbivores but instead cycled into the detritus pathway.

These temperate wetlands are perhaps best known and valued as vital areas for migrant birds and native coastal birds, and as spawning grounds for fish; recently they have tended to be well studied, both for ecological and hunting interests. There are also tropical wetlands, including papyrus swamps in Africa and Asia, and tree swamps such as the Florida Everglades and the wetlands of northern Australia, with areas of open water interspersed with dense stands of swamp-cypress, freshwater mangroves, and sedges, plus a range of insectivorous plants, vast numbers of insects, and turtles and alligators, together with important populations of grazers such as manatee, tapirs and capybara. However, the ecology of many such areas is almost unknown; the hotter swamps are impenetrable and difficult to monitor and are also major centres for diseases such as bilharzia, malaria and yellow fever. In some parts they have been exploited for turtle, caiman and crocodile hunting (for the carapace and skin trades).

The term 'wetlands' also encompasses a range of more specialist categories, notably:
• Riverine forests of the upper Amazon, particularly the 'blackwaters' coloured by high humus contents, with low pH and low nutrient levels, and a highly specialist fauna and flora.
• Floodplains of very large rivers, producing a highly movable littoral zone; the floodplain of the lower Amazon may be up to 100 km across, and is dotted with small lakes.
• Rice paddies in Asia (see Plate 10a, facing p. 310), totally managed communities with monoculture vegetation, but vast enough in many parts of the world to have a strong influence on the fauna, and particularly the birdlife, of surrounding areas.

11.1.4 Freshwater biota

Table 11.7 summarizes the taxa that are abundant in fresh water, and those that are largely or entirely absent (cf. Table 9.1, which also includes a few rare taxa). In all fresh waters, invertebrates tend to dominate the benthos and fish dominate the open water. Rotifers (in huge numbers and with tens or even hundreds of species) and small crustaceans (especially copepods and cladocerans or 'water fleas') are usually numerically dominant in the filtering zooplankton, though in streams blackfly larvae are important specialist filterers. Other insects are common, the majority having only the immature phases in the aquatic habitat, with relatively short-lived aerial adults (mayflies, dragonflies, caddis flies, alder flies, mosquitoes, blackflies, etc.), but amongst the bugs and beetles there are many permanent freshwater species with raptorial diving adults. Also common are planarian, nematode and oligochaete worms (the latter especially in colder regions), amphipod crustaceans, chironomid fly larvae, and bivalve and gastropod molluscs; these groups tend to dominate the deep benthic ('profundal') communities, and tubificid annelid worms and chironomid larvae may reach densities of

Table 11.7 Relative success in freshwater habitats for different taxa (in terms of number of species).

	Very abundant	Moderate	Absent
Algae	Chlorophytes	Other algae	
Plants	Angiosperms	Bryophytes	Conifers
Animals	Crustaceans (ostracods, cladocerans)	Planarians	Echinoderms
	Rotifers	Bryozoans	Cephalopods
	Nematodes	Bivalves	
	Oligochaetes	Tardigrades	
	Gastropods	Other vertebrates	
	Insects		
	Teleosts		

the order of 10^4 per m^2. Fly 'midge' larvae (chironomids and chaoborids) dominate in muddy and sandy/gravelly sediments, with particular species occurring in specific sediment types and at specific depths (varying predictably even between successive riffles and bars in riverine habitats). Some specialist groups also live in the substrata of transient water bodies which may dry up completely; these animals (especially rotifers, tardigrades, nematodes) can undergo 'cryptobiosis' during drought or freezing, and are dealt with as a special case in the next chapter.

The higher trophic levels in fresh water are mainly occupied by fish, especially teleosts; the salmon, pike and perch taxa provide some of the larger species worldwide, with minnows and their kin in smaller size ranges, while cichlids are particularly important in more tropical biomes. Some sharks and rays also penetrate far up rivers, as we saw in Chapter 10. Amphibians routinely live in and around fresh water, as do some reptiles (the alligators and their kin, some water snakes and turtles) and many waterfowl, from flamingos to dippers and kingfishers. Beavers, muskrat, mink, coypu, otters, rats, shrews and voles are the most familiar temperate amphibious freshwater mammals, with less familiar specialists such as desmans and platypus occurring more locally. There are also some riverine specialists, such as hippopotamus, in warmer biomes. All of these are strongly associated with rivers yet they inhabit a macroenvironment effectively remote from the typical river variables such as dissolved oxygen, thermoclines and water turbulence. However, a few mammals are more strictly freshwater in their lifestyles; manatees and river dolphins are perhaps the most truly aquatic freshwater mammals, and Lake Baikal (Siberia) and Lake Iliamna (Alaska) both have species of freshwater seals.

Surprisingly, there is no very clear pattern here; certainly the reasonably large animals endowed with hard coverings (arthropod groups, molluscs, vertebrates) do very well, but then so do some conspicuously soft creatures, such as flatworms, and some very small ones, such as rotifers. There are some obvious freshwater specialists, notably amongst crustaceans, rotifers and teleosts. Only about 3–5% of all insect species are aquatic, but that still represents a considerable abundance and diversity. The main 'failures' are echinoderms (starfish are marginal survivors in estuaries, but no echinoderm seems ever to have lived in fresh waters), tunicates and cephalopods (again never found in fresh or even brackish water throughout their known evolutionary history). Most sponges and cnidarians are excluded, but there are a few successful species such as *Spongilla* and the hydroids *Craspedocusta* and *Hydra*. It is quite hard to find good general reasons for these exclusions (see section 11.9).

11.1.5 *Overall strategies for life in fresh water*

In principle, fresh water poses the same kinds of problems as brackish water but with the hazards more constant through time (lacking a tidal influence) and somewhat accentuated in degree. The medium is always very dilute and puts a great strain on osmoregulatory mechanisms; animals can no longer be true osmoconformers, since cell contents can *never* be as dilute as fresh water or the cells will die. Nor will behavioural and avoidance techniques be of much use, since there is nowhere to go where the osmotic stress is reduced, and no possibility of

short-term shut down until 'better' osmotic conditions return. However, on the positive side the limited need to adapt to osmotic *variation* means that enzyme polymorphisms can be reduced, and a single suite of adaptive responses to give adequate efficiency at ion uptake, volume regulation, and cellular osmoregulation may suffice.

In large lakes, and especially for animals living within the hypolimnion, other factors are likely to be relatively constant within the lifetime of an individual. But in smaller bodies of lentic water, and in flowing waters, variation in other parameters does become a real problem. Obviously the temperature can be highly variable with the season, and there may be strong thermoclines. Transience of the liquid water as a habitat therefore also becomes a problem; temperatures may vary to the point of evaporating away the water body completely in summer and freezing it solid in winter. Oxygen levels can similarly be very variable, and the water may go completely stagnant and anoxic in parts. Ion levels and pH will vary drastically with run-off patterns and with freak inputs. Light levels and hence photosynthetic rate and food availability can vary drastically with the season (especially due to shading from plants on the margins).

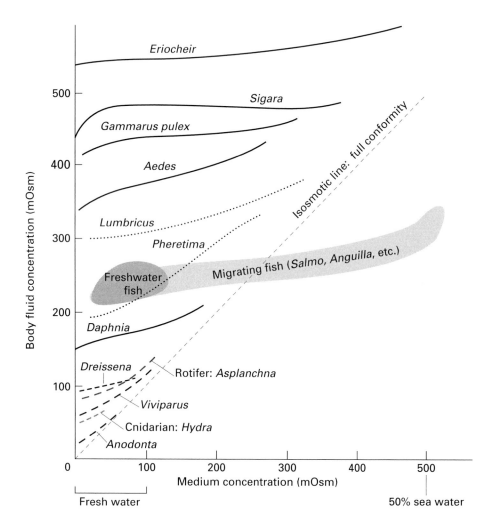

Fig. 11.9 The relationship between internal body fluid concentration and external medium, for various freshwater animals. Arthropods are shown by solid lines, molluscs by dashed lines and worms by dotted lines, with teleost fish represented by the tinted area. Cf. Fig. 10.13.

For all these reasons, reproduction also becomes very tricky; the gametes and young stages, with higher surface area to volume (SA/V) ratios, will find life even harder than adults. It is perhaps not very surprising that freshwater life is dominated by relatively few and rather specialist taxa.

11.2 Osmotic and ionic adaptation and water balance

Freshwater animals face the central problem of a permanently dilute external medium, with sodium, potassium and calcium levels often below 1 mM, with a permanent gradient for ion loss out of their bodies and a net inward osmotic flux of water, so that they must continuously counteract a tendency to become diluted and to swell up. They may also face 'extra' osmotic problems if they live in the littoral lentic or shallow lotic freshwater zones, where periodic emersion is likely on unpredictable timescales (unlike the regular emersion–immersion cycles in the marine littoral zone, dealt with in Chapter 10).

Figure 11.9 shows the classic internal medium vs. external medium plot, this time for a range of fully freshwater invertebrates. Freshwater lake and river dwellers are all capable of osmotic and ionic regulation, as it is impossible to maintain functioning tissues at these continuously low concentrations. Two factors do vary, and tend to be interrelated: the level of body fluid concentration maintained in freshwater (itself commonly 0.1–5 mM), and the tolerance range. Thus the Chinese mitten crab, *Eriocheir*, is very euryhaline, as it can penetrate deeply up rivers but can also live in the sea, to which it must return to breed; even in its freshwater phase it maintains a blood concentration at least two-thirds that of sea water. In contrast, *Anodonta*, a river mussel, is a strictly stenohaline freshwater inhabitant and will not survive above about one-tenth of seawater concentration, having its blood concentrations a little less than one-tenth sea water. The bivalve *Dreissena* is similarly restricted, dying at around 50 mM external NaCl concentrations. Many freshwater bivalve molluscs, rotifers, sponges and cnidarians show this pattern, while gastropods, annelids and insects tend to be intermediate between this and the crustacean mode. Body fluid compositions of typical examples are shown in Table 11.8.

Despite the variation in degrees of hyper-regulation, all freshwater animals (and all terrestrial animals making a secondary return to fresh water) are clearly regulators to some degree, and all have continuous water balance problems because water will always tend to flow into their bodies from the very hyposmotic surrounding fluid.

For all these animals, all the mechanisms described in relation to shores and estuaries in Chapter 10 still apply; i.e. they exhibit:
- reduced permeability;
- ion uptake mechanisms;
- cellular osmoregulation with small osmotic effectors.

But there is also a new mechanism in most freshwater animals:
- regulated hyposmotic urine concentrations.

Table 11.8 Composition of extracellular fluids in freshwater animals.

Animal group/ genus	Concentration (mM)			Osmotic concentration (mOsm)
	Na+	K+	Cl–	
Sponges				
Spongilla				55
Cnidarians				
Chlorohydra				45
Rotifers				
Asplanchna	21	7		81
Molluscs				
Anodonta	16	0.5	12	66
Margaritifera	14	0.3	9	40
Pomacea	56	3	52	139
Viviparus	34	1	31	80
Annelids				
Lumbricus	76	4	43	300
Pheretima	43	7	54	152
Hirudo	136	6	36	200
Crustaceans				
Astacus	208	5	250	477
Eriocheir	309	6	280	636
Potamon	259	8	242	522
Asellus	137	7	125	–
Insects				
Aedes				266
Sialis	109	4	199	436
Ephemera				237
Vertebrates				
Carassius	142	2	107	280
Salmo	161	5	120	320
Rana	92	3	70	210
Alligator	140	4	111	278
Anas	138	3	103	294

11.2.1 Permeability

The permeabilities to water and to sodium (P_w and P_{Na}) for some freshwater animals were included in Table 10.4. The values for both invertebrates and vertebrates tend to be much lower than for marine animals, but there will still be substantial osmotic gain of water and diffusional loss of ions, due to the relatively large gradients involved. The larva of a mosquito (*Aedes*) may gain 3% of its total body water volume per day osmotically, and for the crayfish *Astacus*, with more concentrated blood, the figure may be nearer 5%.

Amphibians are well known for their relatively high permeability, but also show an unusual ability to control P_w, particularly in the region used for water uptake known as the 'pelvic patch', a section of ventral abdominal skin. When dehydrated, or when the bladder is empty, toads in the genus *Bufo* produce the hormone arginine vasotocin, increasing P_w in this area. A second hormone set, the renin-angiotensin system, controls the water uptake rate in the same patch.

11.2.2 *Ion uptake*

Ion uptake is commonly centred on the skin, or on the gills in fish and in invertebrates such as crustaceans and insects (these often having secondary abdominal or anal gills). Rates (as J_{max}) can be quite high (Table 11.9) compared with marine and brackish-water fauna, with affinities (K_m values) substantially lower (cf. Table 10.5). The rate of salt uptake tends to be variable according to the acclimation history; thus the freshwater crustacean *Gammarus zaddachi* is capable of faster uptake after exposure to 0.3 mM Na^+ than to 10 mM Na^+ solutions (Fig. 11.10a). The freshwater crayfishes have been studied in some details, their gill epithelia showing what is now regarded as the typical pattern for freshwater uptake epithelia (Fig. 11.10b). An apical V-ATPase pumps H^+ ions out of the cell, providing a steeper electrochemical gradient for Na^+ entry (so that Na^+/H^+ exchanges are indirectly coupled), and a more direct Cl^-/HCO_3^- exchange also occurs; overall the effect is an apical electroneutral (1 : 1) ion exchange. This is followed by active basal ion transport via the sodium pump, the counter-ions (H^+ and HCO_3^-) being provided by CO_2 from the body fluids. There may be some interrelation with ammonia levels, but this appears to work via acidification effects on gill permeability to NH_3 rather than via direct involvement of NH_4^+ ions.

The essential processes of uptake and regulation in fish gills were considered in Chapter 10, for brackish/freshwater species. The linkage between CO_2 and ammonia excretion by fish gills is also complex (Fig. 11.11). Ammonia probably leaks passively from the gill down favourable blood-to-water gradients, but this is augmented by the carbonic anhydrase-catalysed hydration of CO_2, generating H^+ ions that again help to trap the ammonia as NH_4^+. There are thought to be H^+-ATPases in this epithelium rather than Na^+/H^+ exchangers, and these would assist ammonia excretion by actively extruding more H^+ ions. Note that this is somewhat different from the situation in marine fish (see section 9.10.1), where direct diffusion of NH_4^+ (as Na^+/NH_4^+ exchange) is more likely given the much higher cationic permeability of marine gill surfaces.

Adaptation may involve not only changes in the abundance or rate properties of uptake sites in the membranes, but also, on a grosser level, morphological change in the uptake surfaces, as we saw for the anal gills of brackish-water mosquitoes in Fig. 10.17. Morphological modification to the ion uptake system may also occur 'in reverse', where freshwater invertebrate taxa have some capacity to invade more saline habitats. For example, in *Aedes* mosquito larvae the species that are obligate freshwater dwellers (about 95% of the genus) have a single

Table 11.9 Maximum sodium uptake affinities (K_m) and rates (J_{max}) in freshwater animals.

Animal group/genus	K_m (mM)	J_{max} (mM kg^{-1})
Crustaceans		
Mesidotea entomon	2–3	21
Gammarus duebeni	0.40	20
Gammarus lacustris	0.15	
Gammarus pulex	0.15	
Astacus	0.07	0.05
Annelids		
Lumbricus	1.3	0.1
Hirudo	0.14	
Molluscs		
Lymnaea	0.25	
Margaritana	0.04	
Fish		
Carassius	0.26	0.2
Salmo	0.50	0.4
Amphibians		
Ascaphus	0.07	0.05
Hyla	0.14	0.09
Rana	0.20	0.27
Bufo	0.25	0.48
Xenopus	0.05	
Brackish species general range (see Table 10.5)	0.4–20 (higher)	5–15 (similar range)

rectal segment, whereas euryhaline species have a more complex two-segment pattern. In fact a degree of secondary salt tolerance has evolved at least twice in the mosquitoes: some species osmoregulate using this two-part rectum with both resorptive and secretory cells, while other lineages tolerate salt as osmoconformers.

11.2.3 *Osmolytes and cellular regulation*

Strictly freshwater invertebrate animals cannot afford to accumulate large amounts of intracellular organic osmolytes, in the manner of brackish/estuarine animals (section 10.2.2), so that cellular osmotic regulation by variation of levels of free amino acids is relatively unimportant to them. Instead they tend to regulate their volume using movements of potassium ions from cytoplasm to extracellular fluid, thus reducing osmotic intake and resultant swelling, allowing a suitable electrochemical balance to be maintained. For example, volume regulation in the bivalve *Dreissena* fails if potassium is unavailable, and the ideal ratio of K^+ to Na^+ in the surrounding medium is around 0.01.

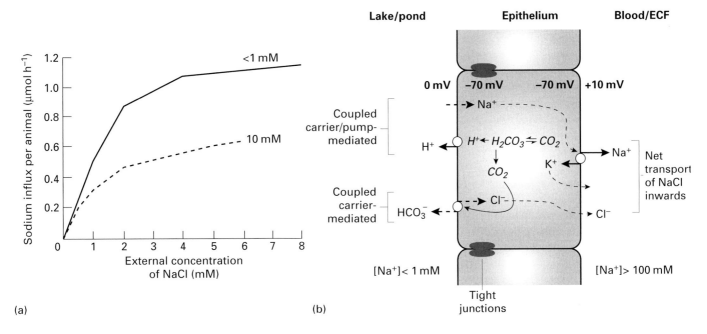

(a)

(b)

Fig. 11.10 Freshwater ion uptake. (a) Acclimation effects: the rate of uptake in the freshwater shrimp, *Gammarus zaddachi*, previously acclimated to very low salinity and to slightly higher salinity. (b) The basic pattern of ion uptake in freshwater gills or skins. ECF, extracellular fluid.

Fig. 11.11 The interaction of ion transport, CO_2 movement and NH_3 excretion across a freshwater fish gill. RBC, red blood cell. (Adapted from Randall & Daxboeck 1984.)

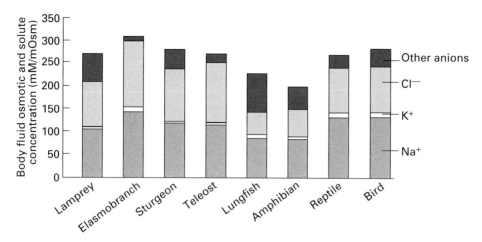

Fig. 11.12 Extracellular fluid composition in a range of freshwater vertebrates. (Adapted from Withers 1992.)

However, freshwater animals that can tolerate a degree of increased salinity do tend to synthesize amino acids as intracellular effectors in brackish media; in the freshwater shrimp *Macrobrachium*, amino acids appear in the haemolymph at 21 or 28 ppt, while gill Na^+/K^+-ATPase levels decrease as the rates of ionic uptake are down-regulated. The levels of amino acids are regulated by endocrine factors from the eyestalks and from neurohormonal cells in the thoracic ganglion. The osmoconforming mosquitoes mentioned earlier also use the trick of accumulating compensatory osmolytes in high-salinity media, in this case incorporating trehalose and proline especially (both these osmolytes also being used in insects for other purposes).

Vertebrates in fresh water routinely have a total osmotic concentration of around 250–400 mOsm, and Fig. 11.12 shows how the extracellular fluid is made up in various taxa of fish. In teleosts and elasmobranchs Na^+ and Cl^- are dominant, while in lungfish and in the 'primitive' fish, such as sturgeon and lampreys, there is a moderate organic component. Intracellularly the patterns are similar but with K^+ inevitably replacing sodium as the main cation. Freshwater reptiles, birds and mammals operate in similar fashion.

11.2.4 Hyposmotic urine

Most marine and brackish-water invertebrates have unmodified, isosmotic urine. Brackish-adapted animals that invade fresh water, such as the crab *Eriocheir*, retain this pattern. But most freshwater forms, with a continuous influx of water through their external surfaces and also usually an unavoidable input of fresh water as they feed, show the additional strategy of recovering ions from their urine to leave it distinctly **hyposmotic** to themselves (urine/blood ratio < 1), as shown in Table 11.10.

The mechanism is essentially the same as ion uptake in gill or skin epithelia, but involves ion resorption from the urine filtrate back into the body. Classic examples of the structures involved are the **flame cells** (or, when collected together in groups, **protonephridia**) shown in Fig. 11.13. These are found in various forms in most 'lower' freshwater invertebrates, including flatworms, rotifers and nemerteans. A single flame cell has an intracellular invagination into which 30–90 flagella project. The surrounding wall has slit-like perforations, where flame cell meets duct cell, and these slits act as the filter. Flagella draw the isosmotic fluid through these slits and then drive it on down the duct, a narrow fast-flow zone where further cells have typical fluid transport structure (see Fig. 4.12), and effect the resorption of ions sufficiently quickly that the water cannot follow proportionately and a progressively hyposmotic fluid is left in the duct lumen.

Other examples of **freshwater kidneys** are shown in Fig. 11.14. The exact classification of these in anatomical or evolutionary terms is still uncertain, but gastropods have a nephridium whose functioning is similar to the basic protonephridial system, though with initial fluid formation by ultrafiltration from the heart cavity (pericardium) or coelom. The crustacean antennal gland (Fig. 11.14b) involves similar filtration and resorption systems, with up to 95% of the initial filtrate being resorbed in the distal tubule in a crayfish (though in amphibious crabs

Table 11.10 Urine concentrations and flow rates in freshwater animals.

Species	Flow rate (ml kg^{-1} h^{-1})	Concentration (mOsm)	Urine/blood ratio
Annelids			
Pheretima		45	0.19
Crustaceans			
Orconnectes	1.3		
Pseudotelphusa	0.13	120	
Eriocheir		800	1.04
Gammarus	21		
Molluscs			
Viviparus	6	30	
Anodonta	19	25	
Fish			
Carassius	14		
Salmo	5		
Lampetra	7		
Esox		30	0.09

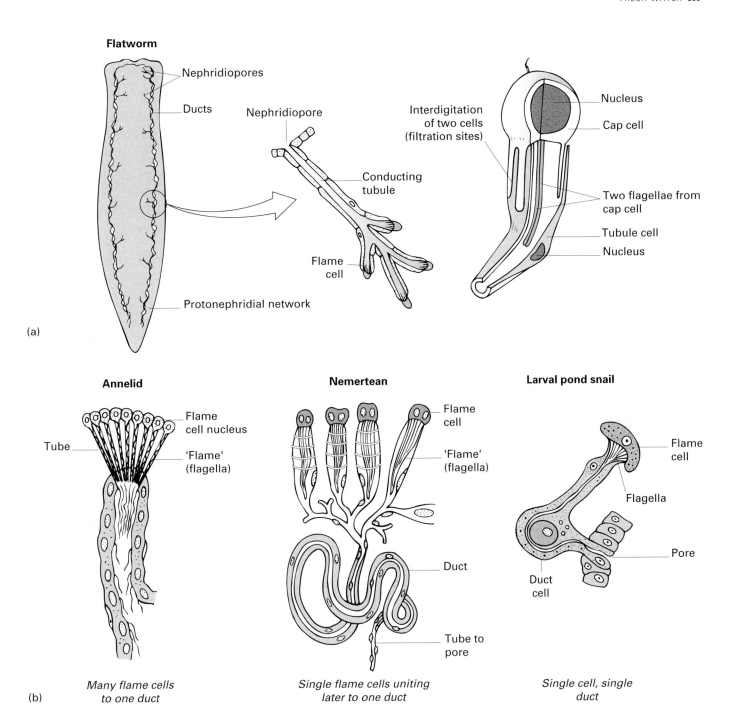

Fig. 11.13 Flame cells, the excretory organs of many smaller freshwater invertebrates: (a) shows the general structure, and (b) shows examples from major freshwater taxa.

further urine reprocessing occurs in the gill chamber when they are living in air—see section 13.2.4). Leeches are an unusual case, living predominantly in fresh water but also occasionally moving around on land and having the additional problem of intermittent blood meals. Their primary urine is formed in multiple paired nephridia tubules by a combination of ultrafiltration from the blood system and a secretory process based on chloride transport from special canalicular cells in the upper tubules (analogous to the secretory urine production system of insects), driving transcellular K^+ flow and paracellular Na^+ flow. The lower tubule then normally resorbs 85% of Na^+ and 97% of K^+, to leave a very hyposmotic urine. However, after a blood meal the flow rate increases due to an up-regulation of paracellular flow, switching the fluid from a K^+-rich to an Na^+-rich balance, and ion resorption is reduced, to give an 80-fold increase in NaCl output and clear excess volume as quickly as possible. Control of this switching is exerted via nephridial nerve cells, releasing a small peptide; these are directly inhibited after blood-feeding by the

Mollusc

Crayfish (crustacean)

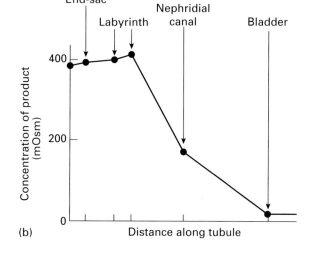

Fig. 11.14 Excretory organs in (a) molluscs and (b) crustaceans from fresh water.

presence of extra Cl⁻, so that the nerves stop firing and peptide release is blocked.

Freshwater insects retain their terrestrially adapted Malpighian tubule system, but often switch to excretion of high percentages of ammonia rather than uric acid. In mosquito larvae (*Aedes*) the secretion rates of fluid, and of sodium and potassium, are controlled by 5-hydroxytryptamine (5HT), and elevation of haemolymph K⁺ also causes increased fluid secretion rates, as do increased salinity and increased feeding rates; but in all these cases urine composition is relatively unaffected.

In vertebrates the main uptake sites are the kidney tubules, and in some cases (e.g. turtles) also the urinary bladder. The urine is invariably hyposmotic to the blood (Table 11.10). In fish and amphibians, the kidney usually contains some nephrons with a glomerulus and some that are aglomerular, gathering fluid from the coelomic cavity (though the proportion of aglomerular nephrons is usually less than in marine fish). In

most cases about half the primary filtrate is resorbed and only half therefore reaches the bladder; but as much as 99% of the filtered ions may be resorbed; Fig. 11.15 shows the functioning of a freshwater fish nephron, and gives an overview of the salt and water balance for the whole animal.

For the relatively few fish that can survive in both sea water and fresh water, such as eels, salmon and blennies, urinary water loss is much increased in fresh water; glomerular nephrons are more abundant and the glomerular filtration rate rises sharply to permit this flow control. For amphibians, filtration rate is highly con-trolled in relation to the hydration state of the animal, largely by the hormone arginine vasotocin. In all animals **urine volume** must be equivalent to the daily osmotic uptake of water, and is therefore high in freshwater fauna. Marine invertebrates may have urine production as low as 0.05 ml kg⁻¹ h⁻¹, but freshwater species (Table 11.10) may achieve 2–20 ml kg⁻¹ h⁻¹ (see above in relation to permeability), although in the relatively few animals that continue to produce isosmotic urine, such as the mitten crab, *Eriocheir*, this volume can be much lower. The same pattern applies in vertebrates, though here the freshwater range is 2–25 ml kg⁻¹ h⁻¹.

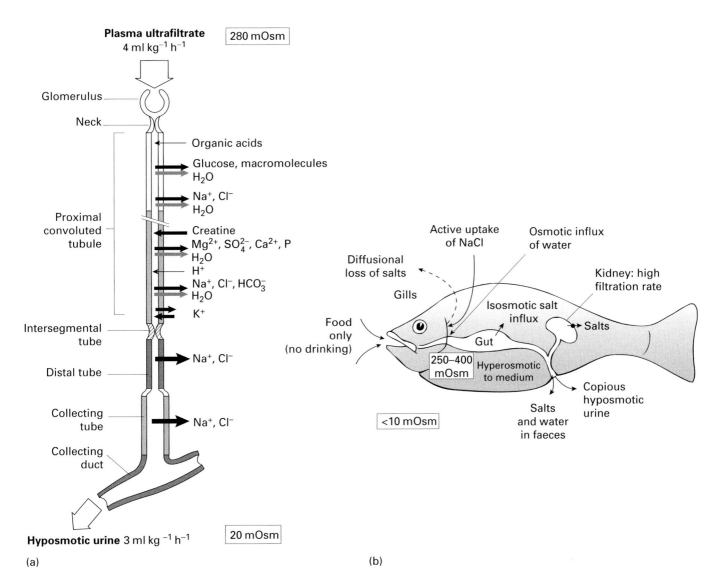

Fig. 11.15 (a) Functioning of a single nephron in a freshwater fish; and (b) an overview of salt and water balance in the whole animal. (a, Adapted from Hickman & Trump 1969.)

11.2.5 *Other ionic problems*

Regulation of ions other than sodium and chloride can be crucial in freshwater bodies which may be individually peculiar in composition. In particular, exposure to high or low pH regimes causes new problems for water and ion regulatory systems, and here analysis has been advanced in view of the problems posed by anthropogenic acidification.

In general, acidic habitats cause problems by disrupting electrolyte balance in both invertebrates and vertebrates, primarily by inhibiting active sodium uptake and increasing diffusional sodium loss. In freshwater clams (*Anodonta*) acute exposure to acidic low pH water produces both low pH and much reduced P_{O_2} in the blood, but this is redressed over a few days, apparently by mobilizing bicarbonate from the shell. Control of

acid–base balance in more active animals can be achieved by varying ventilation patterns to change the availability of CO_2, which provides the counter-ions for NaCl uptake, giving a degree of automatic feedback. The Australian crayfish, or 'yabby', (*Cherax*) can survive in both acid and alkaline water, but becomes hypometabolic in both, with O_2 uptake reduced by at least 40–55%.

In Chapter 10 we noted that in fish, varying salinities lead to changes in the degree of exposure of the branchial 'chloride cells'. Similarly in freshwater species, alkalosis causes the surface area of these cells to increase by withdrawal of the surrounding pavement cells, enhancing the rate of Cl^-/HCO_3^- exchange and so removing basic products from the blood. In acid waters, the chloride cells are covered over by the pavement cells and HCO_3^- is retained.

Magnesium levels may also need particular regulation in freshwater animals, and Mg^{2+} is mainly obtained from food, with gill uptake as a secondary route. Fish can reabsorb Mg^{2+} in the kidney; but in magnesium-deficient fresh waters they can also minimize their losses, increase their intestinal uptake, and mobilize some Mg^{2+} from hard tissues (a reservoir containing around 50% of the body's Mg^{2+} pool).

It should be noted that crustaceans and insects face the additional problem of having to moult and yet maintain osmotic and ionic homeostasis. In the crayfish, freshwater uptake generates the physical force needed to shed the old cuticle and expand the new one, leaving the animal with abnormally dilute haemolymph and calcium deficiency. There is therefore a period of intensive postmoult branchial (gill) NaCl uptake, with additional branchial uptake of calcium and bicarbonate using a Ca^{2+}-ATPase and Ca^{2+}/Na^+ exchange. In insects, the gut helps as a reserve for ions and water during the moult.

11.2.6 Problems of aerial exposure

The remaining water-balance problem for some freshwater animals is that of coping with periodic emersion, which in seasonal droughts may last for weeks or even months. Where water levels merely become lowered, all but the most sessile of animals can retreat down the shore; but where ponds dry up completely, more drastic moves (migration to other ponds, deep burrowing and a period of anaerobic dormancy, or use of a resistant life cycle stage) may be needed. Only rarely do freshwater animals have to face real drought. Freshwater bivalves are one example, where motility is not an option and dispersal is difficult. Many in this taxon have extensive capacities to withstand long emersions, controlling the rate of water loss using valve movements; during emersion Na^+ and Cl^- levels in the blood are tightly regulated. All ammonia production ceases, suggesting they rely on non-protein metabolism, and respiratory acidosis is again compensated using carbonate stores from the shell.

Aerial exposure may also compromise the normal aquatic pattern of nitrogenous excretion using ammonia. Freshwater insects, with a terrestrial ancestry, tend to use allantoin and allantoic acid, apparently unable to switch 'back' to ammonia excretion but at least finding a moderately soluble alternative to uric acid. In lungfish, the African genus *Protopterus* excretes ammonia when in water, and switches to urea when in its terrestrial cocoon phase, while the much more aquatic Australian genus *Neoceratodus* never achieves a high rate of urea production and though it gulps air when necessary it cannot survive in air for any substantial period.

11.3 Thermal adaptation

The patterns of temperature profiles and of heat cycles in lakes are summarized in Fig. 11.16, indicating that problems are very different for animals in the stable hypolimnion compared to those in the seasonally variable epilimnion. In flowing waters thermal problems may be compounded on a short timescale by variation in discharge and volume, hence thermal capacity; and on a longer scale by seasonality and (potentially) latitudinal/altitudinal movement of organisms with the water flow.

Thermal strategies therefore depend critically on the character of a particular freshwater habitat. An additional indirect problem related to thermal variation for all freshwater animals, whether ectotherm or endotherm, is that higher water temperatures promote bacterial action, so there can be drastic and rapid chemical changes in the water, compounding the problems for the rest of the biota. This is considered further in section 11.8.1.

11.3.1 Freshwater ectotherms

The thermal relationships of freshwater ectotherms, as with all aquatic ectotherms, are greatly constrained by the high thermal conductivity and high specific heat capacity of water, the lack of radiative heat gain through much of the habitat, and the impossibility of evaporative heat loss. Heat dissipation at any large surface area such as gills is so rapid that the animal inevitably has a body temperature (T_b) virtually identical to water temperature. Moreover, there can be much more variety of water temperature within the lifetime of a freshwater organism than for organisms in the marine and estuarine habitats considered in earlier chapters.

Biochemical adaptations

Given this variation, the central biochemical adaptations for a freshwater ectotherm, as discussed in Chapter 8, are likely to be: (i) broad-spectrum enzymes whose properties are little affected by thermal change, and/or (ii) several morphs of each enzyme, with a capacity to switch between them seasonally or as appropriate.

In many freshwater animals, different enzyme morphs are expressed in different seasons, and for key enzymes such as the sodium pump Na^+/K^+-ATPase there may be substantial change in protein density in the uptake epithelia and in nerves and muscles; for example, in a teleost fish, the roach, ouabain binding sites increase nearly two-fold in winter. Swimming performance in goldfish and carp is modified following acclimation to both low and high temperatures. Myosin-ATPase activity increases after cold acclimation, and the ratio of different isoforms of the myosin light and heavy chains is also altered, influencing the muscle shortening speed. Thus muscles from cold-acclimated fish can activate, shorten and relax more quickly at low temperatures. In other animals, particularly the salmonid fish, enzyme changes are more limited and control seems to be exercised more by altering membrane fluidity, with an exaggerated homeoviscous response (see Chapter 8).

Behavioural adaptations

The only exceptions to the general rule that T_b matches the water temperature occur for some motile animals that can exploit local thermal niches in shallow waters, using a limited degree of behavioural thermoregulation. Tadpoles may select warmer water at the edges of ponds, and may shuttle in and out of

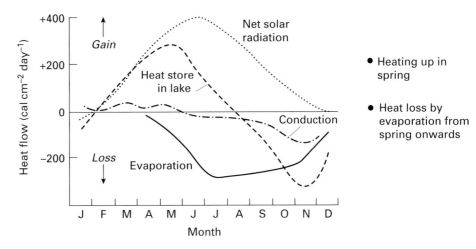

Lake Mendota, Wisconsin (temperate, dimictic)

- Heating up in spring
- Heat loss by evaporation from spring onwards

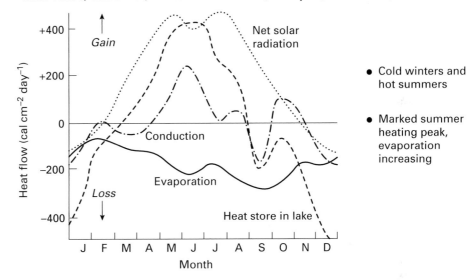

Lake Tahoe, California (Mediterranean climate, subalpine, monomictic)

- Cold winters and hot summers
- Marked summer heating peak, evaporation increasing

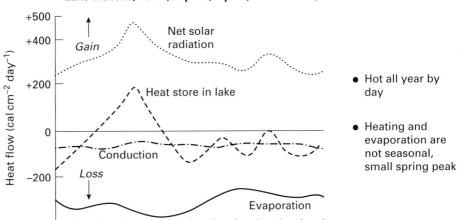

Lake Titicaca, Peru (tropical, alpine, monomictic)

- Hot all year by day
- Heating and evaporation are not seasonal, small spring peak

Fig. 11.16 Heat cycles in three contrasting lakes. The temperate lake has a single annual cycle of temperature, while tropical lakes have more irregular heat storage patterns, and higher and more continuous evaporation. (From Horne & Goldman 1994.)

the warmer waters to maintain a fairly constant T_b. Certain small freshwater fish, and *Hyla* tadpoles, clearly show 'basking' behaviours in these shallow water-edges, orientated to the Sun; and the latter may also aggregate to get warmer. Many semi-aquatic animals also use basking as a sophisticated control of T_b; alligators will generally move into shallow water to warm up, but may also expose part of their backs to the air to allow drying, and if the ambient temperature (T_a) is high enough they will leave the water and bask on the river bank. Water snakes also bask out of the water, especially in the early morning, returning to water around midday; when basking, the species *Nerodia sipedon* has a T_b of 26.3 ± 0.7°C, much lower than expected for this body mass by passive warming in the sun, indicating some thermoregulation to avoid overheating.

Although most aquatic ectotherms cannot have a T_b deviating from the surrounding water temperature, most do still show thermal preferences (T_{pref} or 'eccritic temperatures') wherever there is a gradient of available temperature, for example from deep water to shallow or from midstream to unstirred edges. Preferred body temperatures for a range of freshwater ectotherms appear in Table 11.11, showing clear relations with habitat temperatures. Very subtle changes in preferred temperature occur in some fish, the tropical catfish showing daily patterning (but rather surprisingly to a higher tolerance nocturnally), apparently arising from a control via the light-sensitive pineal gland.

Given the availability of choice of temperatures, and an ability to exploit this, it is not surprising that overall thermal tolerance polygons are generally larger for freshwater animals than for marine ones (cf. Table 8.10). Variations in the size of the polygon for different activities for a freshwater trout are shown in Fig. 11.17. The comparison of upper lethal temperatures against time of exposure in Fig. 11.18 also shows a clear correlation with latitude and habitat for seven species of freshwater fish.

Amphibious animals face varied rates of cooling according to

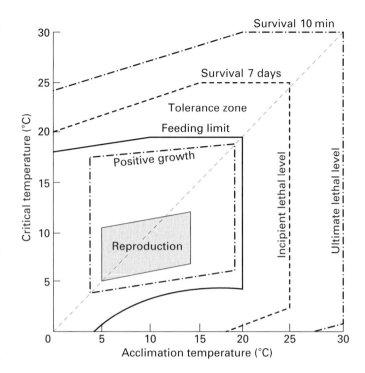

Fig. 11.17 Thermal tolerance polygons for the brown trout (*Salmo trutta*) for various activities. (Adapted from Elliott 1994 and other sources.)

Table 11.11 Preferred body temperatures (T_{pref}) for freshwater ectotherms.

	T_{pref} (°C)
Fish	
New Zealand species	16–27
Bullhead	26
Carp	22–28
East African tilapia	20–32+
Amphibians	
Salamander (juvenile)	25
Bullfrog (juvenile)	24–30
Bullfrog (adult)	22–28
Reptiles	
Box turtle	21–25
Painted turtle	29–35
Alligator	32–35

Fig. 11.18 Upper lethal temperature for varying times of exposure in freshwater fish species. Fish were acclimated at 20°C. (Adapted from Brett 1956.)

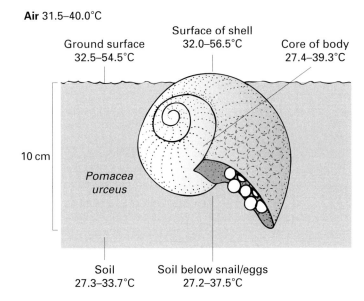

Air 31.5–40.0°C

Ground surface
32.5–54.5°C

Surface of shell
32.0–56.5°C

Core of body
27.4–39.3°C

10 cm

Pomacea urceus

Soil
27.3–33.7°C

Soil below snail/eggs
27.2–37.5°C

Fig. 11.19 Temperature balance in a freshwater swamp snail, during aestivation at the mud surface. The snail's body and its eggs remain below 40°C though the shell surface may reach 56°C. (From Little 1990; adapted from Burky *et al.* 1972; courtesy of Cambridge University Press.)

whether they are in air or water. Small invertebrates that become exposed in summer drought commonly enter aestivating states, and the temperature balance of a snail in this condition is substantially alleviated by partial burial in the pond mud (Fig. 11.19). For larger amphibious animals, there may be variation between rate of heating and cooling within the aerial environment, for example in the freshwater turtle *Mauremys*, suggesting cardiovascular changes similar to those in the iguana (see Fig. 13.25). The turtle raises its T_b by basking in water or in air, and lowers it by immersion.

Coping with freezing

Freezing is a significant problem for temperate freshwater ectotherms, especially in small lotic bodies, though it is worth remembering that even in moderate to high latitude lakes the bottom water probably never freezes and may stay at 4°C where water has its greatest density, protected by insulating snow and ice above. Freshwater animals tend to show little accumulation of colligatively active solutes, and supercool only moderately (–5 to –7°C compared with –10 or –20°C in many terrestrial invertebrates from similar latitude). They are susceptible to inoculative freezing and must therefore rely on not encountering any ice. This is why some species in temperate latitudes seek out larger ponds as winter approaches. Ponds may also be safer than soils in temperate zones; hatchling turtles leave their soil burrows where they nested after birth and seek a pond, since in the soil they are more at risk of inoculative freezing. Certain species of turtle can survive a degree of freezing, however, at least for a few hours.

Freshwater invertebrates in high latitudes may suffer seasonal freezing, and whilst many may migrate away from a freezing front in a stream, or burrow deep enough to avoid actual freezing, whole communities of insect larvae (especially midges) do become frozen within the gravel in Canadian and Alaskan streams, using freeze-tolerance mechanisms to survive (see Chapter 8).

11.3.2 *Freshwater endotherms*

Freshwater endotherms are limited to the secondarily invading birds and mammals. Most of these are amphibious, swimming on the water surface (ducks, etc.) or diving beneath the water in search of food, though the manatees (sea cows) and river dolphins are permanently aquatic. There are no partially endothermic freshwater fish as we met in the marine environment (section 9.3.3), nor are there any known cases of freshwater insect endotherms (heterotherms) as with the terrestrial bees, beetles and moths covered in section 13.3.2.

Endothermic mechanisms for freshwater tetrapods are essentially unmodified from the classical patterns used on land, dealt with in principle in Chapter 8. The main problem for an aquatic endotherm is not with heat generation but with heat retention, the animal being surrounded by a much more conductive medium that is below the body temperature, into which heat is rapidly dissipated. Insulation layers, in the form of fur and feathers, therefore become critical. This raises the problems of wetting, which normally reduces the insulating value of such layers substantially by replacing the trapped air with water as well as compressing the fur or feathers (Fig. 11.20). The common solution in birds is to possess oil glands whose product is spread over the feathers to reduce their wettability, so that an air layer

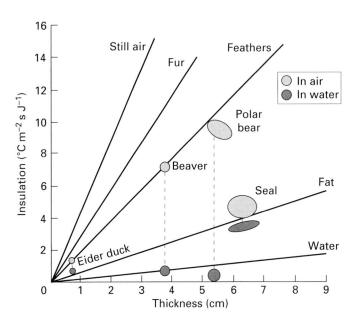

Fig. 11.20 Insulation and the effects of water immersion on fur properties in some aquatic endotherms. (Adapted from Scholander *et al.* 1950.)

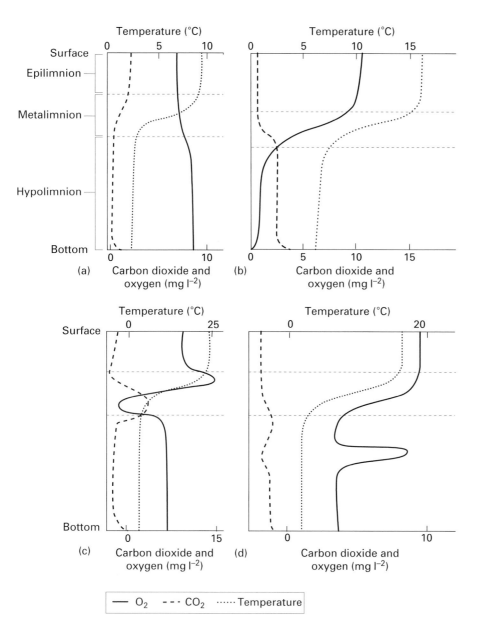

Fig. 11.21 Oxygen and carbon dioxide profiles in different kinds of lake. (a) An oligotrophic lake. (b) A temperate eutrophic lake. (c) A heterograde curve in a relatively unstirred lake where algae just above the thermocline raise oxygen levels and bacteria just below it reduce oxygen. (d) The anomalous pattern that results locally from inflow of dense, cool oxygen-rich stream water that forms a discrete layer. Note that carbon dioxide levels are often inversely related to oxygen levels. (From Horne & Goldman 1994.)

remains trapped against the skin; this has a substantial buoyant effect, useful during surface swimming but increasing the energetic cost of diving somewhat. Mammals may achieve a similar insulating effect with an underfur that retains an air layer, or may use the normal dense fur to trap a layer of stagnant water that at least reduces convective heat loss (the 'wet suit' effect). However, insulation does almost invariably decrease during a dive: in the eider duck, insulation due to feathers in air is about 1.6°C m^{-2} s J^{-1}, but only 0.83°C m^{-2} s J^{-1} in water.

Most freshwater endotherms use countercurrent exchangers in their extremities (feet or flippers) to reduce heat loss, and in cold climates the membranes of tissues in these extremities in animals such as ducks and waders incorporate lipids of short chain length to maintain fluidity—these have melting points up to 30°C lower than the lipids in the body core. Because of the cooling effect of a dive, some semi-aquatic mammals, such as muskrats, also use brief periods of non-shivering thermogenesis in the body core (NST; see section 8.7.2) to warm up after diving or prolonged swimming.

11.4 Respiratory adaptation

Although water contains relatively little oxygen compared with air, in principle fresh waters hold more oxygen than salty ones (see Table 7.2). Often, therefore, small freshwater animals have little problem in gaining oxygen. Equally, they can lose carbon dioxide readily because of its high solubility. It is common to find oxygen and carbon dioxide levels as mirror images of each other in lakes (Fig. 11.21), the pattern of each depending strongly on temperature.

Table 11.12 Composition of gases in freshwater bodies.

	O_2 (ppm)	CO_2 (ppm)
Normal fresh water (surface)	8–10	0.02–0.04
Normal fresh water (100 m depth)	8–10	0.02–0.04
Reed swamp	1–6	3–18
Guinea-grass swamp	0.2–1.2	8–9
Water-hyacinth swamp	2–7	0–9

But the difficulties for respiration arise from the variability of freshwater habitats. Whilst large oligotrophic lakes may stay nearly saturated, several factors can reduce the available oxygen in the water column of rivers and eutrophic lakes, including:

1 Seasonal cycles of productivity and vegetation decay, producing anoxia in the hypolimnion and potentially supersaturation in the epilimnion. For streams, ponds and swamps there may even be a diel pattern of productivity-related oxygen levels.

2 Raised temperature (although somewhat compensated for by increased speed of diffusion at high temperature); in spring and summer lakes may lose up to 50% of their oxygen simply due to temperature change.

3 Prolonged freezing: below the ice all the oxygen may get used up and it is not renewed until the spring melt.

4 Flow patterns where rainfall is highly seasonal.

5 Human interference.

In addition, conditions in the benthic muds of still waters, and in most wetlands such as marshes and swamps, tend to be rather anaerobic (Table 11.12), so that there are whole communities of specially adapted animals there.

Just as with estuarine animals, coping with the variation in oxygen levels can be done at several levels; a simple response shown by many invertebrates and by amphibians is to move to cooler water when hypoxia threatens (the 'behavioural hypothermia' response). Thus frogs preacclimated to 4°C shift their preferred temperature from 6.8 to 1.9°C when also exposed to hypoxia. If hypoxia cannot be avoided by behaviour, animals may respond by either **conforming** or **regulating**. Figure 11.22 includes examples from the mayflies: *Baetis* is a classic conformer, and *Cloeon* a near-perfect regulator over most of the Po_2 range, largely achieving this by variations in the frequency of gill movement. Both it and the crayfish represent the common condition of regulation above some fixed point (P_c; see Chapter 7) and conformity below it, and the values of P_c are lower for most freshwater animals than for either open-water marine or terrestrial species (Table 7.1; cf. Fig. 7.3).

The most obvious and widespread solutions to variable oxygen supply are three-fold:

1 to expand or elaborate the respiratory surface;

2 to use higher-affinity oxygen-storing pigments;

3 to modulate the ventilatory and/or circulatory rates.

Some of these features may be varied on an ontogenetic,

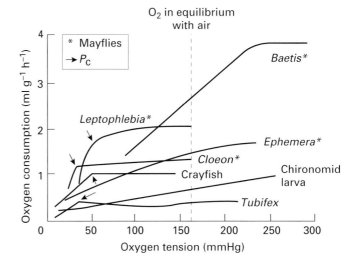

Fig. 11.22 Oxygen consumption in relation to environmental Po_2; conformers and regulators amongst freshwater annelids, insects and crustaceans.

seasonal or daily basis. Alternatively, some freshwater animals opt out of the problem and instead (either intermittently, or permanently if they have terrestrial origins) breathe from the more reliable aerial oxygen supply. A relatively small number of freshwater animals are able to withstand quite prolonged periods of anoxia, as might be experienced when ponds dry out or freeze over.

11.4.1 *Respiratory surfaces*

In freshwater invertebrates gills are the most common adaptation to ensure oxygen uptake, and these can be elaborated from almost any part of the body. Most freshwater soft-bodied invertebrates are derived from marine ancestors already endowed with aquatic respiratory systems, and most annelids and molluscs retain the cutaneous exchange surfaces, or serially repeated filamentous gills or tentacular crown gills, or enclosed lamellate gills, of their relatives. In specialist benthic animals, such as oligochaetes (e.g. *Tubifex* worms), the head is buried in mud and the skin of the posterior half of the body is particularly well vascularized. The wetland oligochaete *Alma* has a particularly deeply grooved tail with dense vascularization. Use of different exchange sites in different conditions also occurs. For example, the common pond snail, *Lymnaea stagnalis*, has 25% cutaneous uptake at low Po_2, but this can rise to 50% as Po_2 rises. The snail *Biomphalaria* has both gills and lungs.

Freshwater crustaceans have equally inherited an efficient gill system from marine ancestors, and the branchial surfaces are generally borne on the legs, either free in more primitive forms, where they may be ventilated by leg movements (as in cladocerans and amphipods), or as complex feathered structures enclosed within a branchial chamber and ventilated by the scaphognathite ('baler') in the decapods (crabs and crayfish).

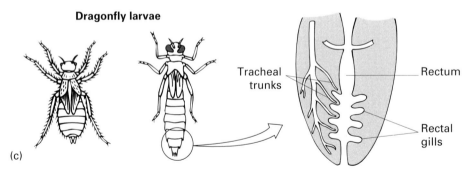

Fig. 11.23 Respiration in freshwater insects. (a) Tracheal gills in damselfly and stonefly. (b) Spiracular gills in an aquatic fly larva. (c) Rectal gills in dragonfly. (Adapted from Hinton 1957 and other sources.)

For freshwater insects that are fully aquatic (as larvae or throughout the life cycle), some modification to the aerially adapted tracheal system is needed, and this usually means the addition of cuticular gills. Many larval forms have tracheal gills (Fig. 11.23a) where extensions of the body surface contain a dense network of tracheae (seen on the abdomens of stonefly, mayfly and caddis fly larvae, and as long tails on many damselfly larvae). Other aquatic insects, particularly fly larvae, have

spiracular gills (Fig. 11.23b), where a tube-like structure extends from a spiracle, or rectal gills (Fig. 11.23c), as in dragonfly larvae, with tracheoles investing folds of the rectal surface. Other insects, especially those that are amphibious, have achieved yet other respiratory devices to breathe from air stores when under water (see section 11.4.4).

Freshwater vertebrates may use cutaneous or gill-based respiration, or be reliant on aerial supplies via lungs. Skin-breathing

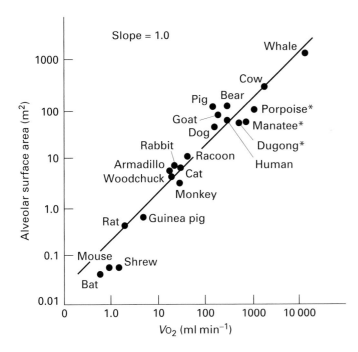

Fig. 11.24 Relationship between alveolar surface area and oxygen uptake in mammals; note that the three freshwater species (*) have somewhat higher uptake rates than predicted from their lung structure.

is quite common in freshwater fish larvae, and many eels and catfish rely on it extensively to sustain their resting metabolic rate. Swamp-dwelling fish often develop a form of air-breathing; the bichir (*Polypterus*) and the true lungfish in the genera *Protopterus*, *Lepidosiren* and *Neoceratodus* all show the development of lungs from the swim-bladder, while less specialized developments of vascularized swim-bladders or gut regions occur in some catfish and bowfins, and in the giant Amazonian fish *Arapaima*. Some salamanders also show a high dependence on cutaneous oxygen uptake, and certain adult frogs have highly vascularized and folded areas of skin, or even areas of 'hairy' skin, that contribute a large part of their oxygen requirement, only using their lungs when active or if exposed to hypoxic surroundings. But most fish, and juvenile or neotenous amphibians, use gills, and nearly all adult amphibians and reptiles, birds and mammals (being secondarily aquatic) rely on air-breathing through lungs. Figure 11.24 shows that while alveolar surface area for mammals is strongly related to oxygen uptake rate (scaling with a mass exponent of 1.0, for all but the smallest mammals), the surface area in freshwater mammals (porpoise, manatee, dugong) produces a rather higher uptake rate than predicted.

11.4.2 *Respiratory pigments*

Increased exploitation of pigments as oxygen carriers and stores is very apparent in common freshwater invertebrate communities, where many more of the species are red or green in general colour than is evident in marine communities. Obvious

examples are the tubificid annelids and the chironomid midge larvae ('bloodworms') living in the muddy benthos. Pigment characteristics, as always, vary with habitat and lifestyle. For example, in the surface-active amphipod *Gammarus*, oxygen consumption rises linearly with oxygen availability, whilst for benthic *Chironomus* worms the shift to full O_2 consumption (full activity) occurs at less than 20% O_2 saturation, its haemoglobin having a characteristically left-shifted saturation curve (Fig. 11.25). In the water flea (*Daphnia*) hypoxia leads to an increase in Hb concentration, which is directly correlated with its swimming ability in oxygen-depleted waters.

Pigment adaptation is particularly evident in the haemoglobins of larger crustaceans and fish with active lifestyles, as might be predicted. In crayfish regulation is achieved by increased haemocyanin affinity. In the blue crab, *Callinectes*, hypoxia is accompanied by an increase in the amount of circulating haemocyanin, but also a shift in the haemocyanins in the blood, with more of the high-affinity 1×6-meric oligomer and less of the normal 2×6-meric oligomer. The properties of the haemocyanin also vary in relation to both temperature and ionic strength of the blood (see Chapter 7). Fish species from rivers where seasonal flow changes may result in very different oxygenation regimes often show rather similar patterns, with seasonal adjustments in the expression of different haemoglobin fractions, and also in haematocrit—red blood cell (RBC) abundance. Goldfish acclimated to different temperature regimes show complex responses involving erythropoiesis (formation of new RBCs), loss of existing RBCs, and division of circulating juvenile RBCs, thus adjusting the abundance of haemoglobin isomorphs without greatly affecting overall haematocrit and blood viscosity. Mobilization of stored RBCs from the spleen (or in the case of some cave-dwelling fish from the liver as well) provides another safety valve.

11.4.3 *Ventilation and circulation*

Increased ventilation rate is usually the first and quickest response to changing oxygen demand in freshwater animals, whether this arises from environmental hypoxia or from metabolic activity. In small animals, such as sponges and rotifers, ventilation is mainly by cilia or flagella, whose activity increases as needed. In invertebrate bottom-dwellers it may involve wiggling of the whole body, as in tubificid worms, or localized bursts of gill activity, in many insect larvae. Molluscs and crustaceans tend to have baling systems and use either an increase in rate or in stroke volume, or both. Crayfish show clear hyperventilation initially in low Po_2, but as hypoxia sets in the animals show bradycardia, and the circulatory flow patterns alter to give increased flow anteriorly over the brain.

Freshwater vertebrates show relatively sophisticated ventilatory responses to hypoxia. Fish employ modulation of both stroke volume and frequency (by varying the buccal and opercular pumping patterns) to match oxygen uptake at gills to demand, as detected by oxygen receptors, usually sited in the

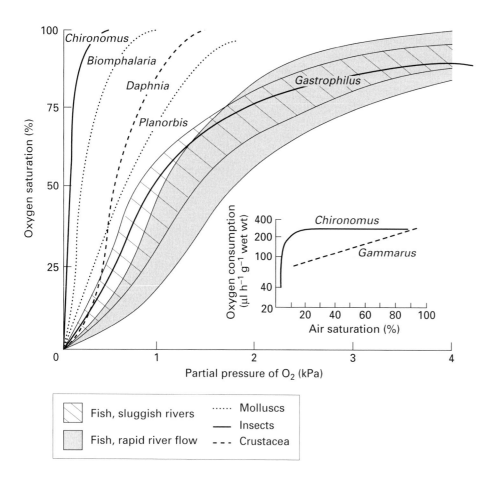

Fig. 11.25 Haemoglobin oxygen-binding curves in freshwater invertebrates and riverine fish; inset shows the effect on oxygen consumption, the chironomid from the stagnant benthos achieving maximum oxygen uptake at very low P_{O_2}. (Data from Weber 1980; and other sources.)

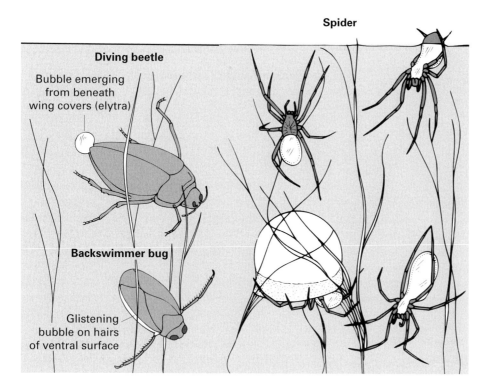

Fig. 11.26 Air bubbles used in freshwater air-breathing insects and spiders (as a diving bell in the spider *Argyroneuta*).

brain and the aorta. In resting snakes and turtles ventilation may be intermittent (with short bursts of breaths between periods of complete rest), but becomes continuous during steady swimming; the green turtle shows a seven-fold increase in mean ventilation frequency and increases in both pulmonary and aortic blood flow.

11.4.4 Respiration in air-breathers and bimodal breathers

Many aquatic animals are secondary invaders from land, using strictly *aerial* supplies. In arthropods two main air-supply systems are used underwater: air bubbles (defined as where the gas phase is compressible) and plastrons (where the gaseous phase is somehow rendered incompressible). With air bubbles (Fig. 11.26), as used in many beetles and bugs, the pressure rises as the animal dives and the gases tend to diffuse from air to water. As the animal consumes the oxygen, though, the Po_2 in the bubble becomes less than in the surrounding water, and from this point on oxygen enters the bubble, so that the total dive time is greater than that allowed by the original oxygen content of the bubble, by about eight times. However, the system cannot continue indefinitely, as the other gases in the bubble increase in partial pressure and diffuse faster to the surrounding water, reducing the bubble's lifespan. In the water spiders, such as *Argyroneuta*, an air bubble is held under a silken framework strung across submerged vegetation, and this may help to prevent its shrinking somewhat. The spider returns to its 'diving bell' intermittently between foraging expeditions. There are also a few species of insect that live in fast-flowing streams where the water flow around the bubble decreases the pressure in the bubble and forces it into an elliptical shape with increased SA/V, so that the bubble does survive indefinitely as a good permanent 'gill'.

Plastrons, found in water bugs and water beetles, are essentially a specialist adaptation on the air-bubble theme, where an air supply is maintained at constant volume and reduced pressure by being contained in a dense mat of fine water-repellent cuticular hairs (Fig. 11.27) that prevent bubble collapse and counteract the tendency for nitrogen to leave. The cuticular hairs that trap the film of air are about 5–10 µm long, usually with a pronounced kink near the tip; in the bug *Aphelocheirus* they are about 0.2 µm in diameter and 0.6 µm apart, giving an amazing 2–3 million hairs mm^{-2}. As the volume of air beneath the hairs tends to contract (as oxygen is used up and nitrogen begins to leave), it is resisted by a high surface tension effect from the increasingly curved meniscus at the air–water junction between each pair of hairs. Further nitrogen loss is prevented and the system reaches a constant volume/reduced pressure equilibrium at a particular depth. The hairs are dense enough in most diving insects to resist a pressure of 3 atmospheres (a depth of 40–50 m) without collapsing. This mat of hair may cover just the ventral surface, or most of the body; but it always covers the areas where the spiracular openings occur.

Special diving adaptations are also found in freshwater reptiles, birds and mammals, and in lentic fresh water the problem of diving may be exacerbated by the presence of a substantial unstirred hypoxic boundary layer; hence, for example, diving times for frogs may be 2–3 times shorter in still waters compared to running waters. For the endotherms, specific diving adaptations differ little from those described in marine tetrapods (section 9.10.3), including blood shunting tricks during submersion, enhanced oxygen-carrying capacity, and use of bradycardia, etc. But for the ectothermic freshwater tetrapods, such as amphibians, snakes and turtles, there is an additional complication in that the water temperatures and hence their own body temperatures are much more variable. To some extent they can solve this problem by their pronounced ability to undergo long periods of apnoea interspersed with bursts of ventilation, but the ventilation rate during such bursts does vary with temperature. For example, in the turtle *Mauremys*, lung ventilation per unit of

Fig. 11.27 Cuticular plastrons in freshwater insects. (a) The principle of plastron design, and (b) the structure and dimensions of cuticular plastron hairs in three species. (a, From *Animal Physiology: Mechanisms and Adaptions* by Randall *et al.* © 1997, 1988, 1983, 1978 by W.H. Freeman and Company. Used with permission.)

O_2 uptake declines linearly with increasing temperature, and alveolar P_{CO_2} therefore increases with temperature. However, at very high temperatures (40°C) there is an increased breathing response, possibly related to evaporative heat loss.

Many essentially freshwater animals are able to use both aerial and aquatic oxygen, and may switch to air-breathing when their aquatic habitat begins to dry up or overheat and become hypoxic, or if it becomes too rich in H_2S due to decomposition. This bimodal breathing represents a complex physiological problem, since air has nearly 30 times as much oxygen as water, whereas water can contain about 28 times as much carbon dioxide as air, giving two very different respiratory environments. Crustaceans, fish and amphibians are most notable as bimodal breathers, though some soft-shell turtles also achieve substantial (cutaneous) aquatic oxygen uptake. All experience the problem that while oxygen becomes easier to acquire on land, carbon dioxide excretion is problematic.

In crustaceans the presence of a high-affinity haemocyanin is common, and switching between sites of oxygen uptake may occur. Many bimodal crabs show the reduced gill area typical of air-breathers (see Chapter 13) but also an elaboration of the branchial chamber epithelia where gas exchange can be effected. The chamber normally contains a limited store of water such that gas exchange can take place from both air and water simultaneously, and in some species gas exchange seems to be partitioned so that oxygen uptake occurs from air into the chamber lining while carbon dioxide excretion occurs mainly across the gills into the stored water. (Note that fully terrestrial crustaceans lose this mixed pattern and can excrete carbon dioxide directly into air from the gills, perhaps due to the incorporation of extra carbonic anhydrase enzyme into their gill epithelia.)

In fish and tetrapods, bimodal breathing obviously requires a marked degree of remodelling of the vascular plumbing, to allow perfusion of gills, lungs or cutaneous sites in various parallel or series arrangements. But at the physiological end it is mainly related to strategic pigment use. Two adaptations are common during the transition to air; a decrease in oxygen affinity and an increase in pigment concentration. The change in pigment affinity may arise intrinsically in some amphibians, or it may be achieved by increases in the organic phosphate : haemoglobin ratio. Behavioural effects also occur, for example in the papyrus swamp fish *Barbus neumayeri* where surface respiration increases in frequency as hypoxia increases and dispersal within the swamp only occurs in the wet season. In the amphibian *Xenopus*, surfacing behaviour depends on temperature, the frequency increasing sharply between 5 and 20°C, but with surfacing duration kept very short (10–40 s) in daylight and only increasing with temperature at night, when the animal may stay at the surface for up to 60 min (probably due to lower predation pressure rather than physiological need). Amphibians suffer from a substantial hypoxic boundary layer in still waters, and can dive for significantly longer periods where the water is well stirred; when diving in still waters they are observed to undergo more voluntary movements, which may serve mainly to stir up the water.

Certain water snakes, such as the garter snake, *Thamnophis*, have an acute seasonal problem with respiration in that they overwinter for several months in submerged water-filled hibernacula at around 5°C. In this period they switch to non-pulmonary ventilation, and their oxygen uptake is only around 50% of that in snakes induced to hibernate in air. At such low temperatures they can survive without anaerobicity using just cutaneous uptake. Turtles that overwinter for 3 months or more under water with persistent anoxia do become anaerobic but cope largely by accumulating some of the resultant lactate in the shell; in the genus *Chrysemys* about 44% of the lactate is buffered in this way. These turtles produce very little of the normal oxidative damage products such as dienes, having high levels of key enzymes such as superoxide dismutase and catalase (Chapter 7), which decline rapidly during anoxia and rise again during reoxygenation. Thus they are minimizing the potential damage by reactive oxygen species (free radicals) during the reoxygenation of organs after anoxic bouts. They also reduce the Na^+/K^+-ATPase activity in their brain during anoxia, with an associated reduction of electrical activity, concomitant with a substantial drop in metabolism and a complete lack of movement.

11.5 Reproductive and life-cycle adaptation

In freshwater habitats there are several key problems that are bound to influence reproductive strategies:

1 The transience and changeability of the habitat, which may require opportunistic breeding carefully tied to seasonal change, and a protected stage in the life cycle.

2 The physiological difficulty of maintaining ion and water balance, made much more difficult in small animals such as larvae and juveniles with a high SA/V ratio.

3 In streams and rivers there is the additional problem, especially for small or immature individuals, of countering continuous downstream flow.

There is therefore a general tendency in freshwater invertebrates to have very short life cycles with a rapid turnover of generations, and to reduce larval forms, with more direct development involving larger and yolkier eggs (many molluscs) or brood pouches (e.g. the crustacean water flea *Daphnia*). Where larvae do persist they tend to be either unusually large, or to adopt a crawling rather than a pelagic habit (e.g. the planula larva of some cnidarians), or to have protected surfaces (e.g. a chitinous covering in that same planula), or in a few special cases to become parasitic on larger animals such as fish (examples include the cnidarian *Polypodium* living on sturgeon, and the bivalve *Unio* whose hooked veliger larvae grow in the gills of various fish). Also rather more species use direct insemination, or protected spermatophores, to protect their gametes prior to fertilization. Protective 'dormant' stages other than shelled eggs include sponge gemmules, bryozoan statoblasts, the durable encysted eggs or larvae of flatworms and nematodes, and the cocoons of certain oligochaete worms, all of which are resistant to drought or freezing or any other environmental stress. Dormancy in all its manifestations is much commoner in fresh-

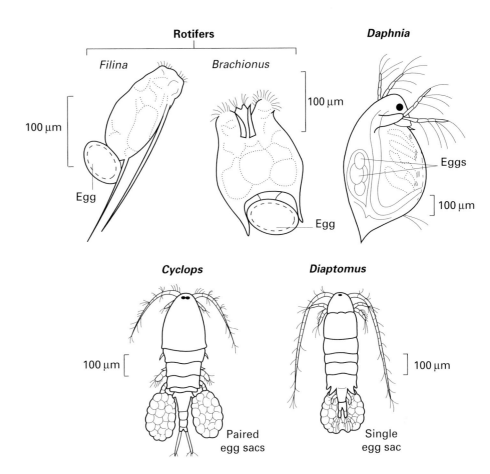

Fig. 11.28 Brooding in freshwater invertebrates.

water than in marine species, though less frequent than on land. These resting stages get widely distributed by winds or water spray or by animal means (such as on the feet of birds), ready to colonize further ponds when conditions are favourable again. In one unusual case a non-resting ostracod actively ensures its own dispersal to better ponds by hitching a lift on migrating toads.

Commonly there is very high annual reproductive output in freshwater invertebrates. One well-studied case is the zebra mussel, *Dreissena*, in which one female may release over a million oocytes per year. The mussel is dioecious and fertilizes its eggs externally, therefore using coordinated maturation and spawning behaviours. Spawning occurs in late spring, with the hormone serotonin acting as a trigger, causing oocytes to undergo accelerated meiosis. Spawned oocytes then contain species-specific sperm attractants, increasing the chances of rapid fertilization.

11.5.1 Short life-cycle strategies

Many of the zooplankton exemplify the short life cycle strategy, allowing them to take best advantage of seasonal algal blooms. Rotifers and cladocerans may complete their lives in just a few days, producing many generations per year (multivoltine); they mature quickly and put most of their assimilated energy into gamete production. In these taxa the time-wasting process of finding a mate is commonly avoided by the use of partheno-

genesis, where the ova develop without fertilization and only females are produced (often brooded in sacs within the carapace; Fig. 11.28). Generations of the asexual female morphs can thus be produced in vast numbers by one colonizing female finding a suitable pond away from predation pressure, and laying many thin-shelled rapidly developing eggs, with a generation time of only 1–4 weeks. A brief phase of production of sexual morphs usually occurs in autumn, or as drought sets in; these mate with each other and produce a further generation, often with heavily shelled eggs that are resistant to desiccation or to low temperatures. This alternation ('cyclical parthenogenesis') combines the rapid population expansion of asexual reproduction with the evolutionary novelty generated by sexual reproduction, and is ideal in an uncertain transient habitat.

11.5.2 Long life-cycle strategies

In contrast to the above, most copepods and insects, and most of the benthic invertebrates, are univoltine and grow relatively slowly through many moult cycles, and they usually do not show parthenogenesis and rather less frequently use dormant stages. Indeed a long adult lifespan is usually a direct alternative to the long dormant or diapausing phases described earlier; out of 167 crustacean species surveyed, not one had an adult life of greater than 1 year *and* a diapausing stage exceeding 1 year. Some copepods use a period of summer diapause, either avoiding fish predation or avoiding food bottlenecks in competition with other grazers; this usually occurs at the egg stage, and the diapausing egg may be capable of extended cryptobiosis (see Chapter 12).

Annual reproductive output in these longer-lived animals is lower (though it may still be high compared to other habitats). It often involves iteroparity, and each egg or embryo is more carefully protected. Freshwater crabs and crayfish have larger eggs than marine/brackish relatives, and higher C : N ratios in the eggs and young larvae, indicative of high lipid content. Copepods carry their egg sacs around on the abdomen, while insects tend to lay them in protected sites stuck to vegetation or buried in mud. In tropical and temperate habitats most of the insect species have an annual life cycle, emerging as adults in summer after the spring burst of food availability, mating and laying eggs within a few days. In higher latitude ponds and lakes the life cycle may be greatly prolonged though, with 5–7 years between egg production and adult emergence. In all cases there may be mass emergence, with clouds of midges or mayflies issuing from the lake or river surfaces on just one or a few days of the year.

Lotic waters are particularly hazardous for free gametes, and for pelagic larval stages, which would move downstream freely, so these are *very* rarely found; instead lotic invertebrates use internal fertilization, and either the young are then retained in the body, or large yolky eggs are firmly cemented to the substrate.

11.5.3 *Phenotypic plasticity and polymorphism*

Either of these broad life-history strategies may be accompanied by a pronounced degree of phenotypic plasticity, allowing the reproductive output to vary with environmental conditions. In *Daphnia*, water temperature has a clear effect, with larger size at maturity (measured as the age when eggs are first laid into the brood pouch) when water temperatures are lower. This operates via an effect on maturation threshold body length two instars before eggs are actually produced, since ovarian development is not initiated until a threshold body size is reached. This may be because higher temperatures and rapid metabolism produce an energy deficit for growth and moulting, or because there is selection against larger size in warmer weather via an external agency such as predation.

The well-known phenomenon of **cyclomorphosis** occurs in many of the parthenogenetic freshwater animals, where generations are polymorphic in form, physiology or behaviour, or all of these. A classic case is again the water flea *Daphnia* (Fig. 11.29a), and something similar occurs in some rotifers. Explanations for this are controversial (see section 11.7).

There are also many examples of resource-based or trophic polymorphisms involving differences in life-history strategy (as well as behaviour and morphology) in freshwater animals. The ecological conditions that promote resource polymorphisms probably include the relaxation of interspecific competition pro-

viding unfilled niches that can be exploited; a striking example occurs with Arctic charr (*Salvelinus alpinus*), which have invaded 'new' freshwater lakes within the last 15 000 years since the retreat of northern ice sheets. In a single volcanic lake in Iceland, four morphs of these charr are found (Fig. 11.29b), differing in size, fin shape, body and head depth, and jaw structure. Two are benthic, the smaller morph exploiting laval tubes; a third is planktivorous and a fourth piscivorous. The phenotypic differences between the morphs include age at maturity, egg size, and reproductive investment; they are partly genetic, but also have a strong environmental component, with triggers acting during early embryonic development.

11.5.4 *Freshwater vertebrates*

Reproduction in freshwater fish shows similar patterns to those found in invertebrates, often with the eggs serving as a protected stage for seasonal endurance. Freshwater fish eggs are generally larger (1–30 mm diameter) than marine fish eggs (0.8–2 mm), and may be protected in mucus froth 'nests', or in vegetation; where a pond or stream dries up the eggs may survive in the bottom mud. Some of the lungfish from swampy habitats can survive as adults encased in a cocoon of hardened mud reinforced with secreted slime, with a narrow air-tube to the surface. Lake species of fish, especially the many endemic cichlid species in the Rift Valley lakes of eastern Africa, have often evolved particularly strange reproductive habits that provide isolating mechanisms between species, including elaborate nest constructions and even 'mouth brooding' of the juveniles. Some lotic fish have also acquired careful nesting patterns, as in sticklebacks and guppies, to avoid the young being swept downstream.

Secondary freshwater vertebrates that are essentially land animals usually resort to a terrestrial site for reproduction. For freshwater reptiles, such as turtles, this is usually the upper shore. Some turtles do manage to lay eggs in mud under shallow water, but these stay in developmental arrest until the water recedes and oxygen is able to diffuse into the embryo. This phase may last up to 12 weeks, and the egg albumen loses up to 90% of its Na^+, while its water diffuses slowly into the yolk but at a very slow rate (the vitelline membrane having a very low P_w) compared with other turtles. Freshwater amphibious mammals also choose a birth site very close to the water, which may be an air-filled burrow in the river bank with an underwater access.

However, some extra problems arise, particularly with the rare ectothermic species that are viviparous and are at extra risk while carrying the embryos. For example, tropical water snakes, when pregnant, have a higher preferred T_b (27–28°C rather than the normal 22–24°C), and thus choose to spend longer at higher temperatures, losing their normal diel pattern of T_b and so both increasing and stabilizing the embryonic developmental rate. Endothermic viviparous species (manatees, lake seals, river dolphins) also have some problems in giving birth whether onshore or under water, paralleling those of the marine cetaceans and pinnipeds.

Fig. 11.29 Examples of polymorphism in freshwater animals. (a) Cyclomorphosis in *Daphnia*; young and adult morphology at different dates from a temperate pond, showing the progressive growth of the 'helmet' after successive moults; water temperatures are also shown. (Adapted from Brooks 1947.) (b) The four morphs of Arctic charr (*Salvelinus alpinus*) from Thingavallvatn, an Icelandic lake: (i) large benthivore; (ii) small benthivore; (iii) piscivore; and (iv) planktivore. (Adapted from Skulason & Smith 1995.)

11.6 Mechanical, locomotory and sensory adaptations

11.6.1 *Depth, buoyancy and locomotion*

Fresh water is rarely very deep, the very large lakes (Baikal, Tanganyika) being only 1000–1500 m, so that freshwater animals do not normally have to cope with great pressures. However, buoyancy is more difficult in fresh water due to the reduced specific gravity of the medium giving very little lift. Hence invertebrates need more exaggerated versions of the buoyancy mechanisms discussed in Chapter 9; either the body must be even more spiky or frilly to increase the drag forces, or more oil or gas must be included within the tissues. Cephalopods do not occur in fresh water, so the problem of achieving enough salt transport from the fluids of the rigid float system does not arise. In fish the swim-bladder is present in a much higher percentage of species and is larger, about 6–9% of body volume in freshwater teleosts compared with only 4–5% in marine species.

Fresh water provides few problems or opportunities for locomotion not found in the marine world (see Chapter 9), except in lotic systems where animals may need to locomote continuously to counter the tendency to flow down river and ultimately out to sea. Station-keeping is a continuous problem for any species that cannot cope with slightly salty conditions. Suckers and hooks for adhesion may be appropriate for bottom-dwellers; certain fish have ventral suckers, and blackfly (*Simulium*) have hooks to hold on to rocks while filter-feeding with elaborate fans of bristles. Adhesive systems, such as the byssus threads of freshwater bivalves, may also be used. In moderately fast flow, a streamlined shape flattened against the substratum may be enough to ensure stability, as the fast water flow over the upper surface effectively presses the animal down and keeps it in place; flatworms often rely on this, and many freshwater bivalve shells are very flattened for similar reasons.

11.6.2 Senses

Vision

In general lentic environments are well lit through most of their depth, and only deep lakes extend beyond the photic zone. However, in both rivers and eutrophic lakes light is depleted by vegetation in summer, so that the light environment can be very variable. Eyes may therefore need to work in very dim conditions, especially for predators in weed-filled ponds or in stretches of river overhung by dense trees. Freshwater beetles and bugs, and fish such as the pike (*Esox*) have eyes with properties similar to those of terrestrial nocturnal species, with large facets or large pupils. Rhodopsins are sensitive to green wavelengths in particular, filtering through the floating or shading foliage.

Amphibious animals share the same problem as some littoral species (Chapter 10), in that they need to be able to focus in both air and water. An unusual solution is found in certain surface-swimmers, that of having two sets of eyes; the four-eyed fish *Anableps* can focus simultaneously on a terrestrial and an aquatic image, having in fact only two eyes but with a pear-shaped lens and two sets of pupils above and below the water meniscus. Any animal that hunts across the air–water interface also has to be able to make allowance for the refraction at the surface; this applies to archer fish shooting water out at insect prey in the air, as much as to herons or kingfishers or bears dipping or diving in to spear fish or invertebrates spotted from above. Diving mammals and birds that hunt by swimming beneath the water, such as otters or dippers, must be able to accommodate their eyes by altering the lens-cornea relationship using the ciliary muscles.

Some teleosts, and the crayfish *Procambarus*, can detect polarized light, an ability that seems to be rare in aquatic animals; as in other (terrestrial) arthropods, this is due to the precisely orientated microvilli in adjacent retinular cells in a single ommatidium being interdigitated at 90° to each other.

Chemoreception

Chemoreception is important to many freshwater animals for prey location, predator avoidance, and location of hosts or mates; in rivers it is particularly favoured because the currents will help to keep chemical plumes directional and reasonably intact. Thus both general and specific sensitivity are appropriate. For example, teleosts have good acid, alkali and salt receptivity, and the minnow can detect 10^{-5} M sucrose or 4×10^{-5} M NaCl. But in these and many other shoaling fish there are also specific sensors for intraspecific cues, such as the alarm substance ('schreck-stoff') released from the skin of damaged individuals; these chemicals are commonly polypeptides or proteins.

Mechanoreception

Many aquatic insects have water pressure (= depth) receptors. For example, the bug *Aphelocheirus* has a group of long cuticular hairs that trap an air bubble, the hairs then compressing around the bubble and detecting its shape change with depth. Other bugs with relatively long bodies use pressure sensors adjacent to their abdominal air spaces to detect differences in tracheal pressure between each segment, so getting information on whether they are swimming head-up or head-down.

Flow receptors are also critical for lotic species, and are designed from deformable neuroepithelial cells or from hair cells (Fig. 9.31) as in other aquatic animals. In larger animals, such as crustaceans and fish, the receptors may be associated with cephalic appendages such as antennae or vibrissae.

Other senses

Electroreception occurs in a range of freshwater fish (Fig. 11.30), with both weak and strong field producers (see section 9.7.5). Weak electric fish sometimes use the system for mate recognition, the male having a discharge frequency markedly lower than the female. The electric eel is one of the most forceful field generators, and can both detect and stun its prey in very murky waters. Some electric elasmobranch fish, including rays and sawfish, penetrate up large rivers. Freshwater amphibians retain the electroreceptors of their ancestors, and the system even persists into a few diving mammals, notably the platypus. Echolocation by contrast is largely lost in freshwater species, although it persists in freshwater dolphins.

11.7 Feeding and being fed on

An overview of the trophic cascade in a typical temperate lake is shown in Fig. 11.31, emphasizing the importance of planktonic and microphagous feeding systems in most freshwater systems.

11.7.1 Microphagous feeding

Most of the freshwater zooplankton are filter-feeders, exploiting the algae, bacteria and detritus in the water column. Some are generalists, but many are highly selective, taking particular kinds of green algae, diatoms or flagellates. Cyanobacteria are generally avoided. Rotifers feed mainly on particles of 1–20 µm, using their ciliated corona. Cladoceran 'water fleas' (including the extremely common herbivorous genera *Daphnia* and *Bosmina*) take slightly larger algae up to 50 µm, and the larger crustacean copepods feed on a larger range again, taking phytoplankton and zooplankton of 5–100 µm. Within each group, some prefer live food, others detritus, while yet others take both types of food material. Surprisingly large numbers of freshwater animals, including cnidarians and some flatworms, have found an additional way of 'exploiting' algae for feeding purposes, adopting a symbiotic association with photosynthetic algae termed zoochlorellae; in *Hydra*, for example, green algae within

---

Fig. 11.30 (*Opposite*) Electric fish in fresh water. (a) The types of electric fish and their pulse patterns. (b) The large electroplaque organ in the trunk of the electric eel, formed from modified myotomes (muscles).

(a)

Freshwater electric fish

Morymrids	(weak pulse 2–5 V)
Gymnarchids and gymnotids	(weak pulse ~1 V)
Catfish	(strong pulse 300 V)
Electric eels	(strong pulse >500 V)

1 V

0.5 ms

0.5 V — Weakly electric, constant low-frequency wave

20 ms

0.1 V — Weakly electric, constant high-frequency wave

1 ms

0.5 V — Weakly electric, variable frequency pulse

100 ms

20 V — Strongly electric pulses

10 ms

Posterior lateral-line nerve

Dorsal branch

Brain

Electric organ

Receptor areas

(b)

Spinal cord

Swim-bladder

Electroplaque organ

Electroplaque (modified muscle)

Detail

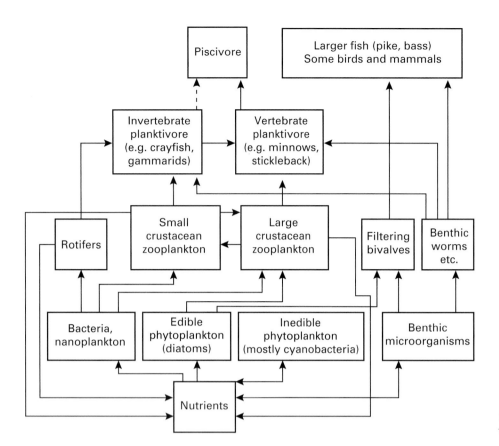

Fig. 11.31 The trophic web in a typical temperate lake.

the animal's cells provide a proportion of the animal's carbon needs (and probably receive some nitrogen in return). Tropical medusae living in crater lakes migrate around the lake daily to keep their symbionts in the sun for as long as possible.

Other benthic invertebrates feed on the sinking remains of the algal blooms, particularly the diatoms, which remain fairly intact as they sink and are a rich source of fatty acids. True detritus feeding (swallowing the mud, in effect) is practised by oligochaete worms (*Tubifex*, etc.). Thus all the vegetation that dies within a freshwater body eventually becomes a food source, although this occurs relatively slowly compared with land habitats due to the absence of fungi and other key decomposers. In the benthic communities grazers are also abundant, scraping microbial communities (the epiphyton) off rocks and plants fairly non-selectively. Mayfly and caddis larvae, gammarid shrimps and most snails are classic microbial grazers.

11.7.2 *Herbivory*

Herbivory on a scale larger than algal grazers is relatively rare in temperate systems, since there are few macrophytes in many lakes, and they are relatively non-diverse in lotic waters. Herbivory is commoner in parts of the tropics, where papyrus and water hyacinth can provide high rewards; large grazers such as hippopotamus, water deer, capybara and dugong are important, and some fish are fruit-eaters.

Where temperate macrophytes do occur, they may be very heavily exploited as living space; a single plant of the pondweed

Potamogeton has been recorded as a habitat for 500–600 individuals, including chironomid midge larvae, oligochaete worms, water mites, ostracods, *Hydra* and small snails. Animals that actually eat the weed are much rarer; pondweeds and water lilies do get attacked by some insects (especially small leaf- and stem-mining larvae) and many molluscs, as well as amphipods, crayfish and some fish. There are also a few herbivorous turtles, and some of the freshwater mammals take aquatic plants as part of their diet—the dugongs and manatees are entirely herbivorous. Defences against herbivory in freshwater plants are relatively poorly studied.

11.7.3 *Carnivory*

Carnivores operating in fresh water may be usefully subdivided into lurkers and hunters. Stationary lurking is exemplified by *Hydra*, taking any small prey that brush against and trigger the stinging cells (cnidoblasts) in its tentacles. Leeches also tend to operate by lurking in the concealing weeds, as do stonefly and alder-fly larvae. Predatory hunting is also common amongst insects (bugs, beetles, dragonfly larvae with their extensible 'mask', and some midge larvae) and almost ubiquitous at some point in the life cycle of fish, since larger zooplankton form an important part of most fish diets. The larger insects, crustaceans, worms and snails, and also many fish, in turn are eaten by hunting reptiles (e.g. crocodiles, anaconda), mammals (e.g. otters, river dolphins) and birds (e.g. herons, dippers).

There is an additional possibility for carnivores in and around

Fig. 11.32 Patterns of diurnal plankton migration in Lake Tahoe, showing abundance of various stages of four species by day and by night. Note that some copepods migrate to the surface at night, and that there are clear sexual differences in distribution in the copepod *Diaptomus*. The thermocline in this lake is at about 20 m. (From Horne & Goldman 1994.)

fresh water, taking advantage of the surface film. Many insects have a specialist lifestyle relying on the surface tension of the water film for support, preying upon other small insects that become entrapped by that same surface tension when they fall into the water, or catching small crustaceans just below the surface. The water striders (gerrid bugs) and the pirate spiders live above the surface, standing on tiptoe on the water film, while the water boatmen or backswimmers (notonectid bugs) live in the water suspended upside down from the surface film by the tip of their abdomens. Exploiting the surface film fauna is rarely an option in marine or estuarine habitats where the film is constantly disrupted by wave actions.

11.7.4 Feeding and growth patterns

Diel and seasonal patterns of food abundance in fresh water have a major effect on growth patterns, which in some parts of a season may be strongly negative. The freshwater flatworms can undergo extraordinary degrees of starvation, reducing their body mass to as little as 1/300 of its maximum and resorbing most of their gut and parenchyma. Seasonality is also very strongly influential on patterns of feeding activities. Many components of the zooplankton show vertical diurnal migrations (Fig. 11.32), and planktivorous carnivores must move with them. In temperate lakes there are commonly 'spring blooms'

Fig. 11.33 Seasonal bloom patterns in Lake Erken (Sweden); in spring for various rotifers, followed by herbivorous cladocerans and carnivorous copepods as algae build up. (From Horne & Goldman 1994.)

(Fig. 11.33), sometimes in two phases, where build-up of diatoms and flagellates allows bursts of herbivorous cladocerans and copepods, accompanied by carnivorous copepods. Fish populations inevitably are trophically linked to these blooms.

11.7.5 *Avoiding predation*

Diurnal movements may be partly related to limiting losses to predators. Other predator avoidance strategies are much in evidence in freshwater plankton. Here cyclomorphosis may be relevant. The small and very transparent summer generations of many cladocerans may be selected for by intense pressure from predatory fish populations; this view is supported by their greater abundance inshore where the fish take refuge from their own predators, compared with greater numbers of the larger and more conspicuous forms in the open water. Similarly the spines, helmets and other protuberances produced seasonally by cyclomorphic rotifers and cladocerans (see Fig. 11.29) may be a defence against invertebrate predation, allowing extra food reserves to be stored in a transparent structure not easily seen by visual hunters and making the animal less easy to swallow. In larger animals, good attachment systems, strong molluscan shells with appropriate resistance to boring mouthparts, and surface prickliness, as seen in many smaller insects, may also serve as antipredator devices.

Predation may also be linked to physiological constraints, perhaps especially for arthropods that must undergo periodic moulting. In the genus *Gammarus*, the species *G. pulex* is strictly freshwater, while *G. tigrinus* prefers water of somewhat higher ionic concentration; where the two species co-occur they prey upon each other's moulting stages. In fresh water the predation is differentially in favour of *G. pulex*, whereas in slightly more saline conditions it is balanced; thus the better physiological adaptation of *G. pulex* to very dilute media is probably the reason for its exclusion of its congener. Snails also have a problem with predators, since in very fresh waters they cannot accumulate enough minerals to make dense calcareous shells and must rely on rather less resistant proteinaceous coverings.

11.8 Anthropogenic problems

11.8.1 *Wastes and pollution*

Human settlements have always tended to develop on river banks, giving natural defences, natural drinking-water, potential power for mills, and above all channels of communication; most of the big cities of today are still straddling the world's major rivers. Unfortunately fresh waters are among the most fragile of all the Earth's habitats and this juxtaposition does not enhance freshwater communities (Plate 4f). These riverside cities have always tended to discharge their sewage into the river; more recently they have discharged all their other wastes

Table 11.13 Categories of wastes discharged into fresh waters.

Type	Source
Acids and alkalis	Industrial
Anions (S^{2-}, SO_3^{2-}, CN^-, etc.)	Industrial, mining
Detergents	Industrial, domestic
Sewage	Domestic
Silage and farm manures	Agricultural
Food wastes	Domestic, agricultural
Gases (Cl_2, NH_3, etc.)	Industrial
Heat	Industrial, power generation
Metals (Cd, Zn, Pb, Hg)	Mining, industrial
Nutrients (nitrates, phosphates)	Agricultural
Oil and oil dispersants	Industrial
Organic toxins (C_6 residues)	Industrial
Pathogens	Various
Pesticides, herbicides	Agricultural
Polychlorinated biphenyls	Shipping, tourism
Radionuclides	Industrial, power generation

as well (Table 11.13). Five main categories of anthropogenic waste can be recognized.

Sewage and silage

Whether from human settlements, intensive farming units, or fish farms, untreated or poorly treated effluents produce a burst of biological oxygen demand in freshwaters, creating an 'oxygen sag' downstream in lotic systems. The **biological oxygen demand** (BOD) of raw sewage may be as high as 120 mg l^{-1} day^{-1}, whilst that of well-treated sewage may reduce to 6 mg l^{-1} day^{-1} whereas clean fresh water is only around 1 mg l^{-1} day^{-1}. This oxygen usage encourages fungal bacterial communities, and animals such as tubificid worms that tolerate low oxygen; it may take a long stretch of river (and/or time) for recovery (Fig. 11.34). Indeed, accelerated (or cultural) **eutrophication** is a well-recognized process in both rivers and lakes (Fig. 11.35). The high algal growth may produce 'blooms', so that the water becomes unpleasant, potentially toxic, and often smelly. Reduced oxygen levels may alter the solubility of nutrients and of some metals. Submerged plants gradually disappear as light is cut out, and animal casualties inevitably follow. First the submerged animals change (*Tubifex* worms replace crustaceans, salmonids lose out to coarse fish like pike and perch), and then surface dwellers decline. Table 11.14 summarizes the major categories of pollution susceptibility in freshwater animals, showing the key role of highly susceptible (anoxia-intolerant) groups, such as mayflies, stoneflies and caddis, as indicator species for environmental damage.

Damage can be reversed by managing the effluent, but this has normally only been done where the lake is a valuable drinking-water source, or valuable as a leisure resource. It can be reduced by putting animal wastes back on the land rather than

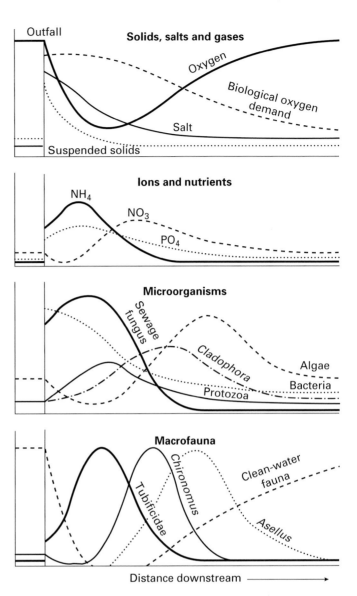

Fig. 11.34 Effects of effluent discharge in a river; downstream patterns of physical and chemical change, and the associated changes in microorganisms and macroinvertebrates. (Adapted from Hynes 1960.)

into rivers; and improved treatment and recycling of human waste has also got to be part of the long-term solution. However, sewage treatment commonly involves surfactants and detergents, and the breakdown products of these may include oestrogenic substances that are known to affect fertility (especially of fish) and which are environmentally persistent.

Mining wastes

Usually there are acid drainage waters from mining activities, resulting from sulphides in ores being converted to sulphuric acid, partly by bacterial action. The acid water then dissolves out metals, like copper, iron and zinc, adding to the pollution

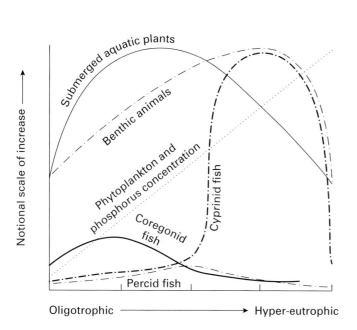

Fig. 11.35 Changes in a temperate lake as it becomes eutrophicated (as indicated by the steady build-up of phosphorus and phytoplankton). Yields of fish may increase overall until the lake is highly eutrophicated, but the cyprinid fish are much less desirable as a commercial resource than are coregonids and perch. (From Moss 1980.)

Table 11.14 Pollution susceptibility in freshwater animals (1 = resistant, 10 = highly susceptible).

Category	Animal taxa
1	Oligochaeta
2	Chironomid larvae
3	Most pond snails; most leeches; water louse *Asellus*
4	*Baetis* mayflies; alder flies; fish leeches
5	Most bugs and beetles; crane flies; blackflies; flatworms
6	*Viviparus* pond snails; mussels; gammarids; some dragonflies
7	Some mayflies; some stoneflies; most caddis
8	Crayfish; most dragonflies
9	Some dragonflies
10	Most mayflies and stoneflies; most caddis; river bug *Aphelocheirus*

problem; copper is especially toxic to fish. A little way downstream, minerals such as iron may subsequently redeposit as red-brown sludge, killing stream biota completely. Mineral deposits may become particularly serious in subtropical rivers in the dry season. This again needs on-site remedial technology, which is expensive but not 'difficult'.

Thermal waste

This arises particularly from electricity generation, where power stations discharge hot water into rivers. Some animal species die directly due to their own low thermal tolerance, others succumb because bacteria proliferate rapidly and reduce the oxygen levels. Heat is a valuable resource, and it is certainly wasteful to offload it in rivers. Some cities are therefore developing recycling systems whereby the heat is piped to buildings. In the shorter term, it is better to use the heat for some purpose near to the generator, perhaps for farming some freshwater species that tolerates or prefers warmer water, such as eels or carp.

Agricultural chemicals

The most important discharges reaching rivers and lakes are nitrates and phosphates. Levels of these ions may be five-fold greater in streams flowing through agricultural land compared to forest. When these ions enter a watercourse, the water tends to get overgrown with algae, whose decay may kill fish by reducing oxygen and/or by increasing alkalinity. This is principally a problem in causing long-term eutrophication, of slow-flowing rivers and especially of lakes. In the UK, East Anglia and parts of the West Country fruit-growing areas are badly affected; many lakes of Europe and in the crop belts of North America are also threatened; and in China one-quarter of all lakes are thus contaminated. On a large scale Lake Erie and the Aral Sea are the most notable, near terminal, casualties, though the latter has been exacerbated by diverting inflows for irrigation leading to massive loss of water volume and salinification (Fig. 11.36).

A second agriculture-related problem is inputs of herbicides and pesticides, such as organochlorines, all directly toxic to freshwater animals, even ones in the least susceptible categories; levels of all these are now controlled by international directives. DDT, which was once widely used and is still applied in some countries, is directly toxic and also has an oestrogenic effect, causing sexual dysfunction in many organisms but especially noticeable as feminization and reduced fertility in male fish.

Industrial outflows

Industrial discharges commonly involve organic and inorganic chemicals, including oils, bleaches, dyes, detergents, formaldehydes and phenols, and a variety of heavy metals, in varying degrees of pretreatment (see Plate 10b, facing p. 310). From older style industrial plants in Eastern Europe and parts of the Third World these factory effluents may be almost untreated, and again tend to be directly toxic. For heavy metals this toxicity can occur at very low doses, often because metals such as cadmium, mercury and zinc are inhibitors of key enzyme activities, while copper is an inhibitor of haemoglobin function.

A particular hazard to freshwater areas used for boating and recreation are the tributyltins (TBTs) and polychlorinated biphenyls (PCBs), used to clean boat hulls and other underwater structures and render them unsuited to larval settlement, but lethal to other animal life where they leak into the lake or river. PCBs also get into the atmosphere, and have recently been recorded precipitating out in the cold air of the polar regions,

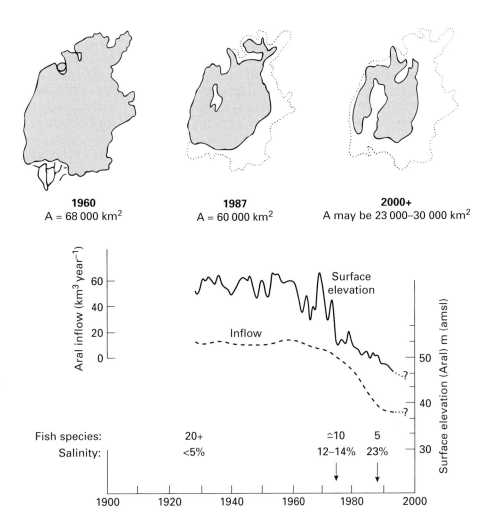

Fig. 11.36 Progressive loss of surface area in the Aral Sea, largely resulting from diversion of inflows for agricultural irrigation in adjacent cotton-growing areas. An elevation drop of just 10 m reduced the area by 60%, and what was formerly the fourth largest freshwater lake in the world is now rising in salinity and much reduced in diversity (with brackish-water species out-competing freshwater species). (Adapted from Micklin 1988 and other sources.)

where once ingested they may be seriously affecting the growth and fertility of animals as large as polar bears.

These last two categories of contaminant are interactive, with eutrophication impacting on concentrations of industrial pollutants. PCB and DDT concentrations in pike, in lakes of southern Sweden, have been shown to be inversely correlated with total phosphorus, chlorophyll and humic content of the water. Otter populations are also reported to persist in eutrophic PCB-contaminated lakes but not in oligotrophic ones. One suggestion is that increased productivity in the lakes reduces the pollution burden in fish and mammals due to higher growth rates, quicker turnover times, and more rapid sedimentation of the pollutants into an inactive form adsorbed to particles (Fig. 11.37).

Agricultural and industrial chemicals are also longer lasting than other pollutants, and many affect the groundwater system more than the rivers. Once there they are nearly impossible to eliminate, as all the cleansing microbial agents need some oxygen source not available underground. However, the rock through which the ground water filters does actually slow the cycling sufficiently so that most organic pollution degrades into

a relatively harmless state. Eventually though, these groundwater aquifers recycle into the drinking-water, and long-lived organic pollutants can sometimes get through the system.

Polluted waters and disease

Contaminated natural fresh waters are the single biggest source of human disease in the world today. Some of the disease results from parasites and pathogens naturally present in fresh water, such as river blindness (carried by blackflies, and caused by the nematode *Onchocerca*), and malaria (due to the protozoan *Plasmodium*, carried by mosquitoes). But a huge additional hazard is added by pathogens deriving from sewage and animal wastes, made worse by erosion/silting after deforestation. Infantile diarrhoea is the single largest killer in areas where these anthropogenic effects are greatest. In the Western world pathogenic contents of water are usually controlled by chlorination, which is not ideal but certainly a relatively acceptable risk compared to the pathogens themselves. The cleaning up of rivers is also well under way in the west, with the Thames and the Rhine now both in reasonable condition compared to the

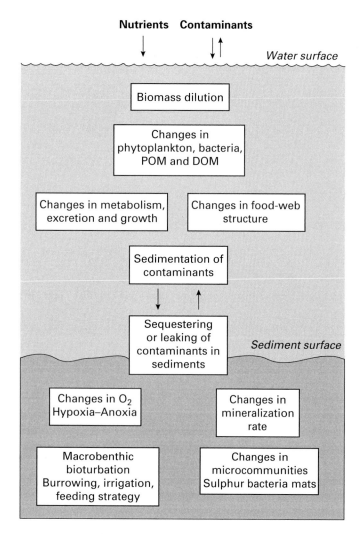

Nutrients   Contaminants

Water surface

Biomass dilution

Changes in phytoplankton, bacteria, POM and DOM

Changes in metabolism, excretion and growth

Changes in food-web structure

Sedimentation of contaminants

Sequestering or leaking of contaminants in sediments

Sediment surface

Changes in O₂ Hypoxia–Anoxia

Changes in mineralization rate

Macrobenthic bioturbation Burrowing, irrigation, feeding strategy

Changes in microcommunities Sulphur bacteria mats

Fig. 11.37 Interactions between contaminants and eutrophication in freshwater environments. Nutrients may increase the primary productivity and thus growth rates of larger animals, increasing turnover rates and sedimentation of contaminants. DOM, dissolved organic matter; POM, particulate organic matter. (Adapted from Park 1991.)

1960s and early 1970s. However, the Elbe, running through the Czech Republic and former East Germany, is still very heavily polluted and lacking in fauna and flora, as little effort was formerly put into ecological planning in major industrial centres such as Leipzig and Dresden.

11.8.2 Acidification

The problem of acid deposition was first noted at least a hundred years ago, with the observation that rainfall was sometimes acidic and that this did harm to vegetation, stonework and metals. But in the 1960s Scandinavian lakes became seriously acidified, with fish and invertebrates dying, and the damage was linked with pollution blown from Britain and north Europe; similar

problems began to appear in North America. By the early 1980s, 8–10% of lakes in Sweden, and a full 20% in the USA, were too acid to support any fish life. In 1997 16 000 out of a total of 85 000 Swedish lakes were recorded as anthropogenically acidified.

Causes

Rain is naturally acidic, because the CO_2 in air dissolves to give weak carbonic acid, having a pH usually about 5.5. But rain in Europe and the USA can now be regularly recorded at 4.0–5.0, with even lower figures in central Europe. Some rainstorms have scored 3.0, and some smogs lower still, with a pH equivalent to vinegar. The obvious clue to the source of this problem is that acid rainfall is a regional rather than a global problem (Fig. 11.38); northern Europe, northeastern USA, and southeastern Canada are much the worst areas, all lying to windward of heavy industry. Sulphur dioxide emissions, either unconverted as a gas or dissolving as sulphuric acid, are therefore targeted as the prime cause of the acidity; 60% of this gas comes from power stations (especially coal-fired ones), and 20% from other industrial plants.

In fact, recent evidence from analysis of foraminiferan shells in lake sediments dates the onset of acidification to around 1800, when coal burning became more common; particles of soot appear in the sediments at the same time. This was really the conclusive evidence as to the cause of acid rain. However, emissions from coal-burning have been going down in the UK, USA and most of western Europe since the early 1970s, and SO_2 has therefore been declining, yet damage has still been worsening. This is suggestive of long-term cumulative effects, and/or interactive effects with other pollutants or stress factors. Several gases are therefore now implicated, but the evidence is especially damning against interactive effects of sulphur dioxide and nitrogen oxides (both from power plants, and the latter also from car exhausts) and ozone (mainly from car exhausts). There are some recent suggestions that climate modification is making matters worse; for example, lakes in the US state of Michigan have become rapidly and seriously acidified only during recent drought summers. Thus if global warming really sets in it may hasten changes in lake chemistry.

Consequences for animals

There are a number of consequences of acidification for freshwater animals:

1 Reduced fecundity and increased mortality occur, operating through a variety of different physiological effects on ion and acid–base balance with knock-on behavioural implications. Animals show reduced body weight in acid waters; this has particularly been recorded for fish and for amphibians, and in the latter its proximate cause is mainly reduced rates of prey capture.

2 Acid rain solubilizes minerals from the river or lake sediment. Of these, aluminium is most important, solubilizing below pH 4.5.

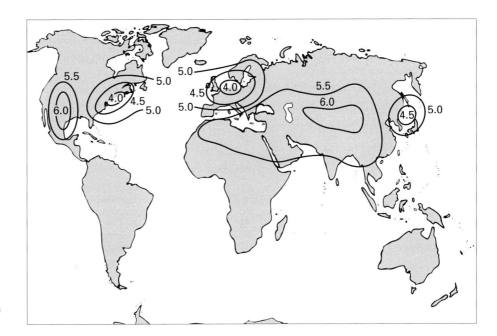

Fig. 11.38 Regional patterns of acidification of rainfall and of lakes.

In many acidified lakes, the death of fish is now clearly attributable to aluminium poisoning, affecting mucus production and clogging the fish gills (mucus has a similar oxygen diffusion coefficient to water, but its presence creates a non-convective and thus effectively stagnant boundary layer around the gill surfaces). It can be reasonably assumed that mucus production in other (non-economic and therefore less-studied) freshwater animals, such as flatworms, snails and annelids, is also seriously affected, causing mortality. Mobilized aluminium also leads to precipitation of phosphate and thus an oligotrophication effect, resulting in reduced algal growth and very clear water.

3 There are direct effects on physiological processes such as cuticle formation in insects and crustaceans (where calcium uptake is especially affected), so reducing these animals' subsequent osmoregulatory abilities. Their resultant mortality upsets the food chain for fish, and these food-chain effects also impact on birds (e.g. dippers and other stream dwellers)—one particular result is thinning of their eggshells and thus increased egg mortality in some species.

4 Particularly acid conditions often occur in spring at snowmelt, coinciding with the hatching time of many salmonid fish. Many fail to hatch at low pH, apparently because the enzyme chorionase, which allows the larva to escape from the chorionic membrane, can only function fully at a pH of 6.5–8.5. Thus fish stocks suffer very considerably.

Abundance of many species therefore declines, particularly snails, amphibians, small crustaceans contributing to the zooplankton, and salmonid fish. Other fish such as eels, and insects and rotifers, are relatively unaffected at first; but in the long term pH levels below 5.0 will kill off most of the fauna, with only some green algae and sphagnum moss flourishing in the littoral fringes.

Dealing with acidification

Some of the 'curative measures' proposed to deal with acidification are also treated with some scepticism by environmental biologists. In the 1970s it was thought that the situation could be helped by dumping ground-up lime into lakes. Norway and Sweden have been attempting to promote lake recovery in this way, until their neighbours reduce their emissions sufficiently; about half of all Sweden's acidified lakes are now treated with lime powder, and some significant increase of species richness has been recorded. The amount of lime needed is very large and very hard to calculate, as it depends on flow patterns and the turnover time of each lake, and liming must be repeated regularly; it also does not affect the inflow streams, which is where fish often breed. 'Target liming', going for streams and bogs in the catchment area, may be the best compromise, and this has been done recently with reasonable success in some lochs in southwest Scotland. But liming tends to kill off sphagnum and other mosses in the catchment zone, which may have other long-term effects on bog formation and stability, and which certainly badly affects birds and invertebrates associated with upland moors and bogs.

There are some recent and encouraging indications that lake and water acidifications can be naturally reversed reasonably quickly when acid inputs stop. Many Scottish lochs have been regaining a higher pH in the 1990s, though no fish are returning yet. A particularly detailed study has been made at Whitepine Lake in Canada. In 1980 the lake was showing severe acid stress, with many acid-resistant perch but a loss of numbers and/or reproductive ability in trout and burbot, plus a decline in diversity of most other taxa. Over the next 10 years, SO_2 emissions were halved and the pH rose from 5.4 to 5.9, with aluminium levels dropping; the trout recovered very effectively, as did the diversity of benthic invertebrates. At least two species of formerly common fish did not recover though, and must be regarded as extinct in that lake. It is also apparent that as yet there is not much improvement of lake quality in Norway or Sweden.

11.8.3 Dams and irrigation systems

The construction of dams and irrigation systems drastically alters flow regimes in river systems, and while this must not be seen as necessarily 'bad', since there may be proliferation of some new and valuable species downstream of management areas, nevertheless it is very difficult to predict the outcome of damming a watercourse, and it is therefore biologically speaking a 'chancy business'. Damming directly affects all migratory species, especially anadromous fish, and both damming and other management schemes tend to remove rapids, waterfalls and shallows, which can greatly reduce a river's ability to re-aerate following any unusual oxygen demand. On the other hand, water in the spills below high dams can become supersaturated with oxygen at up to 140% saturation, and this too can be damaging, causing lethal 'gas bubble disease' in migrating fish.

Damming a river can also have drastic knock-on effects on downstream communities, especially by causing silt to be deposited in the lake above the dam, so being lost to the downstream systems. Both fisheries and natural silt fertilizations of the Nile lowlands were seriously disrupted after construction of the Aswan High Dam. Downstream effects on human communities are also potentially serious, particularly where rivers cross national boundaries and countries compete for the water supply.

11.8.4 Invasions

Human activities have often brought new species (non-indigenous species, NIS) into freshwater ecosystems, which as with all biological invasions can have complicated and far-reaching effects. A number of spectacular invasions of the Great Lakes system in North America have been recorded. In the 1830s the sea lamprey invaded the Lakes via the Hudson river, and seriously disrupted fish stocks. More recently (1986) the European freshwater zebra mussel, *Dreissena polymorpha*, arrived, probably in a ship's ballast water, and has now spread to every one of the Lakes, occurring at densities of at least half a million per square metre in some areas, displacing local clam beds and seriously affecting fish spawning areas, as well as fouling many man-made structures. These *Dreissena* apparently came from populations in the Black Sea with a rather high upper thermal limit, so that in America the species is spreading into much warmer lakes and rivers than it normally tolerates in northern Europe.

11.8.5 Effects of global warming on freshwater systems

Direct effects of carbon dioxide build-up are expected to be limited, in that aquatic animals usually respond more to P_{O_2} than P_{CO_2} (see Chapter 7), and changes in external P_{CO_2} in the water will in any case be very small due to the high solubility of this gas. Temperature effects are of more concern, but will perhaps be less drastic than for the less buffered terrestrial environments (see Chapter 13).

However, the timing and distribution of rainfall and run-off will certainly alter, according to most of the global circulation models attempting to predict future weather changes. There may be 3–15% more rain overall on the latest IPCC (Intergovernmental Panel on Climate Change) model. This is likely to result in increased river currents and discharges, with potential bouts of flooding in vulnerable river delta areas such as Bangladesh. It will also be likely to alter animal zonation patterns in many key rivers; and it could again be especially critical for humans where resources are shared between nations.

11.8.6 Deforestation and afforestation

The presence of forests or other dense vegetation around both lakes and rivers affects the light received and hence flora and fauna patterns, and resultant water chemistry, as well as affecting the allochthonous inputs. Thus whole river systems can be greatly affected by deforestation, with effects even out to sea on vulnerable communities such as coral reefs. There may be a complete loss of the humus layer in the tropics, for example, which runs into the rivers and creates transient 'blackwaters'. On the other hand, replanting naturally or anthropogenically deforested areas with dense stands of conifers, as is occurring in many regions of European and North American uplands, can greatly alter the acidity of river systems running through the area.

11.8.7 Uses of fresh water

The worldwide use of freshwater resources by humans is such that an amazing 73% goes on crop irrigation, usually being used extremely inefficiently; 21% of the usage is industrial; and just 6% involves domestic use. However, these figures are somewhat misleading; most of the irrigation water gets recycled through the agricultural system many times, so it is not actually 'lost'; and much of the industrial water is purified within the plant now, in western countries at least, and so is returned to the cycle.

11.9 Conclusions

It should be very clear that freshwater life is so varied that it is virtually impossible to generalize. Given the problems and the many axes of environmental variations, it is perhaps quite surprising that so many animals live so successfully in these habitats. In fact, most animal taxa achieve some freshwater representation, with many very abundant, and it is not particularly easy to account for this.

Can we end instead by deciding why some groups do not

succeed? We can recall (Table 11.7) that the chief exclusions are echinoderms, tunicates, most sponges and cnidarians, and the cephalopods. For all but the last of these, the answer may be reasonably simple. All these animals are:

- built with large, soft exposed surfaces that could not be made impermeable, either because they are used as food filters (most) or as uptake surfaces (echinoderms and cnidarians);
- from groups that are not already endowed with osmoregulatory organs.

Where an occasional sponge (*Spongilla*) or cnidarian (*Hydra*) does live in fresh water, we know very little about how it is achieved, but it is evident that energy-consuming pumping across the exposed epithelial surfaces must be going on.

For the cephalopods, however, these arguments do not really work. This taxon is entirely marine now, and furthermore always has been. There is no immediately obvious reason why cephalopod skin could not be more impermeable, and the animals have perfectly good kidneys; after all, other molluscs do well in fresh water with the same basic apparatus. Several suggestions have been advanced to explain the cephalopod's restriction to saline waters:

1 They cannot toughen the skin to give greater impermeability because they use this surface so much for delicate, rapid and complex chromatophore signalling, and thicker impermeable layers would mask or slow down this system.

2 Being fast active predators, with quite high metabolic rates, they cannot cope with variation of oxygen because they have a relatively poor pigment (haemocyanin, like nearly all molluscs) with little ability for potentiated loading.

3 Their buoyancy mechanisms depend on active pumping of ions, and this is harder in fresh water against increased gradients.

4 In combination with reasons **2** and **3**, the extra cost of osmoregulation would be just too much for them.

5 Their central nervous system complexity and neuromuscular coordination needs constant conditions to work properly.

It may well be that a mixture of all these—again stressing the interaction of environmental and physiological factors—makes freshwater life just too difficult for cephalopods; perhaps they could cope with any one problem, but not with all of them at once.

However, there is an additional possibility. Cephalopod history has been a continuing saga of coevolution with fish; the cephalopods were dominant in the early Palaeozoic, but when teleosts returned to the sea, the cephalopods slumped, and recovery has only really occurred in those groups (squid) that are most like fish and could compete with them. So perhaps with teleosts already established in fresh water with low blood concentrations, the cephalopods just could not compete in potentially similar predatory niches, so leaving the fish (and subsequently a few other vertebrates) as the supreme actively swimming big predators of fresh waters.

This point neatly underlines some of the messages that should have emerged from this chapter, as a summary of freshwater biology. Firstly, it is crucially important to understand the inter-play of *all* factors in environmental adaptation, including their effects at physiological and behavioural levels and on morphology, biochemistry and life history. Studying the effects of single environmental variables in isolation is particularly inappropriate. Secondly, freshwater organisms give us yet more insight into the critical role of coevolution and convergence between groups of organisms in shaping present environments.

Further reading

Books

Adams, S.M. (1990) *Biological Indicators of Stress in Fish.* American Fisheries Society Bethesda, MD.

Horne, A.J. & Goldman, C.R. (1994) *Limnology.* McGraw-Hill, New York.

Mason, B.J. (1992) *Acid Rain: Its Causes and its Effects on Inland Waters.* Oxford University Press, New York.

Mason, C.F. (1991) *Biology of Freshwater Pollution.* Longman, London.

Morris, R., Taylor, E.W., Brown, D.J.A. & Brown, J.A. (1988) *Acid Toxicity and Aquatic Animals.* Cambridge University Press, Cambridge.

Moss, B. (1998) *Ecology of Fresh Waters: Man & Medium*, 3rd edn. Blackwell Science, Oxford.

Rankin, J.C. & Jensen, F.B. (eds) (1993) *Fish Ecophysiology.* Chapman & Hall, London.

Reviews and scientific papers

Byrne, R.A. & Dietz, T.H. (1997) Ion transport and acid–base balance in freshwater bivalves. *Journal of Experimental Biology* **200**, 457–465.

Byrne, R.A. & McMahon, R.F. (1994) Behavioural and physiological responses to emersion in freshwater bivalves. *American Zoologist* **34**, 194–204.

Chaui-Berlilnck, J.G. & Bicudo, J.E.P.W. (1994) Factors affecting oxygen gain in diving insects. *Journal of Insect Physiology* **40**, 617–622.

Deaton, L.E. & Greenberg, M.J. (1991) The adaptation of bivalve molluscs to oligohaline and fresh waters: phylogenetic and physiological aspects. *Malacological Review* **24**, 1–19.

Dietz, T.H., Neufeld, D.H., Silverman, H. & Wright, S.H. (1998) Cellular volume regulation in freshwater bivalves. *Journal of Comparative Physiology B* **168**, 87–95.

Feder, M.E. & Booth, D.T. (1992) Hypoxic boundary layers surrounding skin-breathing aquatic amphibians: occurrence, consequences and organismal responses. *Journal of Experimental Biology* **166**, 237–251.

Fryer, G. (1996) Diapause: a potent force in the evolution of freshwater crustaceans. *Hydrobiologia* **320**, 1–14.

Goss, G.G., Perry, S.F., Wood, C.M. & Laurent, P. (1992) Relationships between ion and acid–base regulation in freshwater fish. *Journal of Experimental Zoology* **263**, 143–159.

Henry, R.P. (1994) Morphological, behavioural and physiological characterisation of bimodal breathing crustaceans. *American Zoologist* **34**, 205–215.

Kirschner, L.B. (1995) Energetics of osmoregulation in fresh water vertebrates. *Journal of Experimental Zoology* **271**, 243–252.

McKee, D. & Ebert, D. (1996) The effect of temperature on maturation threshold body length in *Daphnia magna. Oecologia* **108**, 627–630.

McMahon, R.F. (1996) The physiological ecology of the zebra mussel *Dreissena polymorpha* in North America and Europe. *American Zoologist* **36**, 339–363.

Morris, S. & Bridges, C.R. (1994) Properties of respiratory pigments in bimodal breathing animals: air and water breathing by fish and crustaceans. *American Zoologist* **34**, 216–228.

Okland, J. & Okland, K.A. (1986) The effects of acid deposition on benthic animals in lakes and streams. *Experientia* **42**, 471–486.

Olson, K. (1994) Circulatory anatomy in bimodally breathing fish. *American Zoologist* **34**, 280–288.

Perry, S.F. (1997) The chloride cell: structure and function in the gills of freshwater fish. *Annual Review of Physiology* **59**, 325–347.

Pynnönen, K. (1994) Hemolymph gases, acid–base status and electrolyte concentration in the freshwater clams *Anodonta anatina* and *Unio tumidus* during exposure to and recovery from acidic conditions. *Physiological Zoology* **67**, 1544–1559.

Ram, J.L., Fong, P.P. & Garton, D.W. (1996) Physiological aspects of zebra mussel reproduction: maturation, spawning and fertilization. *American Zoologist* **36**, 326–338.

Reiber, C.J. (1995) Physiological adaptations of crayfish to the hypoxic environment. *American Zoologist* **35**, 1–11.

Schindler, D.W. (1988) Effects of acid rain on freshwater ecosystems. *Science* **239**, 149–156.

Schwarzbaum, P.J., Wieser, W. & Cossins, A.R. (1992) Species-specific responses of membranes and the $Na^+ + K^+$ pump to temperature change in the kidney of two species of freshwater fish, roach (*Rutilus rutilus*) and Arctic char (*Salvellinus alpinus*). *Physiological Zoology* **65**, 17–34.

Wenning, A. (1996) Managing high salt loads—from neuron to urine in the leech. *Physiological Zoology* **69**, 719–745.

Wheatley, M.G. & Gannon, A.T. (1995) Ion regulation in crayfish: freshwater adaptations and the problem of molting. *American Zoologist* **35**, 49–59.

Wilkie, M.P. (1997) Mechanisms of ammonia excretion across fish gills. *Comparative Biochemistry and Physiology A* **118**, 39–50.

Zerbst-Boroffka, I., Bazin, B. & Wenning, A. (1997) Chloride secretion drives urine formation in leech nephridia. *Journal of Experimental Biology* **200**, 2217–2227.

12 *Special Aquatic Habitats*

In this chapter we cover a range of 'aquatic' habitats that are in various ways not strictly within the definitions of marine, littoral, estuarine or freshwater habitats. Most of these are 'extreme' on one or more of the normal scales: so transient as to share more characteristics with the terrestrial world; or so osmotically or thermally or barically extreme, or so lacking in aqueous content, that they demand very special adaptations. In each case, the animals to be found there are real specialists, and are part of very strange communities often excluded from consideration in classical physiological textbooks.

12.1 Transient water bodies

A range of very specialist habitats occur where fresh water is only intermittently present, whether as puddles or small pools, or in interstitial habitats such as moss cushions and crevices in rocks, or as temporary pools of water trapped within vegetation or animal remains.

12.1.1 *Puddles*

Puddles occur in shallow depressions on slow-draining soils and are rapidly colonized by inocula of phytoplankton such as the spores of *Euglena* and *Chlamydomonas* (flagellate protistans), and by adult insects with aquatic larvae, most notoriously the mosquitoes and other flies. Larval stages of chironomids and ceratopogonids (non-biting and biting midges, respectively) are often hugely abundant, usually achieving success by having very short larval life cycles, often lasting only a few days. Amongst the chironomids, a famous puddle specialist is *Polypedilum vanderplanki*, whose larvae are longer-lived and can survive virtually complete drying out and baking by the Sun by using anhydrobiotic mechanisms as described below. In the biting midge *Dasyhelea* an alternative strategy occurs with the larvae constructing protective cocoons as their puddle dries out. Some water fleas also occur in puddles, from eggs dispersed on birds' feet, etc., and are known to absorb organic nutrients from the water directly through their cuticle. The same may be true of other small invertebrates.

Ground puddles in some parts of Africa and tropical America are also invaded by certain frogs as a tadpole habitat. With luck the tadpoles metamorphose before the puddle dries out. In east Africa the savanna ridge frog (*Ptychadena*) lays its eggs in vast numbers in small pools after rain, and tadpoles occur at densities of up to 1000 m^{-2} of water surface.

Where tadpoles occur, their excretions enrich the puddles and help to speed up development of the insect larvae they share the habitat with.

Puddle specialists must be relatively eurythermal and also tolerant of anoxia, but we have little detailed knowledge of their physiological adaptations.

12.1.2 *Water pools in plants and animals*

Water may be trapped within the leaf bracts of many kinds of plants, but the most conspicuous examples occur in the strong spiky bromeliad plants of the tropics, where the base of each leaf may trap a small tank of water for relatively long periods (Fig. 12.1a). Water is also retained in tree holes produced by natural damage or decay or by certain kinds of burrowing or nesting animals. Water pools also occur in various carnivorous pitcher plants, in the capsules of various nuts and seeds, and even in shelly animal remains. Where such habitats occur in living plants they may be well oxygenated by day but have periods with high levels of CO_2 and low pH; in animals or plant remains the water is commonly strongly eutrophicated and may be anoxic.

In these kinds of habitat a range of insects and other invertebrates can be found, together with larval treefrogs in the tropics. Frogs may deposit eggs and hence tadpoles individually in bromeliad leaf bracts. In some species the tadpoles are non-feeding, living off their yolk reserves; others are predatory, consuming larval insects and even other tadpoles, and in a few species the adults revisit their offspring and provide food for them. Some treefrogs that routinely use bromeliad pools or tree-hole pools have strangely ossified cranial structures, which they use to seal up the entrance to their shelters, and these ossified patches may have unusually low evaporative water loss (EWL)

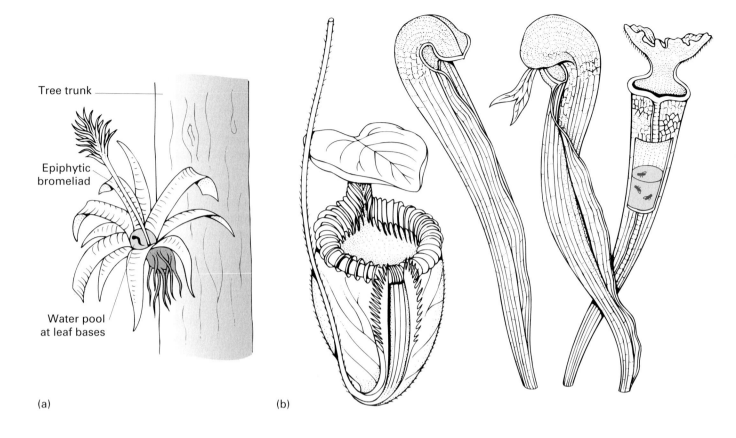

Tree trunk

Epiphytic
bromeliad

Water pool
at leaf bases

(a)   (b)

Fig. 12.1 Transient water pools used by various invertebrates and frogs in (a) leaf bracts of bromeliads and (b) pitcher plants.

rates so that once sealed into its hole the frog is resistant to drying out. Jamaican bromeliads also constitute a home for a specialist genus of crabs (*Metopaulias*), which undergo abbreviated larval development within the tanks provided between the leaf bracts. The larval stages are completed in the low pH water of the tanks (usually pH < 5–6), but not in normal river water from the same habitat (pH 8). Recent evidence indicates that the crabs can achieve some degree of pH and Ca^{2+} control by importing fragments of calcareous snail shells into the tanks.

Pitcher plants provide another kind of puddle (Fig. 12.1b; see Plate 6a, facing p. 310), though here the organisms living in the water may be even more specialist, often being carnivores exploiting the trapped animals that form the plant's 'prey', and thus perhaps best seen as inquilines. Pitcher plant mosquitoes are a well-studied group, able to live and breed in the fluid.

The vacated shells of land snails provide yet another small enclosure that can trap water and provide a temporary aquatic home. In Jamaica the crab *Sesarma* uses these; females carrying eggs find a snail shell of suitable size and fill it with up to 5 ml of water using a hairy patch of cuticle to carry the liquid by capillarity. Juveniles hatch and develop within the snail shell, guarded by the female for up to 3 months.

In these small aquatic habitats within plant tissues or shells, drying out is unlikely but temperatures may still fluctuate markedly. This leads to substantial differences in larval mortality and developmental rate according to the size and exposure of the initial water pool, so that the degree of physiological stress on the larval animal is largely determined by pre-emptive maternal choices.

12.1.3 *Extreme transience: moss cushions, cracks and crevices*

Some of the most transient of all aquatic habitats are sites such as moss cushions, cracks and crevices. Such habitats attract a unique fauna dominated by very small and mostly unfamiliar invertebrates (Fig. 12.2), such as nematodes, rotifers, collembolans, tiny crustaceans, and also the tardigrades (commonly known as 'water bears'). When active these moss-dwellers are covered by fresh well-oxygenated water (either bulk water or at least a covering film of water) and operate much like other freshwater life. But they must also be able to survive when the habitat dries up; this may occur in summer on an hourly or daily basis and with no very predictable pattern, relating more to sporadic rainfall than to clear-cut seasonality, and it may also occur in winter if the microhabitat suddenly freezes.

For these animals, many of which also survive in other extreme habitats such as high-latitude gravels, small crevices on glaciers,

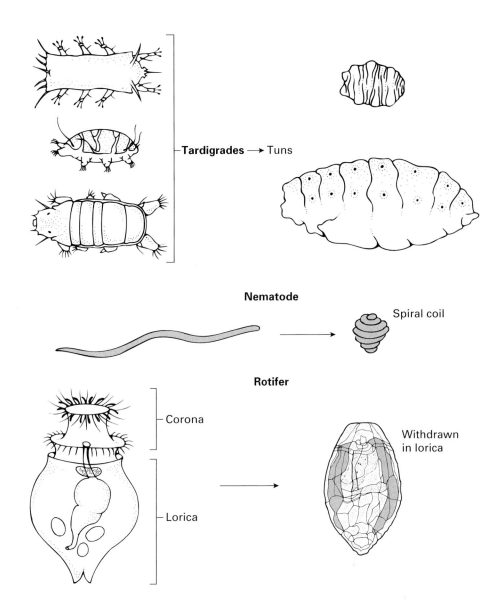

Tardigrades → Tuns

Nematode

Spiral coil

Rotifer

Corona

Lorica

Withdrawn in lorica

Fig. 12.2 Fauna that use cryptobiosis as a survival strategy in transient aquatic habitats.

and sands of desert wadis, the only possible survival strategy is to be able to cope with almost total desiccation, and all therefore exhibit the phenomenon of **cryptobiosis**, or 'hidden life', also known as **anhydrobiosis** (life without water), appearing to be dead for long periods and then apparently miraculously coming back to life days, weeks or even years later when liquid water is supplied.

A number of groups can achieve this in their juvenile stages as eggs or cysts, but others (the rotifers, nematodes and protists) do it as adult organisms. Perhaps the most famous and peculiar examples are the tardigrades, amongst the most endearing of all animals. Figure 12.3 shows an active tardigrade in its hydrated state, and then two further stages showing what happens as its habitat dries out until eventually it is in the form called a 'tun' (meaning barrel), with all the limbs withdrawn, and the inter-segmental cuticle tucked in, leaving just a smooth surface of the

thicker cuticular plates exposed. The animal can be reduced to just a small percentage of its initial water content and yet it stays alive.

Three main mechanisms are known to be involved in the process of entering cryptobiosis, whether it is initiated by high temperature, by drought or by freezing, and the mechanisms involved seem to be the same in all the groups that can achieve this 'playing dead' trick:

1 Changing morphology of the surface, by tucking in more permeable areas of cuticle and changing the relative thickness or chemistry of cuticle layers. Nematodes cannot contract into a tun, so they either spiral themselves up into a neat ball instead, sometimes in aggregations, or they enter cryptobiosis within the shed but not discarded skin from the previous moult. Rotifers withdraw into their lorica (the casing that protects them), and tuck in their high-surface-area feeding structure, the ciliated

(a)

(b)

(c)

Fig. 12.3 SEMs of a tardigrade (a) in its normal hydrated state, (b) shortly after beginning entry to the cryptobiotic state, and (c) when fully into this state (the 'tun'). (Courtesy of J.C. Wright.)

corona. In nematodes the hyaline layer of the cuticle is visibly altered, and in tardigrades wax extrusion may occur.

2 Reducing the permeability of the cuticle, by lipid phase changes and/or wax production, so ensuring that a minimum amount of water is retained. The 'permeability slump' has a time course of tens of minutes in the relatively complex tardigrades such as *Echiniscus*, and only about two minutes in the anhydrobiotic nematode *Ditylenchus*. The final water content of the animal may be only 5–15% of the initial value.

3 Metabolic production of a large quantity of particular sugars, such as trehalose and glycerol, that protect cell membrane phospholipids and proteins from damage during drought by binding into the membranes in place of the water and stabilizing their structures. This process is accompanied by 'cytoplasmic vitrification', turning the intracellular fluid into a glass-like material. Trehalose is an α-1-linked non-reducing disaccharide of glucose, formed in many microorganisms as well as in cryptobiotic animals and in insects. It is probably a stress-protectant for two main reasons: firstly, because it interacts with and directly protects both lipid membranes and proteins, probably by hydrogen bonding to phosphates and other polar residues within the dried out macromolecular assemblages; and secondly, because it naturally forms a 'glass' material, resulting in vitrification of the whole cytoplasm. In a cryptobiotic nematode subjected to either freezing or desiccation it occurs in all tissues, but especially in muscle and in reproductive organs. Its production is catalysed by two key enzymes, trehalose-6-phosphate (T6P) synthase and T6P phosphatase, and its degradation to glucose is mainly controlled by trehalase. However, the regulation of these enzymes is largely unknown. Likewise the reasons for the particular success of trehalose as a stabilizer relative to other sugars are not fully elucidated; in part, it may be that trehalose has a high glass-transition temperature that is relatively unaffected by small amounts of water. Cytoplasmic vitrification due to trehalose has to be maintained throughout the cryptobiotic phase, and if it is lost then phase separations occur in membranes, free radical oxidation begins, and cytoplasm may instead undergo a process akin to crystallization, which irreversibly damages the cells.

A summary of the control of anhydrobiosis is shown in Fig. 12.4; note that there is a need for relatively slow drying in moist air to give a good chance of survival, so that these three mechanisms have time to be set in place, and that if resuscitation begins in dry air it can be reversed. The moisture-retaining properties of mosses tend to ensure humid air in the phases before and after rainfall, so that the animals in this habitat have a reasonable guarantee of successful entry to and exit from cryptobiosis.

The ability to enter cryptobiosis seems to be age dependent, and it cannot be achieved in very young eggs or embryos in most species where it occurs. However, when it occurs in more mature animals it modifies the timing of subsequent life-cycle events but not the reproductive output (the pattern of age-specific fecundity), indicating merely a resetting of an internal clock system.

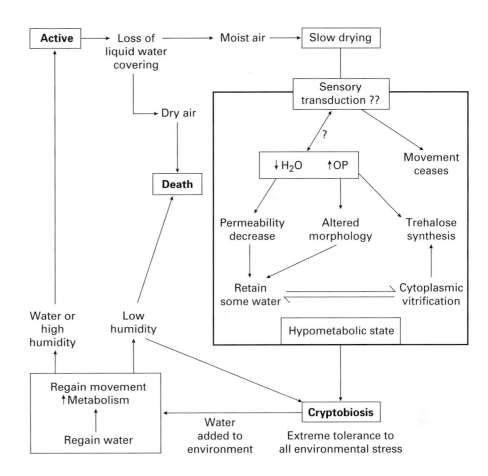

Fig. 12.4 The control of cryptobiosis; the effects of water presence, and humid or dry air externally, and of endogenous controls within the animal. OP, osmotic pressure.

12.2 Osmotically peculiar habitats

12.2.1 High-salinity lakes and brine ponds

There is almost as much water contained in salt lakes (including inland 'seas') in the present world as there is in freshwater lakes (see Table 11.1). Wherever freshwater lakes occur without an outflow they tend over time to turn into salt lakes (Plate 6b), or **endorheic lakes**, and often eventually dry up completely into salt flats (Plate 6c). The resulting inland brackish waters will normally have a salt composition similar to that of the sea, but where they are produced mainly by evaporation from springs they can have a very different ionic composition, and are described as **athalassic** (meaning not sea-like). Compositions of some salt-lake water are shown in Table 12.1; note that in extreme cases the total osmotic concentration may be 6–8 times that of sea water.

Such lakes form particularly in three conditions (Fig. 12.5):
1 In the rain-shadow areas of high mountains—to the north of the Himalayas, in Nevada and Utah and central Canada beyond the Rockies, and in parts of Chile and eastern Australia. Lake Mono in California, and the Great Salt Lake of Utah, are particularly well studied.
2 As the relatively undrained and long-lived lakes formed

Table 12.1 Compositions of some salt lakes.

Lake (location)	Osmotic concentration (mOsm)	Ionic concentrations (mM)						Salinity (ppt)
		Na	K	Mg	Ca	Cl	SO$_4$	
Great Salt Lake (USA)	~6000	3000	90	230	9	3100	150	210
Mono Lake (USA)	~2000							70
Dead Sea (Middle East)	6200–6500							226
Lake Koombekine (Australia)	~7500	4070	17	110	32	1670	72	263

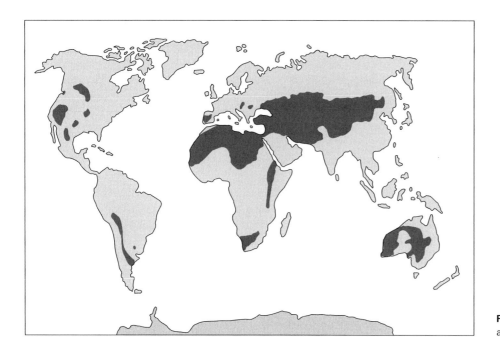

Fig. 12.5 Areas in which endorheic salt lakes are commonly found. (From Moss 1980.)

Table 12.2 East Africa soda lake characteristics.

Lake	Area (km²)	Maximum depth (m)	Mean pH	Alkalinity (mEq l⁻¹)
Turkana	8000	120	9.6	19–24
Manyara	400	1.5	9.5	78–800
Baringo	150	8	9.0	4–10
Magadi	95	0.6	10.1	400–3600
Nakuru	43	1.3	10.4	120–1500

by geological rifting in hot climates where evaporation is intense, notably in East Africa (Table 12.2) and the eastern Mediterranean. The Dead Sea has formed primarily by this process of long-standing evaporative concentration, but receives a small surface input from spring waters and river water, so that periodically it undergoes a substantial mixing event (most recently in 1979).

3 At higher latitudes where lakes forming at the end of the last Ice Age have become isolated from the sea by isostatic rebound (rising land levels as the heavy ice retreats). These lakes are quite common in Iceland, Greenland and parts of Canada, and have very variable degrees of hypersalinity. Often they have become freshened from the top down, and have saline and hypersaline water only at depth.

High-salinity conditions also occur as smaller 'brine ponds' in many areas, often where water accumulates temporarily on top of dried salt flats (Plate 6c) or where pools and coastal lagoons become isolated just above the littoral zone, especially on low-latitude coasts. Here concentrations of up to four times sea water are not uncommon.

Salt lakes and ponds obviously pose a major physiological problem to animals in terms of osmotic and water balance. However, they also tend to be very strongly stratified, due to high water viscosity, needing much stronger wind action to bring about an overturn (i.e. they are usually amictic or monomictic). Thus marked oxygen differences and thermal gradients must also be dealt with for mobile animals to be successful colonizers.

Biota in the lake water

The biota of endorheic lakes and ponds is usually very low in diversity. Extreme cases are dominated by just one or a few species of cyanobacteria, flagellates or halobacteria; but where water is reasonably persistent above the salt pans, the brine shrimps *Artemia* and *Parartemia* are almost ubiquitous colonizers. The species *Artemia salina* may achieve several generations per year in salt lakes as these commonly occur at lower latitudes. In Lake Mono there is a spring generation and a much faster-developing summer generation, when warmer temperature outweighs reduced food availability (but mortality is rather higher with many failures to develop). Other crustaceans also do well in high-salinity pools above tropical shores; the fiddler crab, *Uca rapax*, is common in South American lagoons. In the most extreme salinities all fish are absent, but in salty (up to 140 ppt) high-temperature lakes the desert pupfish (*Cyprinodon*) occurs. The larvae of several species of brine fly (*Ephydra*, *Ephydrella* and *Hydropyrus*) also occur in vast numbers in brine pools in many parts of the world, together with some chironomid midges, such as *Cricotopus*, and mosquitoes, such as *Aedes*. All the insects and crustaceans are extremely good osmoregula-

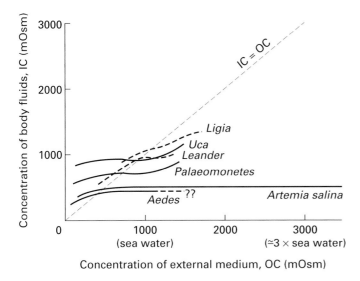

Fig. 12.6 The patterns of hypo-hyper-regulating in various brine-dwelling animals, comparing internal and external media.

tors (Fig. 12.6), maintaining rather constant blood concentrations in both dilute and very salty media; they are sometimes termed 'hypo-hyper-regulators'.

For the various invertebrates faced with these problems, the main response seems to be to increase the drinking rate several fold, this being the only way to acquire the essential water; adult *Artemia* can drink 5–8% of their own body mass per day, with water and ions then gained across the gut epithelium. Surface permeability is relatively low compared to most crustaceans (P_{osm} only 0.1 µm s^{-1}, cf. Table 4.1), so that outward leak of the water thus gained is reduced, but the inevitable consequence of drinking is a salt load, which must be excreted. In adult *Artemia* (Fig. 12.7b) the salt glands on the gills perform this function, though in the larvae it is achieved by a specialized area of tissue at the back of the head, known as the 'neck organ' (Fig. 12.7a). At both sites the epithelium concerned has very high sodium pump (Na$^+$/K$^+$-ATPase) activity, and chloride pumps basally or apically (or both) are also proposed. In the very small larvae of mosquitoes in the genus *Aedes*, hyporegulation is achieved by extraordinary levels of drinking (up to one-third of the body volume per day in *A. campestris* and even double the body volume per day in *A. taeniorhynchus*), with rapid and strongly hyperosmotic salt secretion from the anal papillae projecting from the rectum (the normal site for ion and water regulation in insects).

Many brine-dwelling animals (especially the insects) are able when necessary to excrete urine strongly hyperosmotic to their own blood, with concentrations of up to 8700 mOsm in *Ephydrella*. The crustaceans achieve this via their antennal glands, which can resorb or excrete ions into the urine as necessary, though to only a limited extent. Similarly in the insect's rectum there are large numbers of specialized secretory cells, as

well as the more normal resorptive cells that produce dilute urine in dilute media. The Malpighian tubules of insects from hypersaline lakes rich in Mg^{2+} and SO$_4^{2-}$ are also unusual in being able to pump these ions very rapidly to the lumen. It is worth noting that hyposmotic regulation in the brine shrimp and mosquito larva is relatively much more costly (22–33% of the metabolic budget) than hyperosmotic regulation in any animal yet investigated (cf. Table 5.10), which may help explain why it is a strategy adopted by so few animals.

In very salty water the ion levels may be enough to reduce dissolved oxygen content markedly. It is therefore not surprising that in *Artemia* the concentration of haemoglobin (Hb) and the proportions of different Hb morphs vary with salinity; hypoxia and high salinity both cause an increase in levels of Hb$_3$, a beta homodimeric form of the pigment that is structurally relatively unaffected by high ionic strength.

Artemia also links us back to the issue of cryptobiosis, since the brine shrimp uses this strategy to withstand drying out in the shallow fringes of salt lakes. The resistant stages are usually called 'eggs' but are in fact encysted gastrula embryos; they may have a water content of less than 0.15 ml g^{-1} dry matter, and have no measurable metabolism.

Fauna associated with salt lakes

There are a number of larger animals found in association with salt lakes though not living entirely within the salty water. Most obvious of these are the flamingos, which are spectacularly abundant on the salt lakes in central Africa, such as Lake Nakuru, in some Andean lakes, and in some salt lagoons of southern Europe. They survive by filtering out the unicellular algae. It might be supposed that they would thus be bound to accumulate a high salt load and should therefore require special kidney adaptations, but in fact they solve the problem behaviourally rather than physiologically, by flying off to freshwater lakes regularly for a drink.

12.2.2 *Acid and alkaline waters*

Figure 12.8 summarizes the range of pH found in naturally occurring lakes, in comparison with normal lakes and various other natural substances. Both high and low pH pose particular problems for animal life.

Alkalinity: soda lakes and crater lakes

Moderately alkaline waters occur in chalk and limestone areas worldwide, and lakes with inputs of high pH and few outlets may become increasingly alkaline through their life. This poses real problems for animals' acid–base balance, given the linkage between H$^+$ ions and key transport processes in many tissues.

Problems arise especially in soft-water areas (where run-off is from non-calcareous rocks), where mortality in non-specialists can be very high. Where the water is hard, Ca^{2+} from the

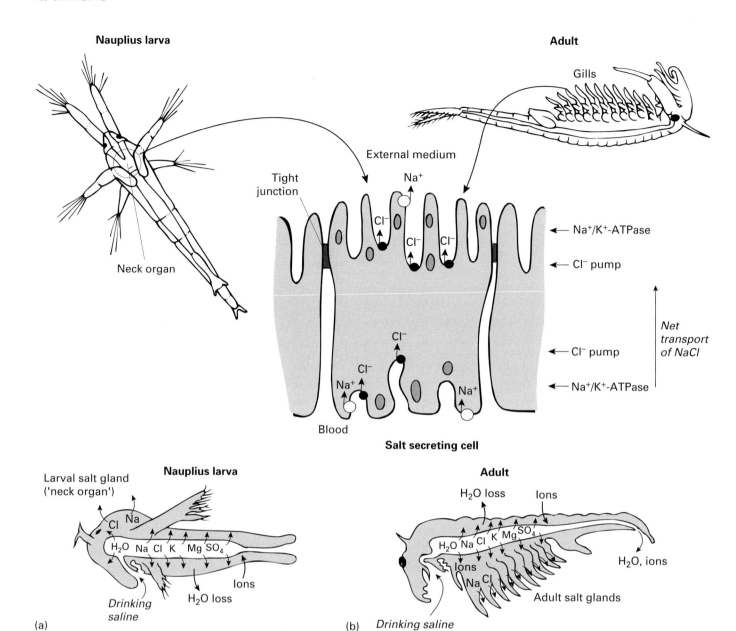

Fig. 12.7 The structures that regulate salt elimination during hypo-regulation in the brine shrimp, *Artemia salina*.

environment (chalk and limestone) appears to protect animals against excessive disruption of their ionic and acid–base balance. The main effect operating may be a calcium stimulation of ammonia excretion, via a Na^+/NH_4^+ exchange mechanism, helping to keep the blood from experiencing raised pH. Certainly fish die if there is no calcium present, and do better in hard waters, which are perhaps fortunately the commonest kind of alkaline aquatic systems.

Moderately alkaline lakes have similar fauna to normal freshwater lakes at similar latitudes, but usually with greatly reduced species diversity. For example, there are just 16 species of chironomid flies in the Neusiedler See in Austria and Hungary, one

of Europe's largest shallow alkaline lakes, compared with 40–50 in many other central European lakes. The chironomid ('bloodworm') biomass is also markedly reduced.

More extreme alkaline conditions occur in 'soda lakes' in various parts of the world, especially in areas of rifting and volcanic activity. One of the better studied examples is Lake Magadi in Kenya (Plate 6d), routinely having a pH of 10 and a temperature of 30–42°C with a 'salinity' equivalent to about 50–65% sea water. Pyramid Lake in Nevada, with a pH of 9.4, and Lake Van in Turkey (pH 9.8), are also homes to well-known alkaline-adapted fish. Certain alkaline crater lakes in Central America lack fish species but are dominated by chironomid larvae and oligochaetes; they are often warmed by peripheral 'geysers' (Plate 6e).

One of the more successful species from African alkaline lakes (and the only fish to survive in Lake Magadi) is *Oreochromis*

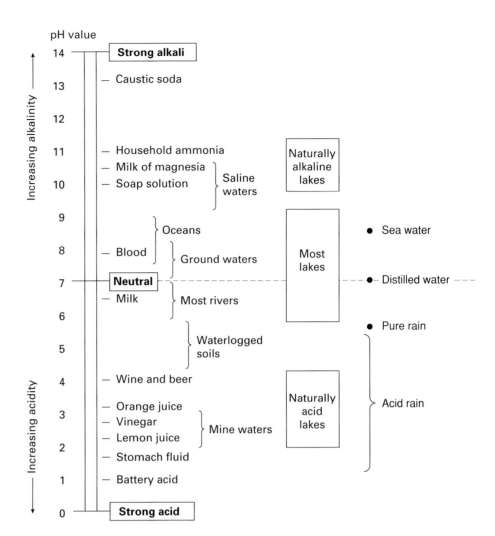

Fig. 12.8 The range of pH found naturally in lakes, in comparison with other natural fluids. Remember that the pH scale is logarithmic.

grahami, a cichlid that excretes at least 90% of its nitrogen as urea so avoiding toxic ammonia accumulation. Trout from Pyramid Lake and the unusual tarek fish from Lake Van achieve a lesser degree of switching to ureotely, and tolerate relatively high levels of ammonia in their tissues.

Oreochromis suffers acidosis and ion-balance disruption if transferred to water of normal pH. This species has apparently standard seawater-type chloride cells in its gills that can be activated or covered over by pavement cells as in their estuarine and freshwater relatives (see section 10.2.2), and which presumably carry out 'salt' excretion as in seawater teleosts, with carbonate/bicarbonate replacing chloride in the main outward transport system. Exposure to non-alkaline waters tends to deactivate these cells in the short term (2–3 h), though this is reversed within about one day of acclimation. However, in the other alkaline lake fish, exposed to relatively dilute media, the gill chloride cells are probably more like the freshwater teleost type (Chapter 11).

As with other osmotically peculiar habitats, the raised salt concentration in these lakes tends to produce low oxygen concentrations, and this may explain the relatively thin diffusion barrier in the gills of *Oreochromis*, rather than a thickened protective gill surface that might have been expected. This fish will perform supplementary surface breathing on occasions.

Various species of mosquito larvae are also found in soda lakes, with *Aedes campestris* (Fig. 12.9) being typical; here again there is excretion of a hyperosmotic urine with HCO_3^- accumulated from the environment replacing the normal chloride.

Lake Natron in East Africa is an extreme example of a large soda lake, fed by hot springs from nearby volcanoes and having a solid soda crust. This particular lake has almost no signs of life, except that it is used by lesser and greater flamingos for breeding, and the young are hatched on raised nesting platforms above the almost crystalline soda. The fledglings rapidly die of heat exhaustion if they fall off the nest; but so long as they stay on the nest platform they survive, fed on regurgitated crop 'milk' from the parents, which has been desalted from their own food. It is presumed that the flamingos migrate in from other lakes such as Nakuru to breed at Lake Natron because there is no predation pressure on the fledglings.

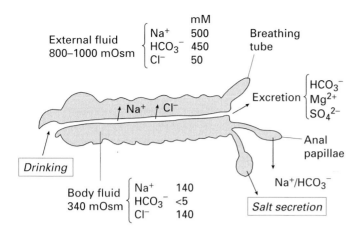

Fig. 12.9 Regulation of body fluids by *Aedes campestris* in strongly alkaline water. (Data from Phillips *et al.* 1971.)

Acidic waters

Acid waters occur as lakes, but also in semi-terrestrial conditions such as peat bogs, and even in some man-made situations such as vinegars. All of these may support animal life, with varying degrees of specialization to regulate the organisms' acid–base balance in extreme H+ surroundings. We looked at some of the consequences of anthropogenic acidification in Chapter 11, and the deleterious effects on species not adapted to cope with low pH. It follows that the same problems must have been solved by acid-water residents. The adaptations needed can therefore be summarized as follows:
- control of the physiological effects of low pH on ion uptake and internal acid–base balance;
- tolerance or detoxification of solubilized minerals, of which aluminium is most important, aluminium poisoning causing mucus production and clogging of gills, etc. (with iron and copper also potentially important depending on the local geology);
- limitation of effects on physiological processes, such as cuticle formation in insects and crustaceans, which would otherwise reduce osmoregulatory abilities;
- changes to the control of hatching in many invertebrates, and especially in vertebrate eggs, because the enzyme chorionase, which allows the larva to escape from the chorionic membrane, can normally only function fully at a pH of 6.5–8.5.

Adaptation of key membrane and enzyme parameters must underlie most of these issues, with proteins functioning under unusual ionic conditions. Many of the acid-water dwellers appear to have slightly to moderately acidic blood, representing some degree of hydrogen ion regulation at gills, kidney or gut that will limit the stress on the cellular constituents. The crustacean *Allanaspides,* from Tasmanian peat bogs at pH as low as 3–4, can keep its blood at neutrality (pH 7), which represents very considerable hydrogen ion regulatory powers. There is

probably net H+ secretion at the gills, and also at the 'neck organ', by modified Na+/H+ exchange processes.

12.2.3 *Living in oils*

In certain parts of the world natural waters become readily contaminated by seepage from underground oil deposits, and the resultant 'hydrocarbon seeps' still support an array of metazoan life (the same may be true of the brine seeps accumulating above undersea oil deposits, though these are scarcely explored). The fluid in these seeps may contain dissolved sulphide and volatile, potentially toxic hydrocarbon compounds, but also provides an enriched source of carbon and nitrogen. Nematodes are particularly well represented, and their numbers often increase with oil content, whereas other taxa such as polychaetes, oligochaetes, gammarids and copepods tend to reduce in number in the more oil-rich seeps. A peculiarly resistant cuticle with a dense inert hyaline layer, and a pronounced tolerance of anoxia, are assumed to contribute to the success of nematodes in such strange habitats.

12.3 Thermally extreme waters

12.3.1 *Deep-sea thermal vents*

Throughout the world's ocean depths, where tectonic plates meet, a mid-ocean ridge is formed from material rising up from the mantle as the continental plates move apart. Sea water seeps downwards through numerous crevices and fractures and becomes heated by the hot mantle magma. The superheated pressurized water (up to 350°C) then forces up through the crust to emerge as 'hydrothermal vents' (Fig. 12.10). The water is rich in silicates, hydrogen sulphide and various sulphide minerals incorporating iron and magnesium, which crystallize as the hot water meets the cold oceanic water, forming 'black smokers', 'chimneys' and other mineral deposits. At the typical depth of 2000–3000 m the ambient water is at about 2°C, with an ammonium concentration of less than 0.5 μM, a nitrate concentration of 40 μM, silicate at 150 μM and no hydrogen sulphide. In contrast the rising vent water contains zero oxygen due to reaction with high sulphide levels, ammonium concentrations can be high, and there may be natural radioactivity. In the zone of mixing, conditions become such that life is possible. Prokaryotes can occur in water almost up to boiling point (which at depth may mean 120°C); the enzymes in such organisms ('extremophiles') are highly specialized, and stabilized with many S–S bonds. However, eukaryotes do not appear until the water is cooled to 55°C and below, while metazoan animals generally need temperatures below 50°C. In practice, most animal life is associated with areas where the vents interact with the surrounding water to produce temperatures below 30°C and hydrogen sulphide concentrations of less than 400 μM. Vent animals are particularly abundant in and on the 'lava' fields that surround the vents, immediately above which the water temperatures may still be

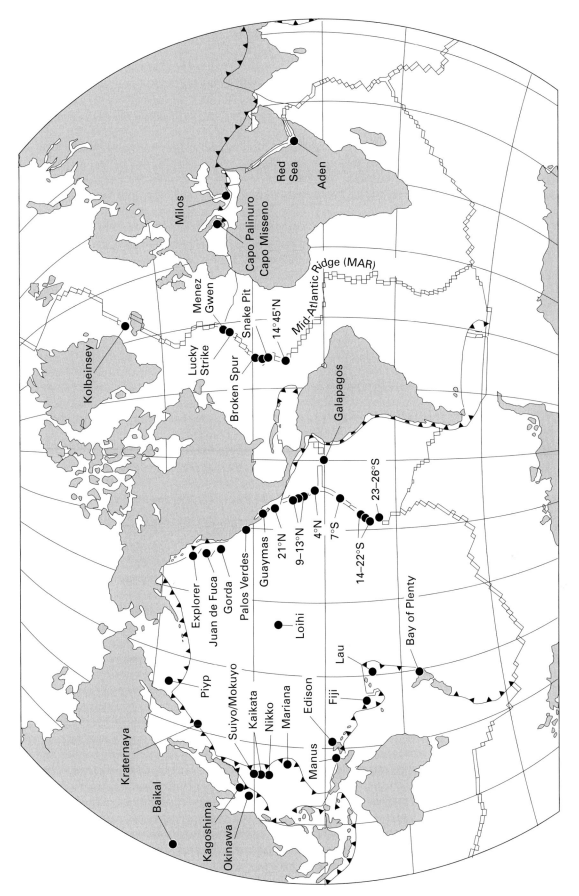

Fig. 12.10 Distribution of hydrothermal vents, primarily around mid-ocean ridges. (From Desbruyeres & Segonzac 1997; copyright Pierre Chevaldonné, IFREMER.)

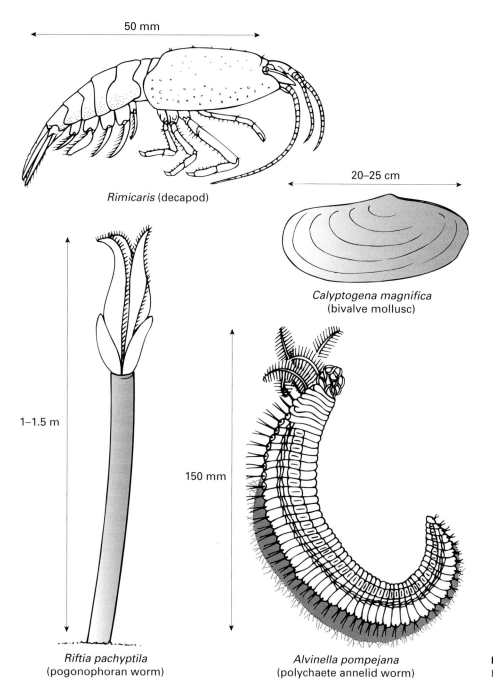

50 mm

Rimicaris (decapod)

20–25 cm

Calyptogena magnifica
(bivalve mollusc)

1–1.5 m

150 mm

Riftia pachyptila
(pogonophoran worm)

Alvinella pompejana
(polychaete annelid worm)

Fig. 12.11 Fauna found at deep
hydrothermal vents in the oceans.

12°C or more; their very presence alters the mineralogy of the chimney structures.

Discovered in 1977, this fauna is dominated by annelid and pogonophoran tubeworms and bivalve molluscs, together with some crustaceans (Fig. 12.11). Some 250–300 species of vent macrofauna have been identified, of which 200 or more are new to science. The fauna is endemic and taxonomically diverse; some vent faunas are also very ancient relative to the surrounding deep sea.

Some vent structures pose an additional problem since they are very unstable, showing major changes in structure and water flow over weeks or months. *In situ* recordings have shown that the temperature and water chemistry at sites where tubeworms and other fauna live are also highly variable as the waters are not well mixed. The vents at 13°N in the east Pacific typically last for only a few years. Furthermore, vent fields are often separated by tens to hundreds of kilometres of 'normal' ocean. Thus the composition of the fauna at particular vents is not fixed, but may be predictably dictated by their age and site in relation to recruitment and larval dispersal.

Coping with high sea temperatures and pressures

Vent animals live in temperatures that may range from nearly ambient (2°C) to 50°C and above. One of the most spectacular cases is the aptly named 'Pompeii worm' (*Alvinella pompejana*), a tube-dwelling polychaete that lives on vent chimneys. The water in its tubes has been recorded at 68°C, with occasional peaks up to 81°C. At the tube mouth the water is only around 22°C, giving a possible gradient of 60°C along the length of the worm's body. This makes *Alvinella* potentially the most eurythermal animal on record. As yet we have no real information on the temperature of its tissues, or whether the proteins and membranes of these worms can function under such conditions, though there are clear differences in the thermostabilities of some of the key enzymes in different species.

Detecting high temperatures may also be an issue for more mobile animals in these habitats, and the deep-sea shrimp *Rimicaris exoculata*, which occurs around vents, lacks normal eyes on its head but has a thoracic 'eye' adapted for vision in very low light and probably mainly used to detect the dim blackbody radiation emitted by the very hot vent water emissions.

High pressures also create problems, with pressures of around 30 MPa being commonplace at the vents of the Pacific Ridge. Studies of pressure effects on vent crabs reveal, as expected from physical principles, a reduced level of damage when high pressures are accompanied by high temperatures (the former causing reduced membrane fluidity, the latter increasing this fluidity); thus effects on key enzymes and on membrane bilayers are relatively mild when the conditions at the vent are reproduced in laboratory studies.

Nutrition and respiration: coping with hydrogen sulphide

At the Galapagos hydrothermal vents, hydrogen sulphide concentrations vary from 0 to 300 µM. Hydrogen sulphide can only accumulate in those environments which are low in oxygen, as it is readily oxidized. Sulphide is potentially toxic, binding to the haem site of cytochrome *c* enzymes to inhibit cellular respiration. Sulphide can also potentially reduce the disulphide bridges in proteins, inhibiting normal protein function. Passive resistance to sulphide (having insensitive enzymes) appears to be rare, and the tubes or mucous layers of the fauna provide no real protection, so most animals living in high sulphide environments have one or more detoxification mechanisms. In addition, nutritional specializations are required to cope with the environment, which is dominated by sulphide rather than by carbon and oxygen.

One possibility is to use sulphide-binding proteins, as found in the pogonophoran tubeworm, *Riftia pachyptila* (Plate 6f), and the hydrothermal vent clam, *Calyptogena magnifica*. These proteins draw free sulphide into the general circulation where it is strongly bound, and thus achieve three important functions:

1 the preservation of aerobic respiration;
2 the prevention of sulphide precipitation in the blood (sulphide granules might impede circulation);
3 the transport of sulphide to internal centres where symbionts' enzymes can exploit it for energy production.

These blood-borne binding proteins have probably evolved separately in each group. The sulphide-binding protein in *R. pachyptila* is the haemoglobin itself, present in high concentration in both vascular and coelomic blood; it can bind oxygen and sulphide at different sites. The same is true for the polychaete *Paralvinella*. In contrast, in the clam *C. magnifica* the haemoglobin occurs in erythrocytes and the sulphide-binding protein occurs quite separately in the serum.

Many of these animals are also symbiotic with bacterial chemoautotrophs that use a sulphur-based metabolism and release carbon compounds to their hosts. The animals provide oxygen and carbon dioxide from the sea water, as well as the necessary dissolved sulphides. Representatives from many phyla incorporate specialized structures containing sulphide-oxidizing bacteria. The nematode *Eubostrichys dianeie* houses sulphur-oxidizing bacteria in a mucus web over the exterior of its body. The worm *A. pompejana* maintains an epithelial 'fur' of fine projections densely packed with sulphur-containing filamentous bacteria. Bivalves have specific intracellular (endosymbiotic) strains of sulphur bacteria housed in their gills in modified cells called bacteriocytes, and gastropods also have bacteriocytes in their gills. In pogonophoran worms, sulphide oxidation occurs in the 'trophosome', an internal organ that fills much of the coelomic cavity and contains extremely high concentrations of sulphide-oxidizing bacteria. It is worth looking at the strategies of some of these animals in more detail.

Bivalves

Vesicomyid bivalves live in many deep-sea reducing environments, including hydrothermal vents, hydrocarbon seeps and anoxic sediments. In the vent clam, *Calyptogena magnifica*, a mouth and gut are present although the feeding and digestive apparatus are reduced. The enlarged gills contain sulphur-oxidizing chemolithoautotrophic bacteria which provide the major energy source of the clams and may average 17% of the animals' wet weight. *C. magnifica* lives on rocks and extends its highly vascularized foot down into cracks, where the hydrothermal fluids rich in hydrogen sulphide (up to 40 µM) are slowly venting. Sulphide appears to be taken up by the foot and is transported bound to the serum sulphide-binding factor via the bloodstream to the gills. The clam's serum can reversibly bind up to 8 mM sulphide, resulting in a significant concentration above ambient seawater levels. The serum-binding component in *C. elongata* has a large molecular mass and contains Zn^{2+} at the active site.

Any unbound sulphide is oxidized to the non-toxic compound thiosulphate in the foot and transported to the gills. The animal therefore has moderate concentrations of circulating thiosulphate but virtually no free sulphide. This suggests that in these high-sulphide environments the clams have evolved the ability to oxidize sulphide to thiosulphate for themselves, and that the incorporation in the gills of chemoautotrophic bacteria that utilize thiosulphate is a secondary adaptation to exploit what would otherwise have been a waste product. The gill bacteriocytes are exposed to the sea water on one side and the blood of the clam on the other, so that nutrients can be obtained from both sources. The siphon of the clam extends into the ambient water such that the animal effectively forms a bridge between the reducing environment of the substrate and the oxygenated sea water. Some of the energy released during sulphur metabolism in the bacteria is trapped as ATP and NADPH, which are used in part to drive net CO_2 fixation. The clams then digest a fraction of the bacteria to satisfy their nutritional needs. As a consequence, the gut can be fairly degenerate (an adaptation found in many of these vent animals).

Other vesicomyid clams, such as *Calyptogena elongata*, which lives on muddy substrates in other parts of the deep sea, appear to have a broadly similar type of symbiosis, with the foot penetrating into the anoxic sulphide-rich sediments and the siphon extending into the water above. Mussels with chemoautotrophic bacteria are also common on hydrothermal vents and in the sulphide-rich oozes. They are much less specialized for symbiosis than the vesicomyid clams, in that they have retained the ability to ingest particulate matter for themselves, and have no significant concentrations of sulphide binding in their gills and haemolymph. The mussel symbionts utilize thiosulphate rather than sulphide. Some species do have methanotrophic (instead of sulphur-oxidizing) symbionts, but must still deal with the potentially toxic sulphide.

Pogonophoran worms

These have a very different strategy for sulphide metabolism. Most of their sulphide oxidation occurs in the 'trophosome'. The worms lack a mouth and gut, and the plume of tentacles serves as the primary site of exchange of gases and solutes with the ambient sea water. The abundant haemoglobin (up to 26% wet mass of worm) in the vascular and coelomic spaces carries both oxygen and sulphide, as we saw earlier, and the sulphide is delivered to the trophosome from the haemoglobin. The bacteria are then able to fix CO_2 by involving the enzyme ribulose 1,5-biphosphate carboxylase, providing reduced carbon compounds, which in *Riftia* sp. supply almost 100% of the worm's requirements. The bacterial symbionts in the pogonophoran trophosome have several adaptations to protect the animal from sulphide toxicity. They have a V_{max} for sulphide that is at least an order of magnitude greater than in free-living sulphur-oxidizing bacteria; and they can shunt excess sulphide into non-toxic elemental sulphur deposits. Elemental sulphur (which can constitute up to 20% dry weight of the trophosome) is further oxidized when blood sulphide levels decrease. The symbionts utilize only sulphide (*not* thiosulphate, as in many other vent animals), and

the worm is highly specialized to minimize the interaction of its own metabolism with sulphide. The use of haemoglobin as the sulphide-binding protein enables the worm to concentrate sulphide from the medium by almost two orders of magnitude. However, because the haemoglobin binds sulphide, and the bacteria use it up, the free concentration of sulphide in the haemolymph is an order of magnitude below external H_2S concentrations. The Hb has a very high affinity for oxygen ($P_{50} = 0.1$–0.3 Torr at pH 7, equivalent to about 2 mM).

Recent evidence suggests that the bacteria in *Riftia pachyptila* can use nitrate in addition to oxygen, producing ammonia and nitrite as end-products. The presence of sulphide stimulates nitrate respiration up to 500 µM sulphide, whereas thiosulphate has no effect, again supporting the idea that the symbionts are sulphide specialists. Nitrate respiration increases with decreasing oxygen concentrations. The use of nitrate respiration presumably enables the symbionts to function in a very low oxygen environment whilst still gaining energy through respiratory pathways.

Crabs

The crab *Bythograea thermydron* also occurs at hydrothermal vents, and shows high tolerance for hydrogen sulphide and for raised temperatures. Like the molluscs described above, it has been shown to oxidize H_2S to thiosulphate, this time in the hepatopancreas. In this species the thiosulphate has an additional adaptive function in increasing the affinity of the respiratory pigment (haemocyanin) for oxygen.

Coping with unusual minerals

Several of the common vent animals exhibit bioaccumulation of unusual elements in their tissues (notably cadmium, copper and in some cases zinc), with levels fluctuating as the vent emissions change. These elements are stored as mineral compounds or bound with insoluble ligands of high chemical stability. Some are also detoxified as more soluble metallothioneins (low molecular weight proteins), whose half-life varies according to the associated cations.

Coping with reproduction and dispersal

Hydrothermal vents are widely separated and often rather short-lived, so that vent organisms face regular bouts of extinction unless they can produce effective dispersive propagules. These may be either larvae or motile adults, which can be carried by seafloor currents supplemented by their own swimming activities. There is little doubt that dispersion is effective and rapid; new vents have been shown to be colonized within a year of their formation. It remains unclear how the colonizers find the vents though; recent evidence indicates that sulphide itself may serve as an attractant, at least to highly motile vent shrimps. Late larval stages of these shrimps have unusual deposits of lipid in their thorax and abdomen, comprising 75–82% wax esters, these same compounds being absent in the adults. These reserves presumably provide the food reserve for a relatively prolonged bathypelagic existence in which opportunities for planktotrophy are rare. A similar explanation may underlie an observation of 'giant' gametes in vent bivalves, the enlarged oocytes having extra yolk to allow an extended lecithotrophic stage.

12.3.2 Hot springs and thermal ponds

Hot springs occur due to geothermal action in terrestrial habitats in areas of volcanic activity. Ground water becomes superheated and forced to the surface through cracks, emerging as steam or as nearly boiling water. In its passage through the laval rocks the water becomes very modified in composition, often heavily charged with hydrogen sulphide, carbonates or silicates. Springs may produce their own outflow streams, or may feed into other natural streams, part of whose flow system is then heavily affected.

Close to the geothermal source only bacteria survive, but downstream of the discharge point algal mats appear and are colonized by specialist protozoans (at about 60°C). As the waters cool below 50°C, some ostracods and larval flies may colonize, and inevitably some nematodes and rotifers appear, albeit with low species diversity. In many hot-spring effluents midge larvae are also early colonizers, with chironomids and ceratopogonids becoming abundant, the latter being especially thermophilic at some Colorado hot springs.

Where geothermal water enters otherwise 'normal' streams, some freshwater animals may drift into the heated and potentially sulphurous regions. Some caddis flies in Californian streams show raised upper critical temperatures (UCT values) in such zones, though unable to penetrate the most strongly affected waters.

Anthropogenic activities also create a new kind of hot thermal environment for animals to exploit in the form of thermal ponds near industrial and nuclear plants. Inevitably, a range of similar highly dispersive opportunist species colonize these. In some areas existing ponds have become heated, and here some fish and invertebrates have proved able to adapt to near-lethal temperatures. The mosquito fish, *Gambusia*, occurs in some thermal ponds, and within about 60 generations has acquired a much higher thermal tolerance (raised UCT), with increased heterozygosity and an apparent heritability as high as 32% for this raised lethal temperature.

12.3.3 Sea-ice dwellers

Polar sea ice

Although polar ice may look completely abiotic, it has fairly recently been appreciated that within the seasonal ice fields there is a maze of fine channels that house a surprisingly complex ecosystem. This is dominated by single-celled algae, but there are also some protozoans and bacteria, and an assortment of animals: amphipods, copepods, nematodes and flatworms in particular.

The ice is colonized by its specialist biota as it forms in the polar autumn, gradually spreading from its summer minimum of 4 million km^2 to its peak of 20 million km^2 around Antarctica. Small ice crystals mix with plankton, which become caught up in the 'frazil' ice as it forms and rises above the sea surface by up to perhaps 2 m. The plankton-containing crystals form 'ice pancakes', stacked on top of each other; as ice is pure water, the sea salt is concentrated in the intervening water, which becomes very hypersaline. Most of the trapped plankton die, and larger animals are crushed by the ice, but the small-bodied, very cold-hardy and euryhaline species can survive, effectively having a habitat that switches from nearly solid to liquid as the annual temperature cycle progresses.

In fact there are separate distinct habitats within the ice. At the upper surface all the water is frozen and tightly packed together at a temperature as low as –15°C; the channels have brine at a concentration at least seven times that of sea water (250 ppt) and little can survive. But lower down the ice sheets contain more abundant pockets and channels, filled with concentrated very cold brine (perhaps –1.5°C) at about 2–4 times sea water; these channels range from a few micrometres to several centimetres across. Here algae are reasonably abundant, accumulating K$^+$ as an osmolyte and producing proline and dimethylsulphoniopropionate (DMSP) as antifreezes. Lower still the sea permeates the ice and it is even more pitted and channelled, though the population of algae is lower as light penetration is much reduced.

In the middle and lower reaches of the ice floes the zooplankton can make a living. Harpacticoid copepods may occur at densities of up to 30 per litre in the lower ice channels, with nematodes also abundant; the occurrence of flatworms, with no protective cuticle, is rather more surprising and their physiological strategies need investigating. These sea ice communities also provide food to the animals in the water beneath the ice, with krill (euphausid shrimps) surviving the winter almost entirely by grazing on the biota at the under surface of sea ice.

In spring as the ice melts, the sea ice communities are released and seed a huge bloom of algal production in the seas around the ice fringe, again affecting the larger marine organisms. The spring surge of diatoms and algae is also a mixed blessing on a larger scale; the organisms release bromine gas, which causes ozone depletion, and dimethyl sulphide (from DMSP), which promotes cloud formation and may help reduce global warming.

Seafloor methane ice

In some parts of the world lumps of 'ice' form under the conditions of low temperature and high pressure that occur in the marine benthos. In fact the ice is a mixture of frozen water, methane (CH_4) and other hydrocarbons, often termed 'methane hydrate'. Lumps of this rock-hard methane ice material are not uncommon in the Gulf of Mexico at depths of 600–800 m and temperatures around 6°C. They form in the relatively anoxic sediment, and can then emerge to lie just above the sea floor. They are extremely unstable, and can evaporate into bubbles almost explosively if the temperature rises even slightly. Recent undersea explorations have shown that even these inhospitable habitats are teeming with life—most commonly a species of polychaete now named *Hesiocaeca methanicola*. The worms appear to 'burrow' into the ice surface by gently wafting water currents at it with their parapodia; then they survive with the assistance of bacteria that colonize their burrows (perhaps finding conditions there particularly amenable) and that can metabolize the methane hydrate into usable organic molecules. As yet it is uncertain whether the worms colonize the 'ice' before or after it emerges from the deoxygenated benthic sludge, but it seems likely that coping with hypoxia is also part of their physiologica repertoire.

Further reading

Book

Desbruyeres, D. & Segonzac, M. (eds) (1997) *Handbook of Deep-Sea Hydrothermal Vent Fauna.* IFREMER, Plouzane.

Reviews and scientific papers

Airriess, C.N. & Childress, J.J. (1994) Homeoviscous properties implicated by the interactive effects of pressure and temperature on the hydrothermal vent crab *Bythograea thermodron. Biological Bulletin* **187**, 208–214.

Brauner, C.J., Ballantyne, C.L., Randall, D.J. & Val, A.L. (1995) Air breathing in the armoured catfish as an adaptation to hypoxic, acidic and hydrogen sulphide rich waters. *Canadian Journal of Zoology* **73**, 739–744.

Cary, S.C., Shank, T. & Stein, J. (1998) Worms bask in extreme temperatures. *Nature* **391**, 545–546.

Cavanaugh, C.M. (1983) Symbiotic autotrophic bacteria in marine environments from sulfide-rich habitats. *Science* **302**, 332–333.

Cosson, R.P. (1996) Bioaccumulation of mineral elements within the vestimentiferan tube worm *Riftia pachyptila.* A review. *Oceanologica Acta* **19**, 163–176.

Cosson, R.P. (1997) Adaptations developed by hydrothermal vent organisms to face the stress of heavy metals. *Bulletin de la Société Zoologique de France* **122**, 109–126.

Crowe, J.H., Hoekstra, F.A. & Crowe, L.M. (1992) Anhydrobiosis. *Annual Review of Physiology* **54**, 579–599.

Crowe, J.H., Carpenter, J.F. & Crowe, L.M. (1998) The role of vitrification in anhydrobiosis. *Annual Review of Physiology* **60**, 73–103.

DeWachter, N., Blust, R. & Decleir, W. (1992) Oxygen bioavailability and haemoglobins in the brine shrimp *Artemia franciscana. Marine Biology* **113**, 193–200.

Ellis, B.A. & Morris, S. (1995) Effects of extreme pH on the physiology of the Australian yabby (*Cherax destructor*). *Journal of Experimental Biology* **198**, 409–418.

Grueber, W.B. & Bradley, T.J. (1994) The evolution of increased salinity tolerance in larvae of *Aedes* mosquitoes: a phylogenetic analysis. *Physiological Zoology* **67**, 566–579.

Juniper, S.K. & Martineau, P. (1995) Alvinellids and sulfides at hydrothermal vents of the eastern Pacific—a review. *American Zoologist* **35**, 174–185.

Laurent, P., Maine, J.N., Bergman, H.L., Narahara, A., Walsh, P.J. & Wood, C.M. (1995) Gill structure of a fish from an alkaline lake: effects of a short-term exposure to neutral conditions. *Canadian Journal of Zoology* **73**, 1170–1181.

Marshall, A.T., Kyriakou, P., Cooper, P.D., Coy, R. & Wright, A. (1995) Osmolality of rectal fluid from two species of osmoregulating brine fly larvae (Diptera: Ephydridae). *Journal of Insect Physiology* **41**, 413–418.

Narahara, A., Bergman, H.L., Laurent, P., Maia, J.N., Walsh, P.J. & Wood, C.M. (1996) Respiratory physiology of the Lake Magadi tilapia (*Oreochromis grahami*), a fish adapted to a hot, alkaline and frequently hypoxic environment. *Physiological Zoology* **69**, 1114–1136.

Nelson, J.A. & Mitchell, G.S. (1992) Blood chemistry response to acid exposure in yellow perch (*Perca flavescens*): comparison of populations from naturally acidic and neutral environments. *Physiological Zoology* **65**, 493–514.

Pond, D., Dixon, D. & Sargent, J. (1997) Wax-ester reserves facilitate dispersal of hydrothermal vent shrimps. *Marine Ecology—Progress Series* **146**, 289–290.

Randall, D.J., Wood, C.M., Perry, S.F. *et al.* (1989) Urea excretion as a strategy for survival in a fish living in a very alkaline environment. *Nature* **337**, 165–166.

Scott, K.M. & Fisher, C.R. (1995) Physiological ecology of sulfide metabolism in hydrothermal vent and cold seep vesicomyid clams and vestimentiferan tube worms. *American Zoologist* **35**, 102–111.

Somme, L. (1996) Anhydrobiosis and cold tolerance in tardigrades. *European Journal of Entomology* **93**, 349–357.

Steichen, D.J., Holbrook, S.J. & Osenberg, C.W. (1996) Distribution and abundance of benthic and demersal macrofauna within a natural hydrocarbon seep. *Marine Ecology—Progress Series* **138**, 71–82.

Sun, W.Q. & Leopold, A.C. (1997) Cytoplasmic vitrification and survival of anhydrobiotic organisms. *Comparative Biochemistry and Physiology A* **117**, 327–333.

Tunnicliffe, V. (1992) Hydrothermal vent communities of the deep sea. *American Scientist* **80**, 336–349.

Tunnicliffe, V. & Fowler, C.M.R. (1996) Influence of sea-floor spreading on the global hydrothermal vent fauna. *Nature* **379**, 531–533.

Wharton, D.A. (1996) Water loss and morphological changes during desiccation of the anhydrobiotic nematode *Ditylenchus dipsaci. Journal of Experimental Biology* **199**, 1085–1093.

Wilke, M.P. & Wood, C.M. (1996) The adaptations of fish to extremely alkaline environments. *Comparative Biochemistry and Physiology B* **113**, 665–673.

Yesaki, T.Y. & Iwama, G.K. (1992) Survival, acid-base regulation, ion regulation and ammonia excretion in rainbow trout in highly alkaline hard water. *Physiological Zoology* **65**, 763–787.

13 Terrestrial Life

13.1 Introduction

The earliest organisms to colonize land were probably prokaryotic cyanobacteria, living in fine silt deposits and screes even back into the Precambrian; certainly more than 550 MYA, and perhaps in excess of 800 MYA. These 'primitive' microorganisms still survive today, as cyanobacterial mats, often in association with terrestrial algae and higher plants. Plants were probably present on land by the Ordovician period, but only begin to appear as fossil deposits from the Silurian (about 430 MYA) when they had become fairly widespread. We can be reasonably sure that animal (metazoan) life on land did not pre-date the appearance of at least some of these multicellular land plants. This is because the plants must first have produced decayed matter to serve as food and as shelter, and once they had a moderately upright growth form they would also give a certain amelioration of conditions at the ground surface, with shade, reduced wind speeds, and a higher, more stable humidity. The presence of burrow traces in Ordovician soils suggests that invertebrates were established by that time, but again fossils are lacking. A range of small myriapod fossils (forerunners of modern millipedes) do appear in the late Silurian, and wingless insects (Apterygota) and some arachnids are present soon after this.

It is nearly impossible to determine what were the first 'real' land animals though. This is because of the complex intergrading of terrestrial habitats with neighbouring semi-aquatic systems. Certainly there would have been ample scope for gradually acquiring characters more appropriate to terrestrial life in a range of small worms, molluscs and arthropodan groups that inhabited the littoral zones, the freshwater marshes and the interstitial spaces of moist fringing soils through the early Palaeozoic. As we will see, the true land environment, where animals essentially live in air rather than in water, presents so many challenges to physiology and lifestyle that successful invasion must have depended on gradually increasing adaptation through a series of intermediate environments; no animal group could have achieved all the necessary modifications of form and function to make the transition to land quickly. However, many animal groups *have* achieved some or all of the necessary adaptations to become partially or fully terrestrial. Some groups did it during the early spurt of terrestrialization that accompanied movements of plant groups onto land in the Silurian, Devonian and Carboniferous. A further range of animals invaded the land later on, when the flowering (angiosperm) plants radiated (perhaps originating in the Cretaceous, around 120 MYA), supplanting many types of conifer and seed fern, and opening up new niches for seed-eaters, pollinators and specialist coevolved herbivores. It is probably no coincidence that these two key phases of land invasion occurred in periods of equable climate across most of the world.

In this chapter we cover the essential strategies of the broad range of animals that live in the majority of terrestrial habitats, particularly in temperate zones and the humid tropics where thermal extremes are rarely encountered and where water balance, though difficult to achieve, is not pushed to the limits for survival. In such habitats, large endothermic animals are more or less always in their thermoneutral zone, and humidities are rarely so low, or food and drinking-water so scarce, as to create grave water-loss problems. Small animals have a huge variety of microclimates to choose from to extend their active periods.

However, there may still be some severe short-term problems. Most of these areas do have significant daily temperature fluctuations, so that small vertebrate endotherms may get chilling problems at night, and tend to exhibit 'heterothermy', as also do large insects (see Chapter 8). There may also be transient droughts or transient waterlogging, so that good avoidance tactics, especially short-range evasive movements and migrations, are a good idea. For similar reasons resistant stages in the life cycle will often be required, especially by smaller animals. Only rarely is excess heat a severe problem in the wet tropics, though a neat example of problematic 'wet heat' does occur in the hot caves of the Neotropics, favoured as roosts by several species of bats, where relative humidity (RH) exceeds 90% and ambient temperature (T_a) levels of up to 40°C year round give rise to basal metabolic rates in the bats that are substantially lower than expected (only 48–66% of the predicted levels). On the whole, though, the inhabitants of tropical areas can be regarded as showing 'normal' terrestrial adaptations.

Property	Water/aquatic	Air/terrestrial	Approximate ratio water : air
Density (g ml⁻¹)	1.00	0.0012	850
Viscosity (kg m⁻¹ s⁻¹)	1.00	0.02	50
Thermal capacity (J ml⁻¹ °C⁻¹)	1.00	0.0003	3300
Velocity of sound (m s⁻¹)	1485	343	4.33
Refractive index	1.33	1.00	1.33
Oxygen content (ml l⁻¹)	4–7	210	1/30
Oxygen diffusion ratio			1/300 000
Carbon dioxide content (ml l⁻¹)	0.4	46	1/115
Salts	Freely available	Not directly available	
Water	Abundant, but may be osmotically unavailable	Rare, always hard to find and keep	
Food for plants	Only at surface	At all levels	
Food for animals	Throughout, all mechanisms possible	Throughout, but often hard to catch or eat	

Table 13.1 Properties of water and air, and of aquatic and terrestrial environments.

Chapter 14 will deal with some special cases of terrestrial life: hot arid deserts, where the hygrothermal endurance limits of animal residents may be severely tested; polar regions, tundra and northern coniferous forests, where extreme cold is superimposed on the generality of terrestrial problems; and montane habitats, where altitude effects may parallel the latitudinal effects at the poles.

13.1.1 *Problems and advantages of life on land*

For any animal, there are a series of distinct problems and equally distinct advantages of life on land and thus in air. Many of these arise directly from the different physical and chemical properties of air and of water (see Table 13.1).

Firstly, water has approximately 1000 times greater **density** than air, and provides a supportive fluid for all the pelagic organisms that float in it. Support of the body (itself essentially watery) is much harder in all terrestrial habitats, and the problems of self-weight and coping with gravity really arise for the first time. The composition and design of skeletons and structures concerned with motility therefore become more critical, and more energy must be expended in maintaining the extra skeletal weight. However, water is also much more **viscous** than air (approximately 50-fold), so that on land the medium actually impedes movement to a much smaller degree and makes acceleration considerably easier. Consequently, on land a specifically adapted animal can achieve faster speeds and an altogether quicker approach to life. The fastest speeds achieved by any animal in water are about 10–20 m s⁻¹, maintained for only very short bursts, whereas animals running on land can achieve around 25 m s⁻¹ in short bursts (66 mph, recorded in a cheetah, is 29 m s⁻¹) and steady speeds of 15 m s⁻¹. Birds may fly in bursts at up to 40 m s⁻¹ and steadily at 20 m s⁻¹. Flying is clearly

the fastest option, and in fact, given the enormous numerical dominance of insects, it would be fair to say that the majority of land animals have adopted flying as their main way of getting around.

This effect of a speeded-up lifestyle is accentuated by the differing **thermal** properties of air. Its thermal capacity is 3000 times less than that of water, so that temperature may change very much more rapidly and on a much finer spatial scale. Terrestrial environmental temperatures also reach substantially higher levels, as the air heats up from direct and reflected sunlight during the daylight hours, so that terrestrial ambient temperatures can exceed 50°C. Since temperature affects so many other aspects of metabolism and physiological activity (Chapters 6–8), we have an added reason why life often proceeds much faster on land. Equally though, polar land habitats can be substantially colder than the neighbouring seas, with no mass of slowly circulating ocean to act as a heat sink; the air over the South Pole can reach temperatures below –70°C towards the end of the long Antarctic winter when the Sun has not been above the horizon for nearly 6 months. This variability of temperature links to a generally much greater instability and unpredictability of terrestrial environments; a great many factors vary much more widely and more quickly for a land animal, so that a faster response and broader sensitivity are likely to be needed.

Two other changing physical properties then become relevant. Watery media have a 4–5-fold higher **velocity of sound** conduction, and most aquatic animals can perceive only low-frequency stimuli, with only the marine mammals using extensive sound communications systems. By contrast, land animals perceive higher-frequency noises, and have a much more widespread and sophisticated capacity for sound production, tying in with their rapid lifestyle. Water also has a higher refractive index (RI) than air, with terrestrial eyes needing more refraction from the cornea to supplement the lens. Whilst this difference in RI affects visual perception across air–water interfaces markedly, it has only limited effects on the visual systems of animals living entirely within one medium.

Then there are differences in the **chemical properties** of air and water, as solvents and as carriers of other molecules. Most obviously, the oxygen which makes up a substantial part (21%) of the atmosphere has only a limited solubility in water, with seas and fresh waters containing only around 4–7 ml l^{-1} compared to about 210 ml l^{-1} of oxygen in air, at least a 30-fold difference. The reduced density of air compared to water also means that the diffusion of gases is very much faster in air, so that in practice oxygen can be supplied to static (non-ventilated) respiring surfaces about 300 000 times faster in air than in water. When ventilation is also taken into account, being much more difficult in viscous water than in air, the business of supplying oxygen to an animal's body is clearly much easier in terrestrial habitats. This again makes a difference to the rate at which aerobic metabolism (and hence many other linked processes, including locomotion) can proceed in the two media, even in the absence of sophisticated respiratory and circulatory systems.

Carbon dioxide is often rather abundant in sea water, though largely in the form of carbonic acid (HCO_3^- ions), the total concentration being about 46 ml l^{-1}; whereas in clean fresh waters and in air it is at quite low concentrations (about 0.3 ml l^{-1}). Simple diffusive loss is nearly always adequate for living organisms in any of these media; however, on land most animals regulate the CO_2 levels in the intermediary circulatory fluid, as this is important in controlling ventilation (see Chapter 7). The availabilities of oxygen and carbon dioxide, together with the much more limited support provided by air for complex respiratory structures, have enormous impact on the design of gas-exchange surfaces in terrestrial animals.

However, the most profound differences in chemical properties between aquatic and terrestrial habitats clearly reside in the **water and salt contents** of the media. Salts are at least theoretically readily available in solution in all watery environments (unless they are exceptionally fresh); on land there are normally no salts in solution, and they must be obtained from food, often requiring complex dietary choices or specific behaviours such as salt-licking or lapping at excreta. But of course those same salts may easily diffuse out of the body of any animal living in an aquatic habit, whilst they are fairly readily retained by terrestrial animals. Similarly, water may seem abundant in all aquatic habitats, but as has repeatedly been seen in earlier chapters it may be osmotically unavailable where the animal's own fluids are markedly hyposmotic (e.g. vertebrates in sea water). Or it may cause problems by a tendency to enter too readily if the animal's tissues are hyperosmotic. On land, for all animals, water loss is always inevitable, but rates may be variable as the ambient water vapour pressure (RH) varies; only when the RH is over 99% does a terrestrial animal come close to achieving water equilibrium.

As a general rule then, for land animals both salts and water have a continuous tendency to leak from the body, so must normally be regulated by careful control of skin, respiratory and excretory surfaces. The smaller the animal, the more critical the control of these regulatory surfaces becomes, making land life for young stages particularly hazardous (and thus entraining a whole complex series of reproductive adaptations). This salt and water regulation could in theory be paid for easily by the greater availability of oxygen, but in practice gaining oxygen always means losing water, so the benefit is at best minimal. Ultimately salt and water balance must be maintained by eating and drinking.

This raises a final category of differences between aquatic and terrestrial habitats relating to feeding habits and community **trophic structures**, linked to the different character of photosynthetic life in the two habitats. In the oceans, where the medium is in constant motion, 'plants' are small and usually short-lived (phytoplankton and benthic macroalgae), whereas on land, with a relatively stable soil to provide anchorage and a degree of longer-term stability, the true plants dominate and may be both larger and more persistent, with specialized multicellular structures both below and above ground level. This leads directly to a difference in the type of organic compound that predominates in plant bodies: in the seas proteins are all-important, allowing rapid growth but little energy storage, whereas on land carbohydrates predominate, allowing a slower-growing organism but one with the ability to store energy over several seasons, decades or even centuries, and with large deposits of purely structural materials. This dichotomy is to a large degree also reflected in the chemistry of the animals of the two environments, and overall there are reckoned to be about 14 000 times greater reserves of carbohydrates on land than in the seas. There may also be a less efficient conversion of carbohydrate to protein by land animals than by their marine counterparts, partly because of the energy requirements of land animals (especially the high proportion of terrestrial birds and mammals); but this is partly offset by the very rapid conversions achieved in terrestrial soils by bacteria and fungi.

There are also inevitably differences in the variety of feeding systems that are practicable in the different media. In the seas nutrients are kept on the move by currents but show a gradual tendency to sink, so that the surface waters where most animals live may become depleted of food, while the benthic zones benefit from a rain of food that can be garnered by a sessile animal. On land the same inevitable effects of gravity mean that foods (of both plant and animal origin) tend to accumulate at the soil surface, which is just where most animals make their living. However, land animals cannot usually adopt a sessile strategy, waiting for their food to come along and be filtered out; the fluid low-density air cannot support a suspended biomass, with only bacteria, very small spores, and the small resistant egg stages of a few animals occurring. Thus the air does not really provide a supply of food to 'filterers' (the obvious exception to this being the extraordinary sophistication shown by orb-web spiders as aerial filter-feeders on flying insects and, as juveniles, on pollen).

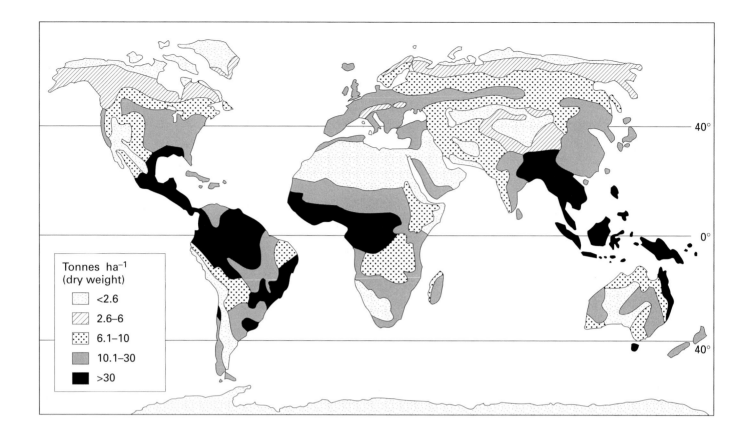

Fig. 13.1 Patterns of primary productivity across the land masses.

Instead, food must generally be actively sought. The layer of dead and decaying detritus is the one easy food source, and probably the starting point for many animal groups in their transition to land. True herbivory (macroherbivory) on land becomes a much more difficult way of life than in water, as land plants are very much tougher (reflecting their own need for support), rather high in carbohydrate, and nutritionally not very well balanced as a diet for animals. However, while herbivorous grazing is rare in aquatic communities, terrestrial grazing is a major way of life for many groups, despite the poor quality of food obtained and the presence of strong anti-herbivore defences (see section 13.8.1). At the same time carnivory may be harder on land because passive sit-and-wait strategies (e.g. suspension-feeding carnivory) are rare, and specific targets must always be involved; most of the prey items are also likely to be moving and responding faster.

Overall, then, the properties of air mean that successful evolution of animal life on land needs more specific adaptation of the skeletal framework and motile systems; more physiological control of hygrothermal balance, with an overriding need for integration of water balance, temperature balance and respiration at all stages of the inevitably more specialized life cycle; and sensory, neurological and behavioural adaptations to permit a more active mode of life for animals generally.

13.1.2 Terrestrial climates and microclimates

The terrestrial environment provides a balance of solids, water and air that is highly amenable to plant growth, with an almost unimpeded supply of light for photosynthesis. The resultant pattern of **primary productivity** over the planet's surface (Fig. 13.1) therefore coincides closely with the supply of light, heat and water; in other words, with meteorological patterns. Ecosystems are generally classified largely in terms of their floral components (Fig. 13.2), giving categories such as tundra, grassland, tropical rain forest, or desert. These patterns of climatically determined plant communities inevitably affect animal distributions, resulting in biomes with distinctive kinds of fauna repeated in different geographical locations around the globe.

In addition, terrestrial **weather** has obvious direct effects on animals' physiology and hence on their behaviour, ecology and distribution. Terrestrial temperatures and rainfall regimes are dramatically variable. Maximum temperatures of 58°C have been recorded, with a record minimum of –89°C. At the same time rainfall may vary from effectively zero in deserts that are in rain-shadow areas, to above 3000 mm per annum in mountain ranges within the tropical rain forests.

The average global surface temperature is only about 9°C, taking all habitats, all seasons and all latitudes into account. However, a more apt figure for terrestrial habitats is a mean ground level temperature of 15°C. There are somewhat higher averages around coasts at all latitudes than in the centre of

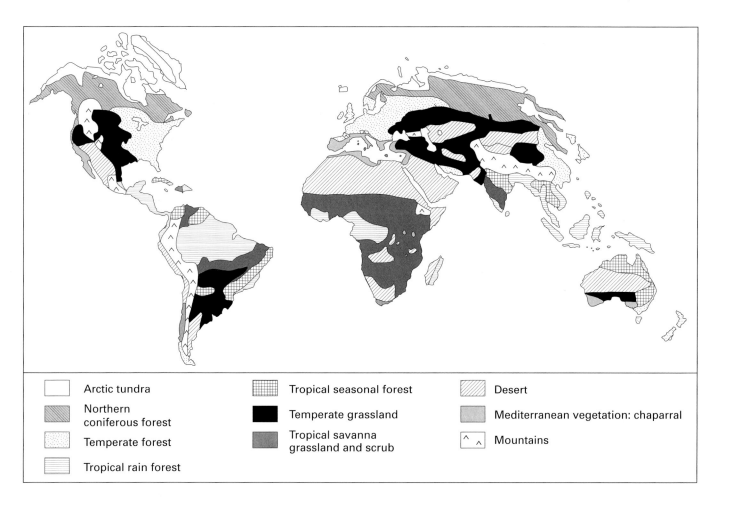

Fig. 13.2 Occurrence of the major terrestrial biomes.

continents, due to the thermally stable sea, giving a 'maritime' rather than a 'continental' weather pattern. However, terrestrial temperatures and climates obviously vary greatly at any one place, largely in relation to all the cyclical patterns outlined in the Introduction to Section C (p. 249).

For most animals with short life cycles it is diurnal temperature change that has the greatest impact on their existence. The most dramatic diurnal changes in terrestrial temperatures, irrespective of habitat, are found close to the ground. This is because during the day solar energy warms this surface faster than the energy can be transmitted downwards or re-radiated away; while during the night the surface loss of infrared radiation is faster than the replacement of heat up through the soil, causing rapid cooling. Since air has a low thermal conductance, heat from the ground will not easily be conducted away from the irradiated surface in the absence of convection, giving rise to steep thermal gradients. Such considerations, plus the presence of vegetation, make terrestrial habitats very varied and complex, particularly for very small animals. Thus it is always more appropriate to talk in terms of **microclimate**, defining the

conditions actually experienced by the animal of interest on an appropriate spatial and temporal scale, rather than the macroclimate that would be experienced by a human (with the head some 2 m above the ground—the height at which our routine meteorological instruments tend to be sited). Most animals live on or in the ground, or closely associated with vegetation emerging from (and intimately connected with) the ground. Microclimatic conditions can be very different 50 mm above a ground surface, where a rodent's body might be, than just 5 mm above that same surface, where an insect or worm might rest. Similarly, conditions 5 mm above the surface may vary from second to second with small-scale air movements and the shifting of shadows, whereas larger animals, with their main body mass and their receptors much further above the ground, will experience a much more stable environment through time.

Effects of soil

The soil provides an important refuge from temperature extremes in terrestrial animals; at depths of 0.2 m all diurnal temperature change is smoothed out in most soil types. Soils are a mixture of minerals derived from the underlying bedrock

Table 13.2 Properties of terrestrial soils.

	Conductivity (W m^{-1} °C^{-1})	Diffusivity (m s^{-1} × 10^{-6})
Soil type		
Sandy		
Dry	0.3	0.24
Wet		0.74
Clay		
Dry		0.18
Wet		0.51
Peat		
Dry	0.06	0.10
Wet		0.12
Water at 10°C		0.14
Air at 10°C		20.50

(and/or by erosion and deposition from other rocks), together with dead organic material (humus). The temperature regime of a particular soil is largely controlled by the thermal conductivity, thermal capacity and diffusivity of its particular mixture of components (Table 13.2). For dry sandy soils **conductivity** may be five times greater than for a dry peat, so the peaty soils heat up much more quickly at their surfaces. However, conductivity alone is misleading, since the presence of water (rather than air) in the soil interstices affects the rate at which soil temperature changes. The most useful measure is therefore the soil **diffusivity**, which indicates more exactly how a particular soil responds to a thermal input, by measuring the time taken for heat to travel within that soil. The lowest values of diffusivity occur for peat soils, whether wet or dry, and the highest values are for saturated sands. Sands and clays permit heat to penetrate quite rapidly, but peats allow only shallow heat penetration and their surfaces are consequently subject to extremes of temperature.

However, soils also vary in their ability to hold on to water, which in turn affects gaseous diffusion rates. Wet sands have an open texture and can be dried out rather easily; wet clays have a high soil moisture tension and dry out only slowly; wet peats dry fairly easily but can be difficult to re-wet. Thus distribution of animals in soils is not solely governed by thermal considerations, and for those animals that rely on soil moisture—either to give a high or stable burrow humidity, or as a source of drinking-water—assessment of soil moisture retention may override thermal diffusivity.

Examples of temperature and humidity profiles in soils, and in burrows within soils, are shown in Fig. 13.3. Burrowing into

(a)

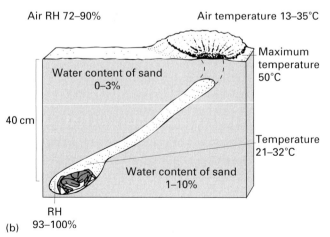

(b)

Fig. 13.3 Profiles of temperature and relative humidity (RH) or water content in terrestrial soils and in burrows. (a) The burrow of a solitary wasp (*Cerceris*) in sandy soils. Dashed lines show ambient conditions. BST, British Summer Time. (From Willmer 1982.) (b) The burrow of the land crab *Gecarcinus* in the upper beach zone. (From Bliss 1979.)

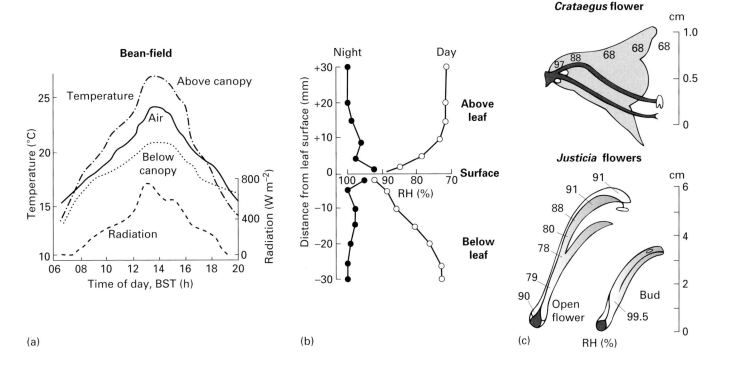

Fig. 13.4 Microclimates associated with plants: (a) the stabilized conditions below the plant canopy; (b) the microclimate around a single leaf in sunshine; and (c) the microclimate within flowers. BST, British Summer Time; RH, relative humidity. (b & c, From Willmer 1982.)

the relatively stable soil layer is an exceptionally useful way to escape, at least temporarily, from many of the problems of life on land. Many land animals spend a considerable part of the day in burrows: this time may coincide with, for example, the hottest and driest hours in summer, with the animal only coming out onto the surface at dawn, dusk or at night; or alternatively in winter the coldest hours may be spent in the burrow, with the animal emerging only in the few warm hours around midday. Whilst in the burrow, the metabolic rate may be reduced or at least may be more constant, with the animal often resting in conditions that are particularly favourable for digestive functions.

Effects of vegetation

Wherever life occurs, the environment changes, and often in ways that make local conditions more suitable for further life. This is particularly true on land, where plants create ecospaces for animals, and both then become habitats for a huge range of symbionts and parasites.

Plant life becomes the focus of **energy exchanges**, raising the

main exchange site above the ground surface. Wherever there is vegetation, the plants reduce the amount of radiation reaching the ground below, and temperature changes at the soil surface and within the plant layer tend to be much reduced. The main temperature gradient occurs instead at the top of the plant 'canopy', whether this is a few millimetres or many tens of metres above the ground (Fig. 13.4a). For all animals living in burrows or in the litter layer, then, plant cover gives a great amelioration of conditions; and a new range of animals can live on and amongst the plants themselves, in a reasonably equable microclimate.

Plants also increase the **humidity** above the ground, however. This effect is quite complex, arising partly from retention of rainwater on the plant surfaces (plants actually reduce the amount of precipitation reaching the ground by anything up to 50%), and partly from the slower evaporative and diffusive dispersion of this precipitation brought about by reduced wind speeds within vegetation. But most importantly, this moisture is continually added to by evaporation and transpiration from the plants themselves, via their stomata, giving a regulation of the water content within the vegetation. In effect, the vegetated area becomes a thick boundary layer of relatively still, stable, moist air.

More drastic effects of plants on the microclimate surrounding them and in the soil below can also be seen. In dry climates where water is scarce plants may deplete the soil moisture around their roots. In wetter climates the opposite effect may occur: plants delay the arrival of rain onto the ground, causing

Table 13.3 Taxa of terrestrial animals.

Worms

'Planarians', or flatworms
  Cryptozoic, interstitial, in soils and damp litter
  Most in humid ecosystems such as tropical rain forest
  Some triclad species in temperate and Mediterranean biomes, under stones and logs
  Nocturnal, carnivorous on other worms and small molluscs
  Genera: *Bipalium*, *Rhynchodemus* and *Microplana*

Nemertean worms, or ribbon worms
  Probably direct from the marine habitat
  Cryptozoic, humid places, supra-littoral zone or in deep litter and among rotting wood
  Nocturnal predators
  Genera: *Geonemertes*, *Argonemertes*

Nematodes, or roundworms
  Strictly interstitial animals, living an essentially freshwater life
  Extraordinarily abundant in soils, with up to 20 million m^{-2} in humic woodland soils

Annelids
  Cryptozoic oligochaetes, commonly known as earthworms
    Deposit-feeding burrowers
    Genera: *Lumbricus*. Tropical and subtropical genera: *Lampito*, *Allolobophora*, also 'giant
      earthworms', more xeric, family Megascolecidae: *Megascolides* and *Pheretima*
  Leeches in tropical forests, damp vegetation
    Predatory or parasitic, usually on vertebrate hosts

Molluscs

Only from class Gastropoda
Order Prosobranchia: some Littorinacea (littoral route), and Architaenioglossa (freshwater route);
  persist only in tropical forests
    Order Pulmonata: variety of shell forms, closure by flaps and ridges; or shell loss (slugs)
    Hygrophiles, active at high humidity, rapid onset of aestivation in drought
    Herbivores, with chitinous radula; or sometimes carnivorous
    Air-breathers with lungs
    Hermaphrodite with complex reproductive systems
    Genera: supra-tidal: *Ovatella* and *Pythia*; slugs and snails: *Helix*, *Pomatia*, *Arion*, *Limax* and in
      deserts *Sphincterochila*

Arthropodan groups

Crustacea
  Copepoda: few, interstitial and cryptozoic in soils
  Amphipoda, family Talitridae: beach-living sandhoppers (*Orchestia*, *Talitrus*), plus true land-
    dwellers only in the tropics and Japan (e.g. *Arcitalitrus*)
  Isopoda, suborder Oniscoidea (woodlice) from damp litter (*Oniscus*), grasslands and soils
    (*Porcellio*, *Armadillidium*), and in desert burrows (*Venezillo*, *Hemilepistus*)
  Decapoda (crabs, lobsters and prawns):
    Anomurans (hermit crabs and their kin): e.g. coconut crab *Birgus* and *Coenobita*
    Brachyuran ('true' crabs): Grapsidae, shingle shores, beaches, mangrove swamps;
      Ocypodidae, beach burrowers, *Ocypode* (fiddler crabs), *Uca* (ghost crabs); Mictyridae,
      soldier crabs; Gecarcinidae, land crabs, but return to water to breed

Chelicerata
  Predominantly terrestrial and xerophilic
  Mainly predatory on insects, using trapping and ambush techniques and venoms
  Araneae (true spiders) and Acari (mites and ticks). Both widespread
  Scorpiones (true scorpions), Uropygi and Amblypygi (whip scorpions and whip spiders), and
    Solpugida (sun spiders)—all in warm and often dry habitats

continued

Table 13.3 (*continued*)

Myriapoda
    Entirely terrestrial
    Chilopoda (centipedes): predators in litter and soils, sometimes in caves
    Diplopoda (millipedes): browsers or detritivores
    Also Pauropoda and Symphyla, tiny and specialist litter-dwellers, cryptozoic

Onychophorans
    All terrestrial, genera *Peripatus* and *Peripatopsis*, 'velvet worms' or 'walking worms'
    Cryptozoic, moist forest habitats, often in rotting wood. Predatory

Insecta
    Wingless insects, Apterygota: litter and cryptozoic habitats
    Winged insects, Pterygota: all possible terrestrial niches (+ fresh water)
    27–29 orders, nine include herbivores, five predominantly carnivores, parasites, parasitoids; also
        omnivores and detritivores
    Many orders into arid zones and true deserts, notably beetles (Coleoptera), ants and bees
        (Hymenoptera), termites (Isoptera) and grasshoppers and locusts (Orthoptera)

Vertebrates
Four classes of tetrapods:
    Amphibia, hygrophilic
    Reptilia (perhaps three separate classes of tortoises, crocodilians, and lizards + snakes)
    Aves
    Mammalia
    (These last three groups all mesophilic and xerophilic)

it to drip steadily through rather than arriving in bursts, so that the run-off from the ground is much reduced and water is more likely to be absorbed into the soil. This effect is especially striking in tropical forests, where run-off is very low and almost no soil erosion occurs; when such an area is deforested, sudden storms can erode the soil very quickly and surrounding areas may experience flash flooding from the run-off.

Plants also produce very local stable microclimates within their own structures due to their own transpiration, with distinct regimes created either side of a leaf (Fig. 13.4b) or within buds and open flowers (Fig. 13.4c); all these zones are exploited by small terrestrial animals.

Effects of animals

Life modifies and very often improves its own environment. Not only do plants modify ecospaces for animals, but the animal itself has effects on its own environment. In fact the presence of an animal in a small enclosed space cannot fail to modify the conditions, as the animal emits both heat and water vapour, and may also produce excreta that will equilibrate with the surrounding air. The nest tunnel of a solitary bee is both warmer and slightly more humid when the female bee is present and digging a new cell than when she is absent foraging. On a larger scale, occupied badger setts have higher levels of CO_2 (up to 0.6%) and lower O_2 levels (around 19.5%) than unoccupied setts. Enclosed bird nests also have their

own microclimate, with humidity rising when the owners are present.

In social animals the effects can be more dramatic, with 'social thermoregulation' occurring. Tropical stingless bees will use endothermic incubation to warm up the brood areas of their nest, and cover cooler areas with insulating secretions; overheating is also controlled, using accelerated and directed wing-fanning. The nests of ants and termites provide more extreme examples of controlled architecture for stable microclimate.

13.1.3 *Categories of terrestrial animals*

'Terrestrial animals' is a rather unhelpful term to a physiologist. Animals that can be broadly considered to live on land cover a very wide range of lifestyles, and it is useful to distinguish between these at the outset. A taxonomic summary of the animals in these various groups is given in Table 13.3 as a reference point for this chapter.

Interstitial fauna

Small animals that are encountered in soils are often experiencing a habitat that is really no different from that of a freshwater animal; they are normally permanently surrounded by a film of fresh water, with which they exchange ions and gases by osmosis and diffusion in a typically aquatic fashion. If the soil dries out too much they become inactive and may enter a cryptobiotic state (see Chapter 12) or may desiccate and die. In this category are all small soil organisms whose dimensions are such that the soil moisture will usually be adequate to give a film of moisture over their bodies; in particular it includes soil nematodes, oligochaetes, mites and rotifers.

Cryptozoic fauna

Soil-dwelling organisms of a larger size will not experience a continuous covering of moisture, and though they may live in exactly the same places as the interstitial fauna described above, these larger animals are more properly terrestrial (e.g. earthworms that burrow in soil are terrestrial in many aspects of their physiology, and may risk drowning in very waterlogged soils). These soil animals, and all others that live in zones of continuously high humidity, are termed **cryptozoic** ('hidden animals'), rarely being seen 'out in the air'. The grouping would include animals from leaf litter and detritus, from burrows and boreholes inside trees and other plants, and from the debris in other animals' nests. Some worms, woodlice, centipedes and many larval insects are familiar animals that belong in this category; so too (stretching a point) do various lizards and snakes, and mammals such as gophers and moles. Many of these 'soil-dwellers' have cylindrical or rather flattened bodies, and are often extremely flexible in three dimensions.

Hygrophilic fauna

There is a further group of animals that still need a water supply and high humidity for activity, but that can tolerate some degree of desiccation. They can best be described as **hygrophilic** ('wet-loving'). This would include snails and slugs, which are active and conspicuous in damp habitats and at night (**nocturnal**) or around dawn and dusk (**crepuscular** for dusk activity, **eocrepuscular** if both dawn and dusk are used). Larger animals with permeable skins, such as amphibians, also belong here. Most of these hygrophilic animals have strategies for survival through water shortages. In temperate and arid-zone species this may involve diapause or aestivation (summer hibernation); in cold habitats it may involve torpor when all water is frozen. Again some of the smaller species use the more extreme form of suspended animation known as cryptobiosis (Chapter 12).

Xerophilic fauna

The final category of terrestrial animals is the most familiar one, and includes all those animals that can be active in dry conditions. These are usually described as **xerophilic** ('dry-loving'), though many do not survive in real drought conditions and might be better termed **mesophilic**, implying a preference for moderate conditions. These are the animals we see above ground and active in all the sorts of terrestrial places where humans live, tolerating the macroclimate of the terrestrial world; the most notable examples are the insects, arachnids, reptiles, birds and mammals.

13.1.4 Littoral and freshwater routes onto land

There are two major alternative routes for the evolution of life on land, and both have certainly occurred several times in different groups. Terrestrial life could arise from animals from the seashore and rock pools moving via the splash zone and onto land from an essentially **marine** source; these animals would be endowed with the kinds of adaptation discussed in Chapter 10. Furthermore, terrestrial life could equally easily arise from animals living in **fresh water**, via the marginal swamps and bogs (see Plates 7a & b, facing p. 310) that inevitably appear as freshwater bodies shrink and dry up, and with the specifically freshwater adaptations outlined in Chapter 11.

In general, the invertebrate animals with a marine littoral evolutionary background could be said to do better as 'true' land animals (i.e. as the xerophilic types discussed above), with a higher ability to tolerate desiccation and osmotic variation. Freshwater groups usually have an ability to produce dilute urine (not helpful to a real land animal) and so they do well as partially terrestrial hygrophiles. Vertebrates could be seen as an exception—the one group that has done well terrestrially from an essentially freshwater background, and which has evolved new ways of coping with the physiological stresses on land.

There is probably a third route onto land, taken by some small soft-bodied worms and perhaps by the insects. This involves an **interstitial** habitat, living entirely within a sediment, where there is buffering against changing salinity and temperature. Sediments can readily serve as a transition zone between aquatic and land life, whether the water within the sediment is salty or fresh, and such habitats probably require less tolerance of osmotic change in their occupants. Any animals adapted to life in sediments are likely to acquire mechanisms for regulating their salt and water balance, and it may be impossible to decide whether this fauna—including the interstitial and cryptozoic animals—moved via intertidal sands or via freshwater muds.

13.1.5 Terrestrial biomes

It is fairly conventional to divide up the terrestrial habitat into the nine 'biomes' shown in Fig. 13.2, roughly determined by their positions along intersecting temperature and rainfall axes (Fig. 13.5). Many other categorizations exist, but this system serves reasonably well in a review of the physiological consequences for animals. There are three major categories of biome that might be considered representative of 'normal' or non-extreme terrestrial life, supporting animals with 'generalist' terrestrial adaptations. These would be (i) the temperate (usually deciduous) forests (Plate 7c); (ii) the temperate and semi-tropical grasslands (steppe, pampas, savanna and prairie; Plates 7d & e) and the not dissimilar scrub forests and 'Mediterranean' zones (Plate 7f); and (iii) the tropical seasonal and monsoon tropical rain forests (see Plates 8a & b, facing p. 310). Each has certain repeatable features in terms of climate, flora and fauna, regardless of which continent is considered.

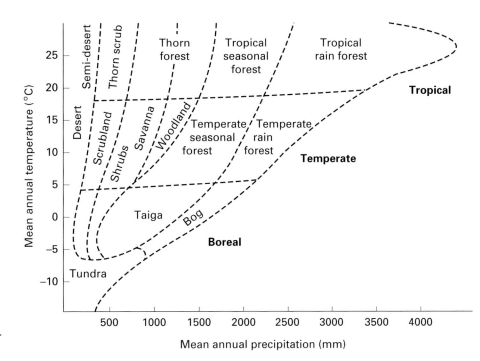

Fig. 13.5 Temperature and rainfall as determinants of the major terrestrial biomes. (Adapted from Holdridge 1947.)

Temperate forests

Forests once covered much of the temperate land mass of the planet, in a wide belt at latitudes between the boreal coniferous forests and the temperate and subtropical grasslands; but they have become increasingly restricted over the last 10 000 years largely through human activities, and in many countries less than 1% of their original area remains. Temperate deciduous forests are rather varied in appearance throughout the globe, with trees from many taxa, shrubs and bulbs, and a dense litter layer. Within such forests, small deer, bears, wild boar, squirrels, chipmunks, rabbits, mice and shrews may thrive, with a great abundance of passerine birds, woodpeckers and owls. Ectothermic vertebrates and many invertebrate groups are also strongly represented in association with the continually renewed supply of dead and dying wood and with the litter; there is also an enormous pterygote insect and molluscan fauna.

Temperate grasslands and scrublands

The grasslands include the steppes of Eurasia, the prairies in North America, the pampas and paramo (Plate 8c) in South America, some parts of Australia, and the savanna and veldt of East and South Africa (Plate 8d). These kinds of habitats may have been partly brought about by human activities (felling and burning forests) over the last few thousand years, and thus may not be really 'natural' at all (though elephants and mammoths may have felled woodland naturally in prehuman times). Today the grasslands tend to exist as flat or rolling landscapes, and most experience periodic drought when the dry soil is eroded by wind. Since they are often sited towards the middle of large land masses, the grasslands can be fairly extreme in terms of climate, with hot summers and cold winters, but the reasonably continuous perennial vegetation provides amelioration of conditions for a wide range of animals. Soils are usually quite shallow and moderately acidic in character. The vegetation is of grasses with many bulbous flowering plants, often with scattered stands of relatively small spiny trees. Grazers are inevitably abundant, with larger animals dominated by equids (horses and zebra), antelopes, and huge numbers of colonial rodents such as marmots (prairie dogs) and gophers; elephants, rhinoceros, buffalo or bison may be less numerous but important in biomass terms. There is frequently a large flightless bird, also evolved to fill the grazing niche; ostriches, rheas and emus occur on different continents. Birds of prey are usually the commonest large carnivores, together with various felids and canids (the cat and dog families). Reptiles and amphibians are fairly sparse. Insects are not hugely diverse, though grasshoppers are common and ants and termites can be incredibly abundant, the ants using grass seeds as food. There is often an ant-eating mammal to exploit these: edentates, pangolins, aardvarks and echidnas have all evolved convergently in the grasslands of different continents. A very large proportion of the fauna is actually based underground, thereby limiting the seasonal temperature stress.

Towards coastlines, and in areas around 40° latitude especially where human influence has been long-standing, the grasslands are commonly replaced by what may be termed 'Mediterranean' habitats, as on European and North African coasts of the Mediterranean Sea (Plate 7f), in California and the western USA ('chaparral' communities) and in parts of South America and Australia. These are characterized by hot summers and mild wet winters, and often have thin and rocky soils. Many of the original grassland animals remain, but the fauna may also become dominated by rodents, small lizards and snakes, and a vast range of insects of which orthopterans (grasshoppers, crickets and cicadas) are often especially conspicuous.

Tropical rain forest

Where rainfall exceeds 200 cm per year and is evenly distributed,

Fig. 13.6 (a) Abundance and (b) biomass of soil animals in temperate and tropical biomes; note that earthworms and winged insects may go off the scale in terms of biomass. Error bars are ±1 standard error; numbers above bars show off-scale values. (Adapted from Little 1990; courtesy of Cambridge University Press.)

evergreen forest results (Plates 8a & b); in areas where there is some variation in rainfall patterns the forest is more seasonal. Mean monthly temperatures may be as high as 24–28°C. Such tropical forests are often said to be more speciose and of higher productivity than any other communities on Earth, though the reality may be more complicated than this. Tropical forest soils are diverse in character on different continents and also within large land masses such as Amazonia. Sometimes they are deep but well weathered, forming clays; but where rainfall is particularly high, and where forests lie over sandstone, the soils may get severe weathering and only a thin layer of organic soil persists over metres of whitened quartz.

Tropical forests are characterized by a pronounced vertical stratification of vegetation from the tallest trees (>50 m) down to

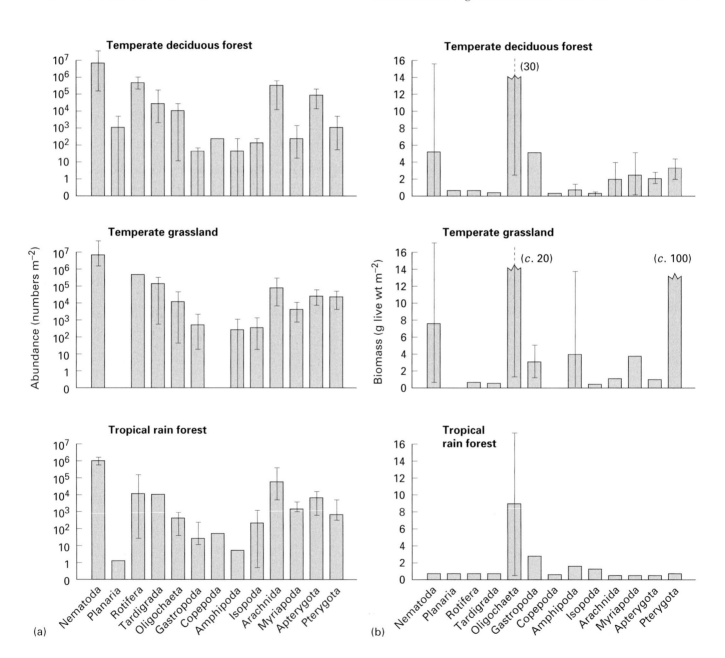

woody shrubs and large herbs and creepers (up to 1000 species of plant per hectare in total). This automatically creates a huge range of habitats; yet the animal biomass in rain forests, at least on the visible scale, is surprisingly low. Many herbivores and frugivores are arboreal and consequently small; primates are especially effective in this role. The main large forest-floor-dwellers are pigs, bovids, insectivores and sometimes elephants. There are few large predators, the main ones being felids (tigers, jaguars and leopards) or canids, such as the dhole in India. However, birds and insects are incredibly abundant throughout the three-dimensional architecture of the forest, as also are reptiles and amphibians. All the marginally terrestrial hygrophilic and cryptozoic litter invertebrates are well represented, though the lack of a deep litter layer in many forests keeps their biomass in check. The preponderance of invertebrates and amphibians underlines the lack of physiological problems; the 'soft-skinned' animals do well.

13.1.6 The success of animals in land habitats

Different groups of animals have very different patterns of success in the various available terrestrial habitats. Quite how these patterns appear depends on whether 'success' is viewed in terms of species diversity, numerical abundance or total biomass. With species diversity it is difficult to disentangle ecosystem effect from latitudinal effect, with species number in virtually all taxa decreasing towards the poles; clearer habitat effects can be seen with abundance and biomass.

On the basis of the relative abundances of the major terrestrial animal groups in the soils of temperate habitats, shown in Fig. 13.6a, it is clear that the small interstitial and cryptic animals dominate the scene, with nematodes occurring by the millions in virtually all sediments and outnumbering other groups in all ecosystems. Tardigrades, rotifers and mites also do well. On a larger scale, oligochaetes and collembolan springtails are also abundant in soils. Planarian flatworms are surprisingly common in cooler ecosystems, while various groups of crustaceans and myriapods generally increase in abundance in warmer soils, as do salamanders. The same is true of gastropod molluscs, with the added complication of more shelled forms in tropical and desert zones and a predominance of slugs in cooler deciduous and coniferous forests. Pterygote insects are common in all soils (particularly as dipteran and beetle larvae in temperate soils and as ants and termites in more tropical zones) but they do not dominate in the spectacular way that they generally do above ground. In general, invertebrate soil communities strongly correlate with vegetation type, with the largest overall abundances in the temperate forests and grassland zones where there is good vegetation cover and a deep protective litter layer. Note that tropical rain-forest soils are *not* especially rich in fauna; in these areas the soil is rapidly leached, often waterlogged and sometimes anoxic, and most of the diversity and abundance of life is above ground here.

Figure 13.6b shows the biomass of different taxa in soils,

Table 13.4 Estimates of number of terrestrial species in major groups of animals.

Animal group	Number of terrestrial species
Platyhelminthes (triclad flatworms)	500
Nemertines	20
Gastropod prosobranchs	4 000
Gastropod pulmonates	20 500
Oligochaetes	2 000
Onychophorans	70
Crustacean isopods	1 000
Crustacean amphipods	50
Crustacean decapods	50+
Arachnids	65 000+
Myriapods	11 000
Insects	5–30 million
Amphibians	2 000
Reptiles	5 000
Birds	9 000
Mammals	4 000

and a rather different picture emerges. The larger body mass of oligochaetes immediately elevates them to dominance. In temperate soils, they are almost always from the family Lumbricidae, the true earthworms, consistent with their well-known role in soil formation and turnover; while in the northern coniferous forests, where the soil is impoverished below the litter layer, the oligochaetes are represented instead by the smaller enchytraeids. Other taxa play only minor roles in most soil habitats; the exceptions are mites in coniferous forests and pterygote insects in temperate grasslands.

If the same plots were to be made including above-ground fauna (the truly terrestrial xerophiles) several major differences would be apparent. Pterygote insects would dominate all the habitats in terms of abundance, while vertebrates, though always low in abundance, would make a significant appearance in the biomass stakes. Looking at species diversity gives us yet another slant on 'success'; Table 13.4 summarizes the species diversity of major land groups and underlines the insects' dominance.

13.2 Ionic and osmotic adaptation and water balance

Maintaining water balance can be seen as the key problem for land animals, interacting with all other aspects of their life, and above all requiring extensive adaptations of the skin and/or cuticle, the excretory systems, and the respiratory systems. The problems are always worse for smaller animals (see Chapter 5), but almost invariably for any truly terrestrial animal water loss must be minimized. In the aquatic habitats described in previous chapters, animals had very variable problems: to take up water and extrude ions (vertebrates in sea water), to try to limit all water exchanges and hold on to ions (in brackish zones), or to take up and conserve ions and try to get rid of water (in freshwater systems). But on land the problem is always clear and constant: to take in and hold on to ions and to take in and hold on to water.

Analysis of how this is achieved is complicated by the existence of the two quite different evolutionary routes onto land. Animal groups that have invaded the land via a littoral route have usually inherited a very high tolerance of changing osmotic conditions and of desiccation, relatively little osmoregulatory ability via skin or kidney, and a respiratory system able to cope with alternating supplies of water or air as the respiratory medium. Animals that arrived from a freshwater ancestry inevitably have pronounced osmoregulation with hyposmotic urine and strong capacities for salt uptake, and essentially aquatic respiratory organs. It could be argued that, because of their high tolerance of change, the animals derived from a marine ancestry have been more successful on land than those from a freshwater background, inheriting an emphasis on expensive osmotic regulation. All of this points to a fundamental and quite surprising point: most land animals are indeed **osmotic tolerators** rather than strong osmotic regulators. It is only some vertebrates (birds and mammals in particular) and perhaps some insects that are particularly good at **osmotic regulation**. Strategies for life on land based on physiological regulation have only been conspicuously successful when the animals concerned have also evolved a relatively impermeable outer covering.

13.2.1 Ionic and osmotic balance

A simple guide to a terrestrial animal's osmotic tolerance can be gained by looking at its ability to withstand changes in **water content** (Table 13.5). Many cryptozoic and hydrophilic animals are fairly intolerant in this respect, especially when they are also small. However, hygrophiles such as earthworms, slugs and snails can often cope with 40–80% water loss, the latter figure being enough to cause very substantial shrinkage of the body. Many terrestrial insects from temperate habitats will survive 30–50% loss of body water, while anurans (frogs and toads) from the same environments can survive around 30% water loss. But most terrestrial birds and mammals, although xerophilic, cannot survive more than 5–10% loss of water; the camel is really exceptional in tolerating 30%.

Examples of the osmotic and ionic concentrations of the blood of land animals are shown in Table 13.6. A number of features stand out. Firstly, most land animals have osmotic concentrations in the range 200–500 mOsm, roughly a quarter to a half that of a marine animal, and more concentrated than most freshwater animals. Secondly, the differences between species are striking, with some invertebrates (mostly those that are thought to have a freshwater background) having osmotic concentrations below 100 mOsm, and some with a clearly marine littoral background

Table 13.5 Tolerance of water loss in temperate land animals.

Animal group	% Water loss tolerated
Flatworm (*Bipalium*)	>50
Earthworm (*Lumbricus*)	70–75
Snail (*Helix*)	50
Slug (*Limax*)	80
Crab (*Gecarcinus*)	15–18
Caterpillar (*Manduca*)	50
Beetle (*Coccinella*)	35
Fly (*Eristalis*)	40
Frogs	28–45
Birds	5–10
Mammals	10–12
Camel	25–30

yielding values of 500–900 mOsm, approaching the typical values for a fully marine animal.

Thirdly, Table 13.6 reveals the variability within species, particularly for relatively permeable land animals, reinforcing the point that land animals are good osmotic tolerators. Where a single figure for osmotic concentration is given in the table, it usually means that only a few individuals have been tested and only under one set of conditions. In practice, when exposed to a range of environmental conditions, or when either deprived of food and water or given a large meal, most terrestrial animals show a wide range of tolerated osmotic concentrations in the blood, reflected in the Table for some groups. In the insect *Trichostetha*, haemolymph concentration may increase from 437 to 742 mOsm during desiccation. Figure 13.7 shows the possibilities for variation in blood concentration in various caterpillars in relation to ambient humidity and desiccation. Other animals' blood may vary drastically in relation to feeding: the housefly, *Musca domestica*, shows a blood concentration increasing from 163 to 762 mOsm after a sugary feed.

13.2.2 Behavioural regulation of water balance

A great deal can be done to alleviate water problems on land by using appropriate behaviour: in other words, as we stressed in the introductory chapters of this book, animals can use behaviour as a first line of defence against most of their environmental problems, and may only need to use complex and expensive physiological adaptations when the behavioural strategy fails.

This is already evident from our categorization of land animals described above. Cryptozoic and hygrophilic fauna are surviving entirely successfully on land because of their ability to find and remain within habitats of high humidity; most of them have inbuilt behavioural patterns that will take them to areas of darkness and high RH, and often also into intimate contact with wet substrates. Once within such zones they may show reduced activity, or activity with increased rates of turning, both

Table 13.6 Blood composition in land animals.

Animal group/species	Osmolarity (mOsm)	Ionic concentrations (mM)					Ancestry
		Na	K	Ca	Mg	Cl	
Nemertea							
Argonemertes dendyi	145						Interstitial
Annelida							
Lumbricus terrestris	150–370	76	4	3		43	Freshwater/ interstitial
Mollusca							
Poteria lineata	74	31	2	5	2	25	Freshwater
Pomatias elegans	254	110	6	16	3	106	Littoral
Helix pomatia	183						Littoral
Agriolimax reticulatus	345						Littoral
Arion ater	97–231						Littoral
Limax maximus	140–200						
Crustacea							
Amphipoda							
Arcitalitrus dorrieni	400						Littoral
Isopoda							
Porcellio scaber	700	227	8	15	11	279	Littoral
Oniscus asellus	577	230	8	17	9	236	Littoral
Decapoda							
Holthuisana transversa	517	270	7	16	5	266	Freshwater
Sudanonautes africanus	~400	207	6	12	11		?
Cardiosoma armatum	744						Littoral
Gecarcinus lateralis	~900	468	12	17	8		Littoral
Insecta	200–510						? Interstitial
Locusta	220–350	60	12	17	25		
Periplaneta	320–380	161	8	4	6	144	
Chelicerata	360–480						? Littoral
Onychophora	180–200						
Diplopoda	150–240						
Chilopoda	320–500						
Vertebrata							Freshwater
Amphibia							
Usual range	180–270						
Bufo bufo	205						
Bufo americanus	310						
Scaphiopus couchii	294–606						
Mammalia							
Homo	295	142	4	5	2	104	
Rattus	320	145	6	3	2	116	

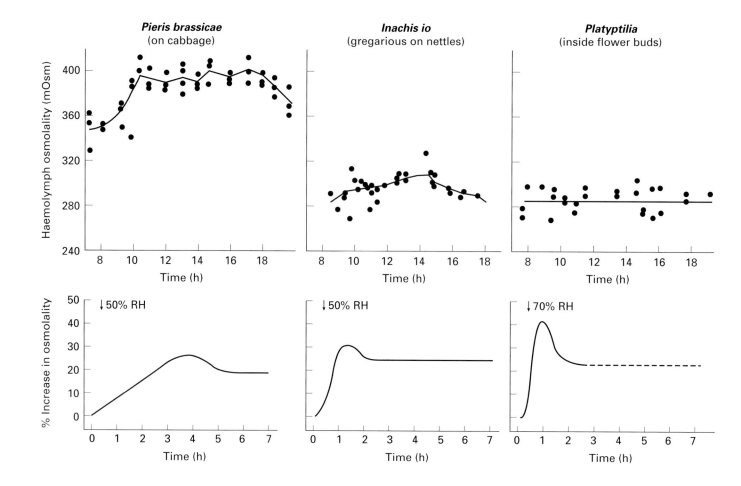

Fig. 13.7 Changes of haemolymph concentration in caterpillars measured within (upper) and measured when removed from (lower) their normal microclimates; species that live in very enclosed spaces (such as *Platyptilia* inside flower buds) suffer excessive water loss and lethal haemolymph concentration increases away from their normal microclimate, whereas species that live in the open (*Pieris* on cabbage leaves) are better able to osmoregulate and survive lower humidities. (Adapted from Willmer 1980.)

behaviours that will tend to keep them in the favoured conditions. Very often they have circadian activity rhythms that peak in the night hours, again to ensure that they encounter only dark (= cooler) and humid conditions. Hygrophiles, such as slugs, may only emerge when they detect a wet substrate and a falling temperature.

All these animals, but also the more xerophilic land animals, also have important behavioural mechanisms that assist their water balance. The choosing of moister rather than drier foods, or of drinking fluids of appropriate osmolarity, are obvious possibilities. Most animals also tend to reduce their activity with prolonged exposure to unfavourable conditions. On a short timescale this may mean retreating to a burrow or other favourable microclimate during the hottest and driest parts

of the day. Taken to extremes on a seasonal scale this leads to winter hibernation or summer aestivation. These responses may be coupled with aggregation, which helps to reduce water loss both in actively feeding larval insects (such as caterpillars) and in diapausing adults, such as fungus beetles and ladybirds; similar effects are to be expected in vertebrates too.

13.2.3 *Skins and cuticles*

Techniques for estimating the permeability of skins were considered in Chapter 5, and the values of permeability to water (P_w) and of cutaneous evaporative water loss (CEWL) were compared across the whole spectrum of animals in Table 5.3; a more detailed terrestrial data set is shown in Table 13.7.

Rates of water loss comparable to those for aquatic animals are found in the terrestrial cryptozoans and hygrophiles— all small and thus with a high surface area to volume ratio— including flatworms, nemerteans and oligochaetes, and also in larger land animals such as the molluscan slugs and snails and the anuran and urodele amphibians. These are all essentially 'soft-skinned' animals, where the skin often retains a respiratory function; most of them have surfaces continuously lubricated with mucus from epidermal glands. In many of these animals, as

Table 13.7 Measures of permeability in temperate terrestrial animals.

Animal group/ species	Resistance (s cm^{-1})	Water flux (mg cm^{-2} h^{-1})	Transpiration rate (µg cm^{-2} h^{-1} mmHg^{-1})	Water turnover (ml g^{-1} day^{-1})
Molluscs				
Helix (active)	1.6		2500	
(inactive)	46			
Limax (slug)				0.96
Otala (aestivating)			16	
Annelids				
Lumbricus (earthworm)	8.8			0.53
Onychophorans				
Peripatopsis				0.91
Crustaceans				
Porcellio (woodlouse)	3.1		110	
Oniscus (woodlouse)			165	
Armadillidium (woodlouse)			20–80	
Ocypode (crab)	32	2.6		
Gecarcinus (crab)	74	1.2		
Arachnids				
Pandinus (scorpion)			80	
Euscorpius (scorpion)			31	
Uroctonus (scorpion)			7.7	
Pinata (spider)	83			
Lycosa (spider)			33	
Ixodes (mite)			60	
Myriapods				
Lithobius (centipede)		3.8	270	
Scolopendra (centipede)			34	
Mastigona (millipede)			234	
Paradesmus (millipede)		1.9		
Glomeris (millipede)			200	
Insects				
Tomocerus (collembolan)			700	
Coptotermes (termite)			37	
Reticulotermes (termite)			28	
Hepialis (caterpillar)			190	
Pringleophaga (caterpillar)				0.24
Bibio (fly)			76	
Eristalis (dronefly)				0.48
Calliphora (blowfly)			51	
Acheta (cricket)			59	0.36
Locusta (locust)		0.8	22	0.13
Periplaneta (cockroach)	199		55	
Diploptera (cockroach)			21	
Blatta (cockroach)			48	
Cicindela (beetle)			24–49	
Rhynchophorus (beetle)			39	
Solenopsis (ant)			25	
Vertebrates				
Caiman (crocodilian)			70	
Gehyra (lizard)	198	0.22		
Amphispiza (sparrow)		1.48		
Columba (dove)	56			
Zenaida (dove)		0.73		
Poephila (finch)		2.69		
Honeyeater				3
Rattus (rat)		0.75		
Cow		9.96		
Merino sheep				0.07

the skin dries out on exposure to low humidities the CEWL decreases quite markedly, so that the figures given in Table 13.7 represent the maximum rate of water loss. But in practice exposure to drought will normally produce a rapid behavioural response, ensuring return to a humid or thoroughly aquatic microhabitat where water reserves can be replenished. If a watery habitat is not located, for example during long, hot summer months, some of these animals go into a period of aestivation and the outer skin dries up almost completely. Here the snails have a particular advantage as they can retreat fully into their shells and may seal off the opening with a calcareous or mucoid lid ('epiphragm'); the small area of exposed mantle collar in a snail such as *Otala* or *Helix* becomes fairly impermeable (see Table 13.7; the transpiration rate in aestivating *Otala* is $16\,\mu g\,cm^{-2}\,h^{-1}\,mmHg^{-1}$). The mechanism involved is not clear, but live aestivating snails lose water less quickly than dead snails, suggesting some active regulation. Certain arboreal frogs also have additional tricks to conserve water; the so-called 'waterproof frogs', in the genera *Phyllomedusa*, *Chiromantis*, *Hyperolius* and *Litoria*, can make themselves much more impermeable in dry spells when they become relatively inactive, due to the addition of a cutaneous lipid layer, derived from a specific posterior lipid gland and 'wiped' over the body using the limbs. Some other treefrogs have a lesser degree of waterproofing related to lipoid secretions from their normal mucous glands.

For the various groups that have supplemented the soft skin with some form of cuticle, rates of water loss are clearly reduced by one or two orders of magnitude. Within each taxon there is still considerable variability, but even the most permeable are always more waterproofed than the soft-bodied animals. However, the terrestrial representatives of some taxa are not necessarily much more impermeable than their aquatic relatives; land crabs and most land isopods, for example, are still subject to rather high rates of water loss.

The waterproofing effect of a cuticle is invariably due to the structure and chemistry of the **epidermis** and/or the **cuticle** itself. The common thread in all the more xerophilic animals is the acquisition of **lipid barriers** at some site (see Fig. 5.4). This is most clearly seen in arthropod cuticles (Fig. 13.8a), where several features contribute to low P_w. Firstly, the epidermis itself exerts some control over P_w; the cuticle of hydrated cockroaches is substantially more permeable than that of desiccated individuals, and the difference is attributed partly to the epidermal layer. Secondly, the bulk exocuticle and endocuticle material of chitin and protein is inherently relatively impermeable, and its water resistance may alter according to the density of its molecular packing as it becomes more or less hydrated, so that an insect under water stress is automatically more impermeable. (These two effects may be related, since the epidermis, influenced by hormones, could be responsible for withdrawing water via dermal gland canals, producing this reduction in P_w.) Thirdly, and usually much the most important of all, the outer layer, or epicuticle, confers very low permeability due to its high lipid content: C_{10}–C_{37} fatty acid derivatives (especially the even-number chain length C_{12}–C_{18} series), together with some branched and unsaturated hydrocarbons. Lipids within the cuticle may be augmented by a thin layer of surface lipid, often predominantly wax esters.

In arachnids the cuticle is traversed by fine ducts from epidermal glands, and the nature of the epicuticular lipids is also often rather different, with a greater sterol component. In fact in these animals the epicuticular lipids may change in nature through the life of an individual, which is probably the reason for glands leading directly to the surface.

Following from this, there is increasing appreciation of the possibilities of adaptive variation in the properties of the arthropod cuticle. Summer-active species may contain more long-chain hydrocarbons than closely related winter-active species, an effect demonstrated both in beetles and in scorpions. The difference occurs even between field-caught winter- and summer-adapted individuals within one species, most clearly demonstrated for the beetle *Eleodes armata*; moreover, pupae of *Drosophila* laboratory-reared from eggs at 17°C have more permeable cuticles than those reared at 24°C, with concomitant changes in constituent lipid chain lengths. The cast-off cuticles (exuviae) of the grasshopper *Melanoplus sanguinipes* contain lipids of significantly higher melting point when the individual is acclimated to higher temperatures, largely due to incorporation of more straight-chain hydrocarbons relative to branched forms; and analysis of siblings shows that the natural genetic variation of their cuticle permeability largely also resides in the straight-chain lipid class. Changes in the cuticle clearly may occur during the lifetime of an individual land arthropod, above and beyond the differences between hydrated and dehydrated individuals referred to earlier. The changes may be related to hormonal control; for example, in cockroaches a hormone from the brain operates at the epidermis to slow desiccation, and water loss rates in the lubber grasshopper (*Romalea*) have been shown to increase with rising humidity, indicating a relaxation of some controlled barrier mechanism. More impressively, in the cattle tick, *Boophilus*, the amount of wax increases when a tick drops off its host, and the thermal properties of the lipids also seem to alter, indicating fine control over the waterproofing properties of the surface layers. Some insects undergo a progressive increase in cuticle permeability through their adult life, perhaps due to increasing abrasion and loss of surface waxes. More spectacularly, some homopteran bugs and coleopteran beetles have superficial wax 'blooms' on their cuticles, which may be replenished seasonally or even daily; and a few have cuticular pores through which they can achieve controlled 'sweating' (see Chapter 14).

In xerophilic tetrapod vertebrates the basic epidermal type, with high concentrations of the protein keratin, is modified by the inclusion of some extra phospholipid in the outer layers of the epidermis (e.g. in the stratum corneum of reptiles). This arrangement works most effectively as a waterproofing layer in the lizards and snakes, where the skin is not interrupted by

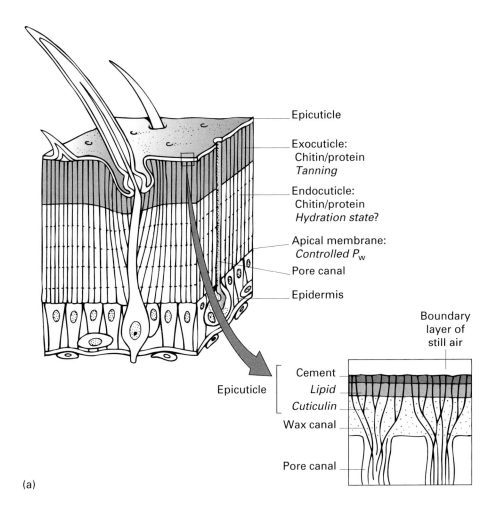

Epicuticle

Exocuticle:
Chitin/protein
Tanning

Endocuticle:
Chitin/protein
Hydration state?

Apical membrane:
Controlled P$_w$

Pore canal

Epidermis

Boundary
layer of
still air

Cement
Lipid
Cuticulin

Epicuticle

Wax canal

Pore canal

(a)

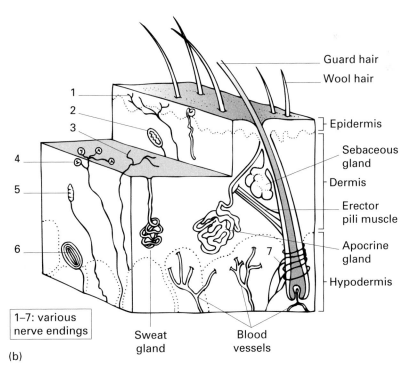

Guard hair
Wool hair

Epidermis

Sebaceous
gland

Dermis

Erector
pili muscle

Apocrine
gland

Hypodermis

1
2
3
4
5
6
7

1–7: various
nerve endings

Sweat
gland

Blood
vessels

(b)

Fig. 13.8 (a) Structure of typical insect cuticle; note that at joints and in growing larval stages it may be much thinner with little sclerotized exocuticle. The main control points of resistance to water loss are shown in *italic*. (b) Structure of mammalian skin (not to scale).

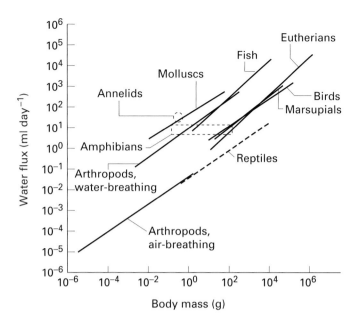

Fig. 13.9 Water flux rates in relation to body size for land animals; all show similar gradients, though arthropods have routinely lower fluxes than vertebrates, and soft-bodied animals much higher fluxes. (Adapted from Nagy & Peterson 1987.)

glandular pores, such as those of sweat glands. Nevertheless, CEWL in reptiles is normally 60–85% of the total evaporative water loss (EWL; higher range in water snakes, and lower in desert tortoises). In the skin of endothermic mammals (Fig. 13.8b), where such glands often have to be present to provide a cooling mechanism, CEWL may be much more variable according to levels of activity and environmental temperatures, though even its basal level is set higher than in a similarly sized reptile.

Reptiles, insects and arachnids, with thick keratinized or sclerotized cuticles having a high lipid content, therefore hold the records for animal impermeability. This is reflected in Fig. 13.9, showing daily water flux always increasing with animal body mass but with the setting lower for arthropods and reptiles, higher for birds and mammals, and of course higher still for annelids and molluscs.

Given reasonably good defences at the external barriers (as in these groups), and/or a moderately large body size to slow the rate of exchanges (as in the vertebrate endotherms), regulation of the internal fluids becomes a viable option.

13.2.4 Regulatory organs

All animals living on land need a site for excretion of waste materials, and most excretory materials, being potentially toxic, have to be lost quickly and diluted in a moderate volume of fluid. Therefore most animals also use this **excretory system** ('kidney') as a site where they can conveniently regulate their water content, and consequently also as a site for regulating salt

balance. In fact most of the so-called excretory systems of animals probably had osmoregulation as their primary function.

Soil animals tend to have osmoregulatory/excretory systems similar to their aquatic relatives, as they will often have an excess of water available and need to excrete it quickly. Animals such as earthworms may need systems that can expel excess water but can also act as water conservers during drought. Xerophilic animals almost always need to conserve as much water as possible, and (with the exception of mammals) most have combined their 'excretory' opening with their gut opening, so that all the excreta are regulated in one site (the rectum) with minimal water loss.

Terrestrial animals therefore show a great variety of mechanisms of excretory water loss. Chapter 5 dealt with the principles of urine formation and regulatory reabsorption; here we look at the form and functioning of the different organs found in land animals.

Cryptozoic and hygrophilic land animals

Most of these retain an essentially aquatic approach to osmoregulation and excretion. Flatworms and nemertines retain the **flame cell** protonephridial system (Fig. 13.10a) discussed in Chapter 11. The lower parts of the flame-cell tubules probably have some resorptive function, but normally allow the primary urine to pass with minimal salt resorption, flushing excess water out of the body as quickly as possible. The product is therefore usually **hyposmotic**, with ammonia as the main nitrogenous waste, both features reflecting the more-or-less freshwater nature of the habitat. However, some terrestrial flatworms are unusual (for invertebrates) in producing a large percentage of their nitrogenous output as urea.

Larger soil-dwellers, such as earthworms, also have a nephridial system, organized segmentally and exiting via nephridiopores, though the organs (Fig. 13.10b) have a more substantial coiled resorptive tubular portion than in the flatworms and nemertines. The product is clearly hyposmotic in most oligochaete species. However, in some of the larger earthworms, such as *Pheretima* from India and the Australasian megascolecids (Fig. 13.10c), the nephridia are 'enteronephric', opening into the gut rather than to the outside world, and here some drying of the faeces + urine is achieved within the rectum. In most terrestrial annelid species the main excretory product is ammonia, though urea levels may increase in worms that are not feeding. In addition a substantial amount of nitrogen is lost as mucoprotein in the copious mucus secretions from the skin.

Land snails and slugs

Terrestrial prosobranch snails have a kidney essentially similar to their littoral and marine kin. Fluid flows to the kidney from the pericardium, and drains directly from there to the outside, with no special absorptive area. Excretion always involves the loss of considerable quantities of water, because there is no

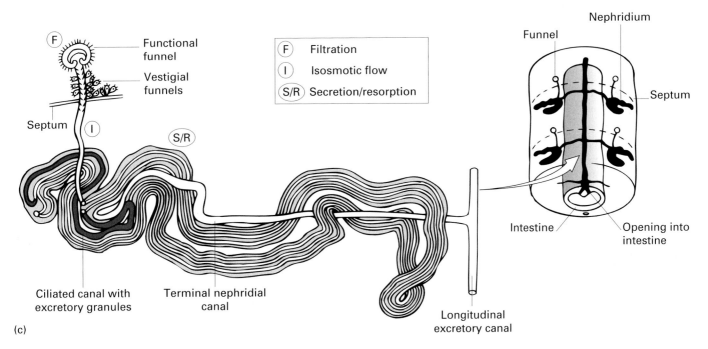

Fig. 13.10 (a) Flame cells from terrestrial nemertines; (b) the more elaborate nephridia of earthworms; and (c) the nephridia of xeric megascolecid earthworms, which empty into the gut allowing greater resorption of water.

spatial separation of the sites of nitrogenous excretion and of water balance regulation.

Terrestrial pulmonates (the more familiar slugs and snails) do rather better, and show a wide range of adaptations to the kidney. The generalized structure from a temperate snail is shown in Fig. 13.11. The kidney has a classical **ultrafiltration** system

to form primary urine, with the fluid supply coming either from the pericardium surrounding the heart or from the blood surrounding the kidney epithelium. The kidney then opens into an elongate ureter, which turns back on itself to give a primary and secondary arm, then opening to the outside at a renal pore in the mantle cavity. Uric acid is secreted into the urine together with some guanine and xanthine, usually within the kidney but sometimes in the upper part of the ureter. The lower ureter is absorptive and regulatory, taking some salts and water back into the body and regulating the urine pH. In damp surroundings, there is a copious flow of hyposmotic urine, but in drought

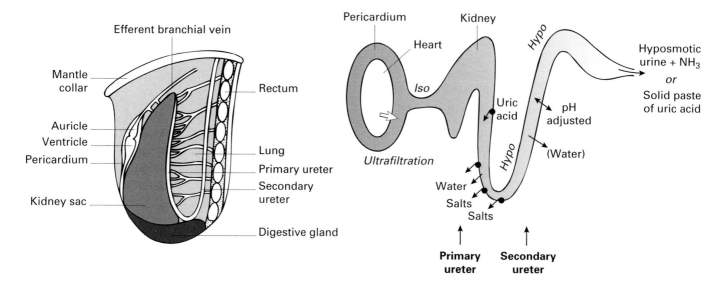

Fig. 13.11 The kidney of pulmonate snails (*Helix*), receiving a filtrate from the pericardium and normally producing a hyposmotic secondary urine in the primary and secondary ureter. (Adapted from Little 1983.)

conditions a dry paste emerges from the renal pore. Curiously, there is no intermediate of a hyperosmotic liquid urine. In other words, either the filtration system is turned on and a dilute watery product emerges, or the filtration is shut down and dry waste is produced independently, from the (spatially separated) uric acid secretion site. In this way pulmonates can safely excrete

Fig. 13.12 (a) Structure and function of the antennal glands in land crabs (note that the branchial chamber often does some 'urine reprocessing'). (b) Coxal gland of a typical arachnid.

their nitrogenous wastes in a wide range of climatic conditions. Indeed, when undergoing aestivation they can survive without any excretion by storing nitrogenous products, especially urea, within the body (though some also lose gaseous ammonia by diffusion through their shell in this phase of their life). Aestivating snails show elevated blood osmolarity, sufficient to depress neuronal activity and reduce the heart rate to around 60% of normal.

Hygrophilic arthropods

Semi-terrestrial crabs, living at the top of beaches and in damp mangrove forest zones of the tropics, show a relatively strong osmoregulatory ability (see Table 13.6). However, this is not achieved by osmoregulation at the **antennal glands** (Fig. 13.12a),

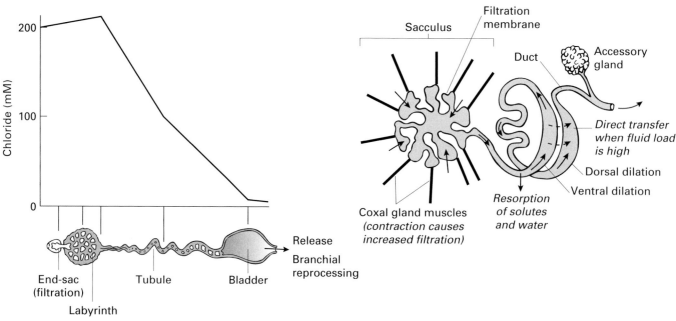

(a)

(b)

the crustacean version of kidneys; these can only produce isosmotic urine. These glands do modify the ionic composition of the filtrate they receive, to regulate blood ion levels, but they cannot produce hyperosmotic urine to aid water conservation. Instead, salt is secreted at other sites, especially at the gill epithelium as in so many aquatic invertebrates (see Chapter 5), but perhaps also in the rectum. The only water balance 'regulation' achieved at the antennal glands concerns the overall rate of **urine flow**, so that in dry conditions flow stops entirely, with no primary ultrafiltration occurring. Land crabs are thus physiologically similar to their aquatic relatives, though the balance between physiological prowess and behavioural strategies to avoid water loss and achieve adequate water input is very variable between species. The only possible extra 'tricks' that some may have are to reingest some of their urine or, quite commonly, to divert some of it into the branchial chamber, from where extrarenal resorption of ions may occur through the gills, resulting in a branchially modified urine known as the 'P product'. This **gill reprocessing** system is under the control of dopamine, which can increase branchial Na^+/K^+-ATPase by around 67%. In the blue crab, *Cardisoma*, around 90% of the urinary ions are reclaimed in the gills, and the two tricks of urine reingestion and reprocessing together allow the robber crab, *Birgus*, to reclaim 70–99% of the ions from its initial urine, especially important when living in particularly dry inland sites where only rainwater is available and ions are in short supply. This phase in urine production also allows the urine to be augmented with extra ammonia.

Systems that are structurally and functionally similar to crustacean antennal glands occur in other hygrophilic arthropods, including flightless insects and myriapods (where the organs involved are maxillary glands, opening near the mouthparts), and in onychophorans (where there are segmentally repeated excretory organs). Most of these glands have two regions (cf. Fig. 13.12b). Firstly, there is a sacculus, where ultrafiltration occurs across thin-walled cells termed podocytes (somewhat similar to cells in the vertebrate kidney glomerulus). The sacculus then leads on to a tubular section in which salts and water are resorbed.

Woodlice (pillbugs) have an extra trick. Urine from their maxillary glands is released via capillary channels to flow over the flat ventral respiratory flaps called pleopods, keeping them moist for gas exchange (and allowing ammonia to diffuse away, disposing of the nitrogenous waste from the woodlouse's body). At least some of this excreted water is then returned to the body via the resorptive rectum, giving an unusual involvement for the gut in that fluid arrives in it via an external route. Species from arid habitats, such as *Hemilepistus*, can resorb significant quantities of water via this rectal route.

Xerophilic arthropods

Two rather different systems occur in truly terrestrial arthropods. Most of the arachnids have a system analogous to the maxillary gland system described above; the structures are termed coxal glands (Fig. 13.12b), and although they have a different origin they function in a similar manner, though in some species ultrafiltration is speeded up by muscular stretching of the walls of the filtration chamber. Unusually, the excretory coxal fluid is often delivered into the prey rather than to the outside world.

All other xerophilic arthropods (insects and many myriapods) have an excretory-osmoregulatory system based on the 'Malpighian tubules plus rectum' composite, where the stages of urine formation and resorption are in widely separated structures. Urine is produced by secretion in the **Malpighian tubule** cells (see Chapter 5); details of this tubule secretion system, which has been very thoroughly studied, are shown in Fig. 13.13. The tubule wall is composed mainly of standard cation-handling cells, with pumps (V-ATPases) effectively moving K^+ in most insects with herbivorous or omnivorous diets, but able to transport Na^+ very rapidly in blood-feeders with high-sodium diets. The entrained flow brings most other solutes passively through into the lumen, though specific urate carriers to ensure excretory clearance are present in many species. There are usually also smaller numbers of 'stellate cells' in the Malpighian tubule wall, which are now known to act mainly as chloride-transporting cells. Malpighian tubules are controlled by hormonal triggers in many insects, with a peptide diuretic hormone from the mesothoracic ganglion operating when a large watery meal is ingested (e.g. a blood meal in ectoparasites), acting synergistically with other hormones such as 5-HT (5-hydroxytryptamine) to give a rapid-onset diuresis. Antidiuretic hormones are also known in some species, as are peptides controlling chloride permeability (Chapter 5), and there are indications that prostaglandins may also have some influence. The net result is a fast and usually isosmotic flow across from the haemolymph into the tubule lumen. Some resorption of salts and water may take place in the lower reaches of the tubules themselves, where P_w is probably reduced; but usually the primary urine is discharged, unmodified, into the gut at the point where the midgut joins the rectum.

The insect hindgut (ileum + rectum) then has a variety of specializations to resorb water effectively and to regulate the combined composition of the faeces and urine. The first part of the hindgut mainly executes a substantial resorption of fluid so that the ileum contents reduce in volume with little change in concentration (except that some molecules specifically secreted into the system earlier on are unable to diffuse out, and so become more concentrated). The **rectum** is the more crucial area, where there occurs substantial changing of concentration of all the urine's components. The first step in this is to draw water from the rectal lumen into the gut cytoplasm; it is assumed that the cells are endowed with osmolytes that draw this water in. Then the rectal cells effectively secrete (to the haemolymph) a hyposmotic fluid. The simplest system that can achieve this is the rectal pad, an intracellular resorption system essentially as described for the model system outlined in Figs 5.14 and 5.15.

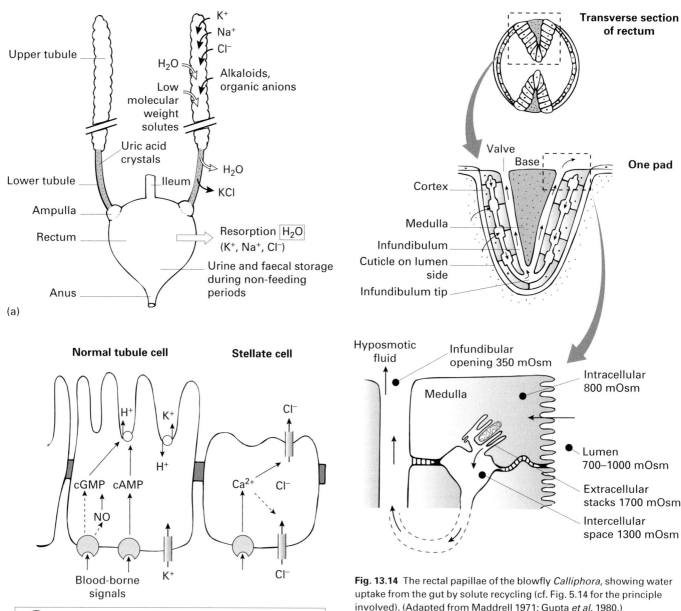

Fig. 13.13 The functioning of Malpighian tubules in insects: (a) an overview of the system; and (b) details of the ion movements in normal tubule cells and stellate cells. (Adapted from Phillips 1981; Maddrell & O'Donnell 1992; O'Donnell *et al.* 1998.)

Fig. 13.14 The rectal papillae of the blowfly *Calliphora*, showing water uptake from the gut by solute recycling (cf. Fig. 5.14 for the principle involved). (Adapted from Maddrell 1971; Gupta *et al.* 1980.)

Slightly larger scale systems operate, using the same principle, in the **rectal papilla** (Fig. 13.14), found in orthopterans and in some flies, such as the common 'greenbottle' or blowfly (*Calliphora*). Spaces between the rectal cells create a compartment separated from both the blood and the gut contents, and ions are transported into these spaces from the cells, with water flowing from the gut across water-permeable membranes, down the osmotic gradient thus created. The fluid that accumulates is then forced out into the haemolymph under the ever-increasing hydrostatic pressure created by the secretory processes, along channels where most of the ions in the fluid can be resorbed and recycled into the cells, but where water cannot readily follow due to a high flow rate and relatively low P_w. This system is a further example of the 'concentration by compartmentalization' mechanism described in Chapter 5; it can also recover useful compounds, such as amino acids and sugars, from gut to blood.

In some beetles and larval lepidopterans a yet more complex system is found, with the blind-endings of the Malpighian tubules reflexed back against the wall of the rectum, to create a **cryptonephridial** system (Fig. 13.15). The spaces around the

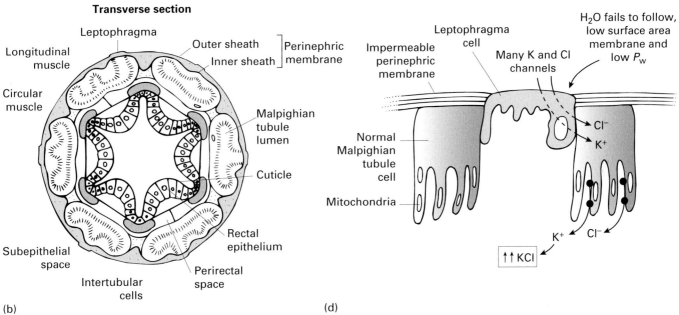

Fig. 13.15 The cryptonephridial system found in a variety of xeric insects, here shown in the flour beetle *Tenebrio*. (a) The whole system with Malpighian tubules folded back against the gut. (b) In transverse section, showing six tubules around the rectum. (c) Ion and water movements and concentration gradients along the system. OC, osmotic concentration. (d) Ion transport at leptophragma cells. (Adapted from Ramsay 1964; Maddrell 1971.)

tubules now form an additional compartment (the perinephric space), which can be used in a similar fashion to the intercellular spaces described above. In addition, the flow forward in the Malpighian tubules is now running in the opposite direction to the flow backward along the rectum, so we have a countercurrent created. The whole complex is enclosed in a relatively impermeable membrane, punctuated with 'leptophragma' cells where intense ion movement from the haemolymph into the tubules occurs (see Fig. 13.15d). At the anterior end of the system concentrations are quite low, though always with a water gradi-

ent from gut to haemolymph; near the anal end of the rectum, there is still a gut-to-blood gradient but by now concentrations in both are much higher and water is still being withdrawn from gut contents, which may reach 4000–5000 mOsm, up to five times seawater concentration. Hence a very concentrated dry paste of uric acid and some faecal waste results.

Tetrapods

The key features of the vertebrate kidney and excretory products were dealt with in Chapter 5; here we need only look at the variation in different land tetrapods. Amphibians certainly inherited an essentially freshwater-adapted kidney from their ancestors, and can only make **hyposmotic** urine. They can decrease the rate of urine formation and increase urine resorption from the bladder by the action of the amphibian antidiuretic hormone (arginine vasotocin), and can also correct their osmotic balance with active salt uptake through the skin.

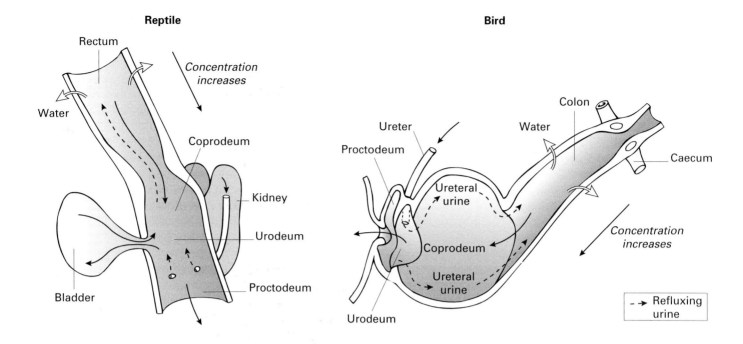

Fig. 13.16 Distal portions of the gut and excretory systems of a reptile and a bird, showing the reflux of urine up into the rectum via the common 'coprodeum'. The hindgut contents are substantially more concentrated than the blood. (Adapted from Skadhauge 1981; Minnich 1982.)

Table 13.8 Urine concentrations and urine/plasma ratios in temperate land animals (values are maxima from dehydrate animals).

Group/species	Urine concentration (mOsm)	Urine/plasma ratio
Insects	~1000–5000	~3–15
Reptiles		
Tuatara	270	1.0
Gecko	325	0.7
Iguana	362	1.0
Birds		
Chicken	538	1.5
Pigeon	655	1.7
Pelican	700	2.0
Mammals		
Rat	2900	9
Domestic cat	3100	10
Vampire bat	4650	14
Beaver	520	1.7
Pig	1100	3
Domestic cow	1160	3.7
Human	1400	4–5

All other tetrapods have rather impermeable skin and cannot use these systems. Reptiles nevertheless remain unable to produce hyperosmotic urine, even when they live in extremely xeric habitats. Their urine is **isosmotic** and excess salts are excreted through their nasal salt glands (Chapter 5). Most birds and mammals also do not normally produce hyperosmotic urine. Instead, when drinking-water is freely available many of them release a hyposmotic fluid; this also occurs where the diet is very watery (as in some fruit bats, which may excrete around 15% of their body mass each day, and in some nectar-feeding hummingbirds). Alternatively, when water-stressed they commonly produce a roughly isosmotic fluid. Some birds (and also some reptiles) discharge this urine into the cloaca but it then flows back up into the rectum where additional (approximately isosmotic) resorption of fluid occurs (Fig. 13.16).

However, some birds and nearly all mammals have a **loop of Henle** interposed between the descending and ascending portions of their nephrons, to give a site of countercurrent multiplication (see Chapter 5), and they can then produce a significantly **hyperosmotic** urine (Table 13.8; U/P ratio > 1) when needed. Here the cells in the hairpin tip of the countercurrent loop are exposed to very high levels of urea and are almost anoxic; they cope by incorporating high levels of betaine and sorbitol into their own cytoplasm as counteracting osmotic effectors (see

Chapter 4). Remember that the absolute length of the loop is not necessarily the key factor in relation to urine concentration. For example, the desert mouse *Notomys* has loops up to 5.2 mm long and can produce urine of 9370 mOsm, while the horse has loops up to 36 mm long and can achieve urine of only (at best) 1900 mOsm. Indeed, in birds urine-concentrating ability commonly decreases with increased loop length, perhaps partly because many of their loops lie crosscurrent rather than countercurrent to blood flow. Urine concentration is determined more by the specific metabolic rate of the tubule tissues (higher in smaller animals, with a much higher density of actively transporting membrane surfaces) than by length of tubule (see Chapter 5).

Most tetrapods excrete urea or (in most birds and reptiles) uric acid, but diet also introduces some variation here; hummingbirds, for example, are able to switch to ammonia secretion when feeding on dilute nectars at low temperatures.

13.2.5 *Acquiring water*

Liquid uptake

Most land animals gain their water by drinking, from streams, rainwater puddles or rain drops on vegetation, or from water vapour condensed as dew. The other important source is the water content of the diet, and a surprising range of terrestrial animals rely on this alone, requiring no free water. Many species will select wetter food sources over drier ones when water-stressed. Some may even take more food than they need just to get enough water, though the evidence for this is controversial. Animals that live in burrows and other enclosed microhabitats may store dry foods, such as seeds, for some time before eating them, and the humid microclimate then allows the foodstuff to hydrate substantially at no cost to the animal. This phenomenon is an important contributor to many desert rodent's water balance (see Chapter 14), eliminating the need for free water in the diet.

However, a number of animals have specific adaptations to acquire water from the environment independently of drinking water or eating food. The possibilities were listed in Chapter 5, and are particularly obvious amongst the various arthropodan groups on land.

Some arthropods have evolved behavioural mechanisms to make use of dawn **dews and fogs**. Simple versions of this involve drinking dew as it forms on plants or on the sand, but some desert species again have more sophisticated ways of using heavy dawn fogs in coastal deserts, which we will deal with in Chapter 14.

Standing water can also be acquired by **osmosis**, in ways that do not involve traditional 'drinking'. This is usually only an option for animals that retain a strong link with water sources anyway, and often retain fairly permeable skin surfaces. Crabs therefore exploit a number of these possibilities. Species with gills that live around high-tide level on beaches can immerse the gill chambers in pools of fresh or slightly brackish water so that water enters the body. The hermit crab, *Coenobita*, can select the osmolarity of the pool it uses according to its own osmotic needs. One species of ghost crab (*Ocypode*) can actually suck water from sand into its gill chamber, creating a pressure of

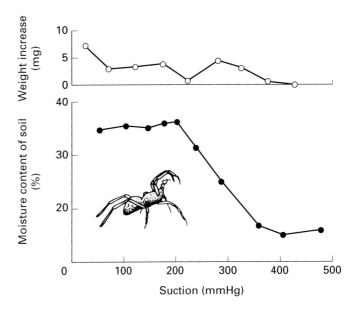

Fig. 13.17 The ability of spiders to suck water from soil of different hydration states; this ability is still present at only 20% soil water content, though weight gain is then very small. (From Little 1990; adapted from Parry 1954; courtesy of Cambridge University Press.)

about 40 mmHg by muscular action on the chamber walls. Many other semi-terrestrial crabs can take up water using a patch of densely hairy cuticle on the ventral body surface; when the crab is squatting down on damp sand, the hairs draw up water by capillarity, and conduct it to the leg-bases, where it is drawn into the gill cavity.

Uptake of water from wet soils and surfaces can also be achieved osmotically in the more terrestrial arthropodan groups. Quite a range of wingless insects and myriapods can absorb water using structures at the top of the legs called **coxal sacs**. These can be everted and placed onto damp substrates, and an active salt-pumping process is used to create a large osmotic gradient that will rapidly take in water. A xeric millipede, *Orthoporus*, can extrude a pad of rectal tissue from its anus that works in a similar fashion. Isopods, such as the littoral *Ligia*, and several of the more terrestrial woodlice, can also absorb water osmotically from wet soils using permeable surfaces at the mouth or anus.

Getting the interstitial liquid water from relatively unsaturated soils generally requires both a very good muscular pump, and modifications of the skin surface to produce a strong **capillarity** that will overcome the capillary attraction of water to soil. Some spiders can extract water from soils (Fig. 13.17) against a suction pressure of 450 mmHg, where the soil has less than 20% water content. They are presumably aided by their strong pharynx muscles (normally used for sucking fluid from prey), so could be said to be preadapted for extracting moisture from soil.

Thus most of the specialist mechanisms for acquiring water are found amongst the arthropodan groups of land animals. Nevertheless, a few vertebrates do achieve similar tricks. Some frogs and toads have patches of abdominal skin ('pelvic patches') specialized to absorb water osmotically; they can sit in puddles or on very damp sand and take water into their bodies, storing a surplus in the bladder. Local sensors within the skin deter them from taking up the water-absorption posture on salty soils or in brackish puddles. This cutaneous drinking response also appears to be under the control of angiotensin II. It can be modified by barometric pressure, so is attuned to the perceived future availability of water. Toads therefore show 'anticipatory cutaneous drinking', before they are dehydrated and with the water-storing bladder still moderately full.

However, most other vertebrates have to rely on drinking liquid water, and may have very much more elaborate physiological controls to regulate their drinking behaviour. Control of **drink-rate** is mainly via osmolarity of the extracellular fluid (ECF), with osmoreceptors feeding information to a 'thirst centre' in the hypothalamus. Decreased ECF volume resulting from water loss also acts as a trigger, causing renin secretion and thus release of angiotensin II, which acts on the 'subfornical organ' close to the brain ventricles, again feeding information to the hypothalamic control centre.

A final possibility sometimes mentioned as a source of water for animals is the **metabolic water** derived from oxidative processes within all the cells of the body, when carbohydrates and other large molecules are broken down. Certainly these processes do produce water as a by-product (see Chapter 6) and animals retain and use this. It is usually only a small component of overall water intake. But in animals from dry habitats, or with very high metabolic rates (e.g. in flight), it may be a significant proportion of the water budget. However, it cannot normally be regarded as a mechanism for gaining 'extra' water. Extra oxidation requires extra ventilation, and in most animals the extra respiratory water loss exceeds the water gained internally. Only in special cases could there be a net gain of water; these may include certain vertebrates with an ability to reduce the water content of their expired air using a nasal countercurrent exchanger, or large flying insects that may also be using countercurrent water saving and that do not use much evaporative cooling in flight (unlike bats and birds).

Vapour uptake

Some representatives of the insects, crustaceans and chelicerates on land can use the technique of **water vapour uptake**, briefly discussed in Chapter 5. This is an unusual and specialist adaptation found only rarely (see Table 5.5): in a few isopods, some mites and ticks, some apterygote insects, and some flightless insects in the orders Anoplura (lice), Siphonaptera (fleas), Dictyoptera (cockroaches) and Coleoptera (beetles, though they show uptake only in the larval stages). An example of uptake rates at different humidities in a flea was shown in Fig. 5.7. Uptake seems to be virtually always associated either with the mouthparts or with the rectum.

In mites and ticks the ability to take up water may vary in different life stages; for example, in deer ticks it is most evident in the larvae, which emerge on the ground and may have to survive for months before encountering a host. In these animals, and also in lice, blocking the mouthparts prevents vapour uptake. A clear **hygroscopic fluid** accumulates near the palps, derived from the salivary glands, which in *Dermatophagoides* (the infamous house-dust mite) produce a secretion that is rich in potassium and chloride, and probably also contains some organic material. At low RH this dries to a crystalline deposit, but at higher humidities water vapour condenses into it and it is then withdrawn by a muscular suction into the pharynx. As yet we do not know how the very high concentrations of solute needed for uptake at low RH can be achieved; for example, uptake at 45% RH (see Table 5.5) would require fluid with a concentration of at least 50 Osm. In the tick *Amblyomma* it is thought that **hydrophilic cuticle** plays a role in water condensation, with the hyperosmotic secreted fluid then altering the water affinity of this cuticle to release the adsorbed water into the sucking pharynx. In the particularly well-studied desert cockroach, *Arenivaga*, the concentration of the salivary fluid alone is certainly inadequate to explain vapour uptake, and other cuticle-based mechanisms are proposed; these are discussed in relation to desert adaptations in Chapter 14.

Other vapour-absorbing arthropods use a rectal rather than a salivary site. The rectum is the normal site for uptake of liquid water from the gut anyway, and its tissues may have very high concentrations of salt to allow osmotic uptake, particularly where the gut has a cryptonephridial arrangement. Calculations show that this is adequate to account for the extraction of water from air above the 88% CEH (critical equilibrium humidity—see Chapter 5) of *Tenebrio* larvae; in effect, during vapour absorption the gut contents have a reversed fluid flow, and ongoing K^+ transport in the Malpighian tubules allows build-up of very high osmotic concentrations there, so extracting water from the rectal tissues and ultimately from the air within the rectum. However, the wingless insect *Thermobia* (commonly known as the firebrat) can take up water anally from air at 45% RH, while the rat flea, *Xenopsylla*, can operate at 65% RH, again via the rectal surfaces; yet neither has a cryptonephridial system. In the firebrat there are three anal sacs, with a much-folded and mitochondria-rich (presumably ion-pump-rich) epithelium, and these may be filled with air and some hygroscopic material. The sacs open to the outside via anal valves, which open and close rhythmically, suggesting a cyclic process perhaps involving **pressure increases**, which would raise the humidity, making uptake easier. This observation links neatly to recent work on water vapour uptake in certain woodlice, where hyperosmotic fluid in the cavity between the pleopods and the abdomen is implicated, with uptake again augmented by pressurizing this cavity cyclically.

13.3 Thermal adaptation

For any animal on land, all the avenues of heat exchange discussed in Chapter 8 are available; a summary is shown in Table 13.9. Heat can readily be gained by conduction and convection from heated surfaces, or by direct radiation from the

Table 13.9 Mechanisms of temperature regulation on land.

	Source	Route	Influences
Heat gain	The Sun	Radiation Conduction Convection	Colour, surface properties Size, and surface area exposed Behaviour (posture, orientation, movement)
	Metabolism		Size and surface area Enzyme concentrations, mitochondrial density Special thermogenic tissues Superficial blood flow, vascular shunts Hormones, nerves Activity
Heat loss		Radiation Conduction Convection	Size/surface area, colour, special exposed surfaces, behaviour (posture, orientation, movement) Superficial blood flow, vascular shunts Hormones, nerves Activity
		Evaporation	Size and surface area Panting, sweating, licking, urinating Superficial blood flow, vascular shunts Hormones, nerves Activity

Sun. Heat can be lost by these same three processes for an animal that moves into cooler air, away from heated surfaces and out of direct sunlight; crucially, it may also be lost by evaporation.

Remember the issues of terminology (Chapter 8). Both **ectothermic** and **endothermic** strategies are favoured in different circumstances. Ectotherms may be thermal **conformers**, but more commonly both strategies on land involve a fairly high degree of more or less expensive **thermoregulation**. Land animals may also be either **stenotherms**, operating over only a narrow range of body temperature (as in most endothermic tetrapods and at the opposite extreme in animals from very protected stable microhabitats) or **eurytherms**, coping with very variable body temperatures as necessary (as in many ectothermic insects and vertebrates). Different strategies and solutions tend to operate in different size ranges; small animals (mostly eurythermic ectotherms) may use a considerable suite of behavioural adaptations to avoid thermal stress, while large animals (mostly stenothermic endotherms) may have elaborate behavioural and physiological techniques to keep their bodies at the preferred constant temperature. Life on land, as far as temperature is concerned, is not just a matter of increasing physiological or biochemical sophistication, as sometimes portrayed; it is more often a case of finding new balances of physiological and behavioural regulation, with the common endpoint being a raised and stable body temperature whenever an animal is active.

13.3.1 Terrestrial ectotherms

As we stressed in Chapter 8, ectotherms are not necessarily 'cold-blooded', or 'at the mercy of their environments' with respect to body temperature, and these traditional terms are particularly inappropriate for many terrestrial ectotherms. Air has a lower thermal conductivity and specific heat than water, therefore it is relatively easy for a land animal to maintain a gradient between body temperature (T_b) and ambient temperature (T_a). In the cryptozoic and hygrophilic fauna a gradient is rarely observed, and the habitats have fairly equable conditions anyway, with an ambient temperature buffered from the more extreme changes in the neighbouring freely moving air; animals living here are moderately eurythermic. However, for the xerophilic animals, T_b may be regulated in a variety of ways, usually to keep it above T_a. This difference is related not only to habitat, but also to **size** (itself linked to habitat, since xerophiles can be, and usually are, much bigger than the cryptozoic and hygrophilic fauna). Small animals inevitably change temperature more rapidly than larger ones, losing and gaining heat on a per gram basis much more quickly; so they cannot maintain a gradient between body and air easily. But a reasonably large xerophilic ectotherm has every chance of regulating its T_b at high and fairly steady levels, principally using behavioural techniques to be active only at appropriate ambient temperatures and to sunbathe or seek shelter; it can be regarded as stenothermic, at least during all periods of actual or imminent activity. Note also that preferred body temperatures, achieved T_b ranges and normal metabolic rate values vary with habitat; for example, tropical species within the lizard genus *Sceloporus* have higher T_b values, metabolic rates about 1.6-fold higher, and higher water flux rates, than do closely related temperate relatives.

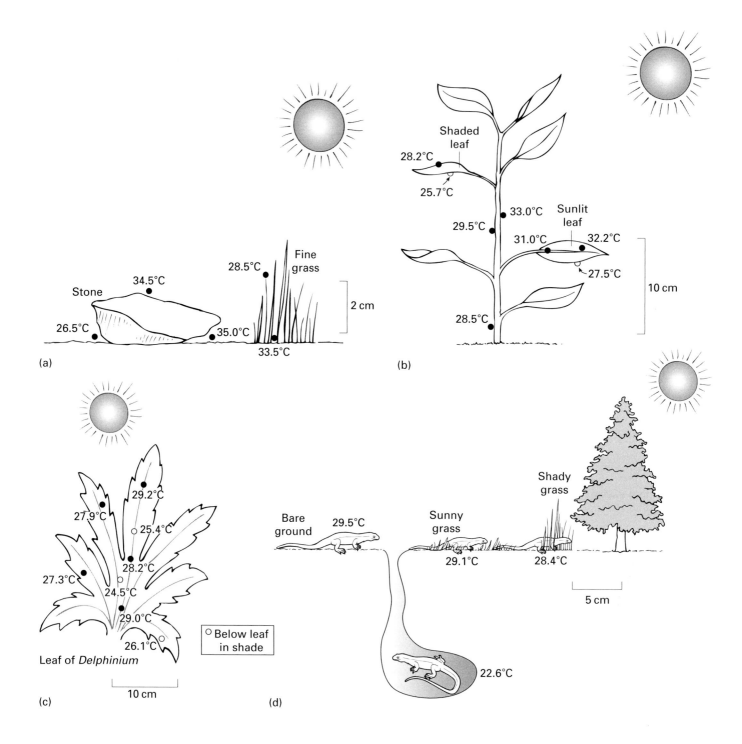

Fig. 13.18 (a–c) Microclimates and body temperatures of small insects (assessed from the temperature of a dead 50 mg insect with an inserted thermocouple); effects of climbing grass blades, moving around small rocks, around plant stems and around leaves.
(d) The effects are still present, though less extreme, for a much larger lizard.

Where ectotherms are concerned, avoidance, tolerance and regulation tend to merge into each other as strategic approaches to varying metabolic rate and achieving temperature balance. Thus animals may use behavioural techniques to avoid very cold microhabitats and avoid the risk of freezing, but could also thus be said to be regulating their T_b above ambient. Nevertheless, it might be useful to deal with avoidance and tolerance first, and then concentrate separately on adaptations that specifically tend to regulate T_b at particular levels.

Thermal avoidance and thermal tolerance

Avoiding the cold

Behavioural choices constitute the most important, and the cheapest, ways of maintaining thermal balance, especially by timing activity appropriately. Here we return to the crucial issue of microclimates. Staying in a warm burrow at night or in winter is the simplest of all strategies. But even when active, relatively small movements can take an animal away from cold spots and into a strikingly different thermal environment, hence bringing about a change in T_b.

Short-term movement effects can be especially striking for small terrestrial ectotherms such as insects, for whom shelter is readily available. Figure 13.18 gives examples of the benefits of climbing up a grass blade, rotating around the stem of a plant as the Sun passes over, or moving around, under or within a shrubby plant. Slightly more substantial movements may be needed for something the size of a lizard to avoid the cold, though choices are still available (Fig. 13.18d). But garter snakes can achieve a better stability of T_b by picking a rock of just the right size and thickness to shelter under than by sun–shade shuttling through a day. This also makes the point that burrows are not the only form of shelter; crevices and gaps within rocks will serve, and in some cases a shelter may even be manufactured, in which case we usually call it a nest. Nests that are not also burrows are relatively uncommon amongst invertebrates or ectotherms generally, but examples may be seen in frog-hopper insects, which produce a frothy ball of foam ('cuckoo spit') around themselves, in 'tent' caterpillars, in the use of rolled up leaves by insects and spiders, and of course in the social insects.

Figure 13.19 shows a few examples of insect behaviour patterns in relation to ambient conditions (temperature, radiation, RH and wind). Not only overall active periods and time above ground, but also orientation, basking postures and even the whole hierarchy of feeding and reproductive behaviours shown are crucially dependent on the hygrothermal environment.

Movement in and out of shelter is usually a very short-range strategy. Longer-range and longer-term movements to avoid cold are generally termed migration, though the immediate stimulus to move may be food supply or photoperiod (see Chapter 8). The alternative way of effectively avoiding cold weather is 'avoidance in time'; ensuring that during the coldest periods the body is 'shut down' or dormant. Many insects achieve this process of dormancy by diapause (see Chapter 8), while other animals may use torpor (or 'hibernation'), dealt with in Chapters 8 and 14. Dormancy in all its forms is conspicuously more common in terrestrial species than in any of the aquatic habitats covered in earlier chapters.

All that said, remember that many terrestrial ectotherms do nevertheless achieve activity and complete their lives with very low T_b values. This can even be true for some flying animals; classic examples are the winter moths (e.g. *Operophtera*), which can fly in freezing air temperatures with their T_b within 1°C of T_a, using a very gentle flapping locomotion, and winter gnats which can maintain high-frequency wingbeats at similar temperatures.

Coping with freezing

Many terrestrial animals can survive freezing temperatures by freeze-tolerance or freeze-avoidance strategies. However, this is largely a phenomenon of colder climates, so we leave consideration of details to Chapter 14. Note, though, that there may be cold winter nights in many temperate habitats, and that invertebrates, amphibians and reptiles in these biomes may have the same kinds of adaptations to cope with freezing as those discussed in section 14.3.2.

Avoiding overheating

Ground-dwelling or plant-surface-dwelling animals are particularly vulnerable to potential overheating since they live in the boundary layer of rapidly heated air. Caterpillars may lift their head and thorax up off the substrate (or allow them to hang down from a leaf under-surface) when overheated. But longer legs are a particular advantage in avoiding boundary layers, and many adult insects and arachnids will raise their trunk well above the surface by 'stilting', effectively standing on 'tiptoe'. Lizards may show similar behaviour, with successively raised legs seen in animals moving over hot sand. Such behaviours not only lift the trunk out of the heated boundary layer but also increase convective cooling by placing most of the body mass in the air moving at a higher velocity above the still surface layer. This can be enhanced by orienting the body side-on to the direction of the wind. On a larger scale, moving to the top of a large rock or up a hillock will increase convective cooling.

Posture may also be used to avoid overheating, usually by orienting the body for minimum exposure to the Sun (commonly 'head on'). Any extremities may also be withdrawn or tucked in, so that heat is not absorbed through their surfaces; in particular, the wings of insects are folded in.

Thermal regulation

The extent to which any animal can thermoregulate is roughly indicated by the slope of the relation between T_b and T_a. With a slope of 1, dependence of T_b on T_a is total, indicating thermoconformity; something approaching this value may be observed in nocturnal ectotherms, and in many cryptozoic animals. However, values of 0.5–0.7 can still indicate a lack of thermoregulation, as such figures can be obtained even in passive systems such as water-filled metal cans lying in the sun. With a slope of 0, though, an animal is clearly showing total independence of environmental temperature, indicating full thermoregulation. Examples of the values for this coefficient are shown in Table 13.10.

However, such data can be misleading for ectotherms. If the body temperatures are collected carefully in the field and only during active periods (usually in daylight) some species will properly appear to be good regulators with a steady high T_b; but where data are gathered over full 24 h cycles the picture will usually be rather different, with very low T_b values at night when the animal is quiescent. Furthermore, very misleading data may be obtained from laboratory studies where animals lack a normal range of choices of microhabitat or of behavioural repertoires. Hypothetical illustrations of these three kinds of assessment are shown in Fig. 13.20. Actual examples of plots of T_b vs. T_a are shown in Fig. 13.21, for a range of reptiles and insects in reasonably natural conditions. Here we consider the ways in which terrestrial ectotherms can gain, lose and conserve heat, to achieve this moderate thermoregulation.

Behavioural regulation to gain heat

Moving to a warmer microclimate is the simplest and most obvious way to warm up, though on its own it only allows T_b to be

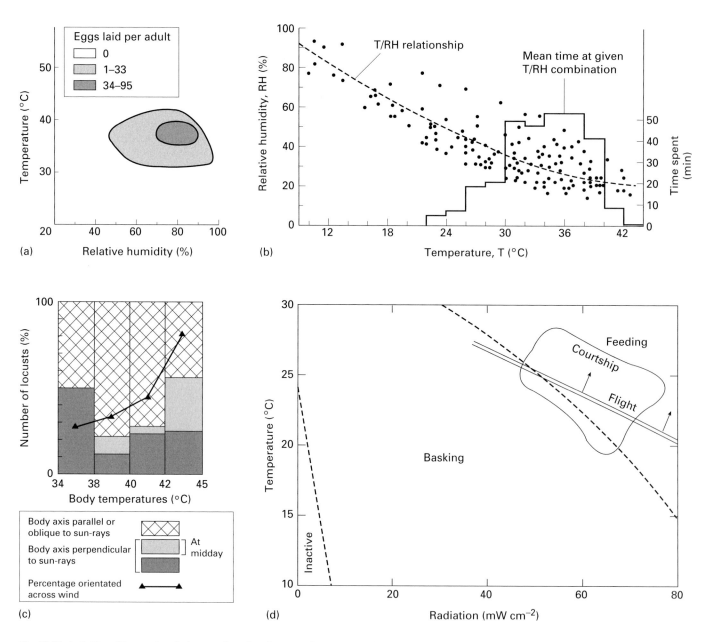

Fig. 13.19 Activities of insects in relation to microclimatic constraints. (a) Oviposition by the apterygote *Thermobia* in relation to temperature and humidity. (b) Active periods of the tiger beetle, *Cicindela*, with the advantages of high temperature overriding the problems of low humidity. (c) Orientation of locusts with respect to Sun and wind at different body temperature (T_b) values. (d) Hierarchy of activities in the butterfly *Heodes* as temperature and radiation increase (note the wide range of conditions for basking). (Data from Sweetman 1938; Waloff 1963; Douwes 1976; Dreisig 1980.)

Table 13.10 Gradients of regression of body temperature (T_b) on ambient temperature (T_a) in terrestrial animals.

Animal	Gradient
Ectotherms	
Anolis lizard (forest)	1.17
Tenebrionid beetle	1.00
Salamander	1.00
Python (non-brooding)	0.48
Cicada (non-singing)	0.44
Anolis lizard (grassland)	0.43
Sphinx moth caterpillar	0.40
Heterotherms	
Honey-bee	0.81
Solitary bee (foraging)	0.65
Honey-bee (foraging)	0.55
Dragonfly	0.43
Naked mole rat	0.41
Bumble-bee (foraging)	0.23
Python (brooding)	0.12
Bumble-bee (brooding)	0.07
Endotherms	
Pocket mouse	0.08
Possum	0.07
Finch	0.05
Parrot	0.05
Weasel	0
Human	0

Fig. 13.20 Hypothetical relationships between body temperature (T_b) and ambient temperature (T_a) for a small ectotherm on land. (a) Field measurements, daytime only, showing good regulation (behavioural). (b) Field measurements for 24 h cycle, with regression line concealing the daytime regulation and night-time conformity. (c) Laboratory studies may give various combinations of these results depending on the choices available to the animals.

raised to the new higher T_a. All the movements described above as part of the strategy of 'avoiding the cold' could also be viewed as ways of gaining heat. In addition, crouching against a heated soil or rock surface, within the still boundary layer, will aid warm-up, and this strategy (sometimes termed **thigmothermy**) is used by reptiles and by surface-dwelling insect types such as ground beetles and tiger beetles.

Basking (heliothermy) is one of the commonest additional mechanisms for gaining radiative heat, to raise T_b well above T_a. It is particularly effective in small ectotherms able to use limited patches of incident sunlight within vegetation, and in animals with wings that can be spread out to gather radiation; in other words, in pterygote insects. Temperatures in moderately sized insects can be raised at least 15°C above ambient by basking alone, and basking behaviour may be carefully controlled by sensors on the wing itself, with a slow wing closure when the local temperature gets too high. The effect of wings as thermal absorbers (and reflectors) may in part explain why adult winged insects are often much better thermoregulators than larval wing-less forms (irrespective of wing muscle activity), even though the latter may have higher body mass (see Fig. 13.21).

Basking postures may also be crucial. Insects exhibit three different basking modes (Fig. 13.22), usually termed dorsal basking, lateral basking and reflectance basking. These are particularly evident in large-winged butterflies, which have a rather narrow thermal window for flight and really exploit basking opportunities. The merits of the three possible postures are somewhat controversial; dorsal basking only works well when the insect is flattened against a warm surface, and the benefit may come principally from the warm trapped air. Body basking, or 'reflectance basking', is supposed to involve the wings reflecting heat down onto the body, and occurs in mainly white-winged insects, but its effectiveness has been seriously doubted, and it may merely involve wings being partially raised to reduce convective cooling. Lateral basking may work partly by trapping a 'greenhouse' of warmed air above the thorax and around the abdomen.

Basking also occurs in vertebrate ectotherms. It is not particularly common in amphibians, since if they achieve a raised T_b

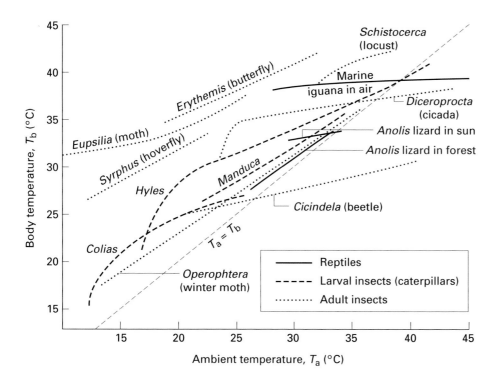

Fig. 13.21 Data collected in the field for insect (larva and adult) and reptile thermoregulation.

they also increase water loss through their permeable skins and cool down again evaporatively—basking achieves no thermal gains. But some species of *Bufo* and *Rana* have preferred T_b values in the range 25–30°C, and as long as water is freely available they will bask to achieve a T_b at least 15°C higher than in the

shade. Reptiles are generally much more regular and predictable in their basking behaviours, and are not so restricted by water availability. The common European lizard, *Lacerta vivipara*, lives in habitats with yearly mean T_a values of around 9°C yet has a preferred T_b of 30°C and an active temperature range of 28–32°C; it can warm from 15°C to 25°C within about 5 min in good sunshine. Many reptile species from fairly open habitats such as grasslands and heaths, or from deserts, are sophisticated

Fig. 13.22 Three postures that may be used for basking in winged insects (see text for more details).

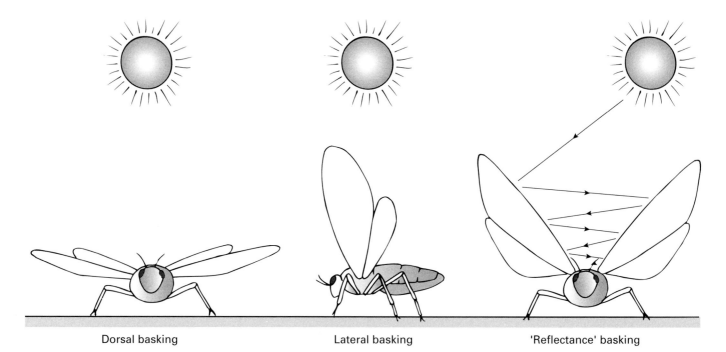

Dorsal basking   Lateral basking   'Reflectance' basking

Fig. 13.23 Patterns of temperature through a day for a *Varanus* monitor lizard, measured by telemetry, showing the rapid rise in body temperature (T_b) (to well above ambient temperatures, T_a) due to early-morning basking and subsequent effective thermoregulation through the daylight hours by sun–shade shuttling. Note that the lizard has high thermal inertia curled in its treehole den at night so that T_b falls quite slowly. (From Cossins & Bowler 1987; with kind permission from Kluwer Academic Publishers.)

basking heliotherms, but forest species rarely find sun-spots large enough to accommodate their lengths. The precision of thermoregulation at a particular preferred temperature can be exceptional in a medium-sized reptile (Fig. 13.23), alternating between side-on basking (often with crests erected) for maximal heat gain and head-on postures to reduce heat uptake. But the preferred temperature is often variable within a species, for example being reduced in poor-quality habitats where food is sparse or predation risks high. The iguanid *Dipsosaurus* maintains $39.1 \pm 2.0°C$ in favourable habitats, but only $32.9 \pm 4.0°C$ (lower and with greater variance) in poor environments where thermoregulation bears a high cost. Within each habitat, reptile T_b values are strongly correlated with success, measured in the short term as locomotory speed or rate of prey capture, or in the longer term as growth rate or reproductive output.

Colour is particularly important for basking species from many taxonomic groups. Since about 50% of the radiant energy from the Sun is in the visible wavelengths of the spectrum, visible reflectance (the colour seen) affects radiative heat gain significantly (see Chapter 8). Insects with white and pale bodies warm more slowly than dark ones; small insects active at dawn and dusk are more likely to be dark, while the middle of the day may be dominated by insects with bright or pale colours (Fig. 13.24). However, insects that live close to the ground and gain heat largely through re-radiated longwave radiation will be less affected by their colour; they are commonly brown or black irrespective of their diurnal activity pattern. Some insects have both white and dark surfaces and may change posture according

to thermal need. Vertebrates may use similar strategies; many savanna antelope have white faces, rumps and ventral surfaces, minimizing absorption of shortwave reflected radiation from the pale sandy soils and when oriented head-on to the Sun.

Given the effects of colour on radiative gain it is no surprise to find many ectotherms able to change colour as a thermoregulatory strategy. Some species of frog can blanch when heat-stressed, changing their reflectance from 35% to 60%. Chameleons are particularly famous for their ability to alter their coloration and indeed their shape, balancing camouflage and thermal need. A cold chameleon darkens and basks, irrespective of other requirements for concealment or aggressive coloration. A warming *Chamaeleo dilepis* can increase its reflectance from 31% at 20°C to 46% at 35°C, with the major changes incurred in the visible and near infrared range (600–1000 nm); this ability is calculated to give a mean change of 0.7°C in its equilibrium temperature averaged over a whole year. Other lizards can also alter between a sandy colour and dark brown, or through green and blue shades; the lizard *Urosaurus ornatus* can change parts of its belly and throat from light green to intense blue using the thermally sensitive iridophore system described in Chapter 8. Some insects too can darken or lighten according to need; for example, certain grasshoppers and dragonflies alter from bright metallic blue or green when warm to almost black when cooled.

Behavioural regulation to conserve heat

Postural effects to conserve heat in terrestrial ectotherms can be

(a) Time, BST (h)

(b) Time, BST (h)

Fig. 13.24 Patterns of insect colour (reflectance) through a day at a basking site (a) and a feeding site on a flower (b); in both cases dark species are commoner towards dawn and dusk (having higher thermal absorption rates) while only light-coloured highly reflective insects are active around midday in full sun. BST, British Summer Time. (From Willmer 1983.)

used to retain the heat gained by basking and heliothermy. General 'balling' postures, with extremities tucked in, are very common, and perhaps best exemplified by the coiling response of snakes, which reduces the effective surface area by 50–70% in the early part of the night, retains heat gained by day, and probably speeds up digestion significantly.

Huddling and aggregation also serve as excellent methods for conserving body heat in ectotherms. Several larger snakes (pythons, anacondas and boas) can be found clustered together at night in the cooler parts of their ranges, slowing down the loss of heat gained by basking. Insects aggregated on their food plants may gain a similar advantage, for example in the tent caterpillars (where the silken tent serves as a reasonably windproof greenhouse) or in various butterflies that mass together in the evening, thereby reducing convection within the group.

Physiological regulation to conserve heat or lose heat

Regulatory conservation or loss of heat in ectotherms by nonbehavioural means is relatively rare, but not impossible; this underlines again the important distinction between the ability to generate heat by endothermy, and thermoregulatory ability to maintain a stable T_b (see Chapter 8). Ectotherms can achieve regulation by vascular adjustments (blood shunting), or by using evaporation. **Blood shunting** to control rates of heat loss occurs quite widely in reptiles, and an example is shown in Fig. 13.25. Rates of cooling are substantially lower than rates of heating, allowing the iguana to retain its acquired heat for

longer, because heart rate is slowed and peripheral circulation decreased, reducing heat loss through the skin. A few ectothermic insects appear to achieve a similar conservation of heat by preventing hot blood in the thorax from losing its heat to the cooler abdomen, using a countercurrent exchanger at the 'waist'.

Evaporative cooling is an excellent way of dissipating heat if, and only if, the consequent loss of water can be tolerated. So it is only common in larger terrestrial ectotherms as a specific thermal strategy. In many small or hygrophilic ectotherms, such as worms, snails and amphibians, it occurs unavoidably whenever the animal gets warmed by the Sun, and in some woodlice the T_b can be depressed by as much as 4–8°C as a result. This could be termed passive evaporation, and the T_b reduction and water loss are often a problem for the animal rather than an advantage. However, some ectotherms, especially reptiles and insects, are able to enhance their evaporative cooling when at risk of overheating, by increasing either cutaneous or respiratory water loss (CEWL or REWL). Insects that can 'sweat' through pores in their cuticles are now reasonably well known, and include desert cicadas such as *Diceroprocta* (see Chapter 14); these have access to ready water supplies from their sap-feeding habit, and can maintain a temperature of 37–38°C at a T_a of 42°C. A few insects also use excretory fluid to achieve cooling; the larvae of the sawfly *Perga* can raise their rear segments vertically and allow excreta to trickle down their exterior surfaces. Two genera of the so-called 'waterproof frogs', *Phyllomedusa* and *Chiromantis*, are also able to increase their normally low CEWL when their body temperatures rise above 37–38°C. In both genera this is due to a rapid onset of watery secretion from epidermal glands, but in *Phyllomedusa* it additionally involves an apparent melting of an epidermal waxy layer. The iguanid lizard *Dipsosaurus* pants whenever T_b exceeds 40°C, and the agamid lizard *Amphibolurus* can use panting to reduce its T_b from about 42 to 39°C. Other lizards use liquids (saliva) regurgitated from the mouth over the throat region to achieve cooling.

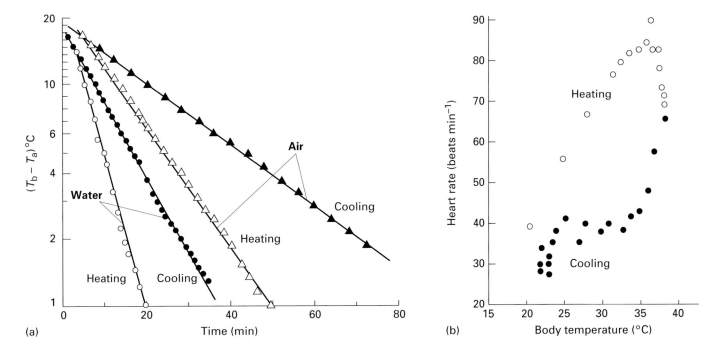

Fig. 13.25 (a) Heating and cooling rates of the marine iguana *Amblyrhynchus cristatus* in water and in air, showing that heating is always faster than cooling. T_a, ambient temperature; T_b, body temperature. (b) This is due to blood shunting and altered heart rate: when cooling there is a marked bradycardia, so blood moves slowly and stays away from the periphery, allowing heat to be retained in the core and reducing the overall cooling rate. (Adapted from Bartholomew & Lasiewski 1965.)

13.3.2 Terrestrial heterotherms and endotherms

Endotherms derive much of their body heat from metabolic processes (endogenous heat) rather than from the Sun or from geological sources. They can therefore maintain a fairly constant T_b by moderate thermoregulation, though they do not necessarily regulate very precisely. Some animals can be endothermic for just part of their lives; they are more properly termed heterothermic. These include both traditional ectotherms, such as some insects, which can turn on endothermic processes (often in just part of their bodies) when they need it to initiate activity; and traditional endotherms, such as hummingbirds and small rodents, which can turn down or turn off endothermic processes when they become too costly to maintain. Other animals, notably the great majority of birds and mammals, are endothermic throughout their lives.

Types and abilities

The basic physiology of a typical endotherm can best be appreciated with a display of the relations between T_a and T_b, and between T_a and metabolic rate (MR), shown in Fig. 13.26 for various mammals. T_b is more or less constant for all ambient temperatures, with a slight upturn occurring as T_a rises beyond about 35°C. At the same time, for placental and marsupial mammals MR is reasonably constant for moderate ambient temperatures but rises at either extreme; at low T_a it rises as the animal compensates for high rates of heat loss, and at high T_a it rises as the ability to thermoregulate begins to break down. Between these two phases of rising MR the animal is in its **thermoneutral zone** (TNZ), and is maintaining its 'normal' T_b. More explanatory details of the relevant terminology were given in Chapter 8.

The values of normally maintained T_b for bird and mammal families were shown in Table 8.11, revealing the generally higher fixed values in birds (about 40°C) compared to mammals (37–38°C), and the rather low values for ratite birds and for monotreme, marsupial and insectivore mammals. Certain taxa also have unusual levels of energy expenditure that affect their thermoregulatory responses; again the Insectivora are notable, where the smallest members (shrews) possess very high mass-specific MRs and the largest members (e.g. hedgehogs) have lower than predicted values, giving an unusually low slope to the classic BMR/M_b (basal metabolic rate/body mass) plot (see Chapter 6). There is no obvious overall trend of T_b with size in the mammals as a whole, however. Nor do mammals and birds from cold climates maintain body temperatures any different from their temperate and tropical relatives, though they have a broader TNZ and may briefly tolerate greater extremes of T_b. The differences between birds and mammals, and between temperate and cold-adapted species, seem to be largely explained by variations in thermal conductance (insulation). Indications of thermoregulation are again best given by the coefficient relating T_b to T_a, and values are included in Table 13.10, showing examples of the very tight regulation in birds and mammals ($k < 0.1$). However, there are also cases of rather loose thermoregulation in a few birds and in the marsupials, and a lack of

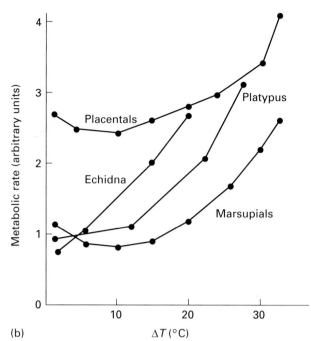

Fig. 13.26 (a) Plots of body temperature (T_b) against ambient temperature (T_a) for endothermic mammals and a lizard in laboratory studies, showing the better regulation in the eutherian cat compared with monotremes and marsupials; and (b) metabolic rate (MR) against ($T_b - T_a$), showing the thermoneutral zone where MR is roughly constant (for marsupials and placentals), rising at either extreme. (Data from Martin 1980.)

thermoregulation in the naked mole rats (the only mammals that hardly use endothermy, living more-or-less permanently in subterranean social nests where T_a may be constant, for example at 26–28°C in *Cryptomys* nests from east Africa). Table 13.10 also shows that some insect heterotherms are nearly as good at thermoregulating as the more strictly endothermic vertebrates.

Avoidance and tolerance

Behavioural avoidance techniques for endotherms are similar to those in ectotherms, for avoiding cold areas, avoiding freezing and avoiding overheating. However, the use of behaviour as a way of achieving equable microclimates and T_b values is limited by the inherently large size of endotherms. With body masses usually in excess of 1 g, and linear dimensions nearly always at least tens of millimetres, it becomes much harder to use the fine-grained variation in climatic conditions. Most of the body is inevitably above the still boundary-layer of air above a soil or rock surface, so that stilting and other postures make little difference to convective exchanges. Endotherms may have a choice between a patch of incident sunlight and the shade under a bush, but they cannot normally choose between the two sides of a leaf. Nevertheless, some aspects of behavioural avoidance are still

very effective. Behaviours that minimize thermal conductance can help animals to limit cooling; assuming a roughly spherical shape, with the body curled up and paws and nose tucked in, is the ideal for a mammal, whilst a fluffed-up squatting posture with the legs covered by feathers and the head drawn down is characteristic of a chilled bird. Many endotherms enhance these insulation/conductance effects either with burrows or by gathering extrinsic materials; nests built by their own efforts or borrowed from other animals may reduce effective conductance and metabolic rate. For example, a lemming surrounded by cotton wool or pieces of collected fur and feathers decreases its own thermal conductance by 40–50%. Similarly a vole maintained in the laboratory at 4°C without nesting material may have a daily energy expenditure of 3.5 kJ g⁻¹, compared with a value of 2.3 kJ g⁻¹ when nest materials are freely available. Huddling behaviours give further benefits; the same vole given no nest but the company of three other voles can lower its daily expenditure to 3.0 kJ g⁻¹. All these effects can be demonstrated under natural conditions as well: for example, shrews in winter have an overall oxygen consumption 20% lower when they have access to a nest than when their nest is sealed off.

Behavioural mechanisms in general tend to keep an animal within its thermoneutral zone and reduce or eliminate the need for raised MRs or more complex physiological regulation.

Thermal regulation

Endothermic heat gain

As we saw in Chapter 8, the principle mechanisms of increasing heat production involve increased activity of skeletal muscles. This may be achieved by exercise, but the drawback is increasing

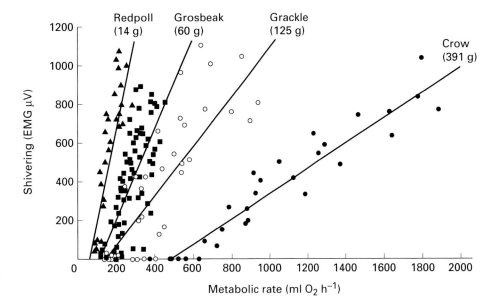

Fig. 13.27 The linear relation between the degree of shivering in various birds and their metabolic rate; the relation is flatter for larger birds, but in all cases shivering substantially elevates metabolic rate and allows warming. (Shivering was quantified from electromyograms, EMGs.) (From West 1965; courtesy of University of Chicago.)

blood flow to the extremities and dishevelling the pelage of birds and mammals, so that heat loss routes are also increased. The better option is to uncouple the muscles from locomotory effectors and simply shiver. Endothermic vertebrates thus show a clear inverse relation between rate of **shivering** (as recorded by an electromyograph) and MR (Fig. 13.27). These animals are regulating the extent of their shivering to compensate for cold and generate internal heat, though the slope of the relation is much shallower for larger animals than for small ones. Shivering tends to have a specific onset temperature (shivering threshold temperature or STT), varying between species and also variable according to acclimation state: in king penguins the STT is about –18°C when cold-acclimated but as high as –9°C after prolonged exposure to 25°C.

A few snakes, such as the Indian pythons (*Python* spp.), use similar shivering mechanisms to achieve endothermic warm-up, during the periods when they are incubating eggs (Fig. 13.28). The female wraps her body tightly round the clutch and produces low-frequency but powerful spasmodic shivering in the trunk muscles for long periods, raising her T_b and thus the egg temperatures to around 30–33°C, perhaps 7–8°C above ambient, allowing faster development.

Certain insects can also shiver. Some moths, beetles, dragonflies, flies, wasps and many bees need to warm their muscles before flying, to achieve an adequate power output (about 100 W) for take-off, i.e. they are obligate heterotherms (see Chapter 8). For example, the saturniid moth *Hyalophora* requires a muscle temperature of at least 35°C to reach the required power output for flight, but once airborne it is able to maintain its thorax at 35°C and fly and feed at air temperatures as low as 10°C. Some winter-flying moths, such as *Eupsilia*, can fly with a thoracic temperature of 30°C at subzero ambient temperatures. Bumble-bees

(*Bombus*) operate best at a muscle temperature around 40°C with a wingbeat frequency of about 50–100 Hz, and must commonly use endothermy to achieve this temperature before take-off. Retention of heat during the warm-up period is aided in all these insects by long scales or dense fur covering the thorax, or internally by elongated air sacs. More details of insect endothermy and other uses for it are given in Table 13.11.

The ability to warm up in insects has been most extensively studied in the bees, where activity patterns must coincide with times of nectar and pollen availability; since these resources often peak in the early morning there may be strong selection for flight at low T_a. There is a significant positive correlation between warm-up rate and body size (Fig. 13.29), probably resulting from the greater surface area to volume (SA/V) ratio and higher heat loss rates of the smaller species. Tropical species may have lower warm-up rates than cool-temperate species of similar size; however, this is complicated by a taxonomic difference between the bees present in temperate and tropical/desert fauna. Species active at low temperatures do have higher MRs, and even within a species this trend can be detected between sea-level and high-altitude (cooler) populations. The ability of individual flying bees to vary thermogenic output is unclear, although honey-bees (*Apis*) certainly seem able to vary metabolic heat production to maintain thermal stability while in flight.

Non-shivering thermogenesis (NST) is another possible way of achieving warm-up (see Chapter 8). It occurs in many placental mammals and some marsupials, especially in young animals, but it is rare in birds. It can produce a two- to four-fold increase in MR in a small mammal such as a rodent, bat or rabbit. NST involves the release of heat from 'futile cycling' of metabolic substrates in the cytoplasm or mitochondria (see Fig. 8.35). NST may occur in the liver and some muscles, but in mammals is

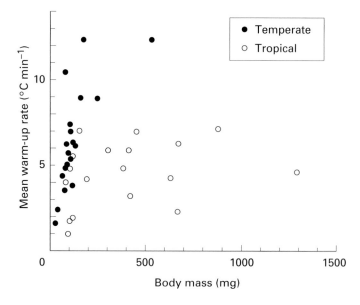

Fig. 13.28 Thermogenic incubation in a python; curled around her eggs, the female contracts her trunk muscles rhythmically to raise metabolic rate and increase T_b. (Adapted from Hutchinson *et al.* 1966; van Mierop & Barnard 1978.)

commonly concentrated into special brown adipose tissue (BAT). This tissue is laid down around the shoulder region of many juvenile and hibernating mammals, where it provides a warm-up site close to the crucial heart and respiratory muscles.

NST may also occur in bumble-bees, though not in other endothermic insects (and some regard it as unproved even in the bumble-bees). It involves futile cycling between fructose 6-phosphate and fructose 1,6-diphosphate, the thoracic muscles having unusually high concentrations of the two necessary enzymes phosphofructokinase (PFK) and fructose biphosphatase (FBPase).

Behavioural heat gain and loss

Endotherms can of course also use behavioural techniques to gain heat as their 'cheapest' option, particularly at times of day when ambient temperatures are low but sunlight is available. They may crouch against east-facing sun-warmed rocks to gain heat by conduction from the ground in the hour after dawn. They may also employ more traditional basking postures, just

Fig. 13.29 Mean warm-up rates in a range of bee species from temperate and tropical habitats, showing an apparently faster warm-up at lower body mass in the temperate species. (In fact this is partly due to a phylogenetic effect, certain families of bee with large bodies being more common in tropical zones.) (Adapted from Stone & Willmer 1989.)

Table 13.11 Insect heterothermy and thermoregulation. Activity after warm-up is flying unless otherwise stated. Blanks indicate information unavailable. Fighting and dung-rolling need a high/controlled body temperature to give a competitive edge; singing needs it to maintain a good output at the correct (species-specific) frequency.

Taxon	Mass range (mg)	Pre-activity thermogenesis?	Biochemical thermogenesis?	Controlled blood shunting?
Odonata				
Libellulids	70–600	Yes		No?
Aeshnids	120–1200	Yes		Yes
Orthoptera				
Tettigonids (katydids)		Yes (sing)		?
Gryllotalpids (mole crickets)		Yes (sing)		?
Neuroptera				
Ascalaphids	60–250	Yes		Yes
Lepidoptera				
Sphingids	200–3000	Yes		Yes
Saturnids	200–2500	Yes		Yes?
Lasiocampids	90–120	Yes		No
Lymantrids		Yes		?
Hesperiids		Yes		?
Coleoptera				
Scarabeids				
Scarabeus	1300–5400	Yes (fight, roll dung)		No
Cotinus	600–2000	Yes		No
Megasoma	10 000–35 000	Yes	No	?
Trichostetha		Yes		?
Diptera				
Syrphus	18–40	Yes		No?
Eristalis	70–150	Yes		Yes?
Sarcophaga	30–130	No		Yes
Gasterophilus	30–240	Yes		No
Calliphora	20–100	Yes		?
Glossina	20–80	Yes		?
Hymenoptera				
Apids (bees) most, e.g.:				
Apis	80–120	Yes (+ brood)	No?	Yes
Bombus	240–600	Yes (+ brood)	Yes?	Yes
Xylocopa	400–2000	Yes		No?
Euglossines	60–700	Yes		?
Centris	100–200	Yes		No?
Andrena	30–50	Yes		?
Anthophora	90–600	Yes		Yes
Vespids (wasps) some, e.g. *Vespa, Vespula*	50–500	Yes (+ brood)	Yes	?

as in ectotherms, to gain heat; mammals as large as camels will bask in the early morning sun, while small rodents and insectivores may get a significant heat input in this way, reducing their need for metabolic thermogenesis. Birds will also bask, and a herring gull basking in the sun may have a measurably lower MR than a nearby gull in the shade. Perhaps most spectac-

ularly, the North American roadrunner, *Geococcyx californianus*, allows its T_b to drop by as much as 4°C overnight and then exposes as much of its body as possible to the sunrise to regain its 'normal' T_b.

Large animals may also use behaviour to dissipate heat. Erection of large and well-vascularized surfaces such as ears

African honey-bee
(738 g colony)

Fig. 13.30 Thermoregulation in honey-bee swarms; profiles are shown at four different ambient temperatures, with the bees much more tightly aggregated as external temperatures fall, giving a much greater thermal gradient across the cluster.

(in elephants or desert foxes) will give a significant heat loss, especially with postures and orientation adjusted to the height and direction of the Sun. The spreading tails of some desert rodents may be used as radiators in some circumstances, but may also have a parasol effect; and birds may spread their wings open in shade and into a breeze to radiate excess heat.

Heat loss also comes from a quick dip in water or wallow in mud, as seen in many African savanna mammals such as hippos and elephants; bouts of temporary submersion followed by moderate sunbathing give evaporative cooling that may be particularly useful for a very large endotherm where internal heat generation may overload the ability of dry surfaces to dissipate heat.

Huddling may be used by endotherms to reduce heat loss, particularly in small mammals and birds, such as bats and wrens, that overwinter in cooler climates where food is scarce; it is particularly useful in flying animals unable to accumulate large lipid reserves to see them through the winter at 'normal' endothermic MRs. Bats may aggregate in groups of several hundred, not to maintain normal body temperatures but to ensure that the T_b values of individuals, all of which are torpid, do not fall below 5–10°C, so avoiding any risk of freezing. Wrens also huddle in winter, with up to 60 entering a single man-made nest box.

Heterotherms also use huddling, conspicuously in the overwintering social honey-bee, *Apis mellifera*. Whereas in most temperate bees only reproductive females overwinter, in *Apis* the whole colony persists, as an almost spherical dense cluster on the central nest combs, with a temperature gradient from core to periphery (Fig. 13.30). The cluster forms when T_a falls to around 13°C, and inserted thermocouples show a steady 20°C within the mass of bees throughout the ensuing winter, rising to over 30°C as brooding begins in spring. This colony homeostasis is probably largely due to the much reduced SA/V ratio of the cluster, rather than to extensive thermogenesis by individual bees, at least in the larger clusters; indeed, in very large groups of perhaps 5000 bees there is little evidence of sustained endothermic muscular activity. Individuals do move around within the cluster, with a positional cycling such that cooler bees migrate in from the periphery; and the temperature of the cluster is controlled by the degree of packing, becoming denser as the ambient temperature falls.

Heat conservation by vascular control

Most endotherms have a capacity to operate vascular shunts to regulate their heat loss rate. In heterothermic insects, endothermic warm-up allows flight and other kinds of activity at low T_a, but once activity begins most of the regulation of T_b is due to blood shunting—in particular the control of heat loss from the thorax—rather than to changes in metabolic rate. The best-studied examples are large sphingid moths and large bees, shown in Fig. 13.31. Bumble-bees have a countercurrent shunt at the narrow petiole between thorax and abdomen, which keeps heat largely within the thorax (where it is generated by the flight muscles). However, this shunt can also be 'turned off', by pulsing the blood flow at the petiole; in an overheated bee, the heated haemolymph from the thorax passes backwards through the petiole in a pulse that is out of phase with cold haemolymph coming forwards from the abdomen. Thus heat is discharged into the abdomen and can be lost from the relatively uninsulated 'thermal window' of the ventral abdominal surface.

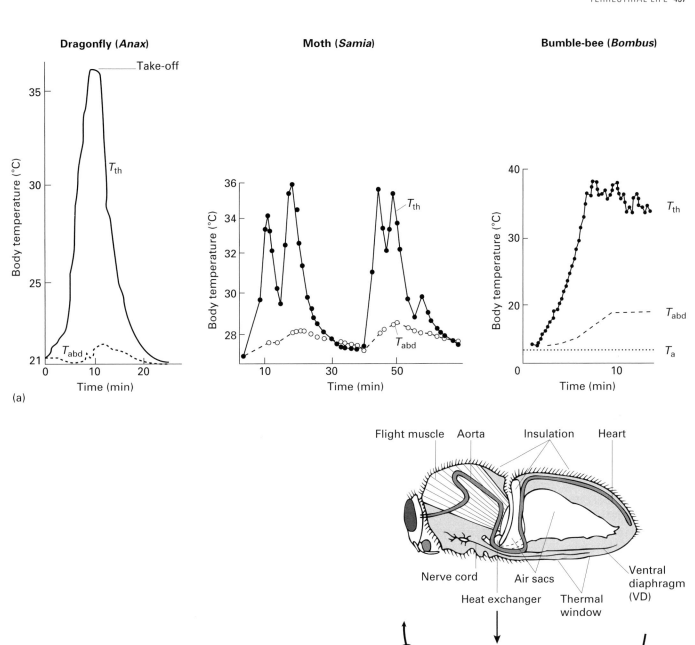

Fig. 13.31 (a) Recordings of warm-up in endothermic bouts in a dragonfly, a sphinx moth and a bumble-bee. T_a, ambient temperature; T_{abd}, abdominal temperature; T_{th}, thoracic temperature. (b) The anatomy of a bumble-bee showing the blood flow (thick arrows) through the petiole ('waist') where countercurrent exchange allows heat (fine arrows) to be retained in the thorax. (Adapted from Oosthuizen 1939; Heinrich 1975, 1976; May 1976.)

In birds and mammals, short-term regulation of heat loss may be achieved by adjustments to the blood flow in peripheral capillary beds; we saw in Chapter 8 that these may be bypassed by arterio-venous shunts. In a normal mammal the blood flows via the skin capillaries, but in a chilled mammal where the maintenance of T_b is under threat, flow to these capillaries is shut off, with all the blood passing through the deeper shunt vessels, so reducing heat loss at the skin surface.

Many endothermic vertebrates from cool climates operate longer-term vascular shunting so that parts of their body are regularly at a lower temperature than the core T_b. Extremities such as feet, flippers, tails, noses or ears, and even keratinized

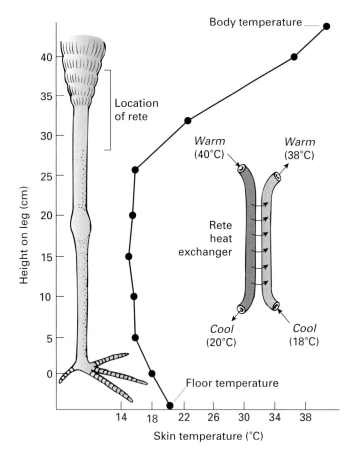

Fig. 13.32 Temperatures recorded from skin surface along the leg of a wood stork, showing the heat-conserving effect of an upper-limb countercurrent rete. (From Kahl 1963; courtesy of University of Chicago.)

areas such as horns, are 'at-risk' sites for an endotherm; they have a high SA/V ratio, are usually poorly insulated, and would allow severe heat loss in cool conditions if their temperatures were maintained. These areas may have an internal vascular countercurrent (rete) associated with them so that warm blood in the afferent arterioles passes close to the efferent venules and heat is transferred across, so that much of the heat never reaches the tip of the extremity. The associated vasculature comes in three different forms (shown in Fig. 8.37).

An example from a temperate bird is shown in Fig. 13.32; a thermal gradient is set up along the bird's limb, reducing the gradient for heat loss to the environment from the foot. Intermittently, pulses of blood are sent to the foot by bringing into play non-rete vessels (e.g. superficial veins or venules) to ensure that just enough warmth is supplied to prevent freezing-damage to the tissues. These bypass routes also mean that the extremities can be used for heat dissipation if the animal ever gets overheated. This is an example of **regional heterothermy** (Chapter 8), and it is common in endotherms from all habitats.

The ability to keep the brain relatively cool using a countercurrent heat-exchanger is also important in many endothermic vertebrates, but as it is particularly well developed in animals from hotter climates we cover it in detail in Chapter 14.

Heat conservation by changing conductance

A decrease in thermal conductance (C) can substantially reduce the lowest ambient temperature that an endotherm can survive, and can also reduce the oxygen consumption (MR). Increased insulation is thus a 'cheap' way to improve survival in cool climates. Larger animals have lower mass-specific C-values than smaller species, because of their lower SA/V ratio, but they usually have a higher absolute thermal conductance. Larger animals therefore have more scope for decreasing their thermal conductance, usually by increasing the density or length of their fur or feathers (pelage). Smaller endotherms tend to be very well insulated anyway, and cannot increase the length of their pelage greatly without hampering the operation of limbs; a small mouse cannot have fur more than a few millimetres in length or its legs would not emerge to meet the ground. Thus in small endotherms, increasing conductance by storing some subcutaneous fat is a better option; this not only insulates the surfaces, but also gives a small increase in body mass with only limited increase in metabolic rate, and simultaneously provides a store of energy for the colder periods when food supplies may be reduced.

Many medium- to large-sized birds and mammals do show acclimatory seasonal adjustments in pelage quality (with a spring moult to lose the thicker winter coat) and sometimes also in pelage colour; this is particularly common in polar and tundra animals and is dealt with in Chapter 14.

The other possible conservation strategy is hypothermia and torpor, also dealt with in detail for animals from cold climates in Chapter 14. But note here that some temperate animals do undergo seasonal torpor, and that many very small endotherms, even in the tropics, undergo transient nocturnal torpidity; Fig. 13.33 shows an example for a Neotropical bat. In bats and in small marsupials this kind of torpor is used to adjust energy expenditure to match food availability, and only occurs when they cannot feed enough to maintain T_b throughout the 24 h cycle.

Heat storage and dissipation

When faced with an excessive heat load, endothermic animals have a number of possible physiological methods for dealing with the heat. If T_b is raised due to extra endogenous heat production (e.g. from bouts of exercise), they may use postural changes, pilo- or ptilo-depression (flattening fur or feathers) and vascular adjustments (extra blood flow to the skin, producing flushing) to lose excess heat. Even if T_b is rising in concert with warm ambient conditions they will still normally be able to conduct, convect or radiate away the heat, this being one of the main

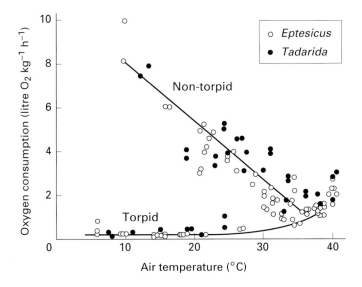

Fig. 13.33 Daily torpor in two species of small bat, with greatly reduced metabolic rate as the temperature drops at night (a very well-fed bat avoids torpor and increases its metabolic rate at night to maintain body temperature, T_b). (Adapted from Herreid & Schmidt-Nielsen 1966.)

advantages of maintaining a high level of body temperature. But if T_a approaches or exceeds T_b the options for regulation become limited. Temporary heat storage (adaptive hyperthermia, with raised T_b) may be worthwhile, to ensure a T_b–T_a gradient, but this can normally only be allowed for short periods; in a small rodent it may be tolerable for just long enough to give time to get

Fig. 13.34 (a) Blood flow rate and water loss rate in an exercising panting dog, also showing increments of rectal and brain temperature (T_{rectal} and T_{brain}) but with T_{brain} much lower due to the cooling effect of panting. (From Baker 1982.) (b) Rates of water loss as percentage body mass in different sized panting mammals; in very small species rates are too high to survive for long, and exercise with panting must be avoided as far as possible in daylight hours. (Adapted from Mitchell *et al.* 1987.)

back to a cool burrow, for example. In larger animals, heat storage has more substantial and longer-term uses, as raising T_b by 1°C takes much longer and represents a much larger quantity of heat stored, so that some large desert animals do use this option on a daily cycle (see Chapter 14). However, for temperate animals it is rare; ultimately the animals must cool down again, and they normally resort to evaporation to achieve this.

Evaporative cooling is, as with ectotherms, an excellent way of dissipating heat if the consequent loss of water can be tolerated; this is more likely to be feasible in endotherms because of their necessarily large size and low SA/V ratio. So it is a common mechanism for regulating T_b in terrestrial endotherms, evolving as a specific thermal strategy.

Evaporative water (and therefore heat) loss may be achieved by either cutaneous (CEWL) or respiratory (REWL) routes. The relative merits of sweating and panting were considered in Chapter 8. In general, panting is a better option, and it is extensively used by birds and by some groups of mammals when exercising (Fig. 13.34); note that in small mammals the EWL can be excessive, and exercise is only possible at night. Birds also have a problem, and for them CEWL may be more important (about 50% of the total water loss) at moderate air temperatures; at higher temperatures REWL usually dominates. In many canids the effectiveness of panting is increased by bypassing the nasal countercurrent system, taking air in at the nose but exhaling entirely through the mouth. Birds and mammals both tend to use a very high frequency of panting, perhaps 10 times the normal respiratory frequency (e.g. from 32 to 320 min⁻¹ in a medium-sized dog); in birds the panting frequency is often close to that of their wingbeat, at around 600 min⁻¹ in pigeons. In each case this also gives a panting frequency that is close to the resonant frequency of the lungs. In many birds REWL is further enhanced by a high frequency 'gular flutter' moving air in and out of the throat region (see Fig. 8.44b).

As with ectotherms, other sources of water may be used when necessary to achieve emergency cooling. Some mammals, particularly marsupials and rodents, salivate and lick their fur to

(a)

(b)

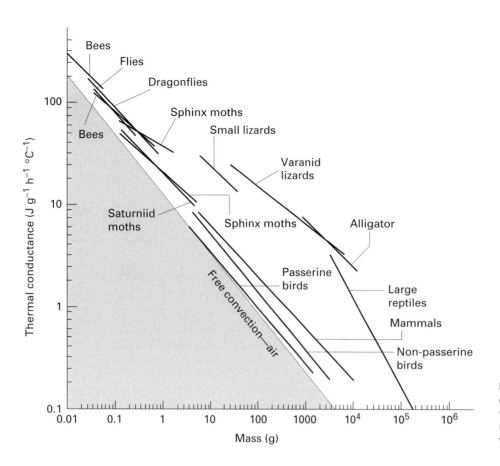

Fig. 13.35 Allometric relations for thermal conductance against mass in insects and vertebrates. Overall the slopes for insects (0.47) and for birds and mammals (0.52) are very similar. (From Withers 1992.)

spread the fluid; a few birds, such as vultures, will urinate over their own legs.

Size limits for terrestrial endotherms

Since the SA/V ratio of animals decreases as they become larger, small endotherms have a relatively large surface area and a very real problem: the metabolic heat generated in the smallest species is lost as fast as it is produced even under the most favourable circumstances. Thus the T_b of small dipterans (weighing, say, 1 mg) is close to the ambient environmental temperature even during long flights with very high rates of heat generation. Insects such as bees require a minimum size of around 20–30 mg before significant heat gains are observed during muscular activity, and in less highly insulated insects the size limit may be much higher, perhaps 100 mg in many taxa. However, once above this size limit, periods of endothermy and temperature regulation with T_b above ambient become a real possibility, as we have described.

The smallest vertebrate endotherms, the hummingbirds and shrews, have a body mass which is at least 10 times greater than that of the smallest endothermic insect. Why are there not even smaller birds and mammals? Are the conditions for heat loss different in insects? The answer to this question is apparently no, since the conductance of large moths and bees falls on a direct extension of the regression line for birds and mammals (Fig. 13.35).

The reason why insects can be successful endotherms at a smaller body size than birds and mammals is therefore probably related to differences in energy output and consequently heat production. In particular, differences in oxygen supply may well account for the constraints on the minimum size of birds and mammals. The cardiovascular system of shrews and hummingbirds is tuned to the limit. The hearts in these animals are two to three times larger than expected from the general scaling of heart size in birds and mammals. The haemoglobin content of the blood (haematocrit) in hummingbirds is also as high as that found in any other animal. It is probably not possible to increase haematocrit further without inordinately raising the viscosity of the blood and therefore the energy needed to circulate it around the body. Even with their large hearts and high haematocrits, the heart rate needed to sustain the maximum activity in shrews and hummingbirds is around 1200–1400 beats min⁻¹. Thus, each heartbeat lasts only 40–50 milliseconds, which probably represents an irreducible minimum for the design of a muscular pump. By contrast, insects are able to sustain higher levels of energy output using their novel and radically different oxygen supply system—the tracheae, described in Chapter 7—allowing oxygen to diffuse directly in the vapour phase to the tissues where it is consumed.

13.4 Respiratory adaptation

Air contains about 21% oxygen, so for the great majority of land animals there is essentially no problem in obtaining the oxygen that they need. Only when constrained within a poorly ventilated space, such as a burrow or cave, or at high altitude (see Chapter 14) is an animal likely to experience a degree of hypoxia.

However, all terrestrial animals must have a respiratory exchange site permeable to oxygen; so they inevitably also have a site permeable to water, from which EWL is bound to occur. Striking the right balance between the conflicting demands of a fast, active lifestyle demanding high aerobic metabolism and the need to conserve water by reducing REWL is a dominant aspect of terrestrial respiratory design. Inevitably the nature of this compromise is affected by the lifestyle and the size of each land animal, as well as by its phylogenetic legacy.

The smallest and soft-bodied land animals are able to breathe cutaneously. A few larger and fairly mesophilic animals retain a system based on the gill, for example, isopod pleopods, though these develop internal 'pseudotracheae' in the more fully terrestrial species. All other land animals have fully internalized gas-exchange sites, either **lung books** or **lungs** or **tracheae**.

In Chapter 5 we considered the general balance of oxygen uptake and water loss from different respiratory surfaces operating in air (see Fig. 5.25), and concluded that for a simple cutaneous surface the ratio of water loss to oxygen uptake might be 2.1 (mg H_2O lost per ml O_2 taken up), whereas for lungs it could theoretically be only 0.2 and for tracheae as low as 0.1. Any terrestrial system of respiration that improves on the poor ratio for skin alone will be strongly selected for, so it is not surprising that both the systems of invaginated surfaces appear to have evolved many times over.

13.4.1 Lung breathing and water loss

Lungs can either be simply **diffusional** or can be **ventilated**, but always involve a substantial **vascularization** of the exchange surface (evident even in the simple terminal 'lung' of some giant earthworms and in the modified branchial chamber of land crabs, where a rather complex double portal system of blood vessels has been described). Diffusional and ventilated lungs represent identical situations in terms of controlling water loss. The lungs of crabs and pulmonate snails, and the lung books of many arachnids (Fig. 13.36), all constitute water-saving devices because by invaginating the exchange surface they help to reduce Po_2 at the point of uptake. This effect is particularly striking in vertebrate alveolar lungs, where the Po_2 at the internal alveolar surfaces is reduced to around half of its atmospheric level, giving the favourable ratio of water loss to oxygen uptake mentioned above. EWL is therefore inherently rather low, around 15–40% of total EWL in a medium-sized reptile.

However, extra savings of water can be made in the nasal exchange systems of some birds and mammals, where, by reducing the temperature of expired air in the turbinals of the 'cold nose', some water recondenses into the nasal mucosa (see Fig. 5.26 for the general principle). Examples of the nasal turbinal systems in vertebrates are shown in Fig. 13.37. The extent of cooling achieved depends on the precise anatomy and length of the system; for example, in humans the expired air has a temperature range of 28–34°C, well below the core temperature of 38°C, and representing a moderate water saving, while in birds the expired air may be as cool as 20°C. However, long noses should not be assumed to be adaptations for water saving during evaporative cooling. A theory that this was the 'reason' for long noses in baboons was largely discarded when it was pointed out that baboons lack the characteristic Steno's gland (permeable tissue producing water for the cooling effect) and lack a valve on the epiglottis to allow the nose to be bypassed during panting; here the long nose probably has more to do with sexual display and flaunting the impressive canines during threat behaviour than it does with physiology!

13.4.2 Tracheal systems and water loss

Tracheal systems by their very nature provide a site where Po_2 can be substantially lowered at the exchange site, giving a favourable oxygen uptake to water loss ratio. In the more hygrophilic land arthropods this is the limit of their water-conserving design features; the spiracles cannot be closed, and REWL must be considerable, but this matters less in the litter habitat of these animals. By contrast, insects have fully closable spiracles, and their respiratory system can have additional specializations to improve ventilation as well as to reduce water loss.

Ventilation and condensation

In at least some insects ventilation of the tracheal system occurs directionally, with most of the intake at the front end of the animal (thoracic spiracles) and most of the exit of spent gases at the rear (Fig. 13.38). In locusts, for example, air enters at spiracles 2 and 3, and leaves mainly through spiracles 5–10; in some beetles only the thoracic spiracles are shaped so as to scoop in air during forward movement. This raises the possibility of operating condenser systems with the expired air (as in vertebrate noses), and some active insects probably exploit this. A locust may have a thoracic temperature of 30°C with the abdomen at only 20°C; and air that is 10°C cooler, while still saturated, holds substantially less water. Certain desert beetles have exploited this option rather more explicitly, as we will see in Chapter 14.

Spiracular control

The onychophorans have open access to their tracheal system, as do most centipedes and millipedes (though a few centipedes do achieve some muscular contraction around the spiracle) and most apterygote insects. In some mites the spiracles have a

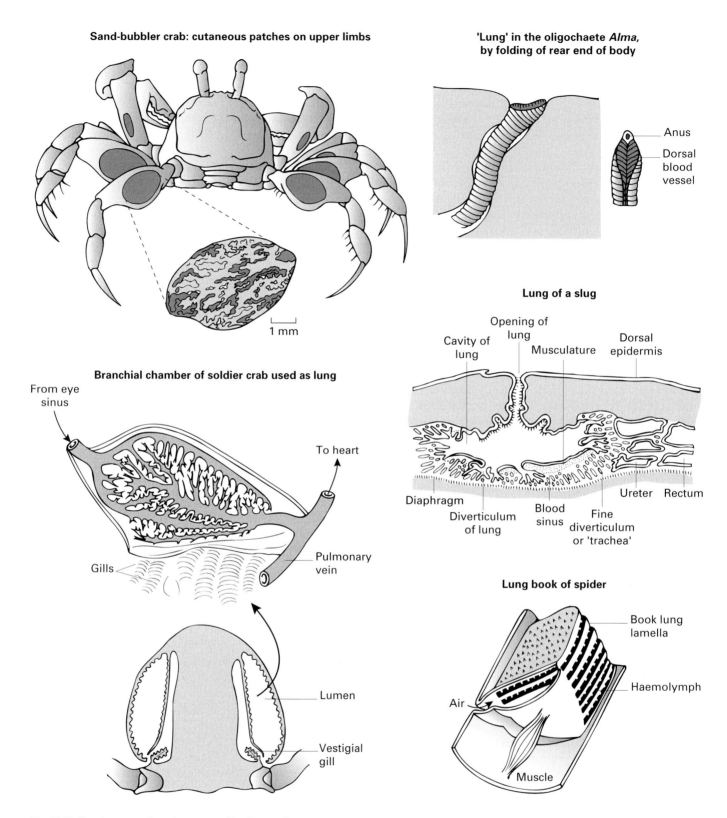

Sand-bubbler crab: cutaneous patches on upper limbs

1 mm

'Lung' in the oligochaete *Alma*, by folding of rear end of body

Anus

Dorsal blood vessel

Lung of a slug

Cavity of lung

Opening of lung

Musculature

Dorsal epidermis

Diaphragm

Diverticulum of lung

Blood sinus

Fine diverticulum or 'trachea'

Ureter

Rectum

Branchial chamber of soldier crab used as lung

From eye sinus

To heart

Pulmonary vein

Gills

Lumen

Vestigial gill

Lung book of spider

Book lung lamella

Haemolymph

Air

Muscle

Fig. 13.36 Respiratory surfaces in a range of land invertebrates: cutaneous patches, lung books and lungs.

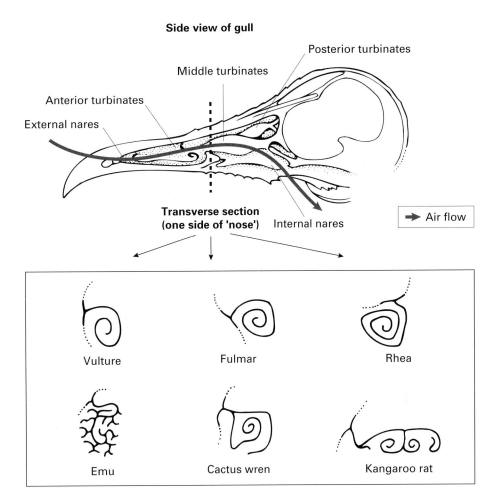

Side view of gull

Posterior turbinates

Middle turbinates

Anterior turbinates

External nares

Transverse section
(one side of 'nose')

Internal nares

→ Air flow

Vulture

Fulmar

Rhea

Emu

Cactus wren

Kangaroo rat

Fig. 13.37 Cross-sections of the nasal passages (turbinates) in various birds and mammals, showing the greater development of the turbinates in more xeric species. (Adapted from Schmidt-Nielsen *et al.* 1970b; Hillenius 1994.)

perforated cuticular covering, and the scale of the perforations is such that they aid water conservation by creating a very humid environment in the enclosed atrium and reducing vapour loss above. But most of the pterygote insects and some arachnids have control systems to open and close cuticular valves over the spiracles and thus manage their water loss. Many have closely interlocking bristles to prevent dust and external water entering the spiracles, and in addition they have two sets of muscles (openers and closers) to control the size of the aperture (see Fig. 7.31). This control may be exercised directly via CO_2 effects on the spiracular closure muscle (Chapter 7), but temperature and humidity, and probably wind speed, also have effects on spiracular closure, and the state of hydration of the insect modulates these effects. The best-known example is the African tsetse fly, *Glossina*, where a direct effect of humidity on spiracular closing has been demonstrated (Fig. 13.39). Thus the respiratory demand and the water balance of the animal can be closely integrated. Fully closed spiracles in a resting insect can reduce REWL by 70–90% in many insects; but in flight the spiracles have to be fully open to oxygenate the thoracic flight muscles, and water balance needs are secondary.

Discontinuous respiration

At the extreme, in some insects, including a range of pupae and some adults, the spiracles remain closed for all but occasional brief bursts of opening. This is termed the **discontinuous ventilation cycle** (DVC), and each cycle is made up of a closed spiracle phase, a flutter-spiracle phase and an open phase. In fact the insect continues to use oxygen at a constant level throughout (Fig. 13.40), but it is able to store carbon dioxide and release it only intermittently, thus also avoiding continuous water loss. When the spiracles are closed, accumulating CO_2 is retained in solution in the haemolymph as bicarbonate, and this means that the air initially taken in and depleted of oxygen has not been replenished with an equivalent gaseous volume. Thus inside the tracheal tubes there is a small but measurable negative pressure as the O_2 is used up. This negative pressure is countered by the spiral thickening of the tracheae, and is enough to ensure that air continues to leak in at the spiracles without significant outward gaseous or water vapour loss, so oxygen continues to be supplied to the animal's tissues ('passive suction ventilation'). After long periods of closed spiracles and gentle inward leak of air,

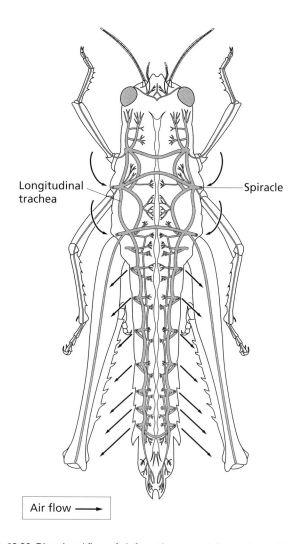

Fig. 13.38 Directional flow of air from thorax to abdomen through the body of a locust, potentially allowing some water conservation by condensation from exhaled air in the cooler abdomen.

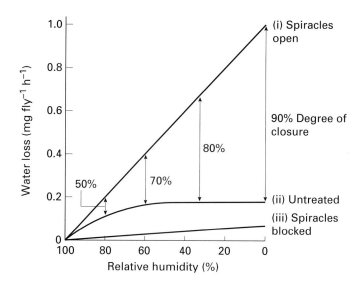

Fig. 13.39 Water loss in a normal tsetse fly, and in one where the spiracles are kept open by raised (15%) CO_2. Respiratory water loss is normally well-controlled and little greater than cutaneous evaporative loss (CEWL; cuticular loss, with all spiracles blocked). (Adapted from Bursell 1957.)

the bicarbonate levels in the haemolymph become too highly elevated and the spiracles open up. Over a period of a few minutes there may be a series of brief spiracular openings (the F phase) where inward oxygenated flow greatly exceeds outward leaks, and then a more substantial open period of real 'flush out', until the haemolymph pH and [HCO_3^-] are back to normal levels (though it is unclear how such massive outward movements of CO_2, from solution, are achieved so quickly). Then the spiracles close again and the cycle repeats, with no apparent 'breathing' occurring for another few minutes or (in pupae) several hours.

This pattern of respiration has now been shown in a range of quiescent adult insects as well as in fairly active ants and various insect pupae, and something very similar occurs in ticks, land snails and even torpid bats. The water-conserving benefits of discontinuous breathing have been disputed on both practical and theoretical grounds, however. Water loss is certainly very low while the spiracles are closed, but it rises greatly at the start of the ventilatory burst. Furthermore, dehydrated grasshoppers tend to reduce rather than increase their use of discontinuous breathing; and some other species exhibiting this pattern have REWL as only a small component of water balance anyway. However, elaborate tests comparing EWL with and without spiracular control in the more xeric animals, such as ants and moth pupae, do indicate a several-fold saving in water loss during DVC, and xeric ants also show lower frequency DVC and higher flutter phase CO_2 emission volumes than mesic relatives. *Drosophila* reared under desiccating stress also show more pronounced cyclic respiration. All these findings suggest that discontinuous breathing is a genuine adaptive response to reduce REWL.

Despite all these adaptations in both lung and tracheal systems, there is ultimately no way of avoiding substantial water loss from respiratory surfaces in a fully active terrestrial animal. The behaviour of the animal in choosing microhabitats and appropriate food and drink, together with its cutaneous and renal adaptations, must compensate for its inevitable respiratory losses.

13.4.3 Oxygen carriage, circulation and respiration

In insects the tracheal system delivers oxygen in the gaseous phase almost to its point of use, and no circulatory fluid is needed, the haemolymph containing no respiratory pigment. In contrast, most other land animals use the blood system to carry the gas to the tissues. Most also use a blood pigment to carry oxygen; molluscs and crustaceans use haemocyanins, while earthworms and tetrapods use haemoglobin. In many vertebrates there is a 'store' of extra pigment available as unused

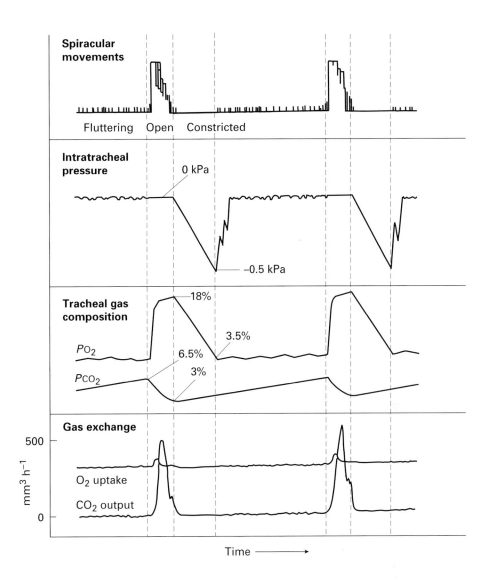

Fig. 13.40 Discontinuous respiration in a silkmoth pupa, showing spiracle movements and the patterns of pressure and gas exchange in the tracheae. (Reprinted from *Journal of Insect Physiology* **12**, Levy, R.I. & Schneiderman, H.A., Discontinuous respiration in insects. IV. Changes in intratracheal pressure during the respiratory cycle of silkworm pupae, pp. 465–492, copyright 1996, with permission from Elsevier Science.)

erythrocytes in the spleen, which can be mobilized if any hypoxia is experienced (or if there is blood loss following wounding). However, the pigments of terrestrial animals are unexceptional—they do not appear to be significantly different in chemistry or in properties from those of closely related aquatic species (see Chapter 7), though they load to a higher percentage oxygen content (see Fig. 7.23).

In contrast, the circulatory patterns of all terrestrial animals except insects do undergo substantial modification relative to aquatic animals. Blood flow has to be organized so that oxygenated blood from the new respiratory sites (in effect, the lungs) is quickly distributed to the most demanding tissues, commonly the locomotory muscles and the brain. In invertebrates the blood always flows from the heart to the systemic organs and then to either the gills or lungs (see Fig. 7.17), so that there is a delay in supplying freshly oxygenated blood where it is needed. However, the rate and pattern of circulation can be modified considerably in relation to demand; even in earthworms, the degree of dilation of cutaneous vessels is under

hormonal control and the O_2 level in the dorsal vessel can rapidly increase on contact with moist oxygenated soils.

The system in vertebrates such as fish is perhaps better organized, with blood flowing from heart to gills to systemic organs (see Figs 7.17 & 7.18). However, in air-breathing vertebrates the situation is complicated by the progressive division of the circulation into two parallel systems, with blood returning to the heart twice in each full circuit of the body; the disadvantage of the 'delay' this introduces in getting freshly oxygenated blood to the respiring tissues is presumably offset by the extra pressure boost given by the second passage through the heart.

13.4.4 Carbon dioxide loss

Carbon dioxide loss in terrestrial animals is much like that in aquatic animals—loss is essentially diffusive, from dissolution in the blood across the permeable surfaces direct into the medium. However, whereas aquatic animals lose most of their CO_2 at the gills, many soft-bodied terrestrial animals use their general body

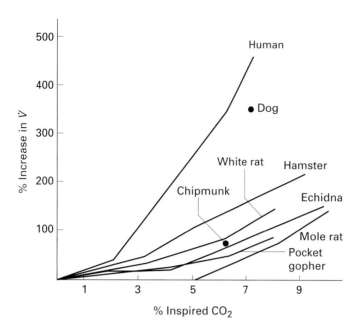

Fig. 13.41 Reduced CO_2 sensitivity in various burrowing animals compared with non-burrowing species.

dangers of hyperventilation) as acid–base balance is again compromised.

Adaptation to land also involves change in the site of CO_2 sensors, from an external site in fish and aquatic invertebrates to internal (usually in the arterial blood system) in the land animals (see Chapter 7). This allows a direct response of increased ventilation when blood Pco_2 builds up. However, this direct response varies somewhat with habitat; for example, CO_2 sensitivity is substantially reduced in most burrowing land animals compared to non-burrowers (Fig. 13.41), reflecting the raised CO_2 levels that must occur within the confines of most burrows and nests. There are a few exceptions to this rule, notably in the burrowing owls (*Athene*) where cardiopulmonary responses are normal but there is a substantially increased blood buffering capacity so the owl can tolerate acidosis.

13.5 Reproductive and life-cycle adaptation

The essential requirement for terrestrial reproduction is to keep eggs and embryos from desiccating (Table 13.12). In terms of types and degrees of reproductive adaptation, we again need to distinguish between animals that achieved terrestriality by a freshwater route and those that invaded across the littoral zone. Littoral animals show a considerable range of reproductive strategies (Chapter 10) reflecting their varied experience of immersion and transient desiccation; many produce small pelagic eggs and larvae, but some have internal fertilization and curtailed larval development, often with parental brooding. Groups such as amphipods and isopods retain this latter habit on land, while other arthropodan groups (insects and myriapods) commonly use sperm droplet or spermatophore transfer systems that may have originated in marine interstitial habitats. Freshwater animals often have a rather different approach, again often fertilizing internally but commonly producing fewer and larger eggs, as dense egg masses attached to the substrate or vegetation, with substantial yolk as a nutritional endowment; they only infrequently adopt brooding habits. Thus the young are released quite early from any association with the mother but are already relatively large to avoid the osmotic problems associated with a very large SA/V ratio in a dilute medium; traditional aquatic

surface. Most worms use this route, as do amphibians. In fact a moderately active amphibian with typical bimodal (cutaneous + lungs) breathing may gain 50–78% of its oxygen from its lungs, but lose 57–84% of its carbon dioxide from its skin.

In land animals with relatively impermeable skins and cuticles (the arthropods and tetrapods) there is little alternative but to lose CO_2 from the respiratory system. Since raised CO_2 levels (hypercapnia) can represent a significant hazard, especially in terms of reduced pH levels and thus acidosis, tracheae and lungs have to be better-ventilated to flush out the carbon dioxide. This is especially important in birds and mammals again, where increases in blood Pco_2 compromise oxygen carriage and delivery, and upset the acid–base balance, with knock-on effects on homeostasis throughout the body. CO_2 becomes an important blood buffer, and there is therefore also a potential problem with losing too much CO_2 too quickly (hence the

Problem	Solution	Examples
Protect the gametes	Return to water	Most crabs, amphibians
	Use sperm droplets	Myriapods
	Use protective spermatophores	Most arachnids, some insects
	Internal insemination	Most insects, spiders, amniote vertebrates
Protect the young	Shelled yolky eggs	Worms, insects, arachnids, snails, amphibians, reptiles, birds, monotremes
	Viviparity	Few insects, all eutherian mammals
	Postnatal parental care	Few insects, some arachnids, many vertebrates

Table 13.12 Problems and solutions for reproduction on land.

larval forms such as trochophores, veligers and nauplius are abandoned. Either route may allow the development of an ovoviviparous or fully viviparous habit, though the marine route with its relatively greater environmental constancy may less often exert selective pressures to abandon oviparity.

13.5.1 *Reproductive strategies*

General life-history patterns

We dealt with the general theories of life-history strategy briefly in Chapter 1. Although the terminology has been criticized and has sometimes been overdone, animals and plants alike can usefully be seen as occupying particular positions on the *r*–*K* continuum (see Table 1.1). From the brief discussion above it will be evident that on land there might inevitably be a general tendency to the *K*-selected end of these possible strategies, with fewer, larger, better-resourced and better protected gametes and young. But the counter-argument is that *r*-selection usually occurs with less predictable habitats and less stable resources, as expected on land. Within the terrestrial animals there is therefore a wide range of possibilities, from the rather profligate egg-laying habits of some amphibians and insects to the extreme of investment in a single offspring once a year or even once every few years in some birds and mammals. It is commonplace to speak of some of the insects as being extremely *r*-selected, but important to bear in mind that within the animal kingdom as a whole their reproductive strategies are actually rather modified towards fewer and higher-investment offspring, albeit produced with great (and sometimes alarming!) abundance and regularity.

Remember that the energy that females invest in reproduction usually scales with a mass exponent (*b*) between 0.5 and 0.9 (see Chapter 6), which means that larger species tend to invest relatively less in their offspring per unit time. But larger animals also tend to live longer, and lifespan scales with a mass exponent of 0.15–0.29. Thus over a female's lifetime, these two factors together mean that the energy invested in reproduction varies more or less isometrically with respect to body mass (*b* = 1.0). Larger animals on land do tend to invest proportionally less of their resources in each breeding cycle, but will have a longer reproductive life and produce more broods; however, this effect is much more evident within a taxon (where reproductive constraints and techniques are comparable) than when comparing across widely differing taxa such as insects and vertebrates.

A further generalization about terrestrial animals is their greater tendency towards seasonal breeding. In most invertebrates breeding occurs annually (more rarely there are multiple times at roughly fixed intervals, in a particular season of the year) but its precise timing is very dependent on climatic conditions; for example, a spring breeding norm may be delayed after a cold winter. However, in many vertebrates there are one or a few breeding seasons each year which are fairly rigidly fixed by hormonal triggers, dependent either on the invariant signal of the photoperiod or perhaps on endogenous timing systems.

Terrestrial invertebrates

Soft-bodied animals living on land are often highly dependent on mucus at many stages of their reproduction. Both flatworms and nemertines may use mucoid sheaths to enclose a copulating pair, and in flatworms there are often multiple eversible cirri each acting as a penis to inject sperm into the general body tissues of any other worm encountered (each being hermaphrodite). In nemertines the gametes may be shed into the mucoid sheath, where fertilization then occurs and the eggs remain (though at least one species of nemertine is ovoviviparous). However, most flatworms and a few nemertines achieve internal fertilization. Flatworms then lay rather well-protected shelled eggs, which hatch directly as small worms with no larval stage.

Earthworms too secrete a mucoid sheath during copulation. Two worms come close together ventrally, lying in opposite directions; each (hermaphrodite) worm makes its own sheath and deposits into it a package of its spermatozoa. The sperm then pass backwards along a seminal groove and enter the sheath of the other worm, migrating into that worm's spermathecae (sperm storage organs). Each worm sloughs off its sheath and they separate. Some time later the worm secretes a further sheath of protein-based material, and some albumen proteins and the eggs are deposited into it, together with some of the stored sperm, at which point fertilization occurs; the sheath slips off the worm, and later hardens and darkens to form the cocoon protecting the fertilized eggs. The young earthworms hatch from the eggs within the cocoon and feed on the stored albumens.

In terrestrial molluscs, many of which are simultaneous hermaphrodites (see Box 9.1), more complex reproductive behaviours and anatomies come into play. Courtship is well developed, with two snails or slugs coming together in an entwining 'dance' of tentacular and oral stroking. In some genera, after a short period of dancing each partner everts a structure called the dart sac and projects a calcareous dart into the other's body, where it stimulates the recipient into the next phase of courtship, culminating in penis protrusion and cross-copulation. Slugs are particularly adept at conducting their copulatory exchanges in mid-air, suspended from vegetation on the end of a mucus thread. Many of the slugs and snails then lay small groups of rather large eggs in a sheltered damp place; even the very large and strongly calcified eggs of the genus *Achatina* (the giant African land snail) are still relatively permeable and desiccate easily. There are even a few pulmonate species that are ovoviviparous and retain the eggs in the oviduct prior to hatching. Note that in both annelids and molluscs there is a trend from jelly-covered eggs in aquatic species to shelled protected eggs in terrestrial species, paralleling the amphibian–reptile transition in the vertebrates.

Crustaceans on land generally show brooding and maternal care. In isopods and amphipods brood-care is present in marine species, and is simply somewhat extended in terrestrial species so that the brooded eggs hatch as miniature adults. In the amphipods another aspect of reproduction is carried over from the marine to the land species, that of precopulatory mate carriage. Aquatic amphipod males carry the female around for some time, then copulate with her and release her very quickly. In the terrestrial forms (e.g. sandhoppers) the male carries the female around only briefly, and copulation is often a lengthier process, but again the male then departs. The female retains the fertilized eggs in her brood pouch for a variable time, depending on her stage in the moult cycle.

Land crabs show rather more sophisticated behaviours associated with mating, but with their essentially marine ancestry they are actually more reliant on water than the two groups described above. In many aquatic crabs mating has to occur at the point when the female has just moulted and her carapace is soft, since at other times the genital operculum is inconveniently hardened in the closed position. The newly moulted female is therefore transiently very vulnerable and may have little choice in who she mates with. In more terrestrial species, though, the operculum may be hinged, so that both copulation and oviposition can occur at any stage of adult life. So in land crabs mating behaviours become much more complex, and the male may have to perform a prolonged courtship ritual to secure a female. The fiddler crabs (*Uca*) and ghost crabs (*Ocypode*) are well known for elaborate mating dances, with visual and acoustic signals, though the complexity and duration of these varies widely. Copulation often occurs inside a burrow (sometimes built specially for the occasion by the male), and the female crab may remain in that burrow until the eggs are mature. Eggs are large, as are any young larval stages that persist, and their carbon/nitrogen ratio is particularly high due to a high content of lipid in the yolk. In some species hatching of young juveniles occurs within the burrow and the young are held against the maternal abdomen, below the pleopods, prior to release. However, most of the grapsid and ocypodid females return to the sea to release their young for a brief marine larval phase, and even the highly terrestrial gecarcinids have to return briefly to sea, often from sites many miles inland.

In the insects and myriapods a different approach to terrestrial reproduction appears, probably derived from an interstitial lifestyle. Male chilopods (centipedes) deposit sperm droplets; at its simplest the drop is laid directly onto the soil, but in many species a pad or web of silk, or a small pillar of gelatinous material, is laid down first. The female centipede then takes up the sperm straight into her genital opening. In another group of myriapods, the symphylans, a sperm droplet is laid by the male and then eaten by the female; as with the flatworms mentioned earlier, the sperm must then migrate within the female body to her sperm-storage organ. The more primitive apterygote insects again use sperm droplets, without copulation, but in many species there is a behavioural ritual whereby the male leads the female into position over his sperm. In all these cases the sperm droplet is an appropriate means of transfer in the humid and protected habitat of cryptozoic fauna.

In most of the arachnids and winged (pterygote) insects, which have emerged fully from the interstitial and litter zones

into the drier, fully terrestrial world, droplets are replaced with more elaborate spermatophores (Fig. 13.42), transferred to the external female genitalia during normal mating behaviour (most arachnids and many insects), or directly inseminated through an intromittent organ into the female (most pterygote insects). Spermatophores may become extremely elaborate, and in some orthopteran groups, such as katydids, the basic sperm packaging is augmented with a gelatinous mass, the spermatophylax (Fig. 13.42c), which the female eats after the male has left. This is thought to have a nutritive function and thus forms a kind of 'courtship feeding'.

Other kinds of nuptial gift come into play with insects during precopulatory behaviour in the higher orders, where direct insemination is the norm; for example, many male flies bring small packages of food to a female. Other complex behaviours are also common; males may wait for (or even dig for) females at emergence sites, or show territorial defence at known feeding sites such as flowers or carrion, or they may form 'leks', male swarms to which females are attracted visually and by scent (pheromones) and where they may choose the most competitive male as a partner. In many insects the males show alternative mating tactics, with some males being territorial and others adopting a 'sneaky' tactic, for example; in many bees and wasps this behavioural variation has been shown to be due to size differences and to the directly related differences in thermal tolerance of larger and smaller (or paler and darker) males. Other female and male behaviours may also affect the reproductive process of course; female insects commonly attract males with pheromones, while either sex may attract the other acoustically with song, or visually with bioluminescent systems, as in fireflies.

Some insects have adopted unusual reproductive modes incorporating asexual phases, and these can usually be related to the particular characters of their terrestrial habitat. Many aphids (Homoptera) reproduce parthenogenetically for much of the year, and some of them produce nymphs directly rather than eggs, with the adults being wingless. Only in the late summer are sexual (usually winged) males and females produced, and these may be non-feeding stages without mouthparts; they mate in protected crevices on the host plant, and the female then disperses to a new habitat and lays just one or a few giant fertilized eggs that overwinter to produce the foundress of the spring (parthenogenetic) generation. Even more remarkably, some midges and some parasitoids regularly omit an adult phase entirely, with larvae becoming sexually mature and laying eggs, or retaining cells that develop internally into new larvae and eventually kill the mother larva. These kinds of life cycles are probably largely determined by food (host) availability rather than by temperature or humidity *per se*.

Insects and most other arthropods produce eggs with a stiff chitinous cuticular covering, often with a waxy layer incorporated to give low permeability (see Chapter 5). This 'eggshell' always bears pores to allow gas exchange, though the total area of such pores may be only 1% of the total egg surface. In many insects the outer surface becomes very complex, with an

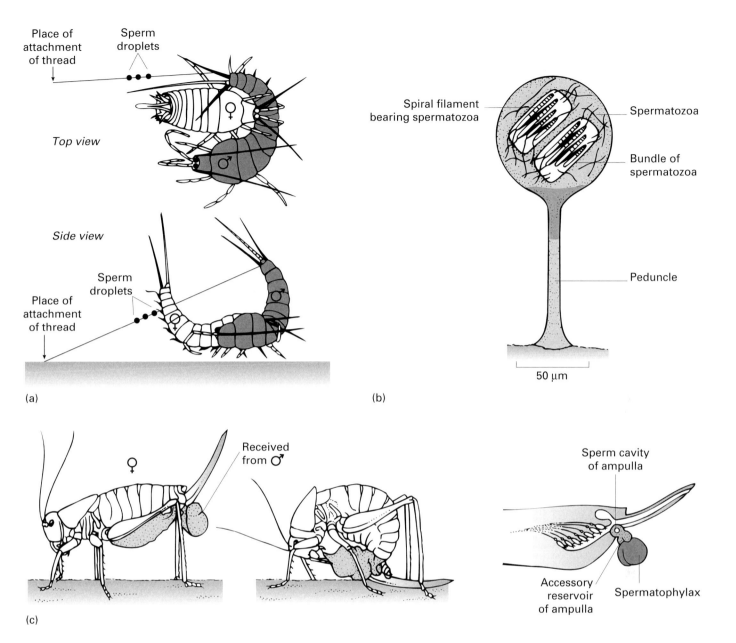

Fig. 13.42 (a) Sperm droplets in an apterygote insect, (b) spermatophores in a myriapod and (c) the elaborate nutritive spermatophylax of an orthopteran, all serving as ways of transferring male gametes on land. (a, From Little 1990, courtesy of Cambridge University Press.)

aeropyle layer of reticulate sponge-like tissue (see Fig. 5.27), so that the eggshell acts as an air-filled plastron in the manner already described for some freshwater insects (see Chapter 11). There is also evidence that air spaces in the eggs of some insect species are structured so that they act as reflectors, keeping the eggs cool in sunlight.

Most insects lay their eggs in specific protected microclimates where desiccation will be minimized. This often involves a site associated with the food source used by the developing larvae or nymphs; eggs may be laid on or in the leaves or flowers or bark of the host plant, or in the flesh of another animal that will be a host for the parasitic or parasitoid juvenile stages. Many terrestrial insects lay their eggs underwater and have an aquatic larval stage; adult dragonflies and mayflies, and some beetles and true flies, must struggle to pierce the surface meniscus of ponds and streams to permit oviposition in a suitable place. Relatively few insects provide any after-care to their young; those that do include a few bugs ('parent bugs'), together with the social insects. However, there are also some species that retain the eggs within the body, usually in the lower ovarioles, and provide nutrition to the developing offspring; for example, tsetse flies (*Glossina*) 'lay' very advanced larvae which pupate without feeding.

In contrast to insects, many arachnids lay their eggs in prepared nests or carry them around with them. Scorpions are typically ovoviviparous or fully viviparous, and these and other arachnids commonly provide brood pouches for their developing embryos or young. Certain spiders also effectively brood their eggs to speed up the development rate; in the genus *Pirata*, the preferred T_b of the female rises when she is carrying eggs, from around 21°C to around 27°C, and she will bask on the upper surface of her moss habitat to achieve this T_b. Other spider genera lay the eggs in a silken nest but will move them around the nest to keep their temperature constant, or even out into the cooler moving air if they get too hot. Arachnids also quite commonly exhibit maternal care after hatching, with juvenile scorpions riding on the parental back and young spiderlings living on within the parental nest for some time after birth.

Vertebrates

Vertebrates, with their freshwater ancestry, inherited typically large eggs and larger offspring size, with more advanced precocious hatchlings that are at such an advantage in osmotically stressful and variably provisioned ponds and streams. Hence freshwater amphibians have distinctly larger eggs than estuarine species, and terrestrial amphibians lay eggs that are even larger—in some species each egg approaches 10% of the mother's linear dimensions. But most amphibians still have unprotected, essentially aquatic eggs, and lay a great many of them (commonly hundreds at a time, and up to 20 000 throughout the lifetime of the average female in some species). The jelly that surrounds each egg expands enormously by taking up water, and the resultant capsule may help to trap heat and keep eggs warm; however, the maximum diameter of modern amphibian eggs is restricted to about 9 mm, as beyond this diffusion alone would not permit adequate oxygenation. The main problem is that a large egg mass of soft jelly-like material inevitably suffers a high mortality to fish, insect larvae, ducks, etc.

In urodele amphibians (salamanders and their kin) the female generally picks up a package of sperm into her cloaca and fertilization occurs internally just before laying. But fertilization is virtually always external in anurans, even though each male may have courted the female carefully and may grasp her tightly during egg laying. These common amphibians need to ensure hydration of the sperm, and many therefore mate within a pond.

An additional obvious disadvantage in amphibian reproduction is that since the eggs are not waterproof, and the male and female both normally have to return to fresh water to lay them and to ensure sperm safety, the amphibians have to be metamorphic, with aquatic young preceding the more terrestrial adults. In practice the young of apodans and urodeles are relatively like the adults except for having gills, and the juveniles are only very different in modern anurans, for whose young the term 'tadpole' is usually reserved. These young stages have to cope with a range of freshwater predators throughout their development.

Faced with all these problems, some frogs have adopted different strategies; they lay far fewer eggs and protect them better, keeping them away from open water for as long as possible. In some cases this involves forms of brooding behaviour, with

egg mortality cut down or even eliminated. For example, the Surinam toads (*Pipa*) have extraordinary broad, flat bodies, and the male spreads newly fertilized eggs over the female's back, where the skin swells to embed the eggs, which disappear entirely within 30 h; the mother remains in water, and about 3 weeks later the young tadpoles break out and swim off. Alternatively, gravid female frogs may find their own predator-free swimming pool, and in tropical forests this commonly involves pools of water that collect within the leaf bases of bromeliad plants (see Chapter 12). At least one Brazilian species makes its own 'birthing pools' with raised mud walls. However, some anurans come closer to a truly terrestrial mode of reproduction. The midwife toad, *Alytes*, is a well-known example occurring in Europe, Majorca and Morocco, where the male wraps up the eggs in stringy mucus around his own back legs for several weeks while he pursues a terrestrial lifestyle; then he stands with his back end in water when the tadpoles are ready to emerge. In some of the South American dendrobatid (arrow poison) frogs, eggs are laid in moist earth and guarded by the male, and after hatching the tadpoles climb onto his mucus-covered back. Some African and Asian treefrogs mate in tree canopies, and then choose a branch overhanging water on which they make a ball of mucus; this is lathered up into an airy froth by both male and female, after which the female lays her eggs in it. Either it then hardens to form a protective desiccation-resistant nest, or in a few species it is kept moistened by the female's urine, until the tadpoles drop out when mature into the pond below. Anuran foam nests reduce desiccation, but also provide insulation, the eggs being up to 8°C warmer (and therefore developing faster) than if unprotected.

A few species of anuran have gone a little further and produce large yolky eggs where the whole development can occur. In the Caribbean whistling frog there may be only 20–30 eggs, each with 50 mm^3 of yolk, and the froglets emerge in about 3 weeks. Occasionally the whole process can be retained inside the adult body. In *Gastrotheca* (the 'marsupial frog') the females have brood pouches on the back, from which emerge either tadpoles or froglets, depending on the species. In *Rhinoderma* ('Darwin's frog' from Chile and Argentina) the male waits for the eggs to hatch and then 'eats' the young, which develop in his enlarged vocal sacs and are subsequently spat out as froglets. The extreme case of terrestrialization is *Nectophrynoides*, a West African frog, where males introduce sperm into the female vent using their cloacal 'tail', so that fertilization occurs inside, and eggs are retained in the oviduct; the tadpoles hatch there and feed off flaky material from the walls of this duct, which are also very well oxygenated by an arterial supply.

All other vertebrates—collectively the Amniota—have overcome the limitation of egg size and egg water balance in a quite different way, with the evolution of the **cleidoic egg** (see Fig. 5.27). This has accessory breathing structures in the egg surface, linked via the chorioallantoic membrane to the embryonic blood system inside the egg, so that reptile and bird eggs can be very large indeed and the young can be highly advanced before

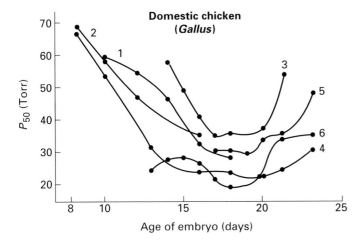

Fig. 13.43 The value of P_{50} for haemoglobin in maturing bird embryos. For six separate individuals greatest oxygen affinity occurs at about 15–20 days (and is due to changing levels of inositol pentaphosphate, IPP—see text). (From Lutz 1980.)

hatching. Reptiles are commonly oviparous, with a few excursions into viviparity, while birds are exclusively oviparous. The eggs produced may be leathery and quite flexible, as in many turtles and lizards, or calcified, as in crocodiles and birds (which often have an outer coating over the eggshell to reduce bacterial invasion). Both types are still relatively permeable to water, but the calcified eggs are sufficiently rigid to resist volume change, and water lost during development, which is around 14–16% of egg mass in birds, is compensated by a growing air space within the shell. This air space is important in facilitating gas exchange in the developing embryos. However, eggs do increase their metabolic rate as they mature, and run into oxygen shortage; they compensate by using inositol pentaphosphate (IPP) to shift the haemoglobin dissociation curve to the right (decreased P_{50}), with a rapid return shift to normal after hatching (Fig. 13.43). The egg air space also allows for some accumulation of wastes, as the ageing embryo switches progressively from producing ammonia to producing urea and then uric acid. The eggshell itself is perforated by pores (see Fig. 5.27) through which both gases and water vapour can diffuse, these pores having a complex anatomy usually constricted towards the inner surface. Though numerous, the pores are tiny and commonly give perhaps only 1 mm² of 'open' exchange surface, though this is adequate to supply the egg's oxygen demand by simple diffusion. These pores inevitably allow rather substantial water loss during incubation, particularly in the latter stages when the shell becomes increasingly eroded.

The cleidoic egg frees its parents from any dependence on liquid water supplies at the site of mating or oviposition, and the amniotes can therefore become truly emancipated terrestrial animals. However, their eggs still require careful siting, either in a nest in a protected environment or retained within the parental body. These techniques serve several functions: for example, preservation against predators (some monitor lizards visit their nest periodically after laying, primarily for defensive reasons) or maintenance of equable humidities (to limit the egg water loss). But their commonest function is control of egg temperature, which must be high enough to achieve a reasonably rapid development. Tuataras in New Zealand provide a nice example, leaving their residential burrows in woodland and producing special nest burrows in open pasture where it is warmer.

There may also be other reasons why temperature control of the egg is important. In many semi-aquatic and terrestrial reptiles a new phenomenon appears, that of **temperature-dependent sex determination**, or TSD. In three of the five major reptilian lineages sex is determined by environmental temperature, and many of these reptiles lack heteromorphic sex chromosomes (i.e. X and Y versions, or some equivalent). TSD is found in all crocodiles and many turtles, but is rare in lizards, and is absent in the tuatara, the amphisbaenians and all snakes which have been examined so far. In general, TSD is more prevalent in long-lived species.

Three general patterns of TSD have been found. In Type A (most crocodilians and lizards), females are produced at low temperatures and males at high temperatures ('FM'). Type B (many turtles) gives males at low temperatures and females at high temperatures ('MF'). Finally, some crocodiles, one lizard and three turtles have been reported to produce females at high and low temperature extremes and males at intermediate temperatures (Type C or 'FMF'). More recent evidence, though, suggests that some cases of FM are actually FMF when sufficiently high temperatures are tested. Patterns tend to be associated with the direction of sexual dimorphism, i.e. which sex is larger.

Figure 13.44 shows the effects of constant temperature on the sex ratio of hatchling turtles. The gonadal sex of the embryo is determined not by the temperature at any particular moment but by the cumulative effects of temperature through a critical phase in development, which in turtles commonly begins shortly after laying and extends through the first half of development. In crocodiles the thermosensitive period is somewhat later, roughly in the third quarter of development (days 30–45), and it coincides with the timing of gonadal differentiation. In the field, the sex ratio of hatchlings is therefore affected by the location of the nests, and the resultant thermal microclimate of the nest interior. Some nests produce both sexes whilst the majority produce only one sex.

Species with TSD are found in thermally patchy environments that allow the production of both sexes. Various hypotheses have been advanced to explain the mechanisms of TSD. The gonadal ratio of androgenic to oestrogenic steroids is known to be important; females are produced by a low ratio and males by a high one. This ratio is controlled by the enzyme P-450 aromatase, which converts testosterone to oestrogen. One idea is that aromatase itself exhibits temperature-dependent activity, or is produced by temperature-dependent gene transcription. Another hypothesis is that incubation temperature affects the secretion of luteinizing hormone (LH) or the density of LH receptors on the gonads; this could involve the regulation of

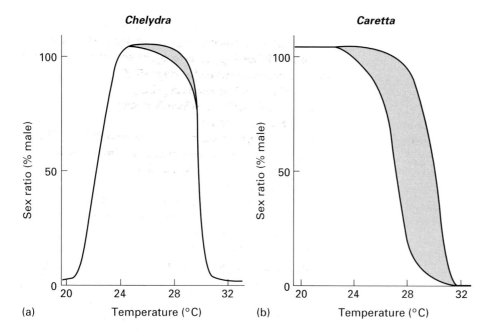

Chelydra

Caretta

Fig. 13.44 The phenomenon of temperature-dependent sex determination (TSD). Sex ratios produced in two different reptiles hatched at constant temperatures. (a) In *Chelydra*, the sex ratio is 100% male at intermediate temperatures but switches to 100% female at lower or higher temperatures (Type C, or FMF). (b) In *Caretta*, higher temperatures produce a shift from all male to all female but with much more scatter (Type B, or MF). Shaded area shows scatter. (Adapted from Janzen & Paukstis 1991.)

gene expression by a temperature-sensitive effector molecule (a DNA-binding protein) which is the product of a sex-linked gene. Some of these ideas are illustrated in Fig. 13.45.

In birds and mammals, sex determination is controlled independently of the environment, with heteromorphic sex chromosomes. In mammals two similar chromosomes (XX) makes a female and the heteromorphic (heterogametic) condition (XY) makes a male, due to the effects of a gene (*SRY*) on the Y chromosome that initiates testis formation. In birds the situation is reversed and the heteromorphs are females. In both groups temperature has no effect on genetic sex, although both do share some later-operating sexual-character-determining genes with the reptiles.

Nevertheless, in birds there is still a need to control the egg temperature, to achieve a smooth and rapid development so that nestlings appear at times when the adult can both feed

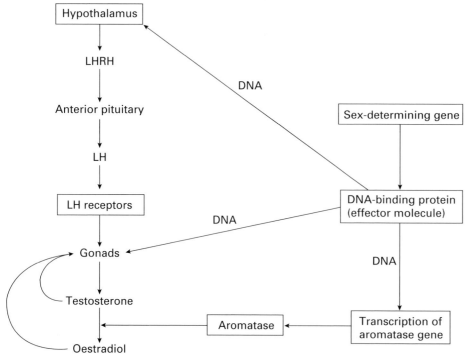

Fig. 13.45 Possible mechanisms controlling temperature-dependent sex determination (TSD). LH, luteinizing hormone; LHRH, LH-releasing hormone. (From Janzen & Paukstis 1991.)

Table 13.13 Milk composition in mammals.

Species	Percentage composition				Energy content (kJ g^{-1})
	Water	Carbohydrate	Protein	Fat	
Human	88	6.5–8	1–2	3–5	
Cow			3.4	4	
Seal				58	
Rhino	95		1.2	0.3	
Arctic fox	72	3	8	16–18	8
Red fox	82	3	6	9	4–5
Polar bear	50–70			20–35	12–20
Sun bear	73				7
Canids	80–95			2–6	4–9
Felids					5–10
Marsupials (early)	9	3–8	~1		
(late)	~1	11–15	17–20		

them adequately and ensure that their body temperatures are maintained. Hence nearly all birds use nests, and prolonged brooding behaviour by one or both parents occurs. All birds also lay eggs whose shape is tailored for particular nesting sites and whose size reflects different degrees of development (precocious or altricial) at hatching time. Birds have never switched to viviparity, and this is presumed to be mainly because the female could not carry the weight of developing young around in her body during flight.

About 90% of all birds are monogamous (at least in theory, though molecular genetic studies are increasingly revealing a surprising degree of infidelity in supposedly lifelong partners!). This is unusual in any animal group, but links to the needs of the young birds before and after hatching. The eggs need to be kept warm by incubation, as the embryo is also 'warm-blooded' but incapable of endothermy and thermoregulation. Then there is a long period of feeding and protecting the young, one or preferably both parents foraging intensively, giving time for behavioural patterns to be learnt, endothermic regulation to be established and stabilized, and flight systems to mature and become manageable.

Mammals have adopted a different solution to reproduction on land from their reptilian and avian relatives. They almost exclusively use viviparity, with very much smaller non-shelled eggs (the only exceptions being the egg-laying monotremes of Australasia, which are reproductively very similar to reptiles until after birth, when mammalian-style lactation occurs). The strategies of retaining the embryo internally and feeding it first from the maternal bloodstream and then from special mammary glands could be seen as the ultimate reproductive adaptations to life on land. Lactation may be one of the most important components of this strategy in terms of selective value. It allows the provision of a highly nutritious and digestible fluid even in mammals with poor-quality diets (e.g. coarse fibrous plants), where the mother may store reserves as fat in times of plenty and then breed in times of dearth; it allows delivery of large meals in a short time, freeing the mother to leave the nest and forage;

it permits the feeding of a litter of many altricial young with under-developed jaws and no teeth; and it promotes a strong bond between parent and offspring, allowing the development of complex learned behaviours. Mammalian milk can be delivered at a concentration suited to the particular habit and lifestyle (Table 13.13); very concentrated in marine mammals, as we saw, but very dilute in many large savanna-dwelling mammals where water balance in the young may be hard to maintain.

Birds and mammals share the problem of newborn young that may not yet be fully capable of endothermy and thermoregulation. In both groups a spectrum of abilities occurs, related to whether the young are altricial or precocial. Many altricial young are virtually naked (lacking fur or feathers) and effectively ectothermic at first, their T_b being dependent on brooding by the mother with a minor assistance from the neonate's own brown fat (BAT; see section 8.7.2). Chicks of boobies are good examples, having no inherent thermoregulation until they grow to about 200 g, but with a constant T_b of 38°C due to parental behaviour. The same is true of most marsupial young, and of some eutherian offspring such as rabbits. It may take days or weeks for normal endothermy and regulation to be achieved. In essence these young animals are ectothermic behavioural regulators, but it is someone else's behaviour that does the work!

However, the altricial strategy has its advantages; it allows shorter gestations and smaller birth sizes, putting less physiological strain on the mother, and it also normally allows larger clutch or litter sizes. In marsupials, and perhaps in other groups living in relatively variable and risky habitats, it allows the mother some choice in whether to continue investing in a youngster; in very harsh conditions, it may pay her to abandon a weak 1 cm long juvenile and invest in a new pregnancy instead, rather than spending up to 7 months (in kangaroos) with an ill-fated offspring in her pouch. Many marsupials can even 'choose' to abort a fetus within the womb if conditions and food availability deteriorate. Average investment in reproduction for a marsupial is therefore less, lactation being a little cheaper and safer to the adult than placentation; in hard times, a eutherian fetus feeds

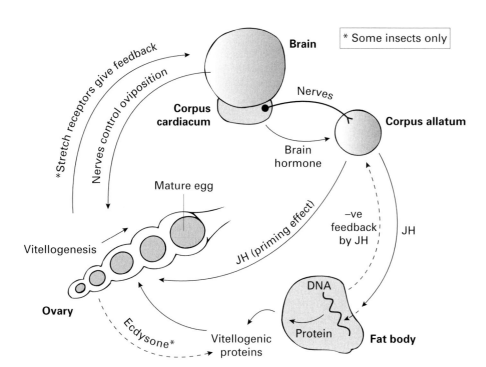

* Some insects only

Fig. 13.46 Key systems for control of reproduction in insects. JH, juvenile hormone.

through the placenta even at the expense of the mother's health. Within the eutherians, altricial young usually occur in taxa that have very short seasonal 'windows' for successful breeding, so that more than one litter can be squeezed into the time available (e.g. desert rodents) or in taxa that have dens or protected home-bases to return to (e.g. canids, primates, felids); precocial young are the rule in antelopes and many other large mammals that breed in the open.

13.5.2 *Reproductive control systems*

With land animals in general requiring more careful production and packaging of gametes, more complex mating and maternal behaviours, and much longer maturation times, there are inevitable consequences for the complexity of control of all stages of reproduction. The simple strategies of a gonadotropic hormone and a spawning hormone or pheromone in many aquatic animals become entirely inadequate; each stage of the reproductive process must be regulated both intrinsically and in relation to the changing environment to ensure that young are produced at a time when their survival chances are optimal. However, many groups have been very inadequately studied and most of our knowledge comes from the insects and the vertebrates (especially birds and mammals); the control systems in these two groups are summarized in Figs 13.46 and 13.47, respectively.

Control of gamete production

The controlling influence of gonadotropic hormones is one of relatively few cases where parallels between terrestrial and aquatic animals can be drawn. Most land animals have a female **gonadotropin**, and many also have a male one (though a few species are born in a relatively advanced state with eggs and sperm already maturing). Land animals use different cues, however; lunar triggers are rare, while a photoperiodic response is very common, often interacting with temperature or nutritional state, and sometimes with specific auditory, visual or even tactile cues from potential mates.

In insects the hormone controlling oogenesis is usually the multipurpose hormone from the corpora allata lying behind the brain; it is termed **juvenile hormone** (JH) because of its role in maturation (see below). Its primary role in reproduction is to initiate vitellogenesis (yolk production) in the ovarioles, and hence egg development. However, in some insects a hormone from the brain operates instead, under the influence of temperature, photoperiod or a specific feeding input (such as a blood meal in mosquitoes). The brain hormone stimulates release of ecdysone ('moulting hormone') from the prothoracic gland, and the ecdysone triggers vitellogenesis and egg maturation. This in turn sets in train an automatic sequence of events through eggshell production, mating and oviposition, probably involving secondary secretions from the ovariole tissues.

In most terrestrial vertebrates gametes need to be produced at precise times, either in a regular cycle or at specific points in the seasonal cycle. Gonadotropic hormones originate in the pituitary, and are released under the influence of small peptide releasing hormones from the hypothalamus. The triggers for production of the gonadotropin are very variable, depending on the timing of reproduction in each species. In annual breeders photoperiod is the most commonly used cue; many medium-

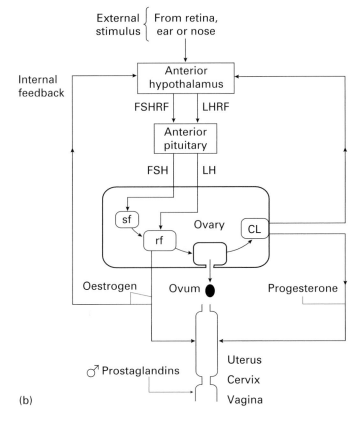

Fig. 13.47 Hormonal control of reproduction in female mammals; (a) the oestrous cycle and (b) ovarian function. CL, corpus luteum; FSH, follicle-stimulating hormone; FSHRF, FSH-releasing factor; LH, luteinizing hormone; LHRF, LH-releasing factor; PGF$_{2\alpha}$, prostaglandin F$_{2\alpha}$; sf, small follicle; rf, ripe follicle. (a, From Goldsworthy *et al.* 1981.)

However, in many social and herd mammals seasonal breeding is abandoned to avoid the production of many vulnerable young at the same time, so females become receptive at random; though in highly seasonal habitats food availability may force seasonal breeding even on herd animals.

The main gonadotropin in female vertebrates is **follicle-stimulating hormone** (FSH), which causes mobilization and growth of follicles within the ovary. In males both FSH and another hormone, **luteinizing hormone** (LH), are released. The gonadotropic hormones then regulate the release of further hormones (**androgens**) from the gonads, which in the female are **oestrogens** and which act to prepare the reproductive tract for further stages in the reproductive cycle (eggshell production, or placentation). In males the LH stimulates nutritive interstitial cells in the testes to produce **testosterone**, while FSH stimulates spermatogenesis.

Control of mating behaviour

In soft-bodied terrestrial animals mating behaviour tends to be limited, though we have seen that snails and slugs do have complex courtships, probably initiated at a distance by the scent of a potential partner and maintained by sense organs primarily in the tentacles. In insects mating behaviour is often triggered seasonally by hormones but is very much controlled by pheromones in most species that have been studied. The most important controls are female pheromones, often serving to attract a male from downwind (see later). Further short-range pheromones may be released by the female to arrest and guide an incoming male, sometimes eliciting particular male behaviours, and contact chemicals from the female and male genitalia may be required to ensure full copulatory success.

In birds and mammals, female mating behaviour (termed **oestrus** in mammals) is initiated by the effects of oestrogen on the central nervous system; it is normally a short period of intense sexual receptivity. In males circulating testosterone brings about changes in external genitalia and development of secondary characters such as breeding coloration or breeding plumage, and it may also stimulate specific territorial behaviours and mate-attracting behaviours, with visual components and/or acoustic song components. In addition pheromones come into play; these are often released in the male urine to attract a female, or vice versa, or both. Some of the complex behaviours and their controls in breeding birds are shown in Fig. 13.48.

Control of gamete release and fertilization

The stimulus for gamete release in insects is often indirect, because of the ability of many insects to store sperm and release them as required when an egg is mature and an oviposition site has been located. Many insects, notably dragonflies and fleas, have highly complex sperm storage organs, where the sperm donations of several different males may be kept. In some cases the last sperm to enter are the first to be used in fertilizations,

sized mammals rely on declining photoperiod in autumn to initiate gamete maturation and sexual activity, so that young are born in the spring as food becomes abundant, whereas large species with gestation lasting nearly a year commonly have phases of sexual activity in the spring for the same reason.

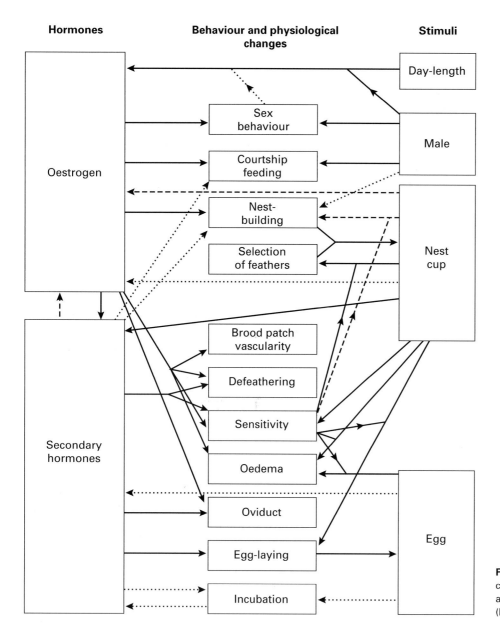

Fig. 13.48 Effects of hormones in controlling bird breeding behaviours and accompanying physiological changes. (From Hinde 1982.)

and in these species males may try to scoop existing sperm out of the female before inseminating her themselves. In other cases the first sperm to enter take precedence. But recent evidence from DNA fingerprinting suggests that certain female dragonflies may exercise some control over which sperm fertilize which eggs, and they will behave rather differently according to which sperm they have 'chosen'. Most of the control here is probably nervous rather than hormonal; stretch receptors may signal when a mature egg is present in the lower ovariole, and nerves will control the release of sperm from the spermatheca.

There is also some evidence of control of sperm release in the hermaphroditic land snails, and in genera such as *Helix* this may be related to the stimulus from the calcareous 'love darts' that each partner fires into the other.

In land vertebrates, ovulation is triggered when the level of circulating oestrogens interacts with the pituitary to affect the response to the 'releasing factor' hormones from the hypothalamus. This may be achieved cyclically by complex internal feedbacks as in humans, or may occur in response to an external stimulus such as scent or the sound of a male calling; in some cases the act of copulation itself triggers ovulation. Instead of continuing to produce FSH, the pituitary then produces a surge of LH. This acts on the ovary to release the ovum from its follicle (the remnants of which become the corpus luteum and take over production of the second major female gonadotropin, **progesterone**). One or more eggs are released into the fallopian tubes, and normally encounter spermatozoa from the male ejaculate in the upper portion of the uterus. In mammals, prostaglandin hormones in the male ejaculate may facilitate this by causing contractions in the uterine muscles.

Fertilization is controlled in a similar manner in most animals, but details at the molecular level are still somewhat unclear. Mammals are inevitably best known; here the spermatozoa, as many as 250 million ml^{-1} of seminal fluid, begin to approach the egg by dissolving the sticky cumulus cells that surround it, using hyaluronic acid derived from the sperm tip or acrosome. They then penetrate the zona pellucens and the vitellogenic membrane of the ovum itself. Normally only one sperm achieves complete penetration and fertilizes the ovum nucleus, all other sperm then being prevented from entry.

Internal fertilization poses some extra problems for land animals. Firstly, infective agents such as bacteria and fungi may gain access to the body through this route, encountering relatively unprotected surfaces. In higher vertebrates this problem is partly solved by the presence of specific commensal bacteria (*Lactobacillus vaginalis*) within the vagina that react with glycogen (common in cells in the walls of the reproductive tract, especially at the high levels of oestrogen preceding ovulation) to produce a highly acidic environment (pH 5 in humans), in which most foreign bacteria and fungi cannot grow. Seminal fluid therefore has to be well-buffered and alkaline (pH 7.2–7.8). A further problem is that immunologically foreign material is introduced deep within the body, where it may contact highly vascularized surfaces and will provoke an immune response. In mammals there is an exudation of neutrophil white cells from the uterine wall, which quickly agglutinate and begin to destroy the sperm; fertilization must therefore occur quickly and efficiently, and the male gamete of land animals has only a very short life outside its owner's body.

Control of egg maturation and development

Oviparity: shelled eggs and hatching

In insects and arachnids, hatching of the eggs (whether as nymphs, larvae or tiny adults) is usually highly attuned to environmental conditions. Where the eggs have overwintered or diapaused, temperature is the commonest cue for hatching, though cumulative 'day-degrees' above a particular temperature are often needed rather than a simple threshold effect. If the eggs have been laid in water, for example in dragonflies, mosquitoes and some hoverflies, rising temperature and decreasing oxygen levels together may serve as a cue.

After hatching, time to maturity (the final moult to adult form and/or size) is determined in insects by the interplay of two main hormones, JH and ecdysone. As the titre of JH in the haemolymph falls progressively, each moulting episode triggered by ecdysone results in a less juvenile morphology until finally the adult genes are fully expressed and the adult body architecture appears. However, there may also be maternal effects on maturation determined across generations via the physiology of the egg-laying adult, so that eggs laid as winter approaches have an inherently greater tendency to enter diapause (see Chapter 8).

Viviparous development

In mammals the initial control of fetal development is largely brought about by the gonadal androgens, oestrogen and progesterone. Oestrogen causes the uterine wall to prepare for implantation by the fertilized embryo, with thickening, mucus production and enhanced vascularization. Once fertilization has occurred the presence of the embryo brings about persistence of the corpus luteum, probably by blocking production or reception of a female prostaglandin hormone from the uterine wall. From this point, the fetus has therefore begun to exert some control over the maternal physiology, a process that will continue for the weeks or months of pregnancy. Progesterone is secreted throughout this period, either from corpora lutea in the ovary or from the uterine placental tissues themselves. Oestrogen is also present at low levels, and begins to function in preparing the mammary glands.

When the fetus reaches maturity, in most species it triggers its own birth with a spurt of corticotropic releasing hormone from its hypothalamus, triggering adrenocorticotropin release from its pituitary and thus cortisone release from its adrenal cortex. This affects enzyme activity in the placenta and brings about conversion of progesterones to oestrogens, and production of prostaglandin in the placenta, causing smooth muscle contraction and cervical dilation in the mother. The oestrogens in turn stimulate more prostaglandin production, giving a positive-feedback loop and a surge of hormones that initiates a release of oxytocin from the maternal pituitary, bringing about maximal uterine contractions and expulsion of the fetus. Thus pregnancy and birth involve a complex array of interactions between fetal and maternal hormones, with little intervention from environmental stimuli, though there may be induced abortions if environmental stresses become extreme.

Post-hatching care and maternal behaviour

Little is known about the control of parental care in most insects, but the exceptions which have been extensively studied are the social insects, particularly the honey-bees. Here pheromones released by the queen (usually the sole egg-layer) control the behaviour of all the other females, the sterile workers, who are full- or half-sisters of the new eggs. Workers construct and provision the egg cells, tend and clean the larvae, and gather provisions from outside the nest to maintain the young and the queen, all under the influence of 'queen substance'. In most other insects and in the arachnids exhibiting parental care it must be assumed that one or more of the fairly small number of arthropod hormones act to influence the female behaviour.

In mammals, birth (parturition) is followed by a rapid decline of both oestrogen and progesterone in the mother once the placenta is shed, and when the young then begin to suckle the physical stimuli promote a surge of prolactin-releasing hormone from the hypothalamus and then prolactin from the pituitary, which together with oxytocin initiates and maintains milk production and ejection.

Overall, we can see that reproductive problems on land are solved by increased investment of time or energy or both, by one or both parents; whether in finding a mate and achieving efficient sperm transfer, in protecting and provisioning the eggs, or in either preparing for or directly overseeing the postnatal needs of the juveniles. All of these changes necessitate a far more elaborate behavioural repertoire, and substantially greater sophistication of the neural and hormonal control systems.

13.6 Locomotion and mechanical adaptations

13.6.1 *Soft-bodied animals*

Soft-bodied animals, with essentially hydrostatic support systems, are at a serious disadvantage on land. They will have less support from the medium and a tendency to flatten under their own weight; thus they need relatively thicker muscle layers to maintain tone and shape compared to related aquatic forms. Their entire anatomy and functional integrity relies on maintained hydration, and any degree of desiccation will compromise their normal activity—resisting water loss is crucial. They will also suffer severely from the effects of friction in crawling over or burrowing through the relatively hard substrates prevalent on land, and may be damaged by such contacts; so they require more copious production of mucus as a lubricant and as a protectant, which entails extra water loss. Clearly this design will only work adequately for small animals in the humid microhabitats of soil or litter layers.

Most nematodes are so small (20–50 µm diameter) that they remain within the water films around soil particles; indeed, their design is such that they locomote most efficiently when within such films. Thus they cannot be considered terrestrial in terms of locomotion, any more than in terms of their physiological relations with their environment. However, land flatworms and nemertines are somewhat larger and do experience a less than saturated world, and have certain locomotory adaptations to ease their passage through it. Land flatworms move with contractile muscular waves, which are of longer wavelength than their aquatic kin, so that the body touches the ground only intermittently (Fig. 13.49a), leaving broad mucoid 'footprints' rather than a continuous trail. Some nemertines probably use the same trick of limited ventral contact points, but many of them have an additional 'looping' strategy to achieve fast escape movements: the proboscis is shot out ahead of the worm and its tip adheres to the ground, so that the rest of the body can be rapidly drawn up to it using longitudinal muscle contractions (Fig. 13.49c).

Earthworms are distinctly larger, and move with more obvious longitudinal and circular waves of contractions, accentuated by their pronounced segmentation (Fig. 13.49b). Their movements occur almost entirely within the soil, the expanded segments (plus chaetae) acting as anchors; fundamentally their locomotion differs little from that of marine annelids (Chapter 9). However, their near relatives the leeches (Hirudinea) are often found moving about well away from water in tropical forests, using the suckers to achieve a looping motion (Fig. 13.49c) that reduces contact with the substrate, thus reducing frictional costs and potential damage.

Gastropod molluscs are 'soft-bodied' in terms of locomotion (and remember that many of them spend a large part of their life underground and burrowing, with the shell reduced and flattened to give a more worm-like profile). The gastropod foot operates in essentially the same way as the body of a flatworm. Here the properties of the mucus are critical, as it undergoes rapid phase changes that alternately stick the foot to the substrate and allow it to slide freely along. The wave-like motion of the foot can again be modified to limit ground contact, though; terrestrial snails and slugs often leave a discontinuous 'footprint' trail indicating the reduced points of contact, and in the genus *Pomatia* this trend is taken one stage further with the foot divided longitudinally and the right and left halves taking alternate 'steps' forward, a phenomenon sometimes optimistically termed the 'galloping' gait of snails!

In evolutionary terms the real innovators in soft-bodied land locomotion have been the tropical onychophorans (velvet worms), with the body wall protruded into many pairs of stumpy legs. The body operates essentially hydrostatically, undergoing tremendous shape changes (up to 10-fold variation in cross-sectional area), and dehydration causes loss of tone and a disrupted gait. But the legs act as 'real' limbs: protrusions and retractions due to extrinsic leg muscles give a genuine lever action and greatly reduce frictional contact with the ground.

13.6.2 *Arthropodan systems*

First of all, remember that a great many land arthropods spend most of their life cycle as soft larval stages (the maggots and caterpillars of many endopterygote insects), often completely lacking legs, and moving very similarly to the hydrostatic animals already discussed. Peristaltic muscular waves drive the haemolymph forward to dilate the anterior end and penetrate the substratum, be it soil or the tissues of another animal or of a plant. The cuticle is usually completely unsclerotized, excepting only the head capsule in forms such as soil beetles. Only in the adults of such insects do we find innovative kinds of land locomotion.

Walking and running

Fully functioning legs are present in myriapods, arachnids, juvenile and adult exopterygote insects, and adult endopterygotes. In all of these, the same principles are used for crawling and walking as their aquatic ancestors used for crawling and swimming. They exploit to the full the merits of a cuticle that can be secreted in a rather fluid state and then 'set' (with very variable

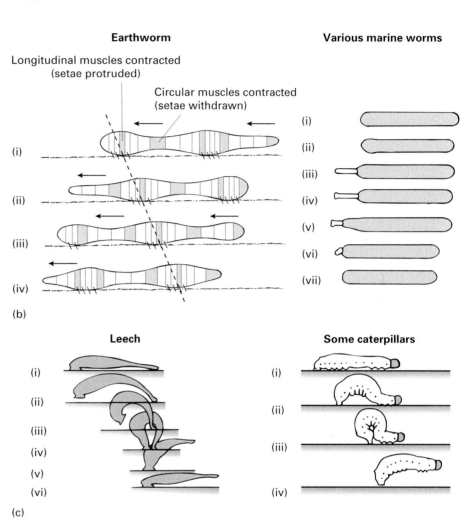

Fig. 13.49 (a) Ciliary gliding, (b) peristaltic waves and (c) looping as ways of moving on land for soft-bodied animals.

degrees of hardness) into any form that is needed. The evolution of a proper leg, with flexors and extensors at each joint, brings immediate advantages on land (Table 13.14). For arthropods, the particular benefits include a lever action, reduced frictional contact with the substrate, and a smooth forward action for the main body mass with only the legs themselves alternately accelerating and decelerating. Efficient locomotion is therefore achieved with reduced number of legs, each containing minimal weight. The most agile land arthropods have only five pairs (decapod crustaceans) or four pairs (arachnids) or three pairs (insects) of legs; and in many decapods and arachnids one pair of legs is not used in the fastest gaits. Muscles are moved to the top of the limbs, with only the attached apodemes (the

arthropod equivalent of tendons) stretching down into the lower limbs (Fig. 13.50). Thus the major changes in locomotion in walking land arthropods as compared with aquatic arthropods are reduced numbers of legs engaged in locomotion, and altered proportions of parts of the legs and leg muscles; there are few fundamental changes in actual mechanisms.

Speeds in land arthropods can be substantially higher than in marine species, largely due to the reduced viscosity of the medium. Land crabs can achieve 1–2 m s^{-1}, with only eight of the ten legs in use and with only three in contact with the ground at any time to form a tripod support, where two legs are thrusting and the third is merely providing balance. Centipedes such as *Scutigera* can also run fast, up to about 0.5 m s^{-1}, each leg having

Table 13.14 Advantages of legs (contrasted with soft-bodied locomotion).

Lever action—small movement of muscle gives large movement at limb-tip

Limited ground contact—reduced friction

Legs stop and start, body moves forward smoothly—reduced acceleration and deceleration of large masses, energy costs reduced. Aided by having main mass of leg (muscle, etc.) at the top, close to body

Muscles can be small instead of sheet-like, so with less connective tissue strapping and increasing speed of contraction

No lateral sinusoidal components that waste energy

Largely independent of hydration state

Effects of muscle contraction are localized, do not affect other body-wall muscles

Increases number of gaits and gait/speed/energy trade-offs

Permit good proprioception

Legs can be diversified for other uses

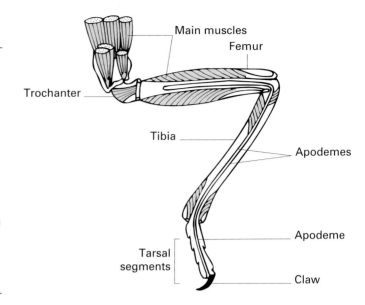

Fig. 13.50 Design of the basic arthropodan leg with muscles confined largely to the coxa and femur and linked to apodemes extending distally. Each element is stronger (per unit mass) and lighter in a terrestrial arthropod than in a similarly sized marine arthropod.

a quick backstroke and a slow recovery phase. These faster speeds tend to be enhanced by longer limbs, but to avoid these tripping over each other their lengths are staggered (Fig. 13.51).

Equally, the forward force of a land arthropod can exceed that of a marine species of the same size; millipedes in particular are renowned for their 'pushing power', moving slowly but very forcefully when burrowing, and often described as 'bulldozing' in their action, with the blunt head and shield-like collar region

thrusting into the soil. Burrowing beetles, with heavy limbs, a narrow anterior and a very heavily sclerotized cuticle, achieve similar forces.

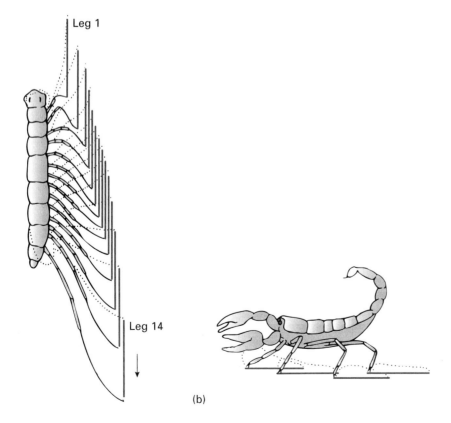

Fig. 13.51 (a) Top view of the gait and stance of a running centipede and (b) side view of a scorpion, both organized such that multiple pairs of legs are different lengths and do not step on each other. Solid bars show stride length; dotted lines show leg positions at end of stride. (Adapted from Manton 1952; Herreid & Fourtner 1981.)

Fig. 13.52 Jumping in a grasshopper and a flea; in both cases the jump is preceded by contraction of muscles which store energy in an elastic cuticular hinge (containing resilin) at the 'knee', which is then explosively released.

Jumping

More significant changes in design and control of locomotory systems occur in arthropods that jump, a rather more radical solution to reducing friction with the substrate. A variety of jumping (saltatory) systems have arisen that are unlike anything seen in aquatic species; they are favoured because they generally use energy-storing systems that can work even at cold temperatures, allowing escape from fast-moving (possibly endothermic) predators. Some spiders are the exception, jumping effectively using a rapid hydrostatic extension of the fourth pair of legs. Collembolans (springtails) are apterygote insects that jump using a forked catapult-like springing organ (furca) on the ventral surface of their abdomen. Fleas and many orthopterans can jump to relatively enormous heights, using energy storage systems in the 'knee' joints of the specialized metathoracic legs, where a rubber-like cuticle protein (resilin) can store energy from muscle contraction relatively slowly and then release it very suddenly to get remarkable acceleration (Fig. 13.52). The elastic stores of a locust leg weigh only about 4 mg but can store and release the energy of a 70 mg muscle. (Note that some marine mantis and pistol shrimps also use energy-storing proteins, but to 'punch' at prey rather than to jump.)

Flying

Flight is restricted to the pterygote insects among the arthropods (though aerial locomotion is achieved by parachuting on silken threads in spiderlings). The wings appear to have evolved initially for purposes other than flight; perhaps as stabilizers, or as areas of increased surface area to aid in thermoregulation, or possibly as 'sails' to assist propulsion over water surface films. The initially stubby wing-buds may have expanded enough to become useful in gliding, and only then became jointed to the thorax and capable of being flapped. Once evolved, and making full use of the advantages of the arthropod chitinous cuticle as a light but immensely strong material, wings became a huge part of the insect success story from the Carboniferous onwards, giving access to new habitats within the canopy of the radiating angiosperm forests and allowing insects to disperse at speeds of up to 10 m s^{-1}. Potentially this allows them to speciate ever more quickly.

Insect flight primitively involved two pairs of large wings with a complex net-like venation, veins being formed where the two apposed cuticular surfaces (with their epidermal cells jettisoned at the moult) are separated by thickened channels carrying haemolymph, nerves and tracheae. These wings have tended to become smaller and/or more easily folded away, allowing winged insects to climb through vegetation, and in some cases to burrow. The wings have also tended to become coupled so that the two wings on each side of the body operate together, efficiently supported by just a few strut-like veins, especially along the leading edge. Strong terrestrial fliers therefore have rather rigid short wings, with mainly longitudinal veins, levering against a dorsal process on the insect thorax like an off-centre see-saw. Two different systems of wing control are found in modern insects (Fig. 13.53). In some, such as dragonflies and cockroaches, the downstroke is produced by **direct muscles** pulling against the flexible and elastic wing-base. But in more advanced orders **indirect muscles** pull on the walls of the thorax instead. The side walls of the thorax are strengthened with cuticle ridges, while the dorsal 'roof' (tergum) is quite flexible. Downward movement of the wings occurs when the longitudinal muscles in the thorax contract, causing the thoracic roof to distort upwards and force the wings down. The upstroke occurs when the vertical dorso-ventral muscles contract and pull the tergum downwards, flicking the wings back up against the fulcrum of the pleural process. Only the up or down positions tend to be stable, and the wing hinge area is again very elastic with a high proportion of resilin in the cuticle, so that the wing clicks very rapidly between its stable states and much of the energy used to power the upstroke is recycled into the next downstroke. In these fliers, including bees, flies and many beetles, the two sets of indirect flight muscle take up most of the volume of the thorax; flight muscle may be 60% of total mass in males of the dragonfly *Plathemis lydia*. These highly developed wing muscles are the main source of internal heat generation in the active insect (see section 13.3), so that locomotory mode is strongly interlinked with other aspects of insect physiology.

Wings must do more than move up and down. They must also

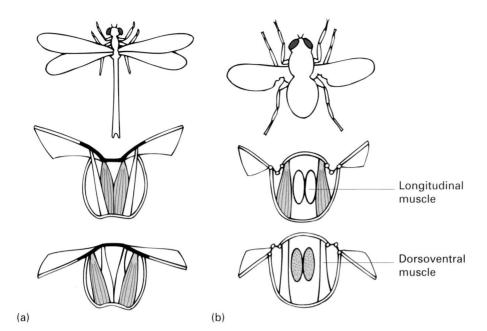

(a)　　　　　　　　　(b)

Longitudinal muscle

Dorsoventral muscle

Fig. 13.53 The two alternative wing muscle systems in insects: (a) direct and (b) indirect.

be moved at an angle (pitch) to give forward thrust, so that a complete wing cycle involves the wing-tip performing an elliptical or figure-of-eight motion. The wing angles and contours must be varied to achieve directional control. These processes are controlled largely automatically in many flying insects by the mechanics of the wing hinge itself, with minor adjustments by smaller direct muscles acting directly on the complex sclerites of the wing-base.

Despite all this, using conventional aerodynamic calculations, many insects should not be able to generate enough lift to fly. They must have ways of generating further downforce. Part of the answer is that they can use the vortices of moving air around the wing; the most recent evidence invokes the idea of 'dynamic stall', showing that insects use their wings with a large angle of attack so that the leading edge vortex (Fig. 13.54a) is prevented from spinning away from the wing and instead helps to draw air faster over the upper side of the wing giving massive lift. Some very small insects also use a 'clap, fling and peel' system, whereby the two wings clap together in the midline at the top of the upstroke and then fling and peel away from each other so that air rushes between them and sets up a new vortex of air circulating around and under each wing (Fig. 13.54b).

Insect flight involves wingbeat frequencies varying from around 4 cycles per second (cps or Hz) in large-winged butterflies through 100–200 Hz in bees and medium-sized flies, to above 1000 Hz in tiny midges. In species that flap their wings at relatively low frequencies each cycle is **neurogenic**, controlled directly by one or more nerve impulses, and the muscle contracts synchronously. For insects with wingbeat frequencies above about 25–30 Hz there is less than one action potential to one contraction; action potentials from the nerves are needed to start the muscle contracting, but once started it continues to oscillate at high frequency. The muscle becomes **myogenic**,

each contraction resulting automatically from the stretching of a muscle as its antagonist contracts ('shortening deactivation' and 'stretch activation'). There is no direct correspondence between nerve and muscle activity, so the muscle is termed asynchronous. This allows very fast wingbeats, facilitated by the lever system (so that a very small decrease in muscle length produces a large movement at the wing-tip) and by the elasticity of the thoracic skeleton. Some flight muscles contract by only 1–2% of their own length, calculated as the equivalent of just one actin-myosin cross-bridge forming, at the limits of what can be achieved with normal biological materials. These muscles may also use fuel exceptionally quickly, usually being supplied with haemolymph containing 150 mM trehalose (equivalent to 300 mM glucose) to produce the massive inward gradient required to keep the system going. Fast wingbeats allow rapid and forceful flight, with a honey-bee able to cruise at 15 mph (around 7 m s^{-1}). Manoeuvrability can also be exceptional, with flies able to hover, fly upside down and backwards, and turn within one body length.

13.6.3 Vertebrate locomotion

Walking and running

Proper 'walking' appeared rather gradually amongst the first land vertebrates. The early labyrinthodont amphibians used a fish-like belly crawl based on the trunk muscles, only slightly augmented by a pulling action from the front limbs and hardly affected by rear-limb action. As skeletal elements fused to form the pectoral and pelvic girdles, each forming strong links to the limbs and spine and becoming freed anteriorly from the back of the skull, so the force provided by the short amphibian limbs could increase, and they became rotated somewhat inwards

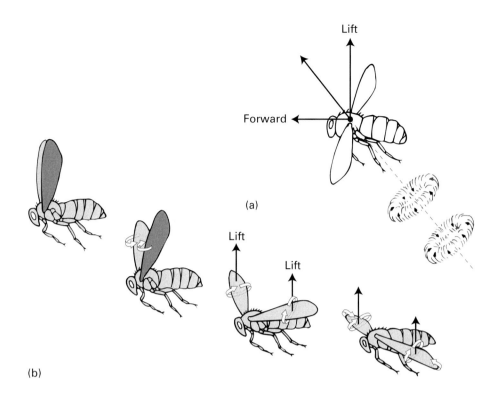

Fig. 13.54 (a) The spinning vortex of air that is shed from an insect in flight, due to the air movements over the wing surfaces, which then helps to give it lift. (b) In small insects the wings clap together on the upstroke and when peeled apart create an extra vortex effect.

to lie below the animal rather than sprawled out to the side (Fig. 13.55). Modern salamanders still use sinusoidal movements of the body to complement their limb action, but in most other tetrapods, the limbs have become greatly modified to support the body well off the ground, giving all the advantages of limbs discussed above. Inevitably there have been concomitant increases in speed, agility and manoeuvrability in three dimensions.

Walking and running therefore become the primary locomotory modes for land vertebrates. Most tetrapods have a slow gait and a fast gait. On four legs the common **gaits** are usually walking (slow) and running (fast), but for animals larger than cats there may be one or more additional gaits, usually termed the trot and (for horses and similar animals) the canter, with the fastest gait then termed galloping. For bipedal animals there is a slow walk and either a hop or a run to go faster.

Quadripedal walking maintains each foot on the ground for more than half the time so that there is always at least two-footed contact with the ground, whereas faster (cursorial) gaits allow successively less prolonged contact until in the gallop all four feet may be off the ground at once and speeds of up to 25–29 m s^{-1} are possible. The most important point about

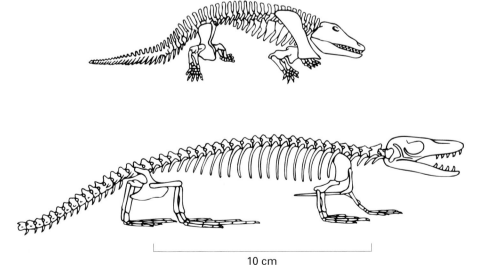

Fig. 13.55 The sprawled gait of early land vertebrates (here a labyrinthodont amphibian from the Permian) contrasted with the upright gait of most modern forms, the legs rotating to lie under the body, holding the belly aloft with much reduced friction.

10 cm

Fig. 13.56 (a) The cursorial leg of a horse, and (b) the three gaits that it can produce at different speeds each with its own optimum speed for minimal oxygen consumption. In (c) we see that a horse in practice shows an almost linear relation between speed and oxygen use, since it switches gait appropriately. (a, From Pough *et al.* 1999; b & c, adapted from Hoyt & Taylor 1981.)

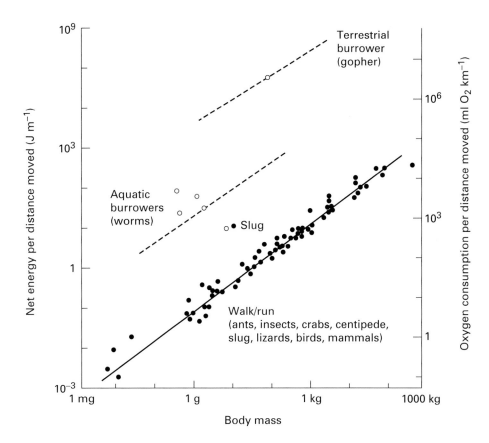

Fig. 13.57 Cost of terrestrial burrowing is substantially higher than either aquatic burrowing in softer substrates, or terrestrial walking and running. (Adapted from Alexander 1982.)

changing gaits is the energy saving incurred; Fig. 13.56b shows that each has its own optimal speed for economy in terms of energy per distance moved, and many vertebrates do in practice use each at the speed where it offers greatest savings. Faster locomotion is aided by longer legs and hence a long stride length, with the leg operating at close to the natural oscillation frequency of a similar length pendulum. As in arthropods, the legs are thin with most of the muscle mass concentrated near the trunk and with tendons and elastic ligaments running down from each muscle (Fig. 13.56a). Speed is also improved by progressively reducing the area of foot contact with the ground so that the posture becomes unguligrade (weight borne on a hoof covering just one or two toes) instead of the primitive plantigrade state (weight borne on the whole foot flat on the ground).

Burrowing and digging in soil or litter remain important locomotory modes even for quite large vertebrates, with some toads, lizards, birds and many mammals dwelling underground in burrows they have excavated. Figure 13.57 shows that terrestrial burrowing is an expensive occupation, but of course it has huge benefits in terms of hygrothermal physiology and concealment. In amphibians and reptiles a subterranean or litter-centred life may be accompanied by loss of limbs, as in salamanders, caecilians, skinks and amphisbaenians. However, birds and mammals always retain girdles and limbs and use the limbs as the digging tools. Life may be permanently centred on a burrow system, with marmots, mole rats, moles and badgers

exemplifying the required adaptations. Forelimbs are generally short and very heavy with stout claws, heads are broad and rounded with the nostrils well protected, and prominent incisor teeth may play a considerable role in burrow excavation.

Jumping

Jumping has evolved several times in vertebrates. It is used as a continuous gait in anuran amphibians, in many birds when on the ground, in many small rodents, and in some marsupial mammals. Moreover, intermittent leaping between trees is common in many primates, and jumping is employed as a transient escape or attack response in many other taxa. Above all, jumping to any significant height again requires speed at take-off, which in vertebrates involves a very rapid straightening of the 'knee' joints, usually with a preparatory bending of the legs, and built-in elasticity.

Anurans are spectacularly specialized for jumping (Fig. 13.58), with several key features:
• long back legs, with an elongated ilium bone in the pelvis plus elongate tarsals in the foot effectively adding two joints to the hind leg for the actual jump;
• the second elements in both limbs fused, as radio-ulna and tibio-fibula, giving greater strength and reduced twisting;
• loss of the tail, since a symmetrical hind kick leaves the tail with reduced function as a stabilizer as in walkers and runners;

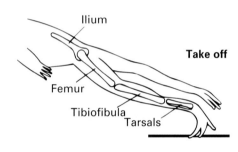

Fig. 13.58 The highly modified skeleton of a frog, showing specializations for jumping.

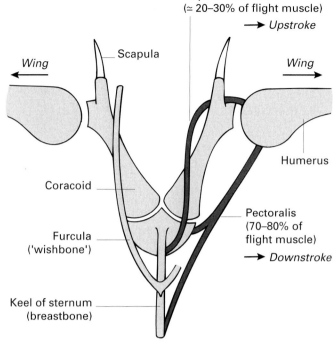

Fig. 13.59 The major flight muscles in a bird, pulling on the sternum and upper forelimb; the supracoracoideus provides the upstroke by acting over a pulley formed by the coracoid and scapula bones.

• a shortened backbone, comprising only about eight vertebrae, with the last few fused as the rod-like urostyle;
• a shoulder girdle modified as a partial shock-absorber.

Jumping in anurans may have evolved primarily to assist escape back to water, since nearly all anurans tend to jump towards higher humidity whenever startled. The Goliath frog of west Africa can jump 3 m, though smaller frogs do proportionately better, possibly using a catapult mechanism to amplify muscle power production. For example, the Cuban treefrog, *Osteopilus septentrionalis*, weighing around 14 g, is able to jump 1.4 m, and calculations show that the power required for take-off is more than 800 W kg^{-1} muscle, with peak power production perhaps twice this, exceeding the maximum output possible from the hindlimb muscles by about seven-fold; thus a substantial part of the work must be performed before take-off and stored in elastic mechanisms. Some treefrogs can jump and then glide 15–18 m, having flaps between fingers and toes and along the body, and sucker-like tips to their toes for safe landing.

Flying

As far as vertebrates are concerned, flying in the strict sense occurs only in birds and in bats (and in extinct pterodactyls), but certain frogs, reptiles and mammals achieve an alternative kind of aerial jumping/gliding. Wing surfaces have been produced in a variety of ways; using membranes between the fingers (bats), or between forelimbs and hindlimbs (flying squirrels, etc.), or feathered extensions from the longbones of the arms (birds). Flight muscles are modifications of the thoracic/forelimb muscles, and in birds two particular muscles become hugely dominant (Fig. 13.59), with the larger pectoralis muscle powering the downstroke and the smaller supracoracoideus muscle producing the recovery upstroke, the two together being up to 35% of the body mass in a strong flier. Both muscles insert on the enlarged keel of the breastbone, and run to the humerus bone in the upper 'arm', the upstroke muscle running over a pulley arrangement so that it can lift the wing. In strong fliers these muscles operate aerobically, with a high capillarity and dense mitochondrial packing, giving red muscle. In birds that fly only as a transient escape response, where stamina is unimportant, the flight muscles are predominantly anaerobic and appear as white muscle.

Flapping flight is extremely expensive, especially for small birds and bats (though recent evidence indicates that it may be somewhat more economical in bats), but of course this is offset by the advantage of an extremely economical means of transportation per unit of mass for a given distance (see Chapter 3). In relatively few very small birds (especially hummingbirds) even more expensive hovering flight is used, normally with fast wingbeats at the resonant frequencies of the wing/thorax system (note the parallel with the insect thorax, using elastic rebound in flight). It seems that when hovering in the cold some hummingbirds can force this system to operate at different frequencies and use the extra heat generated as part of their thermoregulatory system.

Gliding flight is much more economical, using a slow muscle to hold the wing in position; it is widely used for soaring in rising thermals or along the air currents over slopes of land or waves, and can permit cheap travel over great distances, as in petrels and albatrosses.

13.7 Sensory adaptations

Changes in motility and agility, and in the mode and speed of locomotion, have inevitably entailed marked adaptation in the senses and coordinating systems in all the land animals, especially the arthropods and vertebrates.

13.7.1 Mechanoreception and hearing

A sense of touch via mechanoreceptors works well in both

water and air, and mechanoreceptors are not much modified on land as a result. However, they may become more numerous, or concentrated on protrusions around vulnerable areas, to guard against the more forceful impacts that can occur on land. Mammals improve their sense of touch with specific vibrissae (whiskers), particularly around the muzzle and lower legs; arthropod feet and antennae become crowded with mechanoreceptors.

Vibration signals at low frequency work well in aquatic media, and can be used for very long-range signalling (as in the whales); but true aquatic 'hearing' is rare, and most animals use only a localized sense of vibration. In contrast, land animals continue to use a vibrational sense, but also routinely use the sense of hearing as part of their main detection and signalling repertoire, with high-frequency airborne sounds both produced and perceived. Vibrations through the ground are important to many snakes, and various arthropods (e.g. scorpions) also have ventral receptors that pick up vibration through the sand. Land crabs, such as *Uca*, signal by tapping the sand with their claws. However, the related crab *Ocypode* can stridulate using ridges and tubercles on the limbs, producing sounds through the air that attract females to the mating-burrows (see above). Similar systems occur in many insect groups, with crickets and cicadas particularly noted for their species-specific 'calling' songs, produced by serrated areas of the legs, abdomen or wings and often amplified by air-filled chambers or even by constructed resonant burrows (e.g. in the mole crickets). Detection in both crabs and insects usually involves chordotonal organs on the legs, sometimes incorporated in ear-like structures on the legs or abdomen (Fig. 13.60).

Sound production amongst the vertebrates usually involves internalized organs. The calling notes of frogs originate from the

Fig. 13.60 (a) The 'ear' in the locust, and the chordotonal organs that act as sound receptors transducing the soundwave-induced movements of the tympanum into nerve impulses. (From Withers 1992; adapted from Oldfield 1985.) (b) Cuticular sound-producing structures in a cricket. Arrows show movements of the two wings folded over each other above the cricket's back.

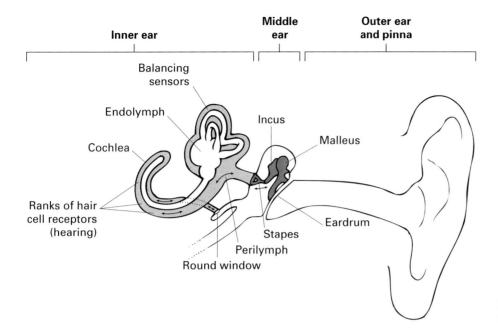

Fig. 13.61 Basic structure of the mammalian ear.

lungs, and are often amplified by throat-sacs; the larynx at the top of the windpipe has a pair of vocal cords and accessory cartilages, to produce the actual noise. From this kind of arrangement also come the intricate songs of passerine birds, and the squeaks, honks, trumpetings (and ultimately speech) of mammals. This is accompanied by increasing sophistication of the ear apparatus in the land tetrapods, with progressive recruitment of 'spare' gill-arch bones and jaw bones into the ear as sound-transmitting ossicles, and extension of the cochlear system of frequency-tuned hair-cell receptors (Fig. 13.61). Sensitivity is improved by variation in the size ratio of the eardrum and the internal oval window; this ratio is commonly 10–25 but reaches 40 in owls, where nocturnal hunting requires acute hearing. Some birds can also hear extremely low-frequency noises (infrasound), which may be used in navigation in pigeons and other species; and elephants are now known to communicate over long distances using infrasound.

13.7.2 *Vision*

Visual communication is more effective over long distances in air than it is in water, light travelling much further without attenuation by absorption. So again it is no surprise that visual signals become more complex and the eyes of land animals more sophisticated. Many land crustaceans respond to visual stimuli over a range of several metres, where their aquatic relatives rarely see more than half a metre; land crabs hunt largely by visual cues whereas aquatic crabs use chemical and tactile cues. Land animals tend to respond to much higher-frequency visual changes, with particularly high 'flicker fusion frequencies' (the frequency above which a photoreceptor cannot detect separate flashes and see a flickering source as a steady light) in the com-

pound eyes of some day-flying insects. Land animals (some insects and some birds, at least) are also able to use the plane of polarization of light as a cue to direction, which very few aquatic animals can do. Most obviously, visual receptors can only respond to those wavelengths that reach an animal's surface unattenuated, so that in the atmosphere there is a wider range of spectral sensitivity than in the hydrosphere and 'colour vision' becomes a key component of many animals' sensory repertoire, with varying pigments in (or overlying) the photoreceptors giving a wide range of possible spectral sensitivities (Fig. 13.62). A few land animals produce and respond to bioluminescence, notably fireflies and glow-worms, and also such unlikely creatures as earthworms; but on the whole it is less common than in the seas.

13.7.3 *Chemical senses*

Chemical communication also becomes more dominant in land animals, with pheromones playing an important part in the lives of most insects and many vertebrates. Chemical signals have relatively poor resolution and their transmission may be rather slow (depending on windspeed), but they operate over considerable range, by day or night, and are not impeded by intervening physical obstacles in the way that vision is. They can also be deposited on soils or vegetation or on other organisms (e.g. a mated female, an already occupied host, or a recently visited food source) where they will persist as markers; and they can potentially form a 'trail', whereas in water a mark or trail would be quickly dissolved or disrupted. Pheromones are best known and best studied in insects, where the receptors on the antennae are very accessible and we can record from individual neurones, allowing detection of thresholds for reception

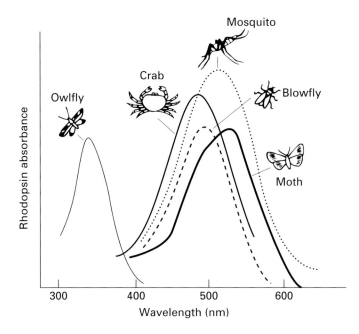

Fig. 13.62 Rhodopsin spectral sensitivities in a range of terrestrial insects and crabs. In man the peak sensitivity is at 498 nm, with cones tuned to 440, 535 and 575 nm. (Data from Hoglund *et al.* 1973.)

Fig. 13.63 Structure of a pheromone plume at different wind speeds; the plume remains coherent in fairly still air and can then travel up to 5000 m. (From Wilson 1970.)

Fig. 13.64 (a) The structure of bombykol, the sex pheromone of the silkmoth; and (b) the three components used in pheromone mixes in bark beetles in the genus *Ips*, each species using a different mixture.

and for response, and recording of very stereotyped behavioural responses.

Pheromones in terrestrial environments have to be picked out from amongst a complex chemical environment, and they have to give unambiguous signals to the receiver. Some degree of specificity may be imparted by the timing of release of the chemical, with many species only emitting their mating pheromones at particular times of day (and/or of season) in still weather where the odour plume will remain reasonably intact and directional (Fig. 13.63), so that male moths and beetles may respond to their females' signals from many hundreds of metres away. But real specificity and non-ambiguity require a novel and volatile chemical. The first of these to be identified was 'bombykol', a complex alcohol, from the female silk moth *Bombyx mori* (Fig. 13.64a). In fact bombykol alone has turned out to be relatively ineffective, and the natural pheromone is a 10 : 1 mixture of bombykol and its aldehyde form, bombykal; only this mix will produce the proper male response. Most pheromones have turned out to be similar kinds of small organic volatiles (acids, aldehydes or ketones), commonly C_5–C_{20}. Most are also multicomponent, like the *B. mori* pheromone, often involving quite a small range of similar chemicals; for example, the oak-roller moth, *Archips*, has an attractant pheromone consisting of 21 components, 17 of which are effective individually, though less so than the complete mixture. Often the species specificity resides in the exact mix of volatile components; in the bark beetle genus *Ips* there are three main components (Fig. 13.64b) of the mating pheromone, with the proportions varying in the different species.

Table 13.15 Functions of pheromones in land animals.

Activity	Specific function
Mating and reproduction	Long-range attraction
	Short-range arrestment
	Courtship behaviour maintenance
	Copulation induction/maintenance
	Postnatal care elicitation
	Fertility suppression (S)
	Nesting behaviour induction (S)
	Larval settling
	Cycle synchronization
Feeding and foraging	Food-source marking (S)
	Trail marking (S)
	Aggregation
Alarm and defence	Conspecific alarm (S)
	Fight induction (S)
Social organization	Hierarchy maintenance (S)
	Caste recognition (S)
	Colony recognition (S)
	Territorial marking
	Dominance marking

S, mainly or solely in social insects, i.e. ants, wasps, bees.

Table 13.16 Frequencies of echolocation sounds in land animals.

Group/species	Frequency (kHz)
Cave-dwelling birds	
Swiftlet	4–8
Oilbird	2–7
Nocturnal insectivores	
Tenrec (*Centetes*)	5–17
Shrew (*Sorex*)	18–60
Shrew (*Crocidura*)	70–110
Nocturnal bats	
Fruit bat (*Rousettus*)	13–40
Greater horseshoe (*Rhinolophus*)	35–90
Mouse-eared (*Myotis*)	25–100
False vampire (*Megaderma*)	30–125
Painted (*Kerivoula*)	230–243

Pheromones in mammals may also achieve specificity from mixing, perhaps even to the level of individual recognition in the scent marks left by territorial species (in these highly olfactory vertebrates the surface of the much-folded nasal olfactory epithelium may be greater than the entire body surface!). These kinds of terrestrial chemical communication systems can therefore apparently evolve and 'speciate' quite rapidly, allowing precise recognition on the basis of relatively few separate small chemicals. Table 13.15 summarizes the functions which pheromones can serve, stressing the prevalence of marking and trail systems.

13.7.4 *Other senses*

As might be expected, land animals generally lack electrical senses as the conductivity of the medium is too low to allow the passage of signals, whether for detecting disturbances of field due to nearby objects (electroreception) or for delivering defensive discharges (electrogenesis). The electrocyte system of fish apparently disappeared very quickly in the first land amphibians (crossopterygians), although use of electrical sensitivity does remain in aquatic vertebrate lineages right up to mammals such as the duck-billed platypus.

Echolocation does have a place on land though, occurring in bats and some shrews active at night and in a few birds that live in caves (Table 13.16), either for prey detection or to avoid obstacles. Bats emit clicks either from their larynx or with their tongues, through the mouth or via an elaborately folded nose structure. They receive the echo with greatly enlarged ear

pinnae, and their ears are highly frequency-tuned. Sounds must be high energy when emitted (up to 120 dB) but on receipt are attenuated down to just a few decibels; how then can a bat avoid deafening itself and still hear the echo? The answer lies with contractions of muscles in the middle ear that dampen the tympanic membrane just as a click is emitted, relaxation occurring exactly at the end of the click so that full sensitivity is restored in time to hear the echo.

One other sense that occurs in a few land animals is the detection of heat; in a sense this is a special case of vision, as it involves picking up wavelengths (infrared radiation) that are outside the normally visible range. It is impractical in water, but useful on land where marked temperature gradients can occur on a small scale. The best known cases are the heat sensors on the heads of pit vipers, though of course many animals have thermal sensors in their skin that will detect and give warning of dangerous heat sources before contact is made.

13.7.5 *Sensory coordination*

Given all these sensory modalities, and their very considerable complexity in many land animals where elaborate intraspecific signalling systems have evolved, it is inevitable that the sensory coordination systems have also tended to become more intricate than those of aquatic animals. Although taxonomic effects are also at work, this increased complexity is clearly expressed in increasing overall brain size (Fig. 13.65a), with particular growth of areas that integrate and pass on the information from the dominant senses (Fig. 13.65b). The antennary lobes of arthropods and the olfactory bulbs of vertebrates are enlarged in terrestrial taxa, as are the optic lobes (and in higher vertebrates the visual cortex). The telencephalon at the front of the brain is the major sensory integrating centre in lower vertebrates, and enlarges markedly to reach its zenith in birds, where it dominates the brain (along with the cerebellum, controlling

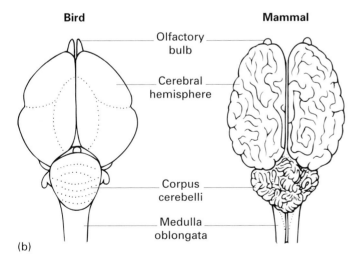

Fig. 13.65 (a) Scaling of brain size with body size for different vertebrate taxa; all show an exponent of about 0.67, but birds and mammals have relatively much larger brains. (Adapted from Jerison 1970.) (b) The increasing size of sensory areas in terrestrial groups compared with fish (with the cerebrum and cerebellum also becoming dominant in the endothermic groups). (From Pough *et al.* 1999.)

motor performance and stability and therefore linked with the problems of flight and balance). Only in mammals is this trend overridden, with the massive growth of the neopallium or cortex dominating the other integrating centres. In birds and mammals, operating at the faster sensory and motor pace demanded by an endothermic lifestyle, the brain is roughly 15 times larger than in an equivalent ectothermic fish, amphibian or reptile (though the main influence here is probably taxonomic rather than habitat-related).

13.8 Feeding and being fed on

In the introduction to this chapter we pointed out that, in comparison with watery habitats, on land the plants are tougher and better defended, while animals are faster. Also the supply of aerial 'plankton' is limited and cannot support many filter-feeding animals (really only the web-building spiders). Land animals can therefore be detritivores, herbivores or active predators, or some omnivorous combination of these; or they can be parasites, though they may perhaps then cease to be really terrestrial (see Chapter 15). In fact, many of the major terrestrial groups are fundamentally predatory, including the flatworms, nemertines, leeches, crabs, centipedes, most arachnid and many insect orders, onychophorans, amphibians and reptiles. Set against this, earthworms, gastropods, amphipod and isopod crustaceans, millipedes, many acarines (mites) and many insect orders are essentially herbivorous. Birds and mammals probably had carnivorous (insectivorous) origins but have diversified into other modes. Indeed amongst the insects, birds and mammals in particular, a huge diversity of trophic roles have evolved, often in concert with radiations in other groups such as the angiosperms. Referring back to Table 13.4, it is noteworthy (though not very surprising) that the most speciose groups are precisely those that have been able to diversify into the whole range of trophic niches rather than sticking to their ancestral feeding mode.

13.8.1 *Herbivores*

Terrestrial plants are not very easy things to eat. Their support systems make them inherently tough, hard to penetrate and hard to digest. These support systems are formed from the carbohydrate cellulose, and the complex and variable chemical lignin, based on a phenyl-propane polymer; Table 13.17 shows

Table 13.17 Fibre contents of different foodstuffs (as % dry matter).

Plant	Cellulose	Lignin
Potato	1.8	0.2
Apple	4.4	0.4
Carrot	8.7	0.3
Wheat bran	9	4
Lettuce	14	2
Alfalfa	27	8

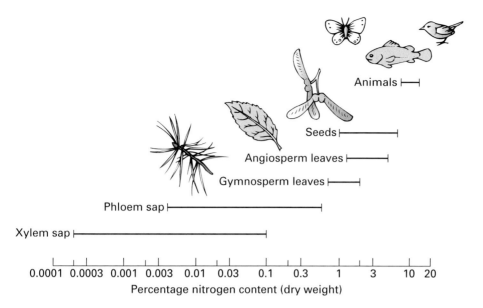

Fig. 13.66 Nitrogen contents of different types of plant tissue; note that almost all are very low compared with animal tissues. (From Strong *et al.* 1984; adapted from Mattson 1980.)

the content of these relatively indigestible components in common terrestrial plants. Vegetation is also hard to hold on to for a small land animal—it is faced with vertical moving wind-blown surfaces, often with a waxy and shiny cuticle. Land plants are also inherently poor nutritionally for animals, having low nitrogen content (Fig. 13.66), the wrong balance of amino acids, insufficient sodium content, and an absence of necessary steroids. There may also be excess water and/or excess sugar for some sap-feeders and fruit-feeders, or excess salt for some desert herbivores. Finally, and again very obviously, land plants are well defended against herbivores, either physically or chemically or both.

Despite being poor food for animals, plants are the primary producers of all trophic systems and are therefore bound to be eaten. Some of the groups mentioned above only eat them once they are dead, decaying, and thus beginning to soften; woodlice, for example, perceive litter as food only when it begins to produce odours of the metabolites that arise from decomposing microorganisms. Others feed only at a cellular level, with stylets used for piercing individually the relatively soft cells of roots and young foliage (many nematodes and some mites). But all groups that eat the tissues of living land plants have to be equipped to get over the first hazard of toughness, so that there is an absolute prerequisite for **hard mouthparts**. The land herbivores are therefore dominated by just three groups:
1 some terrestrial slugs and snails using the chitinous molluscan radula, feeding on leaves;
2 a vast array of insects, using their cuticular mouthparts as chewers for feeding on leaves, roots, seeds, flowers or fruits;
3 a range of vertebrates, using teeth to chew all plant solids.

In addition to hardened mouthparts, the terrestrial herbivores that ingest large quantities of plant material tend to show five other key characteristics. Firstly, they must have **elongated guts**, where the bulky food can be processed adequately; all land

herbivores show markedly greater overall gut length than closely related carnivores. Some birds and mammals can show acclimation by increasing the overall gut length by as much as 50% when fed on particularly poor diets; some birds change gut length seasonally as the diet alters.

Secondly, they must also have guts of unusual structure, as measured by the **coefficient of gut differentiation** (the ratio of stomach and large intestine to small intestine, or of the equivalent areas); in carnivores this is around 0.1–0.4, but in herbivores it may be 2–6, with relatively much reduced absorptive small intestine and much enlarged 'processing' areas.

Thirdly, they nearly all require some form of anaerobic **cellulase-providing symbionts** in their gut to provide the initial fermentation process whereby cellulose is broken down to usable constituents. A fully effective cellulase in fact requires at least three separate kinds of carbohydrase enzyme, and very few animals possess these, the exceptions being a few crustaceans, molluscs and wood-eating beetles.

Fourthly, they need mechanisms to achieve **pH homeostasis** in the gut in face of vegetation with very different pH characteristics (especially if they take in some decomposing leaves as part of their diet); many invertebrate herbivores achieve reasonable pH stability in the midgut with foods ranging between pH 4.0 and 7.5, and in vertebrates homeostasis is even more marked.

Finally, all land herbivores need to supplement their **salt intake**, with sodium being much less available in the diet than for a carnivore or for any aquatic animal. Herbivorous vertebrates usually meet their sodium needs by resorting to salt licks, or selecting some plants of unusually salty composition. Some insects use the alternative of 'puddling' behaviour; moths and butterflies visit small puddles and drink for many minutes or even hours on end, voiding excess water as repetitive anal jets in order to get enough salt. The males of the butterfly *Gluphisia* are especially adapted for puddling, with wide oral slits; much of the acquired sodium is transferred to the female at mating, for incorporation into eggs—a novel form of nuptial gift. Where puddles are contaminated by faecal droppings from other animals they may be particularly favoured as their salt content will be higher; even rain drops that have been in contact with bird droppings on foliage are used in this fashion.

As we will see, most terrestrial herbivores also have special problems with particular plant chemicals which have a defensive function, and therefore need good **detoxification systems**.

Molluscs and other invertebrates

Earthworms eat terrestrial vegetation mainly as already decaying material. They possess a gizzard where food is partially crushed, and a typhlosole (a dorsal longitudinal fold in the gut wall) to increase the absorptive surface area. There are also calciferous glands in the gut epithelium, probably helping to regulate internal pH levels, though it remains unclear whether this is related to a build-up of CO_2 in the burrows or an excess of calcium in the food.

Snails and slugs have inherited the chitinous radula from their aquatic ancestors, and this serves as an effective rasping tool for penetrating the cuticle of leaves and scraping plant tissue into the mouth. Land pulmonates generally have a radula that is wider and flatter than normal, with multiple rows of very similar simple small teeth, functioning rather like sandpaper. They will eat a range of land plants fairly indiscriminately, though they may reject some genera that are heavily chemically defended (see below). They have no problems with calcium, using any excess in shell construction; non-calcareous soils where Ca^{2+} is in short supply may be a major selective force promoting shell loss in slugs.

Some land crabs have become herbivorous, mainly selecting fallen leaves that are beginning to decay but also grazing on living foliage that is within their reach. On some isolated sites, such as Christmas Island, crabs have become the dominant litter recyclers.

Insect herbivory

Within the Insecta, nine out of the 29 orders are entirely or mostly herbivorous, notably the lepidopterans, orthopterans and true bugs (heteropterans and homopterans). Insects are the major terrestrial defoliators in most habitats, with caterpillars in particular eating relatively huge amounts, and in some species growing by a factor of 5000 in just 20 days. Many of the defences found in the plant world are related to deterring these insect herbivores.

Selection and ingestion of food plants

The term 'coevolution' was first coined when it was noted that plants possessing 'odd' chemicals and unusual tastes were only ever fed on by a few genera of relatively specialist insects. Many of the chemicals found in plants have metabolic effects in animals; many, perhaps most of them, have been selected for because they interfere with insect and/or mammalian metabolisms or behaviours, so preventing herbivores from utilizing the plant. Familiar examples include alkaloids, such as atropine in deadly nightshade, solanine in parts of potatoes, and above all nicotine in tobacco; the mustard oils (glucosinolates) that give a 'hot' taste to various crucifers, mild in cabbage but very concentrated in mustard and horseradish; terpenes from many shrubs and trees; and pyrethrum, widely used as a broad-spectrum insecticide. These curious plant chemicals are now generally termed secondary plant compounds or **allelochemicals**. Presumably these plant by-products had serendipitous deterrent effects on insects, and were retained by natural selection. But some insects have subsequently specialized to eat those particular plants, becoming monophagous or oligophagous on them, and leaving the less well-defended plants to be competed for by the mass of generalist polyphagous feeders. Over evolutionary time, the defended plants might increase their investment in chemical defences, so the 'coevolutionary arms race' between herbivorous insects and plants sets in.

It is difficult to classify the chemicals used as defences against insects, a huge range being involved. Table 13.18 gives an overview of chemical types. They can be classified according to effects on the insect herbivore as follows:

1 Growth or development inhibitors—often found in woody plants, and including the tannins, which can be up to 5% dry weight in many trees, and which work by complexing with dietary proteins so the herbivore cannot digest them, and also mimics of insect hormones (JH or ecdysones). These are broad-spectrum dosage-dependent deterrents, present in high quantities—'quantitative defences'—causing decreased growth

Table 13.18 Chemicals used as plant defences against herbivory.

Chemical type	Examples	Sources
Alkaloids	Nicotine	Tobacco
	Atropine	Deadly nightshade
	Solanine	Potato
Thiols, glucosinolates and mustard oil glycosides	Sinigrin	Crucifers
Cardenolides (cardiac glycosides)	Ouabain	Asclepiadaceae, Apocynaceae
Cyanogenic glycosides (releasing cyanide)	Cassavin	Cassava
	Prunacin	Seeds of Rosaceae
Phenolics, terpenes, steroids, coumarins, tannins	Cucurbitacin	Cucumber
	Juvabione	Firs
	Ecdysones	Bracken
	Tannin	Oaks, other trees
Other	Silicates	Grasses
	Odd amino acids	
	Polypeptides	

in a range of herbivores. They are expensive to produce, but hard to counter for the herbivore.

2 Lethal chemicals—alkaloids and glucosinolates are frequently lethal to non-specialist insects. They are produced in only very small doses—'qualitative defences'—and in relatively short-lived weeds; they are cheap to make, and relatively easy to detoxify if eaten by specialists, but most insects never encounter them and are not under selective pressure to develop detoxification systems.

3 Chemicals with physical effects—resins, latexes and gums, oozing from wounded plants and gumming up the feet or mouthparts, or silicates in grasses, which are sharp and painful to eat for mammals (there may be more silicon than nitrogen in a mature maize plant).

4 Feeding or oviposition deterrents—chemicals that modify insect behaviour, e.g. wild potatoes release an aphid alarm pheromone mimic from their trichomes (surface hairs) when bitten.

5 Attractants—bringing in predatory and parasitic insects that feed on the plant's herbivores.

Many toxins are concentrated in the most valuable and susceptible parts of plants, like seeds and young leaves. They can sometimes be induced by damage, and they often occur in higher concentrations in populations more exposed to attack; for example, the wild ginger plant has two morphs, one inhabiting sites where there are few slugs and beetles and another (with more chemical defences) where there are many.

In addition to chemical defences against herbivores, many land plants invest in **physical defences**. Spines, hairs and tri-chomes all make it hard to hold on: sometimes hairs actually puncture and kill invading caterpillars. Alternatively, slippery surfaces, using exudates from glands, can be used to form barriers against attack by insects. As another alternative, some plants can 'cheat' and use **biotic defences**, employing other animals as guards. *Acacia* is the classic case, where the plant provides both nesting space (in expanded thorns) and food (via extra-floral nectaries) to resident ants, who aggressively deter potential insect herbivores and even some large grazing mammals.

Many plants have defences; how then do herbivorous insects eat them at all? **Behavioural strategies** are often the first possibility, allowing the animals to avoid or to limit the effects of the defence, in space or time. They may specifically feed from non-defended parts, such as the phloem, or from inside a leaf-mine (avoiding the most toxic palisade layer of the leaf), or even (for very small larvae) from the leaf surface but avoiding the glandular trichomes in which the toxins are stored; for example, the first instar of *Heliothis* (boll-worm) feeding on cotton can avoid the toxin gossypol by eating around the base of individual hairs. Another way to avoid toxins is shown by grasshoppers that feed on cassava: the animals take a bite and then jump off the leaf, avoiding the sudden release of HCN (cyanide) from the damaged surface. These grasshoppers are gregarious, and many together seem to use the 'bite and retreat' technique to produce wilting of the cassava leaf, so that all can then move in and eat

from a no-longer-defended limp and floppy leaf. Yet another strategy is to cut off the inducible chemical defence supply route: some caterpillars cut a shallow trench in the host-plant leaf, through all the veins, before feeding, so preventing any mobilization of feeding deterrents. A recent further example of behavioural manipulation of the host plant involves leaf-rolling; many insects roll up leaves to make a shelter within which they eat in relative safety and in an equable microclimate, but the rolling also reduces the light falling on the leaf surfaces by 95% in some Neotropical shrubs, which in turn means a 31% decrease in physical toughness and a 15% decrease in leaf tannins.

If behavioural tricks fail, many insects resort to physiological and biochemical techniques against ingested toxins. The most obvious strategy is excretion. For example, the principal defence against nicotine for most tobacco-feeding insects is to excrete this compound very rapidly. *Manduca* (the tobacco hornworm) has a specific active nicotine pump in the walls of its Malpighian tubules. Some such pumps are inducible, for example in milk-weed bugs where a ouabain pump appears after the insects have been exposed to this natural cardiac glycoside in the diet. However, if a compound cannot be excreted, it must be dealt with internally by a detoxification or storage system. Most her-bivores have **mixed function oxidase** (MFO) enzyme systems designed to detoxify a range of unfamiliar chemicals; these are found in the liver of vertebrates and in the midgut and fatbody of insects. The amount of MFO varies with the diet in a pre-dictable manner, with more present in polyphagous species. (This poses a massive problem for insect control, since insect-icide resistance often involves acquiring higher levels of MFO, so that the 'pest' insect tends to become resistant to all insecticides, not just the one used!) Specific detoxification mechanisms also exist, more commonly in the monophagous herbivores that are dealing with small amounts of highly toxic compounds. **Sequestration** is a further option, best understood for the tannins. These tend to be countered in habitual tree-feeders by changing the pH of the foregut or midgut to strong alkalinity, so inhibiting the tannin's complex-forming abilities; the tannin is then tolerated and becomes the equivalent of roughage, accu-mulated in a sequestered inactive form and then either defec-ated or simply retained in the gut and lost at the next moult. But this in turn could make the normal H^+-related uptake systems (see Chapter 4) less rapid, and the processes of normal sugar and amino-acid uptake would be slowed down. In some species the symport proteins that mediate uptake in the gut wall are therefore unusually insensitive to pH, suggesting a long asso-ciation between the tannin-producing trees and the insect's alkaline guts.

Overcoming physical defences principally involves making the mouthparts, whether piercing, sucking, chewing or cutting, sufficiently hardened by strong tanning and cross-linking of the protein/chitinous structures to overcome the basic problem of cellulose-toughened materials. Avoiding penetrating hairs and trichomes is also crucial for softer-bodied arthropods such as caterpillars and aphids. Some caterpillars spin a raft of silk over the leaf and feed from it to get round this defence. Feet adaptations also occur: some aphids have long tarsal tips which allow them to hold on between hairs without damage, while suckers and circlets of fine hairs allow caterpillars to hold on to very slippery surfaces.

As well as circumventing the plant's defences, there are various ways in which the insect can turn the tables back on the plant. One is to use the plant chemical as a feeding attractant, and this has turned out to be extraordinarily common, the supposedly deterrent chemical becoming the insect's **phagostimulant**, initiating and maintaining feeding behaviour. For example, mustard oils are used by *Pieris* butterflies, and gossypol is used by the cotton boll-worm. Even the 'quantitative' defences, such as tannins, are phagostimulants to gypsy moths on oak trees. Alternatively (or often additionally), insects may use the plant chemical as an oviposition cue to identify their preferred host plant, or may use it as a pheromone to attract further insects to feed on the plant and mate with the founder.

There is one further possibility in relation to plant chemicals that is widely exploited by insect herbivores: that of using the plant chemical as the herbivore's own defence. Most caterpillars feed on mildly or moderately poisonous plants, and most 'get by' using either detoxification or excretion. However, their butterflies are largely unprotected, which is a problem as butterflies are conspicuous daytime fliers. Some lepidopterans therefore accumulate significant amounts of plant toxin in larval tissues outside the gut, and pass these chemicals to the adult. They also usually show **aposematism** (warning coloration), so that predators rapidly learn to associate unpleasant taste or illness with the bright (red/yellow/black) colours. The herbivore often adopts an 'aposematic lifestyle', warding off attack before the predator even gets hostile, reinforcing a strong negative search image due to the colours with other obvious visual cues, or conspicuous behaviours such as slow flight or undulating walking. There may be a tendency to aggregate, and perhaps to have tougher body structures to allow handling and learning by predators. The phenomenon of plant-derived chemical defence is particularly common on milkweeds (*Asclepias*), which produce irritant and toxic milky sap but nevertheless attract a range of specialist feeders. Most of these use the plants' cardiac glycosides to achieve unpalatability, and they are yellow/black or red/black. The most familiar examples are the monarch butterflies (*Danaus plexippus*), where larvae store the poison and pass some of it on to the adults, which are very toxic to birds, perhaps explaining why the butterflies can overwinter in such conspicuous aggregations. At least one grasshopper and a few moths actually spray the milkweed poisons out as a defensive secretion when attacked by birds.

A final possibility for insect herbivores is to overcome the plant's defences by releasing chemicals that make the plant form galls. Here plant morphology is altered to the insect's advantage. Galls are pathologically developed cells or tissues, arising by hypertrophy (increased cell size) or hyperplasia (increased cell number). About 2% of all described insects are gall makers

or 'cecidozoa', mainly from the orders Hemiptera, Diptera ('gall midges') and Hymenoptera ('gall wasps'). Gall formation involves two phases, initiation and then growth and maintenance. Initiation is usually in actively growing tissues, young leaves or buds, and involves an exaggerated wounding response sometimes interacting with plant hormones, especially cytokinins. Oral and anal secretions from the insect are also sometimes implicated in gall production. Hemiptera produce saliva containing amino acids and salts, but also low levels of auxins and phenolic compounds probably derived from the plant. There are some recent suggestions that gall formation is also related to a variety of semi-autonomous genetic entities (viruses, plasmids, etc.) transferred from insect to plant during feeding, with the insect thus acting as an RNA or DNA donor. The gall insects get all of their food from the galled tissues, away from potential predators and pathogens, and protected from the weather. But even more important, the gall tissues are sinks for plant assimilates; they tend to be relatively rich in sugars, proteins and lipids. In at least some cases the gall insect also manipulates the plant chemical defences; outer layers of the gall are very heavily endowed with the defensive secondary plant compounds, while the inner nutritive tissues are de-differentiated cells and almost free of toxins.

Digestion of food plants

Once insects have taken plant material into their guts and overcome any toxic effects, they still have the problem of digesting the cellulose and releasing the more nutritious cellular contents. The 'best' insect herbivores achieve 60–65% fibre digestion, as good as that in many mammals. Many do this using a flourishing gut fauna of symbionts, almost always in the hindgut. Some have a relatively small hindgut bacterial fermentation chamber, but primitive termites (having an almost pure cellulose diet) have a very large rectal sac ('paunch') housing flagellate protistans and bacteria. They repeatedly regurgitate and re-swallow the wood they eat to achieve maximum breakdown, which can be up to 89% efficient. The symbiotic gut fauna of arthropods are relatively unstudied, but are known to provide certain key elements in the diet (the B vitamins, and some sterols) in addition to breaking down the cellulose. In termites and some plant-feeding bugs the gut fauna aids in nitrogen balance, metabolizing a proportion of the uric acid waste.

The apparent exceptions to the necessity of symbiotic fauna are certain anobiid and cerambycid wood-eating beetles, and probably also the locust *Schistocerca*, which have low numbers of intestinal bacteria and appear to make their own (endogenous) cellulases, though there are contradictory indications that these derive from intracellular yeasts. Certain woodlice may also have endogenous hindgut cellulases. Another clear exception comes with the more advanced termites, which use a rather special kind of 'extracorporeal symbiont'; these are fungi especially cultivated in their underground nest 'gardens', to provide the initial breakdown of woody material. The termites then consume this partially digested material, and may also produce some foregut cellulase of their own.

Excretory problems for insect herbivores

Some insects feed on plant material that gives them too much water or too much sugar in the diet, and they often have systems to bypass the midgut epithelium (see Chapter 5 & Fig. 5.24). This applies particularly to sap-feeders: xylem (fed on by cicadas, for example) is very dilute with an enormous water content relative to its energetic value, whilst phloem (the food of aphids and many leafhoppers) is excessively rich in sugar and sometimes also nitrogenous foods, but very deficient in inorganic salts. The 'filter chamber' in the gut brings the anterior midgut, Malpighian tubules and anterior hindgut close together, and much of the water taken into the gut thus bypasses the main midgut epithelium and passes directly from anterior midgut to rectum. Phloem-feeders adopt the strategy of rapid excretion instead, with a sugary fluid ('honeydew') emerging from the anus or from dorsal abdominal glands called cornicles, and a certain amount of excess water being lost as salivary secretions. They may also be able to combine simple sugars into oligosaccharides, to reduce the osmotic effect of the latter in drawing water out of the aphids' tissues. Certain fruit-feeding flies, such as the tephritid *Rhagoletis*, exhibit 'bubbling' behaviour after a large dilute meal, repeatedly regurgitating and re-swallowing some of the fluid, presumably to concentrate it by evaporation; many bees do the same thing with floral nectar, though this may also serve as a mode of evaporative cooling.

Vertebrate herbivores

Vertebrates that are predominantly herbivorous include anuran tadpoles (mainly eating relatively soft aquatic vegetation), some reptiles, relatively few birds (except in the sense of specialist feeders on seeds and fruits, which are very common), some marsupials, and many eutherian mammals, especially rodents (specializing on seeds and/or woody tissues), ruminants (leaf-eaters) and primates (leaf- and fruit-eaters). The practice of bulk feeding on green leafy material is rather rare, restricted to the ruminant mammals and a few large species in the other groups (e.g. geese, tortoises), where the necessarily large guts can be accommodated.

Selection and ingestion

Plant defences operate against vertebrates as well as against insects, though the scale of the defence may be different. Smaller scale physical features such as trichomes may become irrelevant and are replaced by sharp and solid thorns, deterring all but the most thick-lipped of browsers. Chemicals are more likely to be irritant or unpalatable rather than toxic, since the larger quantities needed to kill a vertebrate can rarely be achieved. However, there may be enough toxin to have significant physiological effects: for example, ringtail possums feeding on different species of *Eucalyptus*, all with toxic oils, produce urines of different pH, and where the eucalypt is particularly high in terpenes they have to take in more food and then excrete up to 40% of the metabolizable energy, compared with only 10–15% on a low-terpene food plant. Koalas feeding on eucalypts face similar problems and are only just able to detoxify enough leaf to survive, rarely accumulating any fat stores; in droughts, when the gum trees stop producing new leaves, koalas may die of nitrogen deficiency even though they have full stomachs.

Inaccessibility of the most important plant parts is another important aspect of defence against grazing mammals. This is exemplified by the low growth of grasses and of many pasture weeds, where the growing meristems and often also the important reproductive organs may be at ground level; and by the raised canopy of many savanna trees, where the lowest flowering branches are just out of reach of many of the antelope and other herbivorous mammals.

Again plant defences, both physical and chemical, may be 'used' by the herbivore to find and choose its food, producing relatively narrow dietary niches. Primates are particularly fussy about the nature of the plants they ingest, having different and quite specific techniques of manual dexterity to deal with the variety of prickly and stinging vegetation that they encounter. Fruits may be particularly valued, partly because they often rely on being eaten and so are usually sweet and lacking in secondary defences; monkeys and apes often seek out favourites over large distances, revisiting known trees at particular times of year. However, the larger size and necessarily large food volume of most vertebrate herbivores make them fairly unselective in what they eat, and monophagy and oligophagy are rather rare, occurring only in a few (often rather threatened!) cases, such as giant pandas and koalas. Diet selection may be more influenced by what can be found easily and gathered in relatively benign thermal and hygric conditions; small rodents, for example, have been shown to balance nutrition with thermoregulation.

Digestion of food plants

Vertebrates are incapable of digesting cellulose and related plant polymers, but once again they make use of the many species of microorganisms that have the ability to ferment cellulose. The amount of fermentation in vertebrate guts is hindered by a high degree of molecular cross-linking and lignification in ingested plant material. The space available for fermentation in the gastrointestinal tract (GIT) increases in direct proportion to body mass, whereas we know that metabolic rate scales with body mass$^{0.75}$. Thus in small animals the rate of food passage through the GIT is too rapid for extensive microbial growth, and substantial fermentation is impossible. To achieve enough fermentation, several techniques occur. The Australian wood duck, a rare example of obligate foliar herbivory in a bird, has no obvious adaptation to retain or recycle fibre, but selects soft

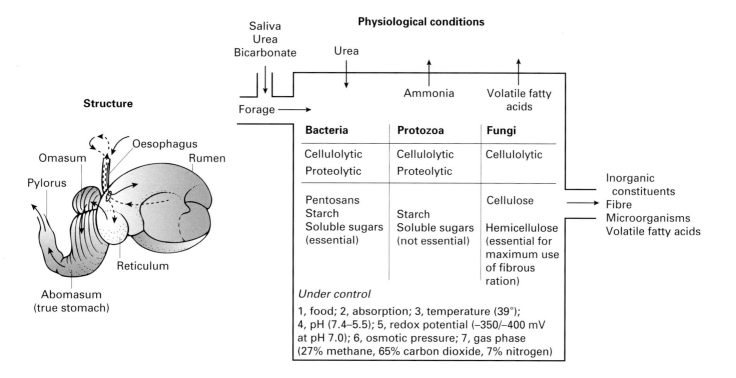

Fig. 13.67 The complex stomachs of a ruminant and their physiological conditions. (From Strong *et al.* 1984; adapted from Mattson 1980.)

grasses and herbs rich in hemicellulose, with little assimilation of lignin or cellulose. Rabbits and other grass-eaters may reingest a proportion of their own faeces (coprophagy) so that the digestion products pass through the GIT several times; the rabbit feeds by night, and while resting by day it produces two kinds of faecal pellets, only reingesting the softer kind (25–65% of total faecal production) for recycling. This gives more time for fermentation and leads to a highly efficient utilization of nitrogen. An alternative approach is to have voluminous intestines, and this adaptation is found in the herbivorous reptiles such as iguanas, with valve-like folds that increase the time food takes to pass through the gut. The iguanas also bask in the sun, which stimulates fermentation but at the expense of higher maintenance costs; in the green iguana, transit time in the gut can be decreased from 10 days to 3 days by raising T_b from 30 to 36°C.

In larger animals there are two additional strategies to achieve good fermentation:

1 Hindgut fermentation: non-ruminant herbivores (many mammals and some birds) have an enlarged area of hindgut, usually a large caecum, where fermentation occurs long after initial gastric digestion. Hindgut fermentation provides the host animal with volatile fatty acids as a source of energy, but the lack of subsequent digestive processes and the limited opportunity for the absorption of other products (especially vitamins) puts the non-ruminants at a relative disadvantage. Efficiency of fibre digestion is variable, between 20 and 65%.

2 Foregut fermentation: usually termed **rumination**, and seen as characteristic of ruminant mammals, though also found in a few birds, such as the hoatzin, which eat leaves. Ruminant herbivores provide accommodation for their microbial partners in a modified stomach, where fermentation precedes gastric digestion. Here efficiency is greater, at 52–80%.

Vertebrate ruminants. The Ruminantia includes bovids (cattle, sheep, goats), cervids (deer, antelope, giraffe, pronghorn) and camelids. Sheep, cattle and goats are the best-known ruminants physiologically, and they share three characteristics. Firstly, they lack upper incisor teeth, having instead a dental pad against which plant material is crushed. Secondly, they have a complex four-chambered stomach (Fig. 13.67). Thirdly, they regurgitate the food from the first stomach chamber and subject it to a lengthy mastication (i.e. they ruminate, or 'chew the cud'). The newly ingested fodder is attacked by a mixed population of anaerobic bacteria, protozoa and fungi. Carbohydrates, proteins and some fats are metabolized (only saturated fatty acids are not subject to fermentation), to produce a foregut mixture of volatile fatty acids, ammonia and microbial cells, containing 50% protein. As the ingesta is reduced to small particles it passes with ruminal fluid to the reticulum (second chamber) and omasum (third chamber), where further physical maceration occurs, together with the absorption of much fluid. True gastric digestion occurs in the abomasum (fourth chamber), which is the equivalent of the stomach of other animals. Since large numbers of microorganisms accompany the ingesta and are killed and digested in the abomasum they represent an appreciable proportion of the solid part of the animal's real food ration.

The particular advantages of pregastric or ruminal digestion are three-fold:

1 Cellulose and other structural plant polymers are solubilized. Of the total ingested energy, about 46% is lost as waste (faeces and methane) and 18% is utilized by the microorganisms for their metabolism and population growth. This extra microbial growth is passed on to the host's gut, however, so that 54% of the total energy is being made available to the ruminant. The 3% that is converted into heat of fermentation is also potentially useful for temperature regulation by the ruminant if it is cold stressed.

2 The microorganisms can use non-protein nitrogen for growth, so that inorganic nitrogen or urea from the host's protein metabolism may be converted to microbial protein, which eventually becomes available to the host animal.

3 The vitamin content of food is increased by microbial synthesis. Ruminants are independent of all dietary vitamins, other than A and D.

Metabolism of the food produces acetic, propionic and butyric acids, roughly in the proportions 70 : 20 : 10, together with some lactate, formate, valerate, isovalerate, ammonia, methane, carbon dioxide and hydrogen. The rumen has very particular and unusual conditions: its temperature is about 39°C, maintained by heat produced during fermentation; its pH varies between 7.4 and 5.5, with a buffer system provided by the bicarbonate from saliva and by the volatile fatty acids; and it has a gas phase consisting of about 27% CH_4, 65% CO_2, 7% N_2 and a trace of H_2, and the redox potential can reach as low as -350 to -400 mV at pH 7, reflecting profoundly anaerobic conditions. Methane present in the rumen represents a sink for the considerable amount of hydrogen produced in fermentation reactions.

Rumen microorganisms consist of bacteria, protists and fungi. At least 12 genera of **bacteria** can be regarded as true rumen organisms, and they act symbiotically such that the endproducts of one class are the substrates of another. Two groups of ciliated **protists** are also found in the rumen, and these can ferment carbohydrates with the production of volatile fatty acids, CO_2 and H_2. In sheep fed a grain diet there are about 2.5 million protists and 10 billion bacteria per ml of rumen fluid, reflecting the ready availability of carbohydrate; whereas on a hay and grass diet with less available carbohydrate, populations of protists are about an order of magnitude lower. Rumen **fungi** are also higher in animals on fibrous diets and are much reduced in animals on a high grain diet. The fungi actively ferment cellulose to acetate, formate, lactate, ethanol, CO_2 and H_2. The latter is converted to methane by bacteria.

Volatile fatty acids (acetate, propionate and butyrate) are absorbed directly from the stomach and utilized immediately. Acetate is especially important: it passes through the rumen epithelium and liver unchanged but is transported in the blood and then oxidized (via acetyl-CoA) in the tissues to produce energy, supplying up to 50% of the carbon expired as CO_2 in a ruminant. The importance of acetate as a source of energy is emphasized by the low blood sugar levels of ruminants (only about half that of adult humans). Ruminants are also unusual in using acetate to fuel milk production and to produce body fat stores.

Other species with ruminant-like digestion of cellulose. A number of non-ruminant placental mammals use ruminant-like digestive processes with a stomach modified for microbial fermentation. These include leaf-eating tree sloths, colobid and langur monkeys, and some rodents. Sloths are noteworthy for their low MR and very large gut volume, which may be up to 37% of body mass; within the foregut fermentation occurs, but at an extremely low rate compared to normal ruminants. The macropod marsupial mammals (kangaroos and wallabies) also have a ruminant-like digestive system. The stomach of the grey kangaroo, *Macropus*, is highly specialized as a fermentation chamber but is more tubular than the four-chambered ruminant stomach, and there is a more continuous flow of material through it. Relatively little material is regurgitated, rechewed and reingested. The fore-stomach again contains bacteria, ciliate protists and fungi, and the overall digestive efficiencies of ruminants and macropods are strikingly similar.

Postgastric fermenters. Postgastric fermentation occurs in many placental mammals (perissodactyls, such as horses; the elephants and hyraxes; some rodents and lagomorphs) and in many marsupials (koala, and many herbivorous possums). The site of fermentation in postgastric fermenters may vary. In koalas the **caecum** predominates ('caecant digestion'), whereas in wombats the colon is crucial and the caecum vestigial. In the horse, bacterial fermentation occurs in the caecum and large intestine, and the volatile fatty acids are absorbed across the wall of the large intestine. In pigs fermentation takes place solely in the large intestine, while starch, sugars, proteins and fat are digested in the small intestine; but fermentation accounts for no more than 10–20% of the feed for pigs. The capybara is particularly interesting, being the largest living rodent at around 50 kg, and possessing both a caecum and a habit of coprophagy, eating its own faeces especially in the early morning.

However, the microorganismal production of the caecant mammals is not normally digested and assimilated, as occurs in ruminants, because the microbial production always occurs after the stomach and small intestine. Consequently, cellulose digestion is less effective in the horse than in a ruminant such as a cow, and horses compensate for this by having a higher rate of food intake. Postgastric fermentation also does not confer the advantage of nitrogen recycling. Nor does it allow the microorganisms to detoxify plant allelochemicals, so that in hindgut fermenters most of these do get into the bloodstream and must be dealt with by the liver. On the other hand it does work well with diets high in fibre or tannins or silica; these pass on through the gut without entering the caecum, whereas in ruminants they cannot leave the rumen until ground to tiny particles, impeding the movement of other materials to the absorptive surfaces.

During the Eocene epoch (57–36 MYA), when forests covered

most of the available land, the dominant herbivores were non-ruminant perissodactyls, now represented by the horses, tapirs and rhinoceroses (a mere six genera of the 158 known to have existed). By the end of the Eocene, however, the artiodactyls (the group that includes the ruminants) had increased, allied to the appearance of grasslands during the Miocene. The ruminants have now come to dominate the artiodactyls, with over 70 genera. Even today, ruminants and horses or zebras often compete in terms of dietary intake and digestive strategies, and horses do better on low-quality forage such as straw. Thus they still do well on grasslands when there is drought, and might perhaps do better in the future in areas of anthropogenic desertification.

The overall effects of terrestrial herbivory

Plants do get eaten, but rarely to excess. They have undergone a prolonged period of coevolution with their herbivores, especially the insects, evolving ever more sophisticated ways of avoiding being eaten, which has resulted in complex chemical niche separation with specialists and generalists on both sides of the competitive arena. Animal–plant trophic relations may have generated much of the present species richness on land, especially in the tropical forest systems where so many of the coevolutionary stories have been worked out. Herbivores can therefore be seen as largely responsible for maintaining genetic diversity and polymorphisms in the land plants, and plant–insect relations have made enormous contributions to shaping present-day environments.

13.8.2 *Carnivores*

Carnivory is generally no more difficult on land than elsewhere, since animal tissues almost by definition provide the perfect balanced diet for other animals. In terms of nutrition and access to the digestible food there may nevertheless be a few difficulties. As with plant-feeders, a major problem lies with skeletal material, though in animals the relatively indigestible parts are discrete, with internal rods or external coverings, rather than forming a protective shield around each individual cell. Land fauna do tend to be dominated by groups with stiff skeletons of course, so most carnivores either have to crush the chitinous exoskeletons of small arthropods or pull the soft flesh off the bones of vertebrates. Chitin is indigestible to many animals, so that the exoskeleton is commonly left behind as a husk; specialist feeders on small arthropods, such as the ant- and termite-eating edentate mammals and the insectivorous bats, tend to have rather low MRs, reflecting their relatively poor-quality diet where a substantial proportion of the captured biomass is discarded. However, some earthworms and crustaceans, and a few vertebrates such as bats and lizards do have a chitinase, and others use bacterial chitinase in the gut. Bone is also fairly indigestible to most carnivores (most merely crushing it to extract the marrow); but there are always a few specialists around to eat even this material, the most famous examples being hyenas.

There are two main strategies for land carnivores: active searchers, and sit-and-wait predators. Either may impose physiological demands on the predator due to the speed of land prey, making any kind of capture difficult and very often intermittent. Many land carnivores therefore have a very high resistance to starvation (notable in some spiders and worms), or if they are active hunters they have unusual stamina, as in various dogs, so that they can pursue prey for long periods. Soft-bodied fauna attack prey with simple jaws and often some extruded enzymes to produce extracellular predigestion, sucking up the resultant soup. Similar tactics are used by most of the arachnids, using powerful sucking pharynx muscles to deal with their liquefied diet. Many of these arachnids use venoms, allowing them to attack prey larger and often more powerful than themselves. Leeches use a modification of this approach in feeding on the blood of much larger animals, using the proboscis to make an incision, then injecting anticoagulants to ensure a steady blood flow. In these animals, and in the abdomens of other blood-feeders such as mosquitoes and certain bugs, the body wall has to be very extensible to permit the intake of large meals at very long intervals. Many blood-feeders have a problem simply of getting too much at once, perhaps feeding only at intervals of weeks or even months. Many leeches, and insects such as the bug *Rhodnius* and the tsetse fly, *Glossina*, have extremely rapid excretory outputs controlled by hormones, to eliminate most of the volume taken in as quickly as possible as dilute urine (up to 50% of the volume being passed within 3 h). *Rhodnius* has also recently been shown to have a haem detoxification system, so that excess haem groups from the prey's blood are polymerized in the midgut into a harmless crystalline form, haemozoin.

Overall, carnivores on land have populations of lower density and there are fewer of them at any particular body mass compared with herbivores (Fig. 13.68), reflecting the inefficiency of transfer of energy through successive trophic levels and the traditional 'ecological pyramid'. Remember, though, that land animals are not only prey for the predators dealt with in this section, but in turn become hosts for a huge range of other animals that feed in and on their bodies; parasitic groups that as endoparasites lead essentially aquatic lives insulated from the vagaries of the terrestrial environment. Many of these animals are from the same groups of terrestrial animals already considered here, notably the insects; they are dealt with in Chapter 15.

13.9 Anthropogenic problems

The effects of human populations on terrestrial ecosystems represent one of the greatest hazards for other animals, a relatively 'new' selective agent determining the survival or extinction of other species. The varied threats that we are imposing on land habitats set complex physiological problems that a few animals survive and even thrive on, where others are stressed beyond their acclimatory or adaptive limits. Here we review some of the major changes being brought about by human activities, and the effects on major groups of land animals.

Fig. 13.68 The relation between body size and abundance for mammalian carnivores compared with herbivores. Note that both absolute numbers and slope are different. (From Peters 1983; courtesy of Cambridge University Press.)

The most obvious effect of humanity is wholesale and direct destruction of habitats, either to make way for human constructions or to provide raw materials. Even where this is on a small scale—for example pollution leaking from a waste site, removal of a few trees to construct a new road, or building a few houses—the effect on the biota is obvious and usually fatal. Large animals may migrate out, but thousands of microhabitats for the small and unglamorous fauna are destroyed. On the much greater scale of mass rain-forest destruction, any potential for moving out may be abolished, and the whole community is extinguished. No adaptation of physiology or life history makes very much difference here. But humans also have more subtle and yet potentially more serious effects on the global ecosystem, and these anthropogenic effects may be more selective between species.

13.9.1 *The greenhouse effect and temperature rise*

Carbon dioxide is in many ways at the root of the atmospheric and climatic regulation problems that have occurred since the beginning of the Holocene period (i.e. since the last Ice Age, 10 000 years ago). If there were no CO_2 the Earth's surface temperature would be only −18°C, instead of the present 15°C

average; we would have a frozen planet, ice-bound and with high reflectivity (albedo) giving no chance of warming and no possibility for the evolution of life. But CO_2 is continually produced by volcanoes erupting, and if unregulated by life it would therefore soon smother and poison all living organisms. In fact, because of life the atmospheric levels of CO_2 have been declining since the Precambrian. Living plants continuously absorb CO_2 from the air, using it as raw material for photosynthesis, releasing it again as carbon compounds underground when they die, and so keeping the atmosphere suitable for other life. The geosphere and the biosphere have interacted, as in so many other ways, maintaining conditions suitable for life.

Some of the CO_2 still persists in the air and acts as a blanket; this kept the Earth warm in the past when the Sun was cooler, but as the Sun grew gradually hotter the CO_2 also became thinner and acted less effectively as a thermal blanket. On balance this has led to rather stable planetary temperatures overall. In a sense we should therefore see the long-term 'problem' for the planet as that of progressive lowering of CO_2 and potential cooling. Looking at the temperature levels through time rather confirms this (Fig. 13.69): the planet has been getting on average colder for many thousands of years.

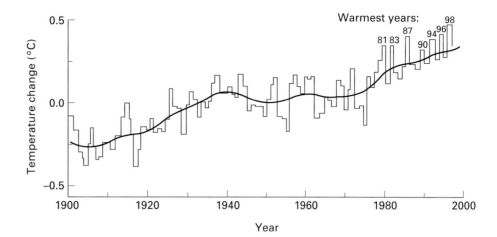

Fig. 13.69 Globally averaged surface (land and ocean) temperatures for the last century, expressed as differences from the mean in 1950–80. (Data from *Climate Change IPCC Report 1995.*)

The problem is that humanity has now caused the CO_2 level to rise by 7–10% in three decades (Fig. 13.70), so increasing its thermal blanket effect. Carbon dioxide levels are naturally cyclic (higher in spring, lower in autumn and winter) but the net levels are undoubtedly rising. The preindustrial average level was 260 parts per million (ppm) and in the 1990s we have reached levels up to 360 ppm, rising by at least 1.5 ppm per year. At the same time the amplitude of the annual cyclical changes are now somewhat greater.

In terms of temperature rise, a doubling of CO_2 levels is calculated to cause a 3°C rise in temperature. So far we have had a clear 0.5–0.7°C temperature rise globally since 1900, with the 1980s and 1990s being exceptionally warm in Europe. The eight hottest years on record (as a global average) have all been in the last two decades. The projections, from the Intergovernmental

Panel on Climate Change (IPCC), are for progressively higher temperatures that go beyond anything seen in the time since man evolved as a species (Fig. 13.71). Thus the biota is having to cope not only with increased CO_2 levels but also with steadily rising average temperatures.

The causes of CO_2 increase are not hard to find. The industrial and domestic burning of fossil fuels currently releases about 5.6 thousand million tonnes (gigatonnes, or Gt) of carbon per year, forming a large part of the problem (Fig. 13.72). Deforestation also has major effects; it leads in the long term to less fixing of CO_2 in vegetation, and the trees themselves are often destroyed by burning which also releases CO_2 directly (up to 2 Gt of carbon per year).

The effect of CO_2 as a pollutant greenhouse gas is probably responsible for about 50% of the total warming effect; however, there are other gases that are increasing due to human activities and that also affect atmospheric temperatures (Fig. 13.73). One major culprit is methane (CH_4), and in many cases deforestation also affects methane levels because forest is replaced by ranches where ruminants are grazed, and ruminants are very good at producing methane from their guts as a by-product of fermentation. Methane is also released from rice paddies in tropical zones (see Plate 10a, facing p. 310) and from bogs and other natural wetlands in temperate zones; perhaps also by dredging up oil, coal and gas deposits. Methane fluxes into the atmosphere are made worse by extensive nitrogenous fertilization as part of human agricultural practice (Plate 10c).

The levels of all of the other gases shown in Fig. 13.73 are also rising; the worst culprits for such emissions tend to be the more industrial countries, with power stations, smelting plants and heavy industrial centres particularly implicated (Plate 10b). Figure 13.74 summarizes the main ways in which human activities have contributed to global warming.

An additional problem in terms of potential temperature rise is that all these kinds of emission to the atmosphere may interact. Interactions may be positive or negative (Table 13.19), but on balance the IPCC group have estimated that positive feedbacks are more likely than negative ones, so that estimates of the

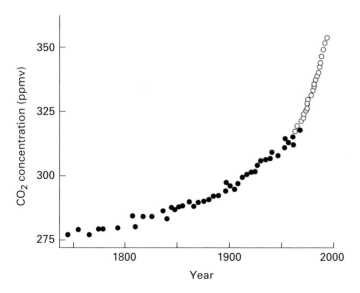

Fig. 13.70 Patterns of atmospheric carbon dioxide recorded since 1750; open symbols are the data from the recording centre at Mauna Loa (Hawaii). (Reprinted with permission from *Nature* (Friedli *et al.* 1986). Copyright 1986 Macmillan Magazines Limited.)

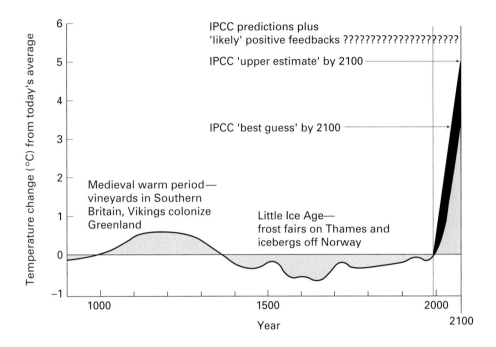

Fig. 13.71 Scenarios for temperature change based on different models of climate sensitivity, from the Intergovernmental Panel on Climate Change (IPCC). (Adapted from *Climate Change IPCC Report 1995*.)

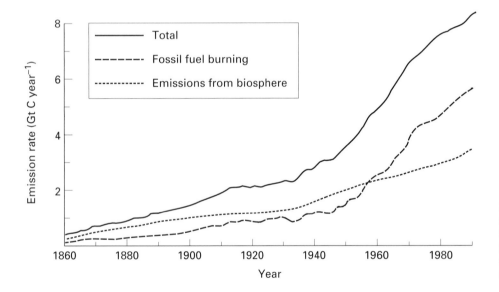

Fig. 13.72 Carbon emission rates increasing through the 20th century, largely paralleling fossil fuel burning but still increasing beyond the 1970s (when fuel use levelled off somewhat), probably reflecting increasing deforestation rates.

magnitude of temperature change and of its timing may actually be on the low and slow side. There is no doubt that CO_2 is rising very fast and abnormally; and while there is some doubt about temperature rise (how much, how fast) there is every likelihood that there will be further changes in the upward direction. Best predictions are for a rise to 600 ppm CO_2 by a date between 2030 and 2080, which might mean a 3–5°C global rise, with worse effects at higher latitudes and in the northern hemisphere, perhaps up to 12°C warmer at the North Pole. This does not sound like a spectacular change, but any rise above the average has enormous implications, especially when it occurs very quickly.

Direct effects of rising carbon dioxide and temperature

Land animals are unlikely to be directly affected at a physiological level by the projected changes in CO_2 levels, since they excrete the gas readily as a waste product of respiration, with very substantial gradients from their tissues to the outside world. Terrestrial animals are usually responsive to P_{CO_2} (more so than aquatic species); but they have their sensors internal to detect the blood concentration, so that direct effects on respiration are unlikely.

However, the effects of rising environmental temperatures on

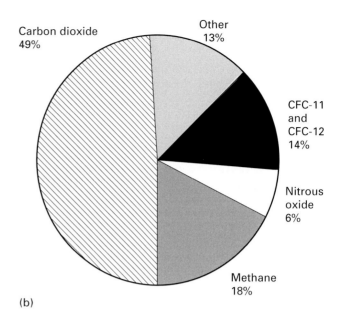

(a)

(b)

Fig. 13.73 Proportional contribution of atmospheric gases (excluding natural water vapour) to greenhouse effect warming: (a) up to 1980 and (b) since 1980. (Adapted from Mintzer 1992.)

physiological processes are potentially more serious. We have explored the various biochemical and physiological mechanisms that underlie thermal adaptation and phenotypic plasticity in animals (Chapter 8), and it is clear that those species with a suite of enzyme and membrane adaptations that render them eurythermal should be able to cope with moderate changes in ambient temperatures. Some eurythermal species may be able to exhibit phenotype plasticity, i.e. switching to slightly different phenotypes as a direct response to different thermal conditions. Others may be able to effect genetic changes in traits with a fast enough time course; this would be especially true for animals with high reproductive outputs and thus rapid evolutionary rates (insects, crustaceans, some worms, but probably only a

few small vertebrates), and for traits that show high heritability. Recent studies indicate that for the common mouse, *Mus domesticus*, body weight and 'nesting score' show high heritability but body temperature and weight of brown fat (BAT) do not, so that more northern mice with larger bodies and an inherent tendency to build good nests might fare better when challenged with climate change. There are yet other kinds of animals that may be able to migrate quickly enough to keep pace with changing climate (after all, some species certainly did live further north in previous interglacial periods and may be able to recolonize lost ground northwards). It is also possible to use behavioural alterations to cope with warming; for example, there is already evidence for 'laying dates' in temperate birds becoming earlier in the spring, probably because food is available earlier, and similar effects are documented for some insects where the earlier start may allow an extra generation to be produced in each annual cycle (potentially serious with 'pest' species such as aphids).

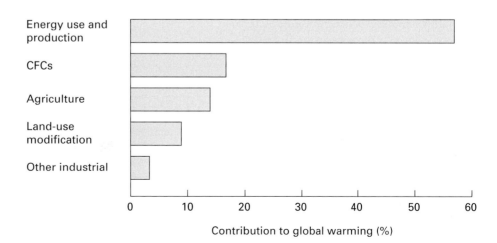

Fig. 13.74 Major influences of human activities on greenhouse emissions and global warming.

Table 13.19 Feedbacks that may alter the greenhouse effect.

Negative feedbacks

Rising temperatures lead to more cloud formation; this increases the reflectivity of the atmosphere and reduces solar inputs

Rising productivity of plants (using high CO_2 levels to fix more carbon) may use up the excess CO_2 and bring things back into balance. (But plant responses are highly variable: weedy plants do well, others are often limited by nutrient resources, perhaps because the enhanced soil microflora are doing better and sequestering the nutrients)

The seas will serve as a sink and mop up excess CO_2. (But present evidence suggests this can only occur where CO_2 levels rise much more slowly; the oceans could perhaps mop up an 'extra' 1–3 Gt per year, whereas 7–8 Gt per year are being released)

Positive feedbacks

Increasing temperature may release more of the carbon held in soils, the dead organic matter being released as CO_2 and CH_4, especially from temperate wetlands and from reserves currently trapped below northern permafrosts

Increasing temperature may enhance sea stratification, with a warm unmixed upper layer having decreased primary production so that it gets depleted of nutrients, giving decreased CO_2 uptake by the seas

Increasing temperatures could cause some polar ice melt, thus exposing new areas of land, giving areas of dark colour with lower albedo and higher heat absorptivity than the ice, and so accelerating the process of warm-up

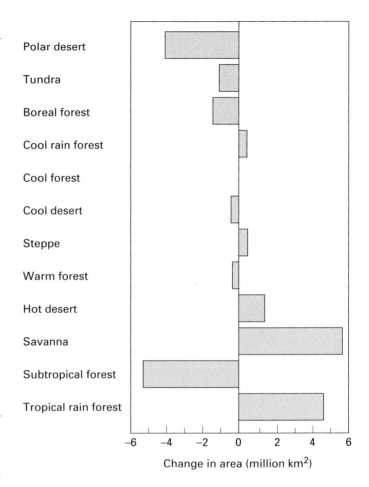

Fig. 13.75 Predicted patterns of change in the land biomes due to global warming.

But stenotherms may be at real risk of experiencing ambient temperatures outside their tolerance ranges. This may be especially problematic towards the poles, where temperature changes may be greatest; remember also that cold-adapted proteins are generally inherently more stenothermal (and likely to denature rapidly) than warm-adapted ones. Thus larger, slowly reproducing, and/or less mobile, polar and high-altitude stenotherms may be at particular risk in a warming world of the future.

Indirect effects

First and foremost a rise in temperature will change the patterns of global climate. There are many alternative computer global circulation models (GCMs), though most are reasonably in agreement and most predict average temperature rises of 1°C already, and up to 5°C in 100 years (varying with latitude). The best guesses for climatic changes accompanying global mean surface warming are large-scale stratospheric cooling, global mean precipitation increases, reduced sea ice with polar winter warming, and continental summer warming and drying with increased cyclones and storms.

Climatic changes of this magnitude must have major effects on vegetation distribution, pushing successive biome types towards the poles. Globally, the current projections of overall changes in biomes (Fig. 13.75) suggest a loss of polar and tundra

systems especially, with some increase in savanna, desert and (if it is not cut down) tropical rain forest. The forest belts have been moving northwards at 1 km per year since the last Ice Age anyway; they probably cannot sustain much faster migration, though they may have hundreds of kilometres of potentially habitable land still to recolonize. More seriously, different maximal migration rates would surely apply to different species of both plants and animals, so that the ecosystems would change and potentially become very unbalanced. Coevolved species could become mismatched in time or in space: pollinators and seed dispersers with the flowering and fruiting of their relevant plants, monophagic predators with their prey, parasites with their normal hosts, and so on; this is especially likely in the tropics, where such relations are often finely tuned.

In general, fast-growing and fast-reproducing (*r*-selected) 'weed' plants and 'pest' animals should almost inevitably do better due to their high dispersal powers, higher reproductive rates, and potential for faster adaptive change. Animal species with eurythermic and euryhaline tolerances should outcompete all other species; polyphagous species should thrive at the expense of monophages; and those with good locomotory/

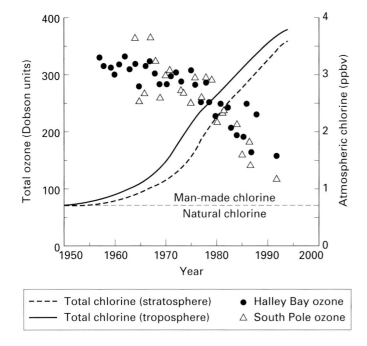

Total ozone (Dobson units)

Atmospheric chlorine (ppbv)

Man-made chlorine
Natural chlorine

Year

- - - - Total chlorine (stratosphere)   ● Halley Bay ozone
——— Total chlorine (troposphere)   △ South Pole ozone

Fig. 13.76 Rising anthropogenic chlorine levels, and associated declines in stratospheric ozone at two sites in Antarctica as the seasonal ozone hole develops.

dispersive powers in at least one stage of the life cycle should do well.

13.9.2 *Ozone depletion*

As a gas, ozone (O_3) is harmful low down in the troposphere; it is an oxidant and a pollutant, and where it collects in cities as part of the 'photochemical smogs' it causes respiratory problems and skin rashes. But ozone is absolutely necessary higher up in the atmosphere, because it helps protect living organisms from potentially harmful ultraviolet B (UV-B) radiation (280–320 nm) in the stratosphere. It is the only atmospheric gas that absorbs strongly around 300 nm. And in the upper layers of the planet's atmosphere, ozone levels have been changing quite drastically in some places and at some times of year.

The first suspicions of a man-made threat to the ozone layer came in the mid-1970s, and a 'hole' was detected between 1977 and 1984 by Antarctic survey teams. The hole now develops each spring in the Antarctic (through September and October) and at its worst grows to be about the size of the USA, spreading over the tip of South America and a moderate area of southern Australia. The chlorofluorocarbon gases (CFCs, especially $CFCl_3$ and CF_2Cl_2) that cause this were 'wonder compounds' when invented in the 1920s, due to their extreme chemostability and low toxicity, so were widely used in refrigeration, bubble-foamed plastics, aerosols and air conditioners. But photodissociation of these CFCs can occur under the influence of short-wavelength

UV light (200–350 nm); they then break down and release Cl^- ions, which react with ozone to produce oxygen. Chlorine is particularly effective because it precipitates a chain reaction with ozone in which the active principle, ClO (chlorine monoxide), is continuously regenerated; a single ClO molecule, once formed, can break down 20 000–100 000 ozone molecules.

The CFCs are very long-lived molecules (15–100 years) and are therefore bound to reach the stratosphere when released at the Earth's surface. The link between the resultant rising levels of man-made chlorine and the seasonal loss of ozone is now quite clear (Fig. 13.76). In the mid-1980s the annual production of CFCs was nearing 1.2 million tons and the total stratospheric content of halogens had risen from 0.6 ppb (parts per billion) to around 3 ppb chlorine and 0.02 ppb bromine. The UN Montreal Protocol of 1987 agreed to halve production of CFCs, and many governments now plan their complete elimination. But chlorine levels are still rising by about 5% per year, the 'hole' is still enlarging each year to cover large parts of the southern continents, and some perturbations are also occurring over the northern polar region. In any event, with the longevity of the molecules involved, recovery of the ozone layer is unlikely until atmospheric chlorine levels fall back below 2 ppb, currently projected for the end of the 21st century if strict controls are adhered to.

Thinning of the ozone layer is potentially serious to life below because organisms become exposed to an excess of UV-B. This has several adverse effects. Firstly, UV-B slows photosynthesis in many plants (not all, but at least 50% of all species tested). It might also decrease productivity substantially in the phytoplankton; marine plants are not well defended against UV-B, and through evolutionary history their main defence has been to live at a depth where it does not penetrate, so they can be harmed by small increases. These effects on land and sea plant productivity would also of course enhance global warming by reducing CO_2 sinks. UV-B may also directly affect some secondary plant chemistry pathways, and there are indications that it reacts directly with DNA in plants. This may relate to its effects in animals, which have mostly been studied in man: a 1% decrease in ozone, by allowing 2% more UV-B through, may cause a 5% increase in benign human skin cancers, of which some unknown percentage will turn malignant. But a 10% decrease in ozone would mean a 50–90% increase in mild skin cancers, and probably a small increase in the much more serious melanomas too. There is also an increasing prevalence of eye cataracts and potential blindness, and some suggestions of effects on the human immune system. Projecting these kinds of effects into the rest of the animal kingdom suggests a serious indirect effect for certain kinds of mono- and oligophagous herbivores, whose food plants could be seriously reduced or out-competed, and a direct risk for those animals that lack a good protective skin or that spend most of their time out in the open. Animals with thicker cuticles or shells, and those that spend much of their life in protected burrows or within the soil or litter layers, may be at little risk.

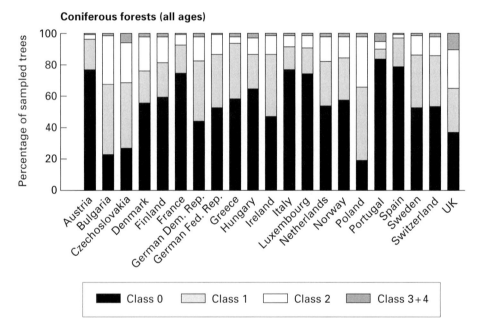

Fig. 13.77 Results of the comprehensive 1989 forest survey in Europe, showing percentage of trees in different damage (defoliation) classes by country; 1–4 = increasing defoliation. Many show less than half the trees with no damage (class 0). (Adapted from *Climate Change IPCC Report 1995*.)

13.9.3 *Acidification of atmosphere and biosphere*

As we noted in Chapter 11 in relation to lake acidification, the problem of acid deposition was first noted at least a hundred years ago, with the observation that rainfall was sometimes acidic and that this did harm to vegetation, stonework and metals. By the 1960s the phenomenon of 'Waldsterben' (tree sickness) was noted, with a general forest decline in southern Germany and Czechoslovakia. Trees, especially the silver fir (*Abies alba*), showed a premature increase in old-age disorders, particularly showing fungal infections and thickening branches. By the late 1970s other trees, including Scots pine, spruce and even beech, were noticeably affected, with bleached leaves, and

by the mid-1980s around a quarter of all trees were affected in western and central Europe. Figure 13.77 summarizes the position throughout Europe from the 1989 tree damage survey: both deciduous and coniferous forests were damaged, the former slightly more so, and matters were especially bad in central Europe (Bulgaria, Czechoslovakia, Poland, Greece and eastern Germany). In 1995–96, 50% of all trees in Germany were assessed as badly damaged. Acid rainfall can be detected very frequently in all these areas, with the rain in Europe and the USA now commonly recorded at pH 4.0–5.0 in areas lying to windward of heavy industry. Sulphur dioxide and nitrogen oxide emissions, either unconverted as gases or dissolving as sulphuric and nitric acids, are the prime causes of the acidity (see Chapter 11). Soils

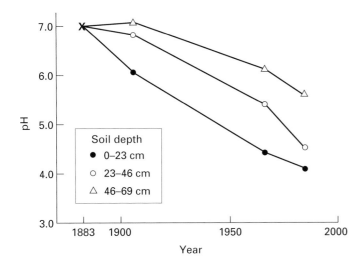

Fig. 13.78 Changes in pH at a site in Britain over the last century, linked to acid precipitation.

are directly affected, even in isolated areas; the average pH of soils in Great Britain over the last 100 years shows a decrease from about 7 to 4.5–6 according to depth (Fig. 13.78).

Acidified rainfall has several effects for land animals:

1 It removes nutrients from the soils. This is relevant to trees, because the acid rain washes out minerals, like magnesium, that are needed for growth, leading to yellowing of conifer needles; thus tree-feeders suffer. However, acid conditions also solubilize aluminium in soil, and aluminium-poisoning affects mucus production in invertebrates such as earthworms and flatworms.

2 The gaseous components of acid rain affect plant growth directly. The gases have interactive negative effects: SO_2 and ozone cause additive damage to leaf growth and synergistic damage to root growth and leaf appearance, while they are antagonistic in their effects on fungal infections. With spruce, SO_2 opens the stomata and lets ozone in, also allowing more drying out and frost susceptibility. Thus the very dry summer of 1976 was one of the worst ever recorded for tree damage in Germany, as the acid gases made drought unusually hard to bear; folivorous animals certainly experienced associated stresses.

3 The dissolved acid destroys useful fungal (mycorrhizal) associations in soils and so indirectly reduces plant growth and affects the associated animals.

4 Reduced diversity in streams and soils (Chapter 11) affects dietary breadth for associated land fauna such as birds and insects.

5 Soil pH directly affects many soft bodied animals, including amphibians; salamanders, for example, suffer disrupted sodium balance and weight loss in acid soils.

6 Very low pH rainfall almost certainly kills small soft-bodied invertebrates in leaf litter directly.

Note that any or all of these effects can leave trees, tree-herbivores and litter animals alive but under stress, and thus more susceptible to other pollutants, to pathogens and to parasites.

13.9.4 *Urbanization*

Wastes and pollutants

Wastes can be divided into four major categories: municipal; industrial; agricultural (including mining and dredging substrates); and nuclear. The amount of each varies for different human communities; agricultural wastes dominate in Europe and most of the USA, whereas in Japan industrial waste is hugely dominant. More than 75% of the waste is recyclable in ways that would pose little threat to the rest of the biota, but *at best* only about 30% does get recycled. Instead, extensive and ever-increasing use of land-fill sites destroys habitats for flora and fauna, while incineration contributes further CO_2 to enhance global warming.

Chemical waste is particularly problematic for the biota. About 80 000 chemicals are now in common use by humans, by far the majority being organic carbon-based compounds, mostly derived ultimately from natural gas and oil. We generate about 250 million tonnes of organic chemical wastes annually. The two main groups are aromatic hydrocarbons used in numerous industrial processes, and organochlorines and organophosphates, especially used as pesticides and herbicides and thus directly threatening to the biota. In addition there are inorganic wastes, with various heavy metals being particularly difficult to dispose of and potentially hazardous.

About 75% of this chemical waste is disposed of in land-fill tips (Plate 10e), mostly around urban areas, but older and/or badly managed tips may threaten the ground water or leak noxious gases, and toxins, gases and microorganisms can all become hazards. The alternative of incineration, which decreased immediately after clean air legislation was put in place in many industrialized countries, is increasing again now. Wastes can generate useful energy, but also greenhouse gases; and the potentially toxic ash still has to go to land-fill. Bacterial remediation has the most promising outlook, in association with modern biotechnology.

Nuclear waste disposal represents a very specific extra problem, because the major effects of radiation are at the level of the genome itself. Reactors, however well managed, contaminate everything nearby with radiation, including clothes, dust and the cooling water, producing low- and intermediate-level wastes (LLW and ILW). They also generate spent fuel, which is still very highly radioactive and constitutes the high-level waste (HLW). This is primarily still uranium, with 1–2% plutonium and 5% other fission products. Plutonium is often described as the most toxic substance on Earth, partly because it has a half-life for decay of 25 000 years; it causes excess mutations in all organisms, potentially leading to teratogenic deformities at birth and carcinogenic effects later in life. A typical reactor may produce

200 kg of it per year, and this is enough in theory to cause cancer in the entire vertebrate population of the world. If it were uncontrolled (or if large quantities of equivalent radiation were released through nuclear warfare or terrorism) it could destroy much of the Earth's animal and plant life; with the somewhat depressing proviso that animals with tough exo-skeletons, and particularly a few species such as cockroaches that have resistant highly dispersed genetic material, might be most able to survive.

Nuclear material cannot be disposed of by controlled burning, which would just generate radioactive smoke instead. Most of the LLW was formerly dumped at sea, but now international conventions ensure it is put into shallow pits on land, together with MLW. The problem of HLW is still quite unresolved, and it is all currently in store at reactor sites. Geologists are not yet agreed on *any* sites, on land or under the sea, as being stable enough for deep burying, even if accompanied by initial stabiliz-ing vitrification.

Man-made habitats (Plate 10)

Man has not only destroyed habitats, but has also, on various scales, created new ones, and certain kinds of animals have exploited these. We mentioned earlier that savanna grasslands are probably a very long-term 'creation' of human activities. An even more obvious recent example is intensively farmed agricultural land (Plate 10c), especially where crops are grown as virtual monocultures. Here the diversity of the endemic fauna and flora is very severely reduced, though there can be 'weed' proliferation and severe plagues of 'pest' insects, largely due to the removal of natural enemies and parasites, but also in part due to the physiological toughness and rapid powers of recovery of such organisms. The native vegetation and all the associated animals get squeezed into tiny islands of hedge-rows and small woods, and they are often at risk of local extinc-tion, having to deal with gross chemical disturbances to the soils and surroundings. This leads to intense ecological stress, with disruption of the normal biotic balance between preda-tors and prey and parasites, and between herbivores, pollinators and plants. It underlines once again the point that for all these relatively equable climatic zones the most important shaper of the community is often the range of competitive and cooper-ative biotic interactions, rather than any particular physiological pressure.

A second kind of man-made habitat is associated with agricul-ture, and involves all the resultant stored products, in grain silos, in processing centres and at points of sale. All provide high concentrations of certain kinds of foods that are by definition attractive to animals, and most also provide very stable (but often very dry) conditions. Grains and seeds are particularly susceptible, and for 'pests' of these the main physiological prob-lem is drought, as these products are stored at low humidity to prevent fungal invasions. The adaptations shown therefore tend to be rather like those of desert organisms, and such habitats

are attractive to many species of mites (often those species capable of water-vapour uptake), to insects such as beetles, and to rodents.

The third and most obvious kind of man-made habitat is of course houses (Plate 10d); in most parts of the world these are deliberately built to have stable mild and reasonably humid interiors, ideal living conditions for a great many insects. Thus our dwellings attract carpet- and clothes-eaters, wood-eaters, and even dry-skin-eaters (the house-dust mites, especially *Dermatophagoides*); and of course generalist detritus-eaters such as cockroaches and rodents.

Notice that all these man-made habitats are attracting species with very high reproductive rates and high evolutionary adap-tation rates, and therefore are stretching our control measures to the limit. Pesticides increasingly do not work, and this situation is certainly not going to get any easier.

13.10 Conclusions

A great many groups of animals have achieved some degree of terrestriality, and it is a mistake to underestimate the success and importance of the small and less 'exciting' soil-dwellers, which may form a huge but invisible part of the land fauna. However, the real success stories are restricted to just a few taxa whose impact on land has been more striking. Molluscs do well in all biomes, but it is nearly always the arachnids and insects and the vertebrates that are dominant, the arthropods taking the small-size niches and vertebrates the larger ones. Above all, this emphasizes the crucial factor of controlling surface properties; the arthropod cuticle and vertebrate dermis have underlain so much of their owners' success. It is worth remembering that there is a real paradox in relation to 'skin'. It has to serve many purposes at once: it is a barrier keeping some things inside the body and keeping others out, but it must also let some things in (notably oxygen and information) and let some out (waste prod-ucts). It is just about impossible to do all these things equally well, and compromises must therefore be made. In the light of this it is perhaps no surprise that the outstanding success story has been that of the arthropod exoskeleton. This represents the best possible compromise for land life: inbuilt defence and protection against unwanted physiological exchanges, coupled with versatile properties residing in the different layers of the structure, which can be varied between and within species and stadia, and a relative ease of making gated entry and exit points. The highly interactive problems of water balance, temperature balance and respiration can be solved in diverse ways using the cuticle; potential modes of locomotion are greatly increased by it; every conceivable trophic possibility can be realized with cuticular mouthparts; and reproduction on land is made easy by protective egg-cases and by the elaboration of the genitalia and of signalling devices to provide species recognition cues. Arachnids and insects, endowed with this material, dominate the terrestrial abundance, biomass and species diversity in almost all habitats.

Further reading

Books

Burggren W.W. & McMahon, B.R. (eds) (1988) *Biology of Land Crabs.* Cambridge University Press, Cambridge.

Cossins, A.R. & Bowler, K. (1987) *Temperature Biology of Animals.* Chapman & Hall, London.

Dejours, P., Bolis, L., Taylor, C.R. & Weibel, E.R. (eds) (1987) *Comparative Physiology: Life in Water and on Land.* Liviana Press, Padova.

Hadley, N.F. (1994) *Water Relations of Terrestrial Arthropods.* Academic Press, San Diego, CA.

Heinrich, B. (1993) *The Hot-Blooded Insects.* Harvard University Press, Cambridge, MA.

Herreid, C.F. & Fourtner, C.R. (1981) *Locomotion and Energetics in Arthropods.* Plenum Press, New York.

Little, C. (1983) *The Colonisation of Land.* Cambridge University Press, Cambridge.

Little, C. (1990) *The Terrestrial Invasion: An Ecophysiological Approach to the Origin of Land Animals.* Cambridge University Press, Cambridge.

Louw, G.N. (1993) *Physiological Animal Ecology.* Longman, Harlow.

Rosenthal, G.A. & Berenbaum, M.R. (1991) *Herbivores: their Interactions with Secondary Plant Metabolites.* Academic Press, San Diego, CA.

Skadhauge, E. (1981) *Osmoregulation in Birds.* Springer-Verlag, Berlin.

Reviews and scientific papers

Alexander, R. McN. (1989) Energy-saving mechanisms in terrestrial locomotion. In: *Energy Transformations in Cells and Organisms* (eds W. Wieser & E. Gnaiger), pp. 170–174. Georg Thieme Verlag.

Beuchat, C.A., Calder, W.A. & Braun, A.J. (1990) The integration of osmoregulation and energy balance in hummingbirds. *Physiological Zoology* **63**, 1059–1081.

Blickhan, R. (1989) Running and hopping. In: *Energy Transformations in Cells and Organisms* (eds W. Wieser & E. Gnaiger), pp. 183–190. Georg Thieme Verlag.

Cazemier, A.E., Op den Camp, H.J.M., Hackstein, J.H.P. & Vogels, G.D. (1997) Fibre digestion in arthropods. *Comparative Biochemistry and Physiology A* **118**, 101–109.

De Vries, M.C. & Wolcott, D.L. (1993) Gaseous ammonia evolution is coupled to reprocessing of urine at the gills of ghost crabs. *Journal of Experimental Zoology* **267**, 97–103.

Full, R.J. (1989) Mechanics and energetics of terrestrial locomotion: bipeds to polypeds. In: *Energy Transformations in Cells and Organisms* (eds W. Wieser & E. Gnaiger), pp. 175–182. Georg Thieme Verlag.

Full, R.J., Zuccarello, D.A. & Tullis, A. (1990) Effect of variation in form on the cost of terrestrial locomotion. *Journal of Experimental Biology* **150**, 233–246.

Geiser, F. (1988) Reduction of metabolism during hibernation and daily torpor in mammals and birds: temperature effect or physiological inhibition? *Journal of Comparative Physiology B* **158**, 25–37.

Hayes, J.P., Speakman, J.R. & Racey, P.A. (1992) The contribution of local heating and reducing exposed surface area to the energetic benefits of huddling by short-tailed field voles. *Physiological Zoology* **65**, 742–762.

Hillard, S.D., von Seckendorff Hoff, K. & Propper, C. (1998) The water absorption response: a behavioural assay for physiological processes in terrestrial amphibians. *Physiological Zoology* **71**, 127–138.

Hoffmann, A.A. & Blows, M.W. (1993) Evolutionary genetics and climate change; will animals adapt to climate change? In: *Biotic Interactions and Global Change* (eds P.M. Kareiva, J.G. Kingsolver & R.B. Huey), pp. 165–178. Sinauer Associates, Sunderland, MA.

Hofman, R.R. (1989) Evolutionary steps of ecophysiological adaptation and diversification of ruminants: a comparative view of their digestive system. *Oecologia* **78**, 443–457.

Lance, V.A. (1997) Sex determination in reptiles: an update. *American Zoologist* **37**, 504–513.

Lovegrove, B.G., Heldmaier, G. & Ruf, T. (1991) Perspectives of endothermy revisited: the endothermic temperature range. *Journal of Thermal Biology* **16**, 185–197.

Minnich, J.E. (1982) The use of water. In: *Biology of the Reptilia*, Vol. 12 (eds C. Gans & F.H. Pough), pp. 325–396. Academic Press, London.

Oliviera, M.F., Silva, J.R., Dansa-Petretski, M. *et al.* (1999) Haem detoxification by an insect. *Nature* **400**, 517–518.

Parsons, P.A. (1990) The metabolic cost of multiple environmental stresses: implications for climatic change and conservation. *Trends in Ecology and Evolution* **5**, 315–317.

Phillips, P.K. & Heath, J.E. (1995) Dependency of surface temperature regulation on body size in terrestrial mammals. *Journal of Thermal Biology* **20**, 281–289.

Quinlan, M.C. & Hadley, N.F. (1993) Gas exchange, ventilatory patterns and water loss in two lubber grasshoppers: quantifying cuticular and respiratory transpiration. *Physiological Zoology* **66**, 628–642.

Schmitz, H. (1994) Thermal characterization of butterfly wings: 1. Absorption in relation to different color, surface structure and basking type. *Journal of Thermal Biology* **19**, 403–412.

Smedley, S.R. & Eisner, T.A. (1995) Sodium uptake by puddling in a moth. *Science* **270**, 1816–1818.

Walton, B.M. & Bennett, A.F. (1993) Temperature-dependent color change in Kenyan chameleons. *Physiological Zoology* **66**, 270–287.

Warburg, M.R. (1994) Review of recent studies on reproduction in terrestrial isopods. *Invertebrate Reproduction and Development* **26**, 45–62.

Willmer P.G. (1980) The effects of a fluctuating environment on the water relations of larval Lepidoptera. *Ecological Entomology* **5**, 271–292.

Willmer, P.G. (1982) Microclimate and the environmental physiology of insects. *Advances in Insect Physiology* **16**, 1–57.

Wright, J.C. & Machin, J. (1993) Energy-dependent water vapor absorption (WVA) in the pleoventral cavity of terrestrial isopods: evidence for pressure cycling as a supplement to the colligative uptake mechanism. *Physiological Zoology* **66**, 193–215.

Wright, J.C. & O'Donnell, M.J. (1992) Osmolality and electrolyte composition of pleon fluid in *Porcellio scaber* (Crustacea, Isopoda): implications for water absorption. *Journal of Experimental Biology* **164**, 189–203.

Zimmerman, L.C. & Tracy, C.R. (1989) Interactions between the environment and ectothermy and herbivory in reptiles. *Physiological Zoology* **62**, 374–409.

14 Extreme Terrestrial Habitats

14.1 Introduction

Extreme terrestrial habitats classically include those that are unusually hot in low latitudes, or those that are unusually cold towards the poles and at high altitude. Hot habitats are normally only a real problem to animals when compounded by aridity, with the high ambient temperatures and low humidity together putting excessive strain on thermal and water (hygric) balance. Where heat is accompanied by high rainfall and freely available water it is less of a problem, as both ectotherms and endotherms can regulate their temperature without compromising their water balance, and many use evaporative cooling to this end. In this chapter we therefore concentrate on the really biotically 'difficult' areas of dry heat, exemplified by deserts. For cold regions, the distinction between arid and humid zones is less important to the fauna, since water loss is much less of a problem at low temperature and evaporative cooling is most unlikely to be an important strategy; cold zones can therefore be treated together in terms of the physiological adaptations needed. All of these habitats tend to be of low productivity as well as climatically extreme, so that animals require specialist adaptations on several fronts, over and above the range of 'normal' terrestrial strategies covered in Chapter 13. For these reasons the extremes of the land masses have always been especially attractive to environmental physiologists, and many of the classic studies come from deserts and ice-fields. This chapter deals with these specialist adaptations, where animals are pushed to their limits; we particularly look at the realms of temperature, water balance, food and energetics, and related effects on life histories. Adaptations to respiratory, reproductive, sensory and loco-motory systems are not covered in detail; essentially these systems are similar in equable and extreme terrestrial habitats.

14.1.1 Common characteristics of life in extreme conditions

Size

It is often said that animals in general are larger towards the poles, and for endothermic animals this idea is formalized as Bergmann's rule. Certainly there are examples that fit well with this rule: penguins, for example, where tropical species are quite small while the much larger king and emperor penguins are found deep in Antarctica; bears, where the large kodiak and polar bears occur further north than other species; and perhaps ospreys, as illustrated in Fig. 14.1. But at the same time many small birds and rodents occur towards the poles, escaping the rigours of their environment by burrowing. In fact if we look at extreme environments of all kinds, whether polar, desert or at high altitude, it might be fairer to say that there is a general rule that being *either* unusually large *or* unusually small is a good strategy. (Inevitably, proving this kind of generalization is difficult as it is confounded by taxonomic artefacts, and a strict phylogenetic comparison has not been attempted; but it is note-worthy that even within a taxon such as the trogid beetles a tendency towards larger and hence more desiccation-resistant species occurs in the Kalahari desert relative to neighbouring more temperate zones.) The logic of being either large or small is obvious: large species get all the advantages of a lower water loss rate and a high thermal inertia so that they can stay warm

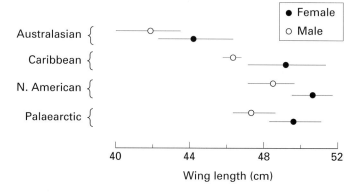

Fig. 14.1 Body size vs. latitude for ospreys (*Pandion haliaetus*) from Australasia to northern Europe, showing mean and standard error for wing length. The data are suggestive of a trend to smaller size at low latitude (though only the Australasian birds are statistically distinct). (Adapted from Prevost 1983.)

longer (at the poles) or stay cool longer (in a desert, as with the camel). The small species get the option, unavailable to larger relatives, of being able to exploit fine-grained microhabitats and escape from the harshest conditions. In deserts in particular, there is a striking paucity of medium-sized animals, which are denied either of these advantages; mammals of the size range that includes the dog, cat and goat only really do well around the fringes of deserts or at oases.

In considering the strategies for effective life in extreme conditions, it is therefore often helpful to split the animals into a small number of (non-taxonomic) groups. It could be said that there are often only two possible broad strategies, which are strongly related to the size of the animals; these are the small 'evaders', and the large 'endurers'. In the particular case of deserts, there is perhaps a third group comprising the rarer middle-sized animals, which might be termed the 'evaporators'.

Lifestyle

In any extreme environment, certain kinds of lifestyle seem to be favoured. Many animals cope using a strategy of **storing energy** during periods of relative abundance to maintain themselves through periods of scarcity, whether in a daily, seasonal or temporally random fashion. Others use a period of reduced metabolic need (dormancy, torpor or aestivation) through periods of scarcity. Others again have a nomadic lifestyle, abandoning each area as it becomes depleted of resources or enters a period of unendurable climatic extreme.

There may also be particular behavioural needs, so that in some extreme habitats there appears to be a trend towards more **gregariousness** or **eusociality**. This may have several advantages: it allows a colony to store food in periods of plenty as a buffer against intermittent and unpredictable scarcity; it allows a relatively small predatory animal, such as an ant, to attack, subdue and store a much larger range of prey sizes; and it may involve shared underground nests that are strongly buffered against climatic extremes. Overall, the population size of a colonial or social species may be much less variable than that of a solitary epigean species.

Range and biogeography

For extreme habitats there appears to be a tendency for animals to have a **greater range** than occurs in more equable biomes. For high-latitude species this has sometimes been formalized as 'Rapoport's rule', indicating that such species range over a broader latitudinal belt than occurs in the tropics. The same seems to be true for elevational range at altitude. These observations are not particularly surprising given that tropical lowland species are likely to be closely adapted to stable and predictable climatic regimes whereas polar or highland species must cope with much greater climatic variation at any one site. The issue is of course compounded by the much greater species richness at low latitude and altitude.

Selective regime

Harsh environments produce populations living on the edge of their abilities to cope, and may involve a particular kind of **adversity selection** (see Chapter 1), species being selected for their tolerance of the physical conditions and for an ability to reproduce successfully even with the small numbers remaining after mass mortalities. Adversity selection allows species to survive through periodic population bottlenecks. At the same time the periodic extinction events keep diversity so low that there are few interactions between the species and little coevolutionary selection.

14.2 Hot and dry habitats: deserts

14.2.1 Types and characteristics of deserts

Occurrence

Arid zones occur between 15 and 40° latitude, either side of the warm and wet equatorial zone. The boundaries of deserts were set quite recently in geological terms; the massive Pleistocene glaciers tied up free water, leading to aridity at low latitudes so that desertification occurred, and in the 10 000 years since the last glaciations these areas have never recovered. We generally think of deserts as exceptionally hot and dry areas with sweeping sand-dunes (see Plate 8e, facing p. 310), but in fact a variety of biotopes are included within this general category and they are surprisingly varied. Their most important shared characteristic is aridity; the three subdivisions of semi-arid (Plate 8f), arid and hyper-arid deserts, as defined by international agreements, cover one-third of the Earth's land surface, about half of this being true desert as traditionally perceived by the layman (Fig. 14.2).

Natural aridity arises for three main reasons, varying in importance for different desert areas. Deserts in parts of North and South America arise from a rain-shadow effect, lying on the leeward side of high mountain ranges (sierras), and where no rainclouds occur as the rising air has precipitated all its water on the mountains (see Plate 9a, facing p. 310). Deserts in the centre of large continents are arid principally due to their sheer distance from the sea; the Gobi and Turkestan deserts are examples. However, the largest deserts on Earth arise from the third factor, occurring in latitudes (25–35°N or S) where there are dry, stable air masses of high pressure, resistant to invasion by storm systems to north or south. The Saharan and Arabian deserts, and the Australian desert, are of this type.

Hyper-arid deserts, where high-pressure weather dominates, normally receive less than 25 mm of rainfall per annum; within this there are areas that effectively receive no rain at all. Even when rain does fall it may be as violent convective showers, causing flash-flooding in dry river beds (wadis; Plate 9b) and very fierce run-off, with little of the rain becoming available to plants or animals in the area; that which does remain local is

Fig. 14.2 The occurrence of deserts on a worldwide basis, with two broad latitudinal belts north and south of the equator.

subject to very rapid evaporation due to the high temperatures. Thus water availability is not only low but highly unpredictable, and all organisms must be opportunistic in responding to its presence. These deserts usually also have high windspeeds, leading to even more fierce evaporation, and they have cloudless skies, so that it can be very cold at night as heat radiates away from the soil. Their soils are hardly leached at all by any passage of water through them, so they are often sandy or stony (leading to high heat penetration and very poor water retention, see section 13.1.2) but also have considerable deposits of salts very near the surface. As this is eroded and stirred by the high winds life becomes even harder for any colonizing plants. Drought alone is difficult, but when combined with hot and cold extremes, and saltiness, and the instability of the terrain, living here is impossible for most kinds of organism. These desert communities are therefore dominated by specialists, usually occurring at low biomass.

Less extreme areas, the arid and semi-arid deserts, may have up to 600 mm of rain per year, more evenly and predictably distributed, and they are often much cooler than the hyper-arid areas. Coastal deserts, such as the Namib (southwest Africa), the Atacama (Chile), parts of southern Israel, and the Baja California coast of USA/Mexico, are characterized by cooling fogs, espe-

cially after dawn. Cooler inland deserts, such as the Gobi and large areas of Patagonia, also occur, with limited fog but prolonged periods of winter cold.

Here we are concerned particularly with the problems faced by the biota in hot arid areas, where terrestrial adaptations are pushed to extremes. We therefore concentrate on animals from the large sand and rock deserts of Saharan north Africa, Arabia and the Middle East, southwestern states of the USA, and Australia. By far the most detailed work to date has derived from fieldwork in the deserts of Arizona and Colorado, the Negev and the Australian outback, all areas where high temperatures and aridity are compounded by high winds and continuously disturbed wind-blown sandy and/or salty soil.

Desert microclimate and vegetation

Microclimatic niches make survival possible for many terrestrial animals, as we saw in Chapter 13. In the Libyan Desert a daytime air temperature of 58°C has been recorded, and other deserts can be below freezing at night; but within desert sands conditions just 100 mm deep are approaching constancy, with diurnal variation virtually eliminated at 200 mm (Fig. 14.3). This degree of 'fine-grained' microclimatic variation becomes paramount in determining animal distributions and lifestyles. Any abiotic irregularity can cause additional microclimatic variation; pebbles and rocks on sand create local shadows and potential shelters, while any vegetation creates not only shade but a local

Fig. 14.3 Patterns of temperature and humidity against depth for dune slopes in a desert, through the 24 h cycle. Notice that there is reasonable constancy at −10 cm and complete stability at −20 and −40 cm depth, while the surface shows a temperature range of nearly 40°C. (From Holm & Edney 1973.)

increase in humidity and a reduction of windspeed. Figure 14.4 shows microclimates recorded around and below a desert shrub in terms of the body temperatures of beetles found in those sites. Desert arthropods commonly occur clustered around vegetation in this fashion.

A great many species also live cryptically, hidden away in microclimatic conditions that allow survival. Burrows 200–300 mm below the surface may be at an almost constant temperature of 35–40°C, and at humidities always above 80–85%, allowing comfortable living for arthropods and small mammals alike even when the surface temperature varies from below freezing to well above 45°C; a specific example for a 200 mm scorpion burrow is shown in Fig. 14.5. Only at dawn and dusk, or after brief periods of rain, are these animals to be seen active on the desert surface.

The particular microclimatic problems of immobile stages such as eggs and pupae have to be solved in other ways. Eggs are laid deep underground or within other organisms by many desert invertebrates. Pupal microclimates can only be regulated

indirectly by choices made in the larval stage, and this again may involve locating a sheltered spot on a host plant, or for parasites a site within a much larger host animal which will itself be maintaining reasonable homeostasis.

Desert vegetation inevitably has characteristic specializations. There is usually a 'crust' of modified sand densely occupied by a community of microorganisms including fungi, lichen and mosses, perhaps only a millimetre in depth and held together by mucilaginous secretions. Hardly any visible perennials exist, except at watercourses (Plate 9b). Small plants that do survive are mostly evaders in time, almost invisible for most of their life as seeds or dried-out prickly husks, undergoing sudden syn-

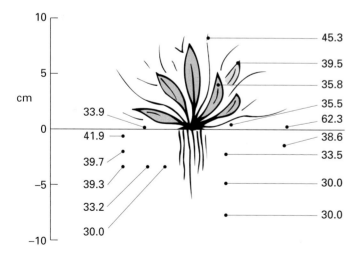

Fig. 14.4 The microclimate (shown by temperatures °C) as experienced by small beetles in, under and around a desert shrub. (Adapted from Hamilton 1975.)

Fig. 14.5 Temperatures in the burrow of the scorpion *Hadrurus*, compared with temperatures at the soil surface and at a depth of 200 mm. (Adapted from Hadley 1970.)

chronous bursts of growth, flowering and seed production after occasional rains. Such plants are often termed ephemerals; they are technically 'arido-passive', being inactive when dried. Some can complete their life cycle in only 14 days. There are also some arido-passive desert plants that endure by virtual cryptobiosis; these are the macroscopic lichens, often called cryptogams. Plants of the genus *Ramalina* from the Negev can survive near-total desiccation for at least a year, with photosynthesis shut right down, and ambient temperatures up to 65°C. They derive their water from dew in the hour or two after dawn, and can even take water from unsaturated air above about 80% relative humidity (RH).

On a larger and more obvious scale, 'normal' plants occur only at oases. These include a variety of palms, a small range of annual flowers, and the few specialist perennial 'arido-active' forms that keep photosynthesizing even in drought. Cacti are the classic examples, with 2000+ species in the Americas; they show loss of leaves, succulent stems, waxy surfaces, deep roots, sunken stomata, C_4 metabolism with reduced temperature sensitivity, and fearsome herbivore deterrents so that all the physiological and morphological specialization is not wasted by surfaces getting damaged. There are also small prickly trees such as creosote bush, mesquite, and many eucalypts and wattles; most of these have enormous roots, penetrating to depths of 50 m or more and vastly in excess of the aerial biomass, and some of them will lose all their leaves and even drop whole branches 'deliberately' when they are particularly severely desiccated.

Desert fauna

Desert faunas have very different taxonomic compositions from other terrestrial habitats, as Fig. 14.6 shows. The microfauna within the soil is as always dominated by nematodes and some groups of microarthropods, especially collembolans (spring-tails) and mites. But in the fauna living some or all of the time above ground differences are striking. Most of the taxa with more permeable surfaces are under-represented, with a very low species richness of earthworms, millipedes, isopods and snails. However, there may be quite a high biomass made up from a few specialist representatives of some of these groups; for example, specialist desert woodlice are surprisingly common in some areas. There are moderate numbers of cockroaches, orthopterans such as locusts and crickets, and surprisingly high numbers of hemipteran bugs such as aphids, hoppers and cicadas, small enough to live in the microclimates around desert plants. Relatively few lepidopterans occur, because the cater-pillar stages are rather large and permeable plant-feeders. However, taxa with grub-like juvenile stages do survive where the larvae are underground root-feeders (e.g. beetle larvae), or are parasitic within other animals (dipterans and some wasps) or are laid in underground nests (most notably bees, ants and termites). There may also be relatively high diversity and biomass for a few particularly adapted groups, notably the beetles, spiders and scorpions from the invertebrate spectrum,

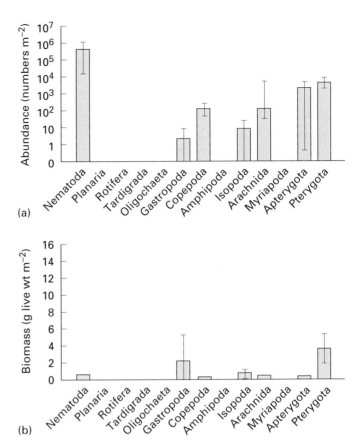

Fig. 14.6 Desert faunal composition (for invertebrates only) in terms of (a) abundance and (b) biomass. (Adapted from Little 1990; courtesy of Cambridge University Press.)

and the reptiles and rodents from the vertebrate classes. In some deserts a single group may be particularly dominant; for example, one family of beetles, the Tenebrionidae, or 'darkling beetles', may account for up to 15% of the species diversity in some deserts, and may be more than 50% of the total insect species diversity in the Namib Desert, while in the western Sahara ants have been calculated to represent 75% of the entire faunal biomass but with the sand cockroach *Heterogamia* being the single most abundant species.

Larger vertebrate animals also of course occur in deserts, and seem most obvious to casual visitors, though they are of low species diversity and low overall biomass. This category would include desert anurans (frogs and toads especially, but also tiger salamanders in American deserts), plus many lizards, snakes and tortoises; and a range of endotherms, dominated by relatively huge numbers of rodents, moderate numbers of antelope, and some specialist birds.

Following the theme outlined in the first section of this chapter, Fig. 14.7 shows how these components of the fauna may be broadly split up into three strategic categories on the basis of their size. We will look at the strategies of each group in turn, though it should be borne in mind that many of the techniques

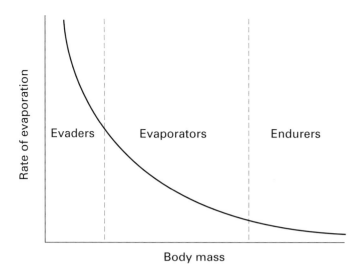

Fig. 14.7 Three main strategies for desert animals, in relation to size and rate of evaporation.

used by the small evaders are also relevant in modified form to many of the larger desert animals.

14.2.2 Desert adaptations in evaders

All the small desert animals, of sheer necessity, come in this category. Four different 'grades' can perhaps be identified:
1 Strict evaders active only at night, otherwise resting deep underground in humid, reasonably equable microclimates—the 'softer' invertebrates like woodlice, apterygote insects, etc., and also nematodes, which may occur at a million per m² even in shallow desert soils. Many of these groups use cryptobiosis (see Chapter 12) when conditions are particularly harsh.
2 Evaders, but rather less restricted; insects, spiders and scorpions, commonly of small size and using daytime evasion with nocturnal/crepuscular behaviour.
3 Small vertebrate ectothermic evaders, the desert amphibians and reptiles, having somewhat different problems.
4 Small vertebrate endotherm evaders, dominated by rodents.

Collectively, all these animals exhibit a wide range of adaptations, most of these being extensions of characteristics found in other terrestrial animals in less severe climates.

Burrowing

A fossorial (digging) lifestyle is characteristic of all members of the first group of animals described above, with woodlice, collembolans, mites and nematodes living within the soil rather than in specific burrows. Many of these migrate through the sandy deposits on a regular cycle, moving up and down each day or through the seasons. The cockroach *Arenivaga* moves up to within 20–50 mm of the soil surface at night, but down to 200–600 mm depth in the day; on a smaller scale many mites move from the shallow litter layer into the soil during the hotter parts of the day.

The use of properly constructed **burrows** combined with fossorial habits is exceptionally common in deserts, being exhibited by virtually all the larger invertebrate desert dwellers and by the ubiquitous rodents. Quite short burrows, even in extreme desert conditions, give reasonably **equable microclimates** (see Fig. 14.5), and burrows may be used by species as large as birds (burrowing seasonally) and even tortoises. Most of these animals show a degree of modification of limbs to allow burrowing, from the modified forelimbs of scorpions to the stout digging front legs of desert tortoises. However, given the ubiquity of burrows, and the fact that constructing one has costs in itself, there are also those who 'cheat' by using a hole created by another animal and are ill-equipped to burrow for themselves.

Burrows become an important component in the regulatory strategies of their owners. For example, desert spiders (which are normally representatives of the burrowing families or are nomadic hunters, not the web-spinners of more temperate habitats) use burrows to shuttle in and out of to keep their body temperature (T_b) reasonably constant. The Australian burrowing species *Geolycosa* emerges soon after sunrise and basks in the sun to raise T_b to 40°C, then basks and burrows in turn to maintain it at this level through much of the day. Desert tortoises in North America use both shallow summer burrows to avoid overheating, and deeper winter dens (often enormous and reused by succeeding generations over decades or centuries) to avoid winter freezing, spending the majority of their life underground. To give a mammalian example, the air within deer mice burrows in Nevada is usually close to 26°C at about 1 m depth throughout the 24 h cycle, where the external temperatures vary between 16 and 44°C. Small rodents such as the antelope ground squirrel (Fig. 14.8a) can thus be briefly active even in the hottest periods of the day, using their cool burrow to unload the heat they pick up on quick forays, so that their T_b never goes above 42–44°C. Most of the nocturnal desert rodents have high and narrow thermoneutral zones, with burrow temperatures always in the range 25–35°C. Indeed, some small desert mammals and birds are gregarious in burrows at night, the clustering habit helping to reduce each individual's metabolic rate (e.g. by 11–22% in mousebirds) while keeping the T_b elevated to normal levels; Fig. 14.8b shows the effect in gerbils.

Burrows may also give high **humidity** that aids osmotic regulatory strategies. For naked mole rats the burrow 700 mm below ground provides not only a constant 26–28°C temperature but also a constant high humidity that helps minimize evaporative water loss (EWL). The spadefoot toads (*Scaphiopus*) only survive in deserts because of an interaction of their permeable skins and their burrowing habit. They live in deep burrows, constructed at the end of the rainy season in moist soil, and initially further water enters the body osmotically from the soil. As the soil dries the gradient is reversed, and the toad then stops secreting any urine and switches to urea-retention (see 'Control and tolerance of water loss' below), raising the concentration of its body fluids to reduce this gradient. The toads can remain buried for 9–10 months without access to further liquid water, and may emerge at the end of the dry season with the blood still only at 600 mOsm.

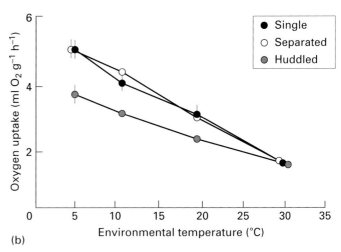

Fig. 14.8 Temperature and activity in small desert mammals. (a) In relation to use of a burrow, for the antelope ground squirrel, where periods of absence from the burrow are brief and heat gained is then unloaded in the cool burrow depths. (Adapted from Bartholomew 1964.) (b) In relation to the presence of conspecifics, in the Mongolian gerbil from relatively cool deserts, where huddling with two other gerbils reduces the need for increased metabolism as temperature falls. (Adapted from Contreras 1984.)

Burrows of course also have advantages beyond the creation of a favourable microclimate. They allow **food storage** (especially for social and gregarious animals), and provide protection from predators and parasites. Use of a burrow also allows a sit-and-wait type of foraging strategy, for example in the mygalomorph spiders, and in insects such as larval ant-lions and tiger beetles, where passing prey fall into or are grabbed from the opening of a small pit in the sandy soil. On a larger scale, sidewinder adders and golden moles also conceal themselves in sand burrows and ambush passing prey.

However, burrows also have their problems. There may be an accumulation of ammonia, less from adult excretion (normally restricted to outside the burrow) than from urination and defecation from juveniles. Furthermore, the restricted entrances of many burrows tend to limit aeration, and when the owner is resident they become zones of reduced Po_2. This is especially important for small mammals and birds with high endothermic metabolic rates; the Po_2 values in the nests of small desert rodents are often only 10–15 kPa, instead of the normal 21 kPa, and this may link to cases of rather low basal metabolic rates (BMRs) in desert rodents discussed below. (Note, however, where animals burrow continuously through sand, rather than constructing a fixed burrow, there is little respiratory problem; the Po_2 near the nose of a sand-swimming desert mole is very similar to that in the free atmosphere, the sand being dry and porous.) We also noted in Chapter 7 that mammals and birds with burrowing habits tend to have reduced sensitivity to the CO_2, allowing them to overcome the related problem of the burrow air having elevated Pco_2 (up to 6 kPa). Some burrowing

rodents even excrete much of their CO_2 as bicarbonate, reducing the build-up of the gas in the burrow.

High mobility

For a surface-active animal in a hot desert, it is particularly useful to be able to locomote at high speeds in short bursts, allowing darting movements between patches of shade or shelter, and rapid retreat when physiological tolerances are pushed close to their limit. Species of *Onymacris* and *Stenocara* (both tenebrionid beetles) utilize rapid sun–shade shuttling as the primary means of maintaining T_b. *Onymacris plana* runs very rapidly between shady spots with the body raised about 15 mm above the substratum, 'stilting' on its long legs; a comparative phylogenetic analysis has confirmed that desert species in this genus do have longer legs than coastal species.

This point emphasizes that there may also be a range of special **locomotory tricks** to move through or over hot sand while minimizing contact of the substrate with the bulk of the body, to reduce conductive heat gain (Fig. 14.9). The sand-swimming moles of the Namib Desert hardly achieve the latter, but do move more rapidly through the hottest sand. Lizards run on tiptoe, and when briefly stationary will raise alternate feet high off the ground in turn to avoid superficial burning. Sidewinder snakes slide rapidly sideways across dune slopes, leaving only an intermittent track as parts of the trunk are always off the substratum. Even more curious is the cartwheeling strategy of a dune spider, adopted during rapid descent of dunes (Fig. 14.9c).

Rapid and darting modes of locomotion are also associated with active foraging, often from a central place such as a nest or burrow. In rodents this commonly involves short, darting forays, using the characteristic hopping gait of the small gerbils and kangaroo rats. In social insects such as ants and bees it may involve predetermined trails, again helping to reduce time on the surface seeking food; the preferred trails lead quickly to known sources of seeds, pollen and nectar, or to prey such as the nests of other ants or termites. These animals often become highly **territorial** and defend their trails and food sources.

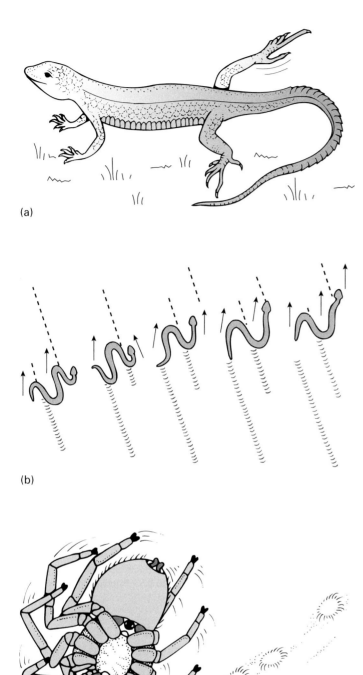

Fig. 14.9 Methods of locomotion on hot sand. (a) Tiptoe stance in lizards (also shown in long-legged insects). (b) Sidewinding snakes, allowing reduced contact with sand. (From Pough *et al.* 1999.) (c) Cartwheeling down dunes in the spider *Carpachne*. (Adapted from Henschel 1990.)

Rhythmic activity patterns

Foraging is rarely continuous in desert animals, but commonly shows a **diel pattern** with numbers of active individuals peaking either at night or at dawn and/or dusk. Scorpions, centipedes and spiders from deserts are far more likely to be nocturnal than are their temperate counterparts, foraging between particular maximum and minimum temperatures. Figure 14.10 shows an example of foraging ants, with the amount of time spent foraging on the surface linked to surface temperature, and the respite length between foraging bouts concomitantly increased as temperature rises. Like many desert species, this ant has an upper tolerable limit of sand surface temperature in the range 47–50°C, and the tight relation between activity and sand surface temperature results in a sharply defined diurnal activity pattern. Other insects may have much lower limits; desert bees may avoid ambient temperatures above about 36°C since their body temperature in flight cannot be greater than around 45°C, while desert caterpillars cease activity above about 34°C and retreat to the shady undersides of leaves or stems.

Many of these desert invertebrates have **endogenous rhythms** that persist under constant laboratory conditions and serve to get the animals away from the harshest surface conditions automatically. Some evidence suggests that the endogenous rhythms of desert animals are particularly likely to be entrained by direct temperature or humidity cues, rather than by photoperiod as is commoner in most aquatic and many terrestrial animals; perhaps this is because in deserts the link between seasonality and productivity is often broken, with spurts of productivity linked instead to sporadic rainfall events. Certainly endogenous rhythms can often be overridden when sudden rainfall produces a period of cool damp surfaces and a subsequent flush of green vegetation, resulting in an abundance of flowers, seeds and ephemeral invertebrate prey items. At the extreme, daily rhythms may be almost entirely suppressed, as in the desert snail *Sphincterochila*, which is active on only a few days each year when the rains have wetted the ground. It may be that endogenous biological clocks have been lost in some desert species, to allow them this more opportunistic approach to organizing their activity.

Where there is a mutualistic relation between food source and feeder, as with pollinating insects such as bees, the plants with which they associate usually also show appropriate diel patterns; many desert shrubs, such as the caper (*Capparis*), open their fragile flowers at dusk and they last only a few hours, during which small nocturnal carpenter bees visit and transfer pollen between individuals.

In some deserts there are also marked seasonal patterns of activity; this depends on the geographical location of each desert, since some experience marked summers and winters and others are nearly aseasonal. **Seasonality** may promote endogenous rhythms entrained by photoperiod. The deserts of Nevada are seasonal with plant growth concentrated in spring and early summer; here desert tortoises show marked seasonal behaviour

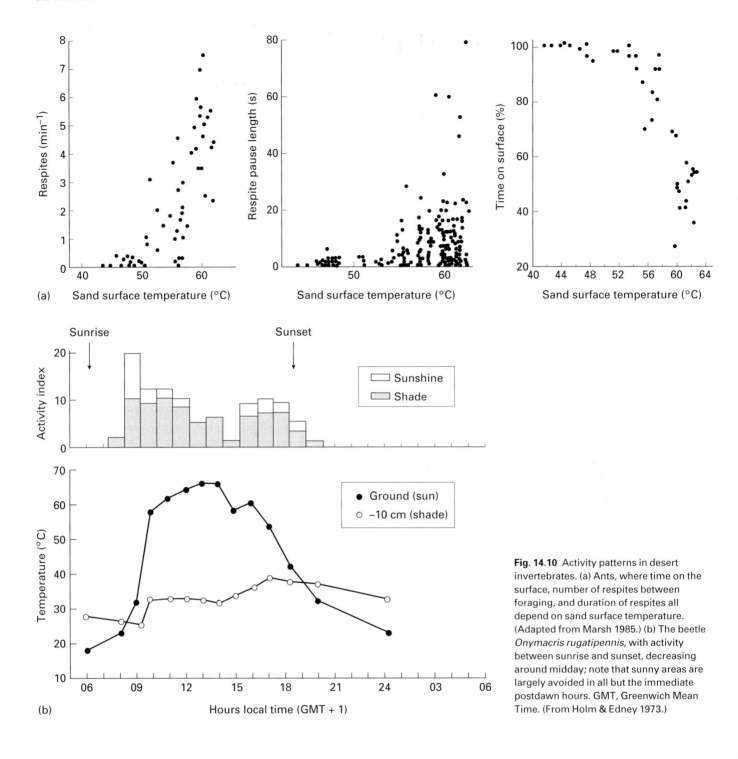

Fig. 14.10 Activity patterns in desert invertebrates. (a) Ants, where time on the surface, number of respites between foraging, and duration of respites all depend on sand surface temperature. (Adapted from Marsh 1985.) (b) The beetle *Onymacris rugatipennis*, with activity between sunrise and sunset, decreasing around midday; note that sunny areas are largely avoided in all but the immediate postdawn hours. GMT, Greenwich Mean Time. (From Holm & Edney 1973.)

patterns (Fig. 14.11a). They hibernate in winter, and in spring are active for only a few hours on some days. At this time the vegetation has too high a water content to give the tortoise a net energy gain; body weight does increase but this is due to water stored in the bladder, and body solids actually decline. In the summer, when the vegetation has dried out somewhat, the tortoises achieve positive energy balance, but incur water loss; there is a depletion of the stores in the bladder coupled with an increase in plasma and urine concentration. In high summer the animals aestivate, emerging briefly only after rainstorms in July; they stay mainly underground until the August/September rains rehydrate the vegetation and give a period of water storage and energy gain prior to hibernation in November. Thus these animals have a seasonal cycle where there is rarely genuine homeostasis; energy and water budgets are balanced over a whole year, but highly varied on a daily or weekly basis.

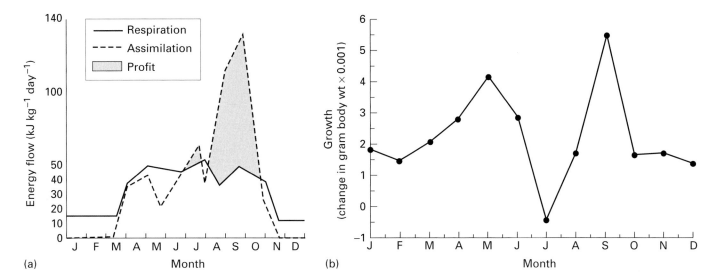

Fig. 14.11 Seasonal energy budgets: (a) for a desert tortoise, with uneven and relatively limited periods of profit when moist vegetation is available; and (b) for a desert scorpion. (a, Adapted from Nagy & Medica 1986; b, adapted from Polis 1988.)

Seasonal patterns of behaviour relating to vegetation quality may also result in complex seasonal patterns of growth rate in invertebrate predators, as shown for a desert scorpion in Fig. 14.11b. Clearly, it is always important to consider the long-term productivity of the whole ecosystem in relation to climate; short-term studies will not do.

Raised thermal tolerance, lowered metabolic rate, and patterns of Q_{10}

Some desert animals are diurnal, and these must inevitably have high thermal tolerances, i.e. **raised upper critical temperature (UCT)**. Desert arthropods commonly have upper limits of 45–47°C. Desert ants, as we have seen, can often tolerate surface temperatures in excess of 50°C for short periods. One species, *Ocymyrmex barbiger*, has a UCT measured at 51.5°C. Possibly even more extreme is an Australian species of *Melophorus*, which can survive for an hour with a T_b of 54°C; it does not become active until surface temperatures rise to 56°C, and has been recorded on sand in excess of 70°C for very brief periods, with no cessation of colony activity even at midday, albeit with long respite periods for individuals amongst cooler vegetation. Some of these animals are clearly on a 'thermal tightrope' in pursuing their activities in a desert, and some do die during foraging treks. Probably such risky tactics are only possible in eusocial animals, where the nest will persist despite some individual losses. There is a trade-off between heat tolerance and 'risk-prone' foraging in Mediterranean ant species, with the more tolerant species foraging 'riskily' closer to their UCT and achieving high foraging efficiency, while less-tolerant species forage in the evening and night for lower rates of return.

Such species may exhibit thickened cuticles and contain lipids with relatively high melting points, but many of them must also have unusually tolerant enzymes and membranes, as yet little studied but assumed to incorporate the kinds of biochemical changes discussed in Chapter 8. There is good comparative evidence that desert species of beetle show a particularly good match between their preferred T_b and their achieved field temperatures, which should allow fine-tuning of membrane and enzyme properties.

It has often been claimed that a **reduced metabolic rate (MR)** is a possible mode of energy conservation for an ectothermic desert animal, either allowing permanently lowered energy expenditure or a temporarily lowered rate during food scarcity, and thus also reducing water losses. The same could be true for small desert endotherms. Some taxa that are pre-eminently desert-adapted, such as scorpions and sun-spiders, have notoriously low MRs. Desert orthopterans, ants and tenebrionid beetles also seem to show a general lowering of MR compared to mesic or hygric relatives; among the tenebrionids, desert species of *Onymacris* show particularly low MRs. Desert lizards (*Uromastyx*) have a lower MR than predicted, as do desert golden moles (only about one-fifth that predicted for their size), probably in both cases linked to low (and poor-quality) food availability. Desert rodents generally have lower resting MRs than mesic species, with indications of somewhat higher rates of non-shivering thermogenesis to compensate for this when necessary. Some small ungulates (e.g. steenbok) and some birds (e.g. pin-tailed sandgrouse) are also reported to have unusually low resting MRs. Perhaps most convincingly, intraspecific effects are well documented. For example, in the brown hare, populations from the Israeli Negev Desert have resting MR values just 61% of those recorded in populations from France. Thus while the data overall are somewhat limited and have not been subject to careful phylogenetic analysis, they do strongly suggest that a general lowering of MR is an adaptive feature. (Nevertheless in some species an opposite effect has been documented; some

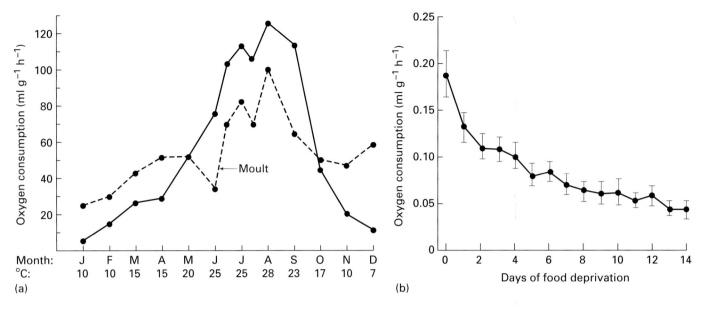

Fig. 14.12 (a) Seasonal differences in metabolic rate for two individuals in the desert millipede *Orthoporus*, measured at the burrow temperature; decreases occur during winter food shortage and at the moult. (From Wooten & Crawford 1975.) (b) The direct effect of food shortage on metabolic rate in the ant *Camponotus*. (From Lighton 1989.)

desert grasshoppers, such as *Taeniopoda*, appear to have unusually high MRs, perhaps compensating for the exceptionally short growing season they experience.)

The ability of desert animals to withstand starvation for months (or even years in a few cases) is also very marked and is accompanied by a further reduction of MR to an extent greater than is possible in mesic congeneric species. Many desert invertebrates that are strongly reliant on plant productivity show a very marked seasonal reduction of MR when food supplies are low (Fig. 14.12a), with a rapid onset of lowered MR when starved of food artificially (Fig. 14.12b). Of course the same effect is often manifest in desert endotherms, where in some cases exceptional reduction in MR is achieved during summer dormancy or aestivation (see below).

Many desert invertebrates also show a particular **pattern of temperature quotients** (Q_{10}; see section 8.2) as indicated in Fig. 14.13 for oxygen consumption (MR) in an ant. The Q_{10} is relatively high at low temperatures, but depressed (i.e. below the expectation of a straight-line fit) in the normal environmental temperatures, rising again at very high temperatures. Remember that a higher Q_{10} means a more substantial response to temperature, so that when ambient temperatures are low and the animals are inactive their MR drops quite substantially with even a small further decrease in T_a. This allows reduced energy expenditure in inactive periods, such as in the desert night. During

Fig. 14.13 Q_{10} adaptation for a desert ant (*Camponotus fulvopilosus*). (a) Declining temperature produces declining oxygen consumption, but the regression line obscures a consistent pattern of relative increases at low temperature and relative reductions close to the normal environmental temperatures (30–35°C). The Q_{10} vs. temperature plot is therefore shown as in (b). (From Lighton 1989.)

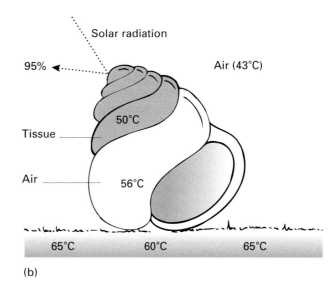

Fig. 14.14 (a) Colour effects in desert beetles, where a species with black elytra has substantially higher temperatures than a similar species with white elytra. (Adapted from Henwood 1975.) (b) The effectiveness of the highly reflective white shell of the snail *Sphincterochila* in reducing heat stress for the soft tissues within. (From Schmidt-Nielsen *et al.* 1971; courtesy of Company of Biologists Ltd.)

normal foraging, in contrast, the Q_{10} is relatively low and there is a more limited response to changing ambient temperatures so that the animals avoid undue 'thermal acceleration'. This kind of pattern has been shown in various ants, orthopterans and beetles, as well as in desert spiders and scorpions; in the active range the values of Q_{10} are commonly only 1.5–2.2.

Colour, shape and posture

Invertebrates encountered in deserts are commonly almost black, almost white, or of a nondescript sandy brown colour. Reasons for all three colour schemes are not hard to find. Black is characteristic of some eocrepuscular species, since dark surfaces allow the body to warm up more quickly in the limited daylight at these otherwise rather cold times of day. White surfaces of very high reflectivity are found in some species active or exposed in the heat of the day, and can be shown to be effective; for example, the beetle *Onymacris brincki*, with white elytra having a reflectance of 35%, has an abdominal temperature of only 37°C in full sunlight when the ground temperature is 47°C, and when the abdomen of a similar beetle with dark elytra is at 43–45°C (Fig. 14.14a). The aestivating snail *Sphincterochila*, which has one of the most highly reflective of all biological surfaces (95%), also demonstrates the efficacy of white surfaces (Fig. 14.14b). However, it is equally true that many insect species active in daytime in deserts are black, which appears to rather spoil the argument. In fact for species with short legs that live very close to the ground colour may be relatively unimportant in thermal terms, since most of their heat gain will come from

longwave radiation reflected off the soil, and absorption of this is unaffected by colour. It may be no accident that the more spectacularly long-legged insect species, gaining heat more from direct shortwave radiation, are more likely to be 'adaptively' white or black, whereas creatures such as scorpions that are flattened closer to the sand are usually drab-brown or dull-black (even those species active by day).

Sandy brown surfaces may also be the best compromise where the animal is essentially burrowing, emerging only briefly and perhaps subject to intense predation pressure, so that the need for crypsis dominates over thermal needs; this is particularly true in the smaller invertebrates. Similar pale, sandy colours are found in most small diurnal mammals in hotter climates (though often with counter-shading for additional camouflage purposes).

Colour change is also reasonably common in desert animals. Quite a number of desert arthropods show a special cuticular feature in having substantial 'wax blooms' on their surface, especially at low humidity. These may be formed from a meshwork of extruded wax filaments (Fig. 14.15), which can be rapidly regenerated if damaged. In some tenebrionid and buprestid beetles they affect the colour of the animal, with *Onymacris rugatipennis* being whitish-blue when the bloom is present but black when it is removed. The amount of wax may vary with the season, being greatest in the hottest months, and it probably helps to protect the animals from solar radiation, having a high reflectivity. Certain desert grasshoppers also change colour, from bright blue-green when warm to almost black when chilled; some desert agamid lizards may use the same trick, being dull dark-brown at dawn and dusk and a bright turquoise around midday. A reed-frog, *Hyperolius viridiflavus*, from very dry African savanna regions, has been shown to change colour dramatically in the dry season as it begins aestivation. Initially the skin turns from yellow-brown to a brilliant white; but as dehydration sets in it turns from white through coppery iridescence and finally to green iridescence when more

Fig. 14.15 Wax blooms extruded on the surface of desert beetles, increasing reflectance and reducing water loss. (From Louw 1993; photographs courtesy of N.F. Hadley.)

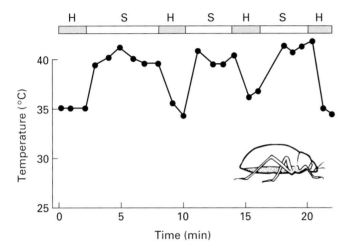

Fig. 14.16 Orientation effects for a desert beetle; thoracic temperatures are reduced when head-on to the Sun (H) and rise in the side-on (S) position. (Adapted from Edney 1971.)

than 25% of the body mass has been lost. This change is brought about by a layer of iridophores in the skin, with crystals causing multilayer interference reflection and hence iridescence; purine deposition in the skin accompanies this process. In dry-adapted skin there are 4–6 layers of iridophores in a layer up to 60 µm thick and with a reflectance of around 65%.

Changes to the fur are also recorded in endotherm evaders, with summer coats usually paler than winter ones. In the Sonoran desert rock squirrels the fur is renewed seasonally and the animals absorb 33–71% less solar heat load in summer as a result (though there is no obvious visible change, suggesting rather subtle changes in structural and optical properties of the hairs).

Colour is highly interactive with shape and with postural adjustment, to get maximum benefit from particular pigmented surfaces and from areas of large surface area. The reed-frog *Hyperolius*, mentioned above, shows 'body shape optimization', with a half-cylinder shape rather than the more common half-sphere, allowing a 'head-on' posture where a large area for conductive and convective heat loss can be shaded by a small area exposed to the direct solar radiation; this frog can survive 3 months of aestivation without water at temperatures up to 45°C. Heat gain in many desert animals is avoided in similar fashion, usually involving orientation into the Sun's rays; the desert beetle shown in Fig. 14.16 illustrates this point. Ideally, and whatever the taxon, the shape should be elongate rather than squat, with the front or rear, and the midline dorsum, as pale as possible.

Control and tolerance of water loss

Many desert animals are exceptionally **impermeable**, and an overview is given in Table 14.1 (see also Tables 5.3 & 13.7). The lowest values of water loss rate have been recorded for certain scorpions and ants, at 0.02 mg cm^{-2} h^{-1} (around 0.03–0.04% fresh weight per hour), with certain tenebrionid beetles (e.g. *Onymacris plana*) only slightly higher at around 0.1 mg cm^{-1} h^{-1}. This impermeability is largely due to the normal cuticular layers described in Chapter 13, which may be enhanced with more epicuticular lipids. However, there are also special cases of 'extra' impermeability; the wax blooms in tenebrionids, described above, certainly help to reduce water loss as well as limiting thermal gain, so that in field-caught *Cauricara* beetles the water loss rate is about 40% higher when the wax bloom is at its minimal thickness in August. In *Onymacris* the occurrence of wax blooms is significantly commoner in desert-interior species than in coastal species. For desert insects in general, cuticular water loss (cutaneous evaporative water loss, CEWL) has been so much reduced that respiratory water loss (respiratory evaporative water loss, REWL) is the dominant component of water balance.

However, while undoubtedly important, relationships between cuticular properties and water loss rates can be oversimplified. For example, the desert fruit fly, *Drosophila mojavensis*, does have more cuticular lipid, and longer hydrocarbon chain lengths in these lipids, than more mesic fruit flies; but since it is also smaller when it develops at high ambient temperatures, its water loss rate per unit body mass is in fact higher than pre-

Table 14.1 Water loss rates and permeabilities in desert animals.

Animal group/ genus	Resistance (s cm^{-2})	Water flux (mg cm^{-2} h^{-1})	Transpiration rate (µg cm^{-2} h^{-1} mmHg^{-1})	Water turnover (ml g^{-1} day^{-1})
Arthropods				
Isopods				
Hemilepistus (woodlouse)			14–23	
Venezillo (woodlouse)			15–32	
Arachnids				
Scorpio (scorpion)	1317			
Androctonus (scorpion)	4167		0.6	
Hadrurus (scorpion)		0.02	1.2	
Buthus (scorpion)			1	
Leiurus (scorpion)			0.8	
Latrodectus (spider)			1.2	
Ornithodorus (mite)			2–4	
Myriapods				
Orthoporus (millipede)	433		8	
Alloporus (millipede)		0.02		
Insects				
Thermobia (apterygote)			15	
Ctenolepisma (apterygote)			0.7	
Rhodnius (bug)			12	
Geocoris (bug)			14	
Arenivaga (cockroach)			12–80	
Diceroprocta (cicada)			100	
Locusta (locust)		0.70		
Onymacris (beetle)	5030	0.10	3.1	
Centrioptera (beetle)			6.3	
Cryptoglossa (beetle)			8.4	
Lepidochora (beetle)			3.1	
Tenebrio (beetle larva)			5	
(beetle pupa)			1	
Eleodes (beetle)		0.20	17	
Glossina (tsetse fly)			8	
(pupa)			0.3	
Manduca (caterpillar)			40	
Anaphes (wasp)			4–8	
Pogonomyrmex (ant)			26	
Messor (ant)			18	
Vertebrates				
Pternohyla (frog, cocooned)	457			
Gopherus (tortoise)	120	1.56		0.003
Sauromalus (lizard)	1360			
Dipsosaurus (lizard)		0.13		0.03
Gehydra (lizard)		0.22		
Pitnophis (snake)		0.23		
Struthio (ostrich)	158	1.73		0.09
Dipodomys (kangaroo rat)		0.53		0.03
Gerbillus (gerbil)				0.13
Peromyscus (cactus mouse)		0.66		
Capra (desert goat)		2.10		0.09
Oryx		3.2–5.9		0.03
Megaleia (red kangaroo)				0.09
Camelus (dromedary)				0.06

dicted, and it has to rely on avoidance of really desiccating conditions. Routinely higher values of cuticular water loss occur in many species from all taxa that are more inclined to seek protected microclimates or moister foods, giving a good inverse correlation between moisture content of the microhabitat and/or diet and the desiccation resistance of a species. For example, the nocturnal honeypot ant, *Myrmecocystus mexicanus*, is more vulnerable to water stress than its diurnal relative *M. mendax*. There is also the complication of EWL ('sweating') in certain insects, as a strategy to avoid lethal overheating; this occurs in some desert cicadas and grasshoppers through cuticular pores, since they have ready access to water in their food (see below). As always, then, it is important not to over-generalize, since there are a range of possible adaptive strategies within and between species, with different balances between physiological and behavioural solutions to problems.

Some desert animals may also be very tolerant of **water loss** (see Table 5.2). Insects from arid habitats may tolerate over 50% water loss, with 75% recorded for some tenebrionids; in such cases dehydration is largely at the expense of the haemolymph, which may be reduced to less than 5% of its normal volume in a very dehydrated beetle. Amino acids contribute to regulation of osmolarity during dehydration (as in estuarine animals, see Chapter 10), with glycerol also playing a role as an osmotic effector. The snail *Sphincterochila* survives 50% loss, and values up to 48% have been recorded in desert toads such as *Scaphiopus*, where again loss of blood volume occurs. Dehydration inevitably accompanies aestivation in many animals, and with no urination occurring there may be a substantial accumulation of the excretory product. For some aestivating Australian frogs, urea levels rise to 100–300 mmol l^{-1}. However, this does not appear to be accompanied by substantial coaccumulation of balancing osmolytes such as methylamines (e.g. trimethylamine oxide, TMAO) or polyols (e.g. sorbitol) as in other urea-accumulators, like elasmobranch fish (see Chapter 9). It may be that the amphibian enzymes are insensitive to perturbation by urea, or even that the perturbing effects are used as an aid to metabolic depression while aestivating. Urea retention linked to having a bladder as a water/urea reservoir was perhaps a 'preadaptation' to becoming terrestrial in amphibians.

Tolerance of water loss in endothermic birds and mammals is a rather different phenomenon, in that the blood is necessarily a much more highly controlled fluid; many of the small desert mammals and birds are characterized by a particular ability to maintain an almost constant plasma volume even when losing substantial body mass and total body water, with the water being lost from other compartments, including the cells, where concomitant adjustments of amino acid and other osmotic effector levels must be involved (see Chapter 5).

Concentration of the excretory product into very **hyperosmotic urine**, with minimal fluid excretion, is an obvious desert adaptation, and may be allied to water-storage systems. Many desert tenebrionid beetles exhibit the cryptonephridial system

for concentrating their urine and faeces (see Chapter 13) and therefore also potentially gain a method of **taking up water vapour**. Desert woodlice (*Hemilepistus*) can resorb extra water from fluid passing up into the rectum after flowing over the pleopods where ammonia is lost. Desert toads use the bladder as a massive water reserve during drought; in contrast, some desert tortoises lack a water-storage bladder (always found in their temperate relatives) and instead have powerful nasal salt glands and cloacal resorptive tissues to regulate their salt and water balance. Desert rodents are famous for having long loops of Henle, in proportion to their size; as we saw in Chapter 5, this is linked to concentrating their urine such that the cells lining the loop are particularly well endowed with mitochondria to support high salt transport rates. The same is true of desert birds, where the resorptive rectal epithelium is also strongly developed. Desert populations of the brown hare form urine up to 4470 mOsm, compared to 2500 mOsm in European populations.

Changing the excretory product also helps; desert toads have switched to excreting urea rather than ammonia, while many other small desert vertebrates (reptiles and birds) excrete most of their nitrogen as **uric acid**. This latter product is ubiquitous among the most successful desert invertebrates, and is commonly excreted as an almost solid paste or crystals of uric acid and urates, with virtually no accompanying water.

The net effect of all these water-retention adaptations can be quite remarkable, and has been best studied in the small desert rodents such as the kangaroo rat (*Dipodomys*) and the hopping mouse (*Notomys*). These animals feed on very dry seeds, and in 1 month may gain less than 1 ml of water from that contained directly in their food; by far the greater gain comes indirectly from the food as metabolic water, perhaps 50 ml. In the same period they lose the same total water, in the ratio of roughly 75% by evaporation, 5% in the faeces and 20% in the urine (at 5000–9000 mOsm). Plasma volume is maintained almost constant under all conditions, and no drinking-water is needed.

Note that water conservation systems cannot be divorced from other problems arising in deserts, notably thermal stress, with similar behavioural strategies alleviating both problems. For example, relatively permeable evaders, such as desert toads, will choose cooler habitats when they are particularly water-stressed, and may show a greater propensity to burrow and to huddle together.

Water uptake and use of condensation

In deserts, the condensation at dawn as the overnight temperature profiles invert can be very substantial, but the dew that forms would normally dissipate into the sand and evaporate very quickly. However, in certain deserts, such as the Namib, dense morning fogs may occur where the water vapour from adjacent cool oceans rolls in over the sands, and condensation may then persist long enough to be really important to the fauna. Some desert tenebrionid beetles have ways to exploit this.

(a)

(b)

Fig. 14.17 The water uptake mechanism in the desert cockroach *Arenivaga*. (a) Fluid is produced in the frontal bodies and flows onto the extruded bladder surface; (b) water vapour is absorbed onto this surface at humidities above 83%, and then drawn back into the oesophagus. (a, Adapted from O'Donnell 1987; b, adapted from Edney 1966.)

Onymacris unguicularis ascends to the tops of dunes during dense dawn fogs, and stands with its hind legs fully stretched up and its head down ('fog-basking'). Water condenses on the body and runs down towards the head where it is drunk, giving up to 34% increase in body weight. *Onymacris plana* selects dunes facing in different directions at different times of day on which to bask (east-facing at dawn, west-facing at dusk, retreating to its burrow at midday; yet other beetles (genus *Lepidochora*) dig short trenches in the sand to gather the condensing dew. It is not only beetles that benefit from desert fogs; water will condense on any surface in these conditions, and many animals lick it off the vegetation. Some geckos and snakes also lick it from their own bodies, and spiders gather it from the hygroscopic silk of their webs.

Water acquisition from sources other than free liquid is the only other option. We have seen that some desert amphibians extract water from soil through their permeable skin when there is a brief rainy period. Some desert arachnids may achieve the same trick at such times, using their suctorial apparatus (see Chapter 13). But this is a rare adaptation in deserts, as soils are normally far too dry for any osmotic or capillary uptake to be possible.

The alternative of using active **water vapour uptake** mechanisms (see Chapter 5) is of obvious utility in deserts. Certainly some species do it via the mouth and salivary glands, including the millipede *Orthoporus*, some desert mites, and the North American desert cockroach, *Arenivaga*. In this well-studied animal uptake is observed above 83% RH (Fig. 14.17b) in association with the protrusion of two salivary sacs (Fig. 14.17a) from below the mouthparts. The concentration of the salivary fluid alone is inadequate to explain vapour uptake, and the secreted fluid is not hygroscopic. Instead a partially physical explanation involving the mat of cuticular hairs on the exposed sacs has been invoked. These hairs appear to undergo cyclical swelling; they take up condensed water when they are exposed and shrunken, releasing this water osmotically to a new flush of salivary fluid which is then resorbed into the mouth along fine capillary channels, allowing the hairs to shrink again. Other desert animals acquire vapour at the alternative site of the rectum, and many tenebrionid beetles achieve vapour uptake using the cryptonephridial chamber. At least one case of vapour uptake by eggs has also been demonstrated, in the Australian desert stick insect *Exatostoma*, which can take up vapour at 40% RH. The 'sneaky' tactic of indirect uptake of water vapour, by storing seeds and consuming them after they have taken in vapour from the air of a humid burrow, could also be mentioned here (it is dealt with fully below).

Respiratory control

By a combination of spiracular control and reduced overall respiratory activity, desert insects are commonly able to maintain a low REWL. Early gravimetric studies were criticized for showing apparently very tight control; for example, at 25°C resting beetles of the genus *Eleodes* appeared to lose only 3% of their total EWL through the spiracles. With more sophisticated techniques using radioactively labelled water most studies have found much variation but in general somewhat higher values; for example, REWL is about 4–5% of total water loss in the

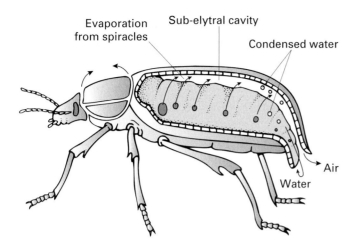

Fig. 14.18 The sub-elytral cavity of a tenebrionid beetle, into which the spiracles open, allowing the cavity to act as a condensation chamber, water being returned to the body via the anus. (Adapted from Ahearn 1970.)

grasshopper *Taeniopoda*, and 8% in the ant *Cataglyphis*, but these values are still notably low (cf. Chapter 13). In larger insects with very impermeable cuticles the value may be higher, so that in the tenebrionid genus *Onymacris* up to 45% of total water is lost through its spiracles. As temperatures rise the control of respiratory water loss also generally becomes less effective, and at 40°C (not uncommon in the Namib Desert) REWL in beetles may be 40–70% of total EWL.

Beetle elytra are thick, and as we noted above in some ten-

Fig. 14.19 Discontinuous respiration in desert species. (a) Oxygen uptake and highly discontinuous carbon dioxide loss in the beetle *Psammodes*. (b) Carbon dioxide loss ($\dot{V}_{CO_2}$) and water loss rate (WLR) in the grasshopper *Romalea*. (Cf. Fig. 13.40.) F, flutter phase; O, open phase; C, closed phase. (Adapted from Lighton 1988; Hadley & Quinlan 1993.)

ebrionid species they are white and highly reflective to reduce radiative heat gain. The air within the **sub-elytral cavity** (below the wing cases, and sheltering the spiracles) is therefore somewhat cooler than ambient air, so that some of the water vapour in the expired air will condense onto the beetle's dorsal abdominal surface, and is recovered via the spiracles or anus (Fig. 14.18). (The system is also found in species with black elytra, however, where the sub-elytral air is actually hotter than ambient; therefore it may be that the major function of the cavity is more mundanely to reduce air flow over the spiracular openings and so limit water loss.)

Many desert insects use a **discontinuous respiration** (DVC; see p. 463), with periods of apnoea, as discussed in Chapter 13. This is particularly common in burrowing species, including many tenebrionids and most ants. Compared with similar mesic species, these ants and beetles show a longer cycle of spiracle closure, and a relatively long flutter phase with a shorter fully open ventilatory phase (Fig. 14.19), presumably giving a saving on water loss. The water saving in cycling animals may be rather limited; it only occurs when the animals are motionless, and the REWL of small desert insects is rather low anyway. Thus savings might be quite limited, but in a desert they may be large enough to make the difference between a positive and a negative net water balance.

Respiratory control of water loss is very difficult in the small endothermic desert vertebrate evaders, with their very much higher metabolic rates. However, **nasal heat exchange** is improved (temperature of expired air is reduced, closer to temperature of inspired air, or T_a) in many species of small desert mammal and bird that have elongated and elaborately coiled nasal passages (turbinals; see Fig. 13.37); and in the kangaroo rat, *Dipodomys*, the expired air is actually cooler than the inspired air (Fig. 14.20). The important point here is that any cooling of the expired air relative to its temperature at the respiratory surface of the lung will cause it to give up moisture as it leaves the body, since saturated cold air holds less water vapour than saturated

(a)

(b)

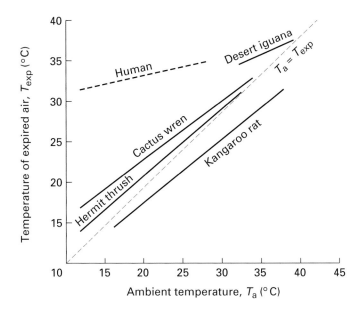

Fig. 14.20 Temperature of expired air (T_{exp}) in desert vertebrates; it is substantially below air temperature, allowing significant water saving (water condensing in the nose). (Data from Murrish & Schmidt-Nielsen 1970; Schmidt-Nielsen *et al.* 1970b; and other sources.)

warm air. Hence water condenses in the elongated nose, which is a site of countercurrent recovery both for heat and water.

Manipulating the microclimate

Many arthropods make their own microclimates, by building burrows (dealt with above) or more specifically by making nests, usually designed to keep the occupant cool. Spiders' nests in the Negev are an interesting exception, being sited on the hottest side of shrubs apparently to increase prey capture while minimizing disturbance from possible browsers. The nest can become very hot indeed, and the adult spiders may move out to the nest entrance during the midday hours; their nests are clearly not a thermal refuge. In fact these spiders are exemplary 'maxitherms', living close to their thermal limits, and not surprisingly the females are a very pale cream colour, cream models being shown to stay cooler than black models.

However, the nests of ants and termites are the most spectacular examples of managed microclimate, with the insulation, ventilation and orientation of the nest all attended to. Termites, for example, are mound-builders in most of Africa and in Australia; in forest areas their nests are compact domes, whereas in semi-arid areas the nests may appear as complex turreted and ridged mounds. But in the hot and/or dry deserts of central Asia termites are mostly subterranean, while in the Sahara there are just a few species and they live almost entirely underground, feeding on remnant semi-fossilized tree trunks from more humid Pleistocene periods. The best-known examples of sophisticated termite nest architecture are the 'compass termites', their nests always being built with the long axis running N/S, so that the structure warms up quickly after dawn but keeps relatively cool in the middle of the day with minimum surface exposed to the Sun. Elaborate patterns of vertical shafts provide ventilation systems (Fig. 14.21a).

Ants' nests are usually underground, and may extend down for several metres (Fig. 14.21b), providing different microclimates at different levels for different functions. For example, *Pogonomyrmex rugosus*, a seed-harvesting species, constructs its storage chambers in the upper 0.7 m of the soil profile, but has its workers (mainly) and larvae (almost entirely) living in galleries 1.5–2 m down.

Getting the right food

Feeding in deserts is necessarily opportunistic, and many species live close to starvation for large parts of their lives. Invertebrates are particularly subject to long periods when there is no food available, between intermittent rains. Desert spiders, for example, may make a capture on fewer than 5% of all nights, and on only 1.5% of all nights during drought periods.

Most of the desert insects are herbivores, many feeding on seeds, which survive for very long periods between rains because the rate of decomposition in a desert is very low. Rodents and some reptiles use the same food resource. Relatively few insects (aphids and hoppers, orthopterans and a few caterpillars) and a few small mammals use the other more succulent parts of plants (often below ground). A small guild of nectarivores usually occurs, often dominated by solitary or semi-social bees from the family Anthophoridae in particular ('flower bees' and 'carpenter bees', particularly diverse in deserts) with some bee-flies (Bombylidae) and small sphinx moths (Sphingidae).

Selection of plant food in deserts may perhaps be more relaxed than in other terrestrial habitats. Granivores certainly collect seeds that are higher in protein content and lower in secondary metabolites by preference, but may be forced to take a wider range in xeric climates than in mesic zones. Foliar feeders may face particularly well-defended plants, since no desert plant can easily afford the desiccating effects of being damaged; physical defences such as spines are ubiquitous and require careful manipulation by any intending herbivore. Chemical defences are also very common, and it is perhaps no coincidence that so many desert plants are used by local human populations as medicines. It might be a fair generalization to say that desert herbivores are more likely to be *either* very generalist, taking whatever is available at any one time, *or* very specialist, having a coevolved dependence on a very few species of relatively abundant well-defended plants; it is the oligophagous species that are quite rare.

Generalist adaptations for desert herbivory probably involve large-capacity but relatively simple guts, able to provide flexibility in dealing with whatever may be available. This is exemplified by the desert tortoise, *Xerobates agassizii*, which eats very fibrous grasses or much softer herbaceous plants according to

Fig. 14.21 (a) Termite and (b) ant nest architecture and temperatures.

the season, utilizing colonic fermentation (see Chapter 13). Its gut constitutes up to 21% of body mass, and is heavily endowed with mucus glands to provide some protection from abrasion, but is otherwise relatively unmodified.

Certain desert plants are halophytic (growing on very salty soils) and attract only specialist feeders because of their high electrolyte content; a classic case is the saltbush, *Atriplex*, common in deserts of the Middle East. Fat sand rats (*Psammomys obesus*) are diurnal rodents that feed specifically on saltbush tissues, but they always scrape off the outermost saltiest layers of the plant leaves before ingesting the underlying tissues.

Conspicuous guilds of specialist feeders may also occur on highly toxic plants such as asclepiads (milkweeds) and solanaceous herbs (the deadly nightshade family); these specialist herbivores can be picked out as they are often aposematically coloured, warning of their own acquisition of some of the plant's chemicals. In some cases desert herbivores can get around this problem of secondary plant chemicals by selectively feeding on plants at times when such defences are reduced; for example, in North American deserts the creosote bush (*Larrea*) has reduced

levels of translocated defensive chemicals in the evenings and is then much more strongly grazed. It has also been suggested that after rains many desert plants are bound to invest more in growth than in defence and so may be particularly favoured by herbivores at such times, many animals taking in plants that they would have avoided during droughts when chemical defences were stronger.

Given a range of generalist and specialist herbivores, there are clearly niches for a limited number of predators, and in most deserts the arachnids and the snakes are pre-eminent. They survive largely by eating the plant-feeding insects, but with the use of poisons they can also take on much bigger creatures such as lizards and rodents. By eating animal flesh, they automatically acquire a well-balanced and relatively moist diet (about 70% water content), so may need little or no extra drinking-water. Spiders have very low metabolic rates and can recycle their silk and recoup its energy, giving very high starvation resistance.

For all desert animals, it is advantageous to take in as much food as possible during the transient periods of plenty. An active desert snail (*Sphincterochila*) may take in nearly 50 times as much energy as it uses in one day of foraging activity. Many social insects and also most of the desert rodents will store food at times of plenty to use later. Sometimes this involves storing within the body itself; laying down fat reserves prior to aestivation is one example, but a more extraordinary version of the same principle is found in the honeypot ants (*Myrmecocystus* and other genera), which use a few members of the colony as living underground storage jars, their abdomens hugely distended with gathered nectar.

Extracorporeal stores are more normal though. Here, storing seeds is a particularly sensible strategy given the low rate of attack by fungi and other decomposing agents, and is also favoured because many seeds will take in water hygroscopically when stored in a burrow at raised humidity, so giving the prudent gatherer an extra source of water in the future. For example, dry seeds collected by gerbils above ground at 10% RH contained 3.7 g water per 100 g; stored at 75% RH in a relatively short burrow, their water content increases to 18.0 g per 100 g. Rodents with cheek pouches for gathering seeds should be best equipped to exploit this effect. However, it has recently been realized that a 'normal' rodent with internal pouches actually loses a significant amount of salivary water to dry seeds when these are held in the pouch for just a few minutes, and cannot recover this water as it mainly stays in the indigestible seed coat. Only some specialist desert rodents with external fur-lined cheek pouches can avoid this. This phenomenon allows many desert heteromyid rodents, such as kangaroo rats, pocket mice and kangaroo mice, to survive on little else but seeds; the kangaroo rats (*Dipodomys*) of the Mojave Desert may eat nothing but dry *Larrea* (creosote bush) seeds for 9 months of the year, without needing any drinking-water.

The sheer volume of stored food in a desert can be astounding; a single termite colony can accumulate 0.5 kg of dried grass, and a nest of harvester ants may contain at least 100 g of seeds, more than four times the total biomass of the ants. One nest of *Messor* ants was calculated to contain around 170 000 seeds!

Aestivation

Many semi-arid and desert animals that are essentially evaders use a period of physiological inactivity as part of their survival repertoire, and this is conventionally termed **aestivation** (summer sleep) as it is usually timed to avoid the hottest and driest periods. MRs are reduced, and usually the thermal tolerance limits are expanded. Growth and reproduction cease, and the animal becomes relatively unresponsive to external stimuli.

As with winter dormancy and torpor, terminology is confusing and different terms are used in different fashions. A period of behavioural inactivity without much physiological adjustment is often termed **quiescence**, whereas deeper states of withdrawal including modified physiological states are termed aestivation, and the deepest forms of metabolic suppression (with no visible sign of life) are called **cryptobiosis** (see Chapter 12). Where the dormant state is triggered internally and occurs as a regular and defined part of the life cycle, particularly in arthropods, it is usually termed **diapause**. These states also differ in the method of arousal, since quiescence normally

ends automatically when conditions improve whereas diapause requires specific signals and internal control mechanisms.

Many burrowing desert invertebrates show periods of quiescence, usually plugging up their burrows, both at the individual level as in spiders and sun-spiders, and at the colony level as in ants. More prolonged dormancy (aestivation) is very characteristic of desert snails, which may spend a large part of their life in this state (up to 98% for *Sphincterochila* in the Negev). In snails such as *Helix* MR is suppressed by 84% after a month in aestivation, and in the more drought-resistant snail *Rhagada* this rises to over 90%, with a 97% reduction in evaporative water loss (and thus relatively little lowering of body water content). In the snail *Otala*, aestivation-specific proteins are synthesized in the hepatopancreas during the process of reducing the MR. This state, with the shell sealed off by a calcareous lid (epiphragm), can prolong survival time to several years. Different species of snail arrange themselves with the body suspended in air on the coolest side of a desert shrub, or on the least insolated side of a rock; the air space within the shell serves as extra insulation for the tissues (see Fig. 14.14b). A few species aestivate in large aggregations, with those at the base of the clump having the best survival chances as they are protected from the sun.

Diapause, as we saw in Chapter 8, is a complex phenomenon of arrested growth and development occurring in arthropods in response to environmental adversity, and it may affect physiology, behaviour and even morphology. It is normally triggered well in advance of the onset of really harsh conditions (see Fig. 8.45). Desert species often lack the photoperiodic cues commonly used by mesic species, relying instead on triggering by food scarcity or climatic factors, always acting cumulatively some time in advance of the onset of the diapause state and so allowing extra control of its occurrence. Thus diapause may persist over many months or even several years in the absence of desert rains.

Aestivation in ectothermic vertebrate evaders occurs in many amphibians and reptiles, which burrow into soil and in the case of amphibians may form a cocoon. Aestivation is particularly well studied in the genus *Scaphiopus* and in several frog genera from Australia, with MR declining by 50–67%. *Scaphiopus* can spend 10 months of every year in aestivation utilizing stored lipid reserves, with reduced haemolymph volume and increased haemolymph concentration accompanied by urea accumulation, as we saw above. In this species the antioxidant enzymes (needed to protect the tissues during arousal and reoxygenation, see Chapter 6) are both abundant and unusually insensitive to urea.

Aestivation also occurs in desert endotherms in drought periods, with behavioural lethargy and reduction of T_b to a level close to ambient; it is well known in desert ground squirrels, gerbils and some possums, where T_b values during the 'summer rest' are commonly around 25–30°C. Gerbils and other burrowers may also show circadian patterns of torpor during daylight, allowing T_b to drop to the burrow temperature of about 30°C (occasionally as low as 19–20°C), thereby saving considerably on both food and water.

Reproduction: protecting eggs and young

Life-history strategies consisting of short lifespan, high reproductive output and frequent breeding are usually debarred by
the scarcity of resources in deserts; infrequent but regular cycles
are also difficult given the unpredictability of the habitat. Many
desert animals are therefore long-lived and breed opportunistic-
ally and very quickly whenever there is rain, being effectively
sterile in some years. Certainly the more traditional terrestrial
patterns of seasonal breeding are rare. Different species may be
more to the *r*-selected or *K*-selected end of the spectrum within
these broad generalizations (see Chapter 1), and in some ways
desert animals therefore represent a mixture of the classic *r* and
K characteristics, conflicting with the normal demarcations of
life-history strategies.

Cryptobiotic nematodes living in the sands can become active
and lay eggs within minutes of their first wetting, and certain
collembolans, mites and isopods can complete a generation
within a week or two. At least some invertebrates appear to be
able to oviposit in anticipation of rain, and the juveniles therefore appear just as the flush of green plants and associated
insects occurs. Many of these examples are underground cryptic
species, and they produce very large numbers of offspring with
every onset of rain, most of which do not survive.

Producing relatively few young in each breeding episode,
and investing heavily in those few offspring, is also favoured,
particularly among surface-dwellers, where maternal care is
therefore rather common. In some special cases the mother
may control the microclimate of the juveniles herself: ants move
the brood around the nest to the most favourable microhabitats,
and many spiders carry their eggs around in an egg sac. Spiders
also take care of their spiderlings, and the Namib dune spider
(*Leucorchestris*) is known to feed its young within the burrow.
Scorpions are particularly renowned for prolonged care of juveniles, these being carried around on the mother's back.

Most of the arthropod evaders burrow and therefore also lay
eggs underground. Even the large migratory locust, unable to
burrow, achieves underground eggs with its very long oviposi-
tor at the end of a telescopic abdomen. Alternatively eggs may
be laid on or in the host food-plant, taking advantage of microclimates on the underside of leaves or within stems or flowers.

The spadefoot toads (*Scaphiopus*) are well known for their
specialized tricks for reproducing in deserts, though they retain
the aquatic egg-laying habit of most mesic amphibians. They
depend on careful timing and very rapid development for their
success; during the seasons when rain might be 'expected' their
gonads mature and they move each night close to their burrow
surface, retreating down again by day if there is no rain. But
when rain does come (even a light shower, <0.5 mm of rain) individuals emerge almost instantly and seek out a temporary
surface pool of water, arriving at this 'breeding pond' on the
same day as the first rain and laying eggs that night. By the next
morning the embryos are well developed and have acquired
moderate thermal tolerance (unusual in amphibian eggs); they
complete their metamorphosis in 2–3 weeks instead of the 10–12
weeks of 'normal' frogs and toads. The pools they live in may
start to dry up and become very crowded, and this triggers a further response, as some of the tadpoles then turn cannibalistic
and develop hypertrophied jaws, speeding up their own development even further with a high-quality tadpole diet!

The desert tortoises, whose seasonal cycle was referred to earlier, manage to reproduce every year despite the variation in
water and food availability, reducing their field MR in drought
years and forfeiting body condition to produce a few eggs; this
'bet-hedging' life-history strategy is probably typical for a relatively large and long-lived desert ectotherm.

Desert rodents use a variety of different reproductive patterns. Some are 'pulse averagers', with slow prolonged reproductive effort matching the historical probability of rainfall
timing; some are 'pulse matchers', responding directly to rain
and food availability; some, which store foods or use completely
reliable foods, are 'pulse ignorers'; and yet others are 'pulse
gamblers', producing large litters at the end of their hibernation
period irrespective of conditions, some of which fail completely.

The most striking factor in many desert-breeding species is
their ability to detect and respond to the very smallest quantities
of rain almost instantly. Spadefoot toads appear to use the low-
frequency sound of rain drops on the ground as their cue; some
desert birds use even the distant sight of rain, or the accompanying slight rise of humidity, as their cue for initiating nesting and
egg-laying.

Sociality

We have already noted that social insects are particularly common in deserts, exploiting the advantages of stable nest microclimates in providing thermal refugia, bulk storage of ephemeral resources, and defence against predators and parasites. The
utility of sociality has been realized in other taxonomic groups
as well. Very arid habitats house a few species of social isopods,
of social spiders, and of at least one social tenebrionid beetle.
Southern African deserts are also the home of a now famous
rodent, the truly social mammal known as the naked mole rat,
Heterocephalus glaber. This has a queen and non-reproductive
workers living together in a burrow network, and its lifestyle is
ideal as a strategy for exploiting limited and patchy food
resources and sharing the costs of tunnelling in hard, baked
soils. It is also noteworthy for being the only mammal that
is approximately ectothermic, having non-insulated surfaces
with very high heat loss rates and thus a potentially variable T_b
(though it can produce heat endothermically in the laboratory).
Its preferred T_b is 33°C, which is also the temperature at which
its gut flora, performing the caecal fermentation of the very high
fibre diet, work fastest.

14.2.3 *Evaporators and their strategies*

The evaporators are the 'middle-sized' animals of the deserts, including some birds, plus dogs, cats, smaller antelope, foxes, etc. The group could also be said to include domesticated goats and sheep, and man himself where he lives on desert fringes.

These animals are all dependent on a reasonable water supply to allow them to cool down by the use of evaporation, at ambient temperatures where other cooling modes are impossible. Evaporators are therefore inevitably few and far between, mostly living only at the desert fringes and where there is access to water at oases. They do rather better in the margins of rocky deserts, where there is more microclimatic variation to exploit on a scale that is useful to them (e.g. sheltering behind boulders). Moderately sized creatures, such as the rock wren and the elf owl, can survive by sheltering in crevices in such habitats.

Evaporative cooling

As with all other animals, either **panting** or **sweating** may be used as the means of evaporation; as we saw in Chapter 8, the latter is less controllable and also loses salts, but it uses up less energy. Panting is normally preferable for desert animals, and it can be combined with countercurrent systems in the nose to save water as described for the smaller evader animals. Desert sandgrouse show a very pronounced gular flutter (see Fig. 8.44) to supplement normal panting, and have markedly enhanced evaporative thermoregulation compared with less arid-adapted relatives; the pin-tailed sandgrouse, *Pterocles alchata*, dissipates 89% of its metabolic heat production by evaporation at 40°C, which is 152% of the allometric prediction. Some xeric birds are particularly well suited to be evaporators by virtue of feeding on nectar. The Australian honeyeater *Manorina flavigula* has a very high daily water turnover and a high and variable urine concentration, reflecting the water inputs it gets from plants.

One group of desert invertebrates deserve honourable mention as being part of the 'evaporator' cohort; these are the insects that can use evaporative water loss for cooling, again because of a reliable watery food supply. The desert cicadas are best known, their transpiration rate increasing at least five-fold when exposed to temperatures above a specific setpoint. Species such as *Diceroprocta apache* release water from pores on their dorsal cuticular plates, allowing them to maintain a body temperature of only 39–45°C when (in the case of males) singing to attract a mate in ambient temperatures of up to 48°C. This particular cicada feeds on the abundant mesquite plants of the North America deserts. The cicada *Okanagodes gracilis* is even more impressive, maintaining activity even through the hottest days ($T_a > 50°C$) while feeding and evaporatively cooling. Control of sweating in the cicada *Diceroprocta* is turning out to be rather complex, involving internal and surface sensors and at least two regulatory pathways that use prostaglandin-like chemical signals.

A less familiar example of the evaporative strategy in desert insects is the grasshopper *Poekilocerus bufonius*, which gets enough water from its asclepiad food-plant to be able to afford some evaporative cooling. It seems probable that other desert xylem-feeders may use a similar technique. We should also mention some social bees, particularly in the genus *Apis*, that must cool their bodies and also their whole colony using gathered water when temperatures rise too high, inevitably causing them to be restricted to oases and desert fringes.

Other thermal and water balance strategies

Inevitably the desert evaporators use most of the behavioural and postural strategies already described above for their smaller counterparts; behaviour continues to be a major part of the overall adaptive pattern. **Huddling** is perhaps more common than burrowing, which can be difficult for these larger animals. The wood hoopoes of sub-Saharan savannas huddle in response to low ambient temperatures, with metabolic rates 30–60% lower at night than for solitary birds, saving up to 29% of the daily energy expenditure.

Desert birds such as babblers, finch larks and sandgrouse show some particular tricks to survive quite deep into the deserts; the mobility endowed by flight enables them to gather seeds over a very wide area, and more importantly to get water from oases. In the sandgrouse genus (*Pterocles*) the males are adapted to **carry water** back to the nest and to the young; they wade chest high into the water and return with it soaked into and below their specialized breast feathers, carrying up to 40 ml per trip over astonishing distances of up to 80 km. Other desert birds use a special kind of 'shading behaviour' at the nest, squatting just above the eggs rather than nestling down on them; this may be more important for cooling the parent's lower body by raising it above the boundary layer, than for allowing convective cooling from the eggs themselves.

However, there are some additional physiological components in the repertoire of these animals. The carotid rete system for **selective brain cooling** (see Chapter 8) appears to be relatively common in these medium to large mammals. It is best developed in those that pant, especially artiodactyls, canids and felids, where the necessary cool blood is that returning from the nasal area, cooled by evaporation. The carotid rete linked to the cold nose in these panting animals can cool the blood to the brain by 2°C or more, which in a heat-stressed desert species is very important in that it avoids triggering the brain-based receptors which normally turn on extra panting mechanisms to cool the body down, and which if activated could lead to disastrous evaporative water loss.

Similar brain cooling occurs in other mammals (even in man), where there is no panting mechanism, and then seems to involve cool blood from the facial veins entering the cavernous sinus at the base of the skull where the carotid artery, though not subdivided into a rete, follows a fairly tortuous course so that its blood can be slightly cooled. In desert birds brain cooling has also been demonstrated, with a countercurrent operating in the ophthalmic rete and involving cooled venous blood from the nose, mouth and eye areas. Brain cooling by heat exchange occurs in at least some reptiles, too, where the brain tissues can be 2–3°C lower than the core body temperature or rectal temperature. This is achieved in desert lizards by a rapid and shallow breathing through open mouth, allowing the pharynx area to cool evaporatively. While this does not actually achieve much body cooling in the sense that true panting can do for a mammal or bird, it does provide a brain temperature-regulating system, for the carotid artery runs very superficially through this pharyngeal area and thus the arterial supply into the brain is again cooled.

Some of the desert birds may also demonstrate a lowered MR and an ability to tolerate **hyperthermia**. The double-banded sandgrouse (*Pterocles bicinctus*) has a resting MR 46% less than the allometric prediction, and a high UCT permitting tolerance of somewhat elevated T_b. In the Negev the great grey shrike (*Lanius excubitor*) shows an increase of T_b with rising T_a even within its normal thermoneutral zone (30–36°C), and this controlled hyperthermia gives it a saving of water by allowing extra dry heat loss rather than EWL. However, this and other desert birds also exhibit a rather high rate of increase of EWL with temperature, a puzzling observation not yet adequately explained.

Many of these medium-sized animals from hot climates seem anomalous in having very dark surfaces (e.g. desert goats, sheep and ravens). The phenomenon of black desert mammals and birds may be explained by several factors. The individual black feathers and hairs can become exceptionally hot at the tip (80°C is not uncommon), enough to actually encourage heat loss to cooler air by convection and radiation, while at the same time the trapped insulating air layer prevents the heat of the pelage being relayed to the skin surface. Black colour may also be appropriate in the winter months for animals such as desert goats, when nights are very cold and food is rare; the MR of a black goat may be 25% lower than that of a white one, because they can absorb more heat from the Sun and have less need to shiver. Furthermore, remember that in all these cases we are considering environments where re-radiated longwave radiation from the bare ground, the absorption of which is largely independent of colour, may be a particularly large component of the total heat spectrum.

Desert goats also show an unusual ability to reduce MRs when food is scarce and maintain their body weights on less than half their normal food intake. Unusually, this involves reduced metabolism mainly in the muscles, whereas in most mammals it is the gut that is 'turned down' most conspicuously.

Medium-sized desert animals also show a pronounced ability to produce **hyperosmotic urine** (Table 14.2), well beyond the levels seen in temperate species of similar body mass (cf. Table 13.8). Mammals with more numerous long loops of Henle inevitably succeed best, with high urine/plasma solute ratios.

Migration

Many evaporators are essentially **nomadic**, using migration as part of their strategy to track the patchy resources, moving between water sources as they fluctuate and become exhausted.

Table 14.2 Urine concentrations and urine/plasma ratios in desert vertebrates.

Taxon	Urine concentration (mOsm)	Urine/plasma ratio
Reptiles		
Desert tortoise	337	1.0
Birds		
Struthio (ostrich)	900	2.7
Kookaburra	944	2.7
Zebra finch	1005	2.8
Savanna sparrow	2020	5.8
Mammals		
Eland	1880	6
Bedouin goat	2200	7
Megaleia (kangaroo)	2700	8
Camelus (camel)	3200	8
Dipodomys (kangaroo rat)	5500	16
Notomys (hopping mouse)	9370	25

This may involve regular movements outward from a fairly permanent central refugium, or a more random wandering between chance-found resource foci. However, migration does impose substantial metabolic costs, and is a risky strategy since location of future supplies can rarely be guaranteed.

Following on from this point, it could also be said that the famous examples of migratory desert insects come into the 'evaporator' category rather than being evaders. Many species of locusts and grasshoppers are neither particularly impermeable nor particularly cryptic in lifestyle, by desert standards; instead they move in vast numbers between patches of water or areas of recent rainfall that are sustaining food resources, ranging over grassland and into deserts, their ability to fly allowing them access to the occasional areas of plenty within profoundly arid areas. Indeed, the true desert locust, *Schistocerca gregaria*, is almost entirely restricted to the hot Palaearctic deserts, moving continuously between wadis and oases; its movements within the Sahara are shown in Fig. 14.22, and are largely determined by the seasonal movements of air masses.

14.2.4 *Endurers and their strategies*

Camel, oryx and other large mammals

The camel and oryx are the famous examples of enduring desert animals, being large endothermic mammals. The camel of the Arabian and Sahara deserts is *Camelus dromedarius*, which in some parts of its range routinely endures air temperatures up to 55°C in the daytime and below freezing at night. The oryx (*Oryx beisa*) is almost equally famous as an Arabian desert gazelle with an exceptional ability to withstand aridity. In the

Fig. 14.22 Locust migrations around the Sahara desert. (Adapted from Farrow 1990.)

same functional category, though less highly specialized, come some of the large gazelles of dry grasslands and deserts, including the eland and the springbok, and perhaps also the two species of African rhinoceros and the African elephant.

Because they are large, such mammals have real difficulties in losing excess heat through their surfaces; much of their body is

Fig. 14.23 Daily patterns of body temperature (T_b) in a camel: (a) a single cycle with daytime heat storage and nocturnal heat dissipation, and (b) the effects of hydration, allowing cooling by evaporation and so reducing the need for the heat storage effect. (Adapted from Schmidt-Nielsen 1964.)

conspicuously lacking in thick fur, as insulation would compromise heat loss even further, but many do have large appendages (ears and tails, or long necks and legs) where heat can be dissipated. Elephant ears viewed with infrared thermography show a remarkable rate of heat dissipation of up to 70 W at a T_a of 32°C, allowing them to shed almost 100% of the animal's total heat loss needs when maximally vasodilated and flapped gently.

These large mammals also tend to be very inactive in the heat of the day, avoiding any metabolic overload; and they will use behavioural strategies to exploit any source of cooling, such as spread-eagling against cooler rocks, seeking the shade under any desert scrub vegetation, orienting head-on to the Sun,

(a)

(b)

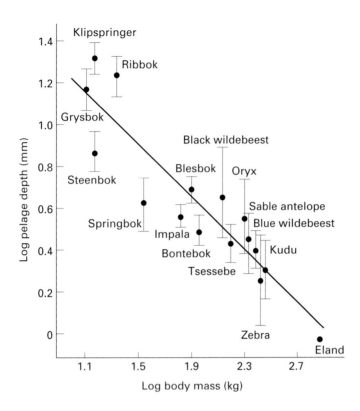

Fig. 14.24 The negative relation between fur depth and body size for various African desert and savanna ungulates (cf. Fig. 8.17). (From Louw 1993.)

Fig. 14.25 Gazelle colour pattern, typical of desert antelopes. (From Louw 1993.)

locating the slightest breeze on a hillock, or wallowing in patches of mud.

However, the great compensation offered by large size is that of high thermal inertia, opening up the possibility of substantial heat storage and the strategy of **adaptive hyperthermia**. Here the body temperature is allowed to fluctuate rather widely, on a daily basis, with heat stored during the day to give a barely tolerable peak T_b just before sunset, and then dissipated during the night so that there is a just tolerable minimum T_b around dawn. Using this technique the camel is able to survive even when dehydrated and in the hottest of deserts, over a 24 h cycle (T_b range 34–41°C, Fig. 14.23; exceptionally up to 45°C briefly). Effectively it is acting as a storage radiator, and this is particularly obvious when the camel T_b is compared with the much more rapidly fluctuating T_b of a small rodent in the same environment (Fig. 14.8a). Storing heat in the daylight hours can reduce the daily water loss of a camel by up to 5 l, since it need not use EWL to regulate T_b tightly.

This type of thermal behaviour is aided by a light colour and thin curly coat, especially dorsally. In stark contrast to the situation in cold climates, the fur of mammals from hot arid habitats reduces in thickness as body size increases (Fig. 14.24; cf. Fig. 8.17). Most desert gazelles are pale brown, but commonly have a darker flank or an almost black strip along the flank (e.g. Thomson's gazelle, Grant's gazelle and springboks; Fig. 14.25),

which is known to facilitate heat gain when the animals stand side-on to the Sun in the early morning. For exactly opposite reasons many also have a white rump and will orientate parallel to the Sun's rays at midday (Fig. 14.26), paralleling the behaviours of ectotherm evaders. Having long ears, long necks and long legs may also be adaptive in providing a range of postural options for controlling heat gain and loss.

A camel is also substantially aided by an ability (unusual amongst large mammals) to endure up to 30% water loss, without losing its ability to behave normally and to feed. For comparison, humans and other mesic mammals cannot tolerate more than 10–12% (cf. Table 13.5). A camel's special ability is that of **drinking** very rapidly to replenish this deficit, up to 200 l or around one-third of its own body weight in one drinking bout lasting just 3 min. It can dehydrate without compromising its blood viscosity; the fluid lost comes from the tissues rather than blood, so that the blood composition and volume remain reasonably constant, the haemoglobin functioning remains normal, and explosive heat death due to reduced blood circulation is avoided. During rehydration, large inputs of drinking-water are stored temporarily (up to 24 h) in the gut to avoid equally dangerous rapid dilution of the blood.

There are many special physiological tricks in the camel and the antelopes to help all this endurance. Some assistance comes from good water conservation mechanisms. They have relatively low water loss rates for their size (around 3–4 mg cm^{-2} h^{-1} in the dehydrated oryx and Grant's gazelle), and can produce very dry faecal pellets by mammalian standards (e.g. about 40–50% water in the camel and the dik-dik, compared with 70–80% in temperate ungulates). As might be expected many of

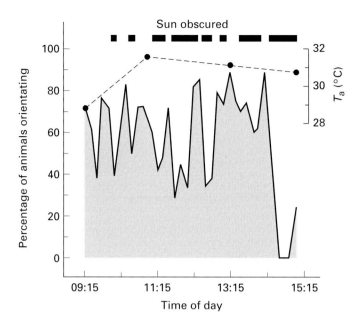

Fig. 14.26 The percentage of gazelles showing an orientation head into or away from the Sun (white face or rump exposed); in cool periods when the Sun is more obscured they are more likely to turn sideways, when the dark flanks absorb heat better. T_a, ambient temperature. (From Louw 1993.)

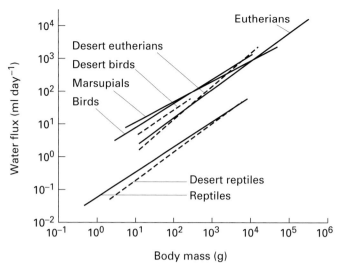

Fig. 14.27 Water flux in desert animals compared with non-desert species. (From Nagy & Petersen 1987.)

them also have very concentrated urine: about 4000 mOsm in the dik-dik, and over 3000 mOsm in the springbok and camel. Of all these large enduring mammals, the dik-dik and the oryx are perhaps the most spectacular in their ability to survive for very long periods with no drinking-water, but of the two only the much larger oryx does so in real desert conditions. The camel, in similar conditions, can survive without drinking for 14 days. These mammals have a series of highly regulated hormonal responses to dehydration that maintain sodium and other ionic balances and water compartmentalization.

In desert antelopes and other ungulates the occasional need to run in the daylight hours may involve a 30–40-fold increase in MR, and thus an enormous heat load. While hyperthermia may be possible for most of the tissues, there is a vital need to keep the brain functioning at a normal temperature. The carotid rete system was first described in desert gazelles, and soon afterwards in the camel, where it is very well developed; this serves the 'brain cooler' function, where cool blood from the nasal heat-exchanger is used to reduce the temperature of the carotid artery blood passing to the brain. Blood cooled in the nasal cavity is diverted to the brain sinuses via the nasal and angular veins. The muscular tone in these veins is particularly temperature sensitive, with relaxation occurring during small thermal increments in the range 33–45°C; under the same conditions the facial veins vasoconstrict, so that as much cool blood as possible is directed to the brain sinuses and carotid rete. A few animals can use the nasal heat-exchange system in a particularly sophisticated manner by expiring non-saturated air, thus achieving even greater water economy in the 'long nose' respiratory system (see Fig. 14.20). The camels are the best-known examples, as they can exhale air at only about 75% RH, thereby saving about 60% of the water unloaded to the air in the lungs.

The net effect of all these features that reduce water loss is a steeper regression between water flux and body mass for larger desert animals as compared to the regressions for all species taken together (Fig. 14.27). This steepened regression is observed especially in eutherian mammals, but also in birds and in reptiles.

Reproduction also produces some constraints in these large desert endurers. Most of them produce rather **dilute milk** (e.g. rhinoceros milk has only 0.3% fat and 1.2% protein, compared with 4% and 3.4%, respectively, in a cow; see Table 13.13), presumably helping to offset evaporative water loss in the juvenile. Also, antelopes and camels need very precocial young as the habitat gives no chance to hide and the young must escape predation and be able to seek shade. Allied with precocity they are often also rather large proportional to the adult at birth, so that they can experience the advantages of thermal inertia as soon as possible.

In addition to physiological features, remember some of the detailed morphological factors that assist the largest mammals in deserts. They have large flattened hooves to avoid sinking in the sand; dense protective eyelashes and strong eyelids to avoid sand damage during sandstorms; pelage that is dense on the back but sparse elsewhere; and a trunk shaped to give a minimal surface area to volume (SA/V) ratio when resting head-on to the Sun in the heat of the day.

Finally, of course, we come to the famous camel's hump. Hopefully, the old wives' tales scarcely need to be refuted: it is not full of water, nor is it full of fat as a source of water (remember that fats do give more energy per gram, and more metabolic water per gram, but produce *less* water per kJ of energy, and with extra evaporative water loss also incurred when providing the respiratory oxygen needed for their full metabolism). The camel's hump is simply fat as a store of food as in most other mammals, but for a large desert animal it is sensible to concentrate the fat only on the dorsal surface since there it also has a localized insulating effect, helping to reduce the uptake of solar radiation from above in the day and allowing the flanks to be used for thermal emission at night.

Ostrich and rhea

The ostrich, *Struthio camelus*, does particularly well in deserts and dry grasslands, as do some other large flightless birds (ratites, including the rhea, emu and cassowary). Their general build—long legs, long necks and often long beaks—in some ways parallels that of the large desert mammals.

Their strategies are rather similar to those outlined above, relying particularly on adaptive hyperthermia and countercurrent cooling. For example, when *Rhea americana* runs for 20 min at 10 km h^{-1} about 75% of the total heat generated is stored, being lost again by radiation and convection after the bird stops running. In these ratites the surface temperatures of all the extremities appear to be well regulated; most obviously, these birds again have the characteristic long noses and elaborate nasal turbinals (see Fig. 13.37). Ostriches are one of the very few animals apart from camels that can exhale unsaturated air (Fig. 14.20); for example, in ambient air at 36°C the exhalant air from an ostrich is at 85% RH, so recovering about 35% of the water initially evaporated into the air during inspiration and saving up to 500 g water per day. And again the cooling effect at the nose can be used to achieve brain cooling; as in other birds this is largely via an elaborate ophthalmic rete just below the brain.

An ostrich at risk of severe overheating also uses a combination of panting (at about 40 breaths per minute) and alterations of the plumage, erecting the rather sparse dorsal feathers to give a thicker insulating layer but one in which air moves freely to give some convective cooling. The birds also orientate face-on to the Sun and droop their wings laterally away from the thorax, giving the rather bare sides of the thorax some shade and allowing them to act as thermal windows.

Ostriches can reduce their urine volume dramatically when dehydrated (Fig. 14.28), with the concentration rising (though only to around 800–1000 mOsm). In such conditions they also become more selective in food choice, taking plants with a higher water content.

Large desert birds face a particular problem in keeping their eggs cool; they cannot build large protective insulated nests in such an environment. The ostrich lays its enormous eggs in a shallow scrape, and relies on its own high thermal inertia and thermoregulatory mechanisms of panting and ptilo-erection to keep the clutch cool beneath it.

The Australian emu (*Dromaius*) is also worth mentioning, as its habitat ranges from high snowy mountains to the central

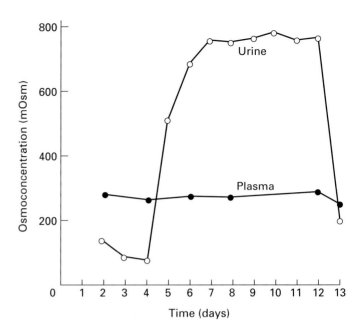

Fig. 14.28 Ostrich urine variability; the bird was dehydrated from day 4 to day 12. Note the minimal variation in blood concentration. (From Louw 1993.)

continental desert, and it can maintain a constant T_b at ambient temperatures ranging from −5 to +45°C despite being too big to exploit any microclimatic variation. Its large thermal inertia certainly helps it; in the deserts it uses techniques similar to those of the ostrich and rhea, while on snow-slopes it maintains T_b largely by reduced conductance and increased heat production, spending much of its time sitting as a huddled ball with all appendages tucked in.

14.2.5 *Desertification and anthropogenic effects*

In a sense a desert is the one type of habitat that is not really fragile and vulnerable to human interference in the way that the rest of the globe is: for deserts are the 'endpoint' of our interference in all low-latitude zones (Table 14.3). Existing deserts are spreading because of human influence, agricultural practices and grazing livestock herds; in some areas entirely new deserts are being created, for example where forests are felled on hillslopes and the land quickly erodes into a dustbowl (Fig. 14.29). These processes are unlikely to be halted; in fact desertification is almost certainly getting worse now, and will continue to get worse into the future, because of pressure from growing human populations, especially in sub-Saharan Africa, and because of projected anthropogenic climatic changes (cf. Fig. 13.75).

Once deserts have formed they are of no use to man, so may suffer relatively little interference, persisting as zones of low but specialized biodiversity. There have been attempts to resurrect certain desert areas by grand irrigation schemes, ongoing particularly in Israel and projected in other parts of the Middle

Table 14.3 Estimated causes of desertification (as percentage of desertified land area).

Regions	Overcultivation	Overstocking	Fuel/wood gathering	Salinization	Urbanization	Other
Northern China	45	16	18	2	3	16
North Africa/Near East	50	26	21	2	1	–
Sahel/East Africa	25	65	10	–	–	–
Middle Asia	10	62	–	9	10	9
Australia	20	75	–	2	1	2
USA	22	73	–	5	–	–

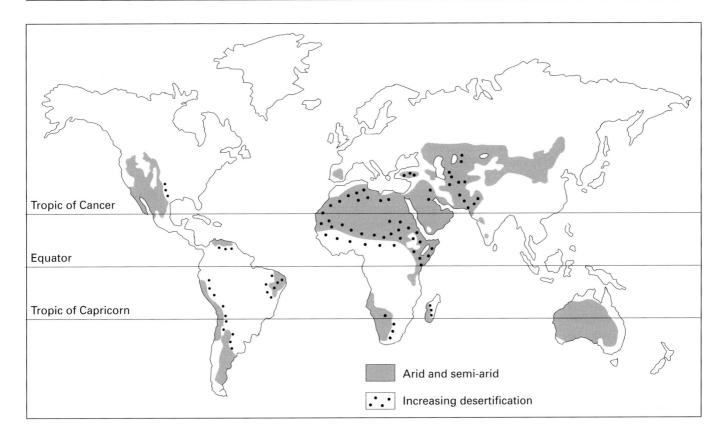

Fig. 14.29 Regions where desertification has increased in the last three decades. (Adapted from Williams 1986.)

East (e.g. diverting water from the Nile to irrigate the Sinai Peninsula); however, such schemes tend to produce localized areas of highly specialized and artificial agricultural productivity rather than returning the land to a natural savanna or subtropical status.

14.2.6 *Overview of hot arid environments*

Before we move on, it is worth reiterating that while we have presented a series of adaptations that desert animals may show, these are very strongly interactive and can never be treated in isolation. To take a simple example, desert insects may have very low values of CEWL such that REWL is a larger proportion of their water balance. They have a number of possible tricks to reduce REWL, but in addition to these they may also show a lowering of MR, which lowers their need for oxygen and hence also lowers REWL. But lowering MR means a reduction in cellular ATP production, which leads to a reduced cellular sodium pumping and a reduced sodium gradient across membranes, in turn leading to a reduced sodium-coupled amino acid transport into cells. The haemolymph of these insects may therefore be low in sodium and high in amino acids. In most insects this would lead to substantial urinary loss of amino acids, but since desert insects also have a very effective rectal resorptive system they are able to sustain their internal conditions. In this fashion many of the key physiological adaptations are linked together. Similar arguments clearly apply in relation to integrated temperature and water balance in camels and other large endurers.

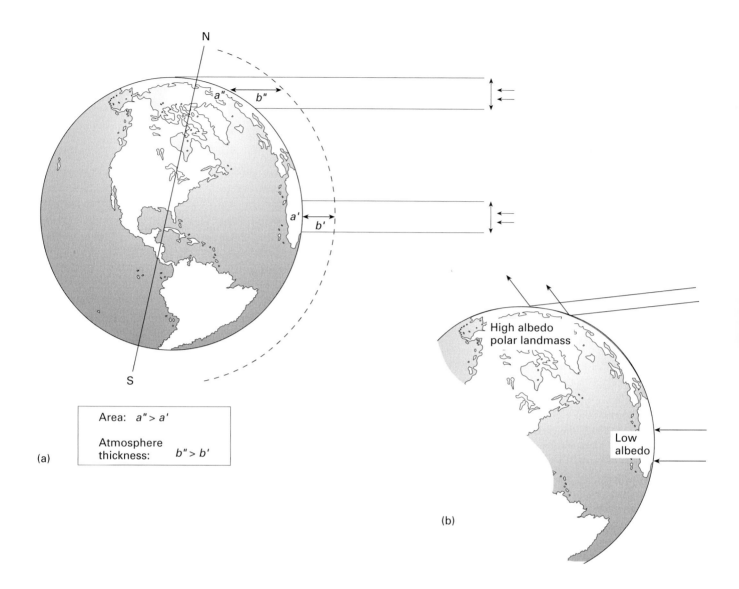

Fig. 14.30 Causes of reduced radiation load at high latitude; (a) input spread over a larger surface area, and (b) reflected by ice (high albedo) over the seas or land masses.

The biology of deserts is strongly dominated by the presence or absence of water, setting the dynamics of resting, foraging and reproducing for all animals and imposing an unusually unpredictable regime requiring opportunistic adaptive strategies. Life appears to proceed as a series of pulses of activity with long periods of relative quiescence in both flora and fauna. For animals that depend on the ephemeral parts of plants—leaves, flowers, sap—these pulses have a direct and immediate impact, and the animals' population densities may thus be directly related to rainfall. For animals that rely on more persistent plant parts such as seeds, or which store foods in nests, and for those that are omnivorous or carnivorous, the pulses can be smoothed out somewhat, and more stable (albeit small) population densities can be maintained.

14.3 Very cold habitats

14.3.1 Types and characteristics of the cold biomes

Most of the living space on this planet is rather cold, since the majority of it is deep cold seas; the animals that live there were dealt with in Chapter 9. But even restricting discussion to terrestrial habitats, the world is predominantly a cold place, especially by human standards as we are essentially a tropical species by origin. However, land-based temperatures that could reasonably be termed 'very cold' are mainly encountered with higher latitude, where the radiant energy from the Sun has to travel through a greater thickness of atmosphere, so that it is partially absorbed, and what does get through must spread out over a larger area of ground (Fig. 14.30a). The circumpolar ecosystems, sometimes collectively termed the boreal systems, are therefore marked by extreme cold; the records for lowest temperatures in overland climates are held by Siberia in the northern hemisphere (−68°C) and by central Antarctica in the southern

Fig. 14.31 Defining the Arctic and Antarctic, and regions of tundra and of taiga. (Adapted from Davenport 1992.)

hemisphere (–89°C). These zones also show a negative winter energy balance (more heat is being lost than is being gained from the Sun), and a very short growing season. They comprise three main biotic types: the polar zones themselves, with continuous ice or snow cover, the tundra zones, with snow and minimal low vegetation in the short summer, and the taiga, where coniferous forests survive. Cold biomes could also be said to include some areas of 'cold desert', often found on relatively high-altitude plateaus in the interior of large continents (e.g. the Tibetan plateau).

The polar zones

Towards the poles, low temperatures lead to snow and ice formation, creating a surface of **high albedo** (high reflectivity), so that some of the sparse incoming solar energy is reflected at a tangent and lost to the system (Fig. 14.30b). Each of the polar regions becomes a giant low-pressure centre, with unmixed cold air at its centre hardly able to mix with the peripheral circulation of milder westerly winds (the 'circumpolar vortices').

Temperatures of course do vary seasonally, with a 'warmer' summer, though the growing season may last only 1 month. The

poles experience periods when the planetary tilt means a complete absence of direct radiation, and by definition beyond the polar circles (roughly 66°N and S) there are days when the Sun never rises. But there is a marked difference between the two hemispheres. The Arctic is merely a frozen sea (Plate 9c), but is fringed by land masses, whereas Antarctica is a real continent, 3000–4000 m above sea level in parts and with another few hundred or thousand metres of ice on top. The south polar region is therefore always colder for a given latitude, because of the presence of a large icy polar continent instead of a thermally buffering polar sea. The coldest northern hemisphere temperatures are recorded around the so-called 'cold pole' of eastern Siberia, and the northern fringes of the land masses of Russia, Canada and Greenland are routinely colder than the true North Pole, again due to the ameliorating effects of the thermally stable Arctic oceanic mass, which is never less than –1.86°C. However, the coldest temperatures in the south occur deep into the continent of Antarctica itself, and may be 20°C lower than in the north.

The polar zones are conventionally defined as the areas within the Arctic and Antarctic Circles (and thus confusingly excluding a small part of the land mass of Antarctica). However, it is more useful to a biologist to define these zones in terms of actual temperatures or in terms of vegetational zones. Figure 14.31 shows the possibilities. The Arctic may be defined by the treeline on the three surrounding land masses, though this has been substantially modified by human activities. The Antarctic may be defined by the extent of pack ice in winter. Alternatively each polar zone could be delimited by the midsummer 10°C isotherm (the line joining all places having a mean air temperature of 10°C in the warmest summer month), which has the advantage of applying to both land and sea.

Polar areas are in many senses best seen as 'cold deserts', suffering in effect from the same kind of aridity that we met earlier in the hot deserts. This is because precipitation rate is low (zero in some parts of Antarctica), and any snow falls largely onto ice, so that water present is unavailable to either plants or animals, which exist in a 'physiological drought'. The Antarctic probably contains 90% of the planet's fresh water, but in a form that is almost entirely useless for supporting life. On the other hand, when snow does fall it creates a covering for overwintering animals where the enclosed air is saturated, in equilibrium with the vapour pressure of the ice and snow; thus animals burrowing within the snow are protected against desiccation.

Polar microclimates and vegetation

Ground temperatures towards the poles are greatly affected by seasonal and daily changes in solar altitude, and small variations of slope can have marked effects on soil temperatures. In the Antarctic the generally colder air and extensive ice cover produce more extreme conditions, where microclimatic amelioration may be irrelevant over large areas. However, the Antarctic fringes do provide areas where soil is protected by a thermal blanket of snow, with relatively stable conditions 50–100 mm below the surface. In the gravelly substrates at the melt line in summer, many small invertebrates do survive, with summer temperatures rising just above freezing for a few hours a day even though T_a may remain below zero. The Antarctic has just two species of flowering plant, together with an abundant fringe of lichens and algae. It might be thought that there is little scope for microclimatic variation in such a shallow layer of vegetation, but in fact thermal gradients on a very small scale can be most striking: for example, the surface of lichens and mosses in the maritime Antarctic may reach 50°C at noon on cloudless summer days even though the air temperature 1 m above the ground is only 5°C.

In the terrestrial Arctic fringes, topography and the influence of the sea have a substantial influence on local climate. Hills and mountains lead to cloud formation and may interrupt the solar insolation since the Sun may be close to the horizon. Hill slopes also give rise to 'katabatic winds', with warmer air flowing down the slopes and giving favourable conditions; by contrast, low temperatures occur where the air is stagnant in valleys. Temperatures are always highest in summer within the reasonably substantial plant boundary layer, where there may be a temperature excess of at least 20°C above ambient amongst small plants. Temperatures may also be raised by several degrees in the soils up to 10 cm below such vegetation and also below lichens and moss turf. Windspeeds may limit this temper-

ature augmentation, however, especially in areas of very patchy discontinuous vegetation.

Polar fauna

In the true Antarctic there are no entirely terrestrial animals except a few mites and two species of midge; the mite *Nanorchestes antarcticus* is the most southerly invertebrate known, occurring within 5° latitude of the South Pole. Even on the fringing maritime Antarctic islands the species diversity is remarkably low, with a limited soil fauna of nematodes, tardigrades and rotifers, and further mites and insects, including some lice and fleas associated with semi-aquatic birds and mammals. This probably reflects the geographical isolation from sources of invading species and the unfavourable direction of winds for dispersal of small species (most of which are wingless). The paucity of animal species is also partly because the Antarctic has nothing like a continuous cover of plants, so that there could be no real herbivores. All the larger animals present are semi-aquatic and instead depend on the sea for food, with seals, whales and penguins dominant; of all animals, penguins are the most profoundly adapted for polar life, the emperor penguins (*Aptenodytes forsteri*) breeding in midwinter in continuous dark at temperatures down to –70°C. Antarctic birds in general, and penguins in particular, are the most important members of their ecosystem, in terms of both biomass and of interactions with other components of the environment; they import organic material from the sea onto the land.

The Arctic, by contrast, having a fringe of plant life on the enclosing land masses, possesses a surprisingly abundant invertebrate fauna, notably spiders, mites, springtails, biting flies, lice, and a reasonable range of butterflies. There is also a substantial burrowing fauna, including enchytraeid oligochaete worms but surprisingly lacking in beetles (at least when compared to their abundance in other habitats). Above ground there is again a wide range of semi-aquatic seals, but only one truly terrestrial carnivore, the polar bear (*Thalarctos maritimus*), though in the Arctic fringes both wolves and Arctic foxes are to be found.

Since in all these habitats the species number is low, the community structure is inevitably very simple, the individual community members above all requiring great capacities for endurance of cold, dark and at times near-starvation.

Tundra zones

Tundra is a biome of low treeless vegetation, occurring in northern Asia and Canada, mostly within the Arctic Circle, often defined as being where temperatures remain below 0°C for at least 7 months of the year. It does not really occur in the southern hemisphere, simply because there are no large land masses at appropriate latitudes. The tundra climate is an extreme version of continental climate, with very short summers between long, cold, dry winters. Tundra in mid-continents is so cold all year that the deep ground is in the condition known as **permafrost**, with no penetration of free liquid water. In the shallow layer of 'active' soil, above the permafrost in summer, there may be a negative thermocline, i.e. temperatures may decline with depth down to the permanently cold permafrost layer; this layer may also be almost permanently waterlogged. Alternating freezing and thawing also make this layer physically unstable, and it may become highly fractured.

Nearer the coasts the soil and climate can be much more variable, depending on the extent of snow cover; here an abundance of snow keeps conditions *less* cold, but also delays the onset of spring growth.

Tundra microclimates and vegetation

The characteristic vegetation is abundant moss and lichen; at best there may also be very sparse and very low-growing trees (especially dwarf willow), and some coarse grasses (Plate 9d). No large trees or shrubs can survive, because the permafrost precludes taproot penetration, and tall plants with shallow roots would inevitably be blown down. The majority of the living plant tissue is underground as roots and extensive rhizomes, and in fact most of the organic carbon is in the form of undecomposed dead remains, with only 2% of it 'alive' (compared with perhaps 50% in most of the forests of the world). Low cushion-like growth helps to trap this dead material and allow local, slow nutrient recycling.

At these high latitudes thick snow may cover the ground to a depth of several metres during most of the winter months, and this provides a whole range of possible habitats. Thick snow has poor thermal conductivity and acts as an excellent insulating layer between cold air and the soil surface, which in many areas is warmed to around 0°C by geothermal energy so that it remains unfrozen; in fact the lowest soil temperatures through most of the tundra zones are typically recorded in autumn rather than in winter. Even in extreme inland regions the minimum winter temperatures at the soil surface under leaf litter and snow are only −4 to −7°C.

Tundra lichens are so well protected biochemically (usually by glycerol) that they never freeze, providing year-round food to anything that can find them. Many of the larger evergreen plants have small leaves with thickened and hairy cuticles, and most have vegetative budding from runners with protected ground-level buds to allow rapid growth in spring.

Tundra fauna

The fauna is inevitably sparse, but there is scope for a reasonable range of rather specialist herbivores. The large herbivores, such as reindeer or caribou (*Rangifera*), migrate south to the timberline in winter and move back to the tundra to breed; medium-sized herbivores such as musk ox (*Ovibos*) and dall sheep and snow sheep (*Ovis*) stay put and survive the winter feeding on lichens. In addition there are small vertebrate herbivores such as

Arctic hares and above all lemmings (microtine rodents). In summer, the tundra is invaded by waterfowl, especially geese, using the long daylight hours to feed. There are relatively few large predators, the commonest being various mustelids and wolves; avian predators include the snowy owl and the gyr falcon. Many of these mammals and birds turn white in winter, giving a high degree of crypsis.

The conspicuous fauna is therefore almost entirely composed of endotherms. Towards the southern edge of the Arctic tundra the species diversity increases, though, with a few reptiles surviving; the common adder (*Vipera berus*) occurs throughout northern Norway well into the Arctic Circle, and lizards of various kinds are also found in Norway, as well as in the southern tip of South America within the Antarctic Circle.

There is of course also a cryptic invertebrate fauna living in and on the vegetation and thin decomposing layer (Fig. 14.32), with mites, springtails and nematodes present in particular abundance, and with a surprisingly high biomass of enchytraeid earthworms. Some gall-forming insects occur on dwarf willow at most latitudes.

The tendency described in the introduction to this chapter, to be large or small, but relatively rarely in the middle size ranges, therefore clearly applies (except perhaps for the special case of migrating bird species). There are also many examples in the tundra illustrating 'Allen's rule', where animals have relatively small extremities compared with their temperate relatives, to avoid undue heat loss. The tundra rodents and mustelids generally have small ears and toes and short tails, as do many polar birds and mammals, particularly in their juvenile stages. The rule is particularly well illustrated by foxes, where the relative sizes of ears, noses and paws decrease progressively from desert species (kit fox and fennec fox) through the ubiquitous temperate red fox to the Arctic fox. The same rule may apply to ectotherms; legs, antennae and mouthparts may be shortened in northern insects, and the incidence of brachyptery (reduced or absent wings) is also higher in Alaskan dipterans than in other faunas.

Boreal forests (taiga)

The more northerly forests of the world are often labelled 'taiga', from a Russian word for 'swamp-forest'. They stretch in a wide belt across Canada (Plate 9e), Scandinavia and northern Europe, and most of Russia, from the northernmost treeline down to a gradual merger with more deciduous temperate forests (see Fig. 14.31). These northern boreal coniferous forests are not really matched in the southern hemisphere, again because land masses are much smaller at the relevant latitudes, though some of the southern beech forests of southern Chile and Argentina do present some substantial parallels.

The climate allows only a short growing period (2–3.5 months), since it is too cold for much of the year and the fairly warm summers have limited rainfall. Soils are nutrient poor and strongly acidic, commonly called 'podzols' and turning with

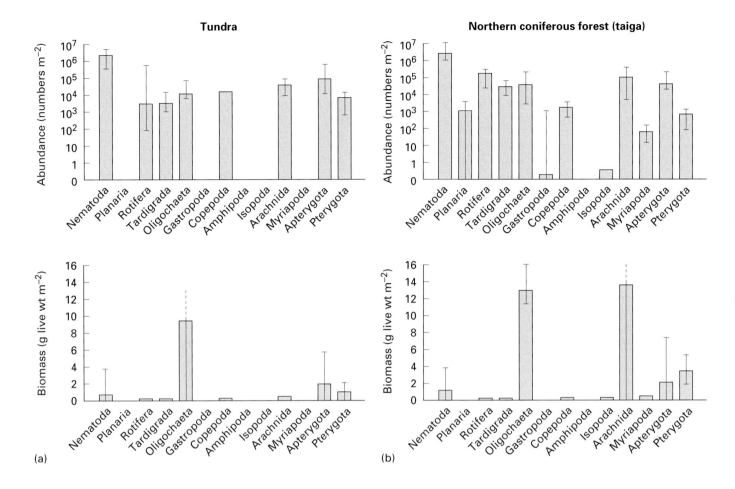

Fig. 14.32 (a) Abundance and (b) biomass of tundra and taiga fauna. (Adapted from Little 1990; courtesy of Cambridge University Press.)

time into peats; remember (section 13.1.2) that peats warm up only slowly but dry out rather easily.

Taiga microclimates and vegetation

Conifers always dominate, usually in low diversity, so that these regions have a dense layer of decomposing conifer needles up to 100 mm deep, producing a very acid effect but providing a thermally buffered zone. There may be larger areas of peaty swampland, dominated by mosses. These swamps are in part created by intermittent bouts of tree-loss in natural wildfires which sweep through the forests, feeding on the resinous undergrowth and litter and sometimes clearing thousands of hectares at a time. Within snow-covered areas a **subnivean** habitat results where temperatures in burrows may be quite mild (Fig. 14.33) especially when gregarious endotherms are present.

Taiga fauna

Apart from the moose, herbivores tend to be of only moderate size (usually smaller than those of the tundra): marmots,

squirrels and voles are common, many relying on seeds and berries and also taking invertebrates as food. Small carnivores are drawn especially from the mustelid group (martens, sables, ermine—these zones are the centre of the fur trade). Larger carnivores include bears, wolves, snow leopard, Siberian tigers and lynx. There are many owls and raptors, and specialist conifer-feeding birds such as grouse and crossbills. Few ectotherm vertebrates live in these forests, though there are rare amphibians such as the wood frog (*Rana sylvatica*) and a few cold-tolerant reptiles (again including the adder) right up to the Arctic Circle.

Invertebrates in the soils are dominated again by nematodes, springtails and a particularly high abundance of mites (Fig. 14.32). Sawflies, gall-midges and weevils (insect groups often specialized in feeding on conifers), and a plethora of biting flies, are evident above ground.

The fauna of cold biomes is clearly dominated by small invertebrate ectotherms and by relatively large bird and mammal endotherms, each of which have their special adaptations. Although we will deal with these separately, it is important to remember that some of the adaptations discussed do apply to both groups. In particular, all the animals that survive at high latitude use behavioural strategies as a first line of defence against the cold, and this is true for both ectotherms and

Fig. 14.33 Microclimate within snow—the subnivean habitat. (Adapted from Davenport 1992.)

endotherms; the major behavioural strategies to be found are exploitation of favourable microhabitats, migration, and the development of gregarious habits. Similarly all of them are likely to share biochemical strategies of enzymes with lower temperature optima and membranes with homeoviscous adaptation. For each category of animal we cover behavioural strategies first and then move on to physiological adaptations.

14.3.2 Small ectotherm evader strategies

Exploitation of microhabitats

As always the utility of behavioural tricks is much enhanced for smaller animals. In the polar zones their choice, and exploitation, of somewhat less severe and more sheltered microhabitats is particularly evident. The winter temperature range beneath snow cover (the subnivean zone) is markedly less than the range of air temperature above snow, and this microhabitat is exploited by a whole range of animals (Fig. 14.33). The advantages of this way of life are greatly enhanced by the penetration of light through snow, allowing plants to stay green and edible. Many invertebrate herbivores exploit this resource, and some are specialized as gall-formers, getting added protection from the modified plant tissue they induce (see Chapter 13).

To create and manipulate microclimate, burrows can be dug into the snow itself, or below it into the litter or soil layers. Some toads in the taiga zones can burrow down more than 1 m to get below the frost line; large numbers of beetles and other ground-dwelling insects also occur at this depth in winter.

Huddling and aggregation

Huddling behaviour is the commonest form of gregariousness used as a thermal strategy, and is very obvious in a whole range of animals from cold habitats. Its effectiveness increases as the difference between air temperature and preferred T_b increases, and with the number of participating individuals. Examples

among the smaller invertebrate ectotherms are not particularly well-studied; the mite *Alaskozetes* aggregates in clusters under rocks, and other mites and some collembolans also aggregate, though it is debatable whether this is real thermally advantageous huddling or just a shared preference for the same microhabitat. Amongst the reptiles several cases are well known, with winter aggregations occurring in many snakes and lizard. The adder is a familiar example, with dens of up to 80 individuals having been found in Finland. Similarly the red-sided garter snake, which occurs in western Canada (where the winter temperature is often –40°C), shelters in huge groups of thousands at a time in sink-holes beneath snow.

Migration

Long-range migration is not a very common strategy for ectotherms from cold climates. There are, however, just a few spectacular examples; the monarch butterflies (*Danaus plexippus*) that move *en masse* from sites in Canada and the northern USA each year to overwinter in Mexico are the best studied case, though only some of them come from really cold habitats. These butterflies then spend the winter in huge aggregations, subsisting largely on their stored lipid reserves (triglycerides, derived from summer feeding on nectar) as there are inadequate nectar supplies to maintain them in the Mexican forests. *Colias* butterflies also migrate seasonally from genuinely arctic habitats in northern Scandinavia to the southern Baltic regions.

Thermoregulation

Small terrestrial animals from cold climates must have highly cold-adapted enzymes and membranes; for example, the mite *Alaskozetes* shows low enzyme activation energies and elevated MRs at low temperatures, relative to temperate species. But they must also be good thermoregulators of necessity, and must make any possible use of the minimal heat inputs available. Most boreal animals are highly adept at **behavioural thermore-**

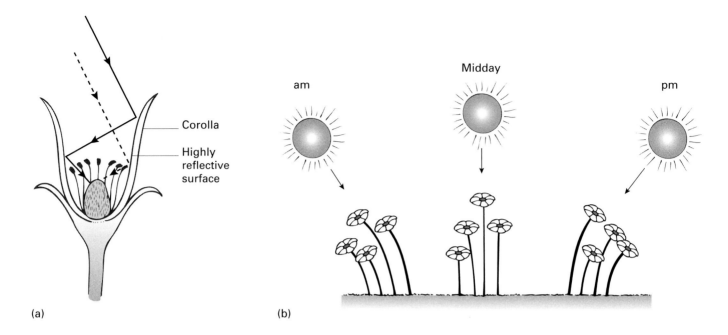

Fig. 14.34 Arctic flowers acting as (a) parabolic reflectors and (b) sun-trackers to give a warm microclimate around their ovules, exploited by many small insects. (Adapted from Kevan 1975.)

gulation by microhabitat choice and aggregation, as described above, but can also utilize sophisticated basking techniques, as with the examples we looked at in temperate habitats (see Chapter 13).

A small ectotherm may raise its T_b at least 25°C above ambient by **basking** in the polar summer. Basking butterflies from Greenland and Canada have been studied in some detail, and they can warm themselves very substantially above ambient using their wings as heat gatherers. Many are melanic, the darker colour improving absorption. Arctic fritillaries are dorsal baskers (see Fig. 13.22), the dorsum of the wings being dark, while *Colias* butterflies are lateral baskers—the Arctic species (e.g. *C. hecla*) are unusually dark in colour especially on the ventral sides of the hind wings; their extra melanism allows them to achieve an excess temperature up to 80% greater than lighter forms. Note that, in addition to basking, contact with the substrate is also used to warm the body in many of these Arctic insects. Butterflies use bare ground for contact warming, seeking warm and sheltered spots.

Basking also occurs in wingless insects, the best example being the 'woolly bear' caterpillars, *Gynaephora groenlandica*, which may spend up to 60% of their active time in basking and only 20% in feeding. This species may spend as long as 14 years in the caterpillar stage, reflecting its very slow growth rate.

There are also some specialist cases where insects use particular species of flowers as basking warm-up sites (e.g. *Dryas integrifolia*, and the Arctic poppy *Papaver radicatum*). These flowers are shaped like bowls, and they rotate by phototropism so that the corolla always points towards the Sun as it passes across the sky each day (solar tracking). Their shape acts as a parabolic reflector to concentrate radiation into the centre of the bowl of the flower, raising it by 5–8°C above ambient (Fig. 14.34). Mosquitoes, hoverflies, blowflies and danceflies all exhibit a raised T_b in association with sitting in these flowers.

Larger polar ectotherms may also use basking as a key component of their thermal strategy. The adders and lizards of northern Norway may bask for large parts of the 20 or more hours of daylight in the summer, feeding between basking bouts to permit lipid stores to be built up for the winter.

In the case of a few large insects, although nominally ectothermic, remember that a component of **endothermy** also comes into play in achieving thermoregulation. The Arctic bumblebee, *Bombus polaris*, found in northern Canada, is able to fly at low temperatures due to its inherent endothermic warm-up (see Chapter 13), and the same almost certainly applies to several *Bombus* species found in northern Scandinavia. These species also use endothermic warm-up at times of egg production, the queen pumping warm blood to her abdomen to speed up egg maturation, and then staying at elevated T_b to brood the larvae within the nest.

Tolerating freezing

Diverse taxa have adopted the 'freeze-tolerance' strategy discussed in Chapter 8, involving some tolerance of extracellular ice formation, with cryoprotectant polyols or sugars to prevent damage. Examples include many terrestrial insects, a few other arthropods such as centipedes, a range of gastropods, annelids and nematodes, and several species of frogs, lizards and turtles that overwinter on land. Recent evidence shows that in polar nematodes freezing may also occur intracellularly, an otherwise unheard of phenomenon; the worms freeze in all body compartments, and can subsequently melt, grow and reproduce apparently normally.

Among the insects, **freeze-tolerance** is the exception rather than the rule, but is widespread in some families from the Coleoptera, Diptera, Hymenoptera and Lepidoptera, especially in larvae and pupae but rarely in adults and apparently never in eggs. In many insects low temperature is the immediate trigger for **polyol synthesis**, in conjunction with other environmental cues, which might include photoperiod, humidity, food availability and to a lesser extent endogenous factors such as hormones and the physical changes of diapause. Two of the best-studied insects are a parasitic wasp, *Bracon cephi*, and the larvae of the goldenrod gall-fly, *Eurosta solidaginis*, which can survive up to 12 weeks in the frozen state. In *Bracon* larvae, which overwinter in frozen Canadian prairies, glycerol is present at up to 25% body weight in winter, giving a supercooling point as low as −47°C. In field populations of *Eurosta*, long-term chilling at the right time of year is required to initiate glycerol production: at least six consecutive days with daily average temperatures of less than 5°C in mid-November. Laboratory experiments show that glycerol is synthesized first, followed by sorbitol as the temperature falls (Fig. 14.35). Glycogen is clearly the source of these polyols, and the glycogen phosphorylase enzyme in the fat body is cold-activated, with low temperature causing a rapid conversion from the inactive to the active form. From September to November the larvae also accumulate mono-unsaturated fatty acids at the expense of saturated forms, suggesting activation of a Δ⁹-desaturase enzyme (see Chapter 8). However, traditional ice-nucleating agents (INAs) have never been identified in

Fig. 14.35 Sorbitol and glycerol synthesis in the gall-fly *Eurosta* at decreasing acclimation temperatures, using up glycogen stores to make the cryoprotectants. (Adapted from Storey *et al.* 1981.)

Eurosta, and recent evidence indicates that calcium phosphate performs this role, existing as small crystalline particles within key tissues such as the Malpighian tubules.

Many animals show **seasonal patterns** of glycerol content, of **supercooling point**, and of melting and freezing point (Fig. 14.36). Examples of the supercooling points (SCPs), and the solutes used to produce them, are shown in Table 14.4 for a

Fig. 14.36 (a) Seasonal changes in glycerol contents and supercooling points in larvae of a caterpillar, and (b) in melting point and freezing point of a larval beetle. (From Hansen 1973.)

(a)

(b)

Table 14.4 Supercooling points (SCPs) in species living at subzero temperatures.

Species	SCP (°C)	Known solutes
Spiders		
Clubiona	−15.4	Glycerol, AFP
Philodromus	−26.2	Glycerol, AFP
Mites		
Alaskozetes	−30	Glycerol, polyols
Lepidoptera		
Pringleophora	−5.0	Glycerol
Isia	−18.2	Glycerol, sorbitol
Nemapogon	−26.1	Trehalose, amino acids
Pieris	−26.2	Trehalose, amino acids
Laspeyresia	−31.5	Glycerol, trehalose
Alsophila	−44.6	Glycerol
Beetles		
Pterostichus	−10.0	Glycerol
Meracantha	−10.3	AFP
Dendroides	−12	Glycerol, sorbitol, AFP
Ips	−32.4	Ethylene glycol
Dendroctonus	−34	Glycerol
Hymenoptera		
Trichiocampus	−8.6	Trehalose
Megachile	−27.7	Glycerol
Camponotus	−28.7	Glycerol
Bracon	−41.2	Glycerol
Eurytoma	−49.2	Glycerol
Flies		
Xylophagus	−6.0	Sugars, amino acids
Tipula	−7	Sorbitol, INP
Eurosta	−10.3	Glycerol, sorbitol, trehalose
Diplolepis	−32.7	Glycerol
Rhabdophaga	−49.1	Glycerol
Vertebrates		
Hyla	−2.0	Glycerol, glucose
Pseudacris	−2.0	Glucose
Rana sylvatica	−3.0	Glucose
Turtle	−3.3	Glucose, amino acids
Northern ground squirrel	−2.9	?None

AFP, antifreeze protein; INP, ice-nucleating protein.

Fig. 14.37 Glycerol contents (means ± standard errors) of the mite *Alaskozetes* at 4°C are moderated in relation to humidity, with glycerol gradually disappearing at high humidities. (From Cannon & Block 1988.)

range of terrestrial ectotherms that are freeze-tolerant; the values are not exceptionally low, with a slow controlled freezing at moderately high subzero temperatures being preferable. Values of −5 to −25°C are clearly quite common, and there is a general correspondence between SCP and the normal site of overwintering: species that survive in vegetation have lower values than those that spend the winter in caves or other more protected locations. Remember that polyols also give protection against

desiccation, which may be an important factor in insects that overwinter in very cold dry air; Fig. 14.37 shows the variation in glycerol content in relation to humidity in the polar mite *Alaskozetes*. Cold and drought are both potential problems for high-latitude animals, but the biochemical and physiological strategies to deal with both problems are often complementary.

Many of the polar arthropods can tolerate substantial ice formation in their extracellular fluids; in some species up to 90% of this water may freeze, particularly when the animal is in contact with external ice and undergoes 'inoculative freezing' from the outside inwards. Precautions against contact with ice are often present to limit the risk; examples include the use of silken cocoons, or galls as in *Eurosta*, or specially impregnated winter-burrow walls in some beetles. Where ice-contact is unavoidable, slow, safe freezing is promoted by ice-nucleating proteins (INPs) in the haemolymph, and such nucleators are fairly common in insects as part of the protective mechanism. For larvae of the crane fly *Tipula trivittata*, both protein and lipoprotein nucleators have been identified in the haemolymph. The INP from the wasp *Vespula maculata* has been purified and is a protein of relative molecular mass 74 000, with an amino acid content that includes 20% glutamate/glutamine, 12% serine and 11% threonine.

At least four species of North American frog and some reptile species can also tolerate freezing. Either glycerol (in the treefrog *Hyla versicolor*) or glucose (in some wood-frogs and in the lizard *Lacerta vivipara*) is used as a cryoprotectant in the few cases that have been studied. The wood-frog *Rana sylvatica*, which is found north of the Arctic Circle, is by far the best-studied vertebrate freeze-tolerator. Here the levels of glucose in the muscle, liver

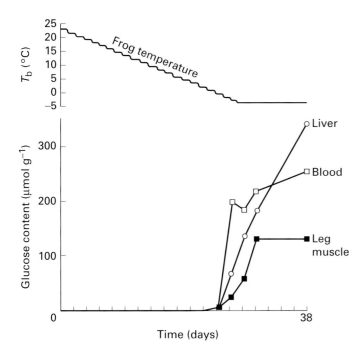

Fig. 14.38 Levels of glucose in the tissues of the freeze-tolerant frog *Rana sylvatica* as its body temperature (T_b) declines. (From Storey & Storey 1984.)

and blood begin to rise about 14 days after T_b reductions are induced in laboratory studies (Fig. 14.38). But at least some of the frogs do not back up their cryoprotectants with nucleating agents, and instead appear to rely on their relatively large size to keep the cooling rate slow enough to ensure that ice only forms extracellularly. Frog skin is highly water permeable, and an important part of the strategy for winter survival is the choice of a well-protected and humid hibernation site where temperatures will fall only slowly.

R. sylvatica is famous for its ability to survive repeated freezing of its whole body, with all breathing and circulation stopped and nerves unresponsive to stimulation. It will survive for 3–14 days in any one freezing cycle, and when frozen a single frog may contain 7–8 g of ice in its coelom and beneath the skin, indicating that 65% of its body water is tied up in the frozen state and all of its organs must be substantially dehydrated. Nevertheless, on thawing it recovers all its functions, with excitability of peripheral nerves and simple reflexes being reinstated in 5–14 h. There is an in-built seasonality in the frog's cryobiology, with autumn frogs better able to survive longer and harsher freezing than summer frogs; the autumn frogs have higher osmotic concentrations and glucose concentrations in their plasma. Glucose is produced from massive glycogen stores (up to 180 mg g^{-1} are built up in the liver before hibernation), and synthesis of the cryoprotectant glucose is triggered by the beginnings of ice nucleation in the body (often when ice first begins to form on the skin of the frog), via an adrenergic and cAMP-mediated signal transduction system. As Fig. 14.38 shows, glucose concentra-

tions may reach 150–300 µmol g wet wt^{-1} in the core organs of frozen wood-frogs (liver, heart and brain) compared with 1–5 µmol g^{-1} in unfrozen frogs. The rapid distribution of glucose from the liver is critically aided by the ability to move this sugar across cell membranes, out of the liver and into other organs via glucose transporters (see Chapter 4); cAMP and protein kinase A are also mobilized during this process. For wood-frog liver the V_{max} value for glucose transport by membrane vesicles is 70 nmol mg^{-1} s^{-1} at 10°C, compared with only 8 nmol mg^{-1} s^{-1} for the common temperate frog *Rana pipiens*. The total number of transporter sites is 4.7-fold higher in the liver of the freeze-tolerant species, and varies seasonally in that species by about eight-fold. There is also a gradient of glucose concentration from the core to the peripheral tissues, which enables the frog to thaw uniformly, thereby avoiding tissue damage in the spring. The thawing of frogs can now be studied by proton magnetic resonance imaging, since the signal undergoes a large change during the transition from ice to water; protons in ice are invisible to this technique and frozen areas therefore show up as black. From such studies it is clear that the liver and heart thaw first during recovery from freezing, so that their vital functions can begin while the rest of the frog is still thawing out. We are also beginning to understand the processes that occur at a gene transcription level during freezing in this species.

Freeze-tolerant animals must survive and maintain some degree of homeostasis despite a lack of circulating fluids. Their enzymes must be able to tolerate very low temperatures and very high salt concentrations. For many of them, in the frozen state energy production switches to the hydrolysis of phosphagens, phosphoarginine (PAr) or phosphocreatine (PCr) (see Chapter 6), followed by anaerobic glycolysis with lactate and alanine as end-products. MR is usually greatly depressed (to 5–10% of normal resting values), giving a typical hypometabolic state.

Avoiding freezing

Many animals that encounter subzero temperatures are freeze-intolerant but are able to survive by virtue of the phenomenon of supercooling alone; such animals must eliminate or mask all INAs as winter approaches. The small terrestrial mites and insects that cannot tolerate ice formation use additional strategies to assist their cold-hardiness: they allow the body to become very dehydrated, to reduce the availability of freezable water, and they may void their guts as winter sets in, to clear any potential ice nucleators. As a result of all these mechanisms, a number of Arctic species have lower critical temperatures of –55°C or even –70°C.

Springtails and mites, among the most important components of polar soil fauna, are almost invariably freeze-intolerant and rely on their supercooling properties for survival. The supercooling points of circulating fluids in these cold-hardy but freeze-intolerant animals are usually (though not invariably) very low compared to those in freeze-tolerant species. Antarctic mites and Arctic insects can generally supercool down to –30 or even –40°C, with eggs and pupae showing even lower values. Some animals can remain almost fully active at subzero temperatures. For example, *Alaskozetes antarcticus* (body mass 150–200 μg) shows peak locomotory activity at 16–24°C but does not enter cold stupor until about –5°C. Another Antarctic mite, *Nanorchestes antarcticus*, has even been observed moving at –11°C, but it has a relatively high temperature requirement for optimum activity, reflecting the high temperatures found at the radiative surface on the mosses, algae and lichen on which this species lives.

In some of the polar insects and mites glycerol acting as an antifreeze can reach concentrations of 25% of fresh body mass during the winter. In addition, due to their small size these animals undergo substantial dehydration, which reduces their content of freezable water and elevates the effective antifreeze concentration. Antifreeze proteins (AFPs) are also now known to occur in some mites, spiders and insects (as well as in the polar fish where they were first identified, described in Chapter 9). The associated antifreeze mRNA has been studied in some north temperate insect species, and its activity is inducible at around 2–4°C in the laboratory, but induction only occurs in winter, and cannot be achieved after about February in these tissues. This suggests that AFP synthesis normally involves an interaction of low temperature priming with an endogenous clock system. Antifreeze proteins tend to be recycled in spring as amino acids for growth and egg production.

Although freeze tolerance may be a 'cheaper' strategy (see Chapter 8), recent evidence suggests that freezing intolerance may be the better strategy in low-temperature but thermally quite variable environments, such as the maritime Antarctic margins. Here small terrestrial springtails and mites are very common, and switching rapidly from inactive to active states may be the ideal way to exploit winter thaws and avoid summer frosts.

Life cycles

Almost all the small 'evader' animals from cold climates have specialized life cycles, in which multivoltinism (having many generations per year) is very rare indeed. Usually the invertebrate fauna show very brief active periods during summer, otherwise being in a resting state as eggs, pupae, etc. They may achieve fairly 'normal' growth rates during the brief summer, but inevitably have lower annual growth increments. This results in two main patterns of life history, each requiring substantial flexibility and opportunism.

The first strategy involves **univoltinism** and adult lifespans that are greatly reduced, so that in some insect species the eggs mature within the pupal phase, with emergence, mating and oviposition squeezed into just a few days of suitable weather. On the other hand, while retaining univoltinism and an opportunistic burst of reproductive activity, there may be a **life-cycle extension** to two or more years in other taxa that normally complete the cycle annually, so that there is much overlap of generations. Mites and springtails have particularly long lifespans (5 years in *Alaskozetes*), as does the caterpillar *Gynaephora* referred to above. *Pardosa* wolf-spiders may have a 6-year life cycle, whereas temperate relatives take only 1 year; the high-latitude populations mature at a much later age. Even within a species the effect can be striking; for example, the springtail *Hypogastrura tullbergi* lives 3–5 years in polar regions and just 2–12 months in temperate regions. Water-striders (bugs living on the surfaces of fresh water) have univoltine cycles with obligate winter diapause in Quebec but show a southwards cline to a bivoltine cycle; interestingly, total reproductive output is similar at all latitudes.

In many Arctic insects 'hibernation' can occur at any stage except the egg, to facilitate this lengthened life; in some cases this may be a true diapause, but more commonly it may be a less specific form of torpor/dormancy, probably because full diapause could be inappropriately triggered by a period of poor summer weather. Torpor may also occur in larger animals, including the relatively rare vertebrate ectotherms in cold climates; in reptiles such as the cobra, living on cold plateaus, it leads to a drop in blood sugar and pyruvate and an increase in blood lactate, but also to a rise in non-protein nitrogen and particularly uric acid levels, indicating protein catabolism. This is in contrast to endothermic vertebrates, where torpor is accompanied mainly by lipid catabolism.

Reproduction itself may also be highly specialized. In insects the females are often wingless and almost immobile (e.g. in crane flies and midges) and males seek them out, presumably using pheromonal cues. The red-sided garter snake provides another example of specialized reproductive behaviour; the males emerge synchronously from the winter dens in spring, then the females come out in small groups and are immediately set upon by males, forming massive 'mating balls'. Each female then bears up to 30 live young. The triggers for emergence are unknown, and androgens in both sexes seem to be at very low levels.

Viviparity in the vertebrate ectotherms is rather common (e.g. *Vipera* and *Lacerta* amongst the reptiles). This may be because retention in the mother's body, which is undergoing typical shuttling heliothermy to maintain its T_b, is the best way of keeping development of the young at a fast and steady rate.

Feeding and predation

As in the extreme desert environments, foodstuffs may pose a problem. Many of the small invertebrates are omnivorous with little diet specialization, taking only detritus for much of the year. Macrovegetation may be unavailable or deeply frozen, and is heavily physically defended, so that animals with food storage habits are favoured. However, lichens do remain unfrozen so that caribou, using their hooves to dig and tough lips to forage, can survive year-round. Green plants may also remain in a minimally active state within the snow layer, with just a little light penetrating, providing enough food for the burrowing rodent populations. Small populations of carnivores are found in tundra and taiga, and large populations of biting ectoparasites are also to be found, benefiting both from blood meals and from the microclimates created by their hosts.

14.3.3 Large endotherm endurer strategies

Use of microclimates and burrows

The subnivean vegetation is a food resource for some small specialist herbivorous mammals such as lemmings and shrews. They in turn become a food source for small carnivores like foxes and raptors. Thus a whole submerged community exists through the winter, and some of the mammals even choose to reproduce in winter because of the benefits of snow cover.

These animals are all exhibiting a form of microhabitat selection, but for the large ones it is more specifically expressed as a **burrowing** behaviour. Perhaps the most famous mammalian examples are the lemmings, found in most of the northern tundra areas throughout the year. They dig winter burrows within the snow, lined with grass, where they live gregariously and do not enter torpor; in fact these winter burrows may be warmer places than the lemmings' summer holes dug in the earth. Rodents in the tundra and taiga biomes have a particular problem in being the major prey for raptorial birds, so their burrows provide an important escape from predation as well as a climatic refuge (though owls can hear lemmings even under a dense snow layer). Some raptors can also detect rodent abundance by perceiving the ultraviolet emission from deposits of urine and faeces. Therefore lemmings in Scandinavia and Canada have elaborated the architecture of their summer burrow, unusually for a rodent, to include a latrine area so that no clues are left above ground. In winter, when predation risk is much lower, their snow burrows have no latrine and they defecate freely away from the nest. The burrows of ground squirrels, used during winter torpor, are also decidedly elaborate, and are thought to act as biological condensing towers, water evaporating from the warm lower levels and recondensing in the upper cold levels, where it may be available as an important water source for the squirrels when they arouse.

Permanent Arctic residents among the birds also use snow burrows, though rarely gregariously; the classic example here is the burrowing rock ptarmigan, rather larger than a lemming. Burrowing is also employed quite extensively by seabirds at high latitudes; petrels and shearwaters dig into rock crevices, while puffins excavate burrows in the turf above sea cliffs in many cold maritime areas of the northern hemisphere. Even the largest terrestrial animal found at the poles, the polar bear, may dig temporary snow dens in the harshest periods of the winter, with suckling females spending many months in deep snow dens.

Polar and tundra endotherms also use **basking** in association with microclimatic variation, which may have little effect on their core temperatures but will certainly help to maintain peripheral temperatures and thus reduce the overall cost of thermoregulation. Semi-aquatic animals such as penguins and seals bask extensively, to assist in the replacement of heat lost during dives; they may rest facing into the Sun with the wings (or in the case of seals the tail flippers) raised and splayed out.

Migration

Migration is the most striking behavioural strategy shown by the bigger polar animals, and could be seen as a larger-scale version of the movements between microhabitats discussed above. It is also perhaps the equivalent of a life cycle involving resistant stages in terrestrial polar invertebrates, providing an escape in space rather than in time. It is infrequently encountered in ectothermic tetrapods, the reptiles and amphibians, since their locomotory speeds and endurance are limited. But there are many endothermic examples, classic cases being the caribou and reindeer, migratory mammals that trek overland to the forested taiga zones as winter approaches, and the lemmings with their autumnal migrations southward. Polar bears also migrate south across the frozen Hudson Bay in winter. Among the birds migration is especially common since movement by flight is energetically favourable. Penguins of course miss out on this benefit; many of the smaller penguin species merely move away from the Antarctic mainland and spend the winter at sea or on small islands such as the Falklands and South Georgia. But high latitudes in summer provide areas of high productivity and thus serve as good feeding grounds for flying birds, with ducks, terns and many waders moving into littoral zones, and martins and swallows moving onto exposed land areas to exploit the copious insect life. Most of the summer tundra and taiga fauna retreats to more temperate areas for the winter months (Fig. 14.39). Some then show quite profound changes of lifestyle; for example, the knot (*Calidris canutus*) switches from being a rufous-feathered arthropod-eating land bird in summer in the high Arctic tundra where it breeds, to being a grey-feathered mollusc-eating temperate shorebird in the non-breeding winter season. But some birds may avoid this need for a changed lifestyle by moving to the opposite hemisphere to enjoy another 'summer'. At the extreme, the Arctic tern (*Sterna paradisaea*) flies from pole to pole every year, spending the northern summer in Greenland or Norway and the southern summer around the shores of Antarctica.

Migration may have multiple functions and advantages, and may be triggered by multiple cues, but it nearly always results in taking the animals to more equable thermal environments. However, migrations in the endothermic birds are probably less to do with avoiding very low temperatures and rather more to do with leaving areas where foraging will become unprofitable due to the imminent snow cover. The same can be said of the large mammals, whose lichen food would be inaccessible under a thick snow cover. In semi-terrestrial mammals such as elephant seals (*Mirounga leonina*), the situation may be rather different in that migration by swimming may be unrelated to food but necessitated by the requirements of breathing through ice holes,

Fig. 14.39 Migration patterns in tundra animals.

so that the seals move north from Antarctica to areas where the ice is thin enough to maintain such a hole; but again temperature itself is not the main problem. It is normally only in small birds and mammals that cold alone would be likely to be lethal.

Migration may be set in train by temperature cues in a few cases, but is probably more commonly brought about by changing **photoperiod** or by food supply. In most cases, the start of migration is inevitably preceded by a period of laying down energy stores as fat, to fuel the journey.

Reproduction and life-cycle adjustment

Endotherms that migrate into high latitudes for the summer commonly breed there, exploiting the transiently highly productive ecosystem. This exposure to prolonged summer day-lengths necessitates a 'resetting' of the response to photoperiod, to ensure that gonadal development is initiated at the correct time, avoiding an inappropriately early stimulation of the reproductive system and associated behavioural repertoires. Primarily, the relation between day-length and gonadotropin secretion must be readjusted. The effect of decreased sensitivity to photoperiod and hence a shorter reproductive cycle can be seen in birds that have been introduced to new latitudes, and in species such as the house sparrow, with a wide latitudinal distribution, where the period of active spermatogenesis is shortened by about 2 days per degree of latitude the further north the birds breed.

High-latitude reproduction may also require a higher degree of **parental investment** from one or both parents. In the hamster *Podopus*, from cold deserts, the reproductive cycle is extremely compressed, requiring rapid milk production, which puts real strain on the mother's water balance, and also maternal hyperthermia. Species from the more arid and cold parts of the desert show biparental care, the male's presence increasing humidity in the burrow and so alleviating thermoregulatory and water balance stresses for the female; whereas species from areas with more predictable rainfall rear the young without male assistance.

Huddling and aggregation

Huddling in endotherms has been a favourite topic for polar researchers, as it is so spectacular in some seals and penguins. Smaller penguins probably huddle largely to ensure group cohesion and aid mating access in the spring; the thermal advantage for these, in the face of horrendous wind-chill factors in the Antarctic winter, may be very limited. But penguins such as the emperor penguin, *Aptenodytes forsteri*, and the king penguin, *A. patagonica*, have the additional advantage of large size and compact shape, lacking protruding 'eartufts' and tails. Emperor males may spend the winter in the Antarctic interior incubating eggs, and fasting for up to 115 days, when they certainly do gain thermal benefit from huddling. At moderately cold temperatures the emperor penguins form fairly loose semi-circular huddles, all with their backs to the wind; when temperatures drop to the typical –50°C of an Antarctic winter, with windspeeds over 160 kph (100 mph), the birds form much denser tight huddles of several thousand individuals, giving a biomass of perhaps 100 tonnes. Most of each bird's surface is then in contact with other penguins rather than with the cold air, reducing weight loss per individual (from metabolism of stored fat) by 50% relative to a non-huddling penguin, and saving as much as

80% of the heat loss that would occur from one bird alone. There is a 25% reduction in MR, perhaps in part because the birds sleep for long periods.

Polar mammals such as seals and walruses exhibit a less dramatic but also thermally effective huddling when hauled out on beaches in cold weather. In many species, individuals that get too cold on land will return to the sea at $-1.86°C$ to 'warm up'. However, there are a few species of pinniped, such as elephant seals, that have moult periods when they have to remain out of water and fairly immobile for several weeks. In colder periods, these seals will huddle together as pods of up to 30 individuals, staying warm enough to melt surrounding snow.

Burrowing combined with huddling may give energetic benefits in cold climates, but it may be effective at least as much due to the local heating of the burrow that a huddle produces, as to the reduced surface area of each participating individual. For example, in short-tailed voles (*Microtus agrestis*) the local heating effect is estimated as 55% of the total energy saving.

Insulation

When active (i.e. in less cold periods) the regulation of T_b in cold-climate endotherms is largely at the heat *loss* end. The larger polar mammals show an obvious tendency to have long, fine and densely packed fur, commonly 30–70 mm thick in the larger species (see Fig. 8.17) and without any particular correlation with body size. However, in the small species such as lemmings or weasels the fur is much shorter, and usually not significantly longer than in an equivalent tropical species, although it is denser, and is spread more completely over the body including the head and extremities.

Fur in boreal mammals also varies with season (Fig. 14.40). Arctic foxes and the stoat (ermine) are also classic examples of sub-Arctic tundra species that grow a denser and highly camouflaged **winter coat**. The advantages of increased pelt thickness are strictly size-related though: mammals of less than 1 kg show negligible improvements in winter pelt depth, whereas larger mammals such as the red fox and the timber wolf show depth and insulation increments almost in proportion to body size. Phenotypic acclimatory responses also occur; domestic dogs, cats and rabbits may all increase their coat length when kept in colder conditions. The fur of large boreal mammals is often so effective that the complete thermal gradient from air temperature to normal T_b is contained across the fur and skin (Fig. 14.41). The polar bear is something of an exception in having coarse fur, probably because it frequently dives in the water between ice floes and must shake water quickly out of the fur again when it surfaces. There is an apparent paradox in animal coloration here, in that the polar bear and many other polar animals are white, when black fur would appear to be beneficial in absorbing maximum solar radiation. For example, the light coat of the Siberian hamster in winter has a thermal reflectance of 43% compared with the summer coat of 18%, and for the snow-shoe hare the relevant figures are 61% (white in winter) and 23%

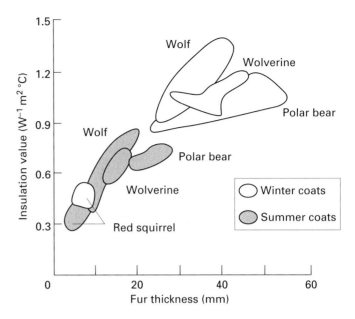

Fig. 14.40 Winter coats of four species of mammals are thicker and better insulating than summer coats; note that changes are greater for larger species. (Adapted from Hart 1956.)

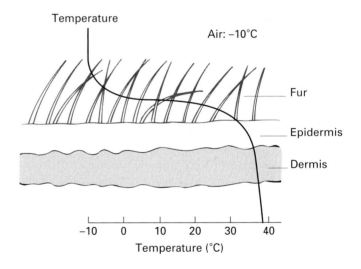

Fig. 14.41 The temperature gradient across the surface of a high-latitude mammal; the gradient is largely across the highly effective fur, and skin surface temperatures are close to core temperatures.

(brown in summer). An obvious explanation is that white pelage makes the animals less conspicuous against a bright ice or snow background, whether they are predators stalking their food or prey trying to avoid being eaten. But this explanation has recently been supplemented for polar bears by a demonstration that the hairs are very unusual in having hollow shafts, which has two effects: firstly, the air space reflects visible light so that the hair appears white (while it is in fact translucent), and secondly, the hairs act like optical fibres, with total internal reflection of shortwave UV radiation so that it passes in from the hair tips to the bear's skin, allowing radiative heating through an apparently 'reflective' pelt.

High-latitude birds such as penguins have dense, smooth layers of feathers, extending over all their extremities, and particularly effective in resisting ruffling and loss of insulation at high windspeeds. Beneath the plumage is a substantial adipose layer in winter, providing insulation during diving and a massive food store for the incubating and often non-feeding adults; king penguin males fast for a month or more at the start of their breeding cycle, and 93% of their energy is then provided by the subcutaneous fat stores. The insulation in penguins is so effective that they can survive in icy water for days on end, with their feet continuously vasodilated to maintain the thermal window effect for heat loss to prevent overheating; and can breed on land in air well below freezing point while maintaining high levels of REWL even when motionless, again to avoid overheating.

The terrestrial high-latitude mammals do not have particularly thick subcutaneous **adipose layers**, as these are too heavy and cumbersome. However, aquatic and amphibious mammals such as whales, seals and polar bears do match the penguins in investing in a massive 'blubber' layer, since for them fur would lose most of its value through being wetted (wet fur is up to 50-fold more prone to heat loss than dry fur). The relative utility of fur to terrestrial and aquatic animals is neatly demonstrated by many of the seals, where the essentially terrestrial pups have much denser fur than the primarily aquatic blubber-bearing adults. Baby seals have similar problems to penguins and can die of overheating rather readily; indeed sea lions in the Galapagos have to breed in cool caves or behind boulders away from the Sun. Some amphibious diving mammals, such as beavers and mink, do have a pelt of dense air-trapping underfur beneath a coarser wettable outer fur, but these animals normally only immerse themselves for short periods. Species that dive in sea water rather than fresh water, such as the sea otter (*Enhydra lutra*), have an extra problem of salt crystals forming in the pelt, which compromises its insulating role both on land and on return to the water; these species spend unusually large proportions of their day in grooming the fur to remove the salt.

Control of heat loss at extremities

Many polar animals suffer from potentially extreme rates of heat loss at their extremities, which usually have less dense fur to aid manoeuvrability and/or sensitivity. In seals and whales, for example, it is calculated that 10–30% of heat production in a resting animal is lost through the flippers, fins and flukes, rising to 70–80% during moderate exercise. Most of these animals are using **countercurrents** at their extremities to prevent these losses being even higher. Seal and dolphin flippers are among the best-studied cases of this regional heterothermy, with elaborate rete systems (see Fig. 9.43). Each artery is surrounded by a ring of small veins, normally allowing a countercurrent exchange to reduce heat loss through the uninsulated flippers. But when the

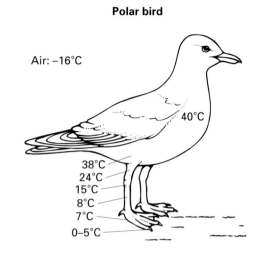

Polar bird

Air: −16°C

40°C

38°C
24°C
15°C
8°C
7°C
0–5°C

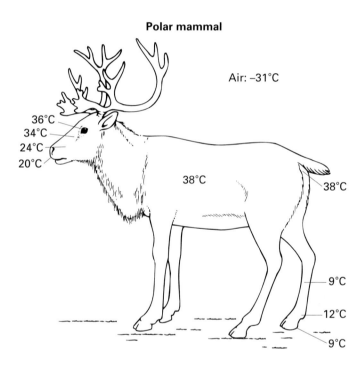

Polar mammal

Air: −31°C

36°C
34°C
24°C
20°C

38°C

38°C

9°C

12°C

9°C

Fig. 14.42 Peripheral countercurrents producing cold extremities in polar mammals and birds. (Adapted from Irving & Krogh 1955.)

animal is overheating due to prolonged exercise, the anatomy of the system is such that the raised blood flow and higher blood pressure cause the central artery to dilate and the surrounding veins to collapse, bypassing the heat-exchanger and making blood return from the flipper by alternative more peripheral veins, losing heat to the surrounding water in the process.

Similar heat-exchangers normally keep hot blood away from gull and penguin feet, from caribou noses and hooves, from beaver tails and probably even from the small ears of many polar mammals (Fig. 14.42). Since so many of these animals are standing on or swimming in ice-cold environments, they particularly

Fig. 14.43 Metabolic rates (MR) vs. declining temperature in three polar endotherms; the thermoneutral zone (TNZ) is relatively broad, especially for the Arctic gull, in which MR does not increase even at −30°C. (From Davenport 1992.)

need to ensure that 'frostbite' (cold-induced tissue damage) does not set in, so they must also supply the peripheral tissues with occasional pulses of warmer blood to give some protection; at very low temperatures the heat transfer to the tip becomes small but continuous (see Fig. 8.40).

Control of heat production

Some high-latitude animals do have **higher basal metabolic rates** than predicted from the simple allometric equation (see Chapter 3). In rodents such as lemmings it is 200–240% greater than predicted. MR can also be modified in many cold-adapted animals if food supplies permit. But note that, on the whole, metabolism is less sensitive to temperature than in tropical animals (Fig. 14.43). Polar animals have a much broader zone of environmental temperature where they keep 'ticking over' at

much the same rate regardless of ambient temperature (i.e. a **broader thermoneutral zone**).

Hypothermia and torpor

Hypothermia involves an adaptive decline in core T_b, to a new value that may be just a few degrees below normal (perhaps 30–35°C) or may be very low indeed, often below 10°C and within 1 or 2°C of T_a. Slight reductions in T_b are best described just as mild hypothermia; many birds and mammals use this in cold spells, but these animals can normally become active again more-or-less instantaneously. The term **torpor** is generally used for the phenomenon of a relatively deep hypothermia involving a pronounced naturally occurring decline in T_b where MR, respiration and circulation are strongly depressed; its occurrence in birds and mammals was outlined in Chapter 8. Torpor requires the specific and energetically expensive phenomenon of **arousal** to permit subsequent activity.

Remember that 'hibernation' is a much looser term, sometimes thought of as synonymous with prolonged and profound torpor but widely misused in relation to such animals as badgers and bears, both of which sleep for long periods in the winter but readily can and do become active when hungry; the descriptive terms 'winter sleep' or 'carnivore lethargy' are often preferred in modern literature. Brown bears (*Ursus arctos arctos*), for example, show 'hibernation' at only 4–5°C below normal T_b, with very minor modifications to circulatory pattern and a slightly increased red blood cell count, probably triggered by an autumn decrease in production of thyroid hormones.

Torpor is not especially common in polar or tundra mammals, because rates of heat loss are too high for it to be sustainable at all, the habitats being too unproductive to allow adequate accumulations of fat reserves; polar species that do exhibit torpor, such as the Arctic ground squirrel (*Spermophilus parryi*), use it for only brief periods. It is the small eutherian mammals of the boreal forests that yield the best examples of prolonged deep torpor.

Torpor normally has three phases; that is to say, the rapid entry phase and the rather slower arousal phase can be physiologically distinguished from the middle period of relatively steady, low MR, when the MR is much reduced, representing a very large energy saving. In these boreal mammals there is no single trigger for going into a torpid hypothermic state; its onset may vary with food supply and with the imposed ambient temperature regime, though where there is a yearly cycle the onset of torpor is generally regulated in part through photoperiodic changes acting via the endocrine system. The onset of torpor may be quite gradual, and it largely results from failure to increase MR as T_a declines, with ventilation slowing rather than increasing. This leads to an inevitable decrease in T_b and a further decrease in metabolism, so that the animal slides rather gently, often over 8–24 h, into a torpid state without initiating any special effects such as increased conductance. Entry into this state sometimes involves a short series of transient drops in T_b

Fig. 14.44 The onset of torpor in a ground squirrel, with progressive lowering of brain temperature (T_{brain}) on successive nights. (Adapted from Strumwasser 1960.)

Fig. 14.45 Temperature and oxygen consumption in a bat during arousal from torpor, showing that brown adipose tissue (BAT) metabolism is important but forms only part of the whole thermogenic response. (From Davenport 1992; adapted from Jansky 1973.)

Table 14.5 Time to enter and arouse from torpor for animals of different body mass. Calculated from allometric equations, for cooling and arousing at 15°C, body temperature (T_b) changing from 17 to 37°C.

Species	Body mass	Entry time	Arousal time
Shrew	2 g	35 min	13 min
Hummingbird	4 g	59 min	17 min
Honey possum	10 g	80 min	24 min
Poorwill	40 g	224 min	41 min
Nightjar	86 g	350 min	55 min
Vulture	230 g	39 h	3.2 h
Echidna	3.5 kg	27 h	3.8 h
Marmot	4.0 kg	29 h	4.0 h
Badger	9.0 kg	45 h	5.4 h
Bear	80 kg	138 h	12.3 h

(Fig. 14.44), with 'test runs', followed by reinstatement of regulation, then a single smooth rapid drop to the new low level during which thermoregulation is temporarily abandoned and the body cools largely passively. For a bird or mammal of about 2 g, the torpid body temperature of 15°C would be reached in only about 40 min, whereas for a 100 kg bear to become fully torpid would take a week (Table 14.5). Simple calculations nevertheless show that for any endotherm, despite the rapid rate of warm-up required for arousal and its cost in energy and food, even short periods of torpor are energetically worthwhile.

In the fully torpid state, the MR may be only 2–20% of its normal resting value, and it is primarily body fat that is catabolized. Metabolism is suppressed mainly by a reversible phosphorylation of key glycolytic enzymes, especially pyruvate dehydrogenase. Enzymes also become relatively temperature-insensitive, and levels of phosphagens are decreased. As a result, for example, the oxygen consumption of a torpid bat is only 2.5% that of the active rate (Fig. 14.45), so that its fat reserves will last 40 times longer then 'normal' under cold conditions. In at least some torpid mammals the degree of saturation of the dietary fats affects torpor, with an unsaturated diet giving lower minimum T_b and longer torpor bouts, possibly via interactions with the energy-regulating protein leptin. But this enhanced torpor is at the expense of greater lipid peroxidation during the torpid period; in ground squirrels the intake of polyunsaturated fatty acids such as linoleic and linolenic acids is therefore reduced during the autumn months.

In full torpor, ventilation may be as low as 1–2 breaths per minute in a small mammal, with periods of total apnoea of up to 5 min. Although torpid animals appear unresponsive and uncoordinated, they still have functioning neural control of their own condition. If T_b is in danger of dropping too low, with the animal approaching freezing, there may be spontaneous arousal or a gradual increase in heat production to keep the body at 4–5°C without actually initiating expensive full warm-up. This is reflected in a rise in oxygen consumption as the air temperature falls too low, and again indicates that the torpid animals are not entirely going over to ectotherm-like thermal biology.

Arousal from torpor may occur quite rapidly (1–6 h) relative to the process of entering torpor. Arousal is also very size-dependent (see Table 14.5). However, arousal rate also inevitably depends on the starting point of T_b, since MR and temperature are so closely related. In fact at a very low body temperature it is impossible for chilled tissues to use oxygen rapidly and MR may not be able to get high enough to raise T_b at all; in other words, for each size of animal there is a theoretical body temperature threshold below which it cannot fall or arousal becomes biochemically impossible. Each species also in practice has a minimum temperature from which it will initiate arousal, known as the **critical arousal temperature**, and this parameter seems relatively unrelated to body size or to phylogeny, perhaps reflecting natural habitat temperatures (e.g. it is 2°C for the boreal pocket mouse *Perognathus longimembris*, but as high as 20°C for the pygmy mouse *Baiomys taylori*). It may also be linked to habitat humidities, since progressive desiccation may be a real problem for small hibernating mammals such as the little brown bat, where EWL during torpor would exceed metabolic water production at any humidity less than 99.3% RH.

Arousal in most boreal mammals is a two-stage process, triggered by day-length and/or rising temperature. At first it is not accompanied by obvious shivering or muscular activity, because a large part of the initial thermogenesis in mammals is achieved not by the muscles but by **non-shivering thermogenesis** (NST; see Chapter 8) in the specific tissue known as brown fat or brown adipose tissue (BAT). The BAT where NST occurs is found particularly around the neck and shoulder areas of torpid mammals (and of many newborn mammals too, where it may be up to 5% of the birthweight), though it is absent in birds. Its activation results in a very rapid warming of the thoracic and head area, including heart, spine and brain. Rectal temperature lags many minutes behind, indicating active control of blood distribution during arousal. Tracer studies show a high blood flow to the heart, to the brown fat itself, and then to skeletal muscles, so that key organs can be supplied with oxygenated blood. Only then are corticosteroids and insulin released into the blood so that the muscles can initiate shivering thermogenesis, to give the second phase of speeded up arousal. Thus in arousing hedgehogs blood insulin secretion does not begin until T_b has risen to about 25°C, showing the importance of NST in the early warming stage. Indeed, NST accounts for about 40% of the heat needed for full arousal in hamsters, and somewhat less in bats (Fig. 14.45).

Figure 14.46 summarizes the role of reduced winter metabolism and torpor in mammals of different sizes. In large species seasonal acclimation (with MR reduced to about 50% of summer levels) is largely achieved by extra fur insulation and torpidity is avoided. In smaller mammals, there may be prolonged deep torpor in some cases, but in other species winter activity is maintained by a combination of weight reduction, improved thermal conductance and a much reduced degree (length and/or depth) of torpor; the winter MR of these species is about 30–40% of that in summer, compared with a drop to 15–20% of summer levels in the small mammals with deep prolonged torpor.

Coping with freezing

Because of their endothermic and thermoregulatory abilities, most of the 'endurer' animals are not required to cope with freezing and cannot do so if very low temperatures are forced

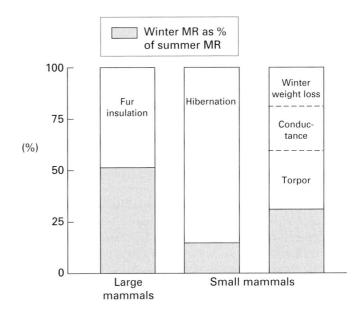

Fig. 14.46 A comparison of the seasonal thermal strategies of large and small mammals; large species mainly respond to winter with improved insulation, while small mammals use either prolonged torpor (hibernation) or a mixture of weight loss, improved insulation and short periods of torpor. MR, metabolic rate. (Adapted from Heldmaier *et al.* 1989.)

upon them in laboratory studies. However, there is at least one endotherm, the Arctic ground squirrel, mentioned above as a true polar user of torpor, that has the ability to supercool to −2.9°C to avoid freezing.

Food supplies

As we saw for ectotherms, foodstuffs may pose a problem. Many tundra and taiga species that do not migrate are food hoarders, hiding caches of seeds or nuts for future use, though they normally combine this with periods of torpor. Herbivores tend to face rather nutrient-poor and high-fibre diets, and gut mass may be larger than normal; it may be subject to acclimation, since it has been shown to increase markedly as a proportion of body mass in lemmings fed on low-quality high-fibre diets. Polar mammals and birds that are herbivorous show more distinct plant preferences than related temperate species (e.g. willow ptarmigans select willow, rock ptarmigans select birch).

Permanent active residents such as polar bears, penguins and pinnipeds tend to be carnivores, or more especially piscivores; fish and other large marine fauna are the only rich dietary items available year-round that are not themselves frozen. Polar mammals have the additional problem of providing adequate nutrition to their offspring, often in the depths of winter; it is no coincidence that cetacean and pinniped milks are unusually rich in fats and proteins, and polar bears too produce a milk richer than that of other Carnivora (see Table 13.13).

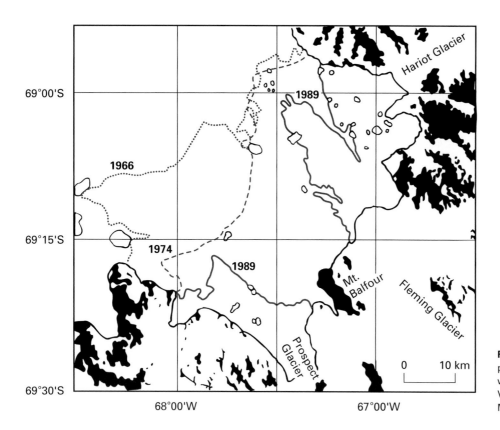

Fig. 14.47 Documented ice-sheet thinning in part of Antarctica since 1966. (Reprinted with permission from *Nature* (Doake & Vaughan 1991). Copyright 1991 Macmillan Magazines Limited.)

14.3.4 *Anthropogenic effects in the cold biomes*

The low overall productivity in boreal zones makes the plants, and hence the communities as a whole, particularly susceptible to human interference. Humans have occupied the Arctic tundra since at least the Neolithic, but had little impact before this century except in reducing some populations of vertebrates through whaling, hunting and fur-trapping activities.

In the last 40 years, however, exploration and accompanying **exploitation** have boomed, with oil, gas and minerals much sought after and with pipelines, roads and railways all invading the landscape. The problem is that localized damage to these environments is only very slowly repaired by regenerative growth, and soil turnover is even slower, so that we can still see extensive surface damage from the early gas and oil pipelines installed across Alaska and Siberia up to 35 years ago, with damage accentuated by activities causing localized scree flows and the creation of deep gullies. Such lines may seriously interrupt animal migrations.

Global warming also poses particular problems for high-latitude systems. Most of the climate models predict substantially faster and greater changes in temperature at the poles, commonly as much as 6–12°C within the next century, leading to warmer polar winters. An early proposal that this might lead to more snowfall and hence a thicker more extensive area of snow and ice-field, with greater albedo and greater reflection of radiation, has not been upheld by the evidence;

there is now extensive documentation of the thinning and **break-up of ice sheets**. In Antarctica several ice-shelves have been monitored carefully and show marked disintegration and retreat of the ice front (Fig. 14.47), while in Greenland ice thinning has also been reported. Since many polar invertebrates appear to have upper lethal temperatures rather close to their current maximum microclimate temperatures, there may be casualties from overheating.

Ozone depletion is currently a phenomenon very much centred on the polar zones, with polar stratospheric clouds implicated in the reactions with atmospheric halogens (chlorine and bromine) that lead to ozone destruction. The seasonal Antarctic ozone 'hole' is still expanding every year, and some annual thinning in the Arctic is also documented. The ozone-depleted air has on occasions moved out to lie over the southern tips of Australia and South America, and routinely affects the Antarctic continent and fringes. The expected effects of increased frequencies of skin cancers, and of cataracts, should certainly be most likely in the polar mammals and birds, with penguins perhaps most profoundly at risk.

All of these problems can be accentuated by increased predation risks for weakened animals, faced by introduced predators; feral mice in the sub-Antarctic are already a problem for insects. Hence in some ways these cold terrestrial environments are every bit as 'fragile' as the tropical forests and temperate wetlands that are more common areas of concern, but as yet they have received much less research effort.

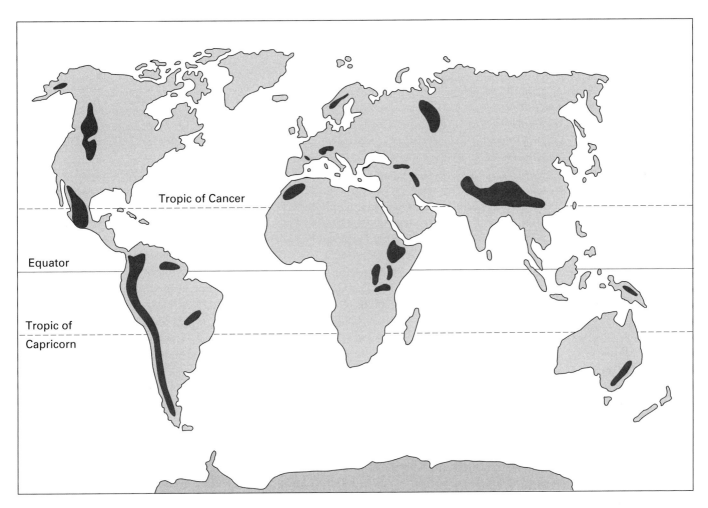

Fig. 14.48 The world's mountain ranges.

14.4 Life at altitude

Life at altitude in many ways presents a special case of low-temperature habitat, where height above sea level parallels the effects of latitudinal distance from the equator. However, there are also additional problems relating to temperature, water availability, respiration and pressure, so these habitats are given a section of their own here.

14.4.1 *Montane habitats*

Occurrence

Mountains occur in all continents (Fig. 14.48), and in a wide range of forms, both as long mountain chains and as relatively isolated peaks. The process of **orogenesis** (mountain building) is primarily related to plate tectonic movements, with the land mass rising up where two plates collide and forming a mountain chain along the perimeter of one plate. This may be either at the edge of a continent where a submerged plate undergoes subduction below a land mass, as with the Andes in South America; or it may be in the centre of a continent formed from two land masses abutting on each other, as with the Ural mountains between Europe and Asia, or the Himalayas between Asia and India. However, every mountain range is unique, since the processes underlying orogenesis are very complex. Furthermore, every existing mountain range is of different age, some still rising and forming, others the result of hundreds of millions of years of fracturing and erosion since they first formed, thus giving a vast range of different types of formations. Some key characteristics of major mountain ranges are given in Table 14.6.

Being areas of marked recent geological upheaval, younger montane areas also tend to suffer from other geological phenomena such as earthquakes and vulcanism. The Andes, for example, are part of the Pacific 'ring of fire', with numerous active volcanoes. Mountainous areas may also be sufficiently elevated to be frozen, thus giving rise to glaciers, and the scouring effects of recent or current glaciation are often very obvious.

From a biological point of view, several of these geological factors have an important impact. Elevation in itself is perhaps the most obvious, with effects on temperature, pressure, and oxygen availability. In mountain ranges daily average temperatures are reduced by about 1°C for every 150 m of altitude. This effect is roughly similar at all latitudes, so that even tropical

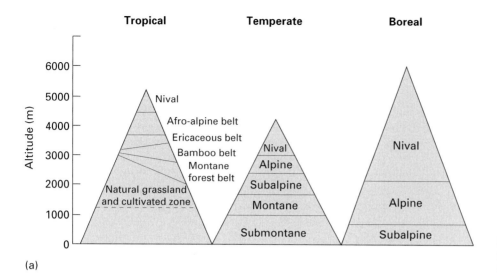

Fig. 14.49 Patterns of floral change with altitude in different mountain ranges. (a) Schematic zonation. (*continued*)

(a)

Table 14.6 Characteristics of mountain ranges.

Mountain range	Highest peak (m(ft))	Latitude
Alps	Blanc 4807 (15 771)	45–47°N
Andes	Ojos del Salado 7034 (23 241)	35°S–7°N
Antarctica	Vinson Massif 5140 (16 863)	
Appalachians	Mitchell 2037 (6684)	34–42°N
Atlas	Toubkal 4165 (13 664)	30–33°N
Australian Alps/ Great Dividing Range	Kosciusko 2230 (7316)	28–38°S
Caucasians	Elbrus 5642 (18 510)	42–43°N
Ethiopian plateau	Ras Dashen 4620 (15 157)	5–15°N
Himalayas	Everest 8848 (29 028)	28–35°N
Iran/Turkish plateau	Ararat (Büyük Agri) 5165 (16 945)	30–42°N
Kenyan plateau	Kilimanjaro 5895 (19 340)	5°S–3°N
Pyrenees	Pico de Aneto 3404 (11 168)	42°N
Rockies	Elbert 4399 (14 431)	35–62°N
Sierra Madre	Citlaltepetl 5699 (18 697)	35–40°N
Sierra Nevada	Whitney 4418 (14 495)	36–39°N
Southern Alps (NZ)	Cook 3764 (12 346)	43–45°S
Urals	Narodnaya 1894 (6213)	53–68°N

mountain ranges sitting astride the equator can be snow-covered (Plate 9f). They are therefore profoundly cold at night even though highly irradiated by day, and these diurnal cycles of extreme heat and cold are accentuated during the dry season. A classic example is the isolated peak of Mt Kilimanjaro in Tanzania, almost on the equator yet with snow and glaciers at the summit all year round even in blazing midday sunshine. Other tropical mountain ranges may be extremely arid, with the northern Andes having exceptionally low-humidity air since the precipitation has been shed from air rising up the west-facing slopes, giving a rain-shadow effect. These dry mountain peaks are again always profoundly cold by night; many animals living there must show expanded thermal tolerances. In more temperate latitudes, mountain ranges may show a lesser diurnal temperature range but a very drastic seasonal variation, so that temperate mountain ranges such as the Alps and Rockies have an extremely variable snowline.

Aside from elevation, mountains also may pose challenges in terms of geological instability, with periodic fracturing, quaking or volcanic emissions. While potentially lethal for local populations in the short term, such effects may in the longer run produce increased diversity (by separating subpopulations on isolated peaks or plateaus, leading to potential allopatric speciation), or in the case of vulcanism to increased local soil fertility and biotic productivity.

Montane vegetation

Vegetational patterns at altitude parallel those with latitude, but may show the same sequences compressed into a much smaller linear dimension (Fig. 14.49a). In ascending from the lower slopes one may pass from deciduous forest and chaparral or grassland into coniferous forest, above which is a distinct treeline; above that again there may be regions of shrubby vegetation or ericaceous plants, sages and junipers, with some

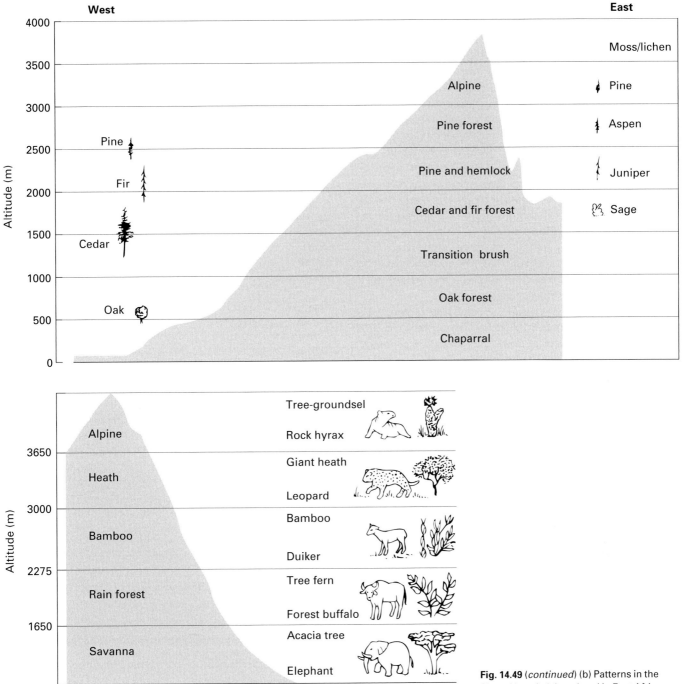

Fig. 14.49 (*continued*) (b) Patterns in the Sierra Nevada (above) and in East Africa (below).

(b)

thinning grass, then bare rock with lichens. The snowline in winter descends downwards through these successive tiers of vegetation.

But the actual vegetation on any particular mountain range is very variable with continent and latitude (Fig. 14.49b), so that it is very hard to generalize beyond this. In addition, mountain ranges have often acted as refugia and as centres of isolation at different times in their history, so that many of them have unique combinations of flora.

Montane fauna

The upper lichen zones of mountains have an invertebrate fauna largely restricted to the soil layers and including springtails, mites and nematodes. Spiders, particularly from the genus *Pardosa*, occur in most mountain ranges, subsisting partly on windblown insect fauna such as aphids; some insects also occur as residents (up to about 6000 m in the Himalayas). In these upper rocky areas, where there is patchy vegetation there is also a range of relatively large and quite specialized vertebrates, particularly camelids, with the bactrian camel occurring in the Himalayas, the dromedary in central Asia, and the llama, alpaca, guanaco and vicuña in South American mountain ranges. There are also various sheep and goats, including the alpine ibex, mouflon and chamoix, the Himalayan goral, tahr and markhor, and the Rocky Mountain bighorn. Amongst the crags there may be resident raptorial birds (eagles, etc.) and a range of passing migrant birds. Just a little lower down, where soil is deeper and vegetation more prolific, there is a wide diversity of burrowing rodents, providing prey for the raptors. Where there are lakes within the mountain ranges, particularly in the long chain of the Andes, quite a range of waterfowl also occurs, especially geese and coots but also including flamingos.

Further down the mountains faunal diversity inevitably increases and largely parallels the fauna of the equivalent latitudinal belt, with a particularly strong representation of insects, such as grasshoppers, and some pollinators, such as bees, butterflies and flies. Amphibians and reptiles become rather common, each species often having a quite substantial altitudinal range. Rodents are again common, as are various large deer and antelope (elk, pronghorn, taruga, klipspringer, etc.) in dense bush and forests, and very bulky animals such as the yak in plateau areas of Asia. Thus many taxa of animals have surprisingly high altitudinal limits (Table 14.7).

To make the point that every mountain range is different, an example of the faunal distribution up a tropical mountain is shown in Fig. 14.49b. Here hyrax are a speciality in the upper slopes, with leopards making homes in isolated crags and only descending at night to feed; lower down the typical African bush fauna appears.

It is generally found that species diversity decreases with altitude, and that species ranges increase with altitude; the habitats thus become dominated by relatively few, relatively specialist, species extending over broad elevational bands. An example of elevation ranges for bats in the Andes is shown in Fig. 14.50.

14.4.2 *Thermal problems*

The main environmental effect of altitude, as far as animals are concerned, is the **declining temperature**. On most mountains there is also a particularly stark contrast between conditions in sun (since the radiative load is very high) and in the shade, and often a stark contrast between day and night. Thermal problems can also be greatly magnified by high winds.

Animals in mountain ranges may therefore have to exhibit a high thermal tolerance, to avoid freezing by night, and to cope with overheating and desiccation by day; in tropical mountains the change between the two conditions may be extremely sudden

Table 14.7 Altitudinal limits for various vertebrates.

Taxon	Altitude (m)	Location
Fish		
Trout	2800	Alps
Trout	3800	Andes
Nemachilus	4700	Asia
Amphibians		
Salamander	3000	Alps
Eleutherodactylus	4500	Andes
Toad	5000	Himalayas
Reptiles		
Lizard	3400	Rockies
Skink	4000	Kilimanjaro
Iguanid	4900	Andes
Lizard	5500	Himalayas
Birds		
Several	4000–6500	Rockies
Lammergeier	5000–6000	Himalayas
Barheaded goose	8800+	Himalayas
Mammals		
Deer mouse	4000	Rockies
Human (permanent)	4500+	Andes, Himalayas
Llama, alpaca	4800–5400	Andes
Chinchilla	5000	Andes
Yak	5800	Asia
Taruca deer	6000	Andes

at sunrise and sunset, with no time to 'prepare' on a daily basis. To make matters worse, since the smaller animals are active and feeding by day they cannot empty their guts and are bound to be full of ice-nucleating agents, so that they are particularly vulnerable to bodily freezing.

Heliothermy: basking, sun–shade shuttling and colour

Amongst ectotherms, the insects, lizards and snakes use basking and shuttling **heliothermy** to a marked degree. There have been classic studies on an Andean lizard species, *Liolaemus multiformis*, living at 4000 m, which emerges from its burrow at dawn (07:00 h) with an air temperature of only −5°C and basks on heaps of vegetation that insulate it from the ice, warming up to a T_b of 35°C within 2 hours (Fig. 14.51). However, life at altitude does complicate the issue of thermal balance and patterns are not straightforward. Within the genus *Liolaemus* there is no clear association between thermal tolerance and elevation, because varying body masses affect thermal inertia; with mass factored out, high-elevation species heat more slowly. High-altitude Andean frogs do have a broader thermal tolerance than

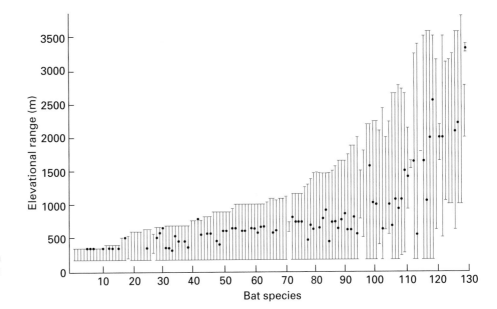

Fig. 14.50 Elevational ranges of 129 bat species in the Andes; species that occur at a higher mean altitude (solid circle) also have substantially greater altitudinal ranges. (From Patterson *et al.* 1996.)

Fig. 14.51 Basking in a Peruvian mountain lizard allows cloacal temperature to rise sharply after dawn and then be maintained at a steady 35–38°C, well above air temperature. (Adapted from Pearson 1954.)

dramatically reduced compared to lower temperatures. Pigment cell sensitivity thus appears to be acclimated at the cellular level, to allow appropriate thermoregulatory responses.

Postural thermoregulation is also very obvious in ectothermic montane fauna; basking and ground-hugging is common in montane butterflies such as *Colias* species, and crouching postures side-on to the Sun are conspicuous in montane grasshoppers such as *Melanoplus sanguinipes*, which can raise T_b up to 20°C above ambient. These high-altitude grasshoppers often exhibit higher resting MR than their low-altitude congeners, which may compensate for lower temperatures and shorter seasons. For many montane animals there are indications that the MR rises and the critical thermal maximum decreases, inter- and intraspecifically, with increasing altitude.

Use of vegetation and microclimate

Because of the combination of freezing and desiccation dangers referred to above, many montane animals from the tropics cannot be either particularly cold-hardy or particularly desiccation resistant, and there is a strong tendency to use behavioural rather than physiological techniques to survive. Hiding in crevices and under rocks, and seeking shelter within vegetation, are almost ubiquitous strategies amongst invertebrates. Thus, for example, insects sampled from above 4000 m on Mt Kenya showed almost no special adaptations to low temperature, but instead sheltered in the rosettes of large *Senecio* plants and in other dry flower-heads. The predominance of large rosette plants in these habitats may be particularly important in providing stable microclimates amongst the bulk of decaying older leaves.

Certain insects are particularly adept at exploiting the sharp

congeners from neighbouring lowlands, and greater aerobic metabolic scopes at low temperatures. The lizard *Sceloporus variabilis* achieves a mean T_b of only 28.9°C at 1000 m in Mexico, whereas the same species at 45 m can achieve 32.4°C, despite the two populations having similar metabolic rates. Another closely related lizard, *S. jarrovii*, is winter-active in mountains, and it can darken rapidly when cold, with the melanophores directly responding to melanophore-stimulating hormone. Skin samples show a differing dose–response curve to this hormone at different temperatures, and at 35°C the darkening response is temperature gradients around mountain vegetation. One example is the checkerspot butterfly, *Euphydryas gillettii*, in the mountains of Wyoming, USA, which preferentially lays its eggs on the southeast-facing leaves of certain honeysuckles, to gain maximum warming from the early morning sun. Eggs laid in such sites mature more quickly than those from other zones of the plant, and so are more likely to produce caterpillars that will have fed enough to survive when the early montane winter sets in.

Microhabitat selection certainly also matters for vertebrates, but the physical nature of the terrain at high elevations may limit the options for these larger animals. For example, the microclimate choices are relatively similar for the Andean and lowland frogs mentioned above, although the times of day (and air temperatures) at which each kind of habitat is sought out do vary markedly.

Burrowing and huddling

Burrowing in general serves as a way of avoiding very cold temperatures for montane animals. The Andean lizard *Liolaemus* spends on average 82% of its time in deep burrows, with 15% basking and crouching on warm surfaces and just 2–3% actively foraging or interacting with other animals. Deer mice in Wyoming mountains similarly use burrows to avoid freezing; with night temperatures of –15°C, the burrow never falls below 0°C. Note that this deer mouse is the same species that can be found in the Nevada desert further south in the USA, where, as we saw, the burrows have the opposite role of providing a buffer against overheating.

Montane animals also frequently combine the benefits of **burrows** and **aggregations**. The alpine marmot (*Marmota marmota*) overwinters in large underground nests, containing 5–8 animals of the same family (whereas its close relative the woodchuck does not live at altitude, and hibernates singly). The alpine marmot undergoes torpor with a T_b of only 7–8°C for periods of about 2 weeks, interspersed with short arousals when T_b rises to normal levels for a day or so. Within the nest all the individuals are synchronized, and huddling occurs both during torpor and during the high T_b phases. When one individual begins to arouse, some of its body heat is transferred to the other marmots, helping them to warm passively; the first one to be fully aroused also grooms its nestmates and covers them with extra nest material to aid their warming. The marmot burrows are commonly around 12°C internally in autumn, but this has dropped close to 0°C by early spring. When burrow temperature falls below 5°C, the marmots raise their MR somewhat, but because of the burrow this is only required for rather less than half the total period of torpor. The more animals there are in the burrow, the longer it takes for the burrow temperature to fall to 5°C, giving a real benefit to 'social hibernation'.

Even in the larger mountain mammals a gregarious habit is common, with sheep and goats living in small herds and huddling together at night and in stormy weather.

Insulation, heat distribution and metabolic rate

Montane mammals show some interesting adaptations of furs. The duck-billed platypus, *Ornithorhynchus anatinus*, lives in mountain streams in eastern Australia and may dive for prolonged periods in water close to freezing point; it has one of the most effective fur insulations known, with a dense underfur and flattened guard hairs to trap air, combined with countercurrent heat-exchangers at the base of its tail and all four limbs. *Oreamnos americanus*, a wild goat from the Rocky Mountains, has evolved the same strategy as we met earlier in the polar bear, with a rather coarse pelt of hollow hairs that appears white (and presumably it gets the same advantages). The large camelids, deer and antelope from mountain areas are also famous for the depth and quality of their pelage. Many are used as providers of high-quality wools (e.g. chinchillas, and alpaca and cashmere goats). The individual hairs may be very long, most spectacularly in the wild yak where the rather coarse fur almost reaches the ground. During the short summer periods, when these animals might risk overheating, blood can be shunted to the less insulated extremities to dissipate excess heat.

In several invertebrates, notably insects, mass-specific **metabolic rates** tend to increase with elevation, and it has been argued that this is an adaptation to accelerate development in species that remain annual (see below). However, in the grasshopper genus *Xanthippes*, population differences in mass explained most of the effects (high-altitude grasshoppers were smaller), and when mass was factored out little effect of elevation remained. In heterothermic bees T_b and warm-up rate increase with altitude, although it is unclear whether this results from decreased thermal conductance or higher mass-specific heat generation rate.

Coping with cold and with freezing

Small ectotherms in mountains may have real problems with achieving activity in continuously cold conditions. In insects this is sometimes circumvented by a degree of endothermy, as for example in the rainbeetles (*Plecoma*) of the Sierra Nevada of the USA, which achieve a T_b of 35°C even in air temperatures around freezing. Bumble-bees in the Himalayas and Andes achieve similar feats.

Many small animals from seasonal temperate mountains have classic physiological adaptations to achieve cold-hardiness in winter, with both freeze-tolerance and freeze-intolerance exhibited. Smaller species (mites, springtails, etc.) tend to be **freeze-intolerant**, voiding their guts to eliminate nucleating agents and accumulating glycerol or other polyols or sugars. Larger invertebrates tend to be **freeze-tolerant**, accumulating both nucleating agents and cryoprotectants, and surviving repeated freezing of their extracellular fluids. As in other habitats, levels of protection against freezing tend to show seasonality; Fig. 14.52 shows the strongly seasonal variation in supercooling points for alpine beetles.

Fig. 14.52 Seasonal variation in supercooling point in three species of alpine ground beetles. (From Sømme 1995.)

Fig. 14.53 Rates of water loss (cutaneous evaporative water loss, CEWL) in montane beetles compared with those from arid lowlands (measured at similar temperature and humidity); there is a lowered evaporative water loss (EWL) in tenebrionids from Pico del Teide (Tenerife), but for carabids (living in the litter layer) there is little effect except in the relatively exposed Norwegian beetles. (From Sømme 1995.)

One of the more spectacular examples of freezing tolerance is the New Zealand weta (*Hemideina maori*), a very large montane grasshopper that survives in the frozen state for much of the winter, and which can be seen freezing and thawing on a daily basis in spring and autumn. The weta may have more than 80% of its body water converted into ice, promoted by an ice-nucleating protein, though intracellular contents remain unfrozen through osmotic dehydration effects and a build-up of proline. A few cases of freeze-tolerance in species from tropical mountains have also been recorded, including beetles from Mt Kenya, *Meridacris* grasshoppers from the northern Andes, and *Agrotis* caterpillars from Mauna Loa on Hawaii.

14.4.3 Dehydration and water balance

Animals suffer an increased tendency to dehydrate at altitude. At low ambient temperatures the water vapour content of the ambient air is reduced, and any animal with a T_b above T_a will lose water very quickly in its warm exhaled saturated breath; furthermore the diffusion constant for all gases, including water vapour, increases with altitude. Increased windspeeds may add to the dehydrating effect. Where there is standing water or snow, larger animals, including most of the vertebrates, may have little problem, but small invertebrates are more at risk.

Invertebrates from mountains may therefore have relatively high **dehydration tolerance**. For example, tenebrionid beetles from exposed habitats on the slopes of Mt Teide (Tenerife, 2000–2500 m) can survive at least 40 days at 5% RH, with only

15–30% weight loss, matching or outdoing some more xeric lowland species for rates of CEWL. However, in groups such as carabid beetles that normally live within the litter layer the montane species are no more desiccation-resistant than similar lowland species; only a few species that survive in rather exposed zones above the treeline in Norway have lowered CEWL values (Fig. 14.53). In addition to cuticular adaptations, there may be **excretory adaptations** to minimize water loss in invertebrates. Montane tenebrionids normally have a cryptonephridial system, and the gut lumen (and therefore the urine/faeces) may reach 2500–2750 mOsm.

For endotherms the issues of water balance are somewhat more complicated. Many mammals are reported to have **lower total body water** values at altitude, and in humans this effect can be marked (29 kg in Andean populations compared with 39 kg in weight- and age-matched lowland humans). High-altitude populations can also clear a drinking-water load much more quickly, mainly because the collecting ducts of the kidney become relatively insensitive to the hormone arginine vasopressin. It is still unclear whether other aspects of water-loss rates are altered in mammals. Careful analysis of deer mice water-loss rates with altitude show a clear increase, but this was an indirect effect, with mass and MR as confounding variables. Differences in the thermal environment are probably the main cause of higher overall water flux rates in small mammals at higher altitude.

14.4.4 *Pressure and respiration*

Another important effect of altitude is the change in barometric pressure, which dramatically alters the **partial pressures of gases** in the atmosphere. The relationship between altitude and pressure is exponential, pressure decreasing by 50% for every 5500 m of elevation. Thus at the top of Mt Everest (8848 m above sea level) the barometric pressure is only 250 Torr or 33 kPa, about one-third of the sea-level value, with Po_2 at 6.9 kPa.

The additional physiological problems of altitude are therefore mainly respiratory, due to low pressure and low oxygen. This is unlikely to be a significant problem for animals in high-altitude ponds and streams, and indeed the amphibious African crab *Potamonautes warreni* shows little specialization in ponds at 1400 m elevation. Nor is it a substantial problem for terrestrial arthropods using a tracheal respiratory system, which is rarely at the limits of its capacity to supply oxygen. But for vertebrates with lungs it represents an additional constraint on performance, and especially for the maintenance of a high BMR in endotherms. Table 14.8 shows the effects of altitude on partial pressures of gases for a human.

For non-adapted birds and mammals, hyperventilation is the normal response, in an attempt to maintain alveolar Po_2, but arterial oxygen levels inevitably decline and for humans loss of consciousness results at about 50% arterial oxygen concentrations, normally occurring at around 7000 m above sea level. **Hyperventilation** also tends to lead to reduced alveolar Pco_2, which in turn inhibits respiration for terrestrial vertebrates (see Chapter 7), giving an automatic brake on the acclimatory response. This can be partly overcome by excreting excess carbonate at the kidneys leading to metabolic acidosis and thus an improved central chemoreceptor response to increase the respiratory rate. The need to oxygenate the tissues conflicts sharply with the need for acid–base homeostasis. Inorganic phosphate and 2,3-diphosphoglycerate in the red blood cells play some role in changing the haemoglobin (Hb) affinity to assist in this balance, overriding the Bohr shift (Fig. 14.54).

For resident adapted animals, the respiratory and circulatory systems become modified; in vertebrates this means that **extra pigment** in the blood (increased haematocrit), with high-affinity

properties, and enhanced lung volume are the commonest responses. **High-affinity haemoglobin** in mammals and birds is particularly striking, with the oxygen dissociation curve well to the left; this is particularly apparent in camelids (Fig. 14.55). It remains unclear, though, whether the camelids have this characteristic due to residence at high altitude over hundreds of generations or whether they have come to inhabit high altitudes because they were already endowed with suitable haemoglobin. A comparison of carnivores from high altitude in Peru shows P_{50} values of 31.1 Torr (puma) and 18.5 Torr (fox) compared with 36.3 and 26.2 Torr, respectively, in the same genus or species from sea level, suggesting that in this group the change in respiratory pigment affinity and structure had occurred after invasion of the montane habitat. A high-altitude rodent from China, the pika, reveals further circulatory adaptations, having reduced pulmonary vasoconstriction responses with a larger right ventricle giving high pulmonary perfusion; in this species there is no increase in haematocrit.

Comparisons of the circulatory systems of high-altitude birds (Andean coots) show further circulatory changes. **Capillarity** is

Table 14.8 Altitude and pressure effects in humans.

Altitude (m)	Atmospheric pressure (kPa)	Ambient Po_2 (kPa)	Alveolar Po_2 (kPa)	Alveolar Pco_2 (kPa)
0	101	21.1	13.8	5.3
3 100	71	14.6	8.9	4.8
4 340	62	12.8	6.0	–
5 300	Ceiling for unacclimated humans			
5 500	Highest human habitation			
6 200	46	9.7	5.3	3.2
6 000–7 000	Ceiling for acclimated humans			
8 500	Human breathing possible for a few hours			
8 848	33	6.9	4.0	1.5
9 200	30	6.3	2.8	–
12 300	19	3.9	1.1	–
14 460	Ceiling for humans with pure oxygen supply			
15 400	12	2.4	0.1	~0
20 000	6	1.3	0	0

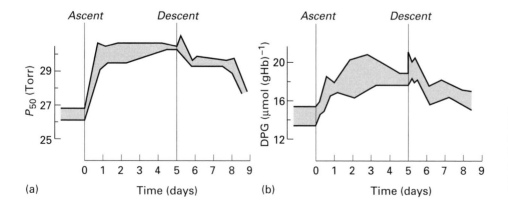

Fig. 14.54 (a) Acclimatory changes in haemoglobin affinity in humans at altitude (4510 m) for 5 days, and (b) the associated change in erythrocyte diphosphoglycerate (DPG) concentrations. (Adapted from Bouverot 1985.)

Fig. 14.55 Oxygen dissociation curves and lowered P_{50} values in high-altitude llama and vicuña compared with sea-level mammals. (Data from Hall *et al.* 1936.)

increased in muscles, and mean muscle fibre diameter is decreased. These changes appear to be sufficient to allow regulated energy production in muscles without increases in enzymatic activity, since oxidative enzyme function shows no significant differences from sea-level coots.

Birds' eggs also have a particular respiratory problem at altitude, the shell needing larger than usual pores to compensate for reduced rates of gas exchange and therefore incurring rapid EWL. Most birds that nest at altitude must compensate by providing more protected nest holes with a humid microclimate.

14.4.5 *Life cycles*

As with polar species, montane invertebrates often have **extended life cycles**, needing two or more seasons to complete a cycle that in related lowland congeners would take only one year. Carabid beetles from the Austrian Alps routinely take 2–3 years; even at the relatively limited heights of Snowdonia in Wales, carabids seem to be biennial whereas their lowland conspecifics are annual. Wolf spiders of the genus *Pardosa* have a 2-year cycle above the timberline in the Rockies, whereas species below the timberline always complete the cycle in one season; the same applies for high-altitude *Pardosa* from Norway (3 years) and from Austria (2–3 years). Similar effects have been reported in mites, springtails, moths and crane flies. **Diapausing** stages may contribute to the lengthened cycle, and as in other extreme habitats this phase is often triggered by temperature rather than by photoperiod, allowing greater phenotypic plasticity. However, there is also evidence that on reaching adulthood high-altitude populations undergo more rapid senescence;

in *Melanoplus* grasshoppers montane populations aged and died more quickly than lowland populations when both were kept in a common equable environment, perhaps due to selection on their reproductive schedules.

An alternative strategy at somewhat lower elevations, or where microclimatic variation can be exploited to speed up development, is to **curtail the life cycle** somewhat so that it can be fitted into one season. A Rocky Mountain grasshopper, *Aeropedellus*, has one fewer nymphal stage than normal, and many insects at similar elevations are rather smaller as adults than in lowland populations. But patterns of change with altitude are not simple. In lizards, growth rates and reproductive potentials can either increase or decrease with elevation, depending on rather subtle balances of behavioural adjustments and local climate patterns.

In mammals, the most conspicuous adaptation is that of extremely **precocial young**, which on high mountain slopes can move around and look after themselves almost from the first day after birth. Birds appear to show an alternative pattern of increased biparental care in montane species (linked with less sexual dimorphism and relatively drabber males). This may be associated with another problem for many high-altitude species, that of attracting a mate for reproductive purposes. Data on this issue are limited, but at least in Andean frogs there is evidence that the more montane species show less energetically demanding vocal behaviours.

14.4.6 *Other problems and adaptations in mountains*

A major problem in high mountain crags is that of locomotion, and of keeping a grip and maintaining balance on unstable difficult terrain that is often very windswept. For the large mammals that make a living in these habitats, some show careful slow gaits, picking their way amongst rocks, while others have a bouncing jumping locomotion achieved at amazing speeds. The latter group have major adaptations, particularly of the feet, to give this sure-footedness. The bighorn sheep of the Rockies have soft elastic pads in their feet that absorb the shock of impacts and provide an effective grip on slippery rock.

Communication across craggy mountain expanses may be difficult, with many obstructions to long-range visual signals, so that auditory communication has become particularly common, in amphibians, rodents and the larger mammals. Honking and barking sounds that carry well in thinner cool air are used to communicate between and among the herds in larger mammals, and to signal danger among marmots and other rodents.

Some higher-altitude populations may be among the most susceptible to potential **global climatic change**. Most obviously, they almost by definition occur in isolated small populations, relatively close to the edge in terms of local extinction. More specifically, analyses of locomotor performance and stamina in a widespread North American lizard *Urosaurus ornatus*, across a range of habitats, indicate that the high-elevation populations in

(a)

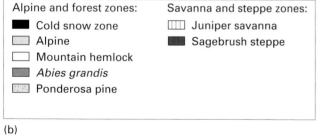

(b)

Fig. 14.56 Effects of global warming on ecosystems at altitude.
(a) A warming of 3.5°C is predicted to move floral zones upwards and severely reduce the high alpine communities in general. (Adapted from Beniston 1994.) (b) Predictions for the east and west faces of a northern hemisphere mountain range, for two different levels of warming.

ponderosa pine forests may be the most at risk, lacking an ability to modify their geographical distribution on an appropriate timescale.

The overall pattern of life on mountain slopes is likely to change markedly in response to continued global warming (Fig. 14.56). An analysis of postulated changes in vegetation distribution on the east and west slopes of a temperate mountain range (Fig. 14.56b) shows the eastern faces lose much of their interesting alpine upper zones and become dominated by sagebrush scrub, while the west loses alpine, hemlock and fir habitats and gains further oak savanna.

14.5 Aerial life

Aerial life is in some senses a special case of life at altitude, where the animal has usually generated the altitude by its own activity. This 'altitude' may be very limited, in small insects, but can exceed that of the highest mountains for some birds, who are able to fly at heights greater than those achieved by human mountaineers using accessory oxygen (see Table 14.7). All of these animals may therefore have altitude-related problems as described above, relating to temperature, water balance, pressure and respiration; but in addition they have to solve the more mechanical problems of keeping a low body weight and good buoyancy, of having appropriately shaped aerodynamic surfaces and, of course, of fuelling. The mechanical and physiological problems often become coupled, since flying entails substantial power output, potentially producing unusually high heat loads and compounding the problems of water balance by increasing convective and evaporative cooling and/or increasing metabolic water production. In flying animals, temperature, water balance and respiratory problems are particularly closely tied together.

Many organisms live continually in the air for short periods, often in dispersive phases and often involuntarily. A great many more use flight and other kinds of airborne locomotion in short bursts, for prey capture, escape from predators, or competition for and acquisition of mates. Just a few insects and some birds (e.g. swifts) stay aloft for extended periods, and could be said to lead truly aerial lives. It is worth remembering that *most* animals fly; the majority of insects of course, but also most birds (of which there are twice as many species as there are of mammals) and even about a quarter of all mammals (the bats).

Flight, whether of short or long duration, needs particularly **high-energy foods** such as nectar, fruit or very good quality flesh, especially in the form of fish for the birds that spend long periods aloft gliding and flying over the sea. Very few leaf-eating herbivores are good fliers, since the diet is too poor and the gut needs to be too large. Fuelling within the body has to be critically adjusted to keep the wing musculature adequately supplied; in insects, proline and lipids are used, while in vertebrate fliers glycogen stores remain the primary source of blood glucose. However, small passerine migratory species have particularly high levels of plasma triglycerides, and this may be an adaptive system of fuel supply.

Wing muscle design also varies according to lifestyle and time spent aloft. Birds that fly only in short bursts have predominantly **white anaerobic muscle** fibres in their flight muscles, while prolonged fliers have predominantly red aerobic fibres in the main pectoralis (downstroke) muscles. Flying birds have the highest mass-specific energy turnover, and the highest core body temperatures, of any animals; the design features of the smallest hummingbirds, which allow this power output, were considered in Chapter 6. Because they are generally small, birds cannot store large amounts of energy and may have fewer options for patterns of **energy allocation** than any other vertebrates. Their power output in sustained flight is 8–12 times that of the resting MR, reflecting a much higher factorial scope than is found in exercising mammals of similar size.

Respiratory cycles are often coupled to wingbeat frequency in flying animals, with thoracic musculature (in both insects and birds) generating both effects and so giving a saving on muscular effort. The insect tracheal system is probably always adequate to cope with supplying oxygen fast enough, but for vertebrates it is a different story, particularly when flying at any height. At very high altitudes, where the partial pressure of oxygen is low, the driving force for diffusion of oxygen across the pulmonary surfaces is very low, and here only the crosscurrent blood flow system of the parabronchial lung of birds (Chapter 7) will suffice, by ensuring that partially depleted air continually encounters new deoxygenated blood. Hence high flying is only possible for birds, and is never seen in bats.

The water balance problems of aerial animals are rather poorly understood. Flapping flight in birds certainly increases **respiratory water loss** and may also increase cutaneous losses leading to rather rapid dehydration. However, all flying birds appear to be excellent plasma-volume regulators, even when undergoing substantial dehydration, an ability that in mammals tends only to occur in desert species. It appears that this goes with an ability to **tolerate high heat loads** during flight. In insects, small species appear to suffer net water loss in flight, which may constrain their flight duration and require modified dietary choices (e.g. more dilute nectar for bees). But large insect species, for example bees of more than 200 mg, appear to generate more metabolic water than they can readily cope with due to their exceptionally high metabolic rates, and they have been observed jettisoning excess water while in flight, both by loss of saliva or crop fluid from the mouth and by copious dilute urination.

Finally it should be noted that flying is an expensive business, with costs ranging from 50 to 120 W kg^{-1} in birds and bats, and up to 300 W kg^{-1} in insects. There may therefore be design trade-offs between flight apparatus and other organ systems. For example, in territorial insect species where flight determines mating success, such as in the dragonfly *Plathemis*, extra investment in flight muscle may mean reduced gut tissues and fat reserves in the males, and perhaps reduced longevity.

14.6 Conclusions

Extreme environments pose a series of linked problems for their inhabitants, and some of these problems are rather similar whether we are looking at deserts, polar regions or mountains. Not only are the problems similar, but the convergence of adaptation between quite different groups is striking. Animals living at these extremes have repeatedly acquired the same kinds of solutions: from behaviour (use of burrows and other refugia, gregariousness) through to features at the gross morphological scale (overall size and shape, and the reduction or elaboration of appendages) down to physiological systems (peripheral counter-current exchangers, altered insulation, changes in MR and periods of hypometabolism) and the fine details of biochemical endowments (altered membrane and enzyme properties, use of osmolytes that protect against freezing and desiccation damage). Convergent evolution is extremely commonplace where there are only limited possible solutions to common physical problems.

Further reading

Books

Bouverot, P. (1985) *Adaptations to Altitude-Hypoxia in Vertebrates*. Springer-Verlag, Berlin.

Bradshaw, D.S. (1986) *Ecophysiology of Desert Reptiles*. Academic Press, Sydney.

Carey, C., Florant, G.L., Wunder, B.A. & Horowitz, B. (1993) *Life in the Cold: Ecological, Physiological and Molecular Mechanisms*. Westview Press, Boulder, CO.

Hadley, N.F. (1994) *Water Relations of Terrestrial Arthropods*. Academic Press, San Diego, CA.

Heller, H.C., Musacchia, X.J. & Wang, L.C.H. (eds) (1986) *Living in the Cold: Physiological and Biochemical Adaptations*. Elsevier, New York.

Hochachka, P.W. & Guppy, M. (1987) *Metabolic Arrest and the Control of Biological Time*. Harvard University Press, Cambridge, MA.

Lee, R.E. & Denlinger, D.L. (1991) *Insects at Low Temperature*. Chapman & Hall, New York.

Louw, G.N. (1993) *Physiological Animal Ecology*. Longman Scientific and Technical, Harlow.

Louw, G.N. & Seely, M.K. (1982) *Ecology of Desert Organisms*. Longman, London.

Reviews and scientific papers

Ancel, A., Visser, H., Handrich, Y., Masman, D. & LeMaho, Y. (1997) Energy saving in huddling penguins. *Nature* **385**, 304–305.

Badyaev, A.V. (1997) Altitudinal variation in sexual dimorphism: a new pattern and alternative hypotheses. *Behavioral Ecology* **8**, 675–690.

Baker, M.A. (1982) Brain cooling in endotherms in heat and exercise. *Annual Review of Physiology* **44**, 85–96.

Bale, J.S. (1993) Classes of insect cold hardiness. *Functional Ecology* **7**, 751–753.

Block, W. (1995) Insects and freezing. *Science Progress* **78**, 349–372.

Butterfield, J. (1996) Carabid life cycle strategies and climate change; a study on an altitude transect. *Ecological Entomology* **21**, 9–16.

Caceres, C.E. (1997) Dormancy in invertebrates. *Invertebrate Biology* **116**, 371–383.

Cerda, X., Retana, J. & Cros, S. (1998) Critical thermal limits in Mediterranean ant species: trade off between mortality risk and foraging performance. *Functional Ecology* **12**, 45–55.

Christian, K.A. & Morton, S.R. (1992) Extreme thermophilia in a central Australian ant *Melophorus bagoti*. *Physiological Zoology* **65**, 885–905.

Convey, P. (1996) Overwintering strategies of terrestrial invertebrates in Antarctica—the significance of flexibility in extreme seasonal environments. *European Journal of Entomology* **93**, 489–505.

Danks, H.V. (1992) Long life cycles in insects. *Canadian Entomologist* **124**, 167–187.

Elkhawad, A.O. (1992) Selective brain cooling in desert animals: the camel (*Camelus dromedarius*). *Comparative Biochemistry and Physiology A* **101**, 195–201.

Faraci, F.M. (1991) Adaptations to hypoxia in birds—how to fly high. *Annual Review of Physiology* **53**, 59–70.

Florant, G.L. (1998) Lipid metabolism in hibernators: the importance of essential fatty acids. *American Zoologist* **38**, 331–340.

Gaede, K. & Knülle, W. (1997) On the mechanism of water vapour sorption from unsaturated atmospheres by ticks. *Journal of Experimental Biology* **200**, 1491–1498.

Gibbs, A.G., Louie, A.K. & Ayala, J.A. (1998) Effects of temperature on cuticular lipids and water balance in a desert *Drosophila*: is thermal acclimation beneficial? *Journal of Experimental Biology* **201**, 71–80.

Haim, A. & Izhaki, I. (1995) Comparative physiology of thermoregulation in rodents: adaptations to arid and mesic environments. *Journal of Arid Environments* **31**, 431–440.

Hainsworth, F.R. (1995) Optimal body temperatures with shuttling—desert antelope ground-squirrels. *Animal Behaviour* **49**, 107–116.

Henen, B.T. (1997) Seasonal and annual energy budgets of female desert tortoises (*Gopherus agassizii*). *Ecology* **78**, 283–296.

Jenni-Eiermann, S. & Lukas, J. (1992) High plasma triglyceride levels in small birds during migratory flight: a new pathway for fuel supply during endurance locomotion at very high mass-specific metabolic rates? *Physiological Zoology* **65**, 112–123.

Klok, C.J. & Chown, S.L. (1997) Critical thermal limits, temperature tolerance and water balance of a sub-Antarctic caterpillar *Pringleophora marioni*. *Journal of Insect Physiology* **43**, 685–694.

Kuhnen, G. (1997) Selective brain cooling reduces respiratory water loss during heat stress. *Comparative Biochemistry and Physiology A* **118**, 891–895.

Machin, J. & O'Donnell, M.J. (1991) Rectal complex ion activities and electrochemical gradients in larvae of the desert beetle *Onymacris*: comparisons with tenebrionids. *Journal of Insect Physiology* **37**, 829–837.

Nagy, K.A. & Gruchacz, M.J. (1994) Seasonal water and energy metabolism of the desert-dwelling kangaroo rat (*Dipodomys merriami*). *Physiological Zoology* **67**, 1461–1478.

Navas, C.A. (1996) Metabolic physiology, locomotor performance and thermal niche breadth in high-elevation anurans. *Physiological Zoology* **69**, 1481–1501.

O'Donnell, M.J. (1982) Hydrophilic cuticle; the basis for water vapour absorption by the desert burrowing cockroach *Arenivaga investigata*. *Journal of Experimental Biology* **99**, 43–60.

O'Donnell, M.J. (1987) Water vapour absorption by arthropods: different sites, different mechanisms. In: *Comparative Physiology: Life in Water and on Land* (eds P. Dejours, L. Bolis, C.R. Taylor & E.R. Weibel). Liviana Press, Padova.

Patterson, B.D., Pacheco, V. & Solari, S. (1996) Distribution of bats along an elevational gradient in the Andes of south-eastern Peru. *Journal of Zoology* **240**, 637–658.

Reinertsen, R.F.E. & Haftorn, S. (1986) Different metabolic strategies of northern birds for nocturnal survival. *Journal of Comparative Physiology* **156**, 655–663.

Samaja, M. (1997) Blood gas transport at altitude. *Respiration* **64**, 422–428.

Schmidt-Nielsen, K. (1981) Counter-current systems in animals. *Scientific American* **244**, 118–128.

Schmidt-Nielsen, K., Kanwisher, J., Lasiewski, R.C., Cohn, J.E. & Bretz, W.L. (1969) Temperature regulation and respiration in the ostrich. *Condor* **71**, 341–352.

Schmidt-Nielsen, K., Schroter, R.C. & Shkolnik, A. (1981) Desaturation of exhaled air in camels. *Proceedings of the Royal Society of London* **211**, 305–319.

Sherbrooke, W.C., Castrucci, A.M.deL. & Hadley, M.E. (1994) Temperature effects on in vitro skin darkening in the mountain spiny lizard *Sceloporus jarrovii*: a thermoregulatory adaptation? *Physiological Zoology* **67**, 659–672.

Sinclair, B.J. & Wharton, D.A. (1997) Avoidance of intracellular freezing by the New Zealand alpine weta *Hemideina maori* (Orthoptera; Stenopelmatidae) *Journal of Insect Physiology* **43**, 621–625.

Stone, G.N. & Purvis, A. (1992) Warm-up rates during arousal from torpor in heterothermic mammals—physiological correlates and a comparison with heterothermic insects. *Journal of Comparative Physiology B* **162**, 284–295.

Storey, K.B. (1997) Metabolic regulation in mammalian hibernation: enzyme and protein adaptations. *Comparative Biochemistry and Physiology A* **118**, 1115–1124.

Storey, K.B. & Storey, J.M. (1990) Metabolic rate depression and biochemical adaptation in anaerobiosis, hibernation and estivation. *Quarterly Review of Biology* **65**, 145–174.

Storey, K.B. & Storey, J.M. (1996) Natural freezing survival in animals. *Annual Review of Ecological Systematics* **27**, 365–386.

Strathdee, A.T. & Bale, J.S. (1998) Life on the edge: insect ecology in arctic environments. *Annual Review of Entomology* **43**, 85–106.

Tatar, M., Gray, D.W. & Carey, J.R. (1997) Altitudinal variation for senescence in *Melanoplus* grasshoppers. *Oecologia* **111**, 357–364.

Walsberg, G.E., Weaver, T. & Wolf, B.O. (1997) Seasonal adjustment of solar heat gain independent of coat coloration in a desert mammal. *Physiological Zoology* **70**, 150–157.

Ward, D. & Seely, M.K. (1996) Adaptation and constraint in the evolution of the physiology and behaviour of Namib Desert tenebrionid beetles. *Evolution* **50**, 1231–1237.

Wharton, D.A. & Ferns, D.J. (1995) Survival of intracellular freezing by the Antarctic nematode *Panagrolaimus davidi*. *Journal of Experimental Biology* **198**, 1381–1387.

Willmer, P.G. & Stone, G.N. (1997) Temperature and water relations in desert bees. *Journal of Thermal Biology* **22**, 453–465.

Withers, P.C. (1983) Energy, water and solute balance of the ostrich *Struthio camelus*. *Physiological Zoology* **56**, 568–579.

Zachariassen, K.E. (1996) The water conserving physiological compromises of desert insects. *European Journal of Entomology* **93**, 359–367.

15 *Parasitic Habitats*

15.1 Introduction

A great many animals live in an intimate association with another organism, either on its outer surface or within its body; these 'commensals' range in their effect on the host from harmful parasites to beneficial symbionts (see Box 15.1), though here we are concerned primarily with parasites having an intimate and negative effect on their host. Living in or on other organisms is phylogenetically extremely widespread in the Metazoa, and it is clear that invasions of the body surfaces and interiors of other species have happened many times. Here we deal mainly with parasites living on other animals, though of course many land animals are notionally 'parasitic' on plants.

Parasites are known in most of the common phyla of animals (Table 15.1), but certain groups are extremely rich in parasitic species, particularly the tapeworms, flukes, acanthocephalans

---

Box 15.1 Interactions between species

It is traditional to view interactions between species in terms of positive and negative impacts for each partner, resulting in an 'interaction grid' as follows:

		Species 1		
		+ve	0	−ve
Species 2	+ve	Mutualism	Commensalism	Contramensalism
	0		Neutralism	Amensalism
	−ve			Competition

However, the terminology is often confused in existing literature, particularly in relation to the terms 'commensalism' and 'symbiosis'. **Commensalism** is often used in a wider sense than is shown in the matrix above, to include all kinds of situations where two species live together (usually one on or within the surfaces of another). **Symbiosis** is sometimes used as a catch-all term for species living and interacting together, but is perhaps better reserved for a more restricted usage where two species are both benefiting from an interaction, making symbiosis synonymous with mutualism.

Thus zooxanthellae within corals are an example of mutualism or symbiosis, where both plant and animal benefit; bromeliad plants that live on a tree trunk but take nothing from the tree are true commensals; and herbivory and carnivory are both 'contramensalisms'. But the boundaries are not always clear-cut, and in almost every case both partners may over evolutionary time gradually change the relationship in ways that benefit themselves.

On such a matrix, **parasitism** is usually classified as a **contramensalism**; the parasite benefits and the host is harmed.

Note, however, that the boundaries between parasitism and predation become rather difficult, and in effect there is a continuum of kinds of relationship. Predators kill and eat many items of prey which are usually smaller than they are; parasites do not normally kill their single much larger host but eat it (or its resources) while it lives. A biting fly, or a vampire bat, that alights on a larger animal and sucks its blood is normally seen as predatory (but sometimes as parasitic, especially if it transmits disease); a fly that punctures the skin to feed and then lays eggs in the wound is more clearly a parasitic species. The term parasitism therefore needs to be reserved for associations that not only have a negative impact on the host but that also involve a relatively lengthy association between the two species (so excluding the 'hit-and-run' biting flies).

There still remain problems with many arthropod 'parasites' (commonly wasps and flies) where one or more eggs are laid on a host (usually a larval insect, not dissimilar in size from its attacker) by the adult female. The resulting offspring hatch and feed on the host, which is usually immobilized by their presence but still alive (hence the relation is parasitic) but which dies when they mature into adults and leave the carcass (hence the outcome is effectively predation). The term '**parasitoid**' is commonly used for these intermediate kinds of relationship.

For our purposes, the parasitoids are living for most of their life in an internal enclosed environment created by the host and with physiological problems like those of a parasite rather than those of a predator; therefore we will include them in this chapter without further distinguishing them from 'true' parasites.

Table 15.1 Occurrence of parasite members in animal groups.

Animal group	Importance as parasites
Poriferans (sponges)	A few 'ectoparasitic', boring into shells, etc.
Cnidarians (sea anemones, hydroids, etc.)	A few ectoparasites on fish, etc.
Ctenophores (comb-jellies)	Very few
Platyhelminths (tapeworms and flukes)	All groups (except most turbellarians) are ecto- and endoparasites on invertebrate and vertebrate hosts
Nemertines	A few parasitic/symbiotic? in other invertebrates
Rotifers	A few ectoparasites
Nematodes (roundworms, filarial worms)	Huge numbers of endoparasites in invertebrate and vertebrate hosts
Nematomorphs (horsehair worms)	All endoparasitic in arthropods as larvae
Acanthocephalans (spiny-headed worms)	All endoparasitic in vertebrate guts
Molluscs	Some larvae are parasitic on fish; a few adult ectoparasites and endoparasites
Annelids	Leeches are ectoparasitic/predators; a few polychaetes are parasitic
Insects	Very numerous ectoparasites (e.g. lice, fleas, tsetse flies, mosquitoes); also parasitoids on other insects (e.g. ichneumon wasps, flies); some larvae endoparasitic in vertebrates (e.g. bot flies)
Crustaceans	Uncommon; some barnacles on crabs, some copepods and branchiurans mostly on fish or on other crustaceans, pentastomids in vertebrate lungs
Arachnids (spiders, mites, etc.)	Mites and ticks are ectoparasites
Chordates	Some jawless fish are ectoparasites/predators

and nematodes (collectively often termed **helminths**), and the insects. Nematodes are extraordinarily abundant and highly speciose, while the insect group contains huge numbers of species (especially parasitic wasps and flies) that develop as larvae inside other insects or inside plants. From consideration of these two groups alone, which dominate the animal kingdom numerically, it is probably fair to conclude that *most* animal species are parasites.

This chapter surveys the departures from a free-living physiology that are associated with a range of types of parasitism. There are two main aspects to cover. Firstly, being a parasite involves coping with the *abiotic* conditions to be found when living in or on a particular part of the host. For many parasites, homeostatic mechanisms within the host provide the parasite with energy and nutrients, regulate temperature and water balance, and supply the dissolved gases for aerobic respiratory exchange while removing metabolic wastes; in this sense, the parasitic life is physiologically 'easy' most of the time. However, some parts of the host environment (e.g. the acidic or anoxic parts of the gut) are less suitable for normal life; thus parasites may show adaptations to unusual pH, redox potential, dis-

solved gas levels and so on. Some of these adaptations parallel features found in free-living organisms that occupy similarly extreme environments—for example, some of the adaptations found in gut-inhabiting nematodes and platyhelminths are similar to those found in metazoans inhabiting anoxic muds.

However, being a parasite also involves interactions with a host, which constitutes a *biotic* environment, and which creates a unique set of problems. Most obviously, the host commonly seeks to evict the parasite, and is therefore an aggressive environment that 'fights back'; and just as hosts are part of the parasite environment, so parasites are an ancient part of the host environment. Many hosts have developed a range of defences against parasite attack, and these in turn have led to an equally diverse array of parasite countermeasures. Whereas physiological solutions to a given abiotic environmental challenge should have long-term selective value in that environment, the same cannot be said of the host–parasite arms race, which is an on-going evolutionary struggle with no predefined optimum or endpoint. Defence and counter-defence are therefore a dominant part of the biology of many parasites, and a large part of this chapter deals with the physiology of host defences and the biochemical, physiological and behavioural tricks used by parasites to cope with these.

There are other ways in which a biotic environment is unusual. Animals normally control their own metabolic resource allocation using hormones as control agents, making 'decisions' on allocations to growth and to reproduction, eventually undergoing programmed senescence followed inevitably by death. But many parasites are not passive bystanders in the control of host resource allocation, but instead actively re-route resources to favour their own development, by releasing hormones and neuromodulators that operate directly on host receptors. Effects include prevention of host reproduction, alteration of host size, induction of entirely new tissues for the parasite's benefit, and considerable changes to host longevity.

To be successful a parasite must also reproduce or be transmitted within its host's lifetime, posing a further set of physiological hurdles to be overcome, and mortality of propagules is often high. Parasites show many adaptations that improve transmission between hosts, including methods to synchronize parasite and host reproduction, and manipulation of host behaviours. These mechanisms operate through intimate chemical interactions between parasite and host, giving an added dimension to parasite environmental physiology.

Clearly, many of the more dramatic metabolic adaptations in the environmental physiology of endoparasites are responses to a host environment that is both hostile and capable of coevolutionary responses. This is what makes parasite environmental physiology unique.

15.2 The parasite environment

The environment of parasitic organisms is unusual in consisting of two components:
1 the macrohabitat in which the host organism lives;
2 the microhabitat, resulting from modification to the free-living environment brought about by living in or on a particular region of the host.

The importance of these two components in determining the environment experienced by a parasite varies enormously across the spectrum of host–parasite interactions, differing between parasite species and between stages in the same parasite life cycle. At one extreme, some parasites living on the external surface of their host (termed **ectoparasites**) may experience much the same environment as their host, and so require similar physiological tolerances and adaptations to free-living organisms in the same habitat. At the other extreme, parasites living within host tissues (termed **endoparasites**) commonly experience a relatively stable but often highly unusual internal microhabitat, and they may have their thermal, osmotic, respiratory and nutritional needs met entirely by the host. The term 'endoparasite' covers a wide range of different niches within a host's body (especially if the host is a vertebrate), and differences between these niches have had major consequences for the environmental physiology of parasites. In addition to the

ectoparasitic environment on the skin, we therefore need to consider three major categories of endoparasitic niche—respiratory passages, gut and blood—in what follows.

15.2.1 *Living on the skin*

For parasites in aquatic environments there is little 'microclimatic' amelioration around the surfaces of the host animal, whereas for terrestrial ectoparasites, especially on birds and mammals, the environment close to the host skin may be much less thermally variable and much less desiccating than the macrohabitat. Skin ectoparasites on a land vertebrate may need only fairly narrow thermal tolerances and many will have relatively little problem with water balance. This will be especially true if the parasite is 'plugged in' permanently or intermittently to the capillaries or tissues of its host, as with many ticks and fleas, and thus endowed with a freely available water supply.

The physiological tolerance of particular groups of skin parasites may mean that they are capable of living in some environments and not others. This is again seen most clearly when comparing terrestrial and aquatic habitats. For example, while platyhelminth flukes are major ectoparasites of vertebrates in marine habitats, they are far less abundant and speciose in terrestrial hosts, presumably being unable to cope with the desiccation associated with a terrestrial environment given their permeable outer surfaces (see Chapter 5). Some monogeneans do survive in terrestrial habitats by generally avoiding the most exposed outer surface of the host, and limiting any free-living phase in their life cycle to periods when the host is in water. A good example of this is the common fluke *Polystoma*, which as an adult inhabits the bladder of frogs.

15.2.2 *Living in respiratory passages*

Many parasites inhabit the buccal cavities or respiratory chambers of larger animals; most notable are the great range of flukes and leeches found in the gills and buccal cavities of marine and freshwater crustaceans and fish. These locations are still essentially ectoparasitic, though less exposed to some aspects of the external conditions. Many parasites have taken this route further, and inhabit deeper respiratory passages as endoparasites. These environments are moist enough to allow essentially aquatic taxa to survive; for example, the lungs of terrestrial vertebrates support crustaceans (e.g. pentastomids in snakes and birds) and platyhelminths (e.g. the human lung fluke). Parasites derived from terrestrial ancestors have also invaded respiratory passageways, for example the mites whose infestation of the tracheae of bees causes the disease varroa, and the maggots of the nostril fly, *Oestrus ovis*, which live within the nasal passages of sheep and goats. Here most aspects of the parasite's physical environment (such as temperature, levels of respiratory gases and water content) are dominated by the host, and physiological problems are relatively limited. Entry and exit of eggs or infective larval stages is also uncomplicated, via the respiratory openings. A dramatic example of this is provided by the larvae of the nostril fly, which when mature migrate towards the nostril openings and are expelled into the environment in an explosive sneeze! However, this also points to a major drawback of life in respiratory passages; many hosts, and especially vertebrates, have defences such as coughing and sneezing that serve to eject parasites.

Many of the parasites in respiratory passages feed by penetrating the wall of the host's body to feed on blood; they then become vulnerable to attack by the various blood-mediated host defences discussed below, and their environment is best regarded as somewhat intermediate between respiratory and blood-based.

15.2.3 *Living in the gut*

Parasites inhabiting the gut are surrounded by host food material, which provides nutrition for many of them (particularly cestode tapeworms). The gut lining is also richly vascularized, providing a further food source for those gut parasites that penetrate the gut wall and feed on blood (nematodes and some trematode flukes) or tissue fluids (acanthocephalan worms). Entry via the mouth is easily achieved, and exit of eggs or infective larval stages is also uncomplicated, via the anus or possibly via the mouth if the parasite can move itself or its propagules upwards against the gut movements.

However, there are severe problems as well. The gut of most potential hosts is a relatively anoxic environment, especially in its lower reaches, and many gut parasites are therefore at least facultatively capable of anaerobic respiration. Also, gut parasites are exposed to the digestive processes of the host, including extremes of pH, high levels of activity of a range of enzymes, and the mechanical action of peristalsis. Conditions are most extreme in the stomach, with a very low pH, abundant proteolytic enzymes, and a high turnover of the stomach lining making attachment difficult; relatively few metazoan parasites survive here. At a given location in the gut, conditions may also change dramatically and cyclically during the processing of a single host meal. Species feeding on blood may be attacked by blood-borne host defences, and some host defences are also known to reach non-blood-feeders in the gut lumen. Furthermore, the gut again has ways of physically removing parasites at either end, by vomiting or by the copious faecal emissions (diarrhoea) induced by some parasitic infections.

15.2.4 *Living in tissues and blood*

Host tissues include not only the muscles, body cavity and organs of the host, but also host circulatory systems containing cells (e.g. blood and lymph). In some ways, these constitute ideal parasite habitats: as long as the parasite tissues have similar general requirements to host tissues, they are provided with nutrients and oxygen, and waste products and carbon dioxide are removed for them. In other ways, inhabiting host tissues is

the most demanding of all parasitic habitats. Firstly, although tissue parasites are protected from physical removal, they are exposed to the full range of host defences based on host recognition of non-self tissue (described below). Such exposure has given rise to a spectacular diversity of parasite countermeasures, including the exploitation of a number of specific parts of the host's body—**immunologically privileged sites**—where host defences are reduced. Secondly, there is often no direct aperture in the host's tissues through which the parasite can infect the host or release eggs of larvae. Tissue-inhabiting parasites thus tend to have complex life cycles and infective strategies; eggs or larvae may escape by release into the gut or lungs (thus encountering, albeit briefly, the problems already outlined for such areas), or may be transmitted by a blood-feeding vector such as a mosquito. Entry into a second host may involve boring into host tissues by larvae, boring out of the host's gut after being swallowed, or being injected by blood-sucking insect vectors.

15.2.5 *Microenvironments in space and time*

A feature of parasites inhabiting each of these habitat types is that the region of the host body they exploit is extremely specific (e.g. a particular part of the respiratory tract, or of the gut, or a particular set of blood vessels). Characteristics of internal host environments that confer this specificity not only include general aspects such as pH, salt concentrations and levels of oxygen and carbon dioxide, but many more intimate fluctuations in host physiology. Recognition of these is important in many aspects of parasite biology, ranging from parasite migration within the host to the timing of parasite reproduction.

Host physiology may alter on a variety of temporal scales; changes may be unidirectional and long-term (such as those associated with sexual maturation), or medium-term and cyclic (such as the changes in levels of mammalian reproductive hormones in the blood, or changes in levels of insect moulting hormones), or operate on even shorter timescales (such as changes in the chemical composition of the gut associated with processing of a meal, or changes in the levels of metabolites associated with host activity–sleep cycles), and these short-term changes can be particularly important in synchronizing release of infective stages in some parasites (e.g. release of nematode filariae into the blood). In addition, many hosts are able to deploy mechanisms that have evolved specifically to impede parasites, including a wide range of relatively specific bloodborne (humoral) and cell-mediated responses, as well as some more widespread defensive systems.

The tapeworm *Hymenolepis diminuta* provides the best-studied example of adult migration between microenvironments within the host gut. The adults show a daily (circadian) migration in response to the position of food along the gut. When the host's stomach is full after feeding, the scolex ('head' region) of the worm is found within the intestine close to its exit from the stomach (the pyloric sphincter). With the passage of food into the intestine, the worm shifts back down the intestine,

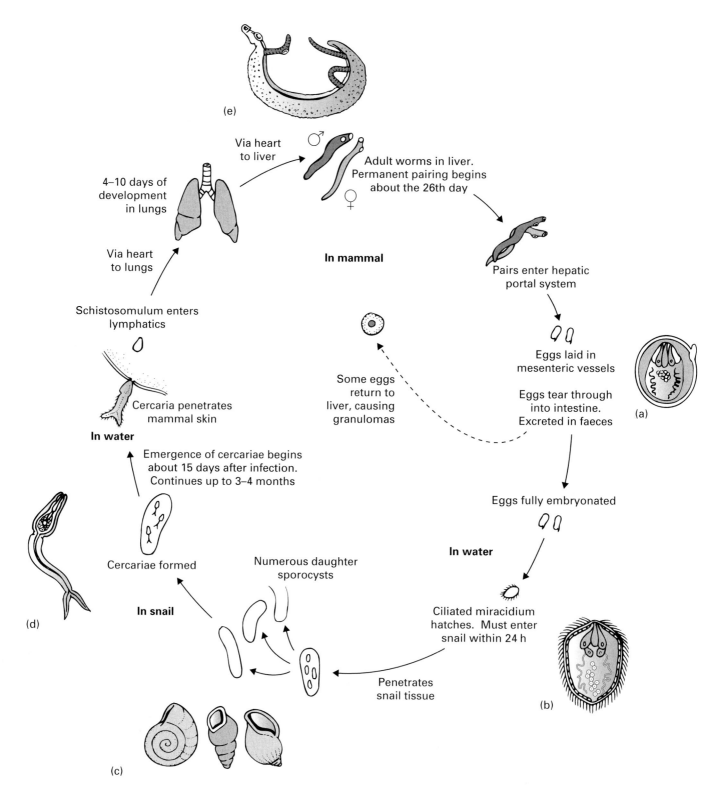

Fig. 15.1 The life cycle of the human blood fluke, *Schistosoma mansoni*.
Enlarged life-cycle stages are: (a) a fertilized egg, (b) a miracidium larva,
(c) alternative snail hosts, (d) a cercaria larva and (e) a pair of adult
flukes, with the female lying in a groove in the male. (Adapted from
Trager 1986.)

and finally moves forward again to await the next host meal. An adaptive explanation for this change in position is that the cestode is using the balance between surrounding host gut contents and physiological environment to achieve adequate absorption of nutrients. Indeed there is evidence that some parasites adjust their position in the host's gut when the host *expects* to be fed, even if the host does not actually feed; this occurs in the digenean fluke *Bunoderina encoliae*, which inhabits the rectum of sticklebacks, and it suggests that parasites can detect stimuli more subtle than the mere presence/absence of gut contents.

It is also important to realize that parasites (particularly endoparasites) may move between very different environments during their life cycle. This is well illustrated by the life stages of the digenean fluke *Schistosoma mansoni* (Fig. 15.1), which causes the disease bilharzia in humans. Schistosomes generally live as adults in the blood system of vertebrates, and adults of *S. mansoni* prefer the fine mesenteric blood vessels surrounding the human gut. Schistosome eggs are spiny, and with the help of the host's immune system (see section 15.7) make their way out of the host's body with faeces or urine. The eggs hatch into free-swimming ciliated miracidium larvae, which locate and penetrate a freshwater snail as the second host organism. The miracidium larvae give rise to a different type of larva, the sporocyst. These reproduce asexually in the snail, producing first a second generation of sporocysts and then a third larval type, the cercaria. The cercariae leave the snail and swim actively until they encounter human skin; they then bore through the skin and shed their ciliated larval epidermis, metamorphosing very rapidly into a smaller version of the adult worm, called a schistosomula. These immature adults migrate to the blood supply, and travel in it until they reach the hepatic portal vein. This life cycle thus involves two distinct dispersive free-living freshwater stages, which must cope with the vagaries and variability of freshwater bodies (see Chapter 11), and also must cope with abrupt changes in their external environment as they penetrate the tissues of their next host. There are also two stages that develop inside host tissues, but the two hosts are extremely disjunct taxonomically, differing in such basic variables as body temperature and levels of dissolved gases and metabolites, and also in the types and magnitudes of defences they are able to mount against invading organisms. The different environments encountered by the life stages of *Schistosoma mansoni* have inevitably selected for stage-specific physiological adaptations; successful completion of the life cycle requires appropriate sense organs at each stage to detect which environment the current life cycle stage is in. These sensory inputs must be linked to switches able to trigger the expression of genes relevant to each particular environment. This example shows that the modifications to the environmental physiology of a free-living organism—even a relatively 'simple' platyhelminth—may be extremely complex.

A feature of any complex life cycle is that the behaviour and environmental physiology of any one life cycle stage may limit the distribution of the parasite. Thus while adult *Schistosoma* can probably survive in human beings anywhere, they also require freshwater habitats containing the snail intermediate host. Furthermore, snail and human hosts must be abundant enough for the larvae involved to find them in their very limited lifetimes—a matter of hours in each case. Considerations of this kind mean that many parasites have a limited distribution even in regions where they are generally endemic.

15.2.6 *The parasite environment and parasite speciation*

The importance of host microenvironments in the evolutionary radiation of parasite groups is encapsulated in what has become known as Fahrenholz's rule, which states that the evolutionary radiations of parasites tend to parallel those of their hosts (Fig. 15.2). Parasite groups that show this pattern tend to be highly adapted to their host environment and to infect only a small diversity of host species; for example, 78% of monogenean parasites of fish and 70% of parasitic copepod crustaceans are restricted to a single host species. However, not all parasite groups follow Fahrenholz's rule, and some parasitic taxa are characteristic of a given habitat rather than a group of closely related hosts; a good example is the fluke *Benedenia hawaiiensis*, which is an exception to the monogenean generalization made above, being able to infect 24 fish species distributed in a range of families but all living in the same coral reef habitat. Study of the relative importance of parasite–host relationships and external macroenvironments in determining speciation patterns in parasites is very much in its early stages, though, because the sampling of host species for their parasite faunas is still very incomplete.

15.3 Basic parasite physiology

15.3.1 *Ionic and osmotic adaptation and water balance*

In general, parasites are either bathed in, or are taking in fairly continuously, a fluid that is already being regulated by their host. In this sense they have fewer problems with osmotic physiology than almost any other kind of animal; once inside the host many of them are highly permeable, are fairly intolerant of water loss (because they never experience it), and have weak or absent kidney function. However, it must always be remembered that the larval or transmission phases may have very different problems, surviving outside the host in what may be very different osmotic conditions. Endoparasitic nematodes that infect insect larvae are known to remain within the host cadaver for variable times dependent on the external humidity; at low relative humidity (RH) they may stay within the corpse for up to 50 days rather than emerge and risk desiccation.

Many aquatic ectoparasites are surprisingly sensitive to variation in the salinity of the environment inhabited by their host,

Hosts **Parasites**

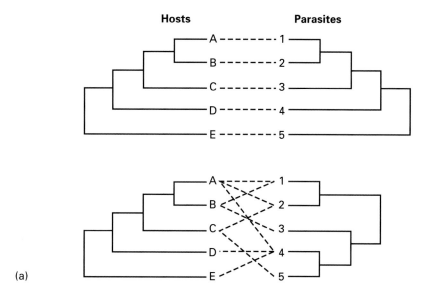

(a)

Pocket gophers **Chewing lice**

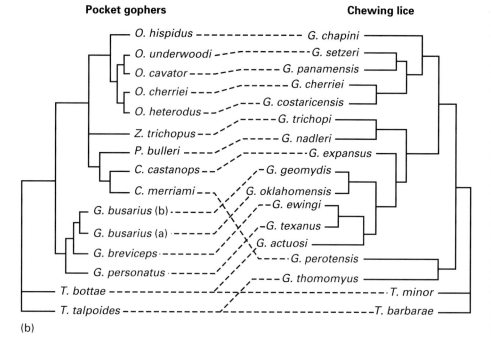

(b)

Fig. 15.2 (a) Alternative patterns of speciation in parasites and their hosts, with host–parasite relationships indicated by dashed lines. In the upper example, speciation of host-specific parasites exactly tracks speciation of hosts. Such parallel evolution results in Fahrenholz's rule: i.e. taxonomically related hosts have taxonomically related parasites. The lower example shows a case in which host and non-specific parasite phylogenies do not match. (Reprinted from *International Journal for Parasitology* **23**, Page, R.D.M., Parasites, phylogeny and cospeciation, pp. 499–506, copyright 1993, with permission from Elsevier Science.). (b) A clear example of parallel evolution, demonstrated between pocket gophers (rodents of the family Geomyidae) and their chewing lice (order Mallophaga, family Trichodectidae). Gophers: O, *Orthogeomys*; Z, *Zygogeomys*; P, *Pappogeomys*; C, *Cratogeomys*; G, *Geomys*; T, *Thomomys*. Lice: G, *Geomydoecus*; T, *Thomomydoecus*. (From Hafner & Page 1995.)

and monogenean and trematode platyhelminths are generally rare in low-salinity environments. Where the host is better able to tolerate variation in salinity than its ectoparasite, the ectoparasite may be absent from part of the host's range; this applies for the monogenean fluke *Benedenia melleni,* which has a lower tolerance of high salinity than its fish hosts.

Terrestrial ectoparasites may have very little amelioration of water balance or ionic problems across their external surfaces relative to a free-living animal in the same environment, but do have the advantage of being plugged in to a reliable regulated fluid source. Some species with terrestrial hosts may have the option of excreting into the host (as in the tick shown in Fig. 5.22) rather than to the outside world.

Endoparasites are probably usually in osmotic balance with their animal host's tissues, though this is rarely studied. For any host of reasonable size, whether in sea water, fresh water or on land, this will give the parasite a stable habitat; in land and freshwater animals, and in marine vertebrates, a stable osmotic concentration of 100–400 mOsm is likely, while in marine invertebrates the concentration is around 1000 mOsm. The interesting experiences will come for parasites on migratory aquatic hosts, for example eels, salmon and crabs such as *Eriocheir* that move between seas and rivers. Little is known of such parasites, but it is very likely that the host does most of the regulatory work and the parasite conforms with the resulting osmotic concentration of the host tissues.

One interesting example of host habitat effects on parasite physiology is provided by some endoparasitic wasps that develop inside larvae of species of fruit fly (*Drosophila*). These fruit flies live in fermenting fruit, and particular species differ in their tolerance of the ethanol generated by fermentation in this habitat. Parallel adaptations are found in the parasitic wasps: strains and species that attack *Drosophila* in alcohol-rich habitats show greater tolerance of ethanol.

15.3.2 *Thermal adaptation*

Small aquatic ectoparasites effectively experience the thermal vagaries of the macrohabitat, since water is usually well stirred and of quite high thermal conductivity. Thus temperature may be important in constraining the environment occupied by these kinds of parasite. For example, the trematode fluke *Derogenes varicus* is absent from hosts in warm marine areas, preferring cool temperate seas; and it is found near the surface at cooler high latitudes but in deeper waters at lower latitudes.

For terrestrial ectoparasites, temperature balance is also usually fairly easy, as they exist mainly within the stable boundary layer created around a reasonably sized host. Ectoparasites are therefore common on land vertebrates, and especially on endothermic hosts where the trapped, still air within fur or feathers may be at a rather constant temperature (close to the host's body temperature, T_b), and at a high humidity. Thus an ectoparasite here may need a rather high but narrow-band thermal tolerance, and can use enzymes and membranes with rather limited acclimatory ranges. In cold climates, ectoparasites tend to congregate in the warmest places; the 'armpit' or 'groin' regions of mammals, or beneath the wings on birds. On marine mammals they may be fairly evenly dispersed in the aquatic phase, but tend to accumulate on the tail flipper when seals and walruses haul out to give birth, this being the warmest spot and used for basking; for example, the lice on elephant-seal flippers stay at 27–34°C throughout a diurnal cycle.

For endoparasites, problems are again effectively handed over to the host's regulatory systems for large parts of the life cycle. But many endoparasite life cycles involve transitions between different thermal environments. This is particularly true for those life cycles (such as the infection of human hosts by schistosome cercariae, or the injection of nematode filaria larvae into mammalian blood by an insect vector) where a free-swimming larva enters an endothermic host. Such abrupt changes in thermal environment can be very damaging (as discussed in Chapter 8), and parasites make extensive use of heat-shock proteins (HSPs) to protect their metabolism. Heat-shock protein manufacture can occupy a major proportion of a parasite's metabolism during such transitions; in adult schistosomes, production of HSP-70 represents more than 1% of the total adult protein production, and levels are higher again in the post-invasion schistosomula larva. The transmission stages of vertebrate endoparasites may also need much broader thermal

tolerance; nematode parasites have a marked degree of cold tolerance that is also susceptible to rapid selection. They have also been shown to change their membrane lipid saturation in similar fashion to other ectotherms exposed to colder conditions. Most nematodes that exploit insect hosts also have a suite of inducible HSPs that can protect them from excess heat (e.g. in a heated decomposing corpse until conditions are favourable for leaving). There are indications that nematode parasites of vertebrate hosts living in extremely arid and hot conditions, such as those of the dromedary of the Sahel, have rather specialist suites of enzymes with reduced isozymic variation compared with nematodes from more mesic hosts.

While intricate temperature regulation may not be a notable feature of parasitic adaptation, sensitivity to temperature certainly is. For many parasites that develop in or on endothermic hosts, temperature is a major cue for larval penetration of the host, or for egg hatching within the host. For example, schistosome cercariae move up a temperature gradient when locating human skin, and fleas locate new hosts by moving towards warmth. Hatching of a range of eggs, including taeniid tapeworms, is triggered by temperature. Nematodes that have snails as intermediate hosts have an unusually high developmental temperature threshold, and this probably helps to ensure that larvae remain in their first instar while the snail undergoes winter hibernation, the first instar being more cold-resistant than later larvae. In general, a very wide range of parasites associated with endothermic hosts show high sensitivity to temperature gradients as infective stages.

While parasites may show adaptation to high body temperatures in endotherms (37–41°C), a range of host organisms react to infection by producing short bursts of exceptionally high body temperature—generally termed fever (see Chapter 8). Fever in vertebrate endotherms is induced in response to chemicals termed **pyrogens**, which can be produced either by parasitic organisms (exogenous pyrogens) or by host blood cells (endogenous pyrogens). In mammals both operate, as endogenous pyrogens are produced by leucocytes (host defensive cells; see section 15.7) in response to exogenous pyrogens, and they act directly on the hypothalamic temperature control centre to elevate the setpoint about which temperature is regulated. Some ectothermic animals (including some caterpillars and some lizards) also react to infection by elevating their body temperature. In these cases elevation is usually achieved by a change in behaviour—increased basking duration, and altered posture to increase the surface over which heat is absorbed. In these ectothermic cases, fever has a demonstrated advantage in conferring protection from bacterial infections, but in endotherms the role of fever is less clear; high temperatures may slow infections down, but can also cause harm or even death to the host. A further intriguing example is provided by bumble-bees infected with the larvae of conopid flies. Such bees forsake the high regulated temperature of the nest and live outside. This may reduce the developmental rate of the parasite, so extending the life of the host.

15.3.3 *Respiratory adaptation*

Ectoparasites

In general, ectoparasites exchange gas with the macroenvironment, whether water or air, and show little specialization. However, some aquatic ectoparasites do show a range of behavioural responses when exposed to hypoxia by the behaviour of their host. For example, *Entobdella soleae*, a monogenean ectoparasite of the sole, shows increased body undulations to maintain movement of water over the body surface through which gas exchange takes place. Similarly fish lice (which are crustaceans) increase the ventilatory movements of their pleopods (modified abdominal limbs). Parasites that occupy the outer respiratory passages of their hosts are supplied with oxygen by the host's ventilatory currents (e.g. flies living in the nasal passages of ruminants, or marine leeches in the branchial cavities of fish).

Some tissue endoparasites are able to be 'ectoparasitic' in terms of respiration, reaching the exterior for gas exchange by causing lesions in the host body wall; for example, the larvae of many tachinid flies have posterior spiracles armed with hooks that they use to penetrate either the skin of their host, or an air sac or major tracheal airway. The host responds by forming a funnel of wound tissue around the parasitoid larva's spiracles, providing permanent access to gas exchange with the exterior. The jigger flea has a similar trick, the adult living beneath the skin as little more than an egg-producing bladder, but with the posterior segments of the flea containing a narrow airway that penetrates the host's skin through a small pore.

Endoparasites

All endoparasites need the physiological scope to cope with the particular levels of oxygen and CO_2 (and the associated pH) of their niche within the host. However, in any one particular spot these conditions should be relatively constant, so in some sense the adaptation is not too difficult. Some parasitic sites, such as the bloodstream and lungs, are unambiguously aerobic, while the lumen of the small intestine and the bile duct may have consistent but extremely low oxygen tensions; the haemoglobin of the fly *Gasterophilus* can apparently take up O_2 even from the stomach contents of the horse it lives in, having a P_{50} of only 0.003 kPa. Parasitic worms such as the nematode *Ascaris lumbricoides* or the fluke *Fasciola hepatica* are oxygen conformers, i.e. the rate of oxygen uptake varies with the partial pressure.

But the greatest physiological problem by far is usually that respiration must necessarily be anaerobic, particularly in the large intestine.

Relative use of anaerobic respiration

In a number of ways, respiration in parasites inhabiting relatively anaerobic habitats (such as the lower gut of vertebrates) parallels pathways seen in free-living metazoans inhabiting environments such as anoxic surface muds in aquatic habitats, and the pathways must often have been inherited from free-living ancestors. Indeed, parasites sometimes show metabolic pathways that seem unsuited to their habitat in the host. For example, anaerobic respiration is common in adult nematodes and platyhelminths, and is found even in species inhabiting relatively aerobic habitats, such as the blood of vertebrates. Use of relatively inefficient anaerobic metabolism even where oxygen is available suggests that efficient exploitation of high-energy metabolites has not been important in these species. Perhaps this is because these materials are so abundant in the host environment that parasites can 'afford' to be wasteful. In explaining the metabolic pathways used for energy generation by parasites, we therefore need to separate genuine adaptations to selective pressures from ancestral mechanisms characteristic of particular taxa which may have little or nothing to do with a parasitic way of life. Remember also that use of entirely anaerobic respiration does not necessarily mean that a particular species can thrive in the total absence of oxygen. Schistosomes, for example, respire largely anaerobically as adults, but require a certain minimal level of oxygen to carry out the synthetic reactions associated with the manufacture of egg shells. In the total absence of oxygen, adult schistosomes survive well but cannot reproduce.

Aerobic respiration is now known to occur at a low level in many adult parasites formerly thought to rely entirely on anaerobic respiration. While the activity of the aerobic respiration pathways is limited, the far greater yield of ATP per glucose molecule generated aerobically means that even low activity can make a significant contribution to the parasite's energy budget. In the filarial nematode *Litomosoides carinii*, adults pass only 2% of their carbon through aerobic respiration, and 98% is fermented; yet starting with 100 mol of glucose, the energy yield of the 98 mol processed via fermentation is 196 mol of ATP, and the 2 mol of glucose respired aerobically generate 72 mol of ATP, or 27% of total energy storage.

The large intestinal nematode *Ascaris lumbricoides* exhibits transitions in energy metabolism during its life cycle that are characteristic of a number of parasites. The eggs leave the host with faeces, and survive in the soil. Larval development within the egg is aerobic, with high activity of enzymes involved in the Krebs cycle. After ingestion of the egg, a second-instar larva emerges within the host's gut. It burrows through the gut mucosa to enter the bloodstream, reaching the liver 2–3 days after infection, where it moults into a third-instar larva. This then switches to anaerobic metabolism as it migrates via the heart to the lungs, reaching them after 7–9 days. The larva then moves up the trachea and is swallowed, reaching the intestine again after 8–10 days where it moults into a fourth instar before finally reaching adulthood. Both the fourth-instar larva and the adult can continue their development in the complete absence of oxygen.

In *Schistosoma*, the transition from aerobic respiration in the cercaria to anaerobic metabolism in the schistosomula (see

Figs 15.1 & 15.8) is very abrupt. When a cercaria penetrates human skin there are two rapid morphological changes: the propulsive tail is lost, and the outer layer of cuticle (the glycocalyx) surrounding the remaining anterior part of the body is shed. The loss of the glycocalyx results in a massive increase of the permeability of the parasite epidermis to host metabolites, some of which are thought to be the cause of the metabolic shift from aerobiosis to anaerobiosis. The key host metabolite may be serotonin (5-hydroxytryptamine, 5-HT), which among other effects in vertebrates activates the enzyme adenyl cyclase and promotes the formation of cyclic AMP. This has a direct effect on respiration through its ability to activate a number of enzymes involved in glycolysis.

In other parasites, shifts between mechanisms may be gradual rather than abrupt. In large parasitic worms, mitochondria with cristae on the inner membrane are concentrated in outer tissues, while mitochondria in deeper tissues lack cristae. Cristate mitochondria generally indicate aerobic respiration, and absence of cristae usually indicates anaerobiosis, leading to the conclusion that deeper tissues predominantly use anaerobic respiration. This has been explained as the result of limited diffusion of oxygen into the tissues of larger parasites; it may also explain in part the general shift from aerobic to anaerobic respiration through the life cycle of parasitic platyhelminths, being an effect of increasing size as well as of changes in habitat.

Anaerobic pathways

Respiration involves the oxidative breakdown of a small organic molecule, and this is most typically glucose, but raw materials can include amino acids, glycerol or fatty acids (see Chapter 6). It seems that parasites have an almost complete dependency on carbohydrate (either as glycogen or exogenous glucose) as their sole energy source. In adult helminths there is no active β-oxidation sequence, so there is no catabolism of fatty acids, and amino acid catabolism is very limited. Similarly, there is no evidence of the co-fermentation of amino acids and carbohydrate or of fatty acids and carbohydrate; this is found in the free-living larvae of some endoparasites, but is absent from the adults.

There are two other important differences relating to parasite anaerobiosis compared with free-living versions. Firstly, for parasites it is a semi-permanent state, not an emergency measure until oxygen returns; thus helminths do not accumulate anaerobic end-products for later resynthesis into glycogen. Secondly, in parasitic helminths anaerobic metabolism persists even if some oxygen is available, and the anaerobic pathways are not inherently inhibited by the presence of oxygen.

In general, anaerobic parasitic helminths can be divided, on the basis of their end-products, into two groups:
1 those that rely essentially on glycolysis alone and produce lactate or some other reduced derivative of pyruvate as end-products of carbohydrate breakdown ('homolactic' fermentation);
2 those that go beyond glycolysis to fix carbon dioxide and have what is often referred to as an '*Ascaris* type metabolism'.

In parasitic helminths where carbohydrate is broken down glycolytically, the amount of lactate varies from 30% under aerobic conditions to 80% under anaerobic conditions. Acetate and acetoin, the other end-products, are produced by the decarboxylation of pyruvate via the pyruvate dehydrogenase complex. In the acanthocephalan *Moniliformis*, the main end-product is ethanol, with small amounts of lactate, succinate and volatile fatty acids. These parasitic helminths do not produce alanine, octopine, alanopine, etc. as anaerobic end-products, unlike free-living anaerobic invertebrates.

Ascaris adults illustrate the alternative anaerobic energy-generation mechanism that is also found in other (unrelated) gut parasites, such as the tapeworm *Hymenolepis*, as well as in some free-living invertebrates. Glycolysis terminates with phosphoenolpyruvate (PEP), rather than pyruvate, and the enzyme PEP carboxykinase then converts the 3-carbon PEP to a 4-carbon product, oxaloacetate, a step that generates 1 ATP. The enzyme malate dehydrogenase is then used to reduce oxaloacetate to malate. In the process, the NADH formed in glycolysis is oxidized to NAD+, maintaining the NADH/NAD+ balance.

The malate produced enters the mitochondrion, and rather than entering the normal Krebs cycle it acts as the substrate for two other linked pathways. These pathways together are termed **malate dismutation**; malate is reduced in one series of reactions, and oxidized in the other. In the oxidation route it is converted to pyruvate and CO_2, while in the reduction route it is converted to fumarate and then succinate (Fig. 15.3). These two reactions are coupled in that the oxidation route generates NADH from NAD, and this NADH is then responsible for the reduction of malate in the second route. As in normal aerobic respiration, the NADH generated is the raw material for an electron transport system. The cytochrome involved in the anaerobic electron transfer system (cytochrome b_{558}) is different to those used in the usual aerobic chain (cytochromes a, a_3, b, c and c_1), and the terminal electron acceptor is fumarate rather than oxygen. The cytochrome and fumarate reductase enzyme make up a complex that is the major component of adult *Ascaris* mitochondria, and rhodoquinone is used as an electron carrier, rather than the ubiquinone characteristic of the aerobic electron transfer chain (see Chapter 6). The redox potential of ubiquinone is more positive (+113 mV) than that of the fumarate/succinate couple (+33 mV), such that electron transfer to fumarate would not be favoured. In contrast, the redox potential of rhodoquinone is strongly negative (–63 mV), favouring electron transfer to fumarate. In *Ascaris* and many other parasites the fumarate, succinate and pyruvate are usually further metabolized to propionate, acetate and volatile fatty acids such as 2-methylvalerate and 2-methylbutyrate.

Energy is thought to be stored in ATP at several stages in the dismutation reaction. The best-known ATP generation point is during the electron transfer chain in the reduction route. The enzyme fumarate reductase, which converts fumarate to succinate, is coupled to an electron transfer system in the mitochondrial inner membrane, phosphorylating ADP to ATP. This is the major known energy storage reaction in *Ascaris*. Three other

Fig. 15.3 The malate dismutation reaction in the mitochondria of the nematode *Ascaris lumbricoides*. Note the principal stage resulting in ATP generation (between fumarate and succinate), together with three other possible ATP- or GTP-generating stages (marked with ?). PEP, phosphoenolpyruvate. (Adapted from Trager 1986.)

Table 15.2 Pathways used in parasite anaerobic respiration.

Respiration type	End-product	mol ATP generated per mol glucose	Organism
Aerobic	CO_2, H_2O	36	Almost all aerobes
Anaerobic	Lactic acid	2	Very common; e.g. *Schistosoma mansoni*, filarial nematodes
Anaerobic	Ethanol	2	Some helminths
Anaerobic	Alanine	2	Many invertebrates
Anaerobic	Acetate + succinate	3.7	Many helminths
Anaerobic	Acetate + propionate	5.4	Many helminths
Anaerobic	Acetate + propionate, methylbutyrate, methylvalerate	2–5	Some nematodes, e.g. *Ascaris lumbricoides*, *Hymenolepis diminuta*

possible stages at which ATP or GTP are generated are also indicated in Fig. 15.3. The endpoint of the metabolism of malate is the fatty acid 2-methylbutyrate. In contrast to the 36–38 ATP molecules generated from a single glucose molecule in traditional aerobic metabolism, the *Ascaris* mechanism generates between three (two in glycolysis, one additionally in the mitochondrion) and five (if the additional ATP-generation steps in the mitochondrion actually occur).

This pathway is unusual in the role played by PEP carboxykinase, which differs radically from its usual role in animals. In mammals this enzyme is present in the liver, and is involved in the manufacture of glucose from lactic acid (gluconeogenesis; see Chapter 6). In contrast to the carbon fixation seen in *Ascaris*, in mammals the enzyme catalyses the same reaction but in the opposite direction! In effect the animal is using the second span of the tricarboxylic acid cycle, from succinate to oxaloacetate, but

operating in the reverse direction; it has what has been described as a partial, reversed tricarboxylic acid cycle. How is this reversal achieved? Three aspects of the situation in *Ascaris* favour this kind of CO_2 fixation:

1 Levels of the substrate PEP are far higher in *Ascaris* tissues than they are in mammalian tissues; pyruvate kinase, which normally breaks PEP down, is present only in very low concentrations in *Ascaris*.

2 *Ascaris* PEP carboxykinase has a seven-fold higher K_m (i.e. lower affinity) for oxaloacetate than for PEP.

3 Oxaloacetate is rapidly metabolized by the enzyme malate dehydrogenase, favouring further CO_2 fixation by the PEP carboxykinase.

Similar transitions in energy metabolism between life-cycle stages are seen in other parasites. *Schistosoma mansoni* has two free-living larval stages, the miracidium that locates the snail intermediate host, and the cercaria that penetrates the human host. Cercariae are definitely aerobic, and miracidia are probably so, but the adults rely to a large extent on anaerobic respiration, even in the presence of oxygen. In contrast to the metabolic pathways used by adult *Ascaris*, schistosome adults generate lactic acid in much the same way as anaerobically functioning invertebrates described in Chapter 6, yielding only two molecules of ATP per molecule of glucose. The same energy storage system is also important in filarial nematodes. Thus across parasites as a whole, a range of end-products of anaerobic glycolysis are found, most yielding only two ATP per molecule of glucose (Table 15.2).

15.3.4 Feeding

Parasite feeding strategies can be broadly divided into two very different types:

1 Parasites feeding on host tissues and fluids via the mouth. This category includes those feeding on blood or haemolymph, and on host exudates such as mucus on the skin.

2 Parasites absorbing nutrients from host body fluids or gut contents, across their body wall.

Feeding via the mouth

Where the parasite takes in food via the mouth, it is possible for the parasite's body to be enclosed in a cuticle or sheath that is metabolically relatively inactive, conferring protection against attack by host immune responses or digestive enzymes. Most nematodes, for example, feed via the mouth and have metabolically inactive cuticles that are thick and multilayered. The same is true of many blood-feeding ectoparasites, whose outer body layers must protect them from conditions in the external environment. The mouthparts of all blood-feeders need to be modified for attachment and/or suction (Fig. 15.4).

To generalize, the main problems for this group, and hence the principal feeding adaptations found in members of this group, are four-fold:

1 coping with large volumes of fluid;

2 coping with the metabolic breakdown products of blood (particularly iron);

3 coping with possible blood-borne infective agents such as bacteria;

4 coping with humoral or cell-based host defence systems active in ingested blood.

Many parasites from a diversity of animal groups feed on blood, including platyhelminths (monogeneans and trematodes), nematodes, leeches, crustacean fish-lice, arachnids such as ticks, and insects such as lice, fleas, mosquitoes and tsetse flies. Problems may result from very intermittent meals of large volume; adaptations are best known in leeches and in biting insects, where in both cases diuretic elimination of water is initiated by hormones (often triggered by body-wall stretch receptors) to promote very rapid fluid loss after a meal. Initial blood digestion occurs in response to enzymes released by pharyngeal

or 'salivary' glands. Homogeneous acellular products of this digestion, including haemoglobin, are usually digested intracellularly. Iron is discarded back into the gut lumen by exocytosis and shed with normal faeces, or in the case of schistosomes iron-rich waste products are voided by regurgitation (flukes usually have a blind-ending gut, with no anus). To cope with the third problem, it is known that antibacterial peptides are up-regulated in response to ingestion of a blood meal in the blood-feeding fly *Stomoxys calcitrans*, and such a system may be widespread.

Some parasites with normal guts and mouths that could feed on blood or other tissues nevertheless use the skin as an uptake surface for key nutrients; for example, the monogenean *Diclidophora merlangi* feeds mainly on blood, but supplements its diet by absorption of amino acids through its outer tegument.

Most parasites can use their metabolic pathways to synthesize the majority of the amino acids they require, often by the described routes relating to carbon dioxide fixation, and most can synthesize pyrimidines; but few if any can manufacture the purine ring, which must be obtained from the host. Salvage pathways therefore exist in many parasites to allow recycling of certain host molecules from the blood before they are broken down in normal catabolic paths.

Direct absorption across cuticles

Where direct absorption occurs, the outer tissues of the parasite are typically highly metabolically active, and experience rapid fluxes of nutrients, ions and water; it is not possible for the parasite to isolate itself from the host by a barrier layer. Direct absorption of host nutrients across the body wall is found in a wide diversity of parasites, particularly among the platyhelminths. It is also found in acanthocephalan worms, in the first larval stage of some insect endoparasitoids, and in a small number of highly specialized nematodes.

Parasitic platyhelminth teguments consist of several layers (Fig. 15.5a & b). The outer layer is composed of syncytial living tissue, the nuclei lying more centrally within the animal below layers of circular and longitudinal muscle. The outer surface of the tegument is covered in a thin glycocalyx of mucopolysaccharides and glycoproteins. This layer has a net negative charge, and is able to bind both inorganic and organic charged molecules, including host enzymes. These may then be used to digest surrounding food material, which is absorbed by the parasite. The surface area of the tegument is enlarged (by up to 10 times) by microvilli (called microtriches in the case of cestodes), which are particularly well developed in cestode tapeworms (Fig. 15.5b), and are presumably adaptations for large-scale secretion and absorption across the tegument. The outer surface of acanthocephalan worms also has an enlarged surface area (Fig. 15.5c), in this case through high densities of surface pores, which lead to blind-ending spaces ('crypts'), enlarging the surface area about 60-fold. The nematodes that absorb nutrients across their body wall lack the usual thick and non-absorptive cuticle, and develop microvilli all over their surface, through which it is assumed nutrient absorption takes place.

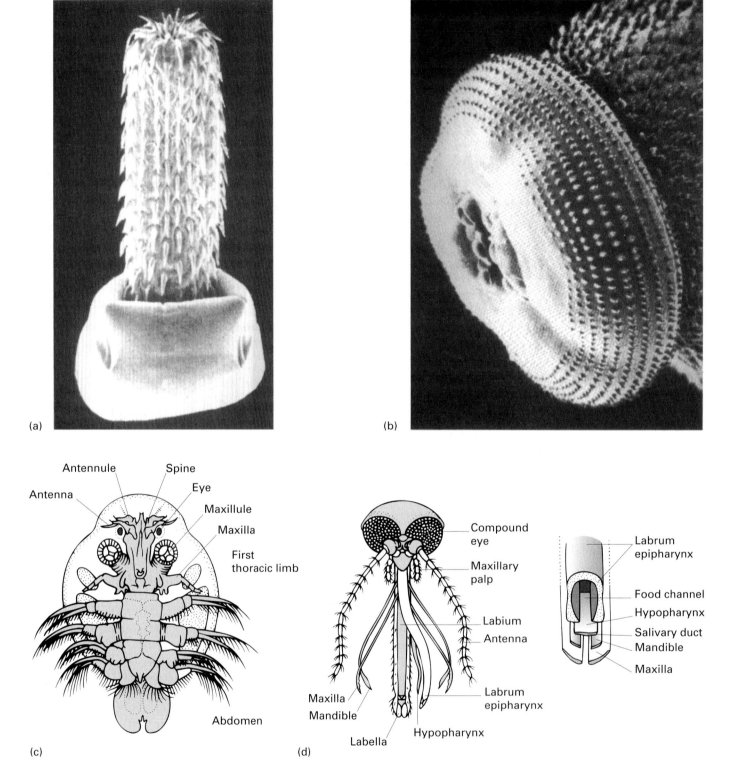

Fig. 15.4 Mouthparts of blood-feeding parasites. (a) The proboscis of an acanthocephalan worm. (From Matthews 1998; photograph by H. Melhorn.) (b) The spiny head collar of the nematode *Gnathostoma*, with mouth at the centre. (From Matthews 1998; photograph by L. Gibbons, CAB International.) (c) Ventral view of a crustacean fish louse, where the enlarged maxillules (limbs modified for feeding) hold the parasite to the fish's skin by suction. (Adapted from Kaestner 1980.) (d) Piercing mouthparts of a mosquito, with the components spread out of their natural alignment; to the right are the components in their functioning arrangement. (From Brusca & Brusca 1990.)

Digenea

(a)

Cestoda

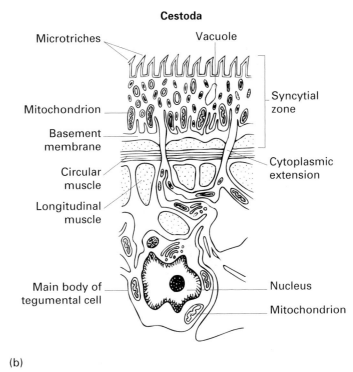

(b)

Acanthocephala

(c)

Fig. 15.5 Highly modified cuticles in gut parasites. (a) Tegument of a digenean fluke; (b) tegument of a cestode tapeworm; (c) cuticle of an acanthocephalan worm, showing internal crypts. (Adapted from Matthews 1998.)

The major physiological adaptations in nutrition are associated with digestion and absorption of host material, both inside and often also immediately outside the parasite's body, while protecting the same tissues from attack by host defences. Cestodes and acanthocephalans especially are in competition with their hosts for nutrients in the gut lumen. If they are to compete successfully for these, their ability to bind and absorb nutrients must be at least as good as or better than the host's. Adult cestodes, for example, are able to absorb all detectable glucose from their surrounding medium, indicating a very high-affinity (low K_m) active uptake for this compound.

Living in the gut clearly has real hazards, and gut parasites show a variety of tricks to enhance their own competitive success. Many gut parasites coat themselves in mucus, which is a good ion-binding matrix, to limit the effects of extreme pH in the host gut. Some intestinal parasites inhibit digestive activity of the hosts and indirectly inhibit vitamin and blood-sugar metabolism. For example, the tapeworm *Diphyllobothrium latum* has very high affinity for vitamin B_{12}, and can cause deficiency disease (pernicious anaemia) in humans. In the nematode *Nippostrongylus*, adults secrete large quantities of acetylcholinesterase from their excretory pore, perhaps to produce local inhibition of peristaltic activity in the host gut (which will also help prevent them being dislodged). Insect parasitoids may disrupt host absorption of materials from the haemolymph, which increases resource availability for themselves.

Many parasites secrete enzymes associated with feeding. A number of gut-inhabiting parasites are able to reduce the risk of digestion by their host by releasing 'anti-enzymes', compounds that neutralize the host enzymes. Survival may also be augmented by tolerance of high levels of toxic compounds, as illustrated by tapeworms in the guts of rabbits. Rabbit bile contains very high levels of deoxycholic acid—36 mg ml⁻¹, in comparison with 1.77 mg ml⁻¹ in dog bile. Rabbit gut parasites that live below the point where the bile duct enters the intestine are able to cope with this level of deoxycholic acid, while larvae of the dog tapeworm, *Echinococcus granulosus* (which lives in a similar position in dogs), are destroyed by the bile acid concentrations found in rabbits.

15.4 Reproduction and transmission

15.4.1 *The goals and challenges of transmission*

All hosts have a finite lifetime, and the long-term survival of parasites requires the transmission of infective stages from one host to another. In marine and freshwater environments, transmission between hosts can take place, as we have seen, via ciliated free-living larval stages or (in the case of ectoparasites in particular) by direct adult migration. In many cases larval dispersal is passive, without direct attraction to the hosts, although larvae may concentrate in areas more often visited by the host. An example is larval concentration in surface waters at times when intermediate hosts feed.

The situation is quite different in terrestrial habitats. Adult tapeworms, nematodes and other parasitic worms, though able to locomote and survive in very moist terrestrial habitats, are unable to survive outside the host in drier environments; movement from one host to another thus cannot be direct. However, direct transmission is often possible for ectoparasites with a more recent terrestrial free-living ancestry, such as biting flies, fleas or ticks. Endoparasites are generally unable to survive as adults in terrestrial habitats and rely on resistant eggs and/or larvae released from the host to achieve infection of further hosts. One parasitic strategy is to produce long-lived desiccation-resistant eggs that are released into the environment in large numbers and taken up by the same host species (usually via the mouth). The host occupied by the adult parasite (usually

termed the **definitive host**) may be extremely patchily distributed in the environment, making the possibility of direct infection of another host very low. Some parasites have evolved life cycles that exploit an **intermediate host** species, usually one which is more abundant in the larval environment than the definitive host. Many parasitic life cycles include **asexual reproduction** by larvae in the intermediate host, increasing the probability that at least one larva will successfully reach either a further intermediate host or the definitive host. In many life cycles, the intermediate host is a prey item for the next intermediate host or the definitive host, so that the limited ability of the parasite to move between hosts is vastly enhanced by the ability of a predator to detect and locomote towards its prey. The same advantage applies for parasites transmitted by vectors. For endoparasites such as tapeworms that infect terrestrial hosts, complex life cycles of this type can be seen as a necessary consequence of the limited ability of the parasite to survive and locomote on land. What a blood-feeding tsetse fly (*Glossina morsitans*) can achieve by flying between two cattle, a tapeworm must achieve by apparently tortuous and indirect means.

Parasite environmental physiology shows adaptations associated with several steps in the transmission process. Firstly, there are parasite traits associated with different methods of movement between hosts. These include adaptations to particular environments, and the dramatic changes that often take place when moving from one to the other. Secondly, a number of parasites have evolved mechanisms allowing reproductive synchronization with their host, and some show daily migrations in the host blood system that enable them to make use of blood-feeding insect vectors. Thirdly, and most elegantly, parasites that reach their definitive host by moving up a food chain are sometimes able to manipulate the behaviour of their intermediate host such that they are more likely to be eaten.

15.4.2 *Movement between hosts*

Reinfection requires that a suitable host is located, and usually a specific part of the host's body. In cases where the parasite habitat is essentially external, movement of individual adults or larvae through the medium may be possible. In marine ectoparasitic platyhelminths such as flukes, detection and settling of larvae on host external surfaces is analogous to detection of settling sites by the planktonic larvae of many other marine organisms. In many endoparasites, the situation is rather more complicated. Reinfection involves not only release of eggs or larvae from adults that may be located deep within host tissues, but also subsequent re-entering of hosts. This is achieved via the mouth and digestive tract, or via the host's skin. In the latter case, location of the adult site in the host usually also involves highly complex larval migration through the host's body.

The physiological tolerances of particular parasite groups have a major bearing on the infection routes possible for free-living adults or larvae. Adults and larvae of many parasites can

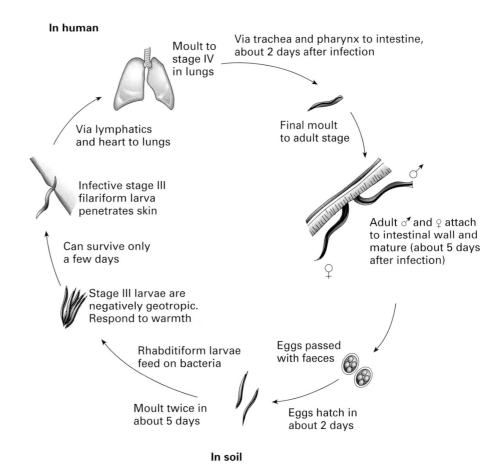

In human

Moult to stage IV in lungs

Via trachea and pharynx to intestine, about 2 days after infection

Via lymphatics and heart to lungs

Final moult to adult stage

Infective stage III filariform larva penetrates skin

Adult ♂ and ♀ attach to intestinal wall and mature (about 5 days after infection)

Can survive only a few days

Stage III larvae are negatively geotropic. Respond to warmth

Rhabditiform larvae feed on bacteria

Eggs passed with faeces

Moult twice in about 5 days

Eggs hatch in about 2 days

In soil

Fig. 15.6 Life cycle of the hookworm nematode *Necator americanus*. (From Trager 1986.)

survive in aquatic environments but are less tolerant of terrestrial ones. In platyhelminth parasites in particular, infection by free-living life stages tends to take place only when the host is aquatic or is a terrestrial animal that intermittently comes to water (as in the life cycle of *Schistosoma*). Some platyhelminth adults can survive for short periods outside the host, even in terrestrial environments. For example, adults of cestode tapeworms release their eggs packed into reproductive segments called proglottids. These leave the host with faeces and can survive and disperse eggs for a while outside the host. Even large proglottids, however, probably do not disperse far. Nematodes are better able to survive in free-living terrestrial environments, and several serious parasites of humans have larvae that spend time in the soil. In the case of the nematode hookworms *Necator americanus* and *Ancyclostoma duodenale*, the parasite life cycles include free-living larval stages that actually feed on bacteria in the soil (Fig. 15.6). Dispersal by larvae in terrestrial environments can also be achieved if they are collected from one host and delivered to another without exposure to the external environment. A number of parasites (particularly nematodes) survive by being spread via blood-feeding vectors, notably insects and ticks. Parasite larvae enter the vector as it feeds from one host, and (sometimes after intermediate larval stages) enter the next host as the vector feeds again.

The alternative to free-swimming larvae or adults is the use of eggs or larvae that are protected from environmental conditions by egg shells or other structures. Parasite eggs leave the host either through a natural aperture (most often with host faeces), or through a hole in the host's body wall induced by the parasite, as in the nematode commonly known as the guinea worm, *Dracunculus medinensis*, or the jigger flea, *Tunga penetrans*. Once released, these resistant stages are unable to locomote themselves, but there is a probability that if they survive long enough they may encounter another host. Sometimes the larva is dispersive for a while, but forms a resistant cyst before reaching the limits of its environmental tolerance. Here the liver fluke *Fasciola hepatica* is a good example; its life cycle is similar to that of *Schistosoma*, except that the cercaria larvae do not actively bore through the skin of their definitive host (a ruminant mammal); instead, they form resistant cysts on grass leaves, and rely on the ruminants to eat them. Dispersive stages of this type typically maintain a very low metabolic rate, are very desiccation-resistant, and able to survive unfavourable osmotic conditions for long periods. Eggs of the gut nematode *Ascaris* possess a lipid layer inside the egg shell that is extremely impermeable to both water loss from the egg and entry of other chemicals from the external environment. In some nematodes, the dispersive stage is a larva that resists desiccation rather differently, by

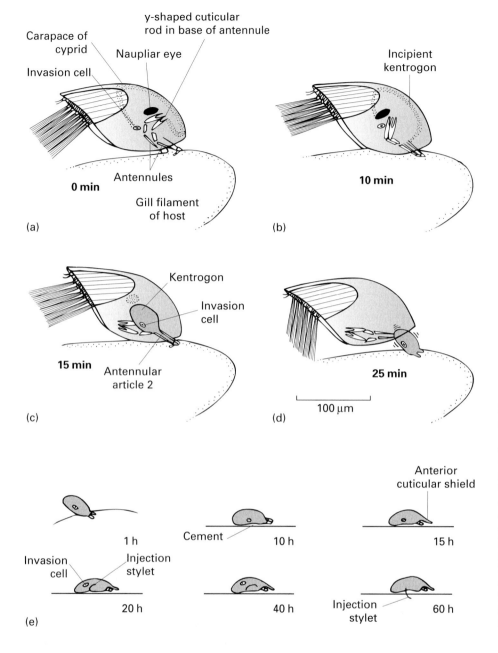

Fig. 15.7 Rapid body reorganization during host infection in the rhizocephalan barnacle *Lernaeodiscus porcellanae*. (a–d) The free-swimming cyprid larva locates and settles on gills of its host, a crab, attaching by secretions from a large cement gland. Within 25 min the cyprid larval body is extensively reorganized, forming a sac-like structure termed the kentrogon. Then the carapace of the cyprid is shed. (e) The kentrogon develops a stylet, and injects a specialized 'invasion cell' into the crab between 40 and 60 h after attachment; this divides to provide a network of parasite tissues, more similar to a fungus in structure than to a typical crustacean. (Adapted from Brusca & Brusca 1990.)

retaining the cuticle of a previous moult as an extra layer around the worm. These adaptations clearly parallel those seen in other essentially aquatic Metazoa—such as rotifers, tardigrades and many free-living nematodes—that disperse through terrestrial environments using a cryptobiotic stage (see Chapter 12) between phases in aquatic habitats.

Infection by free-living adults or larvae

In aquatic ectoparasites, dispersal by free-swimming adults is possible. The adult must already be able to tolerate conditions experienced in the external environment, and needs only to be able to release its attachment to one host and then locate another.

This is certainly possible for some monogenean flukes, but is probably impossible for many species which though ectoparasitic form complex and permanent attachments to the host. This is the case with many monogeneans that grow into the tissues of fish gills.

Free-swimming larvae are found in many parasitic groups, particularly the platyhelminths. The longevity, and sensory and locomotory abilities, of platyhelminth larvae are relatively limited, and without the direct inoculation of another host by a vector, only a tiny proportion reach the next host in the life cycle. Host location by free-swimming larvae involves two general steps. Firstly, responses to general stimuli such as light or gravity allow the larvae to concentrate in regions of the environment

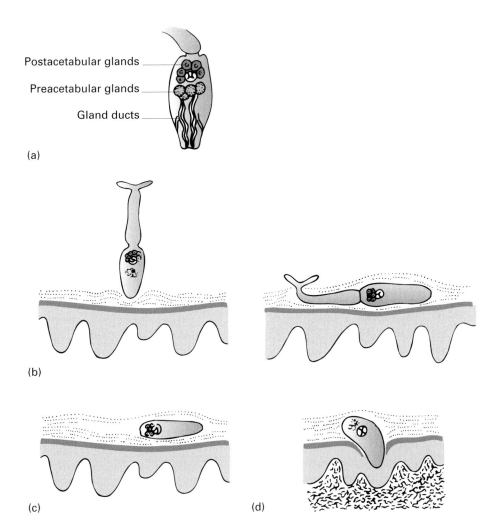

Postacetabular glands
Preacetabular glands
Gland ducts

(a)

(b)

(c)

(d)

Fig. 15.8 Penetration of human skin by a cercaria larva of *Schistosoma*. (a) Enlargement of anterior portion of a cercaria, showing glands and ducts; (b) adhesion to, and penetration of, the skin surface; (c) shedding of the larval tail; (d) migration of the metamorphosed schistosomula larva into deeper tissues and then to the human's circulation. (Adapted from Matthews 1998.)

where contact with the next host has a better than random probability. For example, larvae of flukes that attack bottom-living fish tend to stay near the bottom, while species attacking surface-swimming fish are positively phototropic and maintain themselves in upper water layers. Secondly, responses to characteristic host cues stimulate egg hatching, larval orientation, and attachment behaviours. Thus in the monogenean *Entobdella soleae*, the sole fluke, which parasitizes a bottom-living fish, eggs hatch in response to urea in the mucus of the host's skin, and detection of this cue stimulates release of a proteolytic hatching fluid and emergence of the larva. The larvae show a diurnal rhythm in their hatching activity, which peaks during the middle of the day when their hosts are lying inactive and accessible. Having located a host, the larvae use adhesive glandular secretions to maintain contact until the main attachment organ (the hook-bearing opisthaptor) can be applied to the fish's skin.

The situation becomes more complex in endoparasites, many of which have complex life cycles. The miracidium larvae of schistosomes (see Fig. 15.1) locate the intermediate snail host, and the cercaria larvae re-enter the definitive host. There is a dramatic and rapid reorganization of the larval body on entering a host from the free-living aquatic environment; in the case of the schistosome transitions a motile outer layer (ciliated in the miracidium, and smooth with a muscular swimming tail attached in the cercaria) is rapidly shed to leave a new outer layer. An even more extreme example is shown in Fig. 15.7 for *Lernaeodiscus porcellanae*, a rhizocephalan barnacle that is parasitic on crabs.

In life cycles involving more than one host, stage-specific host location behaviours and recognition cues must be used. Larval recognition of the host is normally only effective over short distances, and for schistosomes it involves movement up a temperature gradient, and response to lipids characteristic of host skin. Once contact is made, the cercaria uses secretions from two specialized acetabular glands whose ducts open into an anterior sucker (Fig. 15.8). Secretions from these glands are sticky, and while holding the larva in place help to soften the outer layers of the host's skin. Once the larva has gained entry into a small break in the host's skin, proteolytic secretions break down host cells and allow the larva to move inwards. Then the larva undergoes a dramatic transformation into the next larval stage, the schistosomula; the free-swimming cuticle and tail are shed (Fig. 15.8), leaving the highly specialized absorptive outer covering (tegument) characteristic of the adult. While the body wall of the cercaria was relatively impermeable, though probably with salt uptake regions to maintain larval metabolism in the hyposmotic freshwater environment, the new tegument is the permeable surface through which all metabolic exchange with the host takes place for the rest of the parasite's life. An important early function of the tegument is the rapid acquisition of host surface antigens, rendering the parasite invisible to the host immune system (see below).

Infection by free-living larvae in non-aquatic terrestrial environments is less common. In the nematode parasites of mammals that have free-living larvae surviving and feeding in the soil, infection of a host is preceded by a moult to a non-feeding stage, which must locate a host within a few days. Host location is brought about by negative geotropism (keeping the larvae at the soil surface) and by rapid movement towards a source of warmth. Once contact is made, the filariae of the nematode hookworm *Necator americanus* also bore through the skin of their vertebrate host, using proteolytic secretions from their oesophageal glands to break down host cells in their path.

Vector-mediated parasites

In a sense, parasites utilizing vectors avoid some of the problems associated with entering the external environment. They do not need sense organs to find the next host, since the vector does that for them, and they are not exposed to risks of desiccation. However, they encounter other challenges. Firstly, the larvae must reach a part of their host in which they become accessible to their vector. For parasites occupying internal tissues, this requires migration towards the host body surface in the peripheral circulation. Although this brings the parasites to within reach of the mouthparts of blood-sucking vectors, travelling in the bloodstream may expose the parasite larvae to intensified host defence. A partial solution to this problem is for the parasite to show a very brief and precisely timed diurnal migration between deeper tissues and the blood vessels just beneath the skin, as observed for the microfilariae of the nematode *Wuchereria bancrofti*. The vector exploited by the parasite varies geographically, and parasites emerge in the peripheral circulation at the right time for their local vector. Thus numbers of *Wuchereria* microfilariae just beneath the skin peak during the night in regions where the main vectors are night-biting *Culex* and *Anopheles* species, and during the day where the main vector is the day-biting *Aedes*. The parasites must detect some stimulus in host metabolic cycles that has a daily rhythm, and experimental confirmation of this comes from studies of *Wuchereria* filarial migrations in people working night shifts, with biorhythms 12 h out of phase in comparison with the normal human cycle, who show filarial migrations that are 12 h out of phase with filariae in people working day shifts (unfortunately for the parasites, this results in larval presence in the peripheral host blood system at a time when their mosquito vectors are not active).

In addition to getting the timing right, vector-borne parasites must also be present in the blood in an area of the body favoured by their vector. *Onchocerca* is another nematode parasite of humans, being the agent of the disease river blindness. It is transmitted by several species of biting blackflies that show species-specific preferences for biting particular parts of human hosts. Thus the nematode larvae aggregate in the legs in regions where their vector preferentially bites legs, and in the upper body when this region of the host is preferred. This clearly requires very specific larval detection of—and response to—host stimuli.

A second set of challenges involves parasite responses to the environments present in the vector. Vector-transmitted parasites are initially swallowed into the gut of the vector, and thence move to the blood and on into the tissues. Survival requires parasite resistance to host digestive processes, and defences in host tissues, which are extensive even in invertebrates (see below). Eventually (sometimes after intermediate larval stages) the larvae reach the vector's salivary glands or proboscis, and are re-injected into another host. This requires the parasite to cope with three very different vector environments, often within a very short time-frame.

Reinfection via the gut

Infection via the gut typically involves eggs or encysted larvae that emerge in response to the conditions in a certain part of the host gut. Protective coats around the egg or larva, coupled with relatively specific hatching or excysting stimuli, allow the parasite to reach the appropriate part of the host gut without being digested on the way. Protective coats include egg shells, cysts formed by larvae, and larval sheaths in some nematodes. They also include cases where the larva is surrounded by tissue laid down by an intermediate host; in the tapeworm *Taenia solium* the cysticercus larvae are surrounded by cysts of pig tissue in the muscles, and emergence is triggered when the flesh of the pig is consumed and then digested by the human definitive host.

A wide range of stimuli are known to trigger emergence from eggs or cysts in the gut. Generally, stimuli in invertebrate hosts are less specific than those in vertebrates, even where these are both present in the same parasite life cycle. For reasons that are not always obvious, gut parasites usually occupy a relatively specific position in the host's alimentary canal (Fig. 15.9). This may represent the region of the gut in which resistance to host defences and levels of nutrient acquisition are optimum for a given parasite. Host guts show characteristic gradients in chemical and physical conditions along their lengths (Fig. 15.10), and there may be triggers corresponding to particular combinations of conditions that allow parasites to emerge at the correct position. Conditions required for emergence of larvae can be divided into two types. Firstly, some stimuli indicate that the infective stage has passed unsuitable regions of the gut;

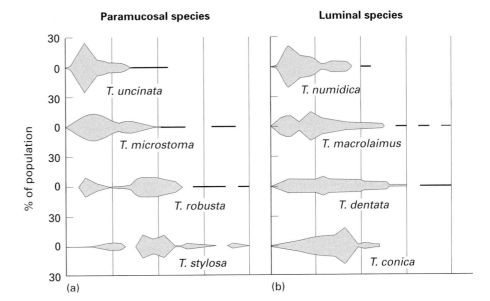

Fig. 15.9 Distribution of nematode species of the genus *Tachygonetria* in the intestine of the Greek tortoise. The eight species are divided into two groups; (a) those living close to the gut wall (paramucosal) and (b) those inhabiting the lumen of the gut. For each group, the distribution of the parasites along the intestine is shown from left to right. (Adapted from Schad 1963.)

thus emergence of several intestinal parasites is initiated by low pH, characteristic of the host stomach, though hatching does not progress further until more favourable conditions are reached.

Fig. 15.10 Gradients in conditions along the mammalian gut, for: (a) water and salts; (b) pH and osmotic pressure; (c) bile salts; and (d) glucose and redox potential (Eh). (From Matthews 1998.)

Secondly, some stimuli indicate more directly that the egg or larva has reached the right part of the gut for hatching or metamorphosis. For example, the increase in temperature associated with entry into an endothermic host provides a very simple trigger for a few parasites. Alternatively, conditions in the host gut may stimulate secretion of enzymes by the egg or larva, leading to rupturing of the egg/cyst wall; *Ascaris* eggs hatch in response

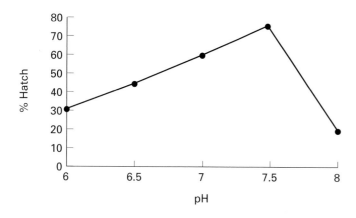

Fig. 15.11 The pH sensitivity of hatching by eggs of the gut nematode *Ascaris lumbricoides*. (From Matthews 1998.)

to a critical concentration of CO_2, an appropriate pH (slightly alkaline—see Fig. 15.11) and a high temperature (close to 37°C). Within the eggs, larvae release lipase and chitinase, and the lipid-rich inner layer of the egg shell increases in permeability. The enzymes diffuse through and dissolve a small area, through which the larva emerges. As another option, digestive enzymes present in the host gut may break down the egg/cyst wall; here examples are provided by hatching of the eggs of tapeworms, and excystment of digeneans and cestodes. Host enzymes remove the outer layers of the egg or cyst, and then bile causes changes in the permeability of the remaining membranes.

15.5 Parasite sensory abilities

It is usually suggested that in comparison with their free-living relatives, many parasites have reduced sense organs—particularly those associated with stimuli absent from the interior of a host's body, such as light. This is true of some parasites, such as the rhizocephalan barnacle *Sacculina*. This parasite, which in the adult stage looks nothing like a barnacle and quite like a mushroom, forms a network of tissue within the body of its crab host, extending externally as a formless mass of tissue beneath the crab's abdomen. Unlike the free-living barnacles from which it is derived, the adult *Sacculina* lacks obvious sense organs.

However, to extend the example of *Sacculina* over all parasites may be unwise. Many parasites are able to successfully locate hosts as larvae, and furthermore to locate amazingly specific microhabitats within their hosts. Within the human blood system, young larvae of *Schistosoma* are able to recognize and congregate in the hepatic portal vein, while other flukes are able to migrate to a wide diversity of highly specific sites in other parts of the body, including the brain, heart and bladder.

It could be that parasites infect hosts at random, but die in the 'wrong' hosts, or in the wrong part of the right host—generating specificity without sensory selection of hosts by the parasite. Alternatively, parasites could possess sense organs allowing specific selection and rejection of alternative hosts and sites.

Site recognition could in part be attributed to a small number of highly specific recognition cues, perhaps mediated by recognition of specific host cell types at a molecular level. Detailed studies are now revealing that parasite sense organs and nervous systems are far more complex than previously appreciated. Superficially simple endoparasites such as trematodes and nematodes have highly complex sensory systems. As an example, adults of the marine aspidogastrid fluke *Lobatosoma manteri* have a probable total of 20 000–40 000 surface sensory receptors of at least 14 different types. Such a complex sensory system may be required to monitor the physiology of the host, and to prevent over-exploitation leading to host death (and so parasite death).

The ability of parasites to find such specific locations in their hosts suggests an ability (across a range of parasite species) to detect extremely precise combinations of pH, osmotic concentration, oxygen tension, patterns of fluid pressure in water currents, and temperature. In addition, the exploitation of different habitats through the life cycle also implies that different subsets of receptors or activation thresholds for given behaviours are present in different life cycle stages. Remember also that parasite behaviours are often very flexible, and able to take into account variation in the parasite's environment. Thus the adults of a number of intestinal worms are able to adjust their location in the gut in response to the digestive cycle of their host, while others adjust their spatial patterning in the host in response to whether it is well fed or is fasting. These alterations of position again suggest a sophisticated ability to detect and respond to changes in environmental stimuli.

15.6 Parasite regulation of host physiology

Parasite manipulation of the way hosts invest their resources is widespread, and is thought to result in greater resource availability for the parasite. The scale of the manipulation can range from a small area of host tissue immediately around the parasite to reorganization of the host's entire body. The mechanism of parasite modification is rarely known, but in at least some cases it is through active secretion of chemical signals.

15.6.1 Modification of host tissue growth

On the most local scale, parasites may modify the physiology and development of individual cells. The nematode *Trichinella spiralis* is able to transform individual cells in the muscles of its pig host, secretory/excretory products of the worm transforming muscle cells into nurse cells, probably through direct modulation of host genomic expression.

Some of the clearest examples of parasite control of host tissue differentiation come from insects that form galls in plants (Fig. 15.12). Galls are plant tissues induced by another organism, providing that organism with food and a measure of physical protection (see Chapter 13). Typically, galls consist of nutritive inner tissues, on which the galling organisms feed, and outer

Fig. 15.12 Galls formed by cynipid gall-wasps. (a) Longitudinal section through a gall induced on a rose, showing the larva (La) in its hollow cell (LC), and surrounding plant tissue layers on which the larva feeds (NT, nutritive tissue; PNT, parenchymatous nutritive tissue; VT, vascular tissue supplying the gall; C, outer plant tissues). (From Brooks & Shorthouse 1998.) (b–f) Examples of galls induced on oak trees by gall-wasps: (b) *Neuroterus quercus baccarum*; (c) *Andricus aestivalis*; (d) *A. quercus tozae*; (e) *A. coronatus*; (f) *A. tomentosa*.

tissues that are often hardened. The gall-forming organism is often able to induce the plant to produce abnormal cell types and structures, and to direct resources such as nitrogen preferentially to gall tissues. A group of small wasps (commonly termed gall-wasps, in the hymenopteran family Cynipidae) induce the most structurally complex and diverse galls known. At the centre of the gall is a layer of endosperm-like tissue on which the gall-wasp larva browses. A series of outer tissue layers give the gall of each species a characteristic form, and show a wide diversity across gall-wasp species, including layers of woody or spongy tissue, complex air spaces within the gall, and surface coats of sticky resins, hairs or spines. It is thought that these outer tissues protect the gall-former from attack by insect parasites and vertebrate predators. In addition to physical protection, gall tissues are often rich in toxic secondary plant com-

pounds (see Chapter 13). Some oak-feeding gall-wasps are even able to induce oak tissues that secrete a sucrose-rich nectar, attracting ants to the surface of the gall, which further protect the gall-wasp from insect enemies.

15.6.2 *Modification of host reproduction and life cycles*

Rather than initiating the development of novel tissues, parasites may prolong existing phases of the host's life cycle to favour their own development. One way of diverting resources into parasite growth is to suppress sexual maturation of the host, such that host resources are never directed into eggs or sperm. This occurs in chaetognaths and snails infected with trematode larvae, and in crabs infected by *Sacculina*. In the snail case, trematode larvae interfere with the endocrine control of host development, causing the host snail's central nervous system to manufacture a protein ('schistosomin') that interferes with several of the snail's usual hormones involved in the control of egg development. A vertebrate example is given by tapeworm larvae that suppress sexual behaviour and egg maturation in many fish, probably by preventing production of the host's gonadotropic hormones.

Sometimes the parasite does not completely inhibit reproduction, but inhibits or delays it. For example, male hamsters infected by schistosomes show decreased testosterone levels and so reduced investment in male behaviour and secondary sexual characters. The change is thought to be under parasite control, through release of chemicals termed opioid peptides. These have many effects on host behaviour and immune responses, and in the hamster suppress immune defences as well as altering host investment in reproductive effort.

A second parasite-induced effect is host gigantism, sometimes associated with host castration. Again, the suggestion is that such a manipulation of resource investment by the host results in greater resource availability for the parasite. Host gigantism in response to infection is best-known for cestode larvae of the genus *Spirometra*, whose larvae live in mice. The larvae secrete a hormone ('plerocercoid growth factor') that mimics the effects of pituitary growth hormone, causing infected mice to reach a spectacular size.

Some parasites are also able to prevent moulting in their host, for example in crabs infected by *Sacculina*. Suppression of moulting is required because the parasite has external parts that disrupt the usual form of its host's exoskeleton, and an attempt to moult would probably kill both host and parasite. In other cases, suppression of host moulting maintains feeding and growth by a particular host life cycle stage attacked by the parasite, so enhancing parasite growth.

The clearest examples of this type of host manipulation are to be found among insect **parasitoids**, representing a special case of parasitism in which the parasite always kills the host. **Idiobiont parasitoids** kill or paralyse their host at the time of egg-laying, and no further host growth takes place. **Koinobiont parasitoids** represent a more subtle strategy in which the parasite allows the host to continue to develop after egg-laying; parasitoid larvae avoid damaging organs essential to host survival and growth, and may also manipulate the host's endocrine physiology, nutritional biochemistry and/or behaviour. The eggs or larvae of koinobiont parasitoids typically delay their own development until the host reaches a large size, and may also prolong the life cycle of the host, both strategies resulting in greater resource availability for the growing parasite. For example, the parasitoid *Cotesia congregatus* is able to prevent the larva of its caterpillar host *Manduca sexta* (the tobacco hornworm) from metamorphosing into a pupa, and can induce up to six extra host larval instars. Metamorphosis of the host is usually brought about by a drop in levels of juvenile hormone and increase in the level of ecdysone in the host's haemolymph. Parasitoid larvae are able to prevent the drop in levels of juvenile

hormone, and some may also block the conversion of inactive ecdysone into the active 20-hydroxyecdysone. However, it is worth noting that delay in parasitoid development as the host grows may come with a cost; most koinobiont insect parasitoids are endoparasites, and they are thus exposed to host defences for rather longer.

Synchronization between host and parasite life cycles is also an option that many parasites exploit, ensuring that parasite young are produced at the same time as host young. Moulting in insect parasitoids appears to occur in response to host titres of juvenile hormone and ecdysone. More spectacularly, rabbit fleas time their own reproduction by sensing the changes in circulating hormones in their pregnant female host's blood.

15.6.3 *Parasite manipulation of host behaviour*

Many host behaviours have direct impacts on parasites, ranging from obvious examples such as grooming behaviour, coughing, sneezing or vomiting, to less obvious cases such as predator avoidance. It is in the interests of parasites to increase the probability of host behaviours that benefit the parasite, and to suppress those which do not. Host behaviours result from the interactions of stimulus detection and assimilation systems (sense organs and central nervous system), communication mechanisms (nervous and endocrine systems) and effector mechanisms (muscles). An increasing number of cases are known in which parasites interfere with one or more of these components. Much of this is achieved by chemical interventions, and thus represents components of the environmental physiology of the parasite.

Parasite-induced host behaviours can be divided into three types:

1 Behaviours that help maintain the parasite in the host. An example is provided by the behaviour of portunid sand crabs infected with *Sacculina* (similar to the parasite in Fig. 15.7), which show increased cleaning behaviour, tending the parasite as if it were a crab egg mass stored in the same location. This behaviour is shown even by male crabs, which do not usually carry egg masses.

2 Behaviours that increase the probability of parasite transmission. For example, healthy individuals of the estuarine snail *Ilyanassa obsoleta* avoid being exposed by tidal flow. In contrast, individuals infected with the trematode *Gynaecotyla adunca* follow high tides and become stranded on beaches and sandbars, increasing the probability that larvae will reach the semi-terrestrial crustaceans that are the next host. In the acanthocephalan worm *Polymorphus paradoxus*, for which surface-feeding ducks are the definitive host and the freshwater amphipod *Gammarus* is an intermediate host, the parasite alters the amphipod behaviour from benthic scavenging well away from ducks to a suicidal pattern of skimming along the surface and clinging to surface weed (Fig. 15.13), where ducks readily eat them. The related parasite *P. marilis* causes the gammarids to swim in the lower waters of the pond, ensuring that they are picked up by a different required host, in this case diving ducks

Polymorphus paradoxus

Polymorphus marilis

Fig. 15.13 Behavioural manipulations by parasites: the interactions of duck predators and gammarid amphipod crustaceans infested by different species of acanthocephalan worms (*Polymorphus*). One species causes gammarids to stay at the surface where they are eaten by dabblers such as mallard (a), while another causes them to swim at depths where divers such as scaup eat them (b).

such as scaup. *Polymorphus* is thought to alter *Gammarus* behaviour by somehow interfering with 5-HT metabolism.

3 Behaviours that favour parasite survival after host death. Insect parasitoids may pupate within or on their dead host, and a number of species manipulate the host so that before it dies it reaches a location suitable for pupal survival. For example, conopid flies that parasitize bumble-bees ensure that just before death the bee burrows into soil (a highly unusual behaviour for a worker bumble-bee), providing the fly larva with a protected location in which to pupate.

15.7 Biotic interactions: host–parasite conflicts

15.7.1 *Host defences*

Host defence mechanisms and parasite responses have presumably been evolving since one species first invaded another, hundreds of millions of years ago. Thus parasites are part of the host environment, requiring the evolution of defence mechanisms. Similarly, as soon as hosts begin to evolve defences, there is strong selection on parasites to evolve countermeasures. Defences and countermeasures thus both fall within the remit of environmental physiology. This is particularly true for endoparasites, although several host defence mechanisms are able to reach ectoparasites on the outer body surface.

It is a common misconception that host defences are essentially limited to vertebrates, but this is certainly not the case. It is now known that a number of host defence systems are widespread phylogenetically, and probably of very ancient origin (Fig. 15.14 and Table 15.3), while others appear to be limited to vertebrates. As well as withstanding host defences, some parasites have evolved mechanisms that directly suppress host defences, or utilize them in their own development. Control of host immune responses involves the complex interplay of a variety of chemical messages, including hormones that influence host behaviour; the secretions of some parasites, perhaps primarily aimed at suppression of host immune responses, thus also modify host behaviours to the benefit of the parasite.

There are a number of defence systems that parasites have to withstand, and what follows is a very brief review of the environmental physiology of host defence.

Distinguishing 'self' from 'non-self', and initiation of defence

Defence in an organism relies on the ability to distinguish self cells from non-self cells. Recognition normally takes place in response to complex glycoproteins, termed **antigens**, present on the surface of cell membranes. A primitive recognition system in all animal cells is mediated by a number of proteins, including agglutinins, cecropins and lectins, in tissue fluids such as the blood. A second recognition system results from detection of non-self antigens by a range of receptors on the surface of host white blood cells (found in invertebrates as well as vertebrates; see Table 7.8). Vertebrates possess an additional highly complex self-recognition system, involving cell surface markers of the **major histocompatibility complex** (MHC). This involves

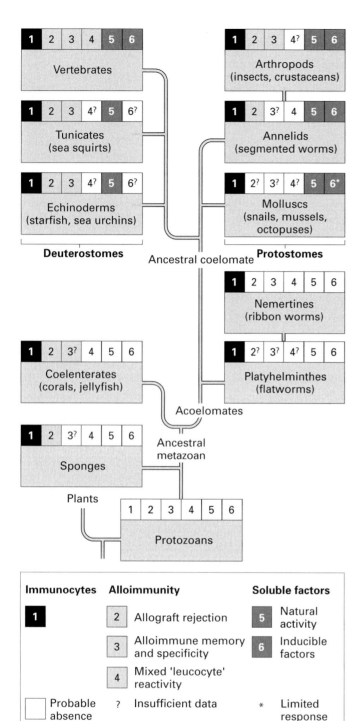

Fig. 15.14 Phylogenetic patterns in immune-type responses and defence systems in the Metazoa. Cells attacking non-self tissues (immunocytes) are ubiquitous, while more complex cellular immune responses and humoral responses (natural and inducible soluble factors) have a more limited distribution. See also Table 15.3. (From Roitt *et al.* 1998.)

incorporation into host cell membranes of gene products coded for by two groups of polymorphic loci—MHC class 1 (on all adult cells), and MHC class 2 (on a subset of adult cells). The presence of these surface markers gives cells a highly characteristic signature recognized by host defences. Either the presence of other surface antigens or the absence of the correct MHC signature can lead to attack by host defence systems.

Once tissues have been recognized as foreign, hosts may mount an **immune response** against them. Such a response can take several forms, broadly divided into those involving blood and other cells (cell-mediated responses), and those relying on the presence of compounds existing freely in host blood and tissue fluids (humoral responses). The initial reaction to foreign tissue is often a local swelling of host tissue, and local aggregation of host defensive cells. This defensive effort is achieved through signalling between host cells; avoidance of host defences by parasites can therefore be brought about by manipulating these signals.

Signalling between host cells during the immune response

Both vertebrates and invertebrates possess tissues, particularly white blood cells, whose primary function is the control of the behaviour and production of particular types of host defensive cell. In vertebrates this control includes instructions for the manufacture of highly specific defensive **antibodies** by host cells that never actually confront parasites directly. On a more local scale, signals between cells dictate the type of response mounted in the immediate vicinity of a particular parasite. Communication is achieved via cell surface receptors to specific messenger molecules, and differential stimulation of sets of receptors can produce very different responses by the same host cell. The most important chemical signals can broadly be divided into two types: cytokines, and hormones. The known diversity of these signals is far higher in vertebrates than in invertebrates.

Cytokines

These are polypeptides released by a variety of activated immune and non-immune cells (including endocrine glands), regulating host defences in both invertebrates and vertebrates. There are four basic sets of cytokines (Table 15.4), with a range of effects summarized in Fig. 15.15. They are best studied in vertebrates, but at least three types are already known in invertebrates, and the invertebrate interleukin IL-1 shows a marked diversity of functions: (i) stimulation of phagocytosis and proliferation in white blood cells; (ii) promotion of leucocyte aggregation leading to encapsulation of non-self tissue; (iii) cytotoxicity to some cell types; and (iv) induction of increased vascular permeability allowing greater accessibility of non-self tissue to host defences.

Other chemical mediators of immune responses

Vertebrate white blood cells possess surface receptors for many

Table 15.3 Evolution of the immune system. (Adapted from Rowley & Ratcliffe 1988.)

Evolutionary step or selection pressure	Immunological implications
Single-celled animals	Recognition and discrimination
Multicellularity (including colonial forms)	Histocompatibility system, recognition and short-term memory
Mesoderm and circulatory system, nutrition and defence as separate functions	Freely circulating and more diverse blood cell types, cellular immunity and erythrocytes
Cancer and viral infections associated with increasing complexity and longevity	Immunosurveillance of own cells for those that are infected or cancerous
Ancestral protovertebrates	Increased recognition and discriminatory powers?
Lower vertebrates: increased size, longer lifespan and reduced reproductive potential compared with invertebrates	True lymphocytes, lymphoid tissue and antibody production (IgM), longer-term memory
Emergence onto land, exposure to irradiation and development of high-pressure blood vascular systems	Bone marrow, additional antibody classes, T and B lymphocytes, lymphoid organs with increased complexity
Amniotes (reptiles, birds, mammals) with loss of free-living larval form	Advanced differentiation of immunocompetent cells allowing increased diversity and efficiency of immune system
Endothermy provides a more favourable environment for pathogens	Increased efficiency of immune system, integrated cellular and humoral responses, germinal centres in secondary lymphoid organs, lymph nodes
Viviparity with maternal–fetal interactions	Additional fine-tuning of immune system to avoid rejection by mother

Table 15.4 Types of cytokines in animals.

Type	Functions
Interferons	Defence of the body against viruses
Interleukins (17 different types)	Directing other cells to divide or differentiate Up- or down-regulation of cellular processes
Colony-stimulating factors	Division and differentiation of bone marrow, hence levels of specific white blood cell types
Tumour necrosis factors α and β, and transforming growth factor	Mediating inflammation and cytotoxic responses

blood cells can elevate or depress the production of cortico-steroids, while interleukins from pituitary and adrenal glands affect the numbers and activities of white blood cells.

hormones, neuropeptides and neurotransmitters, including steroids, epinephrine (adrenaline) and norepinephrine (noradrenaline), enkephalins, endorphins and others. Corticosteroids, endorphins and enkephalins are of particular importance for parasite environmental physiology because they suppress host immune responses. Cytokine- and hormone-mediated systems can interact in complex ways; for example, interleukins from

Defence systems

The defences mobilized following detection of non-self tissue can be broadly divided into two types. Firstly, there are non-specific, innate immune defences (also termed 'natural', 'non-adaptive' and 'non-anticipatory'). These tend to occur in response to a broad spectrum of parasites, with relatively general defensive properties. The defences have little or no 'memory', so prior exposure to a parasite does not result in more effective defence next time the host is attacked.

Secondly, there are induced, adaptive defences (also termed 'specific' or 'anticipatory'). These defences are more specific to particular parasites, and include those with a 'memory' that confer enhanced protection of the host after initial exposure to the parasite.

The division between these types is in part taxonomic. A range of non-specific systems are common to all Metazoa, while long-term memory in the immune system and the manufacture

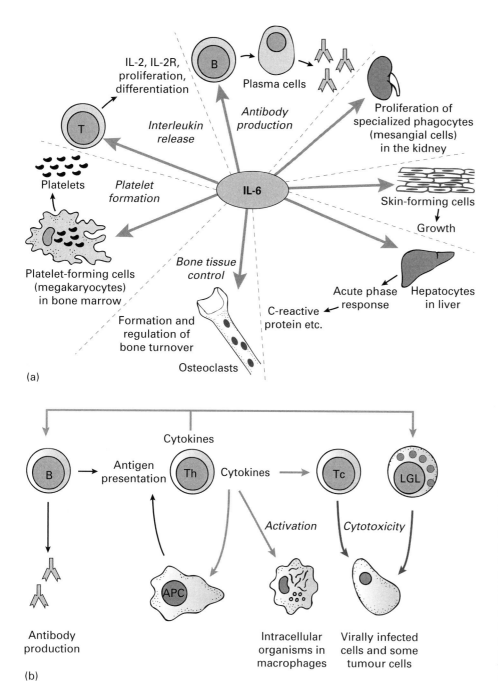

(a)

(b)

Fig. 15.15 (a) The range of effects of a cytokine, interleukin-6 (IL-6). (b) An example of the role of cytokines in mediating interactions between populations of white blood cells: here, between T-helper cells (Th), B cells (B), cytotoxic T cells (Tc) and large granular lymphocytes (LGL). Th cells are stimulated by antigen-presenting cells (APC) and B cells to produce cytokines; macrophages are activated to kill intracellular microorganisms; and Tc cells and large granular lymphocytes recognize and kill target cells. See text for more details. (From Roitt *et al.* 1998.)

of highly antigen-specific antibodies (immunoglobulins) are known only in vertebrates.

Non-specific immunity

Both cell-mediated and humoral non-specific immunity systems are widespread in both invertebrates and vertebrates. Cell-mediated defences include three reactions mediated by host cells—phagocytosis, natural killer cells, and encapsulation—which may be ubiquitous in the Metazoa. All are mediated by

white blood cells (leucocytes—see Fig. 15.16), which may have their evolutionary roots in the phagocytic cells responsible for intracellular digestion and transport of food in basal metazoans. Phagocytosis occurs in invertebrates and vertebrates alike, when phagocytes move towards and engulf invading material, which is digested intracellularly. Natural killer (NK) cells are large granular white blood cells that recognize and destroy non-self cells by releasing cytotoxins, possibly pore-forming proteins that become incorporated in the membrane of the non-self cell so that it either swells and explodes through osmotic inflow of

Leucocytes (white blood cells)　　　　　　　　　　　　**Other**

Lymphocytes　　　　Phagocytes　　　　Auxillary cells

Cell　B cell　T cell　Large granular lymphocyte　Mononuclear phagocyte　Neutrophil　Eosinophil　Basophil　Mast cell　Platelets　Tissue cells

Soluble mediators　Antibodies　Cytokines　Complement　Inflammatory mediators　Interferons Cytokines

Fig. 15.16 Summary of the main cell types involved in the immune system, the majority belonging to one of the classes of white blood cells (leucocytes). (From Roitt *et al.* 1998.)

Table 15.5 Cells and tissues of invertebrate immune systems. (From Roitt *et al.* 1998.)

Cells/tissues	Role(s) in immunity/physiology
Mucus, cuticle, shells, tests and/or gut barrier	Physico-chemical barriers to invasion
Five groups of free and sessile white blood cells	Mediate cellular and many of the humoral defence reactions
Progenitor cells	May act as stem cells for other cell types
Phagocytic cells	Phagocytosis, encapsulation, clotting, wound healing and killing
Haemostatic cells	Plasma gelation and clotting by cell aggregation; non-self recognition, lysozyme and agglutinin production
Nutritive cells	Encapsulation reactions and wound healing? Nutritive role?
Pigmented cells	Role in defence (if any) unknown; respiratory function
Fixed cells such as pericardial cells, nephrocytes or pore cells, etc.	Pinocytose colloids and small particulates; synthesize lysozyme (pericardial cells) and other antimicrobial factors?
Haemopoietic organs—well organized in some invertebrates	Haemopoiesis and phagocytosis; synthesize antimicrobial factors in a few animals
Fat body (insects), midgut and sinus lining cells (molluscs, crustaceans)	Synthesize immune proteins and agglutinins (fat body), phagocytosis (midgut cells), clearance of foreign particles

water, or ceases to function as a result of metabolite leakage. The third process of encapsulation involves surrounding non-self material with tightly adhering layers of host cells. Parasites may be killed when walled in by encapsulation; or this may be merely the first stage in a sequence of cell-mediated host responses resulting in the destruction of the parasite, including the action of lysosomal enzymes such as peroxidases and lysozymes. The cell-mediated defences of invertebrates are now known to involve a wide diversity of cell types (Table 15.5).

Humoral defences include toxins that are permanently present in the tissues of some metazoans as defences against predators and parasites alike; these occur in many sponges and tunicates, and in those herbivorous insects that store toxic secondary plant compounds in their tissues (Chapter 13). More generally, humoral defences common to vertebrates and invertebrates include proteins that adhere to and aggregate foreign material (e.g. agglutinins and lectins), antibacterial peptides (e.g. tachyplesins and defensins) and relatively broad-spectrum digestive enzymes (lysozymes). Although many of these broad-spectrum humoral defences are described as principally 'antibacterial', some confer effective protection against metazoan parasites. Humoral defences can be rapidly elevated in response to invasion (termed an 'induced response'), but the duration of response is generally only a few days. Agglutinins may show some specificity to markers on particular types of invading cell, and short-term memory in humoral immunity (up to several weeks) has been demonstrated in molluscs and insects.

Adaptive immunity

Adaptive immunity involves the development of highly specific host responses to particular parasite antigens. This is coupled with an ability of the host immune system to maintain a small population of cells specific to a given parasite antigen even when

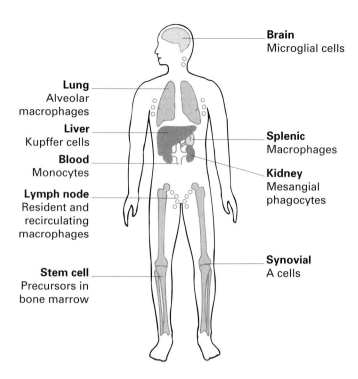

Fig. 15.17 Distribution of different types of phagocytic cells (collectively termed mononuclear phagocytes, or monocytes) in vertebrates. These are manufactured in the bone marrow, and pass out of the blood to form a range of tissue-specific morphs. (From Roitt *et al.* 1998.)

the infection is cleared. This immunological 'memory' allows later reinfections to be met with a rapid and highly specific defensive response.

As far as is known, such adaptive immunity is only present in vertebrates, and it represents a major breakthrough in controlling parasites. If rats are infected with the nematode *Nippostrongylus brasiliensis*, nematode larvae reach the gut and mature, beginning to release eggs about 5 days after infection. Egg production reaches a peak about 9 days after infection, and then decreases rapidly. After 21 days, all the adult worms have been expelled from the intestine. But if these exposed rats are then reinoculated with infective nematode larvae, very few adult worms reach maturity, and infections are cleared in 10 days rather than 21.

The specific leucocyte membrane receptors and antibodies involved contain a common structural motif, the **immunoglobulin domain**. In vertebrates there is a wide diversity of similar genes with this motif, almost certainly derived by multiple gene duplication events and sequence divergence from a single ancestral gene. Several cell types are involved:

1 Mononuclear phagocytes ('monocytes'). A range of different forms of these cells exists in different vertebrate tissues (Fig. 15.17), including mammalian monocytes, which become tissue macrophages. They recognize particular portions of antigen molecules, and if these are not compatible with the MHC signature of host cells, then the structure bearing them is

attacked. In addition to their phagocytic ability, macrophages can destroy other cells by secreting a range of toxic compounds. Cytotoxicity can be through release of proteolytic and other cytolytic enzymes, or through release of toxic compounds termed 'reactive oxygen intermediates' and 'reactive nitrogen intermediates'. Oxygen intermediates are formed by enzymes that reduce oxygen to the superoxide anion, OO^-. This in turn can give rise to toxic hydrogen peroxide and hydroxyl radicals. Monocytes in the blood use peroxidase to create more toxic bleach-like oxidants. Nitrogen intermediates are formed by the enzyme nitric acid synthase, which combines oxygen with nitrogen from the amino acid arginine to produce toxic nitric oxide. Under certain conditions, nitrogen intermediates and oxygen intermediates may interact to produce even more toxic peroxynitrites. These compounds are all highly oxidizing, and their application in defence is sometimes termed the 'oxidative burst'; it can destroy multicellular parasites such as nematode microfilaria larvae and the cercariae of schistosomes.

2 Neutrophils. These are the most common type of leucocyte in the blood, and are shorter-lived since after ingesting target material they die. Neutrophils are able to mount cytotoxic responses using toxic proteins and reactive O_2/N_2 intermediates, against targets labelled with antibodies. This defence is also effective against various metazoan parasites.

3 Eosinophils. These are a specialized group of leucocytes that can recognize and damage large extracellular parasites. Although capable of phagocytosis, they function principally by cytotoxic responses.

4 Lymphocytes. These are wholly responsible for the specific recognition of parasite antigens, central to adaptive immunity. There are two types, derived from bone marrow stem cells: T cells develop in the thymus gland, while B cells develop in the bone marrow in mammals. Both B and T cells accumulate in the lymph nodes and spleen, from where they respond to parasite antigens in the blood. T cells exist in two types with different surface markers; T-helper (Th) cells and cytotoxic T cells (Tc). Tc cells are involved mainly in recognizing antigens resulting from intracellular parasites, and killing infected cells, while Th cells are more important in responses to multicellular parasites. Th cells divide to produce a clonal population of antigen-specific cells, which then release cytokines to activate other cells in the immune system (see Fig. 15.15b). B cells also possess highly specific antigen recognition abilities, and once activated they manufacture and release large quantities of antibody.

Antibodies and complement

Immunoglobulin (Ig) antibodies are proteins composed of four peptides—two heavy (≈ 400 amino acids) and two light (≈ 200 amino acids) chains (Fig. 15.18) bound together to give a Y-shaped structure. The arms of the Y contain highly variable regions on both the heavy and light chains, and this is the part of the immunoglobulin involved in binding to a specific antigen. The base of the Y has a more constant amino acid sequence

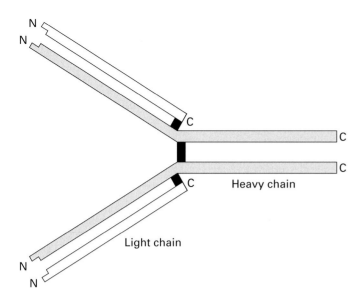

Fig. 15.18 The basic structure of immunoglobulins. The Y-shaped molecule consists of two identical light polypeptide chains and two identical heavy polypeptide chains linked together by disulphide bonds (black). N is the amino terminal and C the carboxy terminal of the chains. The variable regions are at the N-terminal ends of the heavy and light chains. (From Roitt *et al.* 1998.)

characteristic of particular immunoglobulin types, and is the part of the molecule recognized by host cells. In mammals, antibody molecules can be divided into four types, with characteristically different roles in host defence.

1 IgM. This is the first immunoglobulin secreted in response to infection. It is a five-molecule polymer that stimulates phagocytosis and also binds cells bearing the correct antigen together. It is secreted across the wall of the intestine, and is thus active against gut parasites.

2 IgG. After 4–5 days, IgM is replaced by IgG, which plays a major role in activating cytotoxicity by macrophages, neutrophils and eosinophils. It also activates a complex cascade of reactions termed the **complement system**. This consists of at least eight factors that interact to result in a membrane attack complex, forming pores in bacterial cell walls. Complement components also attract macrophages, enhance phagocytosis and increase local vascular permeability.

3 IgA. This is secreted in body fluids such as tears, saliva and milk, and boosts immunity of young infants. It is also secreted into the gut, and is resistant to enzymatic attack.

4 IgE. This is bound to cells termed mast cells. In response to antigens and cytokine messages from activated T cells, it causes the mast cells to release chemicals including histamines into the surrounding tissues and stimulates cytotoxicity by macrophages and eosinophils. The histamines cause direct damage to parasites, or disrupt the usual parasite microhabitat by tissue inflammation, or increase exposure of the parasite to other attacking agents as a result of increased vascular permeability.

Mast cells and IgE are important in defence against helminth infections, particularly for worms that live in contact with the gut lining. Activated mast cells cause shedding of the intestinal epithelium, aiding removal of the worm.

Antibodies are able to reach parasites in a wide diversity of locations in or on the host. While tissue and gut parasites are most obviously exposed to attack, parasites on the body surface can also be reached if they feed on blood. Fish produce antibodies in response to attack by ectoparasitic monogenean flukes, targeting them either via the blood they ingest or via the mucus on the fish skin. Mammals produce antibodies to anticoagulants in the saliva of arthropod blood-feeders such as ticks, leading to infiltration of the bite site by mast cells. Ticks attempting to feed from immune hosts may fail to engorge, fail to moult normally, or die of desiccation.

Coordination of host defences

In parasites with complex life cycles, each host provides a different defensive environment, and a single parasite species may be attacked by several of the immune responses described above. Often particular components of the immune system are of differing importance in different tissues, and where parasite larvae migrate through the host they may be exposed to a sequence of defences. This can be illustrated by the diversity of defences brought to bear by the human immune system on the larvae of *Schistosoma mansoni*. Attacks are directed against the larval tegument, whose antigens induce Th lymphocytes to initiate production of IgG and IgE antibodies by B cells. The antibodies bind to the larval tegument. Direct humoral attack on the worm is carried out by the complement system, both alone and in concert with worm-bound antibodies. A cytokine, interferon, secreted by Th lymphocytes, activates macrophages, which in turn secrete another cytokine, tumour necrosis factor. This activates neutrophils, eosinophils and platelets. All four blood cell types respond to the antibody-labelled worm tegument with cytotoxins. Macrophages and neutrophils attack the host tegument using toxic oxygen and nitrogen metabolites, while eosinophils attack using toxic proteins. Worm antigens also bind to specific IgE receptors on mast cells, and these in turn are involved in activating the eosinophils.

15.7.2 *Parasite countermeasures*

There are two kinds of protection available to parasites in or on their hosts. One type is associated with resisting chemical attack in extreme environments, particularly the gut. The second is associated with avoiding specific antiparasite aspects of host physiology. Some aspects of parasite protection are common to both types of environmental challenge.

Physical protection

Isolation of the parasite from external conditions confers

protection from all unfavourable aspects of the surrounding environment. Some parasite larvae resist host defences in their intermediate hosts by constructing thick-walled protective cysts, as in the nematode *Trichinella spiralis*. These cysts are broken down when the parasite enters the next host in its life cycle (normally by predation of its intermediate host), either by host- or parasite-generated enzymes. Adult parasites in their definitive host may avoid immune responses by inducing host tissues to form a cyst around them. This can perhaps be regarded as parasite adaptation to survive within a cyst following detection of a host encapsulation response.

Immunologically privileged sites

These are regions of the host body in which immune responses are reduced or lacking, usually due to the absence of host white blood cells. In vertebrates such sites include the eye lens (occupied by larvae of some digenean fish flukes) and parts of the central nervous system. In insects certain sites (including nerve ganglia, salivary glands, gonads, muscle fibres and the fat body) show far lower risks on encapsulation than others, and these too are exploited by parasites. Life-cycle stages that migrate through host tissues are more at risk from host immune defences than those inhabiting the gut lumen, and it is possible that exploitation of the gut as an adult site may represent avoidance of host immune responses.

After parasites enter the host, the speed with which they reach immunologically privileged sites may be crucial to their survival. The larvae of some endoparasitic tachinid flies hatch in the gut of their insect host where they are relatively safe from host defences; they then migrate rapidly from the gut wall towards the anterior nervous system of their host, a protected site. On the way they initiate a vigorous encapsulation response from their host, and only the most rapidly moving are able to reach safe sites in the nervous system. Tachinids also illustrate the fact that safe sites are sometimes exploited only for a while immediately after the parasite enters the host. After a period, generation of more effective defences by the tachinid larvae allows them to re-emerge into the better-defended host haemocoel without being encapsulated.

Detoxification of host humoral defences

Parasites are able to neutralize a very wide range of toxic host compounds, including toxins generally present in the body fluids, the complement system, toxic reactive oxygen and nitrogen intermediates, and antibodies. Where hosts sequester compounds to deter attackers, specialist parasites are usually able to detoxify these compounds, although generalist parasites may be susceptible. For example, larvae of moths of the genus *Zygaena* sequester cyanogenic compounds from the food plants and are avoided by most parasitoids, but a few specialist parasitoid wasps possess enzymes that detoxify these compounds.

Parasites show a range of responses to humoral products of the immune system. Metazoan parasites are often able to resist attack by the host complement system through possession of surface glycoproteins that mimic host molecules that inhibit the complement cascade. A number of parasites show enzyme-based defences against the reactive oxygen and nitrogen intermediates from cytotoxic phagocytes. The enzymes are the typical metazoan antioxidants (see Chapter 7) and may either be secreted from the parasite's body or bound to its surface. Thus the nematode *Onchocerca* secretes superoxide dismutase, filarial nematodes inhabiting the lymph system use a surface-bound glutathione peroxidase, and schistosomes use a surface-bound glutathione S-transferase.

Detoxification of antibodies is achieved in some nematodes and trematode flukes by cleaving the antibody molecules with proteases. The antibody molecules are split in such a way that the part of the heavy chains essential for detection of the antibody by host cells is removed.

Hiding and molecular mimicry

Intracellular parasites such as protistan plasmodia or coccidia are able to 'hide' inside host cells, such that their antigens are not exposed to host immune systems. Relatively few metazoan parasites are small enough to hide inside cells, but the filarial nematode *Trichinella* is an example, hiding inside the large cells of pig skeletal muscle.

Parasites are sometimes able to avoid detection by the host immune system in other ways, either coating their surface with host tissues, or disguising their tissues with host antigens (molecular disguise). *Schistosoma mansoni* absorbs host antigens soon after invading the human body, and so effectively disappears from host detectors of 'non-self'. The larva is able to absorb surface antigens corresponding to blood groups and the major histocompatibility complex. An immune response is mounted to the antigens presented immediately after invasion, but the larvae have 'disappeared' by the time the host response becomes effective. However, later invasions by cercariae are often destroyed by the activated defences during penetration of the human skin. Immunity to reinfection established by the first wave of parasites is termed 'concomitant immunity', and may help reduce over-exploitation of hosts by a given parasite species. Schistosomes are able to incorporate host antigens in both their snail and human hosts; their ability to take up host antigens is highly host specific, and cercaria larvae invading the incorrect host are rapidly destroyed. For example, the cercariae of marine schistosomes, normally living in sea birds, occasionally try to penetrate the skin of human swimmers or fishermen; they are halted by host defences and die in the skin, leading to a dermatitis condition known variously as 'swimmer's itch', 'clam-digger's itch', 'seabather's eruption' and 'weed itch'.

There is also some evidence that parasites can synthesize their own host-like molecules (molecular mimicry). For example, a few insect parasitoids appear able to coat their eggs with virus-like particles that mimic host proteins.

Changing the antigens presented to the host

If a particular set of surface antigens on a parasite have been changed by the time the host can produce antibodies to them, then the parasite can stay one step ahead of the host's immune system. Changing of parasite antigens is sometimes associated with moults or metamorphoses in the parasite life cycle. This strategy is illustrated by the nematode *Nippostrongylus* in rats. Infective larvae penetrate the skin, and migrate to the lungs in less than 24 h. In the lungs they moult, presenting a different set of stage-specific antigens. The next larval instar moves from the lungs to the intestine, where it moults again to present yet another set of surface antigens. In each case the change is quick enough to prevent effective host immune responses. However, the effect of memory in Th and B cells in vertebrate immune systems is that antibodies against each larval instar can be produced far more rapidly in subsequent infections, leading to levels of acquired host immunity.

A number of parasites are able to maintain a rapid turnover of surface antigens within a single life stage. The best-known example of this is provided by protozoan trypanosomes, the agents of sleeping sickness. By the time the host has mounted an immune response against one set of antigens, the parasite population consists of individuals with a new type of glycoprotein forming their surface coat. This type of strategy has been suggested to explain the low immune responses mounted against some nematodes, such as *Wuchereria bancrofti* (the agent of elephantiasis), which can survive in the lymph system even though this is the heart of the human immune system.

Active suppression of host defences

Some parasites reduce host defences by interfering with the structural and/or functional integrity of host leucocytes. Other parasites are able to divert the recruitment of attacking host blood cells. For example, the complement system activates neutrophils and macrophages through a protease enzyme, elastase. Tapeworms are able to secrete an elastase inhibitor, which prevents the generation of characteristic complement fragments following elastase digestion of targets, so hiding the parasite from host defensive cells.

Suppression of detection may also be involved in the defence of parasitoid eggs inside insect hosts. Within its egg, the parasitoid embryo is surrounded by a layer of cells called teratocytes, and in many species these continue to surround the young larva when the egg shell is lost. The surface of the teratocytes is covered in microvilli, and the cells are also supplied with extensive endoplasmic reticulum, suggesting a secretory function. Some property of teratocytes (as yet unidentified) suppresses host abilities to encapsulate the parasitoid; embryos from which these cells have been removed are quickly encapsulated.

A fascinating mechanism of host defence suppression in insect parasitoids involves the release into the host of polyDNA viruses by the egg-laying female. These viruses have a characteristic structure, either rod-shaped or spindle-shaped, and always contain double-stranded circular DNA molecules coiled into superhelices. The origin of polyDNA viruses remains a subject of debate. One attractive theory is that viruses are in fact parts of the parasitoid genome which are injected into the host to provide local delivery of venom. Evidence for this is the similarity between polyDNA viral sequence and other parasitoid genes expressed in the venom glands. After injection by the female parasitoid, the viruses rapidly infect a whole range of host cells (transcription products of the viral genome are detectable within 2 h of infection), and suppress the host encapsulation of parasitoid eggs.

Parasites in vertebrates show a further set of immunosuppressant adaptations. Some are able to cause direct damage to leucocytes through the release of toxic secretions. In others, the ability of the immune system to produce suitable antibodies is disrupted by the release of large quantities of materials that interfere with antigen processing by macrophages. Alternatively the activity of defensive cells in the immune system may be reduced by parasite-induced release of immunosuppressants, such as prostaglandins and other hormones. This may be achieved either through parasite manipulation of host secretions or (in the case of filarial nematodes and tapeworms) by secretion by the parasites themselves; *Schistosoma mansoni* secretes endorphins, which probably have immunosuppressant effects in both invertebrate and vertebrate hosts. Similar disruption of host defences may also be achieved by altering the balance of cytokines; filarial nematodes induce human hosts to release high concentrations of an IgG which inhibits protective IgE-mediated responses. Similarly, patients infected with schistosomiasis produce IgM and IgG antibodies that inhibit normal cytotoxic responses by neutrophils and eosinophils.

15.7.3 *Parasite exploitation of host defences*

Parasites are adept at turning the tables on their hosts' attempts at defence. For example, they may feed on the tissues produced as a host defence in response to them, or induce such complex and costly defences that the host is weakened and is more likely to be preyed upon by a predator that is itself a host in the parasite life cycle. Some parasites actually need part of the host immune response to survive; here an example is seen in remoras (marine fish) infected with the fluke *Dionchus remorae*, since the egg bundles of this ectoparasite induce the gill epithelium to which they attach to grow around the egg bundles, thus usefully anchoring them in place. Similar immune-response-based attachment of parasites is seen in the reaction around the head of acanthocephalan worms in vertebrate guts, which holds the worms firmly in place. The nematode *Strongyloides*, parasitic on rats, appears to use the host's immune system as a trigger for its own sexuality, only producing sexual phases from larvae that encounter hosts with acquired immune protection; here sex is perhaps being used as an adaptation to counter the rapid evolution of vertebrate immunity.

15.7.4 *Chemical warfare between parasites*

Parasites not only have to contend with host defences, but also with competition and attack from other parasites. Sometimes this occurs within species (even between siblings) and sometimes between species. Suppression of host defences by one parasite may lead to increased susceptibility to attack by others, an outcome which may well be detrimental to the first parasite.

Perhaps in response to this, parasites may deploy secondary host defences (termed 'parasite-mediated internal defences') to prevent damage to host or to first parasites by later opportunists. Such defences may be very general; the bactericidal anal secretions from some insect parasitoids are thought to protect their hosts (which are in effect living carcasses) against bacterial decay, prolonging the 'shelf-life' of the larval food supply. Other examples are more specific; some insect parasitoids lay eggs in previously parasitized hosts that inhibit the development of the resident parasitoids. Competition among insect parasitoid larvae is often determined by direct combat, but differing physiological tolerances for limiting resources, particularly oxygen, may also influence the outcome of these fights. For example, in ichneumonid wasps the eggs and young larvae have a much lower tolerance of oxygen depletion than older larvae, and older larvae are able to deplete oxygen levels in the host such that younger competitors die. Competition is also recorded for nematode and platyhelminth parasites in the gut. Two parasites that normally occupy the same region of intestine when alone may be spatially separated if both invade one host, the competitively inferior species being displaced to a new region. The acanthocephalan *Moniliformis* causes the cestode *Hymenolepis* to move further back in the rat intestine, but if the acanthocephalans are killed the cestodes migrate anteriorly. Physiological interference by the acanthocephalan, perhaps through depletion of essential nutrients or release of alcohol as a toxic excretory product, may be involved. However, just as so many parasites are able to overcome host defences, some parasites (regarded as specialized opportunists) are able to cope with other parasite species in the same host, and consistently parasitize previously infected hosts.

15.8 Conclusions

Parasites may be the single most important cause of reduced fecundity and/or early mortality in natural populations of animals, thus regulating populations and ecological balances at the most fundamental level. There has been an increasing realization that parasites may be one of the major factors shaping all host populations, both in terms of the ecological distribution (ranges occupied) and the cyclical behaviour of population numbers. This is linked to models of how host–parasite interactions may evolve. Are they in a sense mutualistic, the parasite and the host both seeking to minimize the damage done, or entirely aggressive, both host and parasite always seeking to outdo the other, and selection always operating to maximize individual success?

Early work on parasitism tended to stress mutualisms, giving examples like trypanosomes in Africa that are relatively benign to native cattle (with whom they are coevolved) but lethal to introduced cows and to humans. 'Older' parasites (in the evolutionary sense) do seem to be generally less virulent and have slower population increase. However, more recent studies have concentrated on aggressive models, seeing host–parasite relations as a coevolutionary arms race, in which each organism always tries to keep one step ahead of the opposition and have maximal growth rate and maximal virulence. Perhaps a balanced view now is that either model (or an intermediate model of a 'prudent parasite', doing only as much damage as the host can take) may apply in different circumstances. Whether a parasite evolves towards greater or lesser virulence will depend on its genetics and its epidemiology, including factors such as its lifespan relative to its host, its reproductive rate, its transmission mode, the host's population size and dispersion, and so on. In very simple terms, microparasites that have already infected a few hundred other hosts will not 'care' at all about killing off an original host, and nor will a parasite that achieves transmission by one host being eaten by the next. However, many macroparasites—helminths and insects, generally—may have relatively low reproductive rates, longer lifespans relative to their host, and more secure transmission systems, so that they need and benefit from much lesser virulence.

Given these various possibilities, different types of parasites can clearly have wildly different ecological consequences for their hosts. For highly virulent, strongly 'r-selected' microparasites, epidemics and mass mortalities are always possible, and such organisms invading new habitats or new hosts are often lethal. For more benign parasites, effects may be more subtle. For example, the ranges of hosts may be controlled by their susceptibilities to endemic parasites. One example here is the grazers of North America, where the bighorn sheep range is largely determined by the prevalence of lungworms (*Protostrongylus*). Similarly, rinderpest virus in Africa seems to control distribution of native ungulates: it is usually fatal to kudu and eland, has a moderate effect on giraffe and wildebeest, and is fairly harmless to gazelle and hippopotamus.

Parasitism has almost certainly had widespread ecological and evolutionary effects in terms of speciation; there are many examples now known where host and parasite phylogenies can be fairly precisely matched (see Fig. 15.2), strong indication that speciation in one has led to speciation in the other. There may thus be positive feedbacks whereby overall speciation rates are greatly increased; and the effect is so clear that specific parasite distributions are now sometimes used to trace the phylogenetic relationships of their hosts.

On a different kind of evolutionary scale, parasitism may also be the underlying reason for the evolution of sex. If this is true then parasitism has had one of the most vital of all effects on evolutionary rates and processes. Put simply, it is suggested that it was (and is) to escape the pressures of rapidly evolving parasites that other organisms had to use sexual reproduction, despite its rather heavy costs relative to asexual reproduction, just to maintain enough genetic diversity to ensure some resistant strains could keep up with or ahead of the coevolving parasites. The nematode example *Strongyloides*, referred to in the last section, gives some support to this idea.

Sexual selection and the strong divergence between morphologies and colours in the sexes may be a consequence of parasitism too. At a very basic level, evolution of very small sperm in one sex, where the genetic material is stripped of all cytoplasm, may have been selected for partly as a way of reducing possible parasite transmission. Furthermore, at the organism level it is suggested that males (usually) are bright and often gaudy not just to inform females of their quality, but perhaps specifically to indicate their resistance to parasitic loads and give females a way of assessing this resistance. So female choice is related to selecting good genetic resistance to parasites as a desirable feature in a partner.

Some of these ideas are still controversial, but there is no doubt that parasites are indeed coming to be seen as perhaps the major reason for much of the genetic polymorphism seen in host populations. It is hard to imagine what modern environments would be like *without* parasites, since their effects are immense; every organism in existence probably has a suite of various different kinds of parasites acting upon it, and to which it is reacting in contemporary time and over evolutionary time. Thus each organism's current status—its chemistry, behaviour, physiology and ecology—are all consequent upon its parasitological history, and its very genome is a product of coevolution with other infecting organisms. Host–parasite relations could well be the single most important factor affecting modern environments and their occupants.

Further reading

Books

Brooks, D.R. & McClennan, D.A. (1991) *Phylogeny, Ecology and Behavior.* University of Chicago Press, Chicago.

Bryant, C. & Behm, C. (1989) *Biochemical Adaptation in Parasites.* Chapman & Hall, London.

Cox, F.E.G. (1993) *Modern Parasitology: a Textbook of Parasitology,* 2nd edn. Blackwell Scientific Publications, Oxford.

Douglas, A.E. (1994) *Symbiotic Interactions.* Oxford University Press, Oxford.

Godfray, H.C. (1994) *Parasitoids: Behavioural and Evolutionary Ecology.* Princeton University Press, Princeton, NJ.

Marr, J. & Müller, M. (1995) *Biochemistry and Molecular Biology of Parasites.* Academic Press, London.

Matthews, B.E. (1998) *An Introduction to Parasitology.* Cambridge University Press, Cambridge.

Poulin, R. (1998) *Evolutionary Ecology of Parasites; from Individuals to Communities.* Chapman & Hall, London.

Rohde, K. (1993) *Ecology of Marine Parasites.* CAB International, Wallingford.

Roitt, I., Brostoff, J. & Male, D. (1998) *Immunology.* Mosby, London.

Rollinson, D. & Simpson, A.J.G. (1987) *The Biology of Schistosomes.* Academic Press, London.

Shorthouse, J. & Rohfritsch, O. (1992) *Biology of Insect-induced Galls.* Oxford University Press, New York.

Williams, M.A.J. (1994) *Plant Galls: Organisms, Interactions, Populations.* Clarendon Press, Oxford.

Reviews and scientific papers

Baudouin, M. (1975) Host castration as a parasitic strategy. *Evolution* **29**, 335–352.

Bayne, C.J. (1990) Phagocytosis and non-self recognition in invertebrates. *Bioscience* **40**, 723–731.

Beckage, N.E. (1985) Endocrine interactions between endoparasitic insects and their hosts. *Annual Review of Entomology* **30**, 371–413.

Brown, I.M. & Gaugler, R. (1997) Temperature and humidity influence emergence and survival of entomopathogenic nematodes. *Nematologica* **43**, 363–375.

Dunn, P.E. (1990) Humoral immunity in insects. *Bioscience* **40**, 738–744.

Gemmill, A.W., Viney, M.E. & Read, A.F. (1997) Host immune status determines sexuality in a parasitic nematode. *Evolution* **51**, 393–401.

Lawrence, P.O. (1986) Host–parasite hormonal interactions: an overview. *Journal of Insect Biochemistry and Physiology* **32**, 295–298.

Loker, E.S. (1994) On being a parasite in an invertebrate host—a short survival course. *Journal of Parasitology* **80**, 728–747.

Moore, J. (1984) Parasites that change the behaviour of their host. *Scientific American* **250**, 82–89.

Page, R.D.M. (1993) Parasites, phylogeny and cospeciation. *International Journal of Parasitology* **23**, 499–506.

Slansky, F. (1986) Nutritional ecology of endoparasitic insects and their hosts: an overview. *Journal of Insect Physiology* **32**, 255–261.

References

Agnew, D.J. & Taylor, A.C. (1986) Seasonal and diel variations of some physico-chemical parameters of boulder-shore habitats. *Ophelia* **25**, 83–95.

Ahearn, G.A. (1970) The control of water loss in desert tenebrionid beetles. *Journal of Experimental Biology* **53**, 573–595.

Alberts, B., Bray, D., Lewis, J., Raff, M., Roberts, K. & Watson, J.D. (1994) *Molecular Biology of the Cell*. Garland Publishing, New York.

Aleksiuk, M. (1971) Temperature-dependent shifts in the metabolism of a cool temperate reptile, *Thamnophis sirtalis parietalis*. *Comparative Biochemistry and Physiology A* **39**, 459–503.

Alexander, R.McN. (1982) *Locomotion of Animals*. Blackie, Glasgow.

Alexander, R.McN. (1999) *Energy for Animal Life*. Oxford University Press, Oxford.

Alexander, W.R., Southgate, B.A. & Bassindale, R. (1935) Survey of the River Tees—the estuary, chemical and biological. *DSIR Water Pollution Research Technical Paper 5*.

Baker, M.A. (1982) Brain cooling in endotherms in heat and exercise. *Annual Review of Physiology* **44**, 85–96.

Ballantine, W.J. (1961) A biologically-defined exposure scale for the comparative description of rocky shores. *Field Studies* **1**, 1–19.

Barnes, R.S.K. (1981) An experimental study of the pattern and significance of the climbing behaviour of *Hydrobia ulvae* (Pennant). *Journal of the Marine Biological Association of the UK* **61**, 285–299.

Bartholomew, G.A. (1964) The roles of physiology and behaviour in the maintenance of homeostasis in the desert environment. *Symposia of the Society for Experimental Biology* **18**, 7–29.

Bartholomew, G.A. & Lasiewski, R.C. (1965) Heating and cooling rates, heart rate and simulated diving in the Galapagos marine iguana. *Comparative Biochemistry and Physiology* **16**, 575–582.

Bayne, B.L. & Hawkins, A.J.S. (1997) Protein metabolism, the costs of growth, and genomic heterozygosity: Experiments with the mussel *Mytilus galloprovincialis* Lmk. *Physiological Zoology* **70**, 391–402.

Behan-Martin, M., Bowler, K., Jones, G. & Cossins, A.R. (1993) A near-perfect temperature adaptation of bilayer order in vertebrate brain membranes. *Biochimica et Biophysica Acta* **1151**, 216–222.

Beniston, M. (ed.) (1994) *Mountain Environments in Changing Climates*. Routledge, London.

Benzinger, T.H. (1961) The diminution of thermoregulatory sweating during cold reception at the skin. *Proceedings of the National Academy of Sciences of the USA* **47**, 1683–1688.

Binyon, J. (1961) Salinity tolerance and permeability for water of the starfish *Asteria rubens* L. *Journal of the Marine Biological Association of the UK* **41**, 161–174.

Bliss, D.E. (1979) From sea to tree: saga of a land crab. *American Zoologist* **19**, 385–410.

Bond, T.F., Kelly, C.F., Morrison, S.R. & Pereira, N. (1967) Solar atmospheric and terrestrial radiation received by shaded and unshaded animals. *Transactions of the American Society of Agricultural Engineers* **10**, 622–625.

Bone, Q., Marshall, N.B. & Blaxter, J.H.S. (1982) *Biology of Fishes*. Blackie, Glasgow.

Bosch, I., Beauchamp, K.A., Steele, M.E. & Pearse, J.S. (1987) Development, metamorphosis and seasonal abundance of embryos and larvae of the Antarctic sea-urchin *Sterechinus neumayeri*. *Biological Bulletin* **173**, 126–135.

Bouverot, P. (1985) *Adaptation to Altitude-Hypoxia in Vertebrates*. Springer-Verlag, Berlin.

Brett, J. (1956) Some principles of the thermal requirements of fish. *Quarterly Review of Biology* **31**, 75–87.

Brix, O., Borgund, S., Barnung, T., Colosimo, A. & Giardino, B. (1989) Endothermic oxygenation of haenocyanin in the krill *Neganyctiphanes norbegica*. *FEBS Letters* **247**, 177–180.

Brooks, J.L. (1947) Turbulence as an environmental determinant of relative growth in *Daphnia*. *Proccedings of the National Academy of Sciences of the USA* **33**, 141–148.

Brooks, S.E. & Shorthouse, J.D. (1998) Developmental morphology of stem galls of *Diplolepis nodulosa* (Hymenoptera: Cynipidae) and those modified by the inquiline *Periclistus pirata* (Hymenoptera: Cynipidae) on *Rosa blanda* (Rosaceae). *Canadian Journal of Botany* **76**, 365–381.

Brusca, R.C. & Brusca, G.J. (1990) *Invertebrates*. Sinauer Associates, Sunderland, MA.

Burky, A.J., Pacheco, J. & Pereyra, E. (1972) Temperature, water and respiratory regimes of an amphibious snail *Pomacea urceus* from the Venezuelan savannah. *Biological Bulletin* **143**, 304–316.

Bursell, E. (1957) Spiracular control of water loss in the tsetse fly. *Proceedings of the Royal Entomological Society (London)* **32**, 21–29.

Cannon, R.J.C. & Block, W. (1988) Cold tolerance of microarthropods. *Biological Review* **63**, 23–77.

Carey, F.G., Teal, J.M., Kanwisher, J.W. & Lawson, K.D. (1971) Warm-blooded fish. *American Zoologist* **11**, 137–145.

Carter, R.W.G. (1988) *Coastal Environments*. Academic Press, London.

Chan, S.L., Fletcher, G.L. & Hew, C.L. (1993) Control of antifreeze protein gene expression in winter flounder. In: *Biochemistry and Molecular Biology of Fishes*, Vol. 2 (eds P.W. Hochachka & T.P. Mommsen). Elsevier, Amsterdam.

Cheung, W.W.K. & Marshall, A.T. (1973) Water and ion regulation in cicadas in relation to xylem feeding. *Journal of Insect Physiology* **19**, 1801–1816.

Contreras, L.C. (1984) Bioenergetics of huddling: test of a psycho-physiological hypothesis. *Journal of Mammalogy* **65**, 256–262.

Cossins, A.R. & Bowler, K. (1987) *Temperature Biology of Animals*. Chapman & Hall, London.

Cossins, A.R. & Sheterline, P.S. (1983) *Cellular Acclimatisation to Environmental Change*. Cambridge University Press, Cambridge.

Croghan, P.C. & Lockwood, A.P.M. (1968) Ionic regulation of the Baltic race and freshwater races of the isopod *Mesidotea (Saduria) entomon* (L). *Journal of Experimental Biology* **48**, 141–158.

Crowley, T.J. & North, J.R. (1991) *Palaeoclimatology*. Oxford University Press, New York.

Damuth, J. (1991) Ecology—of size and abundance. *Nature* **351**, 268–269.

Davenport, J., Blackstock, N., Davies, D.A. & Yarrington, M. (1987) Observations on the physiology and integumentary structures of the Antarctic pycnogonid *Decalopoda australis. Journal of Zoology* **211**, 451–465.

Davenport, J.A. (1992) *Animal Life at Low Temperature*. Chapman & Hall, London.

de Fur, P.L. & Mangum, C.P. (1979) The effects of environmental variables on the heart rates of invertebrates. *Comparative Biochemistry and Physiology A* **62**, 288–300.

Dejours, P. (1981) *Principles of Comparative Respiratory Physiology*. Elsevier/Holland Biomedical, Amsterdam.

Dejours, P., Bolis, L., Taylor, C.R. & Weibel, E.R. (eds) (1987) *Comparative Physiology: Life in Water and on Land*. Liviana Press, Padova.

Denlinger, D.L. (1985) Hormonal control of diapause. In: *Comprehensive Insect Physiology, Biochemistry and Pharmacology*, Vol. 8. *Endocrinology II* (eds G.A. Kerkut & L.I. Gilbert). Pergamon Press, Oxford.

Denton, E.J. (1961) The buoyancy of fish and cephalopods. *Progress in Biophysics and Biophysical Chemistry* **11**, 178–233.

Denton, E.J. & Gilpin-Brown, J.B. (1961) The distribution of gas and liquid within the cuttlebone. *Journal of the Marine Biological Association of the UK* **41**, 365–381.

Denton, E.J. & Shaw, T.I. (1961) The buoyancy of gelatinous marine animals. *Journal of Physiology* **162**, 14P–15P.

Desbruyeres, D. & Segonzac, M. (eds) (1997) *Handbook of Deep-Sea Hydrothermal Vent Fauna*. IFREMER, Plouzane.

de Vries, A.L. (1980) Biological antifreeze and survival in freezing environments. In: *Animals and Environmental Fitness* (ed. R. Gilles). Pergamon Press, Oxford.

de Vries, A.L. (1988) The role of antifreeze glycopeptides and peptides in the freezing aboidance of Antarctic fishes. *Comparative Biochemistry and Physiology B* **90**, 611–621.

Doake, C.S.M. & Vaughan, D.G. (1991) Rapid disintegration of the Wordie Ice Shelf in response to atmospheric warming. *Nature* **350**, 328–330.

Douwes, P. (1976) Activity in *Heodes virgaureae* in relation to air temperature, solar radiation and time of day. *Oecologia* **22**, 287–298.

Dreisig, H. (1980) Daily activity, thermoregulation and water loss in the tiger beetle *Cicindela hybrida. Oecologia* **44**, 376–389.

Eastman, J.T. (1993) Antifeeze glycopeptides. In: *Antarctic Fish Biology, Evolution in a Unique Environment*. Academic Press, London.

Eastman, J.T. & De Vries, A.L. (1986) Antarctic fishes. *Scientific American* **255**, 106–115.

Eckert, R., Randall, D. & Augustine, G. (1988) *Animal Physiology*. Freeman, New York.

Edney, E.B. (1947) Laboratory studies on the bionomics of the rat fleas *Xenopsylla brasiliensis* Baker and *X. cheopis* Roths. II Water relations during the cocoon period. *Bulletin of Entomological Research* **38**, 263–280.

Edney, E.B. (1953) The temperature of woodlice in the sun. *Journal of Experimental Biology* **30**, 331–349.

Edney, E.B. (1966) Absorption of water vapour from unsaturated air by *Arenivaga* sp. (Polphagidae, Dictyoptera). *Comparative Biochemistry and Physiology* **19**, 387–408.

Edney, E.B. (1971) The body temperature of tenebrionid beetles in the Namib desert of southern Africa. *Journal of Experimental Biology* **55**, 253–272.

Edney, E.B. (1980) The components of water balance. In: *Insect Biology in the Future* (eds M. Locke & D.A. Smith). Academic Press, New York.

Elliott, J.M. (1994) *Quantitative Ecology and the Brown Trout*. Oxford University Press, Oxford.

Else, P.L. & Hulbert, A.J. (1985) An allometric comparison of the mitochondria of mammalian and reptilian tissues: implications for the evolution of endothermy. *Journal of Comparative Physiology* **156**, 3–11.

Farlow, J.O. (1976) A consideration of the trophic dynamics of a late Cretaceous large dinosaur community. *Ecology* **57**, 841–857.

Farrow, R.A. (1990) Flight and migration in acridoids. In: *Biology of Grasshoppers* (eds R.F. Chapman & A. Joern). Wiley, New York.

Feder, M.E. (1987) The analysis of physiological diversity: the prospects for pattern documentation and general questions in ecological physiology. In: *New Directions in Ecological Physiology* (eds M.E. Feder, A.F. Bennett, W.W. Burggren & R.B. Huey). Cambridge University Press, Cambridge.

Flock, A. (1965) Transducing mechanisms in the lateral line canal organ receptors. *Cold Spring Harbor Laboratory Symposia on Quantitative Biology* **30**, 133–144.

Friedli, H., Lotscher, H., Oeschger, H., Siegenthaler, U. & Stauffer, B. (1986) Ice core record of the $^{13}C/^{12}C$ ratio of atmospheric CO_2 in the past two centuries. *Nature* **324**, 237–238.

Fry, F.E.J. & Hochachka, P.W. (1970) Fish. In: *Comparative Physiology of Thermoregulation* (ed. G.C. Whittow). Academic Press, New York.

Geiser, F. (1988) Reduction of metabolism during hibernation and daily torpor in mammals and birds: temperature effect or physiological inhibition? *Journal of Comparative Physiology B* **158**, 25–37.

Gilles, R. (1979) Intracellular organic osmotic effectors. In: *Mechanisms of Osmoregulation in Animals* (ed. R. Gilles). John Wiley & Sons, New York.

Goldsworthy, G.J., Robinson, J. & Mordue, W. (1981) *Endocrinology*. Blackie, Glasgow.

Gracey, A.Y., Logue, J., Tiku, P.E. & Cossins, A.R. (1996) Adaptation of biological membranes to temperature: biophysical perspectives and molecular mechanisms. In: *Animals and Temperature* (eds I.A. Johnston & A.F. Bennett), pp. 1–21. Cambridge University Press, Cambridge.

Graedel, T.E. & Crutzen, P.J. (1993) *Atmospheric Change—An Earth System Perspective*. Freeman, New York.

Green, J. (1968) *The Biology of Estuarine Environments*. University of Washington Press, Seattle.

Gupta, B.L., Wall, B.J., Oschmann, J.L. & Hall, T.A. (1980) Direct microprobe evidence for local concentration gradients and recycling of electrolytes during fluid absorption in the rectal papillae of *Calliphora. Journal of Experimental Biology* **88**, 21–48.

Hadley, N.F. (1970) Micrometeorology and energy exchange in two desert arthropods. *Ecology* **51**, 434–444.

Hadley, N.F. & Quinlan, M. (1993) Discontinuous carbon dioxide release in the Eastern lubber grasshopper *Romalea guttata* and its effect on respiratory transpiration. *Journal of Experimental Biology* **177**, 169–180.

Haefner, P.A. Jr (1969) Temperature and salinity tolerance of the sand shrimp, *Crangon septemspinosa* Say. *Physiological Zoology* **42**, 388–397.

Haefner, P.A. Jr (1970) The effect of low dissolved oxygen concentration on temperature-salinity tolerance of the sand shrimp *Crangon septemspinosa. Physiological Zoology* **43**, 30–37.

Hafner, M.S. & Page, R.D.M. (1995) Molecular phylogenies and host–parasite co-speciation: gophers and lice as a model system. *Philosophical Transactions of the Royal Society of London B* **349**, 77–83.

Hall, F.G., Dill, D.B. & Barron, E.S.G. (1936) Comparative physiology in high altitudes. *Journal of Cellular and Comparative Physiology* **8**, 301–313.

Hamilton, W.J. (1975) Colouration and its thermal consequences for diurnal desert insects. In: *Environmental Physiology of Desert Organisms* (ed. N.F. Hadley). Dowden, Hutchinson & Ross, New York.

Hansen, T. (1973) Variation in glycerol content in relation to cold-hardiness in the larvae of *Petrova resinella* L. (Lepidoptera, Torticidae). *Eesti Teaduste Akadeemia. Toimetised. Bioloogia.* **22**, 105–112.

Hardy, J.D. (1949) Heat transfer. In: *Physiology of Heat Regulation and the Science of Clothing* (ed. L.H. Newburgh). Saunders, Philadelphia.

Hart, J.S. (1956) Seasonal changes in insulation of the fur. *Canadian Journal of Zoology* **34**, 53–57.

Haugaard, N. & Irving, L. (1943) The influence of temperature upon the oxygen consumption of the cunner (*Tautogolabrus adspersus*) in summer and in winter. *Journal of Cellular and Comparative Physiology* **21**, 19–26.

Hawkins, A.J.S., Bayne, B.L., Day, A.J., Rusin, J. & Worrall, C.M. (1989) Genotype-dependent interrelations between energy metabolism, protein metabolism and fitness. In: *Reproduction, Genetics and Distributions of Marine Organisms* (eds J.S. Ryland & P.A. Tyler). Olsen & Olsen, Fredensborg.

Heinrich, B. (1975) Thermoregulation in bumblebees II. Energetics of warm-up and free flight. *Journal of Comparative Physiology* **96**, 155–166.

Heinrich, B. (1976) Heat exchange in relation to blood flow between thorax and abdomen in bumblebees. *Journal of Experimental Biology* **64**, 561–585.

Heldmaier, G., Steinlechner, S., Ruf, T., Wiesinger, H. & Klingenspor, M. (1989) Photoperiod and thermoregulation in vertebrates—body-temperature rhythms and thermogenic acclimation. *Journal of Biological Rhythms* **4**, 251–265.

Henschel, J.R. (1990) Spiders wheel to escape. *South African Journal of Science* **86**, 151–152.

Henwood, K. (1975) A field-tested thermoregulation model for two diurnal Namib Desert tenebrionid beetles. *Ecology* **56**, 1329–1342.

Herreid, C.F. & Fourtner, C.R. (1981) *Locomotion and Energetics in Arthropods.* Plenum Press, New York.

Herreid, C.F. & Schmidt-Nielsen, K. (1966) Oxygen consumption, temperature and water loss in bats from different environments. *American Journal of Physiology* **211**, 1108–1112.

Herring, P.J. & Martin, J.G. (1978) Bioluminescence in fishes. In: *Bioluminescence in Action* (ed. P.J. Herring). Academic Press, London.

Heusner, A.A. (1982) Energy metabolism and body size: is the 0.75 mass exponent of Kleiber's equation a statistical artefact? *Respiration Physiology* **48**, 1–12.

Hickman, C.P. & Trump, B.F. (1969) The kidney. In: *Fish Physiology*, Vol. 1 (eds W.S. Hoar & B.J. Randall). Academic Press, London.

Hillenius, W.J. (1994) Turbinates in therapsids—evidence for late Permian origins of mammalian endothermy. *Evolution* **48**, 207–229.

Hinde, R.A. (1982) *Ethology.* Fontana Press, London.

Hinton, H.E. (1957) The structure and function of the spiracular gills of the fly *Taphrophila vitripennis. Proceedings of the Royal Society of London B* **147**, 90–120.

Hinton, H.E. (1981) *Biology of Insect Eggs.* Pergamon Press, Oxford.

Hochachka, P.W. & Somero, G.N. (1973) *Strategies of Biochemical Adaptation.* Saunders, Philadelphia.

Hochachka, P.W., Emmett, B. & Suarez, R.K. (1988) Limits and constraints on the scaling of oxidative and glycolytic enzymes. *Canadian Journal of Zoology* **66**, 1128–1138.

Hofmeyr, M.D. & Louw, G.N. (1987) Thermoregulation, pelage conductance and renal function in the desert-adapted springbock *Antidorcas marsupialis. Journal of Arid Environments* **13**, 137–151.

Hoglund, G., Hamdorf, K. & Rosner, G. (1973) Trichromatic visual systems in an insect and its sensitivity control by blue light. *Journal of Comparative Physiology* **86**, 265–279.

Holdridge, L.R. (1947) Determination of the world's plant formations from sample climate data. *Science* **105**, 367–368.

Holm, E. & Edney, E.B. (1973) Daily activity of Namib Desert beetles in relation to climate. *Ecology* **54**, 45–56.

Hoppeler, H. & Weibel, E.R. (1998) Limits for oxygen and substrate transport in mammals. *Journal of Experimental Biology* **201**, 1051–1064.

Horne, J. & Goldman, C.R. (1994) *Limnology.* McGraw Hill, London.

Hoyt, D.F. & Taylor, C.R. (1981) Gait and the energetics of locomotion in horses. *Nature* **292**, 239–240.

Huey, R.B. (1987) Phylogeny, history and the comparative method. In: *New Directions in Ecological Physiology* (eds M.E. Feder, A.F. Bennett, W.W. Burggren & R.B. Huey). Cambridge University Press, Cambridge.

Humphreys, W.F. (1979) Production and respiration in animal populations. *Journal of Animal Ecology* **48**, 427–453.

Hutchinson, V.H., Dowling, H.G. & Vinegar, A. (1966) Thermoregulation in a brooding female Indian python *Python molurus bivittatus. Science* **151**, 694–695.

Hynes, H.B.N. (1960) *The Biology of Polluted Waters.* Liverpool University Press, Liverpool.

Ingmanson, D.E. & Wallace, E.J. (1989) *Oceanography.* Wadsworth Publications Co., Belmont, CA.

Innes, A.J. & Taylor, E.W. (1986) The evolution of air-breathing in crustaceans: a functional analysis of branchial, cutaneous and pulmonary gas exchange. *Comparative Biochemistry and Physiology A* **85**, 621–637.

Irving, L. & Krogh, J. (1955) Temperature of skin in the Arctic as a regulator of heat. *Journal of Applied Physiology* **7**, 355–364.

Janzen, F.J. & Paukstis, G.L. (1991) Environmental sex determination in reptiles: ecology, evolution, and experimental design. *Quarterly Review of Biology* **66**, 149–179.

Jerison, H.J. (1970) Brain evolution: new light on old principles. *Science* **170**, 1224–1225.

Jobling, M. (1993) Bioenergetics: feed intake and energy partitioning. In: *Fish Ecophysiology* (eds J.C. Rankin & F.B. Jensen). Chapman & Hall, London.

Jobling, M. & Davis, P.S. (1980) Effects of feeding on metabolic rate and the specific dynamic action in plaice, *Pleuronectes platessa. Journal of Fish Biology* **16**, 629–638.

Johnston, I.A. (1990) Cold adaptation in marine organisms. *Philosophical Transactions of the Royal Society of London B* **326**, 655–667.

Johnston, I.A. & Battram, J. (1993) Feeding energetics and metabolism in demersal fish species from Antarctic, temperate and tropical environments. *Marine Biology* **115**, 7–14.

Johnston, I.A. & Walesby, N.J. (1977) Molecular mechanisms of temperature adaptation in fish myofibrillar adenosine triphosphatases. *Journal of Comparative Physiology* **119**, 195–206.

Jusiak, R. & Poczopko, P. (1972) The effect of ambient temperature and season on total and tissue metabolism of the frog *Rana esculenta* L. *Bulletin of the Academy Sci Pol Sci Ser Biol* **20**, 523–529.

Kaestner, A. (1980) *Invertebrate Zoology*, Vol. 3, *Crustacea*. Wiley, New York.

Kahl, M.P. (1963) Thermoregulation in the wood stork with special reference to the role of the legs. *Physiological Zoology* **36**, 141–151.

Kaufmann, W.R. & Phillips, J.R. (1973) Ion and water balance in the ixodid tick *Dermacentor andersonii. Journal of Experimental Biology* **58**, 523–547.

Kerr, R.A. (1988) Linking earth, ocean and air. *Science* **239**, 259–260.

Kevan, P.G. (1975) Sun-tracking solar furnaces in high arctic flowers: significance for pollination and insects. *Science* **189**, 723–726.

Kilgore, D.K. & Schmidt-Nielsen, K. (1975) Heat loss from ducks' feet emersed in cold water. *Condor* **77**, 475–478.

Kleiber, M. (1932) Body size and animal metabolism. *Hilgardia* **6**, 315–333.

Kleiber, M. (1961) *The Fire of Life: An Introduction to Animal Energetics.* Wiley, New York.

Kram, R. & Taylor, C.R. (1990) Energetics of running: a new perspective. *Nature* **346**, 265–267.

Lasiewski, R.C. & Snyder, G.K. (1969) Responses to high temperature in nestling double-crested and pelagic cormorants. *Auk* **86**, 529–540.

Lees, A.D. (1947) Transpiration and the structure of the epicuticle in ticks. *Journal of Experimental Biology* **23**, 397–410.

Levinton, J.S. (1995) *Marine Biology*. Oxford University Press, New York.

Levy, R.I. & Schneiderman, H.A. (1966) Discontinuous respiration in insects. IV. Changes in intratracheal pressure during the respiratory cycle of silkworm pupae. *Journal of Insect Physiology* **12**, 465–492.

Lighton, J.R.B. (1988) Discontinuous CO_2 emission in a small insect, the formicine ant *Camponotus vicinus*. *Journal of Experimental Biology* **134**, 363–376.

Lighton, J.R.B. (1989) Individual and whole-colony respiration in an African formicine ant. *Functional Ecology* **3**, 523–530.

Little, C. (1983) *The Colonisation of Land*. Cambridge University Press, Cambridge.

Little, C. (1990) *The Terrestrial Invasion; An Ecophysiological Approach to the Origin of Land Animals*. Cambridge University Press, Cambridge.

Lockwood, A.P.M. (1961) The urine of *G. duebeni* and *G. pulex*. *Journal of Experimental Biology* **38**, 647–658.

Lockwood, A.P.M., Sheader, M. & Williams, J.A. (1996) Life in estuaries, salt marshes and coastal waters. In: *Oceanography* (eds C.P. Summerhayes & S.A. Thorpe). Manson, London.

Louw, G. (1993) *Physiological Animal Ecology*. Longman, Harlow.

Lutz, P.L. (1980) On the oxygen affinity of bird blood. *American Zoologist* **20**, 187–198.

Maddrell, S.H.P. (1971) The mechanisms of insect excretory systems. *Advances in Insect Physiology* **8**, 199–331.

Maddrell, S.H.P. & O'Donnell, M.J. (1992) Insect Malphigian tubules— V-ATPase action in ion and fluid transport. *Journal of Experimental Biology* **172**, 4117–4429.

Maetz, J. (1973) Na^+/NH_4^+, Na^+/H^+ exchanges and NH_3 movements across the gill of *Carassius auratus*. *Journal of Experimental Biology* **58**, 255–275.

Manton, S.M. (1952) The evolution of arthropodan locomotory mechanisms. II. General introduction. *Journal of the Linnean Society of London* **42**, 93–117.

Marsh, A.C. (1985) Thermal responses and temperature balance in a desert ant, *Ocymyrmex barbiger*. *Physiological Zoology* **58**, 629–636.

Marsh, B.A. & Branch, G.M. (1979) Circadian and circatidal rhythms of oxygen consumption in the sandy-beach isopod *Tylos granulatus* Krauss. *Journal of Experimental Marine Biology and Ecology* **37**, 77–89.

Martin, R.D. (1980) Body temperature, activity and energy costs. *Nature* **283**, 335–336.

Matthews, B.E. (1998) *An Introduction to Parasitology*. Cambridge University Press, Cambridge.

Mattson, W.J. Jr (1980) Herbivory in relation to plant nitrogen content. *Annual Review of Ecology and Systematics* **11**, 119–161.

May, M.L. (1976) Warming rates as a function of body size in periodic ectotherms. *Journal of Comparative Physiology* **111**, 55–70.

Micklin, P.P. (1988) Desiccation of the Aral Sea—a water management disaster in the Soviet Union. *Science* **241**, 1170–1175.

Minnich, J.E. (1982) The use of water. In: *Biology of the Reptilia*, Vol. 12 (eds C. Gans & F.H. Hough). Academic Press, London.

Mintzer, C. (1992) *Confronting Climate Change*. Cambridge University Press, Cambridge.

Mitchell, D., Laburn, H.P., Nijland, M.J.M., Zurovsky, Y. & Mitchell, G. (1987) Selective brain cooling and survival. *South African Journal of Sciences* **83**, 598–604.

Moore, P.G. & Francis, C.H. (1985) On the water relations and osmoregulation of the beach-hopper *Orchestia gamarellus*. *Journal of Experimental Marine Biology and Ecology* **94**, 131–150.

Moss, B. (1980) *Ecology of Fresh Waters: Man & Medium*. Blackwell Scientific Publications, Oxford.

Murrish, D.E. & Schmidt-Nielsen, K. (1970) Water transport in the cloaca of lizards: active or passive? *Science* **170**, 324–326.

Nachtigall, W. (1983) The biophysics of locomotion in water. In: *Biophysics*. (eds W. Hoppe, W. Lohmann, H. Markl & H. Ziegler). Springer-Verlag, Berlin.

Nagy, K.A. & Medica, P.A. (1986) Physiological ecology of desert tortoises in southern Nevada. *Herpetologica* **42**, 73–92.

Nagy, K.A. & Petersen, C.C. (1987) Water flux scaling. In: *Comparative Physiology: Life in Water and on Land* (eds P. Dejours, L. Bolis, C.R. Taylor & E.R. Weibel). Liviana Press, Padova.

Newell, R.C. (1979) *Biology of Intertidal Animals*. Marine Ecological Surveys Ltd, Faversham.

O'Brien, J. & Block, B.A. (1996) Effects of Ca^{2+} on oxidative phosphorylation in mitochondria from the thermogenic organ of marlin. *Journal of Experimental Biology* **199**, 2679–2687.

O'Donnell, M.J. (1987) Water vapour absorption by arthropods: different sites, different mechanisms. In: *Comparative Physiology: Life in Water and on Land* (eds P. Dejours, L. Bolis, C.R. Taylor & E.R. Weibel). Liviana Press, Padova.

O'Donnell, M.J., Rheault, M.R., Davies, S.A. *et al.* (1998) Hormonally controlled chloride movement across *Drosophila* tubules is via ion channels in stellate cells. *American Journal of Physiology* **42**, R1039–R1049.

Oglesby (1968) Some osmotic responses of the sipunculid worm *Thermiste dyscritum*. *Comparative Biochemistry and Physiology* **26**, 155–177.

Oldfield, B.P. (1985) The role of the tympanal membranes and the receptor array in the tuning of auditory receptors in bushcrickets. In: *Acoustic and Vibrational Communication in Insects* (eds K. Kalmring & N. Elsner). Paul Parey Verlag.

Oosthuizen, M.J. (1939) The body temperature of *Samia cecrophia* Linn. (Lepidoptera, Saturnidae) as influenced by muscular activity. *Journal of the Entomological Society of South Africa* **2**, 63–73.

Page, R.D.M. (1993) Parasites, phylogeny and cospeciation. *International Journal for Parasitology* **23**, 499–506.

Palmiter, R.D., Gagnon, J., Ericsson, L.H. & Walsh, K.A. (1977) Precursor of egg-white lysozyme: amino acid sequence of NH_2 terminal extension. *Journal of Biological Chemistry* **252**, 6383–6393.

Park, C.C. (1991) Trans-frontier air pollution—some geographical issues. *Geography* **76**, 21–35.

Parry, D.A. (1954) On the drinking of soil capillary water by spiders. *Journal of Experimental Biology* **31**, 218–227.

Patterson, B.D., Pacheco, V. & Solari, S. (1996) Distribution of bats along an elevational gradient in the Andes of south-eastern Peru. *Journal of Zoology* **240**, 637–658.

Peaker, M.J. & Linzell, J.L. (1975) *Salt Glands in Birds and Reptiles*. Cambridge University Press, Cambridge.

Pearse, B. (1980) Coated vesicles. *Trends in Biochemical Sciences* **5**, 131–134.

Pearson, O.P. (1954) Habits of the lizard *Liolaemus multiformis* at high altitudes in southern Peru. *Copeia* **1954**, 111–116.

Peters, R.H. (1983) *The Ecological Implications of Body Size*. Cambridge University Press, Cambridge.

Phillips, J.E. (1981) Comparative physiology of insect renal function. *American Journal of Physiology* **241**, R241–257.

Phillips, J.E., Bradley, T.J. & Maddrell, S.H.P. (1971) Mechanisms of ionic and osmotic regulation in saline-water mosquito larvae. In: *Comparative Physiology: Water, Ions and Fluid Mechaniscs* (eds K. Schmidt-Nielsen, L. Bolis & S.H.P. Maddell). Cambridge University Press, Cambridge.

Phillipson, J. (1981) Bioenergetic options and phylogeny. In: *Physiological Ecology: An Evolutionary Approach to Resource Use* (eds C.R. Townsend & P. Calow). Blackwell Scientific Publications, Oxford.

Pierce, S.K. (1971) Volume regulation and valve movements by marine mussels. *Comparative Biochemistry and Physiology A* **39**, 103–117.

Pierce, S.K., Warren, M.K. & West, H.H. (1983) Non amino acid mediated volume regulation in an extreme osmoconformer. *Physiological Zoology* **56**, 445–454.

Piiper, J. & Scheid, P. (1982) Models for a functional analysis of gas exchange organs in vertebrates. *Journal of Applied Physiology* **53**, 1321–1329.

Polis, G.A. (1988) Foraging and evolutionary responses of desert scorpions to harsh environmental periods of food stress. *Journal of Arid Environments* **14**, 123–134.

Pough, F.H., Janis, C.M. & Heiser, J.B. (1999) *Vertebrate Life*, 5th edn. Prentice Hall.

Prevost, Y.A. (1983) Osprey distribution and subspecies taxonomy. In: *Biology and Management of Bald Eagles and Ospreys* (ed. D.M. Bird). Macdonald Raptor Research Centre of McGill University, Quebec.

Purvis, A. & Harvey, P.H. (1997) The right size for a mammal. *Nature* **386**, 332–333.

Ramsay, J.A. (1964) The rectal complex of the mealworm *Tenebrio molitor* L. *Philosophical Transactions of the Royal Society of London B* **248**, 279–314.

Randall, D.J. & Daxboeck, C. (1984) Oxygen and carbon dioxide transfer across fish gills. In: *Fish Physiology* (eds W. Hoar & D. Randall). Academic Press, Orlando, FL.

Redmond, J.R. (1955) The respiratory function of haemocyanin in Crustacea. *Journal of Cellular and Comparative Physiology* **46**, 209–247.

Reinertsen, R.F.E. & Haftorn, S. (1986) Different metabolic strategies of northern birds for nocturnal survival. *Journal of Comparative Physiology* **156**, 655–663.

Richards, O.W. & Davies, R.G. (1977) *Imm's General Textbook of Entomology*. Chapman & Hall, London.

Roitt, I., Brostoff, J. & Male, D. (1998) *Immunology*, 5th edn. Mosby Publishers, London.

Ross, M.H. & Romrell, L.J. (1989) *Histology*. Williams & Wilkins, Baltimore, MD.

Roughton, F.J.W. (1964) Transport of oxygen and carbon dioxide. In: *Handbook of Physiology*, Vol. 3.1. *Respiration* (eds W.O. Fenn & H. Rahn). American Physiological Society, Bethesda, MD.

Rowley, A.F. & Ratcliffe, N.A. (eds) (1988) *Vertebrate Blood Cells*. Cambridge University Press, Cambridge.

Ruppert, R.F. & Smith, P.R. (1988) The functional organisation of filtration nephridia. *Biological Reviews* **63**, 231–358.

Sanders, H.L. (1969) Benthic marine diversity and the stability-time hypothesis. *Brookhaven Symposium in Biology* **22**, 71–81.

Santos, C.A.Z., Penteado, C.H.S. & Mendes, E.G. (1989) The respiratory responses of an amphibious snail *Pomacea lineata* to temperature and oxygen tension variations. *Comparative Biochemistry and Physiology A* **86**, 409–415.

Schad, G.A. (1963) Niche diversification in a parasitic species flock. *Nature* **198**, 404–406.

Schmidt-Nielsen, K. (1964) *Desert Animals—Physiological Problems of Heat and Water*. Oxford University Press, Oxford.

Schmidt-Nielsen, K. (1972) *How Animals Work*. Cambridge University Press, Cambridge.

Schmidt-Nielsen, K. (1984) *Scaling: Why is Animal Size so Important?* Cambridge University Press, Cambridge.

Schmidt-Nielsen, K. (1997) *Animal Physiology*, 5th edn. Cambridge University Press, Cambridge.

Schmidt-Nielsen, K., Bretz, W.L. & Taylor, C.R. (1970a) Panting in dogs: unidirectional air-flow over evaporative surfaces. *Science* **169**, 1102–1104.

Schmidt-Nielsen, K., Hainsworth, F.R. & Murris, D.E. (1970b) Countercurrent heat exchange in the respiratory passages; effects on water and heat balance. *Respiration Physiology* **9**, 263–276.

Schmidt-Nielsen, K., Taylor, C.R. & Shkolnik, A. (1971) Desert snails: problems of heat, water and food. *Journal of Experimental Biology* **55**, 385–398.

Scholander, P.F. (1964) Animals in aquatic environments: diving mammals and birds. In: *Handbook of Physiology*, Section 4 (eds D.B.

Dill, E.F.V. Adolph & C.G. Wilbur). American Physiological Society, Bethesda, MD.

Scholander, P.F., Walters, V., Hock, R. & Irving, L. (1950) Body insulation of some arctic and tropical mammals and birds. *Biological Bulletin* **99**, 225–236.

Skadhauge, E. (1981) *Osmoregulation in Birds*. Springer-Verlag, Berlin.

Skulason, S. & Smith, T.B. (1995) Resource polymorphisms in vertebrates. *Trends in Ecology and Evolution* **10**, 366–370.

Smith-Gill, S.J. (1983) Developmental plasticity: developmental conversion versus phenotypic modulation. *American Zoology* **23**, 47–55.

Somero, G.N., Dahlhoff, E. & Lin, J.J. (1996) Stenotherms and eurytherms: mechanisms establishing thermal optima and tolerance ranges. In: *Animals and Temperature* (eds I.A. Johnston & A.F. Bennett). Cambridge University Press, Cambridge.

Sømme, L. (1995) *Invertebrates in Hot and Cold Arid Environments*. Springer-Verlag, Berlin.

Staehelin, L.A. (1974) Structure and function of intercellular junctions. *International Review of Cytology* **39**, 191–283.

Stewart, M. (1991) *Animal Physiology*. Hodder & Stoughton Educational, Sevenoaks.

Stone, G.N. & Willmer, P.G. (1989) Warm-up rates and body temperature in bees: the importance of body size, thermal regime and phylogeny. *Journal of Experimental Biology* **147**, 303–328.

Storey, K.B. & Storey, J.M. (1984) Biochemical adaptations for freezing tolerance in the wood frog *Rana sylvatica*. *Journal of Comparative Physiology B* **155**, 29–36.

Storey, K.B., Baust, J.G. & Storey, J.M. (1981) Intermediary metabolism during low temperature acclimation in the overwintering gallfly *Eurosta solidaginis*. *Journal of Comparative Physiology B* **144**, 183–190.

Strong, D.R., Lawton, J.H. & Southwood, T.R.E. (1984) *Insects on Plants*. Blackwell Scientific Publications, Oxford.

Strumwasser, F. (1960) Mammalian hibernation. XV. Some physiological principles governing hibernation in *Citellus beecheyi*. *Bulletin of the Museum of Comparative Zoology* **124**, 285–320.

Sweetman, H.L. (1938) Physical ecology of the firebrat, *Thermobia domestica* (Packard). *Ecological Monographs* **8**, 285–311.

Taylor, C.R., Schmidt-Nielsen, K. & Raab, J.L. (1970) Scaling of the energetic cost of running to body size in mammals. *American Journal of Physiology* **219**, 1104–1107.

Taylor, C.R., Maloiy, G.M.O., Weibel, E.R., Langman, V., Kamau, J.M.Z., Seeherman, J.H. & Heglund, N.C. (1981) Design of the mammalian respiratory system. III. Scaling of maximal aerobic capacity to body mass: wild and domestic animals. *Respiration Physiology* **44**, 25–37.

Tenney, S.M. & Temmers, J.E. (1963) Comparative quantitative morphology of the mammalian lung: diffusing area. *Nature* **197**, 54–57.

Theede, H. (1972) Vergleichende ökologisch-physiologische Untersuchungen zur zellulären Kälteresistenz mariner Invertebraten. *Marine Biology* **15**, 160–191.

Torres, J.J. & Somero, G.N. (1988) Vertical distribution and metabolism in Antarctic mesopelagic fish. *Comparative Biochemistry and Physiology B* **90**, 521–528.

Trager, W. (1986) *Living Together: the Biology of Animal Parasitism*. Plenum Press, New York.

Trueman, E.R. (1975) *The Locomotion of Soft-Bodied Animals*. Edward Arnold, London.

van Mierop, L.H.S. & Barnard, S.M. (1978) Further observation on thermoregulation in the brooding female *Python molurus bivattatus* (Serpentes; Boidae). *Copeia* **1978**, 615–621.

Villee, C.A. (1977) *Biology*, 7th edn. Saunders, Philadelphia.

Virkar, R.A. & Webb, K.L. (1970) Free amino acid composition of the soft-shell clam *Mya arenaria* in relation to salinity of the medium. *Comparative Biochemistry and Physiology* **32**, 775–783.

Wall, B.J. & Oschman, J.L. (1970) Water and solute uptake by rectal pads of *Periplaneta americana*. *American Journal of Physiology* **218**, 1208–1215.

Walliman, T., Wyss, M., Brdiczka, D., Nicolay, K. & Eppenberger, H.M. (1992) Intracellular compartmentation, structure and function of creatine-kinase isoenzymes in tissues with high and fluctuating energy demands—the phosphocreatine circuit for cellular energy homeostasis. *Biochemical Journal* **281**, 21–40.

Waloff, Z. (1963) Field studies on solitary and transient desert locusts in the Red Sea area. *Anti-Locust Bulletin* **40**, 1–93.

Walsberg, G.E. & King, J.R. (1978) The energetic consequences of incubation for two passerine species. *Auk* **95**, 644–655.

Wampler, J.E. (1978) Measurements and physical characteristics of luminescence. In: *Bioluminescence in Action* (ed. P.J. Herring). Academic Press, London.

Wayne, R.P. (1991) *Chemistry of Atmospheres*. Clarendon Press, Oxford.

Weber, R.E. (1980) Functions of invertebrate haemoglobins with special reference to adaptations to environmental hypoxia. *American Zoologist* **20**, 79–101.

Weigert, R.G. (1964) Population energetics of meadow spittlebug (*Philaenus spumarius* L.) as affected by migration and habitat. *Ecological Monographs* **34**, 217–241.

West, G.C. (1965) Shivering and heat production in wild birds. *Physiological Zoology* **38**, 111–120.

Wigglesworth, V.B. (1938) The regulation of osmotic pressure and chloride concentration in the haemolymph of the mosquito larvae. *Journal of Experimental Biology* **15**, 235–247.

Wigglesworth, V.B. (1945) Transpiration through the cuticle of insects. *Journal of Experimental Biology* **21**, 97–114.

Wilkie, D.R. (1977) Metabolism and body size. In: *Scale Effects in Animal Locomotion* (ed. T.J. Pedley). Academic Press, London.

Williams, M.A.J. (1986) The creeping desert: what can be done? *Current Affairs Bulletin* **63**, 24–31.

Willmer, P.G. (1980) The effects of a fluctuating environment on the water relations of larval Lepidoptera. *Ecological Entomology* **5**, 271–292.

Willmer, P.G. (1982) Microclimate and the environmental physiology of insects. *Advances in Insect Physiology* **16**, 1–57.

Willmer, P.G. (1983) Thermal constraints on activity patterns in nectar-feeding insects. *Ecological Entomology* **8**, 455–469.

Wilson, E.O. (1970) Chemical communication within animal species. In: *Chemical Ecology* (eds E. Sondheimer & J.B. Simeone). Academic Press, New York.

Withers, P.C. (1992) *Comparative Animal Physiology*. Saunders College Publishing, Fort Worth.

Wittenberg, J.B. (1958) The secretion of inert gas into the swim bladder of fish. *Journal of General Physiology* **41**, 783–804.

Woledge, R.C., Curtin, N.A. & Homster, E. (1985) Energetic aspects of muscle contraction. *Monographs of the Physiological Society 4*.

Wooten, R.C. & Crawford, C.S. (1975) Food, ingestion rates and assimilation in the desert millipede *Orthoporus ornatus* (Girard). *Oecologia* **20**, 231–236.

Zapol, W.M., Liggins, G.C., Schneider, R.C. *et al.* (1979) Regional blood flow during simulated diving in the conscious Weddel seal. *Journal of Applied Physiology* **47**, 968–973.

Index